Second Edition

MACHINE TOOLS

PROCESSES AND APPLICATIONS

GEORGE W. GENEVRO, Ed.D.

Professor Emeritus
Department of Technology Education
California State University, Long Beach

STEPHEN S. HEINEMAN, Ed.D.

Professor of Engineering and Industrial Technology
School of Engineering
California State University, Long Beach

Prentice Hall, Englewood Cliffs, New Jersey 07632

Library of Congress Cataloging-in-Publication Data

GENEVRO, GEORGE W.
Machine tools : processes and applications / George W. Genevro,
Stephen S. Heineman.—2nd ed.
p. cm.
Heineman's name appears first on the earlier edition.

Includes index.
ISBN 0-13-543455-6
1. Machine-tools. 2. Machine-shop practice. I. Heineman,
Stephen S. II. Title.
TJ1185.G417 1991 90-39892
621.9'02—dc20 CIP

Editorial/production supervision
and interior design: **Adele M. Kupchik**
Cover design: **Ben Santora**
Cover photo: **DoALL Company**
Manufacturing buyers: **Lori Bulwin, Mary McCartney, and Ed O'Dougherty**

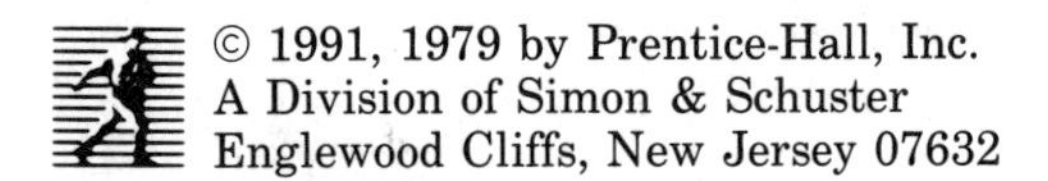

A Division of Simon & Schuster
Englewood Cliffs, New Jersey 07632

Printed in the United States of America

10 9 8 7 6 5 4 3 2 1

ISBN 0-13-543455-6

Prentice-Hall International (UK) Limited, *London*
Prentice-Hall of Australia Pty. Limited, *Sydney*
Prentice-Hall Canada Inc., *Toronto*
Prentice-Hall Hispanoamericana, S.A., *Mexico*
Prentice-Hall of India Private Limited, *New Delhi*
Prentice-Hall of Japan, Inc., *Tokyo*
Simon & Schuster Asia Pte. Ltd., *Singapore*
Editora Prentice-Hall do Brasil, Ltda., *Rio de Janeiro*

Contents

6 SCREW THREADS AND THREADING PROCESSES 100

7 METAL SAWING OPERATIONS AND EQUIPMENT 113

8 DRILLING MACHINES AND OPERATIONS 132

9 ENGINE LATHES AND OPERATIONS 159

10 SPECIAL LATHES 222

11 MILLING MACHINES AND OPERATIONS 240

12 SHAPERS AND OPERATIONS 293

13 GRINDING MACHINES, OPERATIONS, AND ABRASIVE PROCESSES 304

14 BASIC METALLURGY AND HEAT TREATMENT 341

15 METAL-CUTTING THEORY AND CUTTING FLUIDS 364

16 NUMERICAL CONTROL 381

17 ELECTRICAL MACHINING PROCESSES 401

APPENDICES 411

Preface

The continuing evolution and growth of the metalworking industry is heavily dependent on the availability of an ever-broader range of machine tools and on people who can use them effectively. Historically, machine tools have played a dominant role in the development of all industrialized nations. In today's competitive world, the ability to manufacture goods in an economically effective manner is probably even more important than it was in years gone by.

As the metalworking industry expands and the demand for new and more complex products increases, there will be a greater need for imaginative and skilled machinists. There is no doubt that the jobs of future machinists will be more complex—and more interesting—than ever before. New machining methods must be mastered, along with an ever-increasing body of technical information. New specialties may be learned, offering opportunities for advancement into more interesting and challenging jobs with better pay.

The basis for advancement into more complex machining processes or into some of the special areas of the trade is an understanding of machine tools, the processes for which they can be used, and a level of skill in keeping with the tasks to be done. It is with these factors in mind that this book was prepared, drawing on the experience of the authors as machinists, educators, small-business owners, and consultants. Four basic assumptions were developed about the needs of the student who wants to become a machinist, and the seventeen chapters of the text were prepared to meet these needs as effectively as possible.

1. *The student must learn to perform the basic machine tool operations effectively, accurately, and safely.* Since machine tools can be very unforgiving, safety has been stressed throughout the text in addition to Chapter 2, in which general safety practices are covered. Five of the chapters are devoted to basic machine tool operations, such as drilling (Chapter 8), lathe work (Chapters 9 and 10), milling (Chapter 11), and grinding (Chapter 13). Particular care was taken to develop clear, step-by-step operation sequences.

2. *The student must also understand the related processes and theories on which machining processes are heavily dependent.* Since this is of critical importance to those who wish to advance in the metalworking industry, supplementary material in both the text and appendices was carefully selected. Major emphasis was placed on measuring systems (Chapter 5), basic metallurgy (Chapter 14), and metal-cutting theory (Chapter 15), and care was taken to emphasize applications of these and other related materials throughout the book.

3. *The student must be able to relate theories and concepts to the machining operations performed in the shop.* In the text, whenever possible, concepts are introduced in conjunction with machining operations. For example, such factors as the action of cutting fluids, the metallurgical characteristics of the material being machined, cutting tool selec-

tion, and cutting tool geometry are explained and related directly to a specific machining operation.

4. *As the student gains experience, there should be a corresponding increase in decision-making ability.* This is a very important aspect of the machinist's education, so the presentation of material in the text was designed to enhance the student's ability to make effective choices. For example, when more complex operations are introduced, such matters as tool selection, speeds and feeds, chip load computations, and alternative courses of action are discussed. The importance of making the best selection and learning to make carefully considered decisions is stressed.

There is a carefully evaluated balance between material dealing with machining operations and such topics as measuring tools and systems, basic metallurgy, metal-cutting theory, and drawing interpretation (Chapter 3). This balance was developed from the authors' teaching experience at the secondary and college levels and from many discussions with colleagues. The authors' intent is to provide the instructor with a textbook usable at the senior high school, community college, and college levels. The instructor has a number of options in terms of how the book can be used at various levels. For example, the text can be used for short courses or units in such areas as milling machines, lathes, and drill press work and can also be used for a general machine shop course. In addition, it may be used as a basic text in a related theory class for apprentices.

There are more than 900 illustrations in the book. They were selected with two purposes in mind: to show the student a wide range of tools, equipment, and processes, and to help the instructor teach about those machines and processes not usually found in school shops. Important machining and measuring processes are carefully and fully illustrated and described, together with related theory. The authors believe that this is a matter of fundamental importance to the student.

An important aid to the student is a series of review questions at the end of each chapter. These questions cover each main section of a chapter, highlighting the most important terms, concepts, and when applicable, safety precautions. A complete glossary at the end of the book concisely defines each major process, tool, and material and other pertinent points.

Selected tables and appendices are included in the book for the convenience of both the instructor and student. A unique feature of the appendices, for instance, is a section on the use of electronic calculators in the shop. Although the material in the appendices is sufficient for most school shop machining operations, the authors encourage the student to use machinists' handbooks and manufacturers' literature when doing advanced work.

Writing a textbook is a complex task that involves many people directly and indirectly. To our wives we extend heartfelt thanks for patience, encouragement, assistance, and suggestions. The technical skill and thoughtful comments of Michael R. Clancy, who rendered many of the line drawings, are greatly appreciated. Many individuals and companies contributed materials; to all of these we are deeply grateful.

Several machine shop instructors reviewed the original manuscript and provided very valuable comments and suggestions. The opinions of these experienced instructors led to many changes and thoughtful review of the concepts on which the book is based. We extend sincere thanks to Glenn Bettis, East Tennessee State University; Dwight Collins, Elgin Community College; Robert C. Craig, San Francisco State University; Raymond DeKeyser, Ball State University; Donald Hartshorn, Columbus Technical Institute; Joe Merchant, Vincennes University; Ronald A. Witherspoon, San Jose City College; and Albert Zachwieja, Triton College.

George W. Genevro
Stephen S. Heineman

1 Machine Tool Occupations

1.1 INTRODUCTION

The growth and evolution of industry have been characterized by the increasing use of complex machine tools, and this trend appears certain to continue into the foreseeable future. In the early periods of the Industrial Revolution, machine tools were relatively simple and crude. They were, however, capable of performing operations that could not readily be done by hand, and almost invariably they made it possible to manufacture products more accurately and economically.

Machine tools, even in their simplest forms, have always been involved in the manufacture of other machine tools, hand tools, and other shop equipment. The ever-increasing demand for machinery, coupled with the inventive genius of such people as Eli Whitney, Henry Maudsley, and others, resulted in rapid growth of the machine tool industry. These factors, along with the greater availability of metals and the need for machinery in other segments of industry, firmly established machine tools as indispensable to the industrial establishment. Machine tools and machinists have played key roles in bringing the ideas of scientists and engineers into actual use in industrial processes.

In the period following the Revolutionary War in the United States, development of several in-

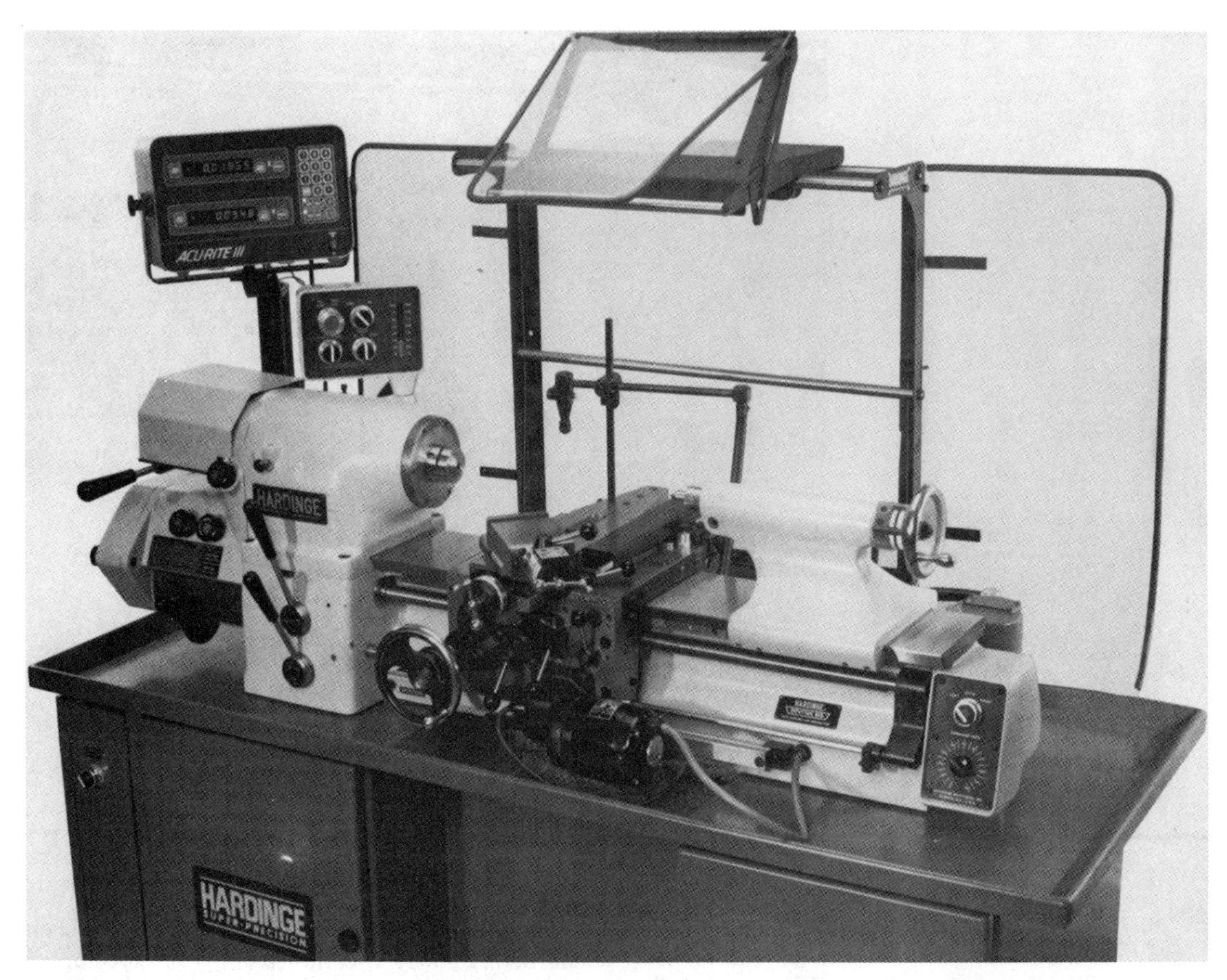

FIGURE 1.1 Precision lathe for manufacturing close tolerance parts. (*Courtesy Hardinge Brothers, Inc.*)

dustries accelerated and the need for machinery expanded. The textile industry, small-arms makers, clock makers, and many others needed machines that could only be produced in machine shops. This made the machine tool industry and machine designers key factors in industrial development. At a somewhat later time, the expansion of railroads and marine transportation broadened the markets for manufactured goods and led to an even greater expansion of the machine trades and related industries.

Although the five basic machine tools—drill press, lathe, milling machine, planer, and grinder—were sufficient for doing most machining jobs when the needs of industry were less complex, it was inevitable that variations of these machines would be developed. In some cases, machines were simply made larger; in other cases, major conceptual changes were incorporated to make the machines more suitable for mass-production purposes. They were also changed to improve accuracy and reduce the effects of such variables as the varying skills of the machine operators.

Several other changes and improvements in machine tools were made possible by developments and inventions that occurred in other segments of industry and science. The invention of the electric motor, development of better cutting tool materials, synthetic abrasives, and better measuring tools, among other things, contributed greatly to the manufacture of more precise, efficient, and durable machine tools. The metallurgist also played an important role in machine tool improvement by providing machine tool builders with better metals for castings, forgings, fasteners, and other parts. Continuing improvements in welding and foundry technology have also been a major factor in the development of better machine tools.

The advent of mass production has had a profound effect on machine tool development. The need for thousands—and in some cases millions—of precisely machined parts led to the development

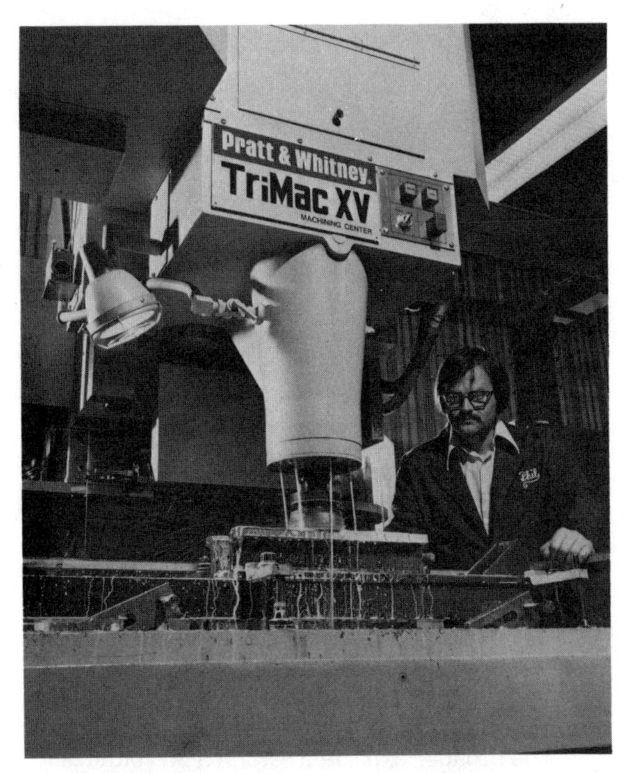

FIGURE 1.2 Skill and experience are required to run complicated machine tools. (*Courtesy Pratt & Whitney Machine Co., Inc.*)

of advanced machine tools such as the automatic screw machine, centerless grinder, multiple-spindle drill press, and others. These efficient machines resulted in lower prices for many industrial and consumer products and helped to accelerate industrial development. They are used extensively in the automotive industry and in other applications where the need for great numbers of parts justifies the large capital outlay.

The increasing need for a wide variety of industrial products is a basic characteristic of any industrial economy. This need can be satisfied only by an active, energetic machine tool industry and sufficient numbers of skilled machine tool builders and machinists trained in the theory and practice of the machine tool occupations.

1.2 THE MACHINIST

The term *machinist* is applied to many people engaged in machine tool occupations. However, the title is not fully descriptive unless it is further defined so that the reader has some indication of the nature, scope, and limitations of the work done by machinists. The subclassifications that follow emphasize the nature of the work and the scope of

FIGURE 1.3 Technicians and machinists work together to solve problems. (*Courtesy Cherry Textron*)

the job. Although the list is by no means complete, it does identify the main segments of the machinist's trade.

1.2.1 Machine Operator

Although the scope of the skills and responsibilities of the *machine operator* may vary, depending on the nature of the manufacturing operation, this type of worker usually operates a production machine and performs a limited number of operations. Depending on the type of product or part being made, the machine operator has little decision-making latitude. Generally, only a little understanding of machine tool theory is needed because the machine is usually prepared by a *setup* or *lead person*. Some skill in the use of measuring tools and preset gages is desirable but not absolutely essential. The operator must be able to make

FIGURE 1.4 Skilled machinists must know how to measure accurately. (*Courtesy L.S. Starrett Co.*)

some minor adjustments and identify any operational problems that occur.

The machine operator requires little training as a rule and is usually paid less than workers in other machine tool occupations. Opportunities for learning machine tool theory and operation in its broader aspects are quite limited because of the nature of the tasks being performed. Specialization in certain mass-production industries and the development of machine tools that allow little operator discretion have resulted in relatively large numbers of machine operator jobs.

1.2.2 Machine Hand

The *machine hand* is a person who has gained considerable skill in the use of one class of machine tool. For example, a skilled milling machine operator is capable of operating all the common types of horizontal and vertical mills. This person can use measuring tools, do layout work, read machine drawings, and make setups of various types. In some industries this type of worker is classified as a "milling machine machinist" because the term "machine hand" is gradually disappearing.

Machine hands are usually found in larger machine shops that allow for a higher degree of specialization among the employees. This type of worker is usually relatively well paid, enjoys a greater degree of job security than does the machine operator, and has greater chances for advancement.

1.2.3 General Machinist

The *general* or *all-around machinist* is the most versatile of machine shop employees. Such a machinist is able to operate all the basic machine tools and usually is capable of learning to use specialized machine tools in a relatively short period. Probably the most important attributes of the general machinist are the abilities to plan work sequences, do the necessary computations, select and use all the necessary measuring tools, and understand the theoretical bases for the operations.

The training of the general machinist includes a specified number of hours of shop work along with classroom work in related theory, usually under the terms of an apprenticeship agreement. Regardless of the type of shop in which the apprenticeship is served, the apprentice is exposed to all the major aspects of the machinist's trade and works on all the basic machine tools. In some situations, the apprentice also works on special machine

FIGURE 1.5 Special machines such as this planer-type contour milling machine are used in heavy industry. (*Courtesy Morey Machinery Co.*)

tools and does various types of bench, assembly, and quality-control work.

The broader experience and background of the general machinist allows this person greater mobility within the trade and can also form the basis for advancement into leadman and foreman positions. Specialization in segments of the machine tool occupations such as tool- and diemaking and experimental machine work may follow.

General machinists who work in large shops may not have the opportunity to learn other metalworking skills, such as welding and heat treating, since there are specialists available to do those jobs. The machinist who works in small shops often has to learn other skills because it is not practical to hire specialists for operations that are done only occasionally. This broadening of one's skills can be very valuable and advantageous to machinists who eventually wish to start their own businesses. In operating a small shop, particularly one that specializes in custom and repair work, the ability to do metalworking operations that are beyond the normal scope of machine work is a very valuable asset.

1.2.4 Experimental Machinist

Increasingly, experienced general machinists work closely with technicians and engineers in the construction of prototype and experimental equipment. The *experimental machinist* usually does no production work but is expected to work to close tolerances and must be able to work from drawings as well as technical descriptions. Some math-

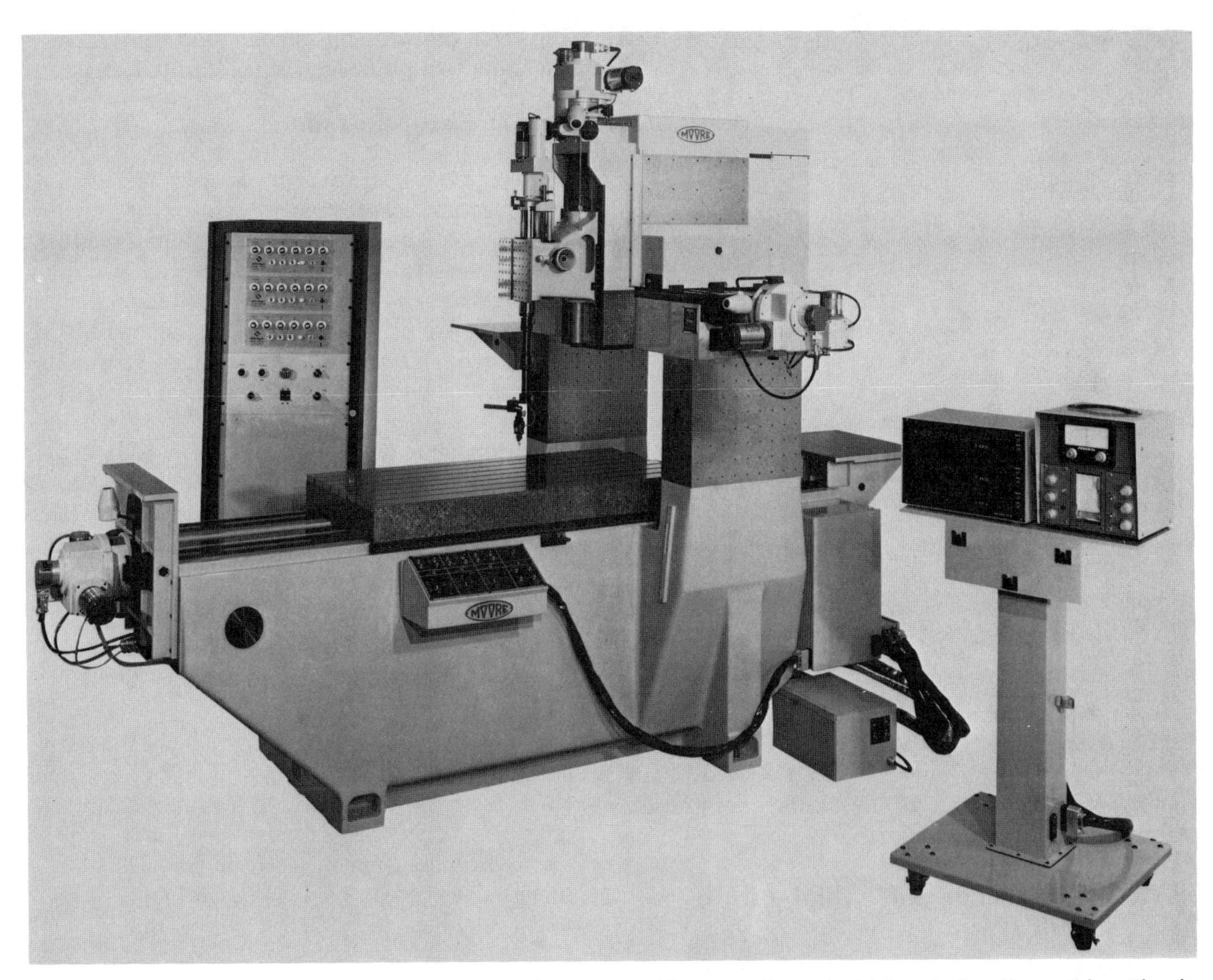

FIGURE 1.6 Machines for very accurate measuring are used in some shops and laboratories. No metal cutting is done on this machine. *(Courtesy Moore Special Tool Co., Inc.)*

ematical ability is highly desirable, and the machinist should be able to perform and specify certain heat-treating and metal-finishing processes. In short, the experimental machinist must not only be a very capable general machinist but must have a thorough understanding of a number of related processes, and in many cases help to solve technical problems.

1.2.5 Tool- and Diemaker

Tool- and diemaking are specialties in the machine tool occupations with high skill requirements. In both cases, the work requires attention to close tolerances and detail. Tool- and diemakers must be able to operate all the basic machine tools along with certain special machines. Knowledge of shop mathematics, basic metallurgy, heat treating, and precision measurement is essential to tool- and diemaking.

There is a considerable amount of diversity within the tool- and diemaking segment of the machinist's trade. In some cases, only a particular type of product, such as injection molding dies, is made in a shop, while in other shops, equipment of many different types and sizes is made. Independent shops, often referred to as *job shops,* produce tools, dies, and other equipment on a contract basis. *Captive shops* are usually part of the facilities of a large industrial organization that needs tooling and related equipment produced, repaired, and altered on a continuing basis.

As industrial products and the processes by which they are produced become more complex, the tool- and diemaker and the engineers who design the tooling are constantly faced with new challenges. In recent years, for example, electrical discharge machining and other nonconventional processes have become major factors in the production of tooling. Use of these processes has made

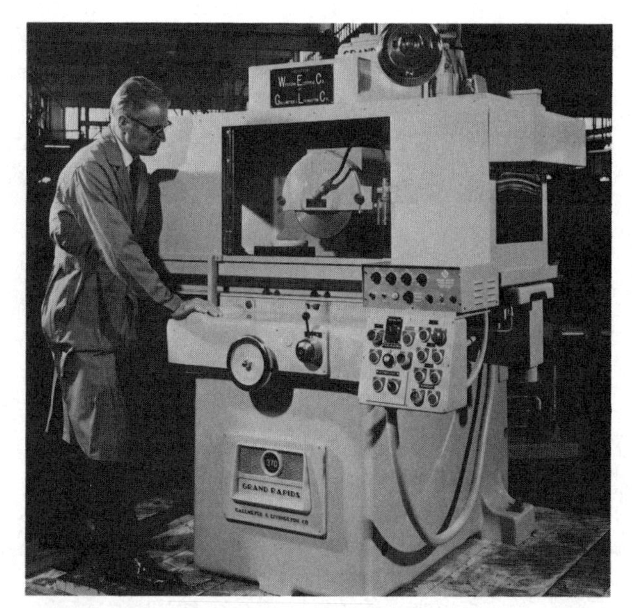

FIGURE 1.7 This precision grinding machine is being operated by a tool- and diemaker. (*Courtesy Gallmeyer & Livingston Co.*)

possible the economical production of dies and related equipment that was not considered feasible previously. It is evident, then, that continued study and development of new skills is very much a part of the tool- and diemakers' job.

1.2.6 Maintenance Machinist and Millwright

The jobs of the *maintenance machinist* and the *millwright* are often similar because in smaller shops and plants one worker may perform both functions at different times. The maintenance machinist is basically concerned with keeping machine tools and other very diverse types of machinery in working order. Occasionally, this person makes or remachines certain replacement parts. This part of the job requires the skills of the general machinist. Under certain circumstances, the maintenance machinist also installs or removes various types of machines.

The primary job of millwrights is the installation and erection of machinery and equipment in factories. The skills of the general machinist are not always required, but millwrights are more versatile when they can do some of the basic machining operations.

Machine tool rebuilding is a specialized variation of the skills just outlined, and the person involved in this type of work has usually served an apprenticeship as a general machinist. The machine rebuilder works to exacting tolerances and must be skilled in the use of precision measuring tools and the equipment used in evaluating machine tool accuracy and performance.

1.2.7 Setup Person

In machine shops where mass-production work is done, most employees are usually machine operators trained to do a limited number of tasks. Since most production machines—such as automatic screw machines—are quite complex, the machines are not set up for production runs by the operators. The *setup person* is usually a skilled machinist who has specialized in the preparation of production machine tools. In larger shops, a setup person may only set up one class of machine, but nevertheless must have a thorough understanding of the theory and practice of mass-produced machined parts. The setup person may also be responsible for preparing the necessary gages for evaluating machined parts and making trial runs of the machines prepared.

1.2.8 Automotive Machinist

Automotive machine work is probably the largest specialized form of machining. It is not a manufacturing activity, as a rule, but it is mainly involved with the machine work necessary to recondition and rebuild automotive components. This type of machine work may be done in job shops, a type of machine shop in which general repair work,

FIGURE 1.8 A setup person prepares an automatic turret lathe for a production run. (*Courtesy Bardons and Oliver, Inc.*)

custom work, and some small-lot manufacturing are usually done. In other cases, garages may have the necessary basic and special machines to do automotive machine work.

The *automotive machinist* is not necessarily trained as a general machinist, but this worker must have many of the same skills and be able to operate at least some basic machine tools as well as special machines. Knowledge of certain aspects of automotive theory and practice and the ability to use precision measuring tools are essential.

Another variation of the machinist's trade that is similar to the automotive machinist's job is the *marine machinist.* The work done by the marine machinist in the process of rebuilding engines or making parts for special installations requires a wide range of skills, along with specialized knowledge of the problems encountered in marine applications. An unusual aspect of the marine machinist's job is that it may involve not only machining the parts but also, installing them. As with almost all other specialties in the machinist's trade, skill in the use of common as well as specialized measuring tools is very important, along with an understanding of the metallurgical problems encountered in marine work.

1.2.9 CNC or NC Machine Operator

The computer numerical control (CNC) machine operator should already be a skilled operator in the category of the CNC machine he or she will operate. With experience in conventional machine tool operation, the CNC operator will already have acquired basic knowledge and skills in the use of measuring tools, blueprint interpretation, machinability of common metals, and identifying correct machining procedures.

The *CNC operator* makes setups with standard and NC tooling, aligns and coordinates machine tables and spindles, installs programmed tapes in electronic consoles, and uses basic inspection tools to check machined parts. A working knowledge of mathematics, including trigonometry and geometry, is highly desirable. Many manually programmed CNC and NC machines require the operator to insert into the control unit data that was calculated from the working drawing.

1.3 LEARNING TO BE A MACHINIST

Machine shop occupations vary in terms of the level of technical skill and knowledge they require. Those who are employed as machine operators usually have very little formal training. In the course of their work they have few opportunities to learn more than the operation of the machine to which they are assigned. On the other hand, general machinists, experimental machinists, and tool- and diemakers must be skilled in the use of both basic and special machine tools. They also must know, and be able to apply, a substantial amount of technical information. Highly skilled employees in the machine trades, therefore, need a substantial, well-planned general education and thorough technical training in the form of apprenticeship and related technical classes.

1.3.1 Apprenticeship

Apprenticeship has been a commonly used means of training craftsmen since ancient times. In its best forms it is a formal agreement that involves planned sequences of work in various machining operations and related technical classes. In a good program, the apprentices are evaluated on a regular basis, and those who are exceptionally able are usually given opportunities to learn additional processes.

The usual term for a machinist's apprenticeship is four years, or approximately 8000 hours. Depending on the nature of the agreement, the apprentice may be required to attend related technical classes for 2 to 4 hours a week. In some cases the apprentice is paid for part or all of the time spent in class. The related classes may be taught in the plant or in technical schools by appropriately certificated instructors.

Unions are often involved in apprenticeship agreements, both in terms of specifying the ratio between journeymen and apprentices and in setting the working conditions and length of apprenticeship. The union also may set performance stan-

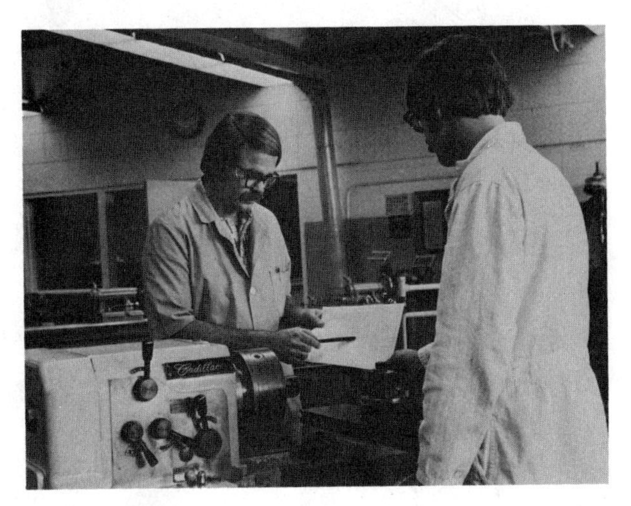

FIGURE 1.9 An apprentice learns by working with a skilled machinist.

dards for the apprentice who is ready to become a journeyman and even conduct formal graduation ceremonies. The minimum age for entry into an apprenticeship program may be specified by the union, and it usually coincides with the age a student may legally leave school. Most future machinists become apprentices at about 18, although it is not unusual for young people between 20 and 25 to enter apprenticeship programs.

In small, nonunion shops apprentices often learn the machinist's trade under relatively informal conditions. In the typical small job shop, the apprentice may have the opportunity to do many jobs without spending specified lengths of time learning any particular operation. Often there is no fixed requirement for attending related technical classes. The apprentice usually learns the necessary technical material by self-study, from the foreman, or from the journeymen in the shop. If the apprentice works in several shops before becoming a journeyman, he or she usually has the opportunity to learn a variety of operations and is exposed to journeymen with different skills and experience. The apprentice who is learning the machinist's trade on an informal basis is strongly advised to take classes in drafting, applied mathematics, basic metallurgy, and other subjects related to machine shop practice.

1.3.2 Trade Schools and Technical Institutes

Throughout the twentieth century, trade schools have played a major role in the training of machinists. They are not a substitute for apprenticeship and direct work experience, but in many cases the machinist's first contact with machine tools takes place in a public school equipped to offer occupational instruction. In some cases, the laboratory instruction can be counted as part of the term of apprenticeship, and the apprentice may return to the same school for additional related technical classroom instruction.

In the public schools, trade instruction is given in comprehensive high schools with vocational education departments, in vocational or polytechnic high schools, in regional occupational schools, or in community colleges. Since 1917 the federal government has provided support for vocational education, and almost all secondary level public schools now offer instruction leading to employment in some segment of the machinist's trade.

Private trade schools generally are of two major types: proprietary schools operated for profit, and schools operated by large companies for the purpose of providing supplementary training for

FIGURE 1.10 Horizontal boring machine. (*Courtesy Summit Machine Tool Manufacturing Corp.*)

apprentices. In either case the training is concentrated and intensive. The main objective is to impart the skills necessary for either initial employment or advancement.

The armed forces also provide opportunities for learning the machinist's trade. The highly mechanized nature of the modern military establishment makes the skills of the machinist indispensable, and almost all the services maintain schools for training machinists. After completing the prescribed training course, the machinist usually works under supervision while gaining actual job experience. The machinist is often required to successfully complete additional related technical classes as a condition for being promoted. In many ways this parallels the advancement of the civilian apprentice toward journeyman status.

Although technical and trade schools are very valuable means of teaching basic skills and advanced related material later in the apprentice's training, they are not a direct substitute for actual work experience. Anyone wishing to learn the machinist's trade well must be prepared to complete the work necessary to meet the apprenticeship requirements and keep skills and related technical information up to date. It is a complex and fascinating trade with many opportunities for advancement and specialization.

1.4 OPPORTUNITIES AND COMPENSATION IN THE MACHINIST'S TRADE

For some individuals, becoming a machine operator or a machinist means that they have achieved their ultimate goal. They may wish to continue working at this occupational level until retirement. Others look upon becoming a machinist as the necessary first step in their career aspirations. With appropriate additional formal education and work experience, machinists can enter a variety of related occupations and professions, including teaching, engineering, and industrial sales. Machinists may also choose to go into business for themselves after learning the necessary business skills.

Regardless of personal aspirations and ambition, machinists will find that they are in a stable, rewarding occupation, particularly if their experience and training qualify them as general machinists, tool- and diemakers, or experimental machinists. Well-trained machinists are seldom out of work, and the diversity of their skills makes them employable in a variety of shops related to machines and machining processes. Employment is generally stable throughout the year since the machinist's work is relatively unaffected by seasonal variations.

Working conditions vary, depending on the type of shop in which the apprentice or machinist works, but the environment is generally pleasant and safe compared with the majority of industrial jobs. In some situations, where very precise work is done, the machinist or toolmaker works in a dust-free, temperature-controlled room. The condition of the working environment is partly the machinist's responsibility, and employers desire personal habits of cleanliness and orderliness. Workers who keep their tools, equipment, and work area in good order are generally superior craftsmen.

The process of becoming a good machinist is not easy. It involves study, work, and the ability and desire to keep on learning after achieving journeyman status. The rewards for learning this interesting and diversified occupation are highly desirable, not only in terms of pay and job satisfaction, but also because of the advancement opportunities open to the versatile machinist.

REVIEW QUESTIONS

1.1 INTRODUCTION

1. What are the five basic machine tools?
2. What part has the metallurgist played in improving machine tools?

1.2 THE MACHINIST

1. How does a machine operator differ from a general or all-around machinist in terms of work experience and skills?
2. List three major skill areas in which the tool- and diemaker must be competent.
3. In what type of machine shop are the skills of the *setup person* usually needed?

1.3 LEARNING TO BE A MACHINIST

1. How long does it usually take, in either hours or years, to complete a machinist's apprenticeship program?
2. Name three areas of related technical study that are useful to the apprentice who wishes to become an all-around machinist.

3. What three different types of schools could a person attend to learn the basic skills in the machine trades?

1.4 OPPORTUNITIES AND COMPENSATION IN THE MACHINIST'S TRADE

1. Identify several specialized areas of machine work that the machinist may enter.
2. Explain why employment in the machine trades is generally stable in comparison to some other technical occupations.
3. Identify two related occupations that the machinist may enter after receiving appropriate additional training.

2

Safety Practices

2.1 INTRODUCTION

The typical machine shop is a relatively safe place in which to work if appropriate work habits are learned early and applicable safety rules are carefully observed. Because machine shops vary in terms of the type and size of machinery used and the nature of the work performed, machinists must learn to identify the hazards in a wide variety of working situations. Once the hazards are known, workers must be able to select the work procedures and safety equipment that will allow them to do their work safely and efficiently.

In the last 50 or 60 years, a number of governmental and private organizations have made substantial contributions to worker safety, and as a result, the number of occupational accidents in practically all fields has dropped. Because of the substantial research that has been done, modern machine tools have effective guards, electrical equipment is safe, and working conditions in most machine shops are more pleasant than in other manufacturing occupations.

The Occupational Safety and Health Act (OSHA), enacted by Congress in late 1970, authorized the establishment and enforcement of work rules and equipment standards designed to lower the number and severity of industrial accidents. Standards were also developed with regard to dust control, noise levels, and a number of other factors related to industrial hygiene. Employers who do not maintain their machinery and facilities

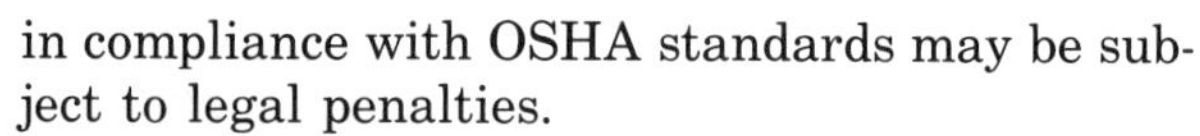

in compliance with OSHA standards may be subject to legal penalties.

In the final analysis, the work habits, knowledge, and attitude of the machinist are the key factors in working safely. Learning to use machine tools safely will make your career as a machinist more pleasant, efficient, and rewarding.

FIGURE 2.1 Eye protection is important. Always wear impact-resistant safety glasses with side shields.

2.2 GENERAL SAFETY PRACTICES

2.2.1 Eye Protection

In the machine shop, eye protection is provided by goggles, face shields, or safety glasses worn by the operator, and the guards or shields that are attached to or placed near the machines. Either way, the machinist's eyes are protected from flying objects such as chips, abrasive particles from grinding wheels, and broken tools (see Fig. 2.1).

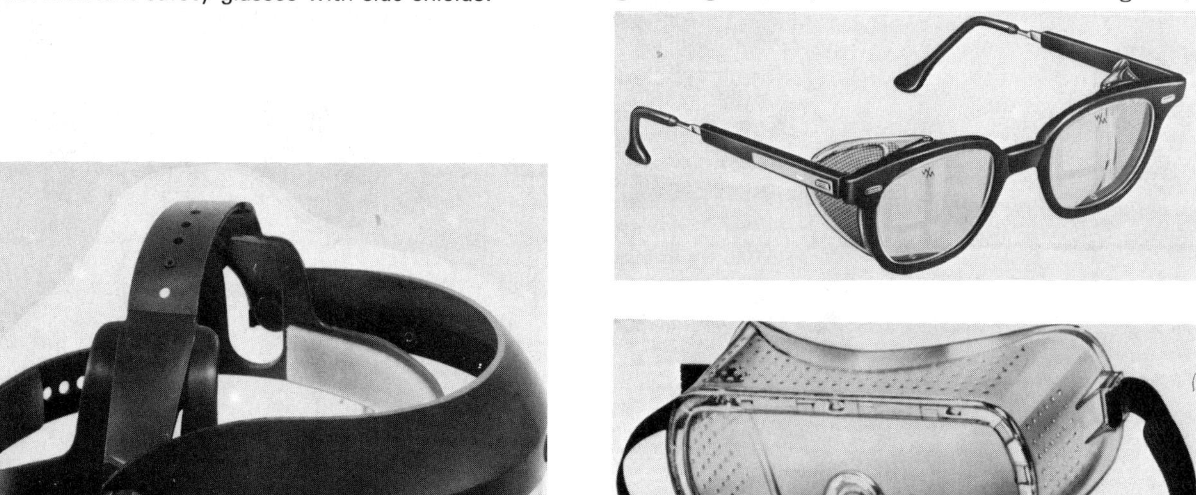

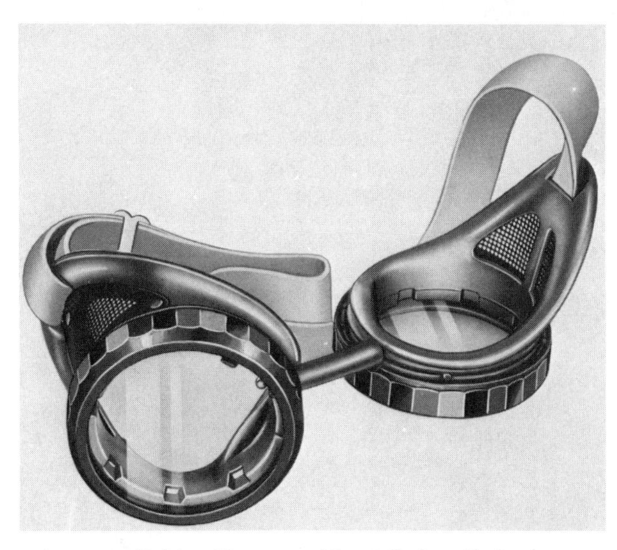

FIGURE 2.2 Many types of eye-protection equipment are available. (*Courtesy Direct Safety Co.*)

Safety glasses are the primary form of eye protection and may be supplemented by the use of face shields. The safety glasses must be OSHA approved and have heavy frames and side shields. Typical safety glasses are shown in Fig. 2.2. Good safety glasses resist impacts without shattering, and the side shields prevent chips or other particles from entering the wearer's eyes.

Goggles may be worn over prescription glasses, but some types of goggles tend to restrict side vision. The all-plastic type of goggles that fit the face contours closely are satisfactory. When corrective glasses must be worn, the ideal solution is to have the prescription lenses ground in safety glass and fitted to strong frames.

Face shields are generally used for supplementary eye protection and to protect the face and neck from chips and other objects (see Fig. 2.3). They are generally made of relatively thin material and are not designed to resist heavy impacts from sharp objects. All eye-protection equipment must be kept clean and free of scratches and blemishes that restrict vision.

It is absolutely necessary that appropriate eye protection be worn at all times in the machine shop. Good vision is indispensable to the machinist.

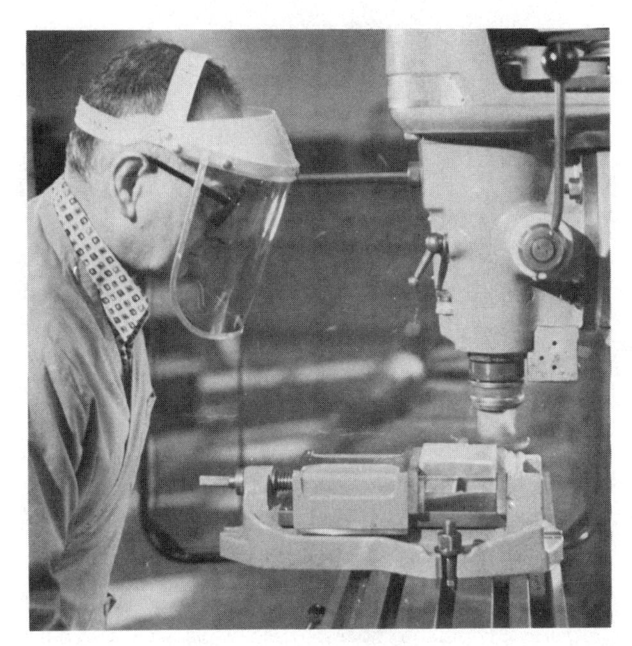

FIGURE 2.3 Always wear a face shield when doing fly-cutting operations.

2.2.2 Industrial Hygiene

In the machine shop, the worker is often exposed to metallic dust from machining operations, cutting fluids, lubricants, solvents, and other materials that may cause discomfort or irritation. Different individuals can tolerate varying amounts of such irritants before discomfort becomes severe. *Any skin or respiratory tract irritations should be reported to the instructor immediately.*

When working with materials that might be irritating, the machinist should be particularly careful about personal cleanliness. Prompt removal of dust, cutting fluids, and other materials from the skin by thorough washing usually prevents irritation. Shop aprons, coveralls, or shop coats must be kept clean.

FIGURE 2.4 Lift properly to avoid back strains.

2.2.3 Lifting

It is often necessary for the machinist to lift fairly heavy objects. If several basic precautions are taken, lifting weights within the person's physical capability is not necessarily hazardous. Lifting should be done with the legs, keeping the back as straight as possible (see Fig. 2.4). Many lifting injuries occur when a person attempts to lift an object that is too heavy. Contact the instructor and secure assistance when lifting extremely heavy objects.

When heavy objects such as lathe chucks must be installed on the machine, blocks or cradles should be used (see Fig. 2.5). This makes it easier to position the chuck and reduces the possibility of back strain or other injury. Objects should be free of oil and grease whenever possible.

FIGURE 2.5 Lathe chucks should be placed in a wooden cradle.

FIGURE 2.6 Always use properly grounded electrical equipment.

FIGURE 2.7 Electrical power panels should have a circuit breaker for each machine and an emergency shutoff switch.

Since floors in the machine shop may be oily, it is possible to slip or lose one's footing even when lifting or carrying light objects. Keeping floors clean and using available oil-absorbent materials will greatly reduce the probability of slips and falls when handling heavy objects.

2.2.4 Electrical Hazards

Electrical devices that are properly installed and serviced are seldom a source of danger in the machine shop. There are, however, cases when misused or damaged power tools can cause serious electrical shocks. Properly grounded electrical power tools will always have a three conductor cord and a three pronged plug as shown in Fig. 2.6. Any electrical equipment that is not properly grounded, has worn cords, or is malfunctioning should be reported immediately to the instructor.

In most shops, there is a power panel with circuit breakers for individual machines and a main switch (Fig. 2.7). Be sure that you know the location of these switches and circuit breakers so they can be used in case of electrical malfunctions.

Individual switches on machines may be of the pushbutton type (see Fig. 2.8) or of the lever type (see Fig. 2.9). Lever-type switches are generally used where the operator must be able to select *forward* or *reverse* rotation of the spindle. In some cases, especially where the machine's motor requires 220- or 440-volt current, an individual breaker panel is used on the machine. This serves to isolate the switch used by the operator from the high-voltage current and also to prevent the machine from starting unexpectedly if there is an electrical service interruption and the switch has been left in the ON position.

2.2.5 Fire Protection

The two major aspects of fire protection in the machine shop are *prevention* and *control.* Fire prevention is the continuing process of keeping electrical equipment, gas appliances, and other equipment clean and in good operating condition. Proper storage of flammable materials in containers and cleanliness of the entire shop are very important factors (see Fig. 2.10). The use of nonflammable solvents and cutting fluids is highly recommended.

A major factor in the control of fires is understanding the burning process and being able to classify fires so that the proper extinguisher can be used. A fire cannot exist unless fuel, oxygen, and a source of ignition all are present (see Fig. 2.11). Removal of one or more of these three elements is necessary to control a fire. In an electrical fire, for example, shutting off the electrical current (the source of ignition) is the essential first step in fighting the fire. Therefore, the location of the main power panels and switches should be well marked and known by those in the shop.

Fires are classified as follows. An understanding of the classifications will help you select the correct extinguisher and fire control procedures:

> *Class A*—fire in combustible materials such as rags, paper, and wood
>
> *Class B*—fire in oils, solvents, greases, waxes, paints and lacquers, and petroleum-based cutting fluids
>
> *Class C*—fires in electrical equipment such as switch boxes, motors, and heating devices
>
> *Class D*—fires in metallic materials

FIGURE 2.8 Pushbutton switch.

FIGURE 2.9 Lever-type switch.

FIGURE 2.10 Solvents should always be kept in safety cans. (*Courtesy Direct Safety Co.*)

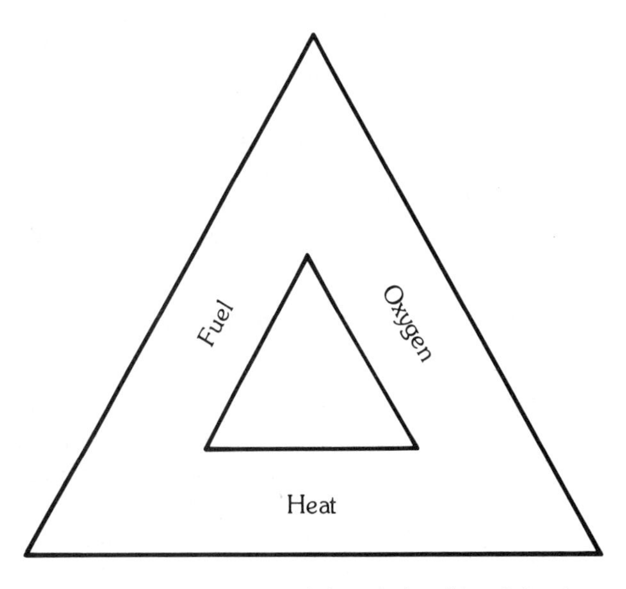

FIGURE 2.11 Understand the relationship of the three elements in any fire.

FIGURE 2.12 Dry powder or CO_2 extinguishers are recommended for controlling fires in machine shops.

Because most fires in machine shops are in class B or C materials, carbon dioxide (CO_2) or dry powder-type extinguishers must be used (see Fig. 2.12). They are effective in controlling class B and C fires by excluding oxygen and are nonconducting, so they can be used on electrical equipment. These extinguishers are also effective on class A fires.

Remember that almost all fires are the result of poor work habits, unclean work areas, poorly maintained equipment, and improper storage of flammable materials (see Fig. 2.13).

2.3 SPECIFIC SAFETY PRACTICES

In the following chapters, safety practices related to specific machines and processes are presented in more detail. There are, however, a number of precautions that must be taken and work habits that must be developed which apply to machine shop work regardless of the processes or machines involved. Here is a brief overview of recommended practices.

FIGURE 2.13 Rags should be kept in safety cans. (*Courtesy Direct Safety Co.*)

2.3.1 First Aid

1. If you are injured, *notify the instructor immediately.* All injuries, even minor cuts and bruises, should receive proper attention.
2. Report to the instructor any irritation caused by dust, cutting fluids, or any other material.

2.3.2 Shop Clothing

1. Wear a *clean* apron, shop coat, or coveralls made of a heavy cotton fabric. Avoid polyester and similar synthetic materials because some of them melt when hit by hot chips (see Fig. 2.14).
2. Do not wear gloves when operating machines.
3. Belts and apron strings must be tied at the back so that they cannot be caught in rotating machinery.
4. Remove wristwatches, rings, and other jewelry when operating machines.
5. Wear shop coats with short sleeves, or with close-fitting long sleeves.

FIGURE 2.14 Wear shop coats with short sleeves or close-fitting long sleeves.

6. Long hair must be controlled so that it will not be caught in rotating machinery. Wear a hat or net.
7. Wear eye protection at *all* times.

2.3.3 Fire Prevention and Control

1. Keep work and storage areas clean.
2. Store combustible materials in appropriate containers.
3. Know the types of fire extinguishers and their locations.

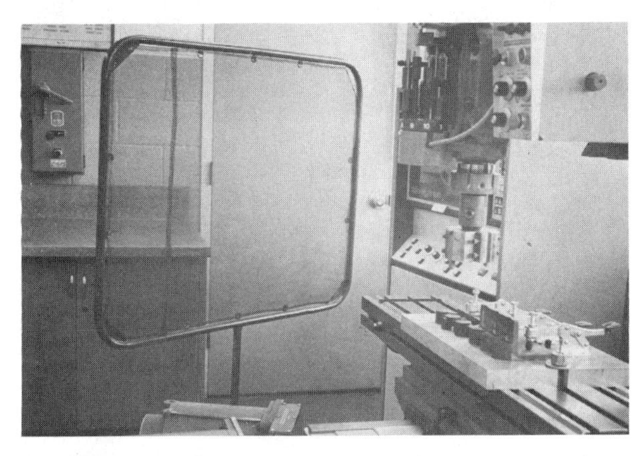

FIGURE 2.15 Chip screens are used to protect workers nearby.

4. Know the location of circuit breakers and switches.

2.3.4 General Housekeeping

1. Keep your work area clean. Cleanliness is a major factor in accident prevention.
2. Report to the instructor equipment that is out of working order.

2.3.5 Concern for Others

Be aware of hazards that you might be causing for others students or workers near you. Select the proper safety equipment to protect them (see Fig. 2.15).

Remember: Your success as a machinist depends on good eyesight, physical dexterity, and the capability for doing a variety of tasks requiring strength and agility. A thorough understanding of safety procedures and development of good work habits will help you achieve a long, injury-free, and satisfying career as a machinist.

Note: In some illustrations in this book machine guards have been removed for clarity. Always use proper machine guards. When in doubt, always consult your instructor.

REVIEW QUESTIONS

2.2 GENERAL SAFETY PRACTICES

1. List three characteristics of good safety glasses.
2. Name one major limitation of a face shield as an eye-protection device.
3. What are the two main hazards to guard against when solvents are used?
4. Briefly describe the best way to lift heavy objects.
5. Identify three common combustible materials in a class B fire.

6. What type(s) of fire extinguisher may be used on a class C fire?

2.3 SPECIFIC SAFETY PRACTICES

1. Why must all injuries be reported to the instructor?
2. What is the recommended fabric for shop aprons or shop coats? Why?
3. What are the recommended types of sleeves on shop coats? Why?
4. List in proper order the two actions you would take in controlling an electrically started fire in a machine shop.
5. Why is it important for portable power tools to be properly grounded?

3 Technical Drawings

3.1 INTRODUCTION

The ability to understand technical drawings and to use the information in machining operations is of utmost importance to the machinist. The drawings used by the machinist may range from simple hand-drawn sketches to complex drawings prepared by professional draftspeople and engineers. In any case, the machinist must be able to read the drawing and derive from it the size, shape, tolerances, material, and other factors about the object to be machined (Fig. 3.1).

FIGURE 3.1 The machinist must understand the technical language of industry.

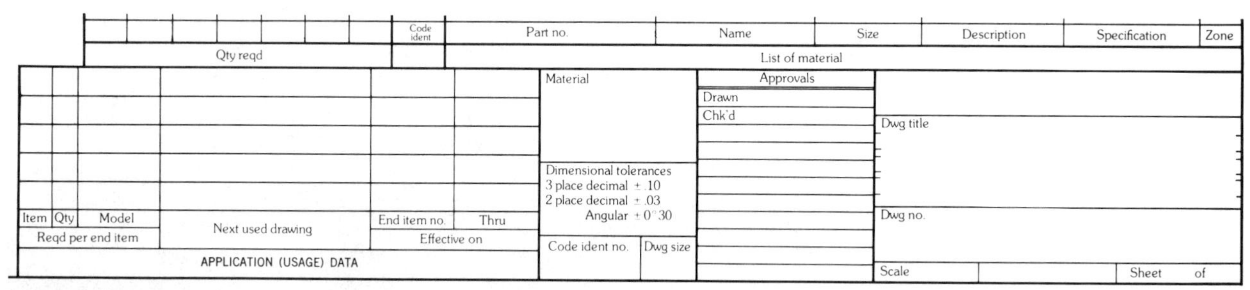

FIGURE 3.2 Title block on typical technical drawing.

Technical drawings are often referred to as the "language of industry." The presentation of information involves the use of words, lines, symbols, numbers, and pictures. In many cases, particular processes are also specified, along with specific materials and alloys to be used. Information on nonmachining processes such as heat treating, shot peening, painting, and plating may also be specified on the drawing (see Fig. 3.2).

From the information on the drawing, machinists often must make a number of decisions: the machine or machines they will use, the auxiliary equipment and measuring tools needed, and the time required to make the part. Because all these decisions affect the accuracy and cost of the part being made, it is evident that the skillful use and interpretation of drawings is critical to the good machinist.

In general, toolmakers, setup people, and prototype machinists must be able to make technical sketches as well as interpret and use drawings. Machinists will use drawings in a variety of working situations throughout their careers. Skill in the use of the language of industry is a major factor in securing better jobs and advancing into more skilled segments of the machinist's trade.

3.2 MACHINIST'S DRAWINGS

Regardless of whether the working drawing is a simple sketch or a complex mechanical drawing, it must provide the information necessary for making the part or object. The machinist must know:

1. The size of all the parts
2. Working tolerances and allowances to provide for interchangeable assembly, press fits, and working clearances
3. The shape of the parts
4. Geometric relationship of the parts
5. The material to be used

This listing is very basic, and working drawings usually contain additional information, depending on the nature of the object and its complexity.

3.2.1 Lines and Symbols

Several different types of lines are used on drawings; each line has a particular meaning and application. Properly used and interpreted, these lines help the machinist visualize the three-dimensional shape of the part.

1. *Visible* or *object* lines show the edges of the part that can be seen in any given view [see (*a*) in Fig. 3.3].
2. *Hidden* lines are necessary to show holes or other cavities in an object [see (*b*) in Fig. 3.3.] Note the relationship between the hidden lines (*b*) and the visible lines (*a*) in the different views of the object (Fig. 3.3).
3. *Section* lines show those areas of an object that were cut in making a *sectional* view (see (*c*) in Fig. 3.3.]. As shown in Fig. 3.14, section lines can be used to indicate the material or materials from which the part is made.
4. *Centerlines* show the axes of symmetrical objects, bolt circles, slots, or paths of motion of a moving part [see (*a*) in Fig. 3.4]. The centerline of a hole or boss may be used as a baseline for other dimensions. Two or more centerlines may also be used to show the dimensional and geometric relationship of holes in a part.
5. *Dimension* lines indicate distance and have an arrowhead at each end, with a break in the middle for the dimension itself (see (*b*) in Fig. 3.4]. Tolerances can be specified in the dimension by placing the largest allowable dimension above and the smallest allowable dimension below the line.
6. *Cutting plane* lines show where an object has been cut when it is necessary to use a section view [see (*c*) in Fig. 3.4]. The arrows

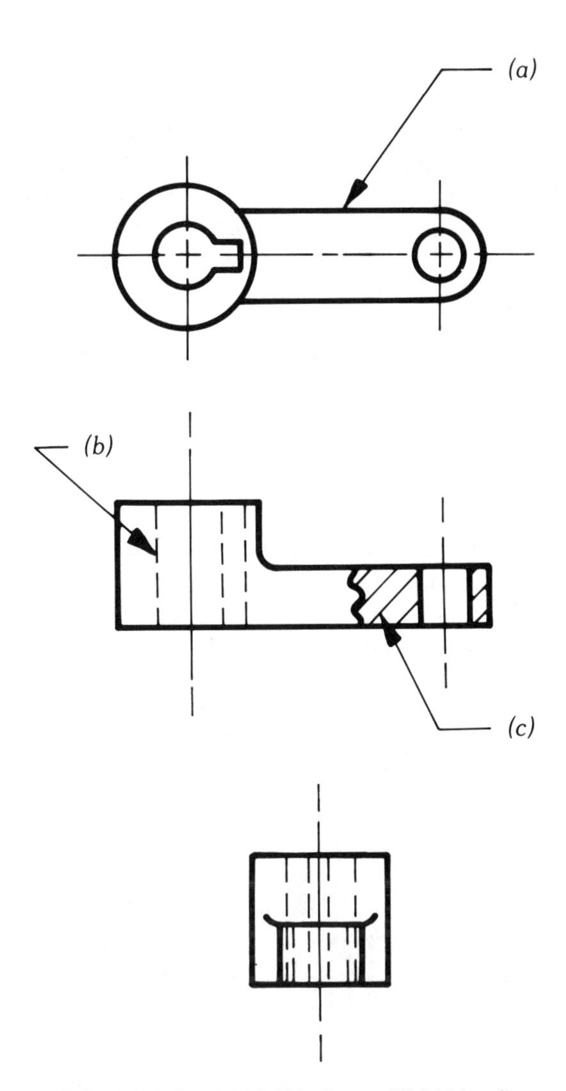

FIGURE 3.3 (*a*) Visible lines; (*b*) hidden lines; (*c*) section lines.

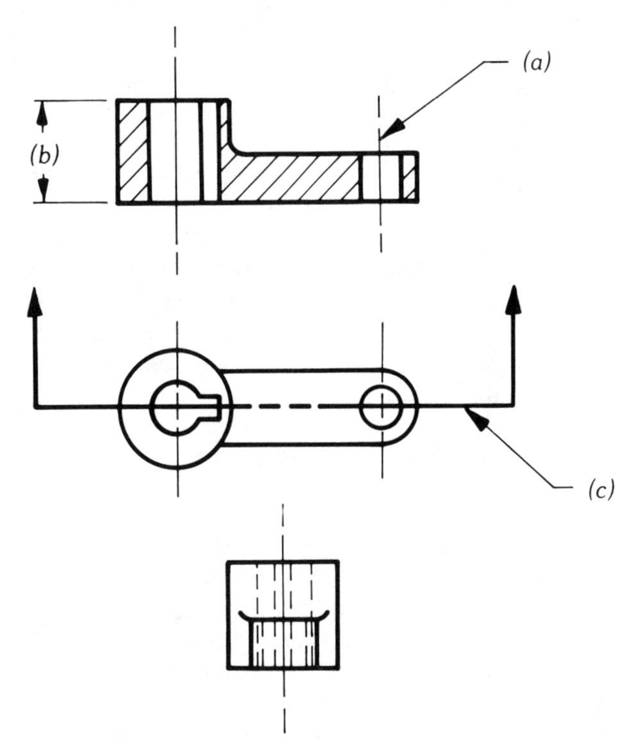

FIGURE 3.4 (*a*) Centerline; (*b*) dimension line; (*c*) cutting plane line.

FIGURE 3.5 (*a*) Short-break line; (*b*) phantom line; (*c*) finish mark.

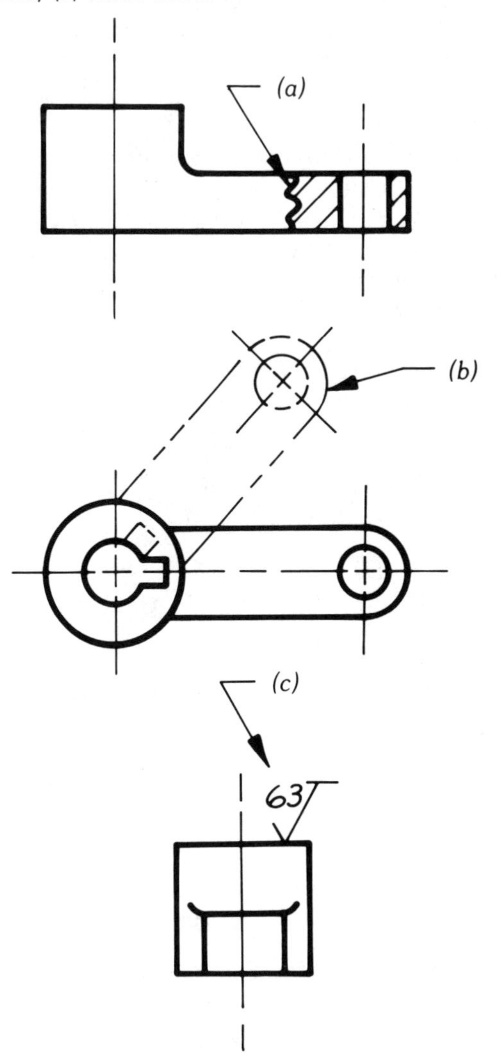

at each end of the cutting plane line indicate the direction in which the viewer is looking.

7. *Short-break* lines indicate a broken-out section of a drawing that is removed to show more clearly some internal feature of the part [see (*a*) in Fig. 3.5]. The broken-out section is indicated by an irregular line of the same thickness as a visible line which is drawn freehand.
8. *Phantom* lines show the range of motion or alternate positions of some parts [see (*b*) in Fig. 3.5]. In some cases, long threads, springs, or other similar items may have all but the ends shown by phantom lines as a time-saving procedure for the draftsperson.
9. *Finish marks* indicate that a surface is to be machined and, in some cases, denote the smoothness of the finish. The number used

with the check mark (see (*c*) in Fig. 3.5] indicates the roughness in *microinches* (millionths of an inch). Finish marks are used on the drawings for parts that are machined from castings or forgings because such parts may not be machined on all surfaces. On parts made from bar stock, the finish may be indicated in the notes on the drawing or specified in a particular machining operation, such as drilling, reaming, or grinding.

3.2.2 Views and Projections

The simplest drawings that machinists use generally have one or two views of the object to be machined. This is usually the minimum necessary to show the size, shape, and geometric arrangement of the part. On more complex parts, three or more views are required, in addition to cross-sectional views of various types. Therefore, it is important that the machinist learns to visualize the objects in three-dimensional form from the information presented in drawings.

It is important to know that there are six basic positions from which an object can be viewed (see Fig. 3.6). The usual directions from which the object is viewed are either perpendicular to or opposite from each other. These views, known as *regular* views, are placed as shown in Fig. 3.6.

Any views that require a line of sight not in alignment with one of the axes of the regular views are known as *auxiliary* views. Auxiliary views are necessary to show the true shape of certain surfaces on objects where the planes of some surfaces are at an angle other than a right angle to the major axes.

Two-view drawings are sufficient when all the necessary information can be presented, as shown in Fig. 3.7. In some cases, sectional views are used as part of the two-view type of presentation. Front and top views of the object are most commonly used, but other arrangements of views may be necessary, depending on the shape of the object.

Three-view drawings are generally necessary when the object has angular, recessed, or irregular surfaces. For objects that can be accurately shown with three views, all lines must be shown in their true length in at least one view, as shown in Fig. 3.8.

Depending on the shape of the object shown in the drawing, the views may be rearranged as in Fig. 3.9. To properly visualize the object in three-dimensional form, it is necessary to look at the object as if it were in a six-sided transparent box (see Fig. 3.10). The relationship of the views to each other can then be more readily understood, regardless of their position on the drawing or sketch.

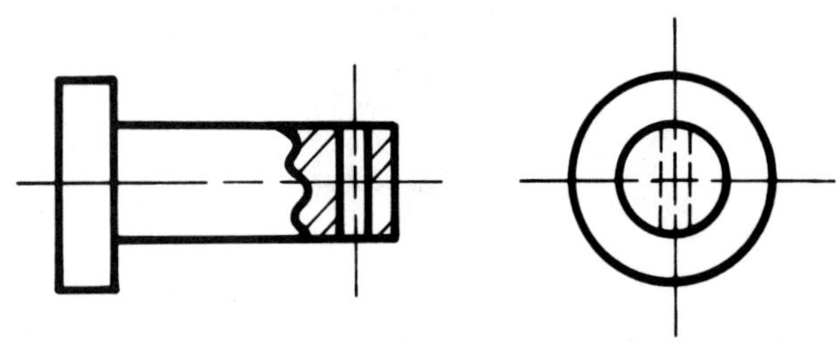

FIGURE 3.7 Only two views are required for this object.

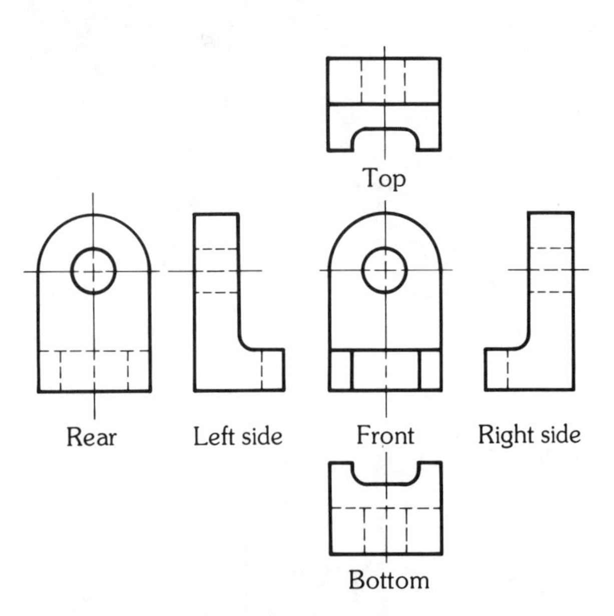

FIGURE 3.6 The six basic projections.

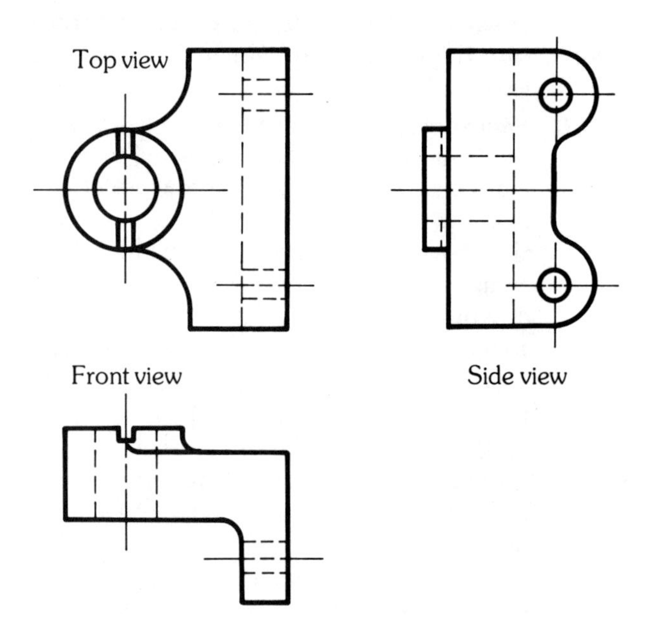

FIGURE 3.8 Typical three-view drawings.

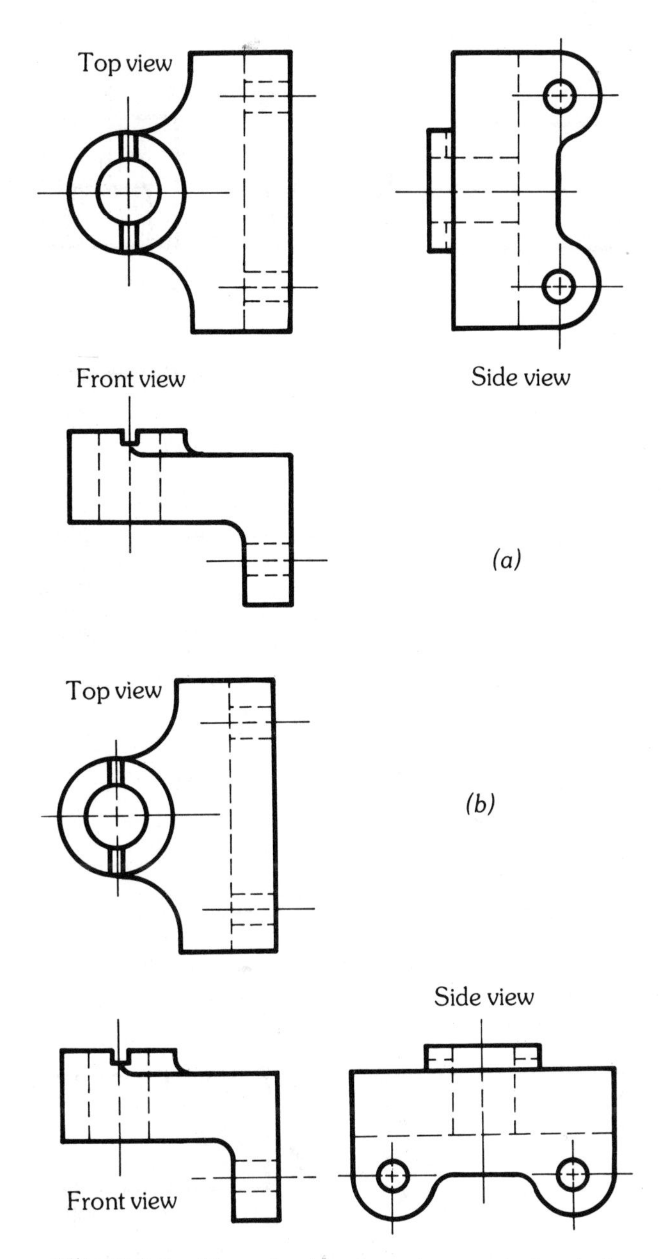

FIGURE 3.9 Alternate side-view placements for clarity.

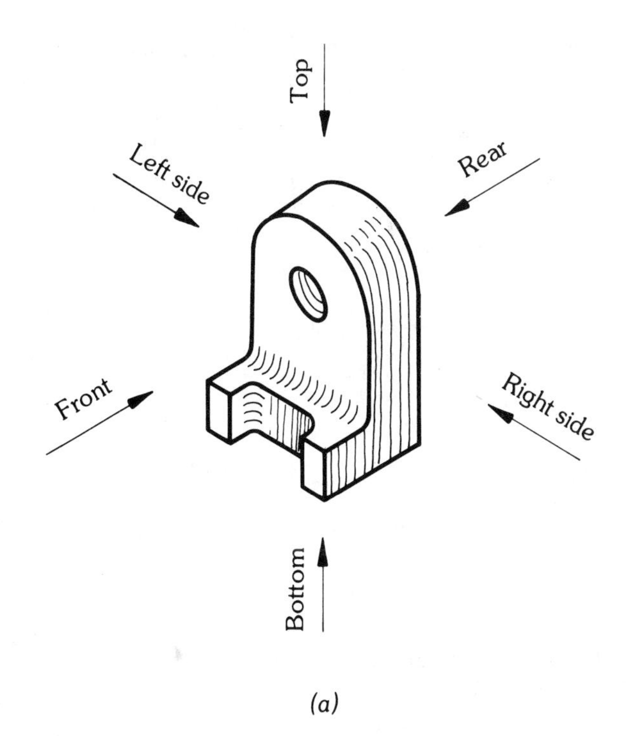

FIGURE 3.10 Visualizing an object from all sides.

Auxiliary views are necessary when parts of an object are not parallel to the regular (horizontal, frontal, and profile) planes of projection. If circular and irregularly curved surfaces are to be shown in their true size and shape, the line of sight must be perpendicular to the plane of those surfaces (see Fig. 3.11). Depending on the complexity of the object, it may be adequately described by one or more regular views and the auxiliary view(s).

Depending on the location of the plane on which it is projected, an auxiliary view is of the *primary* or *secondary* type. A primary auxiliary view shows the actual shape and size of a surface on a plane that is perpendicular to one of the principal planes and at an angle to the other two. A secondary auxiliary view shows the actual shape and size of a surface that is at other than a right angle to all three principal planes. The machinist

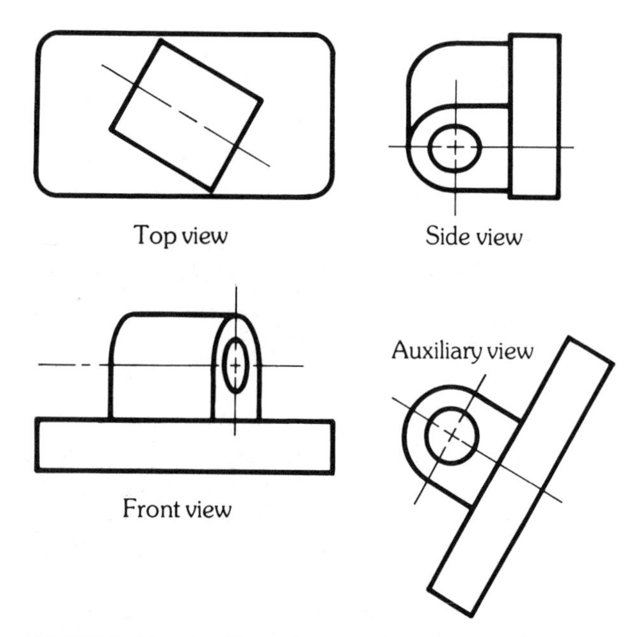

FIGURE 3.11 Auxiliary view used to show angled or irregular surface.

must have a thorough understanding of auxiliary views because they are extensively used in drawings of all types of machinery parts.

Sectional views are often used to show the interior size and shape of parts. Sectional views, either partial or full, are generally clearer and more satisfactory ways to show internal details than hidden lines. To interpret a sectional view properly, the position and direction of the cutting plane must be known (see Fig. 3.12). It usually goes through the centerline and is seen as a line, called the *cutting plane view*, on the main view.

Two arrows placed at the ends of the cutting plane line indicate the direction of sight (see Fig. 3.13). The reader sees the surface that was cut by the cutting plane and other features of the object not touched by the cutting plane. The surface cut by the cutting plane is indicated by *crosshatch lines*.

Crosshatch lines were often used to indicate the material from which the part was made, and this procedure is still used in assembly drawings that require cross-sectional views (see Fig. 3.14). For single parts, the standard crosshatch pattern for cast iron is used for all material, and detailed material specifications are included in the notes.

Exploded views show the relationship of parts in an assembly. They are used in catalogs, manuals, and shop drawings as a guide for the assembly and disassembly of segments of machines. Drawings of this type are seldom dimensioned and are not used by the machinist in actually making parts. The parts shown are usually numbered, and

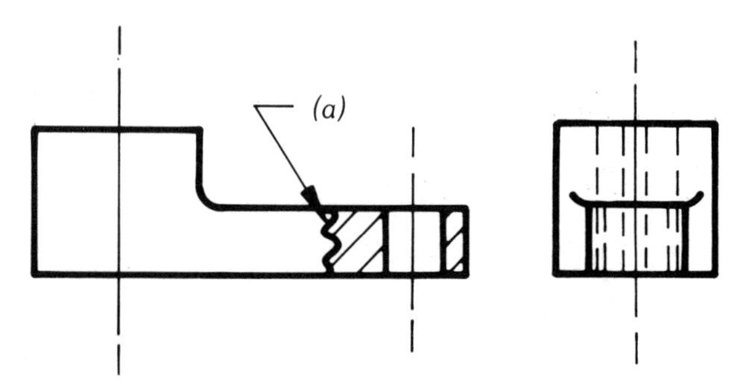

FIGURE 3.12 (*a*) Full sectional view; (*b*) partial sectional view.

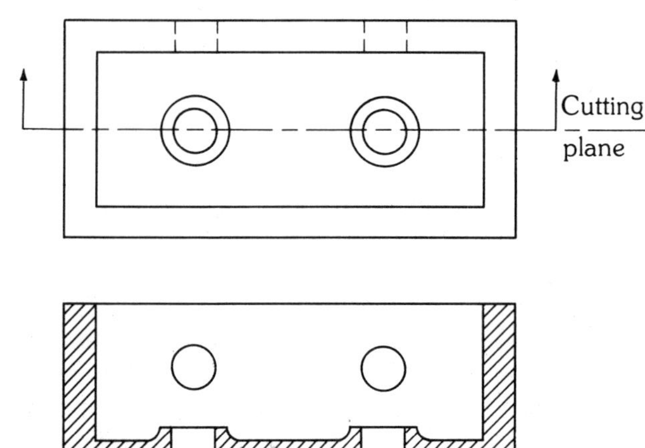

FIGURE 3.13 Position of cutting plane helps viewer visualize the internal shape of the object.

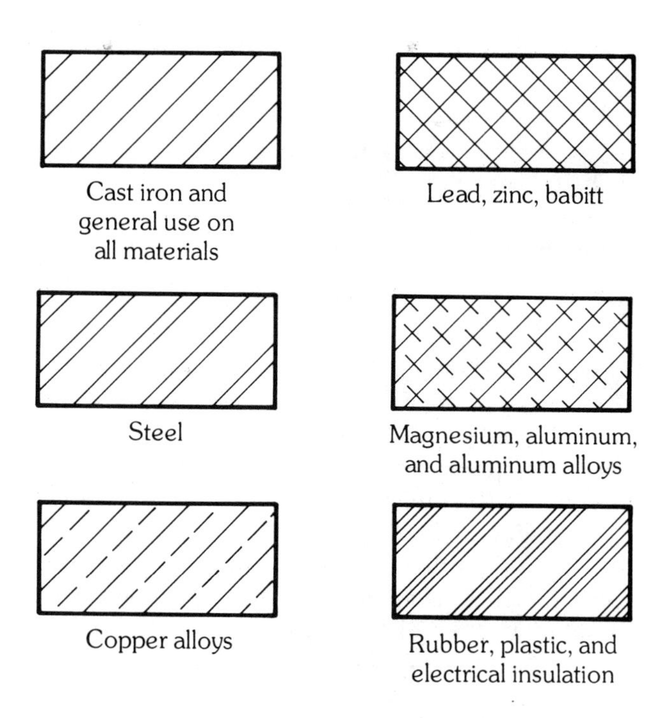

FIGURE 3.14 Type of material indicated by the section line used.

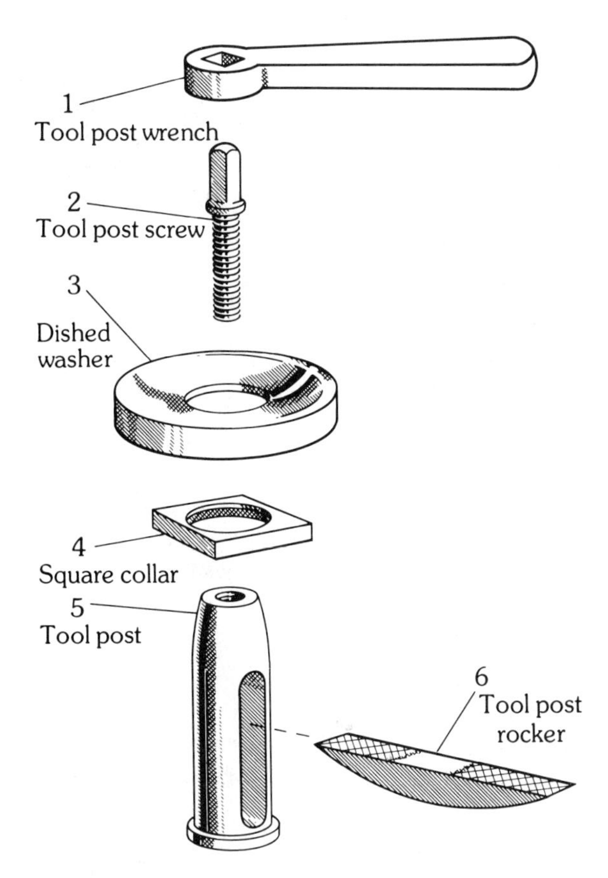

FIGURE 3.15 Exploded view showing the parts of a toolpost.

a parts list and assembly and disassembly procedures may be included (see Fig. 3.15).

More complex machines, such as the belt sander in Fig. 3.16, are often shown by means of exploded views to aid in assembly and in visualizing the relationship of the individual parts. In some cases, sectional views of the more complicated assemblies are also used to help the machinist visualize the assembly. Exploded views are of great value to machinists and millwrights who must disassemble and assemble complicated machines.

3.2.3 Tolerances, Allowances, and Specifications

Dimensions are an absolute necessity on drawings or sketches used by machinists. Because of the relatively close tolerances used in machine work, size of parts, especially those that fit together, must be carefully specified so machinists know exactly what is expected. If machinists understand the

FIGURE 3.16 Exploded view of belt sander.

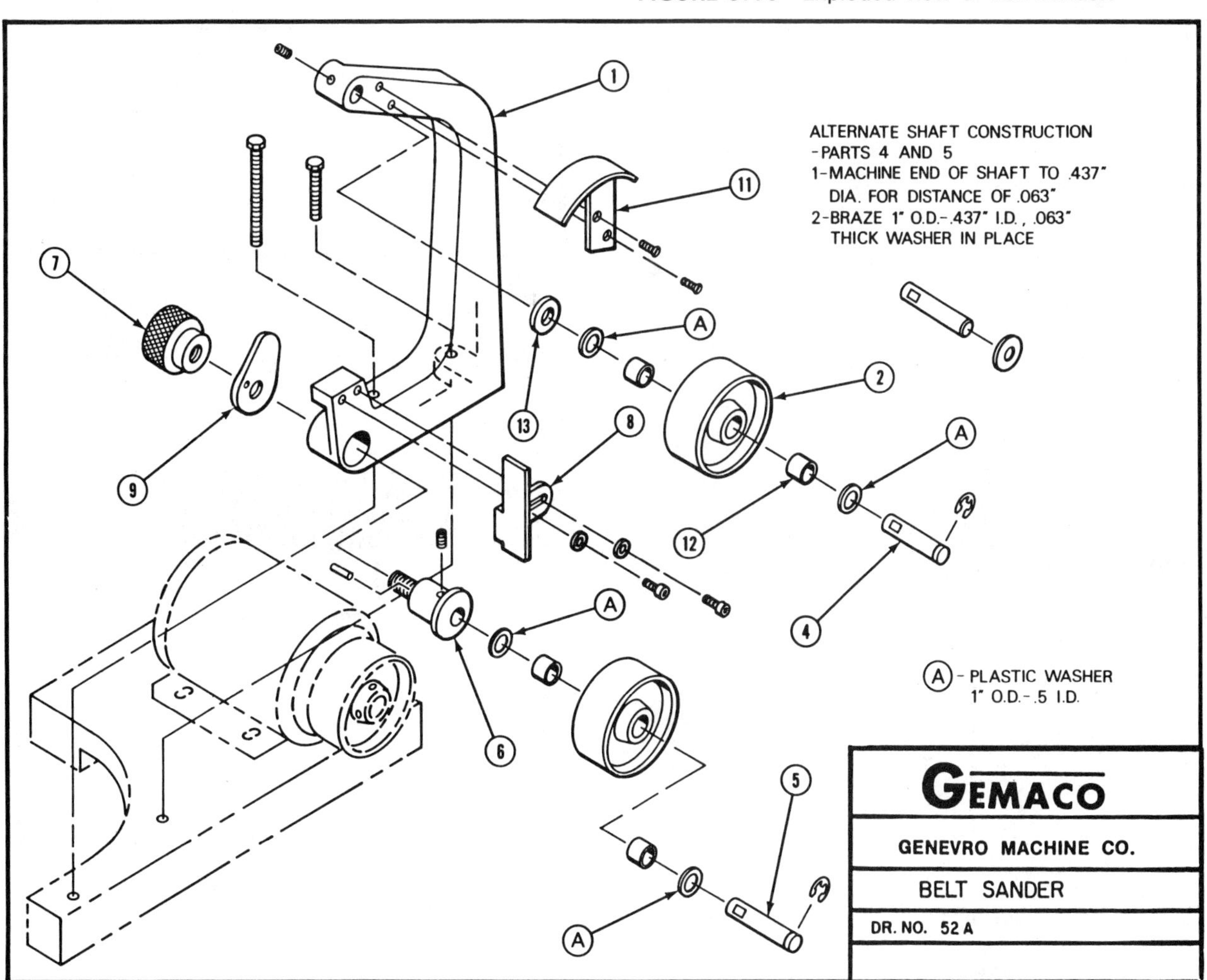

reasons for tolerances and allowances, they can more effectively choose the correct processes and machines for efficiently making the object shown on the drawing. Going beyond the specified tolerances usually results in ruined parts, and holding unnecessarily close tolerances raises costs.

One fundamental concept in mass production is the *interchangeability* of parts. Machinists are responsible for following the specified dimensions and tolerances so that parts manufactured in different shops will fit when brought together for the assembly of new machines or for repair work.

Tolerances should be specified on any dimension appearing on a drawing, either directly on the dimension or in a note. When *limit* dimensioning is used (see Fig. 3.17), the maximum and minimum limits are specified. The usual procedure on a drawing is to place the largest allowable size above the smallest allowable size (for example, 1.502/1.498). When the tolerance is specified in a note, the smaller size is given first (for example, 1.498–1.502).

Tolerances also may be specified by giving a basic size and the amount that the machinist may go above or below that size (for example, 1.500 $\pm$ 0.005) (see Fig. 3.18). In some cases, as shown in Fig. 3.19, the allowed deviation is in only one direction (*unilateral tolerance*) or in both directions (*bilateral tolerance*).

Accumulation of tolerances is a problem that may be largely solved by *baseline* dimensioning (see Fig. 3.20). In this case, one selected surface is used as the base for all measurements. This avoids

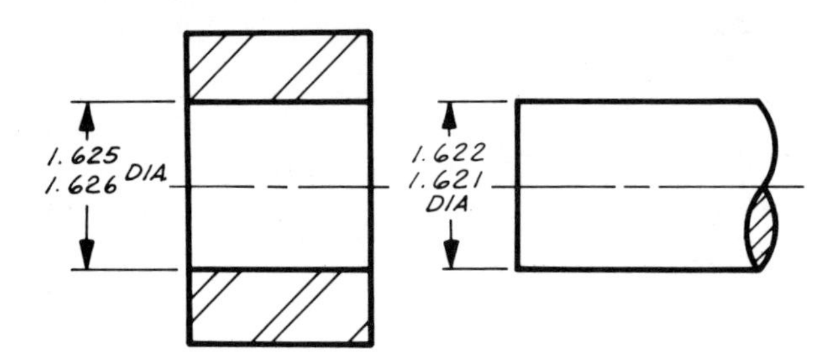

FIGURE 3.17 Example of limit dimensioning.

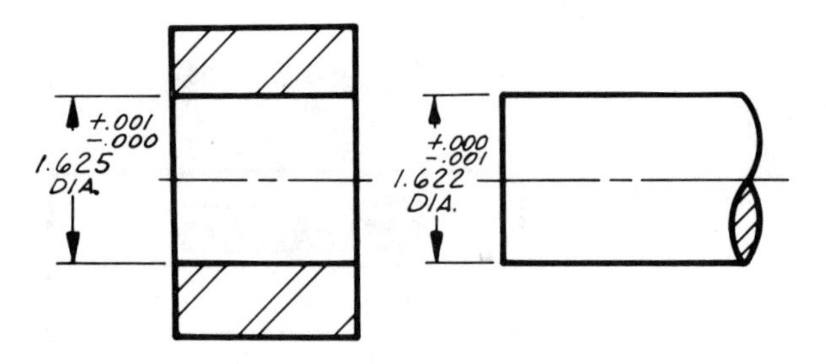

FIGURE 3.18 Example of plus or minus dimensioning.

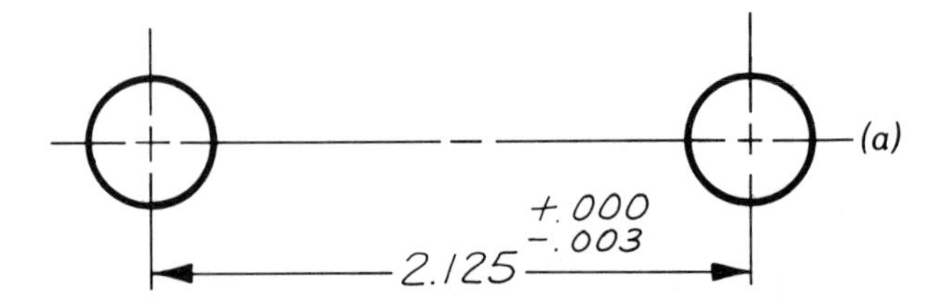

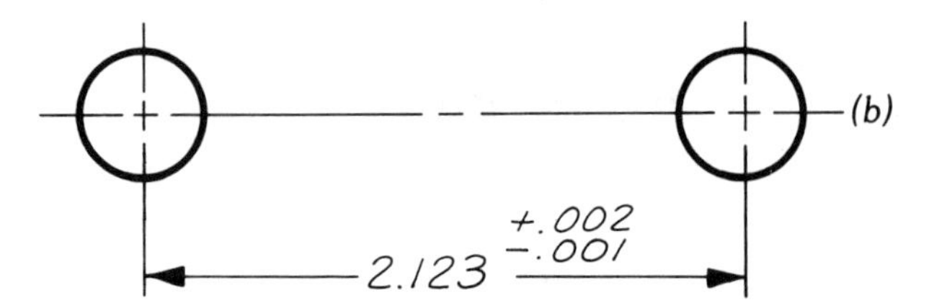

FIGURE 3.19 Tolerances may be (*a*) unilateral or (*b*) bilateral.

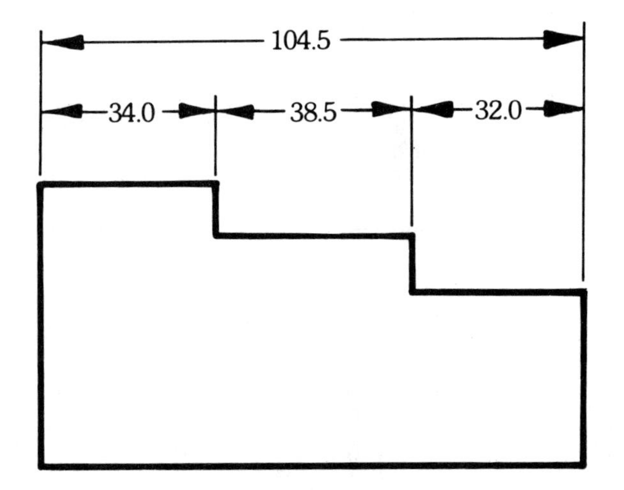

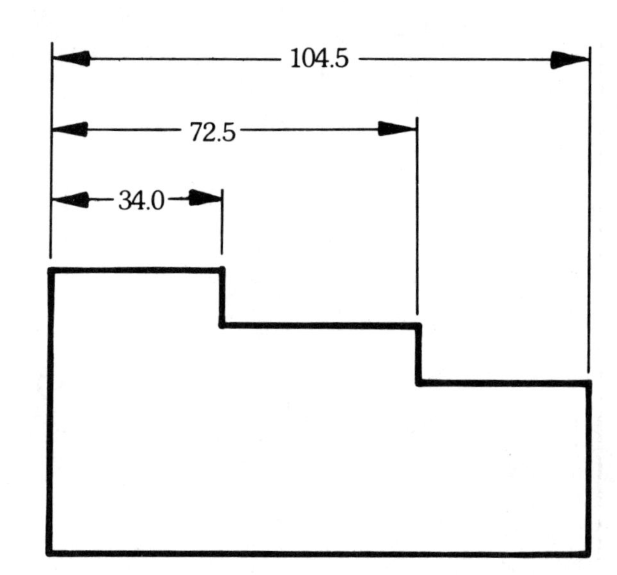

FIGURE 3.20 Cumulative tolerance problems avoided with baseline dimensioning.

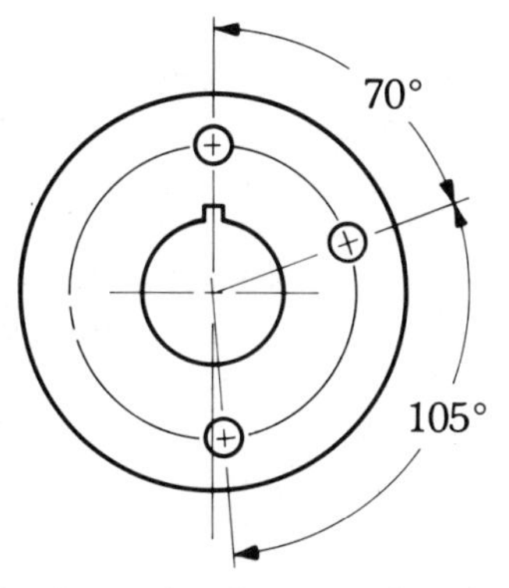

FIGURE 3.21 Example of geometric tolerances.

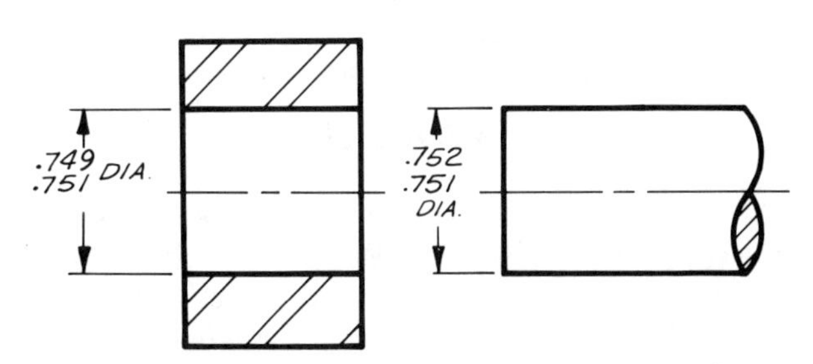

FIGURE 3.22 Negative allowances used for press fit of parts.

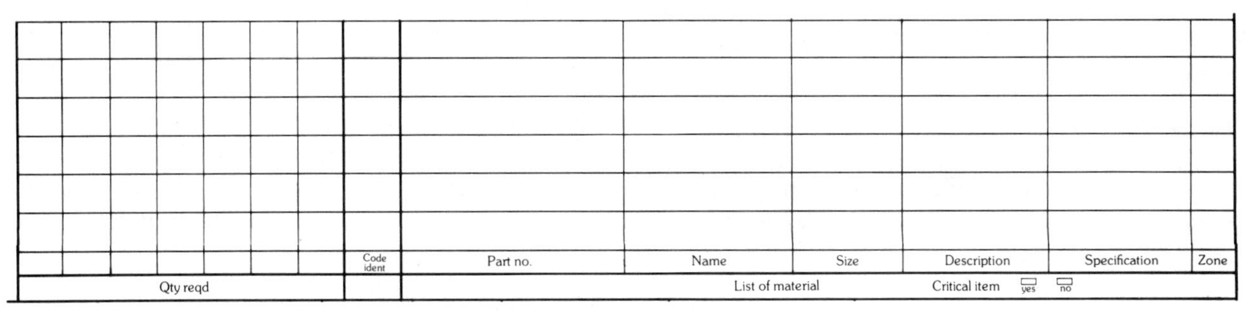

FIGURE 3.23 Bill of materials on a drawing.

buildup of errors either above or below the specified tolerance in relation to the overall dimensions.

Geometric and *angular tolerances* are specified on drawings to let machinists know how far they can deviate from a geometrically perfect condition. The tolerances may be stated as part of the dimension or in separate notes. Geometric tolerances (see Fig. 3.21) are generally used to indicate the accuracy of a part in terms of flatness, angularity, concentricity, and roundness. The angle of a taper may be stated as $25° \pm 0.5°$, for example.

Interference fits are widely used in machinery for the installation of bushings, bearings, and other parts. The allowance is *negative*, with the internal member larger than the hole in the external member, as shown in Fig. 3.22. The parts are assembled using a press or by heating the external member to cause it to expand before assembly. In some cases the internal member is also chilled to cause it to shrink.

A *bill of materials* is usually part of any drawing or technical sketch used by the machinist (see Fig. 3.23). Both metallic and nonmetallic materials may be specified, and standard hardware items such as bolts, nuts, and setscrews are listed. The description of materials must be complete and accurate. Metals, for example, are listed by alloy designation, shape, size, and amount needed for the job. Supplementary notes may be added to the bill of materials when heat treatment or other processes must be specified.

3.3 TECHNICAL SKETCHING

The ability to make and understand technical sketches is very important to the machinist, particularly in general job shop work and experimental machining. Basically, a technical sketch is a freehand drawing of an object, complete with all necessary dimensions, tolerances, notes, and auxiliary views, if necessary (Fig. 3.24). It is made without using drawing instruments, and it is generally done on graph paper, using only a pen-

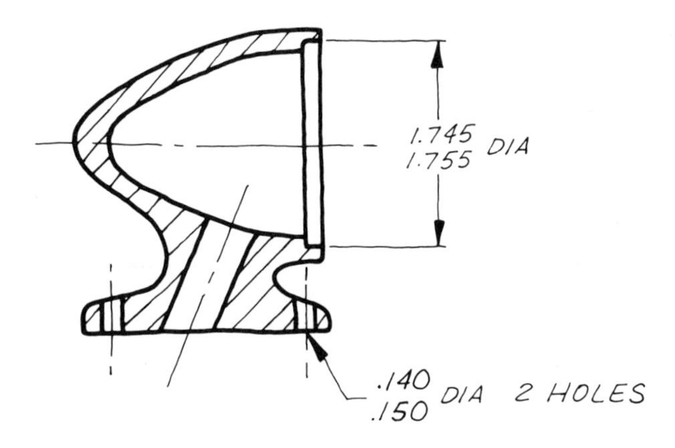

FIGURE 3.24 Typical technical sketch.

cil and an eraser. The technical sketch is usually the first step in presenting design ideas in graphic form. It is a form of the language of the industry that is used by executives, engineers, technicians, shop foremen, and machinists.

3.3.1 Lines and Symbols

Conventional lines and symbols are used in making technical sketches (see Fig. 3.25). Although the drawing is done freehand and the lines will naturally be rough in comparison with a mechanical drawing, the same rules for using different lines apply. For example, *visible* lines are drawn heavily compared to most other lines used, just as in a mechanical drawing. Conventional symbols, such as finish marks, are also used in technical sketches.

3.3.2 Views

Technical sketches of complex objects may require several views, including section and auxiliary views. Being able to visualize the object as if it were suspended in a six-sided glass box is important in making and understanding sketches.

The *three regular views*, as shown in Fig. 3.26, are most often used in technical sketching. In some cases, two views are sufficient, but when complex parts are sketched, auxiliary and section views may also be needed. Section views, either full or partial, are very helpful in showing details of the interior of the object (see Fig. 3.27).

Oblique, isometric, or *perspective* views may also be used in technical sketching to show an object pictorially. *Pictorial* drawings show several faces of an object at the same time and make the object easier to visualize in three-dimensional form (see Fig. 3.28). However, the actual shape cannot be accurately shown as in multiview drawings.

In some cases, a pictorial sketch is made in addition to the multiview drawing to aid in visualizing the object. Pictorial technical sketches may be dimensioned by using conventional dimensioning practice and are made to any desired scale.

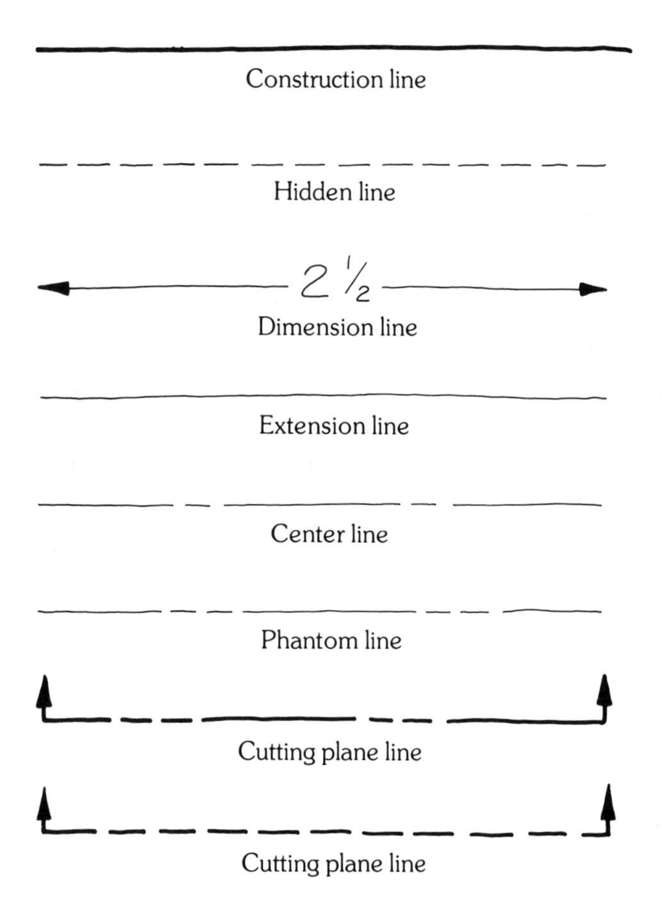

FIGURE 3.25 Lines and symbols used in a technical sketch.

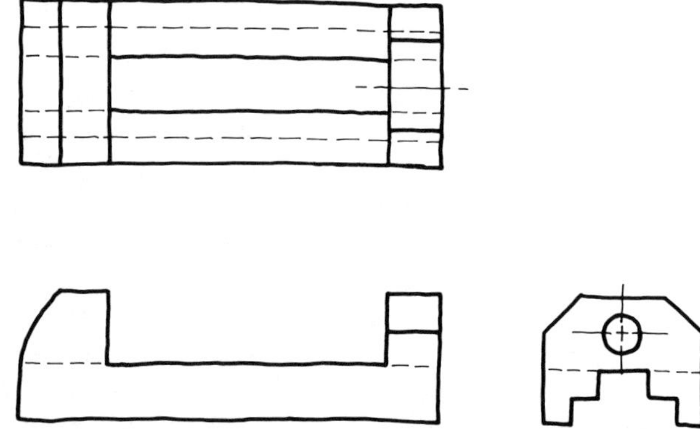

FIGURE 3.26 Three-view technical sketch.

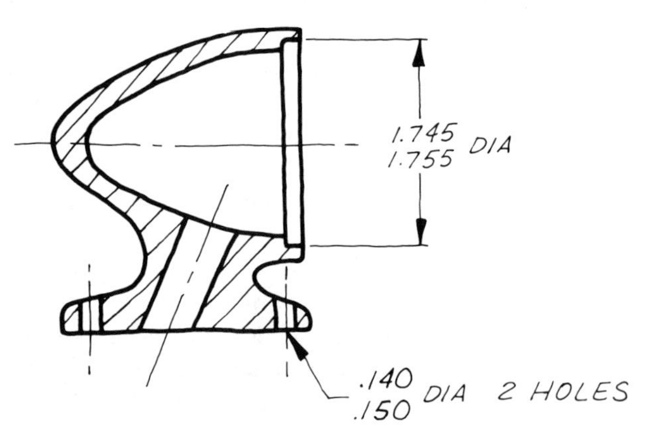

FIGURE 3.27 Section view in a technical sketch.

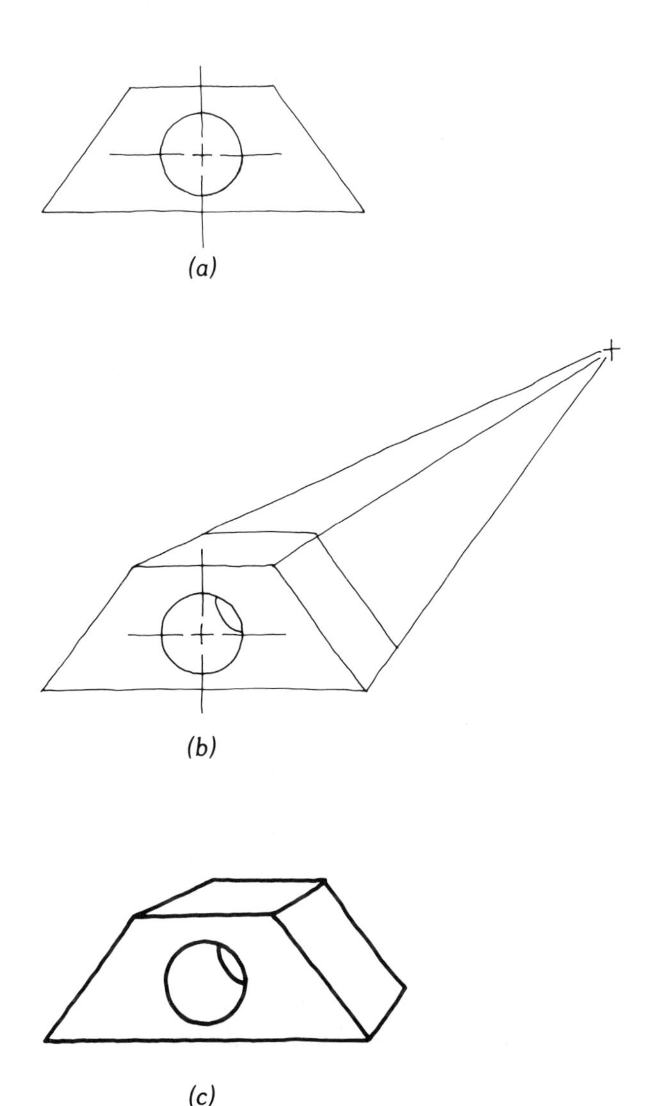
(a)
(b)
(c)

FIGURE 3.28 Isometric technical sketch.

3.3.3 Making Technical Sketches

The basic materials needed for technical sketching are a pencil, eraser, and paper. Graph paper, preferably with four squares to the inch or 5-millimeter (mm) squares, is very helpful in making sketches that are properly proportioned. Isometric graph paper, if it is available, should be used for making isometric sketches.

Almost all technical sketches are composed of arcs, circles, and lines in proper combinations. If the basic techniques shown in Fig. 3.29 and elsewhere in this section are learned well, you will find it easier to make more complex technical sketches. This is a valuable skill for the machinist who wishes to advance into more complex and interesting jobs in the machine shop.

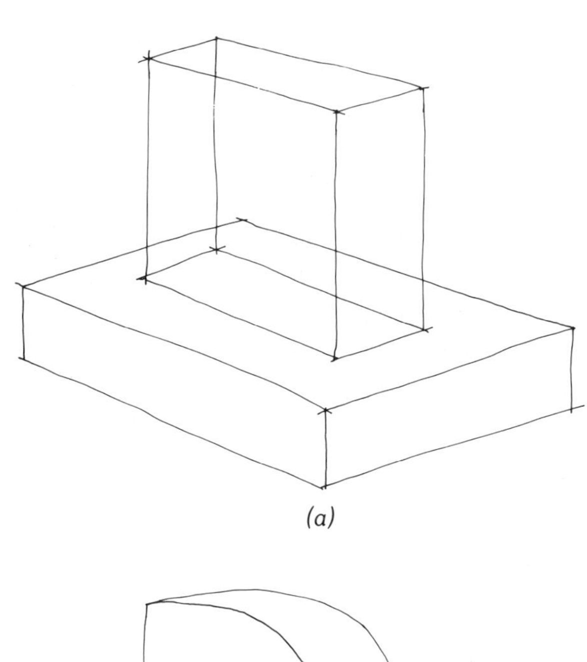
(a)

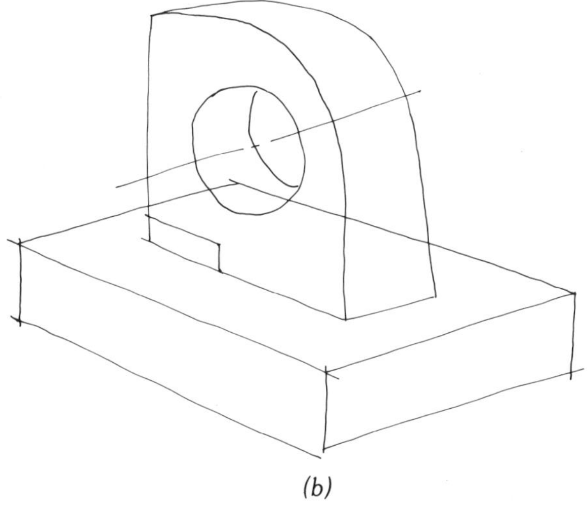
(b)

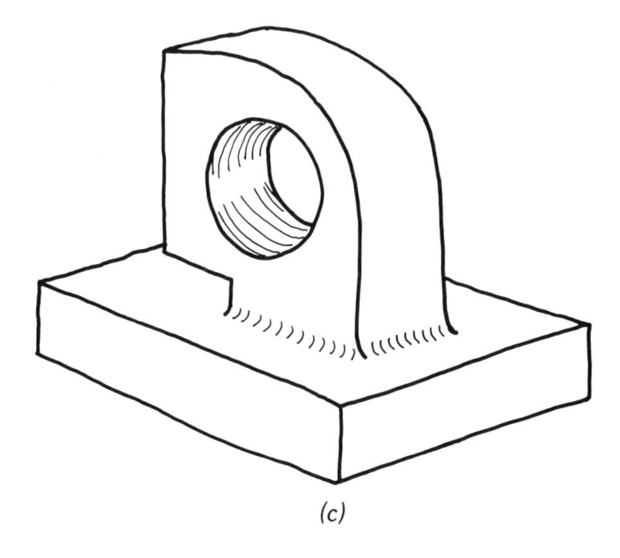
(c)

FIGURE 3.29 Stages in making an isometric technical sketch.

REVIEW QUESTIONS

3.2 MACHINIST'S DRAWINGS

1. What types of lines can be used to show holes or cavities in an object?
2. What two types of section views are used in machine drawings?
3. Explain how the finish on a machined part is specified.
4. When is an auxiliary view necessary on a drawing?
5. What is the main purpose of an exploded view of an object?
6. Briefly explain the process of limit dimensioning.
7. What is meant by *accumulation of tolerances*, and what effect could it have on a machined part?
8. Briefly define baseline dimensioning, and explain its main advantage.
9. What is a negative allowance?

3.3 TECHNICAL SKETCHING

1. Briefly explain why the ability to do technical sketching is valuable to the machinist.
2. What are the basic materials and equipment needed for technical sketching?
3. Briefly list and explain the basic steps in making a simple three-view technical sketch.

STUDENT ACTIVITY

Quite often, the machinist must determine certain dimensions that are missing from a drawing or sketch of parts to be made or reworked. The missing dimension may be needed to establish the working relationship between two parts, such as a press fit for a bushing or the running clearance between a shaft and the bearing in which it rotates. Another common case in which the machinist must calculate or find a dimension is in the selection of a drill for a hole that is to be tapped.

By reference to the drawing (plan set A) of the drill press vise, find dimension A. Please keep in mind that manufacturing tolerances and allowances must be established.

1. What other part of the vise must be used as a basis for calculating dimension A?
2. How will dimension A be expressed?
3. What are the consequences if dimension A is (*a*) too large, or (*b*) too small?
4. Find dimension B either by calculation (refer to Chap. 6) or by reference to the appropriate table at the end of the text.
5. Determine the proper interference fit for the proper assembly of the end caps (part No. 6) onto the sliding handle (part No. 5) and specify the proper drill.

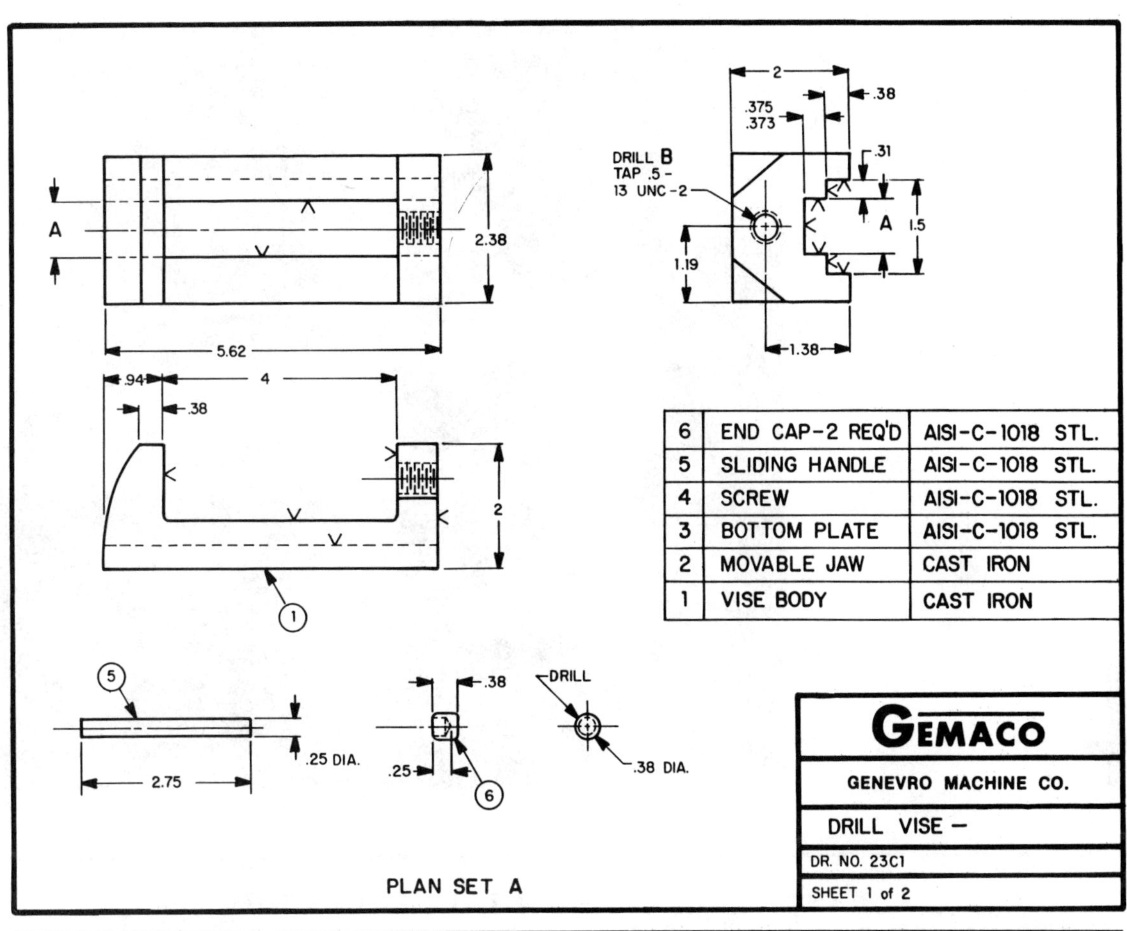

6	END CAP-2 REQ'D	AISI-C-1018 STL.
5	SLIDING HANDLE	AISI-C-1018 STL.
4	SCREW	AISI-C-1018 STL.
3	BOTTOM PLATE	AISI-C-1018 STL.
2	MOVABLE JAW	CAST IRON
1	VISE BODY	CAST IRON

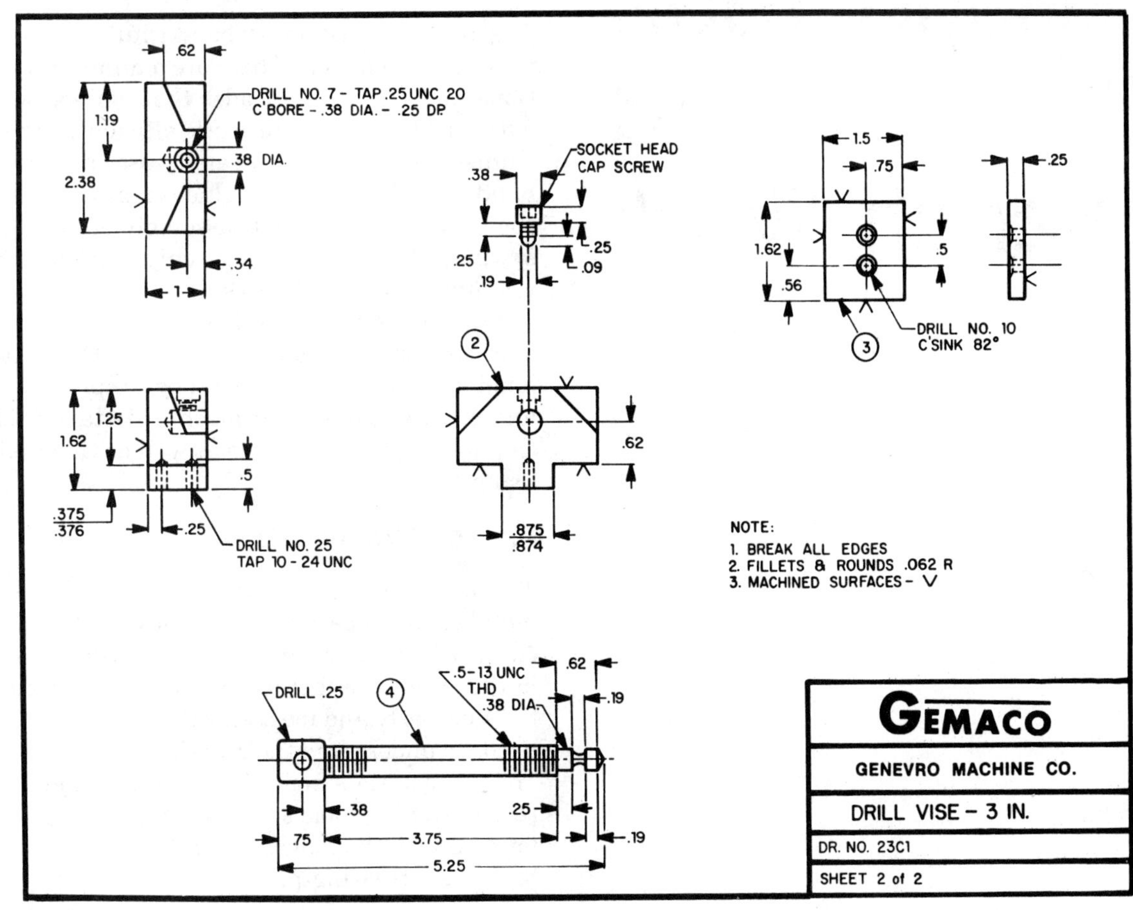

4

Hand Tools and Bench Work

4.1 INTRODUCTION

A good machinist must be skillful with a wide variety of hand tools. There are a number of operations in the machine trades that cannot be done with machines, and to work efficiently the machinist must be able to select, use, and care for hand tools. *Bench work* includes many operations, such as assembly and disassembly of machinery, sawing, threading, filing, chipping, polishing, reaming, and layout work.

In most commercial shops, machinists provide and care for their own hand tools. High-quality hand tools last a long time if they are properly used and maintained. Expert machinists take pride in the condition of their tools as well as their skill in using them.

4.2 CUTTING TOOLS

The machinist uses a variety of hand tools for metal-cutting operations. All these tools have one characteristic in common: their cutting edges are harder than the metal being cut. To use cutting tools properly and most effectively, the machinist needs to understand their working characteristics and the materials from which they are made. When reconditioning tools such as chisels and punches, for example, the machinist must also understand basic heat-treating practices.

4.2.1 Chisels

The *cold chisel* is used in bench work for operations such as chipping, cutting off rivet or bolt heads, and shearing small sheet metal parts. The four most widely used types of cold chisels are the *flat, cape, diamond-point,* and *round-nose chisels* (see Fig. 4.1). Chisels are classified by overall length, type of cutting edge, and width of the cutting edge.

Cold chisels are usually made of medium-carbon tool steel of octagonal cross section. After being forged to shape, the chisel is heat-treated. The cutting end is *hardened* by heating it to a red heat (1450 to 1500 °F) and quenching it in oil or water. It is then *tempered* by reheating to about 375 to 450 °F to reduce brittleness. The head of the chisel is left soft so it will not chip when struck with a hammer.

The cutting edge of a flat or cape chisel is ground to an included angle of 50 to 70 °, depending on the size of the chisel and the operation to be done (see Fig. 4.2). The smaller angle is used for small chisels and for chisels used to shear sheet metal or cut soft materials (see Fig. 4.3). A chisel with the cutting edge ground to a 70 ° included angle is more durable when it is used for jobs such as cutting off the heads of large rivets and bolts, chipping castings, and cutting harder metals. The cutting edge may also be curved slightly, so that it is convex (see Fig. 4.4).

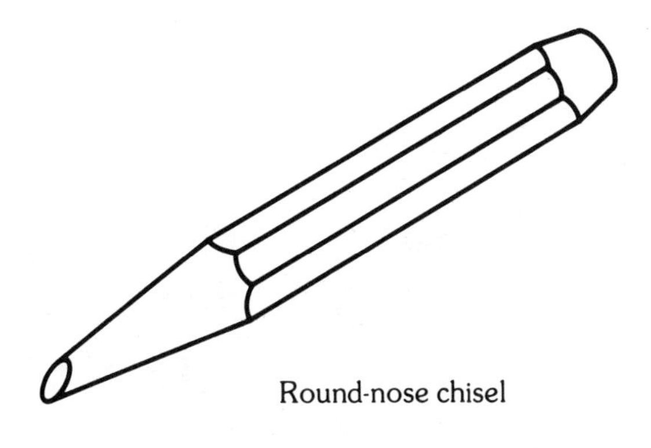

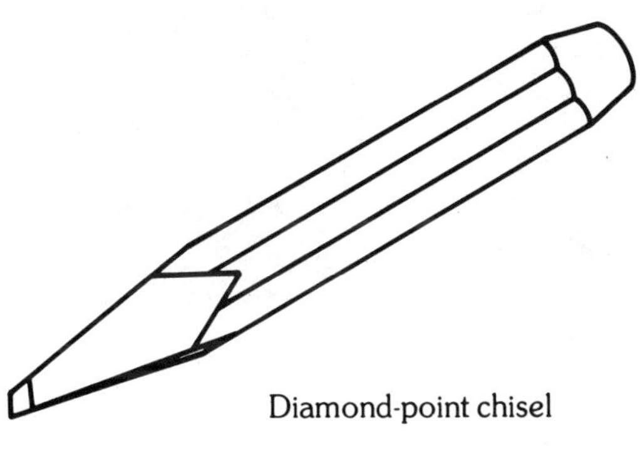

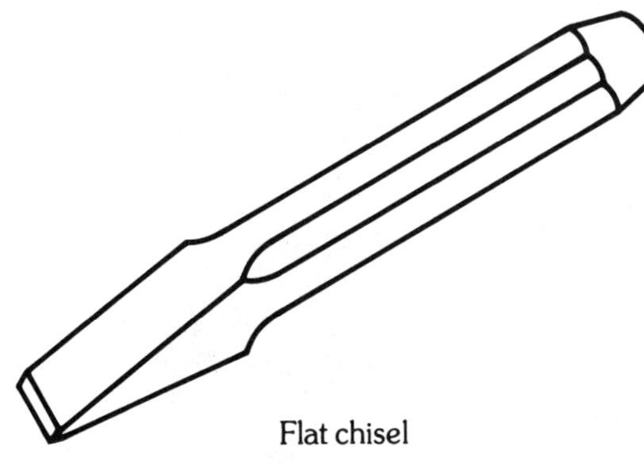

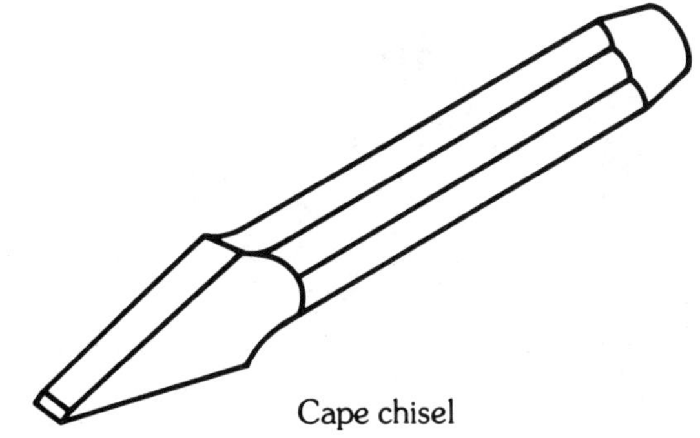

FIGURE 4.1 Chisels.

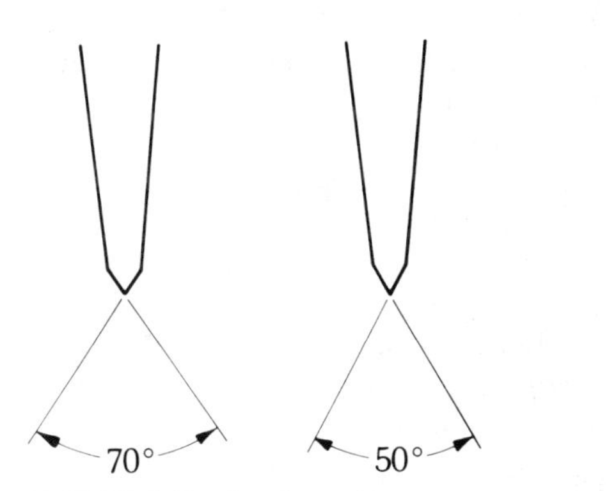

FIGURE 4.2 Cutting-edge angles.

FIGURE 4.3 Shearing sheet metal with a chisel.

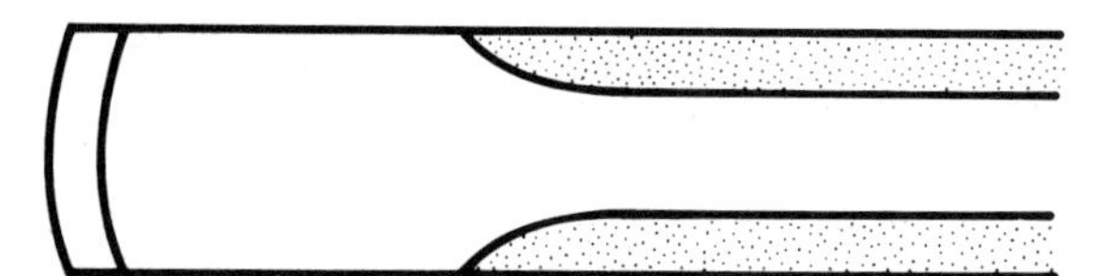

FIGURE 4.4 The cutting edge of a flat chisel should be slightly convex.

Cutting with a Chisel. Selecting the correct chisel and properly sharpening it are important factors in completing a job successfully. Other necessary items are a vise or other means of holding the object to be cut and a machinist's hammer (ball peen hammer) of the correct size.

The chisel should be held firmly enough to position it and struck squarely with only enough force to do the cutting operation (Fig. 4.5). When chipping castings or welds, sand and slag should be removed, if possible, from the areas to be chipped. Wear safety glasses and position yourself so that flying chips will not endanger others.

Sharpening Chisels. Chisels that are properly hardened and tempered must be sharpened with a grinding wheel (Fig. 4.6). The included angle of the cutting edge can be tested with a *center gage* if it is 60°, or a protractor can be used. A simple protractor can be used for checking cutting-edge angles (see Fig. 4.7).

When sharpening chisels, be careful not to overheat the cutting edge. It should be dipped in water frequently to prevent softening of the steel. If the cutting edge is overheated so that a blue color shows on the bright metal (about 375 to 425°F), the chisel may have to be heat-treated to restore its original hardness. The head of the chisel should be ground to a slight taper to eliminate chipping of the mushroomed head (see Fig. 4.8).

FIGURE 4.5 Grip the chisel securely with thumb and fingers.

Safety Tips

1. Wear safety glasses when using or grinding a chisel.
2. Grind head and cutting edge properly.

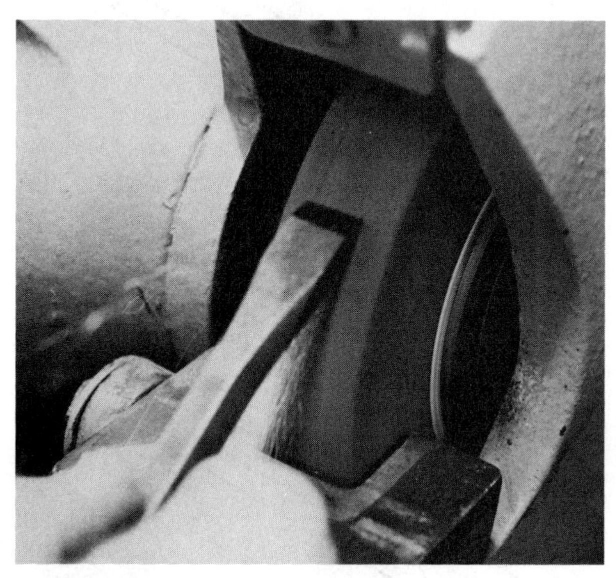

FIGURE 4.6 Sharpening a chisel.

FIGURE 4.7 Checking the cutting edge angle of a chisel.

FIGURE 4.8 Grinding the mushroomed head on a chisel.

4.2.2 Hand Hacksaws

The hand hacksaw is used for sawing many different metals in the machine shop. Although it appears to be an easy tool to use, the machinist must know how to select the proper blade and how fast to operate it when cutting different metals.

Basically, the hand hacksaw consists of a *frame, handle,* and *blade* (see Fig. 4.9). The frame may be adjustable or nonadjustable. The adjustable frames accommodate either a 10- or 12-in. blade. Saws with nonadjustable frames are somewhat more rigid but not as popular. A pistol-type grip is usually provided so the operator can grasp the saw firmly and comfortably. A wing nut is used to apply tension to the blade.

Hacksaw Blades. The blades used in hand hacksaws are classified by length, number of teeth per inch, type of set, and the material from which they are made. Blades may also have flexible or semiflexible backs with only the teeth hardened, or they may be all hard. Blades that are hardened throughout are more brittle but tend to cut straighter when properly used. Typically, a hacksaw blade is 0.5 in. (12.7 mm) wide and about 0.025 in. (0.63 mm) thick, but it will make a cut somewhat wider than the blade because of the set of the teeth.

FIGURE 4.9 Typical hand hacksaw. *(Courtesy Snap-On Tools)*

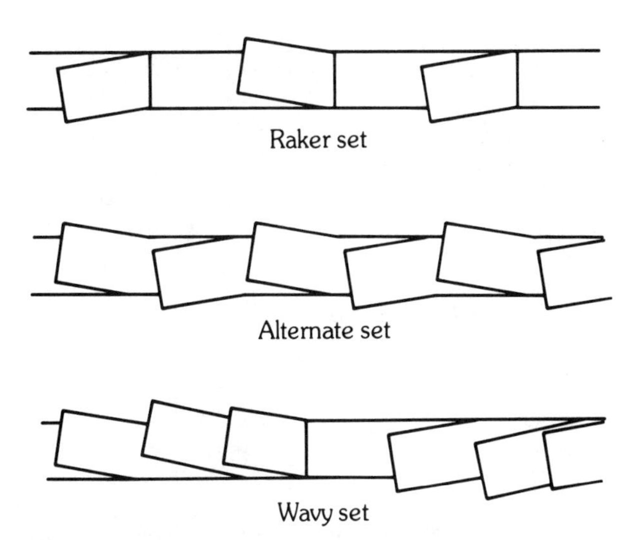

FIGURE 4.10 Wavy, alternate, and raker set hacksaw teeth.

Hand hacksaw blades are made from high-carbon steel or high-speed steel, with the high-speed steel blades costing considerably more. The high-speed steel blades can cut harder materials and last much longer in normal shop use. Blades are available in 10- and 12-in. lengths, with wavy or alternate set (see Fig. 4.10). Blades with 32 teeth per inch usually have *wavy set,* with the teeth offset alternately in groups of three. Blades of coarser pitch have *alternate set* with each tooth displaced either to the right or left. The set of the blade causes the kerf (width of cut) to be wider than the blade and prevents binding.

Selecting Hacksaw Blades. Both the cross-sectional shape of the object to be cut and the material from which it is made must be considered when selecting a hacksaw blade. Generally, a soft material that cuts easily requires a coarse pitch blade so the larger chips will not clog the teeth. When cutting thin tubing or sheet metal, however, use a fine pitch blade because at least two teeth must be in contact with the metal being cut. Otherwise, the blade chatters and dulls easily, causing a ragged cut, or breaks.

Using the Hand Hacksaw. See Figs. 4.11 to 4.13. After the proper blade for the job has been selected, it is installed in the hacksaw frame with the teeth pointing *away* from the handle because the hand hacksaw cuts on the forward stroke. The tension on the blade should be sufficient to keep it straight as cutting pressure is applied. Proper tension also makes the blade last longer.

Operate the hacksaw at about 40 to 60 strokes per minute, depending on the hardness of the material being cut. Remember to go more slowly on the

FIGURE 4.11 Cutting steel plate with a hacksaw.

FIGURE 4.12 Cutting with the blade at 90° to the frame.

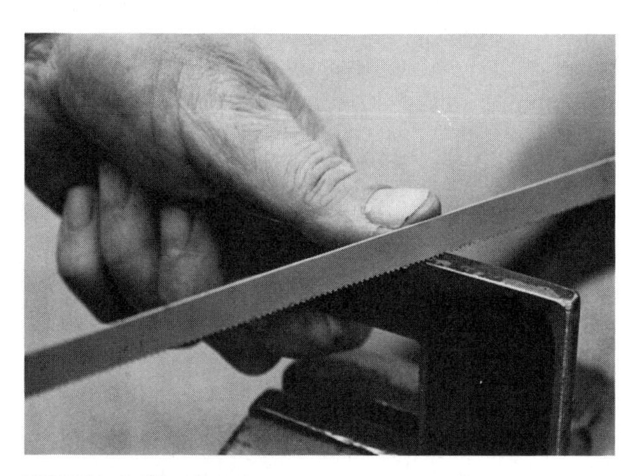

FIGURE 4.13 Starting a cut with the hacksaw.

forward (cutting) stroke. If the hacksaw is operated too rapidly, the set on the teeth wears off and the blade binds in the kerf.

If a blade wears out or breaks before a cut is completed, do not continue the same cut with a new blade if you can avoid doing so. The new blade will bind and wear rapidly if inserted into the original kerf because the outer edges of the teeth will be dulled quickly.

4.2.3 Files

Basically, a file (see Fig. 4.14) is a cutting tool made of hardened high-carbon tool steel. It is capable of cutting all the common metallic materials except hardened steels and is often used by machinists and other metalworkers. Files are generally used for deburring machined parts and for finishing and fitting operations. The skilled machinist must know how to select and use files effectively.

Types of Files. Files are classified by *cross-sectional shape, length, type of cut,* and *coarseness.* All these factors must be considered when selecting a file for a particular job.

1. The *standard cross-sectional shapes* for files are *flat, round, square, half-round,* and *triangular* (sometimes called *three-square*). The half-round file is actually a *segment* of a circle (see Fig. 4.15) in cross section.

   Files vary in outline shapes. Some files, such as hand files, are equal in both thickness and width from the *point* to the *heel.* Others may be tapered in width, thickness, or both. Some flat files have a *safe edge* with no teeth.
2. The *length* of a file is measured from the *heel* to the point. Files are available in lengths of 6, 8, 10, and 12 in. Special types of files, such as jeweler's files, are usually

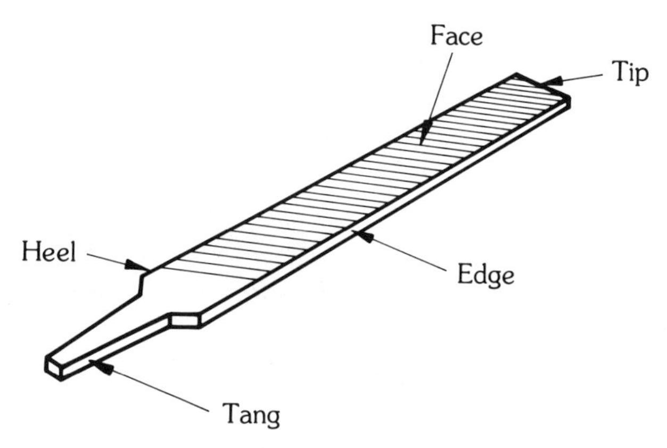

FIGURE 4.14 Main parts of a file.

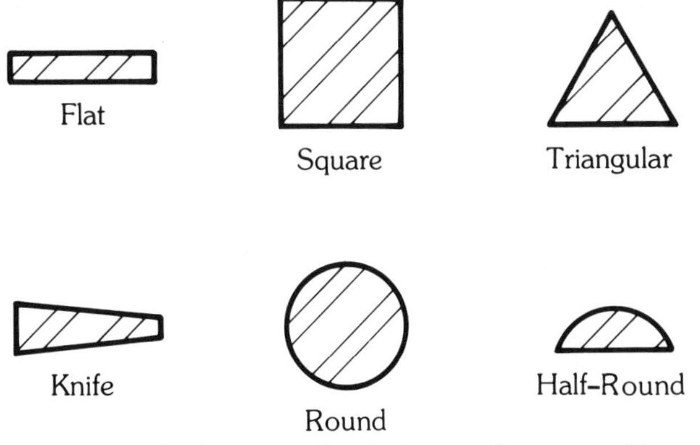

FIGURE 4.15 Cross-sectional shapes of common files.

classified only by cross-sectional shape, with no length specified.

3. The *cut* of a file describes the shape of the teeth. The three basic tooth types are *single-cut, double-cut,* and *rasp-cut.*

   Single-cut files have series of parallel individual teeth that extend at an angle across the face of the file. The angle of the cut to the longitudinal axis may vary between 45 and 85° [see *(a)* in Fig. 4.16]. Files with the 45° angle are generally used for filing work in the lathe. The shearing action of the teeth reduces the tendency for chips to stick to the file, thus producing a better finish.

   Double-cut files have two series of diagonal rows of teeth that cross each other on the face. The deep cut is at a 70 to 80° angle to the longitudinal axis of the file, and the shallower cut is at about a 110° angle to the deep cut [see *(b)* in Fig. 4.16]. The individual teeth are pointed, and the file cuts more rapidly but less smoothly than a single cut file of the same coarseness.

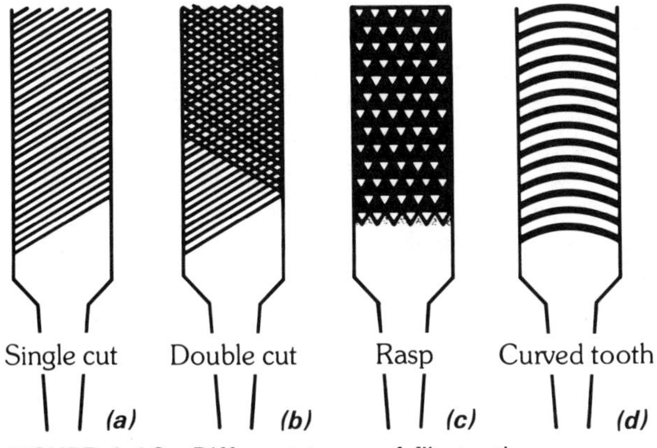

FIGURE 4.16 Different types of file teeth.

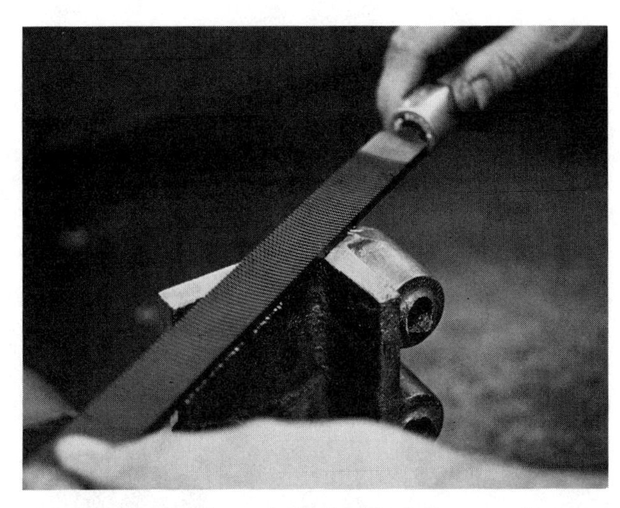
FIGURE 4.17 Curved tooth file being used on an aluminum casting.

   Rasp-cut files have individually formed teeth [see *(c)* in Fig. 4.16] and are seldom used in the machine shop except for rapid removal of soft materials such as aluminum.

   Curved-tooth files (see Fig. 4.17) are a unique type of single-cut file suitable for use on soft, nonferrous metals. The teeth are milled and have a large gullet (opening between the teeth) to accommodate heavy chips. They are ideal for filing aluminum, lead, and most diecasting alloys because of rapid metal removal and the self-cleaning action of the curved teeth.

4. The *coarseness* of a file is an indication of the spacing of the teeth. Both single- and double-cut files are available in *rough, coarse, bastard, second-cut, smooth,* and *dead smooth* grades. For general machine shop work, bastard, second-cut, and smooth files are most frequently used. Machinists doing precision fitting and assembly operations occasionally need dead smooth files.

Special Files. A number of types of files with special characteristics are available for jobs that cannot be done conveniently with standard files. For small, intricate work, the machinist may use *needle files* (Fig. 4.18) or *Swiss pattern files* (Fig. 4.19).

Needle files are generally available in 12 cross-sectional shapes and have round, knurled handles about 1/8 in. (3.17 mm) in diameter. They are about 6 in. (152.4 mm) in length and have very fine teeth.

Swiss pattern files are usually available in 12 shapes, as shown in Fig. 4.19, and with 1/8-in. (3.17-mm) or 1/4-in. (6.35-mm)-diameter round,

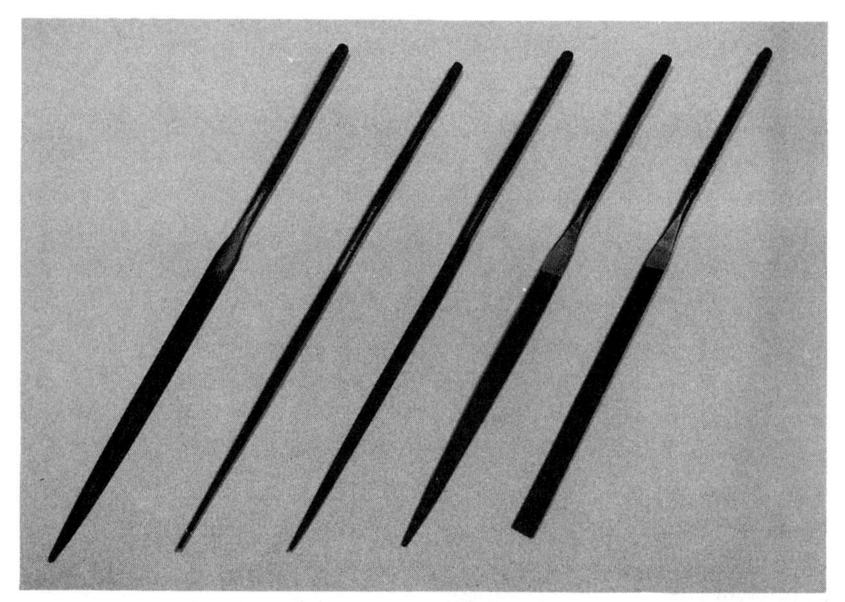

FIGURE 4.18 Needle files for small, delicate work.

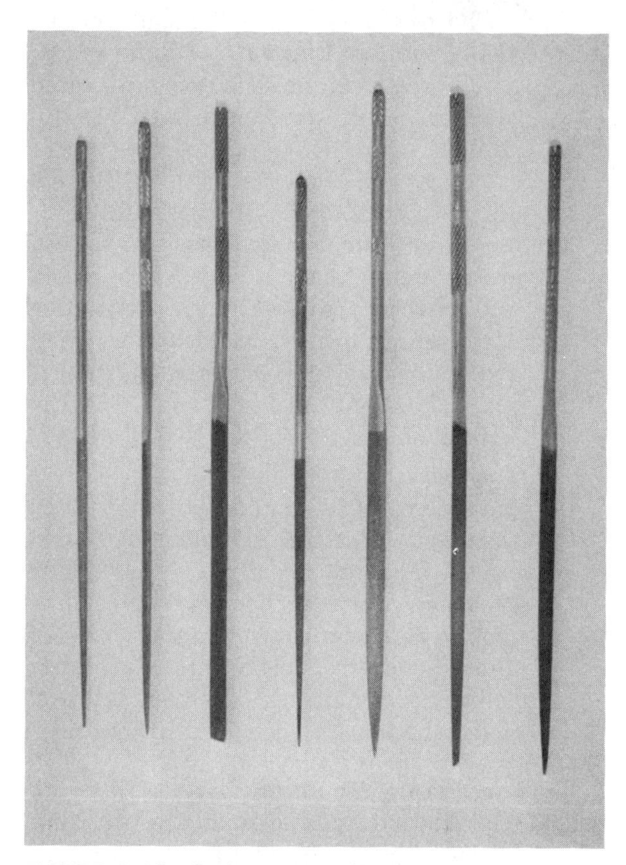

FIGURE 4.19 Swiss pattern file set.

smooth shanks. They are available in grades ranging from the coarsest, No. 00, to No. 6, the finest.

Diesinker rifflers are special files made for finishing cavities in dies and other objects (Fig. 4.20). They are available in about 10 different shapes, depending on the manufacturer, and in coarse, medium, and fine cuts.

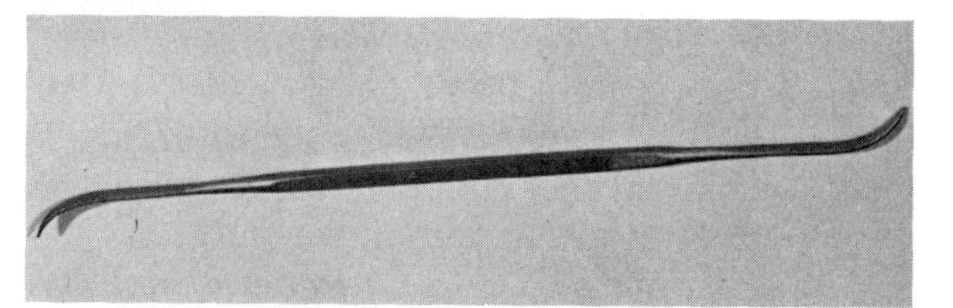

FIGURE 4.20 Riffler files used by die sinkers.

Use and Care of Files. Files must always be used with handles. A handle is necessary for safe use of the file, and a properly installed handle of the correct size improves control during filing (Fig. 4.21).

A *file card* and a *file brush* are used to clean the teeth of the file (Figs. 4.22 and 4.23). Metal chips that become caught in the teeth of a file will scratch the work being filed, especially when a fine-toothed, or *smooth-cut,* file is being used. When cleaning a file, the file card must be moved parallel with the teeth.

When filing nonferrous metals and soft steel, ordinary blackboard chalk can be used on the file to prevent chips from sticking in the teeth and marring the work (Fig. 4.24). The use of chalk results in smoother and more accurately filed surfaces.

Straight filing is one of the most common filing operations. It is used for either rapid metal removal or finishing on many types of metals. Selection of the proper file for the material being filed is very important. When rapid metal removal is necessary and the finish is unimportant, use a double-cut file with coarse tooth spacing. For finishing operations, use a single-cut file with fairly closely spaced teeth.

Remember that pressure is applied to the file

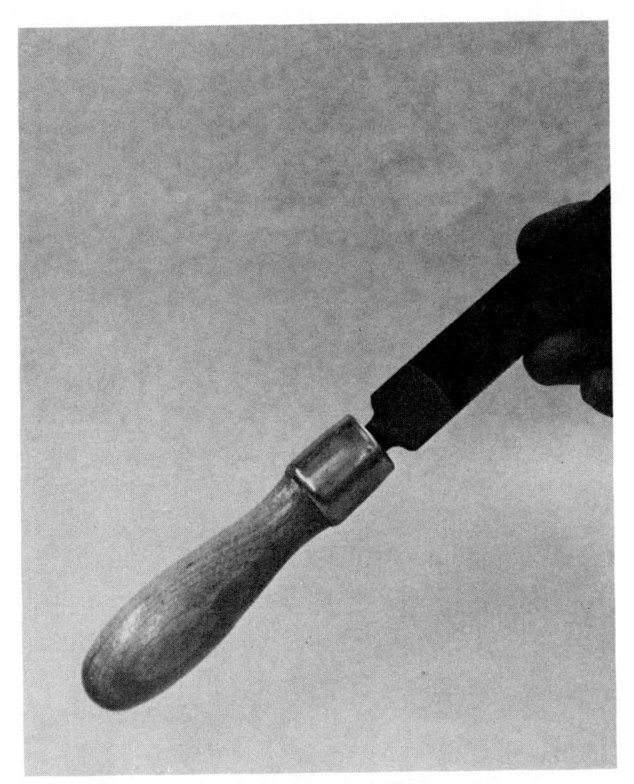

FIGURE 4.21 Properly installed file handle.

FIGURE 4.22 File card and brush.

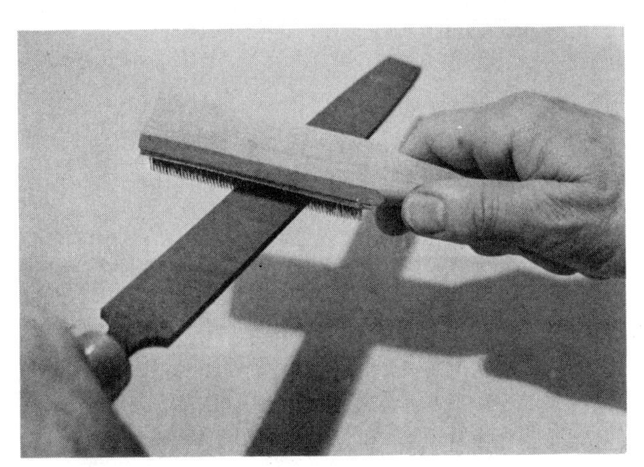

FIGURE 4.23 Cleaning a file.

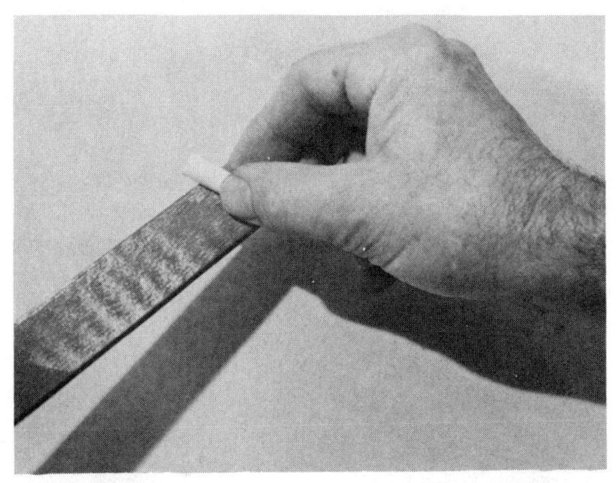

FIGURE 4.24 Chalk may be used as a dry lubricant on fine-toothed files.

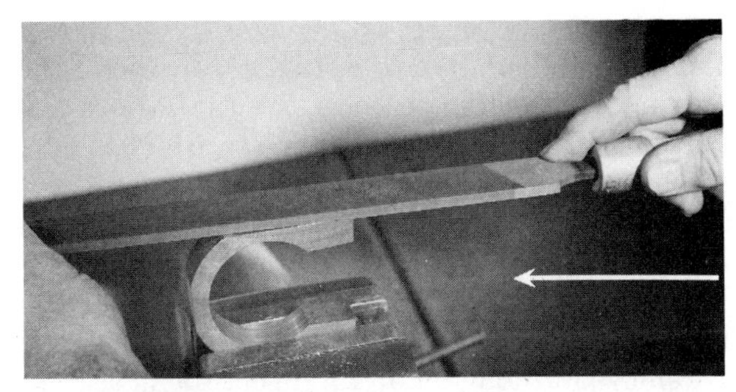

FIGURE 4.25 Straight filing.

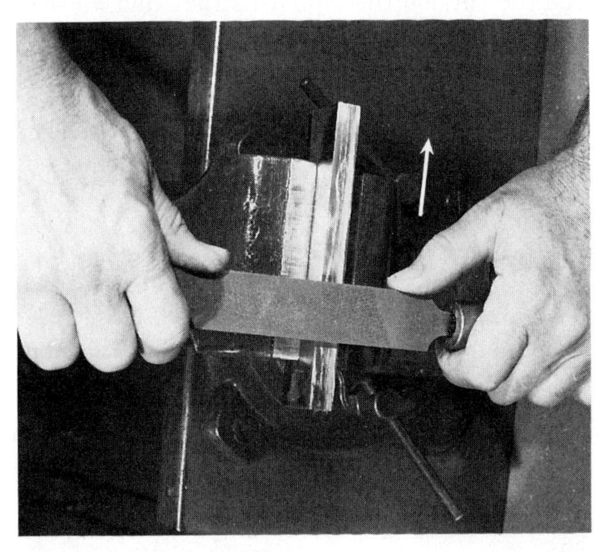

FIGURE 4.26 Draw filing.

only on the forward, or cutting, stroke (see arrow in Figs. 4.25 and 4.26). The teeth of the file point away from the handle, and if pressure is applied during the return stroke, the teeth will be damaged. File with slow, even strokes for best results.

Draw filing is basically a finishing operation almost always done with a single-cut file. Grasp

FIGURE 4.27 Filing in the lathe.

the file with *both* hands (Fig. 4.26), and point the cutting edge of the teeth of the file *toward* the person holding the file. Apply pressure on the *draw* stroke, and relax it on the stroke away from you. In most cases, using chalk on the file helps produce a finer finish and keeps the file clean.

Lathe filing (Fig. 4.27) is usually a deburring or finishing operation in which very little material is removed. A long-angle single-cut file is best for lathe filing because it makes a clean, shearing cut. The file cuts on the forward stroke and must be moved constantly to prevent filling of the teeth with chips. Using chalk helps keep the file clean, especially when filing soft steel and aluminum, and helps produce a better finish. Remember to file left-handed so that you will be clear of the chuck on the lathe.

Rotary Files and Burrs. These cutting tools may be driven by variable-speed flexible-shaft drives or portable electric or air tools (Fig. 4.28)and are used for filing operations that cannot be done with hand files. *Rotary files* are usually made of high-speed steel or tungsten carbide (Fig. 4.29). *Burrs* (Fig. 4.30) are made of high-speed steel or carbide. The more expensive carbide burrs have a much longer life than high-speed steel burrs and can be used on harder materials. The teeth on rotary files are continuous, extending from the tip to the heel, regardless of the shape of the tool.

FIGURE 4.28 Cleaning a cavity with a rotary file.

4.2.4 Hand Reamers

Various *hand reamers* are used in bench and assembly work in the machine shop. They are precision cutting tools and must be used and stored

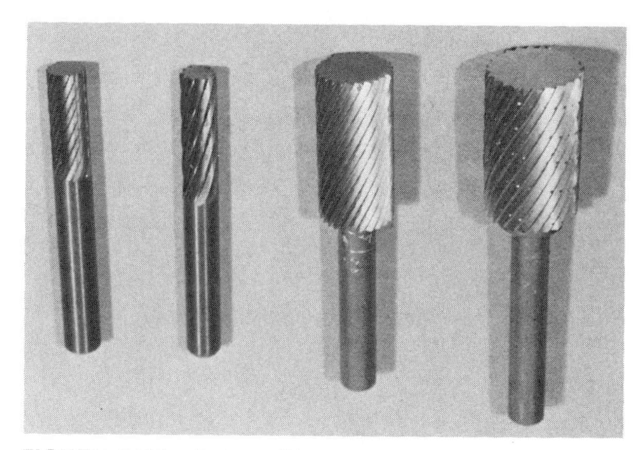

FIGURE 4.29 Rotary files.

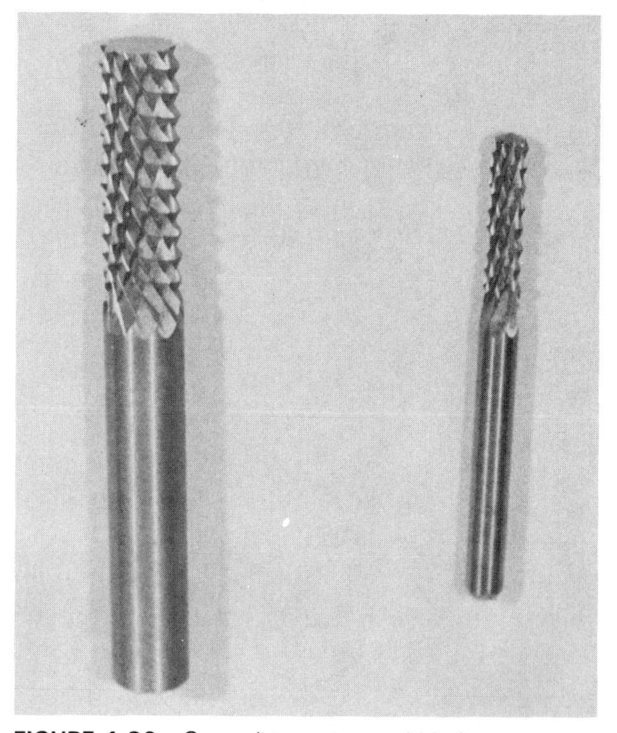

FIGURE 4.30 Ground tungsten carbide burrs.

FIGURE 4.31 Straight hand reamer.

FIGURE 4.33 Expansion reamer. (*Courtesy DoALL Co.*)

carefully to prevent damage. Hand reamers can be identified by the square driving area machined on the end of the shank (Fig. 4.31). The best tool for driving hand reamers is an adjustable tap wrench; smaller hand reamers can be driven with a T-handle tap wrench. Hand reamers are usually made of high-carbon or high-speed steel. On adjustable hand reamers, the body and adjusting mechanism are made of heat-treated medium-carbon steel and the blades are made of high-speed or high-carbon steel.

Solid hand reamers are available with straight or helical flutes, in sizes from 1/8 in. (3.17 mm) to about 1 1/2 in. (38.1 mm). Reamers ground to nonstandard sizes are available from some manufacturers. Helical flute-reamers may have right-hand (leading) or left-hand (trailing) flutes. Hand reamers are turned in the clockwise direction during both the reaming operation and removal of the reamer from the finished hole. Turning the reamer backward can damage the cutting edges.

Hand reamers are intended to remove small amounts of material. Therefore, careful selection of the drill used prior to reaming is necessary so that not more than 0.005 to 0.007 in. (0.13 to 0.18 mm) is left for smaller reamers and 0.012 to 0.015 in. (0.30 to 0.38 mm) for larger reamers. The end of the reamer is ground to a shallow taper for a distance equal to its diameter to allow easier entry. Use of a good petroleum-based cutting oil or wax on all metals except cast iron helps produce a better finish and extends the life of the reamer.

Taper hand reamers are available in standard tapers, such as the Morse and Brown and Sharpe series (Fig. 4.32). Roughing reamers usually have straight flutes with staggered nicks in the cutting edge to provide more rapid cutting action without clogging. Taper reamers for finishing have either straight or left-hand helix flutes and are used to remove only 0.005 to 0.010 in. (0.13 to 0.254 mm) of material from the hole. Use of cutting oil on materials other than cast iron is recommended.

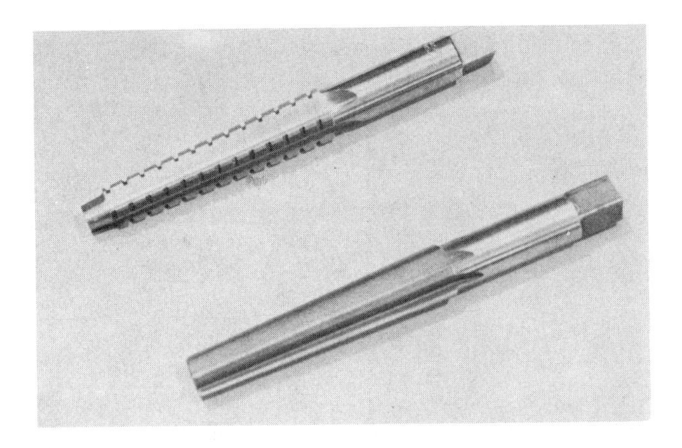

FIGURE 4.32 Tapered reamers: (top) roughing; (bottom) finishing.

Expansion hand reamers (Fig. 4.33) have slots cut next to each flute and an adjusting screw in the hollow center of the reamer body that allows the flutes to be expanded beyond the nominal size. Small expansion reamers (0.25 to 0.5 in. in diameter) can be expanded only 0.004 to 0.006 in. (0.10 to 0.15 mm); larger ones can be expanded up to 0.015 in. (0.38 mm). Because the blades are fragile, the reamer must be used carefully, and cutting oil is recommended on materials other than cast iron.

Adjustable hand reamers (Figs. 4.34 and 4.35) have a threaded body with six or more tapered slots cut in the threaded part. The blades are also tapered on the back side so that the cutting edges are always parallel to each other. The front end of each blade is slightly tapered for easier entry into the hole to be reamed. Two nuts move the blades to the desired setting. Some adjustable hand reamers as well as machine reamers have unequally spaced flutes to reduce the possibility of chatter while reaming.

FIGURE 4.34 Adjustable reamer. (*Courtesy DoALL Co.*)

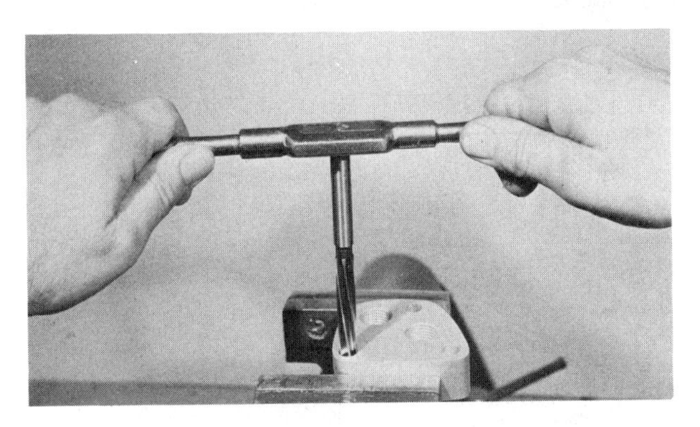

FIGURE 4.35 Hand reaming a hole.

4.3 HAND TOOLS AND VISES

During daily work the machinist uses many tools, such as hammers, wrenches, screwdrivers, punches, and vises. Skillful use and proper maintenance of these tools are signs of good craftsmanship.

4.3.1 Hammers

Of the various types of hammers used by the machinist, the *ball peen hammer* is the most common. It is usually made of medium-carbon alloy steel, and the head is forged to shape. The *peen* and the *face* (see Fig. 4.36) are heat-treated to medium hardness to resist chipping and deformation. Ball peen hammers are classified by the weight of the head and range in size from 4 oz to 2 lb for general machine shop use. The handle is usually made of hickory or ash and held in place with a wedge. Loose handles must be repaired immediately.

Soft hammers are made of lead, plastic, rawhide, brass, or aluminum and are used when the part being struck must not be dented or scratched. Lead hammers (Fig. 4.37) are made by casting the head directly onto a steel handle. When the head becomes deformed or damaged, it is melted off the handle and recast.

Hammers with screw-in plastic or soft metal inserts are very useful in assembly and disassembly of machine parts (Fig. 4.38). Hard metal particles must be kept from becoming embedded in the faces of soft hammers.

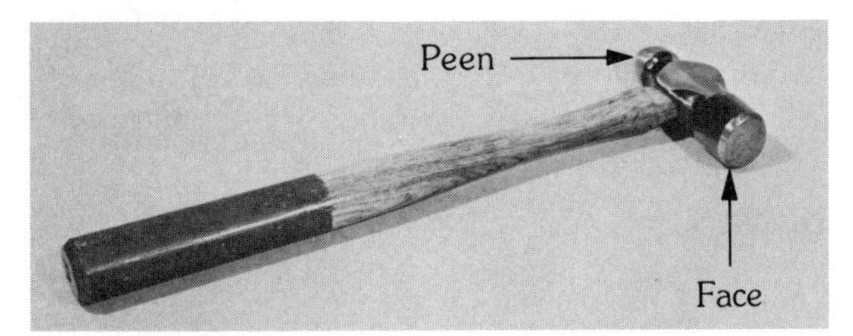

FIGURE 4.36 Parts of a machinist's hammer.

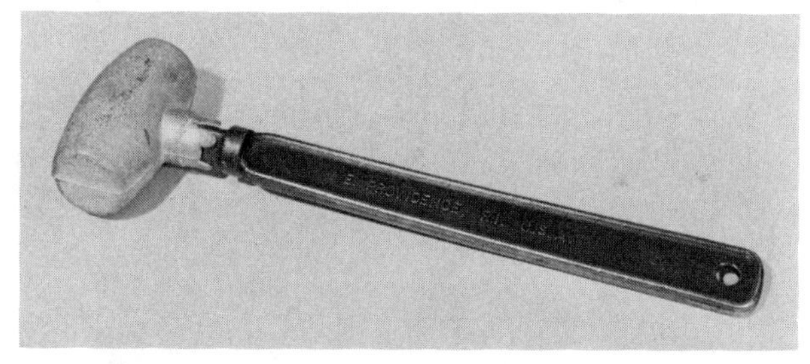

FIGURE 4.37 Lead hammer for setup and assembly.

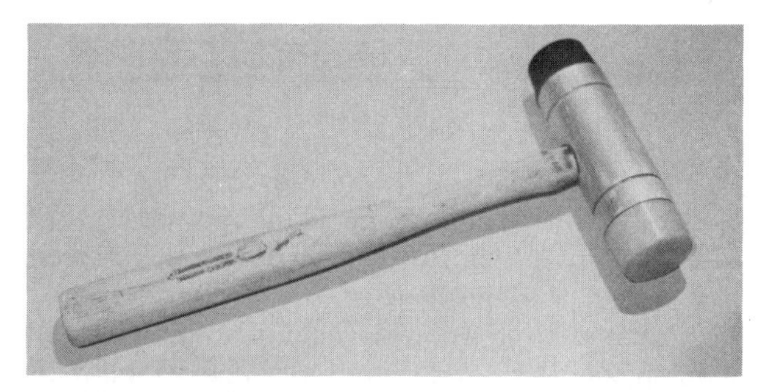

FIGURE 4.38 Plastic-faced hammer.

4.3.2 Wrenches

The machinist uses a variety of solid and adjustable wrenches in daily work. Most wrenches are forged from alloy tool steels and heat-treated to achieve toughness (resistance to breaking) and wear resistance.

Adjustable wrenches usually have one movable and one fixed jaw and are adjusted by a screw mechanism (Fig. 4.39). Adjustable open-end wrenches are classified by overall length, with the opening usually at a 22° angle to the body of the wrench. When using an adjustable open-end wrench, always apply the force in the direction of the adjustable jaw (see Fig. 4.40).

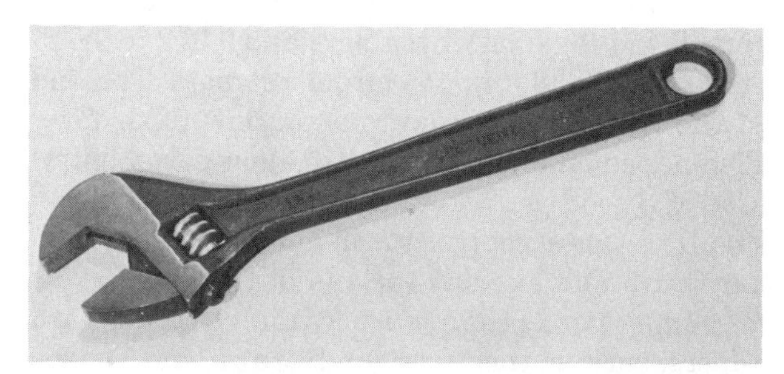

FIGURE 4.39 Adjustable open-end wrench.

FIGURE 4.40 Using an adjustable wrench properly.

Pipe wrenches are another type of adjustable wrench that the machinist may use occasionally. The sharp teeth on the jaws of a pipe wrench will mark the surface being gripped, so never use a pipe wrench on anything that must remain smooth. Pipe wrenches generally have a cast ductile iron or malleable iron handle and movable jaw with hardened medium- or high-carbon steel jaw inserts. Some pipe wrenches have forged aluminum handles. A *strap wrench* (see Fig. 4.41) must be used on round objects that should not be marred or deformed.

Fixed and adjustable *spanner wrenches* of both the hook and pin types are sometimes used in the machine shop (see Fig. 4.42). These wrenches have forged or cast handles and heat-treated steel pins.

Solid wrenches are *box end, open end,* or combination *box and open end* types (see Fig. 4.43). Wrenches of this type are available in a large range of both metric and American sizes and usually forged from heat-treated alloy steel of medium-carbon content. Open-end wrenches have the opening at a 15 or 22 1/2° angle to the body of the wrench and are available in normal and thin patterns. The angled opening of the wrench allows the tools to be used more effectively in confined spaces.

Box end wrenches are made with 6- or 12-sided openings that are pierced during the forging operation and finished by broaching prior to heat treatment. The shank of the wrench is usually angled about 15° to provide clearance, or it is offset (see Fig. 4.44). Box end wrenches are very convenient to use in confined spaces and are generally safer to use than open end or adjustable wrenches because there is less chance of the wrench slipping off the nut or bolt head.

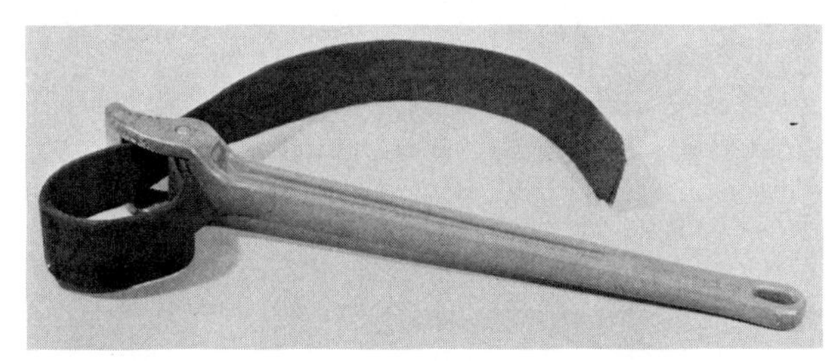

FIGURE 4.41 Strap wrench.

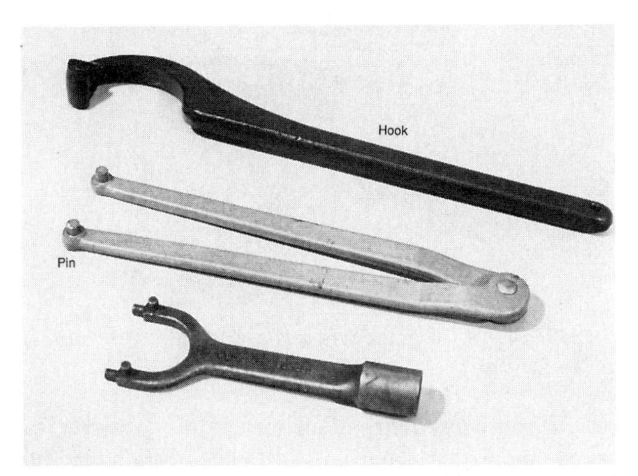

FIGURE 4.42 Spanner wrenches: (top) hook type, and (middle and bottom) pin types.

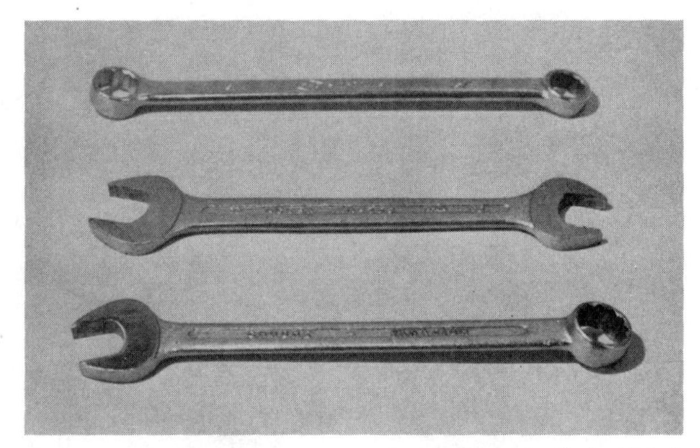

FIGURE 4.43 Solid wrenches: (top) box end, (middle) open end, and (bottom) combination wrenches.

Combination wrenches, as the name implies, are made with an open end wrench on one end and a box end wrench of the same size on the other end. The axis of the open end part is angled at 15° to the centerline of the handle, and the handle is tilted up 15° in relation to the box end part of the wrench.

Socket wrenches are forged and broached from alloy tool steel and available in sizes ranging from 1/4 to 2 in. (6.35 to 50.8 mm), or larger. They are made with either 6- or 12-sided openings and generally are sold in sets. The size of the *drive* (the square opening broached into the back of the socket) and the size nut it will accommodate are the main factors in classifying sockets. Usually, a socket set includes various extensions, a ratchet, speed handles, deep sockets, universal joints, and other accessories (see Fig. 4.45).

Hex wrenches (sometimes called Allen

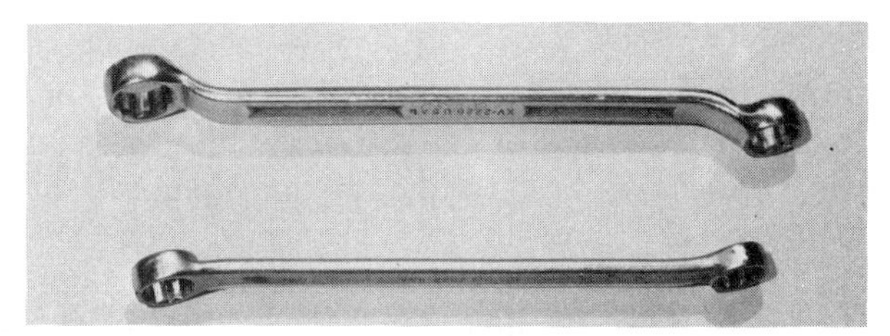

FIGURE 4.44 Angles offset box-end wrenches.

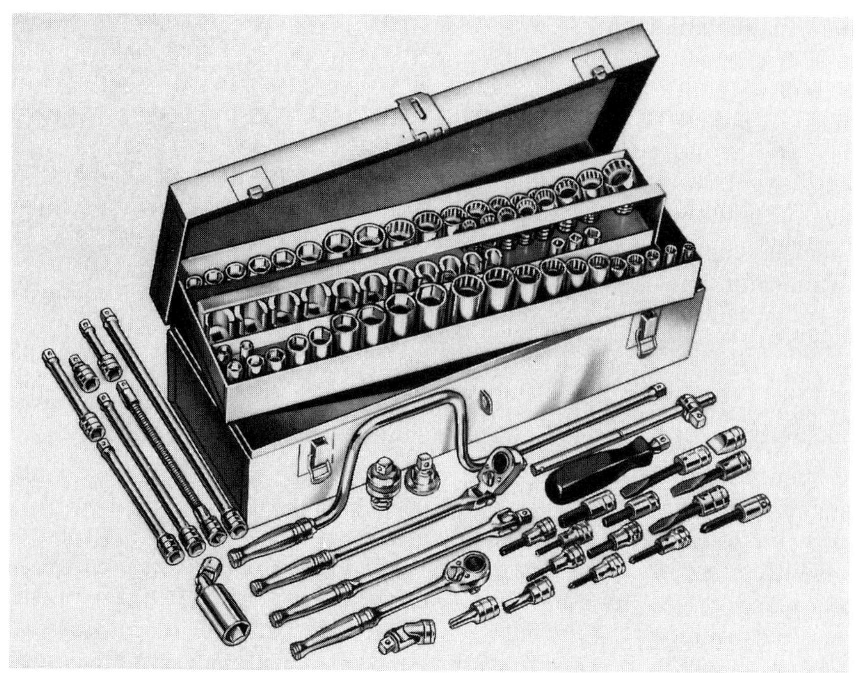

FIGURE 4.45 Set of socket wrenches. (*Courtesy Snap-On Tools*)

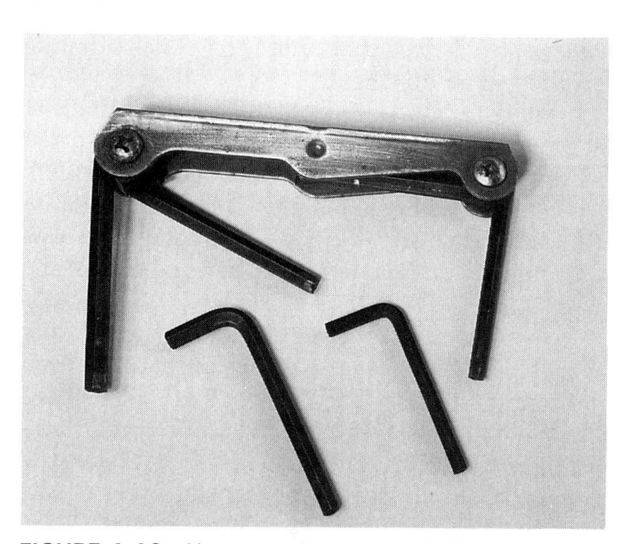

FIGURE 4.46 Hex wrenches are available singly or in sets.

wrenches) are used on headless set screws and socket head capscrews (see Fig. 4.46). They are made of heat-treated tool steel, in sizes ranging from less than 1/16 in. to more than 1 in. across flats. A typical hex wrench set consists of wrenches ranging from 3/32 to 1/2 in. in size.

Torque wrenches (Fig. 4.47) are used for determining the amount of torque (twisting effort) being applied to a nut or capscrew. The indicator on the wrench shows torque in *inch/pounds* or *foot/pounds.* Some torque wrenches show the reading on a gage or scale, others click when a preset torque is reached.

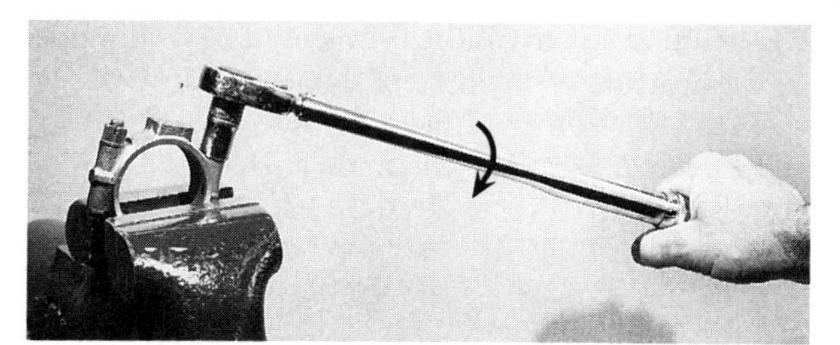

FIGURE 4.47 Torque wrench in use.

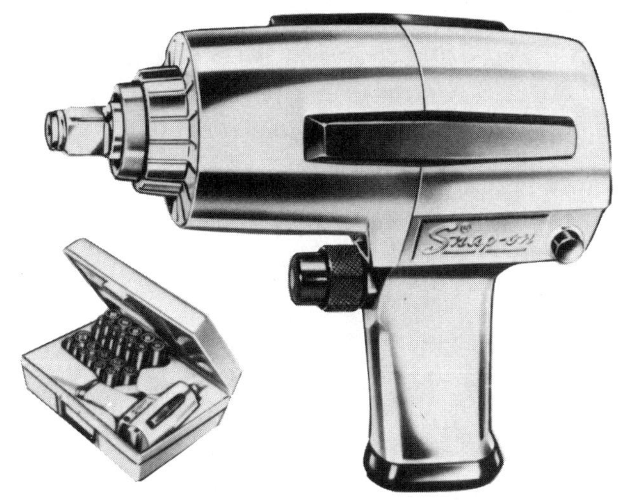

FIGURE 4.48 Impact wrench and sockets. (*Courtesy Snap-On Tools*)

Impact wrenches of either the air or electrical type (see Fig. 4.48) are used to speed up assembly and disassembly operations.

4.3.3 Pliers

Several different types of pliers are used in bench work in the machine shop. Pliers are generally used for gripping, twisting, and cutting operations and should not be used on bolt heads or nuts. Good-quality pliers are forged from medium-carbon tool or alloy steels and are heat-treated for toughness and wear resistance. Some common types are shown in Fig. 4.49.

Combination pliers (also called *slip-joint* pliers) have serrated jaws and a set of cutting edges near the pivot point. They are classified by overall length and are of the *thin pattern* (with thin jaws) or *normal pattern* type.

Long-nose pliers are bent or straight and may have cutting jaws near the pivot point. The closely fitted pivot joint is nonadjustable, and the pliers are intended for light-duty work because of their thin jaws.

Side-cutting pliers are classified by overall length and have a large, closely fitted pivot. The cutting edges are precisely ground and can be used only for cutting wire or rod that is *not* hardened. *Attempting to cut hardened material will ruin the cutting jaws.* Side-cutting pliers also have flat serrated gripping jaws whose surfaces do not quite meet.

Interlocking-joint pliers have a wide range of adjustment and no cutting edges. The gripping jaws are heavily serrated, and the pliers are classified by overall length.

Diagonal cutters have no gripping jaws and are used primarily for cutting wire, cotter pins, and similar objects. The jaws are closely aligned by the main pivot at a slight angle to the handle.

Toggle-clamp pliers (sometimes called *vise-grip* pliers) are a useful tool in the machine shop (see Fig. 4.50). The large mechanical advantage and overcenter locking feature provide heavy holding capacity. There is infinitely variable adjustment of grip within the operating limits of the tool. These pliers are made of a combination of heat-treated medium-carbon alloy steel forgings and stampings. An adjusting screw with a knurled head positions the jaws.

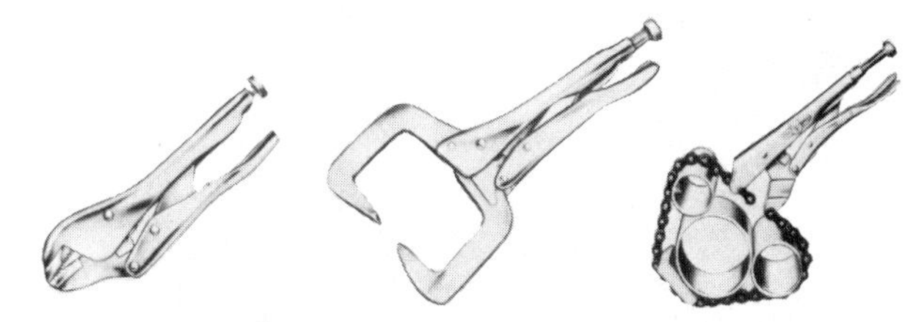

FIGURE 4.50 Toggle-clamp pliers. (*Courtesy Snap-On Tools*)

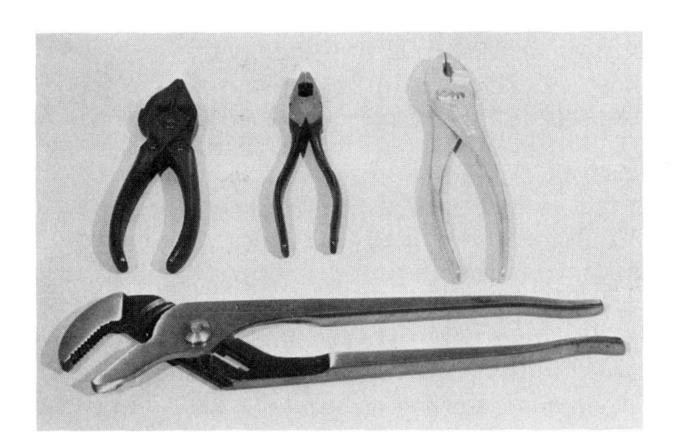

FIGURE 4.49 Common types of pliers.

4.3.4 Screwdrivers

The machinist uses screwdrivers to tighten or loosen fasteners and to make adjustments. The common screwdriver consists of a *handle, shank,* and *blade,* or *tip,* as shown in Fig. 4.51. Screwdrivers are classified by the length of the blade and shank combined and by the type of tip on the blade.

The handle is made of wood or plastic and the shank of medium- or high-carbon tool steel. On large screwdrivers with a square shank, a wrench can be used to increase the twisting force applied (see Fig. 4.52).

The blade of the screwdriver may be *flat* or *crosspoint* (see Fig. 4.53). Flat screwdrivers must closely fit the screw slot to avoid damaging either the screw or the screwdriver. Crosspoint screwdrivers are Reed and Prince or Phillips type. Although they are similar, they should not be interchanged.

Offset screwdrivers (Fig. 4.54) of either the

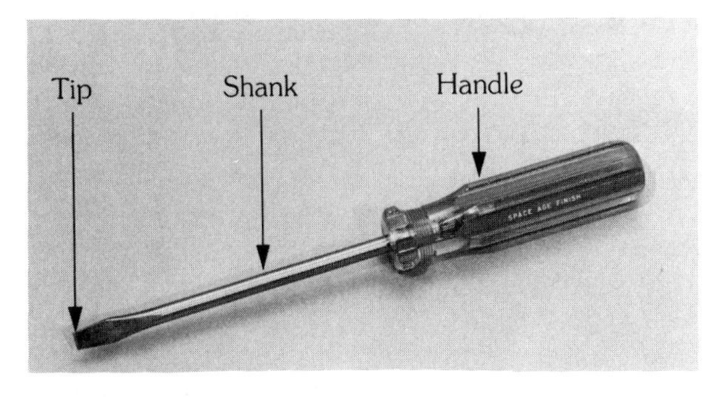

FIGURE 4.51 Parts of a screwdriver.

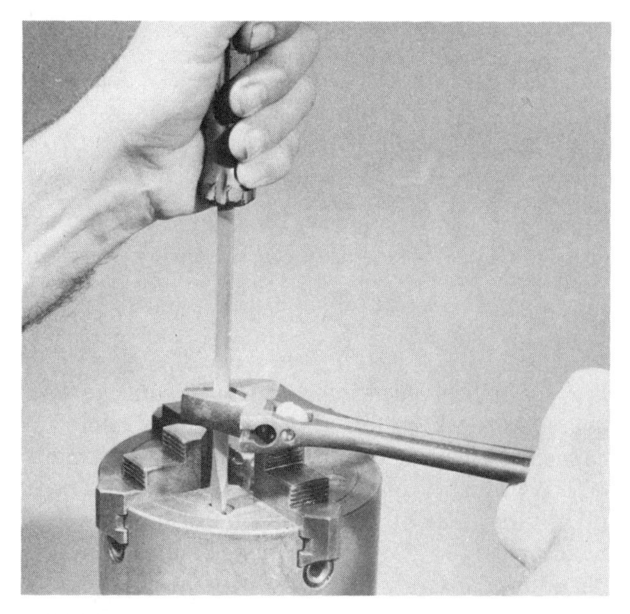

FIGURE 4.52 Heavy-duty screwdriver in use.

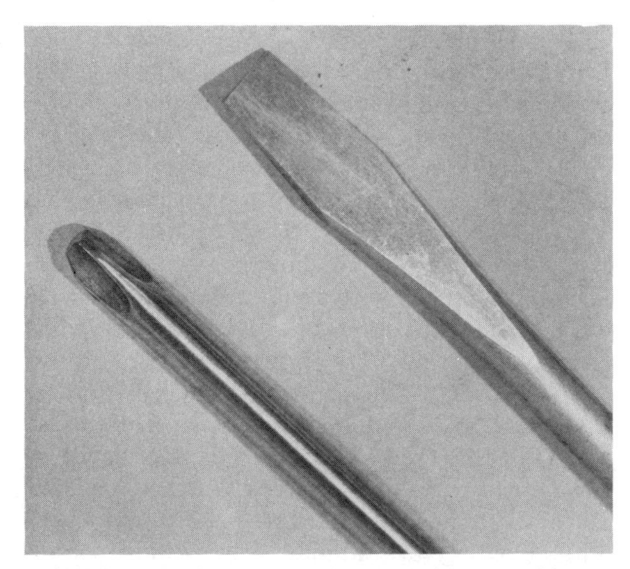

FIGURE 4.53 Standard and cross-point screwdrivers.

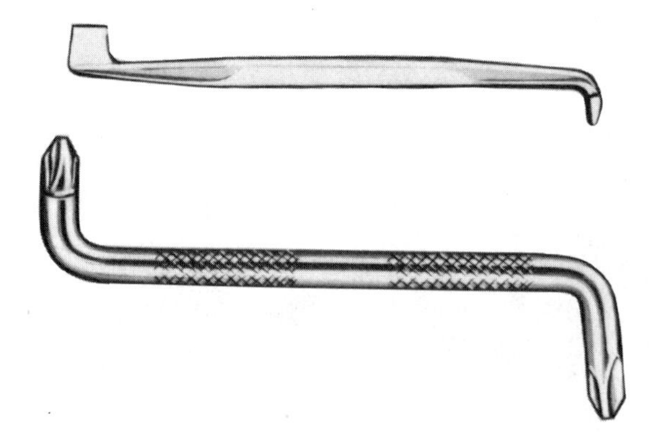

FIGURE 4.54 Offset screwdrivers. *(Courtesy Snap-On Tools)*

FIGURE 4.55 Grinding a screwdriver tip.

FIGURE 4.56 Improper and proper fit of screwdriver tip in screw slot.

plain or ratchet type are useful when working in confined spaces. The blades on the plain offset screwdriver are at a 90° angle to each other.

The blades of common screwdrivers become worn or damaged from use and they can be reconditioned by careful grinding (see Fig. 4.55). The end of the blade must be ground square to the main axis of the screwdriver, and the sides of the blade must be ground parallel to each other at the tip (see Fig. 4.56). The tip of the blade must not be overheated during grinding to prevent it from becoming soft and thus easily damaged or deformed.

4.3.5 Clamps

Various clamps of assorted types and sizes are normally used by the machinist doing bench and assembly work.

C clamps are the type of clamp most widely used. Basically, the C clamp consists of a *frame, screw, handle,* and *swivel pad* (Fig. 4.57). The frame of high-quality heavy-duty clamps is usually forged from medium-carbon steel. Cast frames are made of ductile iron or malleable iron. The screw and

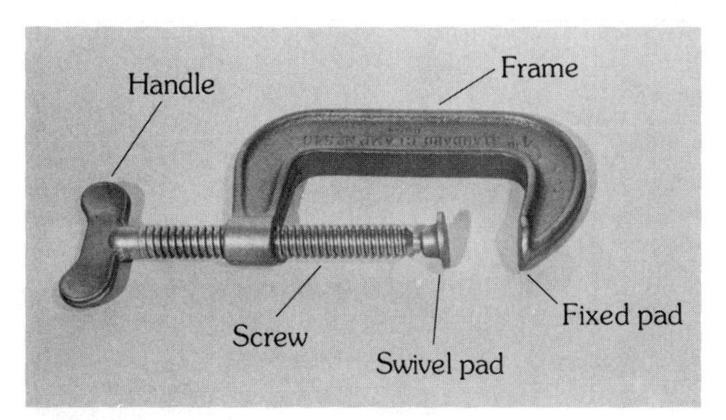

FIGURE 4.57 Parts of a C clamp.

FIGURE 4.58 Tollmaker's parallel clamp. (*Courtesy L.S. Starrett Co.*)

other parts are usually made of steel. C clamps are classified by the maximum opening.

Parallel clamps are often used in assembly operations and layout work. The parallel clamp (see Fig. 4.58) consists of two jaws and two screws. The jaws are usually of heat-treated medium-carbon steel.

4.3.6 Vises

The *bench vise* is the holding device most often used by the machinist for holding objects. The bench vise consists of the *fixed* and *movable jaws,* the *screw* and *nut assembly,* the *handle,* and the *jaw inserts* (see Fig. 4.59). Some vises have swivel bases so that the jaws can be positioned at an angle to the bench front.

The castings on good-quality bench vises are made of ductile iron or malleable iron. The screw, which usually has an Acme thread, is made of steel, and the nut, which is fastened inside the fixed jaw, is generally a bronze casting. The steel jaw inserts may be serrated or smooth and are hardened for maximum wear resistance. Soft jaws made of aluminum, copper, or lead can be placed over the hardened jaws to prevent scarring the surface of

FIGURE 4.59 Major parts of a swivel-base bench vise: (1) body; (2) thread; (3) swivel base; (4) spring screw fastener; (5) steel jaw facings; (6) movable jaw. (*Courtesy L.S. Starrett Co.*)

finished objects held in the vise. Bench vises are classified in terms of size by the width of the jaws.

4.4 HAND THREADING TOOLS

The machinist uses hand *taps* and *dies* to cut internal or external threads in metals and some nonmetallic materials (Fig. 4.60).

4.4.1 Taps

Taps for general shop use are made of high-carbon or high-speed steel. They are heat-treated to achieve a high level of hardness and wear resistance and therefore are quite brittle. The threads on a tap may be either cut or ground. Taps with ground threads are generally considered superior in quality.

The *standard hand tap set* for cutting straight threads consists of a *taper* tap, a *plug* tap, and a *bottoming* tap of the same pitch diameter and pitch (Fig. 4.61). On the taper tap, the first 8 to 10 threads are tapered, and the entering point of the tap is a little smaller than the tap drill size. The taper tap is usually used for starting the threads in a deep hole, but it can be used for the complete threading operation on open-ended holes in thin material. A *plug* tap has the first four or five threads tapered and may be the only tap used in tapping an open-ended hole. It is somewhat more difficult to start in the hole than the taper tap but, it allows completion of the threads in one operation, thus saving time.

A *bottoming* tap has only the first thread tapered and should be used only after a taper or plug tap has been used to thread a blind hole as

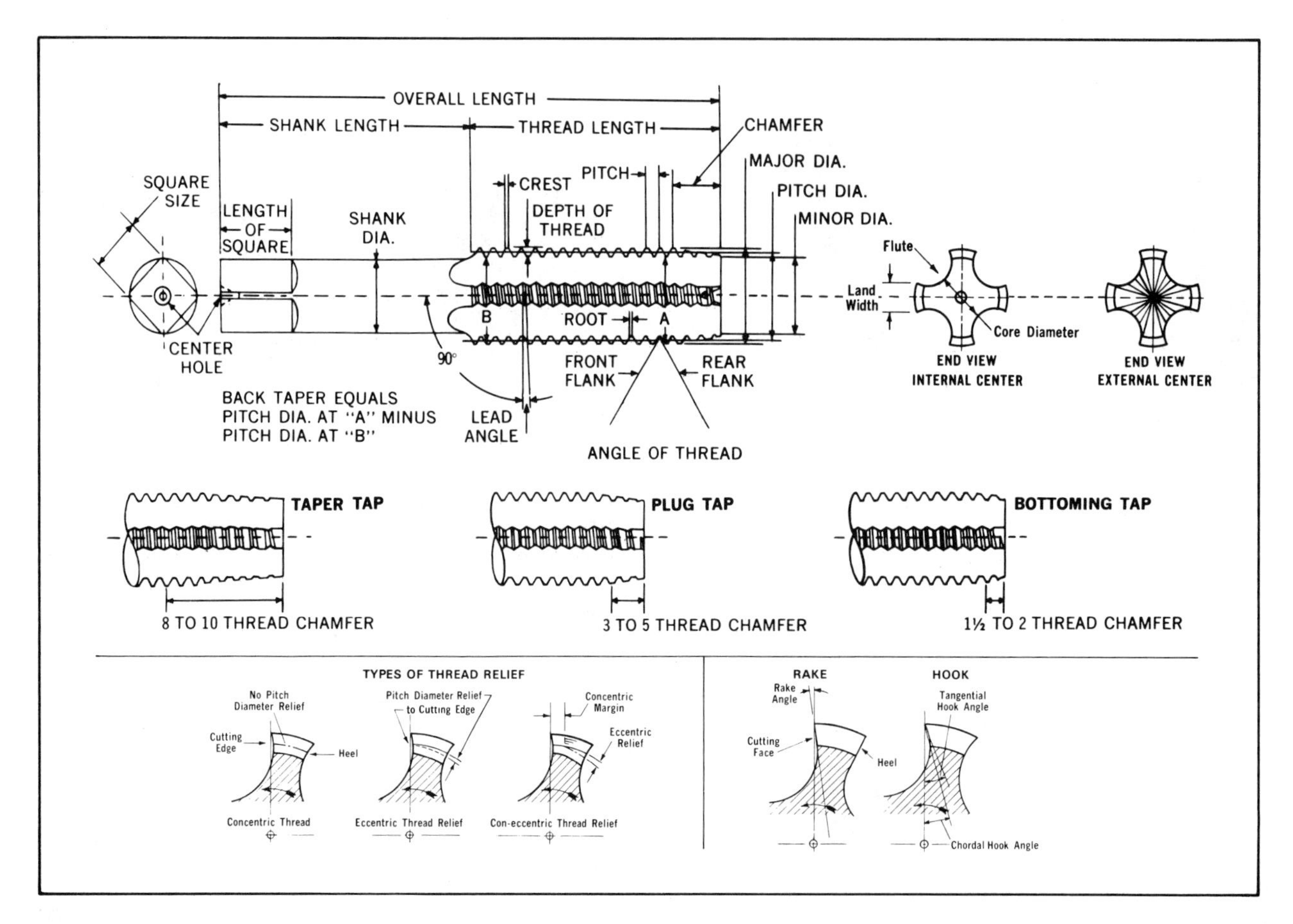

FIGURE 4.60 Characteristics of a hand tap. (*Courtesy Cleveland Twist Drill Co.*)

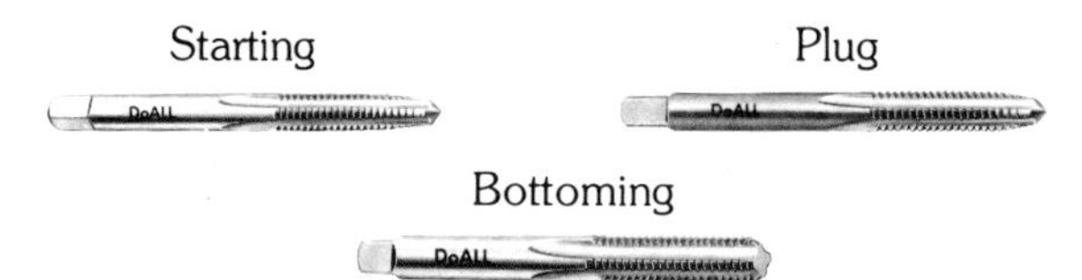

FIGURE 4.61 Set of hand taps. (*Courtesy DoALL Co.*)

far as possible. It is capable of making a complete thread very close to the bottom of a blind hole.

Pipe taps are made with a 3/4-in./ft. taper throughout their length (Fig. 4.62). They are intended only for threading pipe fittings and holes into which a standard external pipe thread will be fitted. Pipe taps are more rugged and have thicker shanks than taps for conventional threads. This is necessary because more effort is required to turn a pipe tap since the pipe tap cuts along its entire surface when making a tapered thread.

Tap Wrenches. Wrenches used to turn taps are T handled for small taps or two-handled and adjustable (see Fig. 4.63). T-handled tap wrenches with long shanks are also available and are useful for working in confined areas. Do not turn a tap with an adjustable end wrench if you can avoid do-

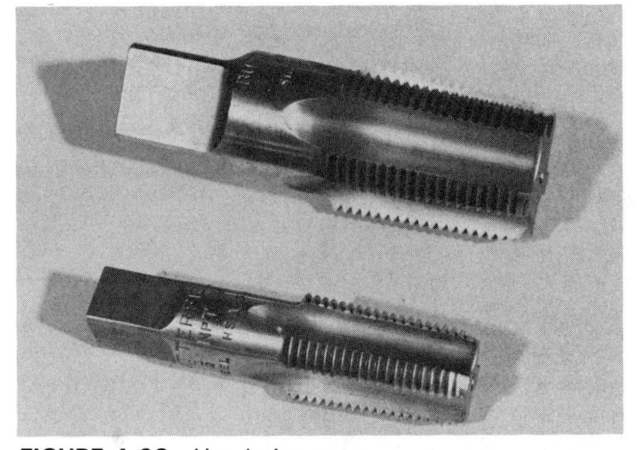

FIGURE 4.62 Hand pipe taps.

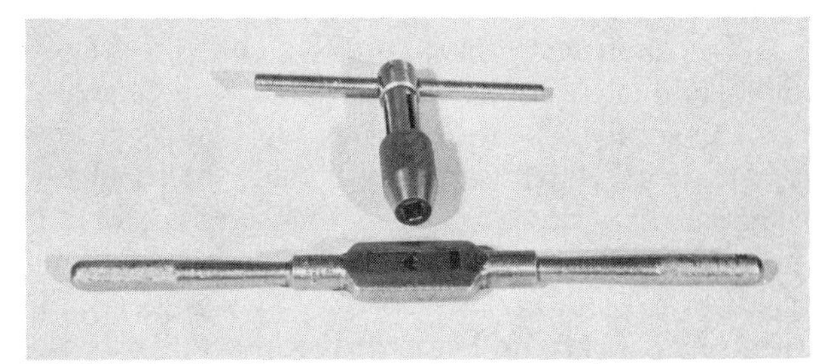

FIGURE 4.63 Two types of tap wrenches.

ing so. The tap is more difficult to start straight and more easily broken because of the uneven twisting effort applied.

Cutting Fluids. Since tapping is an operation that involves high cutting pressures at low speeds, there is a tendency for the tap to bind or gall in the hole. This will cause tap breakage, shortened tap life, or a rough thread unless the proper cutting fluid is used.

In recent years, a number of specialized fluids for tapping have been developed. Although they are very effective, some will cause severe corrosion in the threaded hole and on the tap unless both are thoroughly cleaned with a petroleum-based solvent after use. This is a particularly severe problem when tapping high-strength steels.

When tapping aluminum, magnesium, and certain die-casting alloys, wax compounds, either in liquid or solid form, are quite effective. These materials also work well in tapping copper-based alloys such as brass or bronze.

Gray cast iron is usually tapped without the use of a lubricant or cutting fluid. This is appropriate because of the large amount of free carbon in the material. Ductile iron and malleable iron require the use of cutting fluids for tapping operations.

Tap Drill Selection. A tap drill is specified for every standard thread, and the diameter of the drill is basically the minor diameter of the thread (see Fig. 4.64). The tap drill sizes specified in most tables and charts provide for a 75 percent thread engagement. If a tap drill that is too large is used, the threads are weaker and more easily stripped because of the reduced engagement between the external and internal threads. Using a tap drill that is too small makes the hole hard to tap and increases the chance of tap breakage.

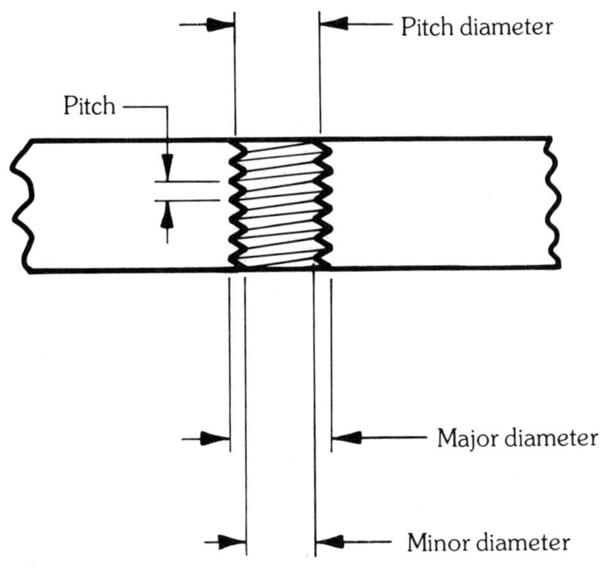

FIGURE 4.64 Characteristics of a tapped thread.

The tap drill size for any Unified thread is calculated from the following formula:

$$\text{TDS} = D - \frac{1}{N}$$

where TDS = tap drill size
D = major diameter of tap
N = number of threads per inch

The tap drill size of 5/8–11 UNC (Unified Coarse; see Chap. 6) thread is found as follows:

$$\begin{aligned}\text{TDS} &= 0.625 - \frac{1}{11} \\ &= 0.625 - 0.09 \\ &= 0.535\end{aligned}$$

The closest standard drill size is 17/32 in. (0.531 in).

Use of Hand Taps. The successful use of hand taps requires an understanding of the characteristics of taps and the materials being threaded. Adhering to the following eight procedures will help you tap threads more effectively:

1. Make sure that you use a tap drill of the right size.
2. If possible, clamp the work to be tapped in a vise.
3. Use the proper taps and tap wrenches. Do not use end wrenches to turn taps.
4. Use cutting fluid where necessary.
5. Start the tap straight and check it as shown in Fig. 4.65.
6. When the tap has been started straight, apply pressure evenly to both arms of the tap wrench.
7. Reverse the tap occasionally to break up chips and allow cutting fluid to enter the cut.
8. When tapping blind holes, do not jam the tap against the bottom of the hole.

Removal of Broken Taps. Breaking a tap generally results from using a dull tap or excessive or unevenly applied force while tapping. Removal of a tap, especially if it is deeply embedded, can be a time-consuming process and may be impossible without ruining the workpiece. Using a tap extrac-

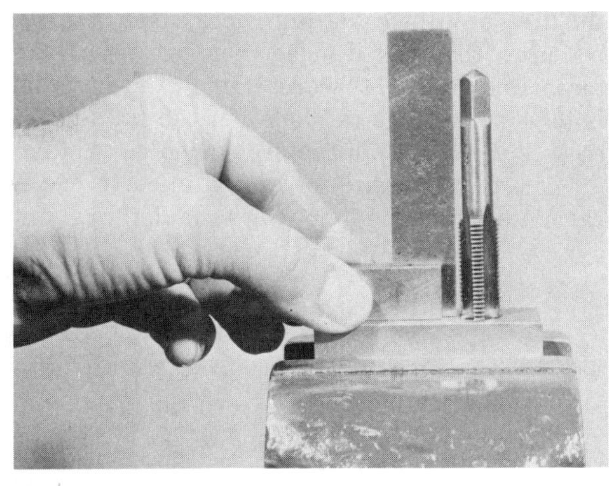

FIGURE 4.65 Checking alignment of tap in partly threaded hole.

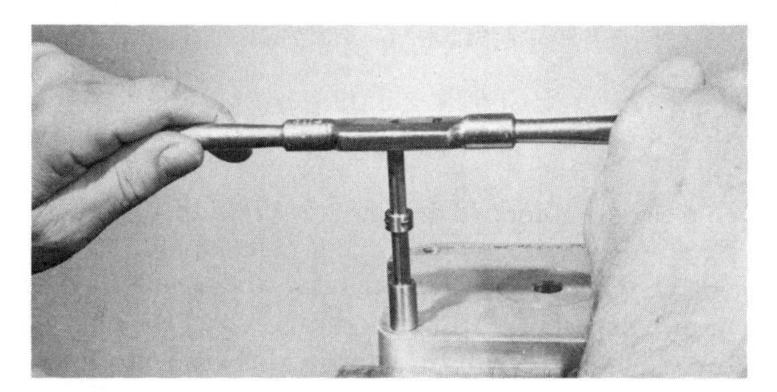

FIGURE 4.66 Using a tap extractor.

tor (see Fig. 4.66) is one possible way of removing the broken tap if it is not too tightly wedged in the hole. Removing broken taps with electric tap disintegrators (electrical discharge machining) is the best procedure if the machinery is available.

4.4.2 Dies

Several different types of dies can be used by the machinist for cutting external threads by hand. Dies are usually made of high-carbon or high-speed steel and are heat-treated for maximum wear resistance. The size-designation system for dies is based on the major diameter of the thread and the pitch. All dies cut straight threads, except American Standard Pipe dies, which cut a thread with a taper of 3/4 in./ft.

Solid Dies. Solid dies are available in hexagonal or square shapes, and are usually made of high-carbon steel. Square dies are available for straight threads in the common sizes and for tapered pipe threads. Hexagonal solid dies (sometimes called *die nuts*) are used for recutting damaged or deformed threads and also can be used for

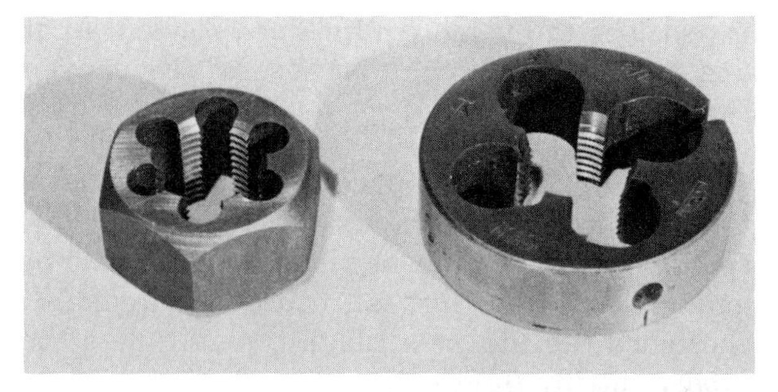

FIGURE 4.67 (Left) solid and (right) split threading dies.

cutting new threads (see Fig. 4.67). Solid dies cut very accurate threads because of their nonmovable cutting edges. The entering side of the die has several incomplete, tapered threads to allow the die to start more easily. The size of the die is stamped on the entering side.

Split Dies. Split dies are available in many sizes in the straight thread series and are made of either high-carbon or high-speed steel. The die can be slightly adjusted above and below the standard depth of the thread by an adjusting screw turned with a small screwdriver (see Fig. 4.68). The first few threads are tapered to allow easier starting of the die on the stock being threaded.

Two-Piece Dies. This type of die has the widest range of adjustment. It also has a guide that helps align the die with the material being threaded. The *die* (see Fig. 4.69), which may be made of high-carbon or high-speed steel, fits into tapered slots in the collet and is held securely in place by the guide. The die is adjusted by two setscrews. The dies are made into matched sets that must never be mixed or inverted.

Diestocks. A diestock consists of a body that is machined to accept the die, and two handles. Some diestocks that are made for split adjustable dies have adjustable guides to help the user align the die with the rod being threaded (see Fig. 4.70).

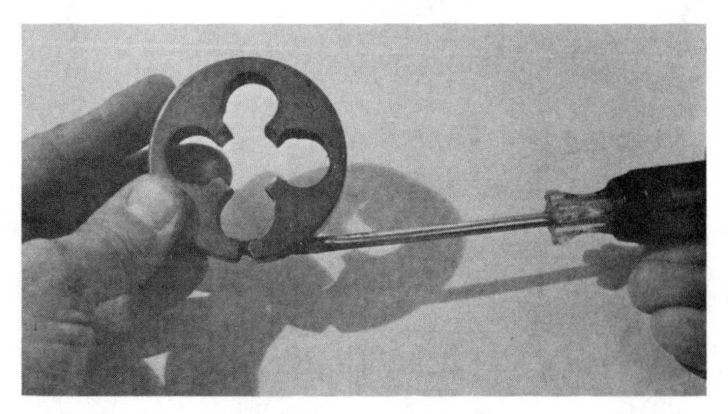

FIGURE 4.68 Adjusting a split die.

FIGURE 4.69 Two-chaser adjustable die.

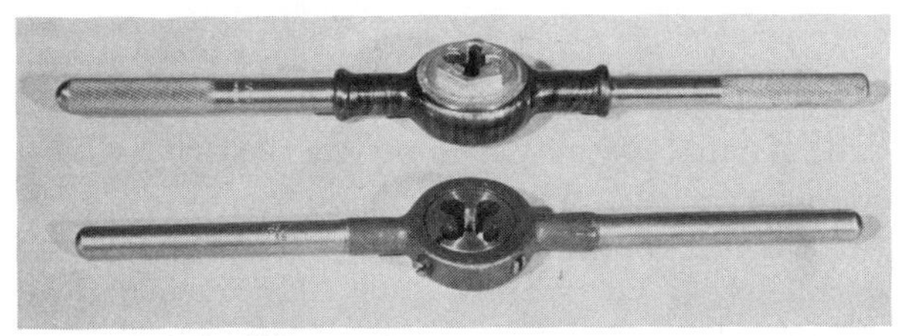

FIGURE 4.70 Diestocks.

Screw-Plate Sets. A typical screw-plate set contains all the taps and dies for a particular range of threads (*example:* 1/4–20 to 1/2–13 UNC), a tap wrench, a diestock, and any adjusting tools that might be needed (see Fig. 4.71).

Use of Threading Dies. Dies can be used to cut external threads on any metallic material that has not been hardened and also on plastic rod (Fig. 4.72). When cutting large-diameter and coarse-pitch threads with an adjustable or two-piece die, the thread can be cut in two or more stages. It should be remembered that threads cut with a die are not as accurate in terms of lead as threads cut on the lathe.

The use of a cutting fluid is strongly recommended on all materials except gray cast iron in order to reduce die wear and help produce smoother threads. When cutting fluids that tend to stain metals or are corrosive are used, both the die and the part being threaded should be washed with a petroleum-based solvent.

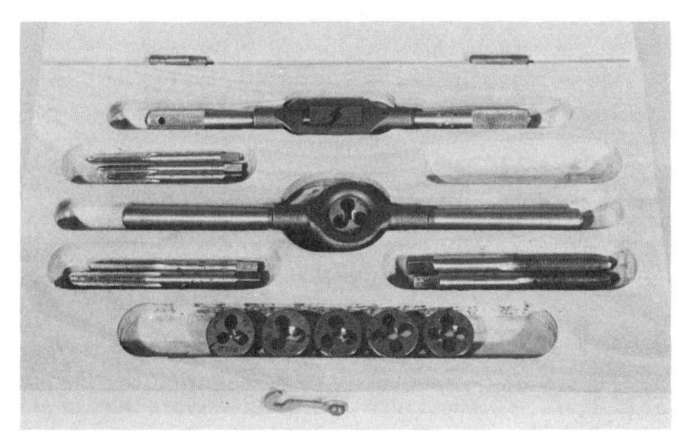

FIGURE 4.71 Screw plate set.

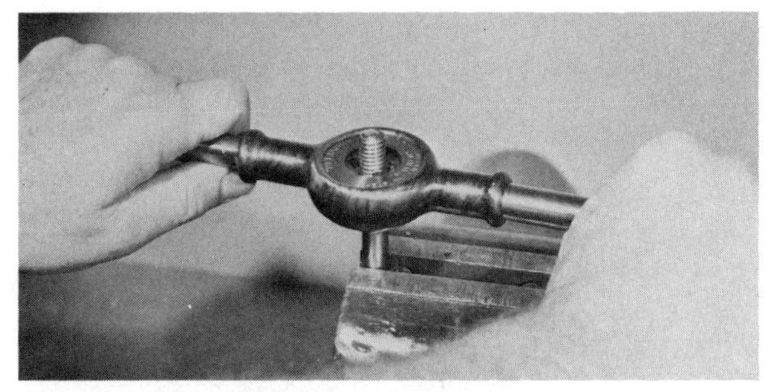

FIGURE 4.72 Cutting a thread with a die.

The machinist must be familiar with all the various thread systems and the means of designating thread sizes and shapes. Refer to Appendices L and M and Chap. 6 for additional information.

4.5 PORTABLE POWER TOOLS

The machinist often has to use hand-held power tools such as drills, impact wrenches, power screwdrivers, die grinders, and power shears. When used with the proper accessories, these versatile tools can be of great help in doing certain tasks quickly and efficiently.

4.5.1 Electrical Tools

Portable electrical power tools can be used wherever a source of appropriate electrical current is available. They are convenient and relatively inexpensive and generally are available in either fixed- or variable-speed versions. For example, the drill shown in Fig. 4.73 is of the variable-speed type and is also reversible. Portable electric drills are usually classified by drill chuck capacity (the larg-

FIGURE 4.73 Electric hand drill.

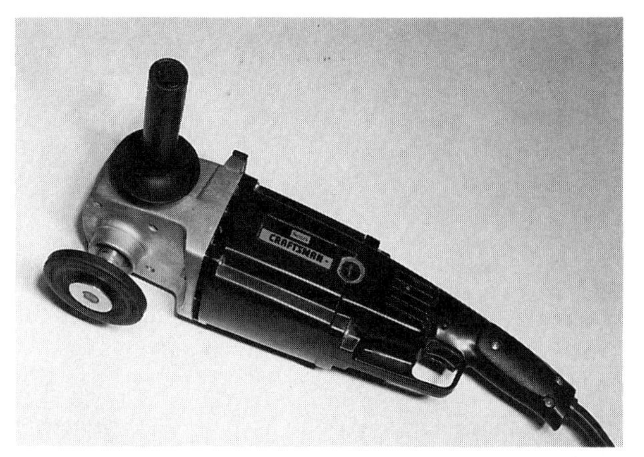

FIGURE 4.74 Portable disk sander/grinder.

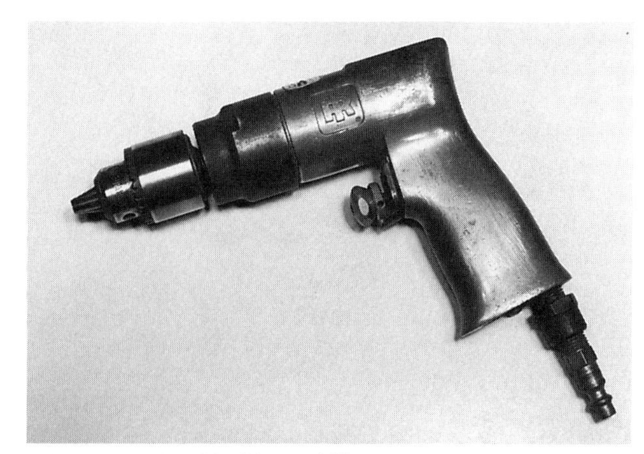

FIGURE 4.75 Air-driven drill.

est drill shank that can be used) and horsepower. For a given chuck capacity, the horsepower rating determines whether the tool is of the light-, medium-, or heavy-duty type.

Disk sanding or grinding operations may be done with the tool shown in Fig. 4.74. Depending on the horsepower rating, such a tool can be used with a 7- or 9-in. sanding disk or with a cup-type grinding wheel. The abrasive used in either disks or wheels is usually aluminum oxide and the coarseness is indicated by a *grit number*. A higher number, such as 100, indicates a relatively fine grit, while a lower number, such as 36, indicates a coarse wheel or disk that leaves a rougher finish.

Eye protection is absolutely necessary when using any portable power tool. When using portable grinders or sanders, it is also important to consider the safety of those working nearby, since particles of metal and abrasive grit travel at high speed when leaving a wheel or disk.

4.5.2 Air Tools

The typical rotary air tool, regardless of whether it is a drill or a grinder, consists of a small air motor that drives a chuck or other tool-holding device either directly or through a gear reduction assembly. Air tools are popular because they are light, compact, and powerful. The speed can be varied by throttling the air supply with a trigger or other control. Another advantage of air tools is that they are spark-free and therefore can be used in areas where explosive fumes are a problem. They are generally somewhat noisier than comparable electric tools.

Air drills may be of the straight type (see Fig. 4.75) or have a head that incorporates a bevel gear so that the chuck is at a 90° angle to the body. A drill of this type may also be used to drive rotary files, burrs, and similar tools.

Air die grinders (see Fig. 4.76) may be used with carbide burrs and rotary files or with abrasives for deburring contouring operations. Tools of this type turn at high speeds, and when used with the proper abrasives, will produce fine finishes.

Larger air grinders are used for operations such as grinding welds or removing the scale or other irregularities from castings in preparation for machining. The grinder shown in Fig. 4.77 may be used with a sanding disk or a grinding wheel and is classified as a medium-duty tool. Much larger portable grinders are available for operations requiring rapid metal removal.

Air tools are fitted with spring-loaded triggers or operating levers so that if the user loses control of the tool, it will stop. As is the case with any tool used for grinding operations, eye protection for the operator and consideration for the safety of others working near the user are essential.

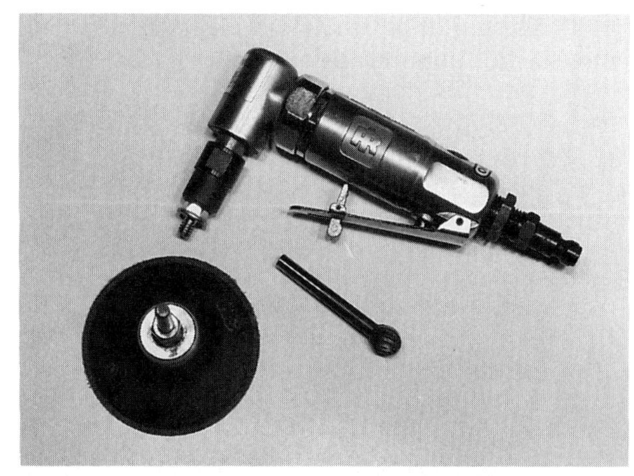

FIGURE 4.76 Air-driven die grinder with 90° angle head.

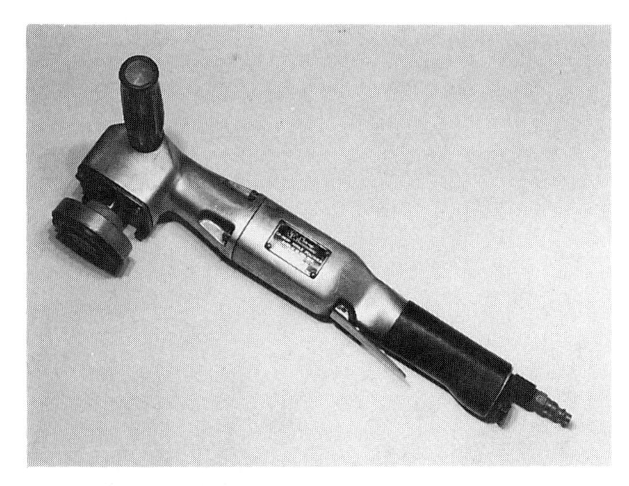

FIGURE 4.77 Air-driven general purpose grinder with a 90° angle head.

REVIEW QUESTIONS

4.2 CUTTING TOOLS

1. Briefly explain how a chisel is hardened and tempered.
2. Explain why the cutting edge of a chisel is ground differently for light and heavy cutting operations.
3. Why must a fine-pitch hacksaw blade be used when cutting sheet metal or thin-walled tubing?
4. What are the effects of operating a hacksaw too rapidly?
5. Of what material are files usually made? Why?
6. What *type* and *cut* of file should be used for rapid removal of metal?
7. Explain the difference between an *expansion* hand reamer and an *adjustable* hand reamer.

4.3 HAND TOOLS AND VISES

1. Of what material and by what process are the heads of ball peen hammers usually made?
2. When using an adjustable open end wrench, why must the force be applied in the direction of the adjustable jaw?
3. Briefly explain how a strap wrench works.
4. Describe the process by which socket wrenches are made.
5. Why is the opening of a typical open end wrench placed at an angle to the body?
6. Briefly explain the function of a torque wrench.
7. How can the tip of a common screwdriver be reconditioned?
8. Identify the processes and materials that can be used to make the frames of C clamps.

4.4 HAND THREADING TOOLS

1. Explain the function of taper, plug, and bottoming taps.
2. Identify and explain two unique features of pipe taps.
3. On what metal should cutting fluids *not* be used when tapping? Why?
4. Of what materials are taps usually made, and what characteristics of these materials make them suitable for taps?
5. How are split dies adjusted?
6. Why are the first three or four threads on the entering side of dies tapered?
7. Explain the size-designation system for taps and dies for both English and metric measuring systems.

4.5 PORTABLE POWER TOOLS

1. How are portable electric drills classified?
2. Why are some electric drills of the variable-speed type?
3. How is the speed controlled on air-driven portable power tools?
4. What safety precautions must be observed when using portable grinders and sanders?

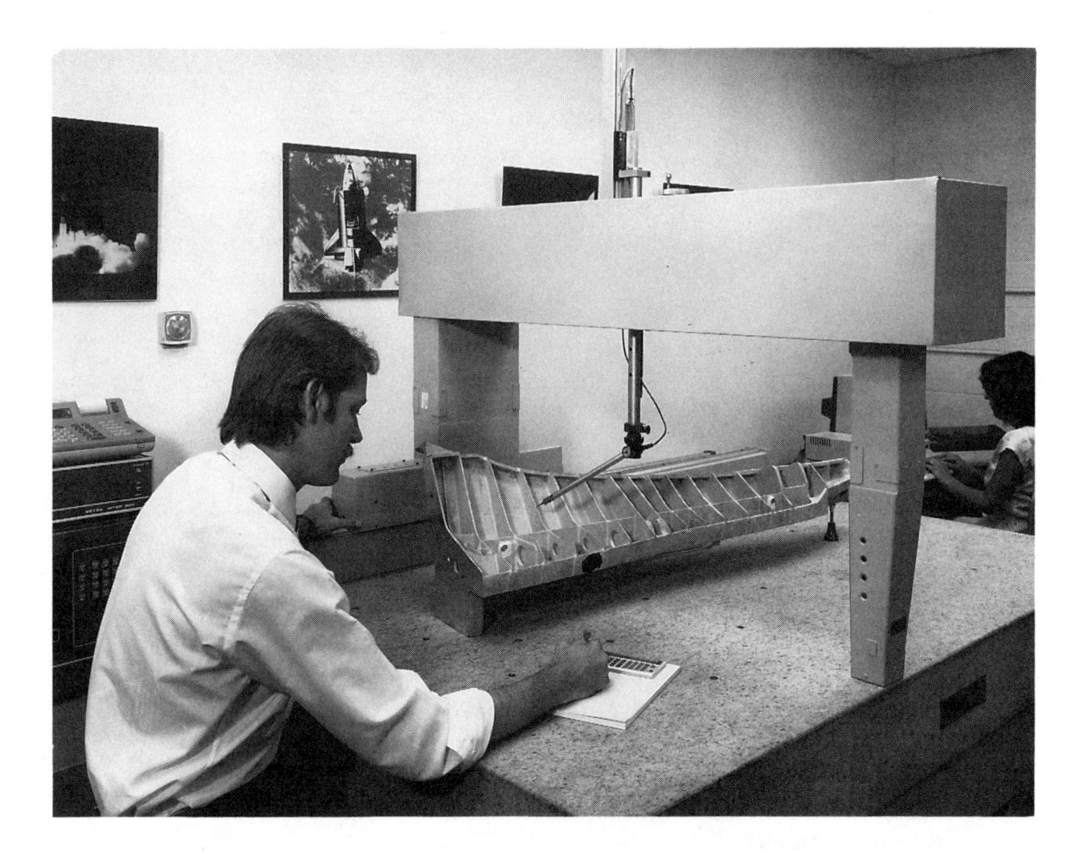

5 Measurement and Layout Procedures

5.1 INTRODUCTION

As industrial products and the machinery to produce them became more complicated, the importance of precise measurement increased. Machinists were required to work to closer tolerances, and better measuring tools were invented and developed. For example, the micrometer, which was invented before 1650, finally came into fairly common use in the United States in the decade just before the Civil War (see Fig. 5.1). As mass-produced machined parts became more common, complex measuring tools were used by machinists as well as by those who inspected parts and finished products.

To be an effective worker in modern machine shops, the machinist must be able to use all the common measuring and layout tools and at least some of the special measuring tools. The machinist must also understand both the English and metric systems of measurement because machinery and measuring tools based on both systems are now commonly used.

Tolerances, limits, allowances, and dimensioning procedures must also be understood. In many cases, selection of which measuring tools will be

Opening photo courtesy of L.S. Starrett Co.

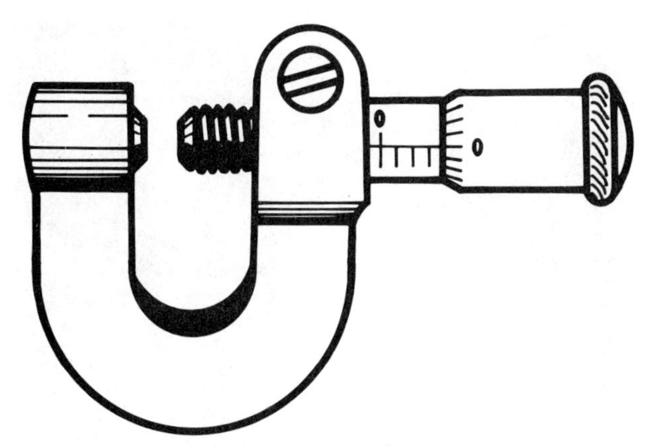

FIGURE 5.1 Micrometer of the Civil War era.

used in a particular situation is determined by specifications that appear on the engineering drawings.

5.2 NONPRECISION MEASURING TOOLS

For work with fairly large tolerances, nonprecision measuring tools such as steel rules, slide calipers without vernier scales, simple protractors, and tape measures are used (Fig. 5.2). A common characteristic of nonprecision measuring tools (sometimes called *semiprecision* tools) is that the *true* scale (the actual length) is read directly. The smallest increment is usually either 1/64 or 1/100 in. Other types of nonprecision measuring tools are used for transferring measurements, comparing profiles of curves or threads, and in layout operations.

FIGURE 5.2 Some measuring tools used by machinists. (*Courtesy L.S. Starrett Co.*)

5.2.1 Steel Rules

Steel rules of a large variety of types and lengths are used in the machine shop (see Fig. 5.3). Good-quality steel rules are made of hardened and tempered alloy steel or stainless steel. Alloy steel rules sometimes have a satin chromium plate finish with black markings. Most measuring tool manufacturers make steel rules in lengths ranging from 1 to 48 in. or more. Rules are *flexible* or *rigid*. Flexible rules can be narrower and are always thinner than a rigid rule of a comparable length.

Steel rules are graduated in English measure, metric measure, or a combination of the two. Rules for English measure are graduated with eights and sixteenths on one side and thirty-seconds and sixty-fourths on the other side (see Fig. 5.4). In some cases, one side of the rule is divided in tenths and fiftieths or tenths and one-hundredths of 1 in. (Fig. 5.5). Metric rules are subdivided into 1- and 0.50-mm graduations (see Fig. 5.6).

Hook Rule. Hook rules are generally available in narrow and regular widths. The hook (see Fig. 5.7) is attached to the rigid rule with a screw. This type of rule is very useful for taking measurements through holes or from nonvisible edges.

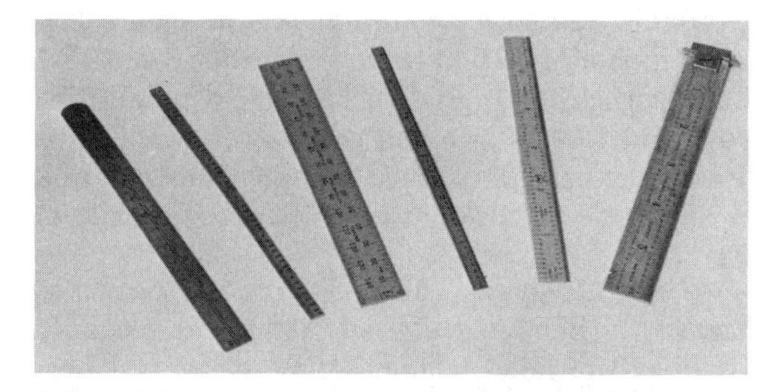

FIGURE 5.3 Steel rules used in machine shops.

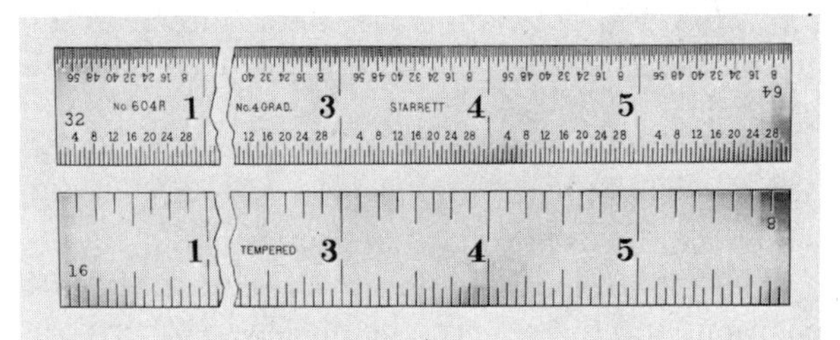

FIGURE 5.4 English measure steel rule. (*Courtesy L.S. Starrett Co.*)

FIGURE 5.5 English measure steel rule with decimal subdivisions. (*Courtesy L.S. Starrett Co.*)

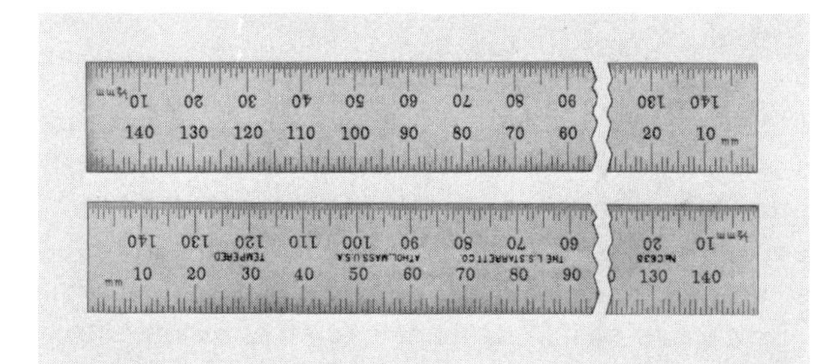

FIGURE 5.6 Metric steel rule. (*Courtesy L.S. Starrett Co.*)

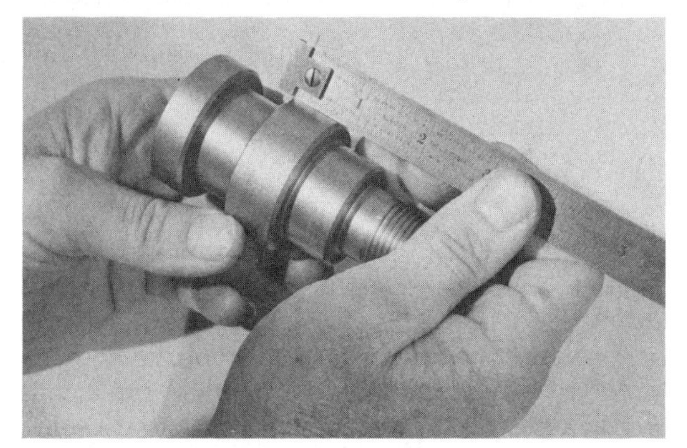

FIGURE 5.7 Hook rule.

Narrow Rule. Narrow rules (Fig. 5.8) are available with or without a hook on the end. The rule is usually 3/16 in. wide and either 6 or 12 in. long, with 1/32-in. graduations on one side of the rule and 1/64-in. graduations on the other side. These types of rules are used for measuring in confined spaces or through small holes.

Depth Gage. This very useful tool, also known as the *rule depth gage* is used to measure the depth of holes, counterbores, slots, and keyways. The rule is clamped to the head by a knurled screw (see Fig. 5.9), and a slot helps interpret the graduations on the scale. A 6-in.-long rule is usually used.

A variation of the rule depth gage is the combination depth and angle gage (see Fig. 5.10), which can be used to lay out angles of 30, 45, 60, and 90°. Two small pins locate the rule in the 90° position, and the rule can be placed in other positions by moving the reference line on the clamp to the desired angular setting.

Caliper Rule. The caliper rule, often called the *slide caliper* (see Fig. 5.11), is used for both internal and external measurements and has 1/32- and 1/64-in. divisions on the true scale. Outside measurements are read at the *out* reference mark; internal measurements are read at the *in* mark.

FIGURE 5.8 Narrow rule in use.

FIGURE 5.9 Measuring with a rule depth gage.

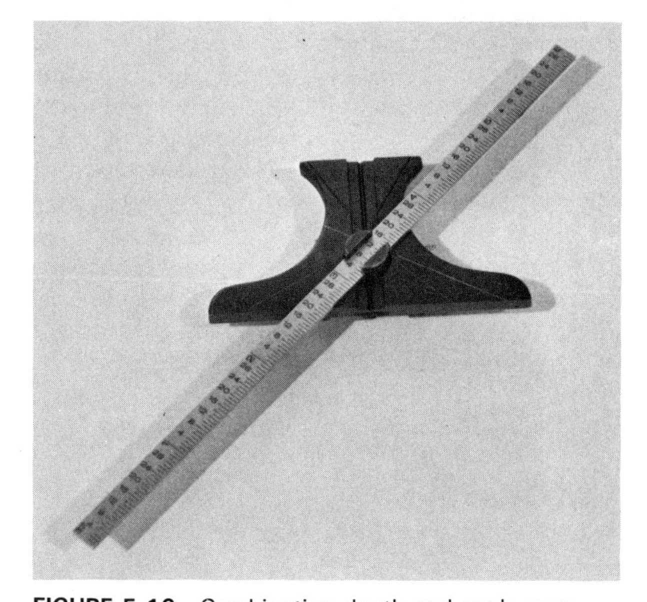

FIGURE 5.10 Combination depth and angle gage.

FIGURE 5.11 Slide caliper rule in use.

There is no vernier scale on this caliper, so it can be read accurately to only 1/64 in. The end of each jaw has an 1/8-in.-thick radius, so the smallest internal measurement that can be taken is 1/4 in. The caliper has a locking mechanism that secures the slide in any position.

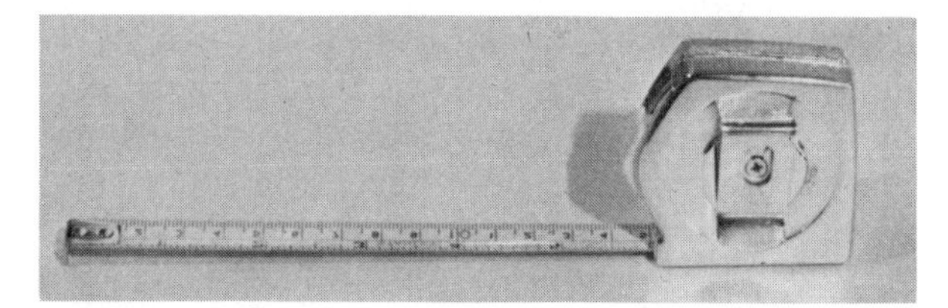

FIGURE 5.12 Tape rules.

Tape Rule. This tool, commonly known as a *steel tape* (Fig. 5.12), is used for measuring larger objects, although it can be used for measuring the diameter of bar stock and similar tasks. The tape rule is usually available in lengths of 6 to 12 ft, with one-thirty-second graduations for the first 6 in. and one-sixteenth graduations after that. It retracts into its case when not in use.

A hook that slides an amount equivalent to its thickness is attached at the end of the tape to provide accurate internal and external measurements. The case of the rule is usually 2 in. wide, so when making internal measurements, 2 in. have to be added to the reading on the tape. Some steel tapes are graduated in both English and metric measure.

5.2.2 The Combination Set

The combination set (see Fig. 5.13) consists of a rigid *steel rule* with a groove in it, a *square head,* a *protractor head,* and a *center head.* The steel rule,

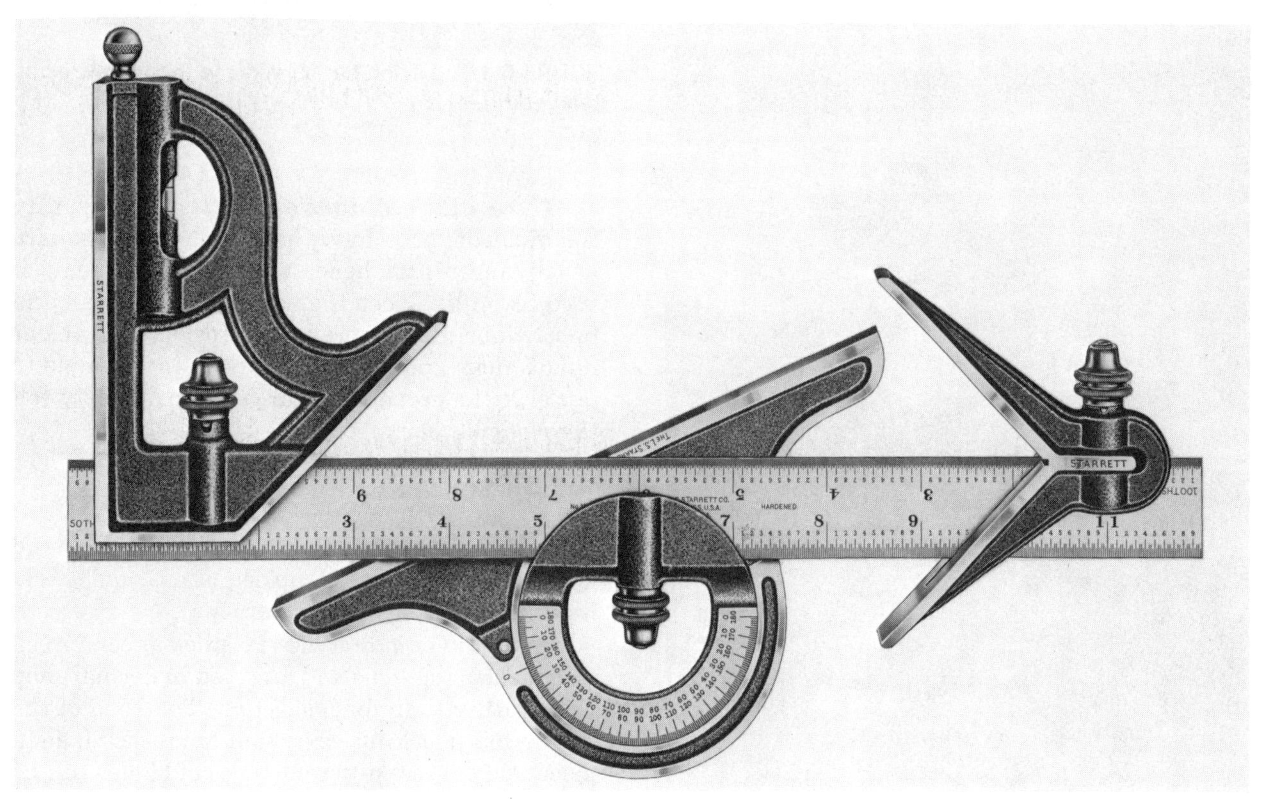

FIGURE 5.13 Combination set. (*Courtesy L. S. Starrett Co.*)

or *blade,* is usually available in 6-, 12-, or 24-in. lengths. Combination sets with 6 in. rules usually have smaller center heads and square heads and may not have a protractor head. The square and protractor heads usually have spirit levels built in. Although these are not precision levels, they are useful for measuring angles where tolerances of 1° or more are permissible. The square head may also have a scriber attached.

The square head, when fitted to the rule, can be used for both measurement and layout work and for checking parts for squareness. It can be used to lay out lines parallel to an edge, lay out 45 or 90° angles, or measure the depth of cavities or slots (see Fig. 5.14).

The *center head* arms are at an angle of 90° to each other, and the edge of the rule bisects this angle when it is clamped in place. The main functions of the center head are locating the center of squares and circles and laying out lines at a 45° angle from the corners of plate or sheet stock (see Fig. 5.15).

The *protractor head* (see Fig. 5.16) is used in conjunction with the rule to measure or lay out angles and to set up parts for machining operations. Most protractor heads are conveniently graduated from 0 to 180° in either direction and have clamping screws for holding the protractor to the blade and for holding angular settings on the degree scale.

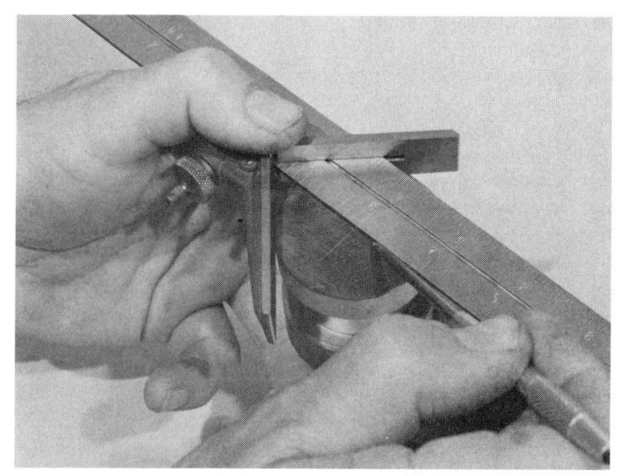

FIGURE 5.15 Finding the center of a round object with center head.

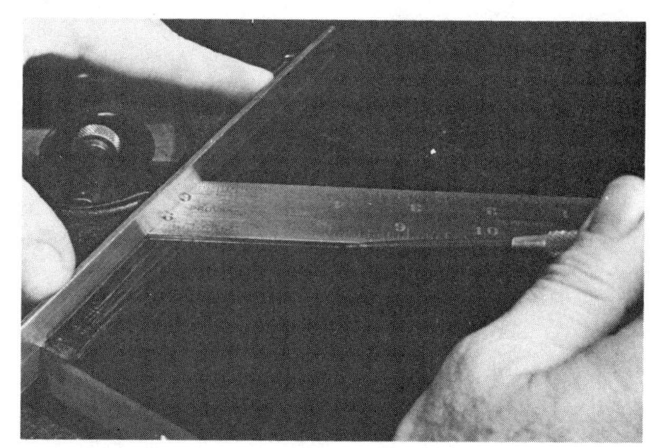

FIGURE 5.16 Laying out an angle with the protractor head and rule.

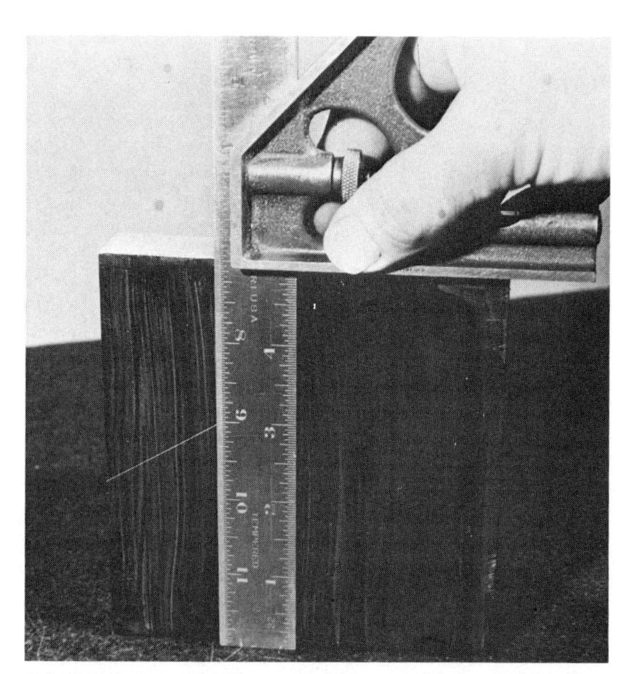

FIGURE 5.14 Height measurement with a combination square and blade.

Care of Combination Sets. Good-quality combination sets have hardened rules, square heads, and center heads and last many years if used carefully. Keep the working parts clean and lightly lubricated to prevent corrosion. Smooth out minor nicks and scratches with a fine oilstone to preserve the proper angular relationship between the working parts.

5.2.3 Squares

The machinist uses various squares for bench and layout work, machine setup operations, and checking machined parts for accuracy.

Adjustable squares are classified as semiprecision measuring tools and are used in normal shop work rather than for inspection. The head of the square has a locking screw that holds the blade in place and several different types of blades can be used. The blades are graduated or plain and have square or angled ends (see Fig. 5.17). The working

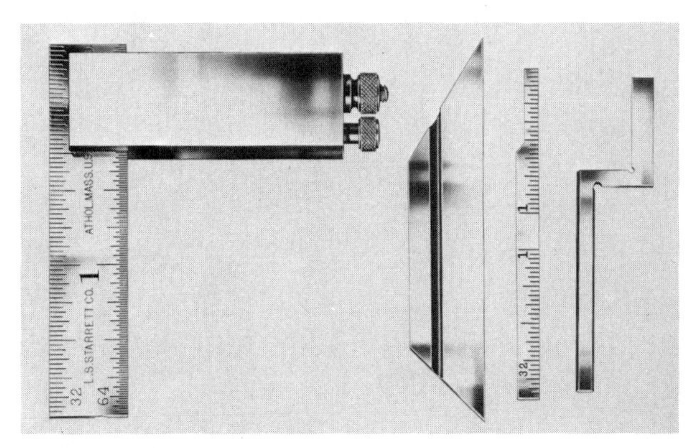

FIGURE 5.17 Adjustable square. (*Courtesy L.S. Starrett Co.*)

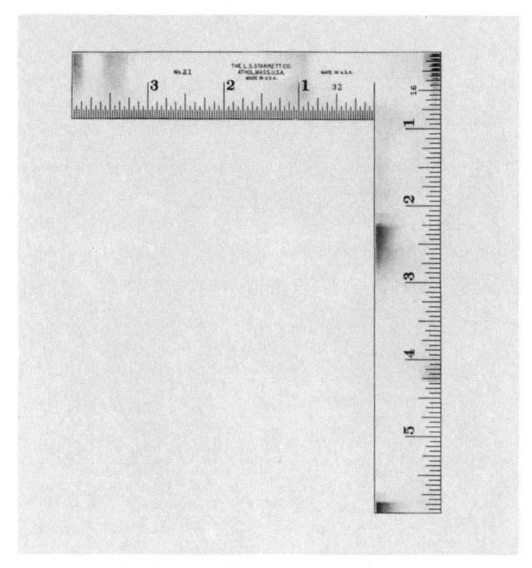

FIGURE 5.19 Millwright's square. (*Courtesy L.S. Starrett Co.*)

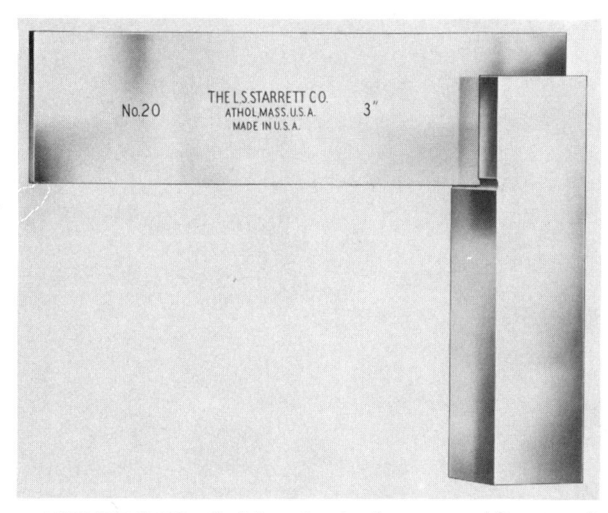

FIGURE 5.20 Solid toolmaker's square. (*Courtesy L.S. Starrett Co.*)

parts of the square are usually made of hardened alloy steel. The blade and slot of any adjustable square must be clean and free of nicks and burrs.

Diemakers' squares are adjustable and give a direct reading of angles varying up to 10° from a 90° angle (see Fig. 5.18). They are available with several types of nongraduated and graduated blades and are very useful for measuring angles and bevels in confined spaces.

Solid squares may be semiprecision or precision. The millwright's square (see Fig. 5.19) is a semiprecision tool used for some types of layout work. Assorted precision solid squares are used by the machinist for inspection and layout work. Solid squares of the type shown in Fig. 5.20 are made from hardened alloy steel parts joined at a precise 90° angle.

The cylindrical square, which is used on a surface plate (see Fig. 5.21), is a true cylinder, with the numbered end lapped square and the other end lapped at an angle with the sides. Out-of-squareness can be read directly in increments of 0.0002 to 0.0012 in. by placing the angled end of the square on the surface plate and bringing the part to be checked in contact.

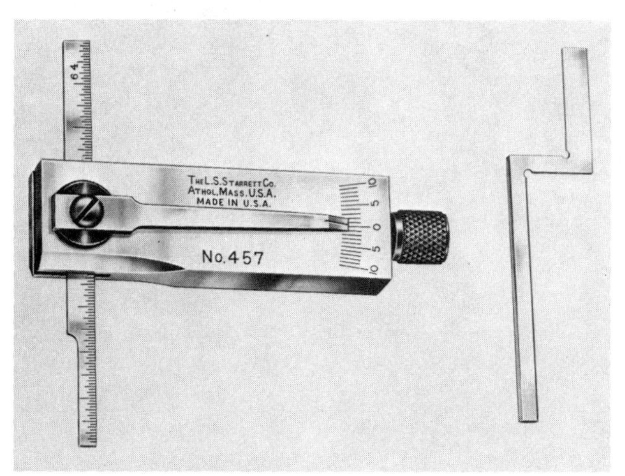

FIGURE 5.18 Diemaker's adjustable square. (*Courtesy L.S. Starrett Co.*)

If the part makes full contact with the square at either of the two 0 marks, it is square. If it makes full contact elsewhere on the square, find the amount of out-of-squareness by reading up to the number from the uppermost dotted line in contact. One side of the square shows out-of-squareness above 90°; the other side is used for angles of less than 90°.

5.2.4 Calipers and Dividers

A variety of calipers are used by the machinist for transferring measurements and doing setup and layout work. Calipers and dividers are classified by the length of the leg, and the fully open capacity

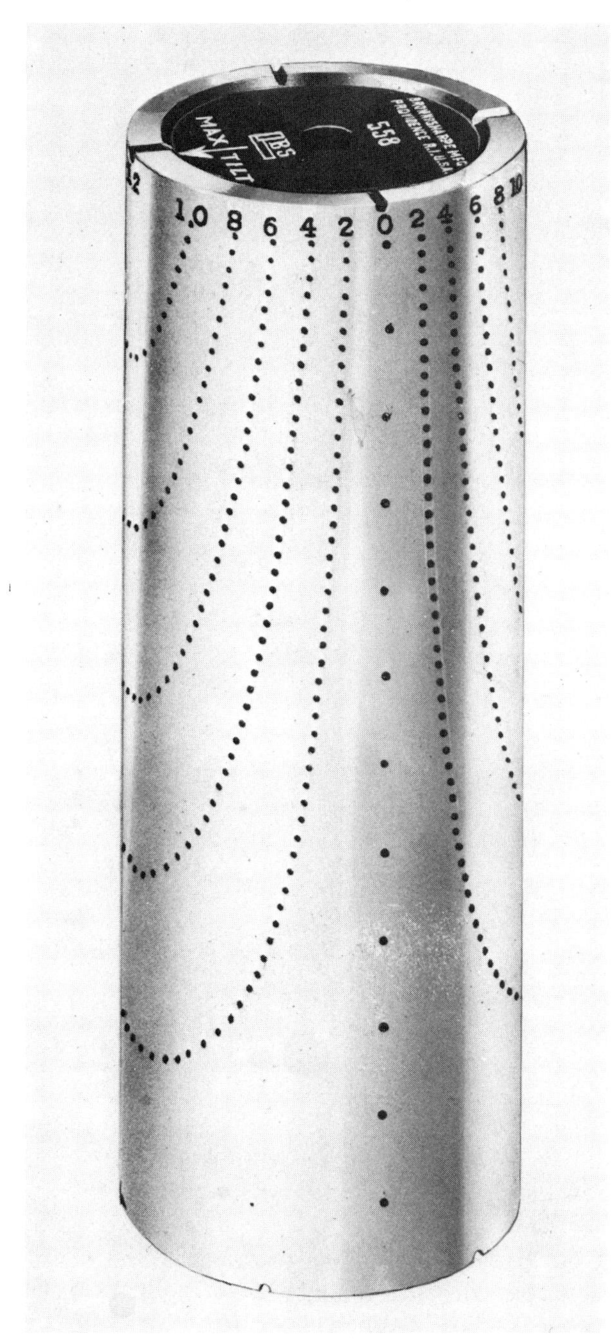

FIGURE 5.21 Cylindrical square. *(Courtesy Brown & Sharpe Manufacturing Co.)*

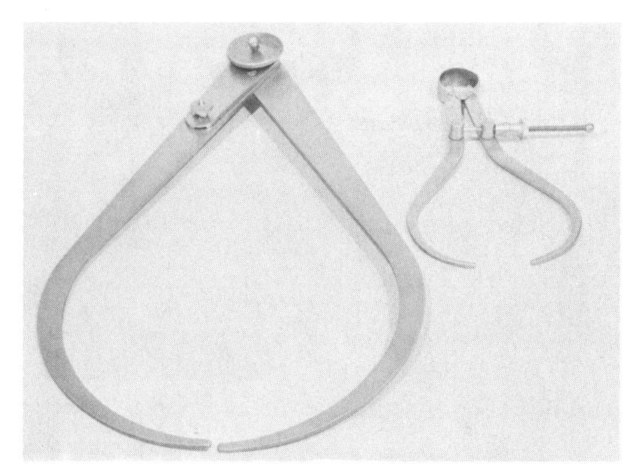

FIGURE 5.22 Spring-joint and firm-joint calipers.

of the tool is about the same as the leg length. For example, a 6-in. outside caliper goes over a workpiece 6 in. in diameter when fully extended.

Calipers are *spring-joint* or *firm-joint* (see Fig. 5.22). *Spring-joint* calipers have a circular spring above the pivot point that tends to move the legs outward. A screw and nut mechanism adjusts the calipers to the desired setting. Some calipers have a split nut that permits large adjustments to be made rapidly. Spring calipers and dividers are usually available in 2- to 8-in. sizes. Spring-joint calipers can be set as shown in Fig. 5.23. Measurements may be transferred as shown in Fig. 5.24.

Firm-joint calipers range in size from 6 to 36 in., and the capacity of the caliper is about 30 percent greater than the nominal size (see Fig. 5.25). *Lock-joint* calipers are a variation of firm joint calipers and have a large knurled nut that locks the caliper at the desired setting. Minor adjustments are then made by turning the small knurled nut when setting the calipers or transferring a measurement (see Fig. 5.26).

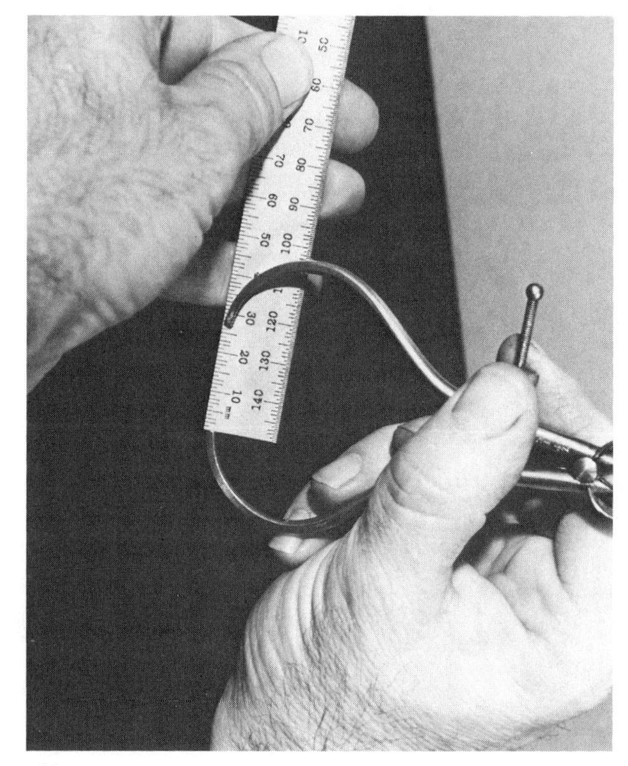

FIGURE 5.23 Setting a spring-joint caliper.

FIGURE 5.24 Inside caliper being set to transfer a measurement.

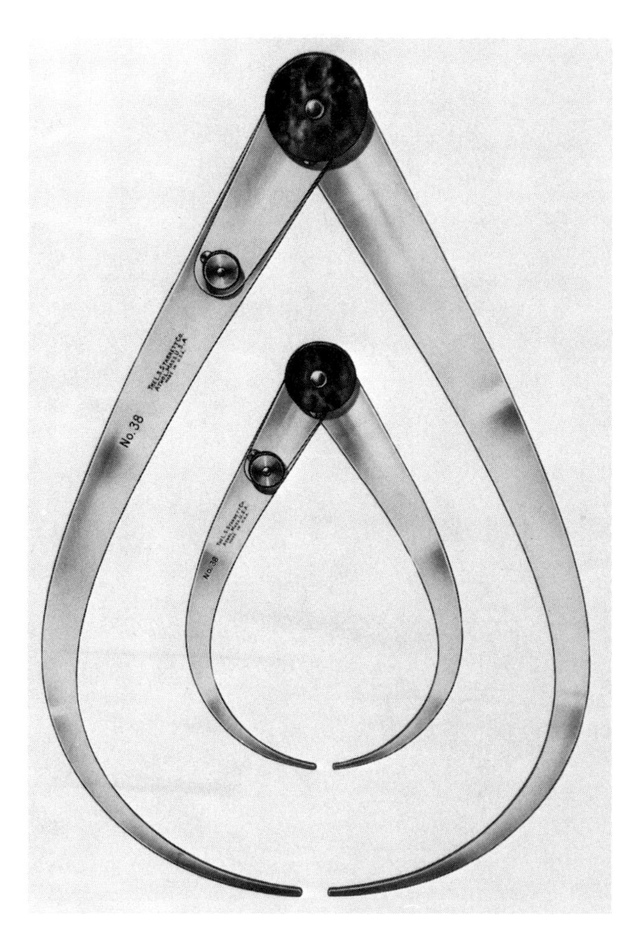

FIGURE 5.26 Lock-joint calipers. (*Courtesy L.S. Starrett Co.*)

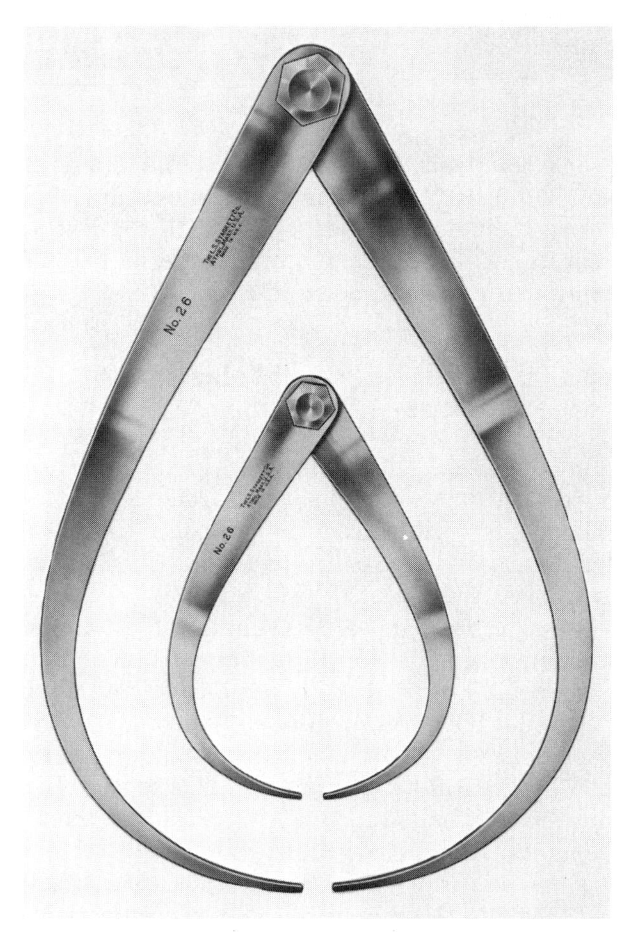

FIGURE 5.25 Firm-joint outside calipers. (*Courtesy L.S. Starrett Co.*)

Lock-joint transfer calipers take measurements in grooves or around edges where it impossible to remove the caliper after it is set. When the transfer caliper is used, one leg and the transfer arm (the short arm with the slotted opening shown in Fig. 5.27) are locked with the large knurled nut. The other leg is then released by loosening the small knurled nut. With one leg free, the caliper can be removed from the work-piece being measured. The leg is then moved back into place and locked with the small knurled nut. The measurement can then be checked with a steel rule.

Hermaphrodite calipers are used in layout work for locating centers and scribing lines along edges (see Fig. 5.28). They are usually available only in firm- or lock-joint types, and the pointed leg is fixed or adjustable.

Dividers (see Fig. 5.29) and *trammels* (see Fig. 5.30) are used for laying out and scribing circles and arcs, stepping off distances on straight lines or circles, and transferring measurements. Dividers may be of the spring-joint or firm-joint type, and the size of both types is determined by the length of the leg.

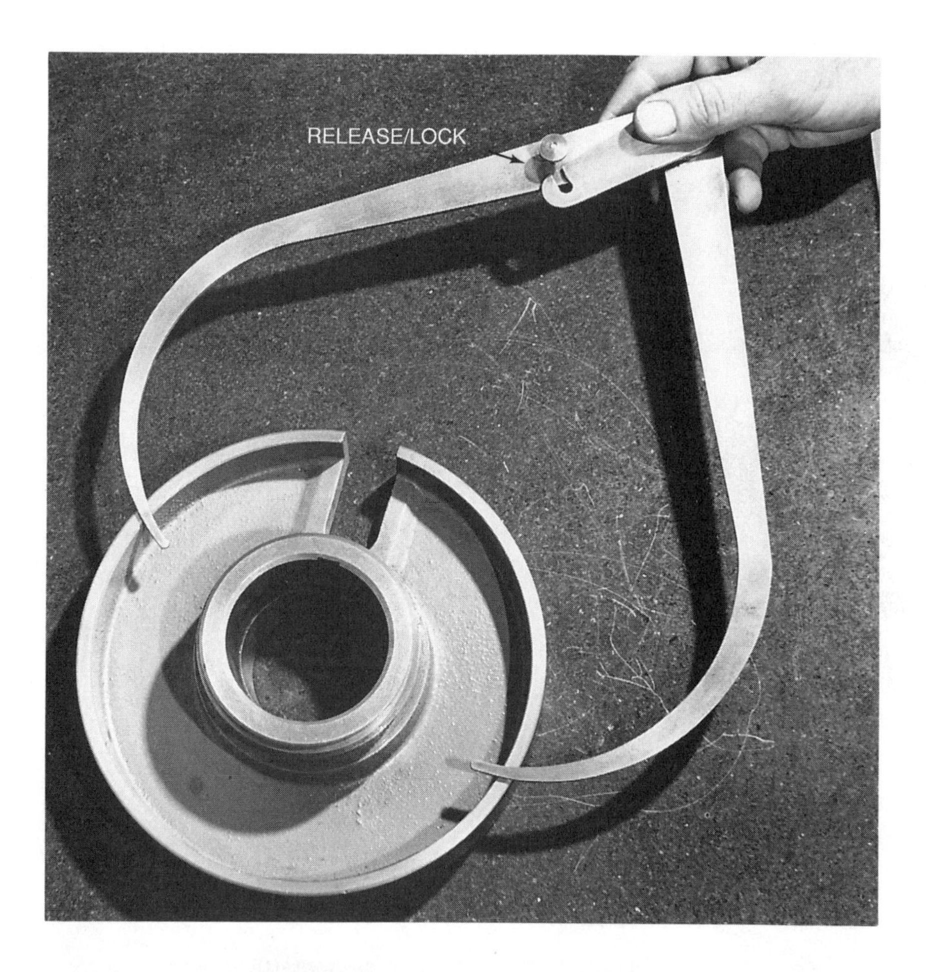

FIGURE 5.27 Transfer calipers.

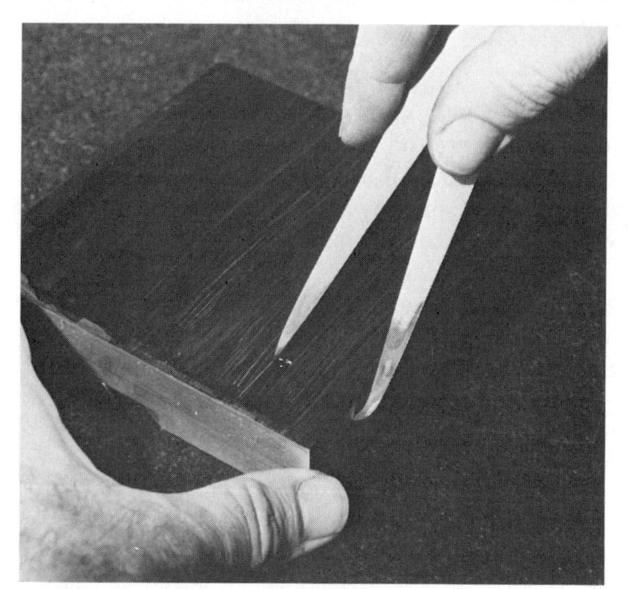

FIGURE 5.28 Scribing a line with hermaphrodite calipers.

The trammel is used for layout and measurement jobs that are beyond the capacity of dividers. The main part of the trammel is the *beam,* which is round or rectangular. The two *heads* are clamped to the beam by screws with knurled heads. On some trammels, one head is adjustable by a screw and nut mechanism. Many attachments are available, including ball attachments that allow accurate measurements from the center of holes. A trammel in use is shown in Fig. 5.31.

5.2.5 Small-Hole and Telescoping Gages

Small-hole and telescoping gages are *transfer* tools used to transfer the dimension from the part being measured. A micrometer or other precision tool is then used to measure the setting on the gage.

Small-hole gages (see Fig. 5.32) are usually sold in sets of four and are available in ball or flat-bottom types. The operating range of a set of gages is from 1/8 to 1/2 in. The basic operating mechanism of the gage consists of (1) the body, with a split ball on the end; (2) a threaded draw screw, with a conical portion that fits in a tapered hole in the split ball; and (3) a knurled nut at the top end. As the knurled nut is tightened, the conical segment is drawn into the ball, thus expanding it.

Ball gages are used in holes and deep slots, and the flat-bottomed gage is used to measure the width of shallow grooves. Do not overtighten the tools,

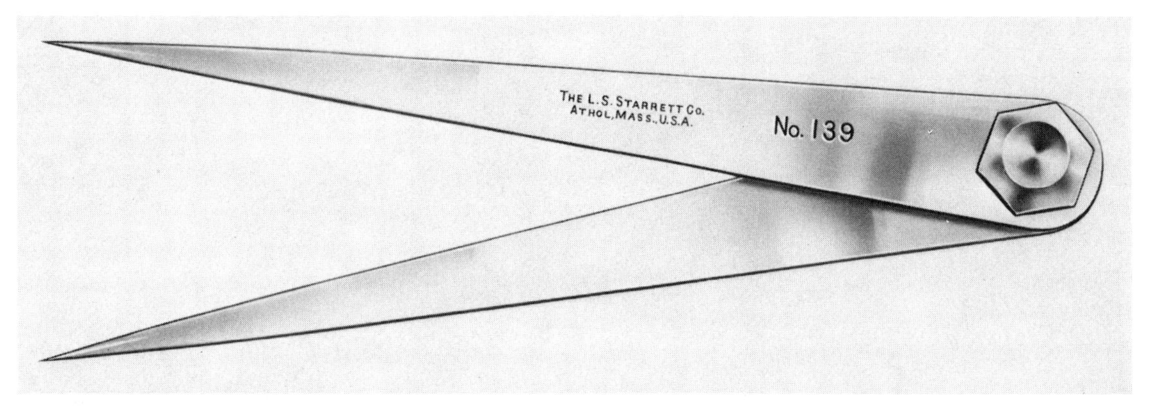

FIGURE 5.29 Firm-joint dividers. (*Courtesy L.S. Starrett Co.*)

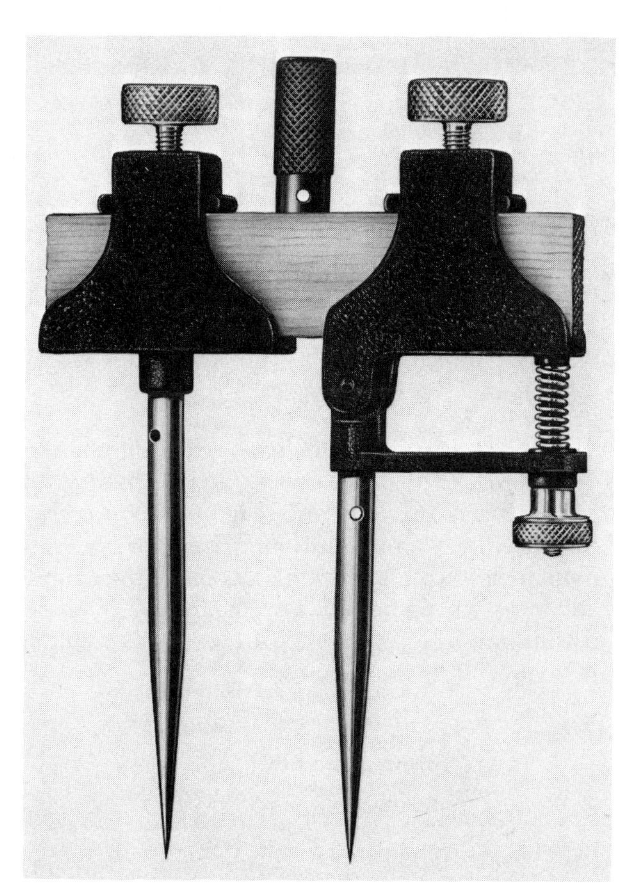

FIGURE 5.30 Trammels. (*Courtesy L.S. Starrett Co.*)

and make sure that the tool and part being measured are clean.

Telescoping gages (see Fig. 5.33) consist of two tubular plungers that are forced apart by internal springs, a handle, and a locking device at the end of the handle. A set of telescoping gages ranges from about 3/8 to 6 in. in capacity. They can be used to transfer measurements from holes or the distance between two surfaces. Once it has been set, the telescope gage is measured with a micrometer or vernier caliper, as shown in Fig. 5.34.

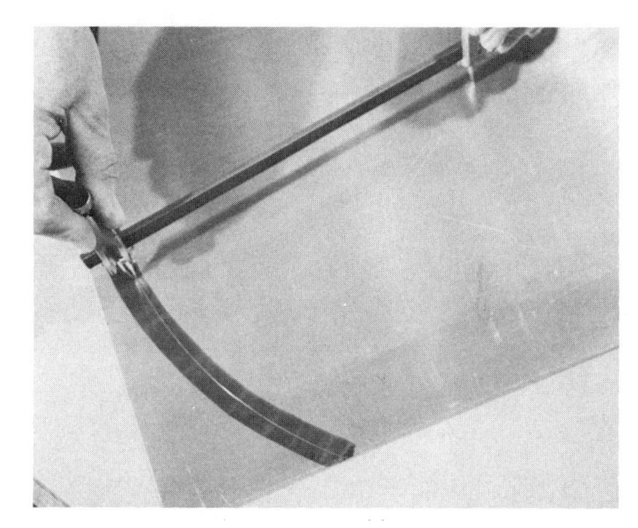

FIGURE 5.31 Scribing an arc with a trammel.

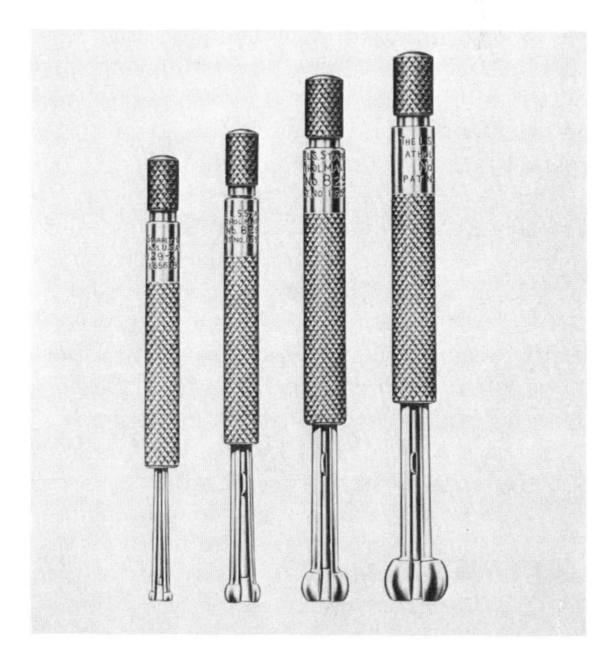

FIGURE 5.32 Small-hole gages. (*Courtesy L.S. Starrett Co.*)

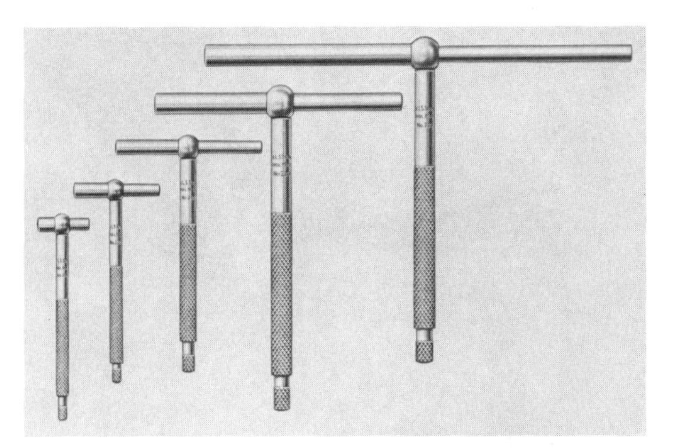

FIGURE 5.33 Telescoping gages. (*Courtesy L.S. Starrett Co.*)

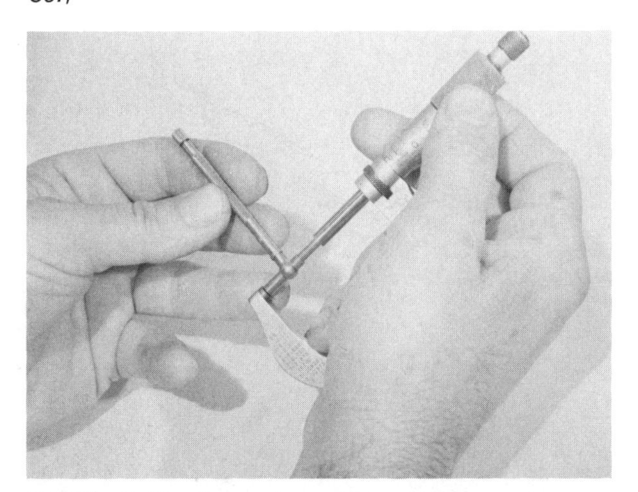

FIGURE 5.34 Measuring a telescope-gage setting.

5.3 PRECISION MEASURING TOOLS

The ability to use many precision measuring tools is an absolute necessity for the machinist. These types of tools are used when doing bench and machine operations and when inspecting completed work. In many cases the accuracy of layout work and machine setups is heavily dependent on the expert use of precision measuring tools.

Generally speaking, the term *precision* applies to those measuring tools that require the use of a screw or other device to amplify the motion of the part that is read by the user. A micrometer and a dial indicator are examples of this principle. A differential scale, such as is used in vernier calipers, height gages, and protractors, may also be used to enhance accuracy by providing larger increments for the user to read. In all cases, the larger increments *represent* smaller increments that cannot be read accurately in their true length.

5.3.1 Micrometers

The *micrometer caliper,* in its various forms, is the most widely used precision measuring tool (Fig. 5.35). Although the operating principle of the

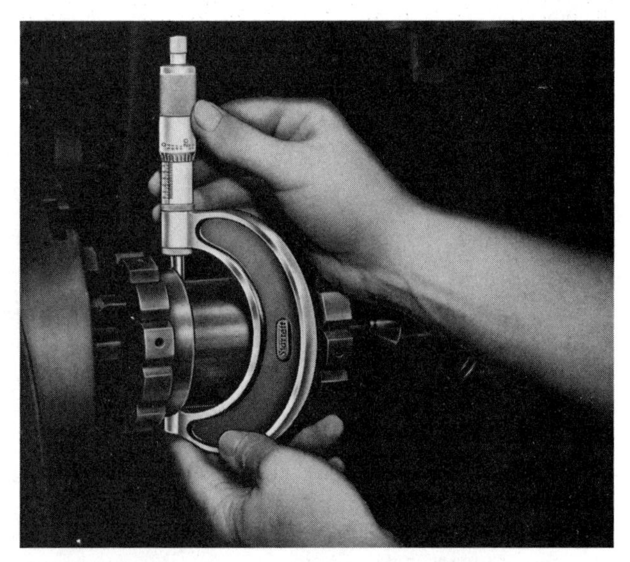

FIGURE 5.35 The micrometer is the basic precision measuring tool. (*L.S. Starrett Co.*)

micrometer was developed almost 350 years ago, micrometers were not produced in significant quantities until just before the U.S. Civil War. As the need for precisely machined products increased rapidly, many variations of the basic outside micrometer were developed.

Operating Principles. All micrometers operate on the principle that an accurate screw will advance the *spindle* (see Fig. 5.36) a precise distance for each revolution. When a screw with a pitch of 40 threads per inch is used, the spindle advances one-fortieth of an inch, or 0.025 in. per revolution. The pitch *(p)* of a thread is computed in thousandths, as follows:

$$p = \frac{1}{\text{number of threads per inch}} = \frac{1}{40} = 0.025 \text{ in.}$$

The *thimble* (Fig. 5.36) is graduated at its lower end into 25 divisions, each division representing one thousandth, or 0.001 in. of spindle travel. On almost all modern micrometers, each division on the thimble is numbered, with each fifth number offset and etched larger as shown in Fig. 5.37.

Regardless of whether the micrometer is used for inside, outside, or depth measurements, the operating range of the spindle, thimble, and sleeve graduations is usually 1 in. On some inside micrometers, the operating range is 1/2 in.

A *vernier* scale can be added to the sleeve of a micrometer to allow direct readings in increments of 1/10,000 in. The vernier scale is etched

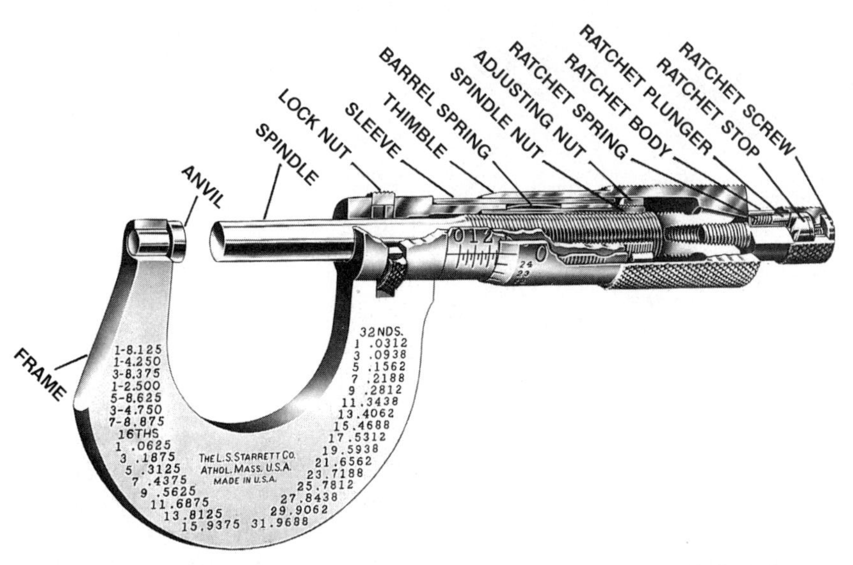

FIGURE 5.36 Outside micrometer. (*Courtesy L.S. Starrett Co.*)

FIGURE 5.37 Vernier scale on a sleeve.

into the sleeve in a convenient position near the true scale (see Fig. 5.38) and has 10 numbered divisions. The 10 divisions on the vernier scale are equal in length to 9 divisions on the thimble, which represents 0.009 in. Each division on the vernier represents 0.0009 in., and because each division on the thimble represents 0.001 in., the *difference* between one vernier scale division and one thimble division represents 0.0001 in. Therefore, movement of the thimble can be accurately read in increments of 1/10,000 in.

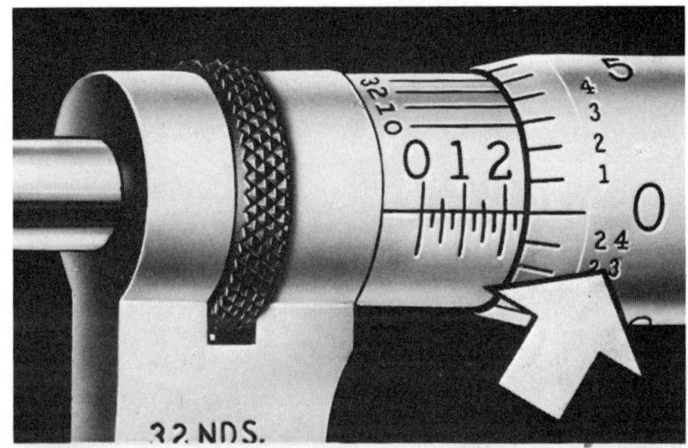

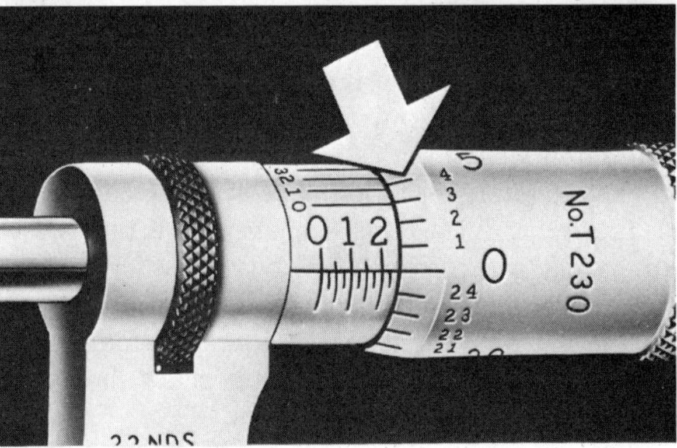

FIGURE 5.38 Micrometer reading of 0.250 in. (*Courtesy L.S. Starrett Co.*)

Reading Micrometers. Because the micrometer spindle has 40 threads per inch, one revolution of the thimble advances the spindle 0.025 in., or the length of one mark on the sleeve. Each mark on the *true scale* on the sleeve of the micrometer is actually 0.025 in. long. Every fourth mark is made longer than the others and numbered. The numbered marks are 0.100 in. apart. Basically, reading the micrometer consists of noting and adding the readings, starting with the largest increments and progressing to the smallest. For example, the micrometer reading shown in Fig. 5.39 is read as follows:

Division 6 is visible on the sleeve. (Remember that each numbered division is 0.100 in. in length.)	**6 × 0.100 = 0.600**
Three small divisions, each 0.025 in. long, are visible on the sleeve.	**3 × 0.025 = 0.075**
Number 13 on thimble is aligned with the index line on the sleeve. Each division on the thimble represents 0.001 in.	**13 × 0.001 = 0.013**
When all the readings are added, the total is 0.688 in.	**0.688**

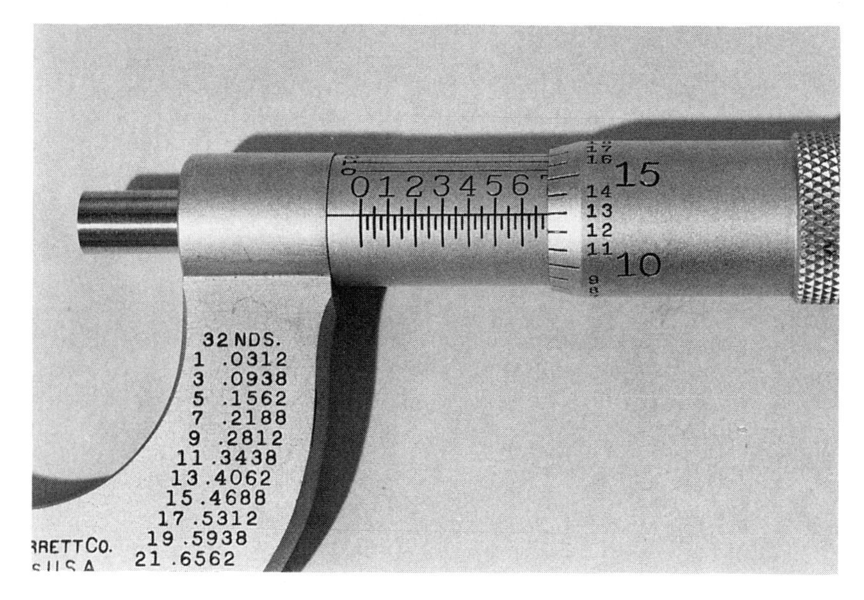

FIGURE 5.39 Micrometer reading of 0.688 in.

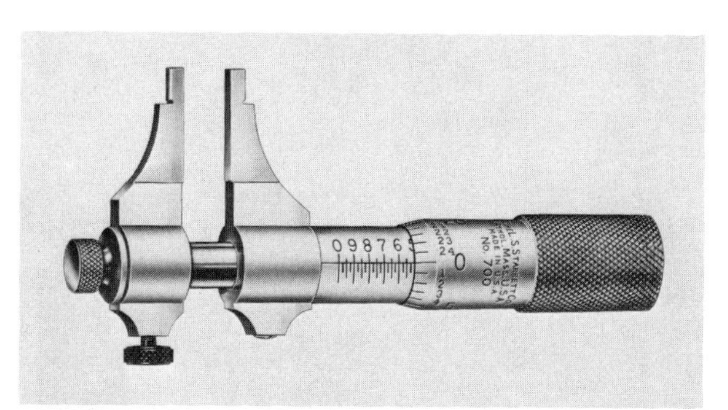

FIGURE 5.40 Inside micrometer caliper. (*Courtesy L.S. Starrett Co.*)

The type of inside micrometer shown in Fig. 5.40 and the depth micrometer in Fig. 5.41 have graduated sleeve scales with the numbers rising in value from right to left. The scale on the thimble is also graduated in the opposite direction when compared to a conventional outside micrometer. Micrometers of this type are also read by noting the largest increment and adding the smaller increments. The depth micrometer scale shown in Fig. 5.42 is thus read as follows:

The edge of the thimble has gone past division 7 on the sleeve.	7 × 0.100 = 0.700
The edge of the thimble has also gone past two *small divisions, and the number* 19 *on the thimble is aligned with the index line on the hub.*	2 × 0.025 = 0.050
When all the readings are added, the total is 0.769 in.	19 × 0.001 = 0.019
	0.769

Reading Vernier Micrometers. The procedure for reading a vernier micrometer is basically the same as for reading a micrometer with

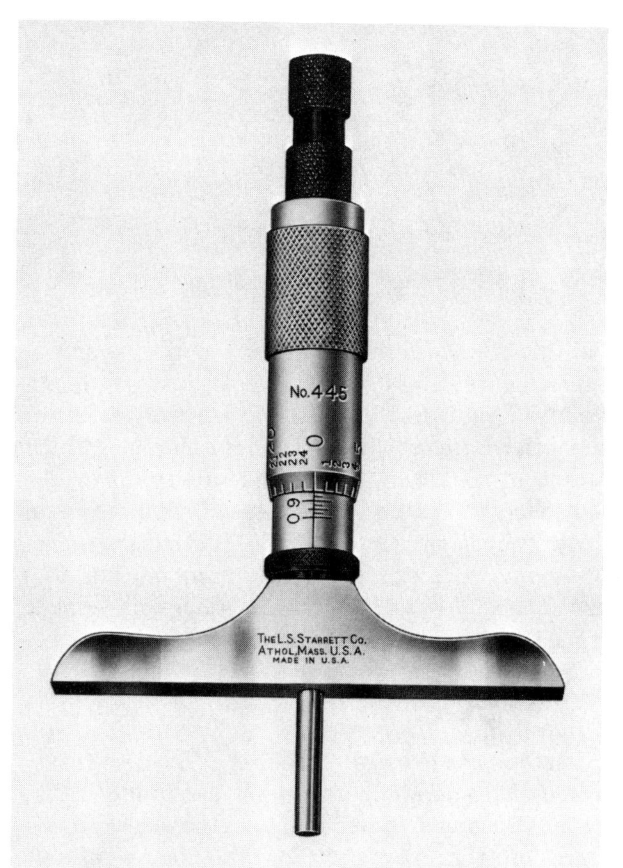

FIGURE 5.41 Depth micrometer. (*Courtesy L.S. Starrett Co.*)

thousandth graduations, except that the reading in ten-thousandths from the vernier scale must be added to the reading in thousandths. The reading shown in Fig. 5.43 is handled as follows:

Division 6 shows on the sleeve.	6 × 0.1000 = 0.6000
Two small divisions are visible. The index line on the sleeve is near the mid-	2 × 0.0250 = 0.0500

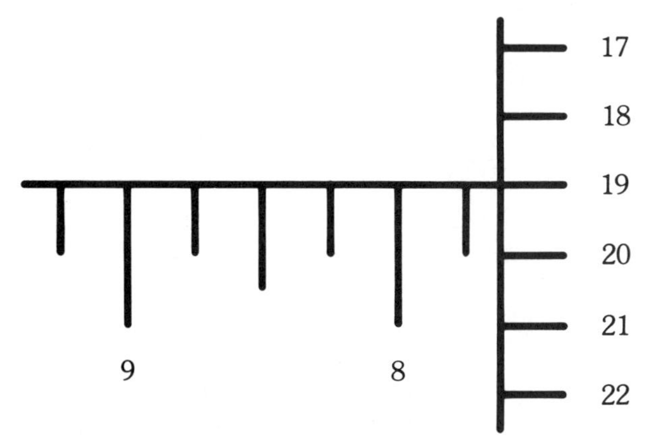

FIGURE 5.42 Depth micrometer reading of 0.769 in.

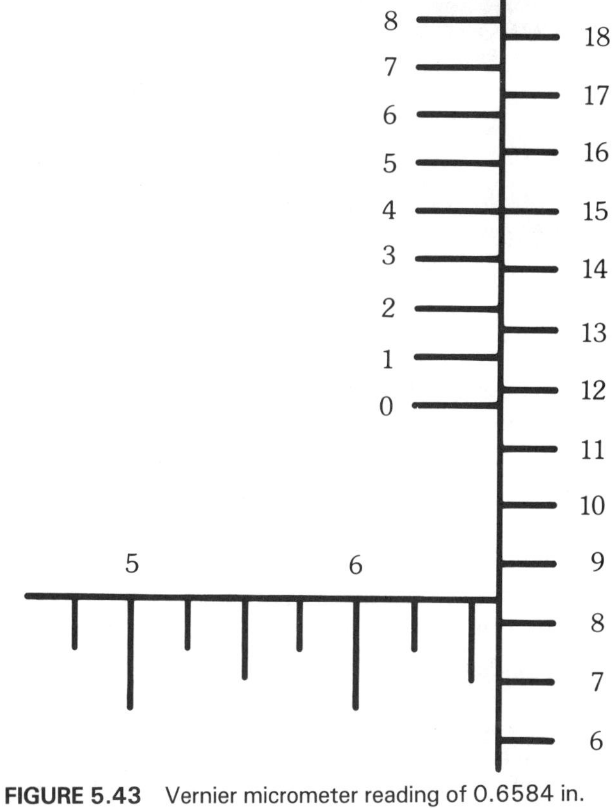

FIGURE 5.43 Vernier micrometer reading of 0.6584 in.

point between the eighth and ninth divisions. Therefore, the smaller reading (0.0080) is used.	**8 × 0.0010 = 0.0080**
Line 4 on the vernier scale lines up with a line on the thimble. This represents 0.0004 and should be added to the reading already noted. A final reading of 0.6584 is achieved by adding.	**= 0.0004**
	0.6584

Reading Metric Micrometers. Metric micrometers are similar in appearance to micrometers graduated for the English system of measurement (see Fig. 5.44). They are available in 0- to 25- and 25- to 50-mm sizes, with a vernier scale that allows readings in increments of 0.002 mm. Metric micrometers without vernier scales are usually available in sizes ranging from 0 to 25 to 125 to 150 mm in increments of 25 mm. Metric micrometers without vernier scales can be read in increments as small as 0.01 mm.

Because the pitch of the spindle screw on metric micrometers is 0.5 mm, the spindle advances or retracts that amount for each revolution. The sleeve of the micrometer has graduations both above and below the index line. The 25 upper graduations are 1 mm apart, and each fifth graduation is numbered. Two turns of the thimble moves the spindle one graduation, or 1 mm.

The lower graduations are midway between the upper graduations and therefore subdivide the upper graduations into 0.5-mm increments. The circumference of the thimble is divided into 50 equal increments, each representing 0.01 mm. Each fifth line is longer than the others and is numbered.

The procedure for reading a metric micrometer

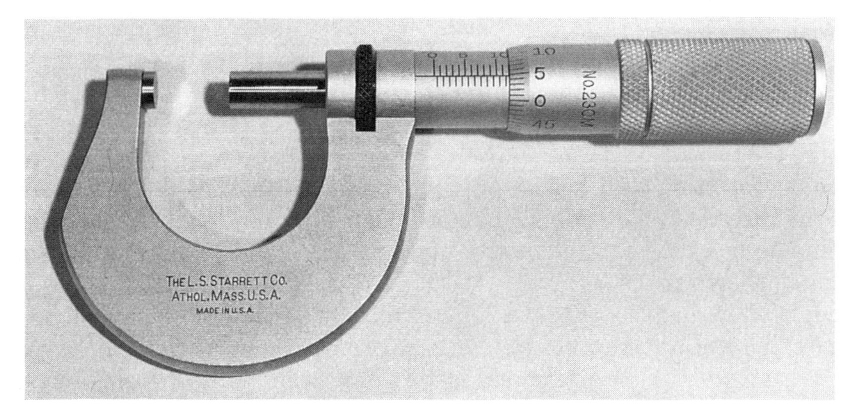

FIGURE 5.44 Graduations on metric micrometer.

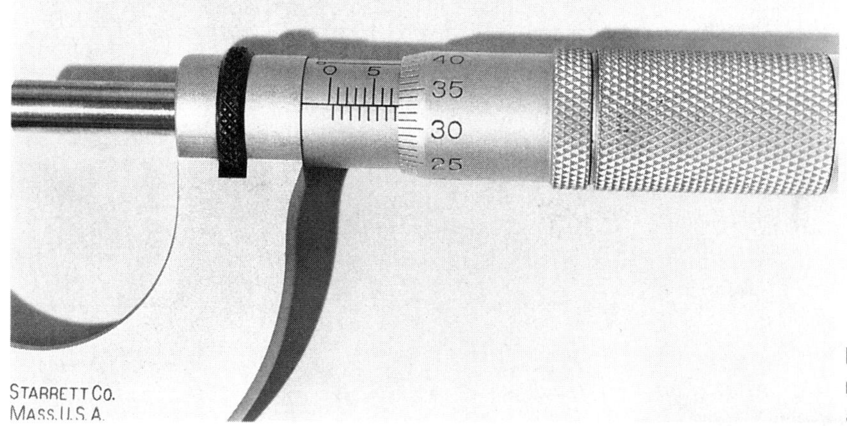

FIGURE 5.45 Metric micrometer reading of 7.83 mm. (*Courtesy L.S. Starrett Co.*)

is basically the same as for reading English measure micrometers. The largest increment is read first, and the smaller increments are added as shown (see Fig. 5.45):

Division 7 is visible on the upper scale.	**7 × 1 mm = 7.00 mm**
The 0.5 millimeter line may be seen on the lower scale.	**1 × 0.5 mm = 0.50 mm**
On the thimble, line 33 is aligned with the reference line, so since each division on the thimble represents 0.01 mm, 0.33 is added to the reading. The final reading is 7.83 mm.	**33 × 0.01 mm = 0.33 mm**
	7.83

Care of Micrometers. If a micrometer is to retain its accuracy in normal shop use, it must be handled carefully and kept clean. Any grit or small chips on the anvil and spindle end of the micrometer cause wear and result in inaccurate readings. Therefore, always clean the anvil and spindle end with a clean cloth before using the micrometer.

Micrometers never should be excessively tightened or forced over a workpiece. This usually results in damage to the spindle threads and nut and inaccurate readings. Most micrometers are fitted with a friction sleeve or a ratchet on the thimble to allow precise contact with the object being measured. These must always be used when taking a measurement.

Adjusting Micrometers. There are two adjustments that can be made to keep micrometers in good working condition. First, if the threads in the spindle nut wear and cause spindle play, adjust the nut with a spanner wrench, as shown in Fig. 5.46. The spindle nut must be adjusted so that there is no play in the threads, but the spindle must turn freely.

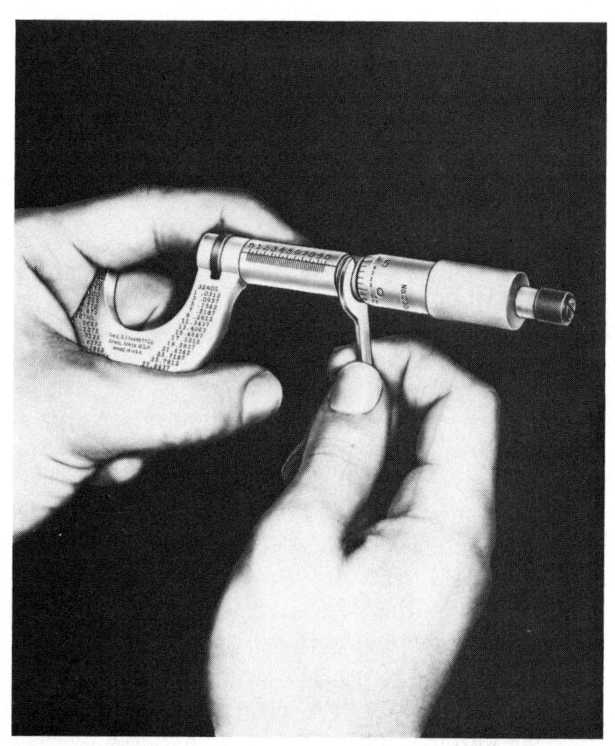

FIGURE 5.46 Adjusting the spindle nut. (*Courtesy L.S. Starrett Co.*)

The second adjustment involves the accuracy of the micrometer. On some micrometers, the sleeve is moved with a spanner, as shown in Fig. 5.47, to bring the index into alignment with the zero on the thimble. On other types, the mechanism holding the spindle to the thimble is loosened and the thimble is brought into alignment. The locking mechanism is then secured. Be sure that the end of the spindle and the anvil are clean and free of nicks before making any adjustments.

Larger micrometers are adjusted in the same manner, except that standards like those shown in Fig. 5.48 are used. The spindle, anvil, and ends of the standards must be clean and free of nicks. Adjust micrometers only when both the micrometer

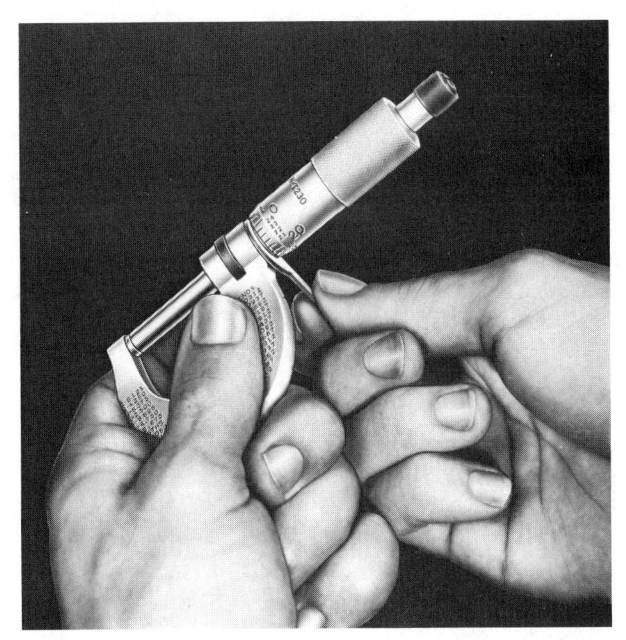

FIGURE 5.47 Adjusting the sleeve. (*Courtesy L.S. Starrett Co.*)

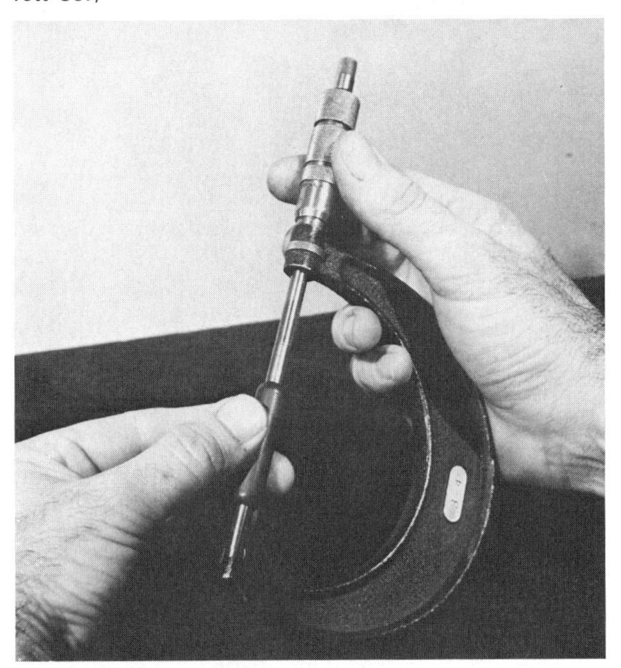

FIGURE 5.48 Checking a micrometer.

and the standards are at normal temperature (approximately 70 °F or 21 °C). This is particularly important when working with large micrometers.

5.3.2 Outside Micrometers

The outside micrometer is the most widely used type of micrometer (Fig. 5.49). The capacity of the micrometer varies from 0.5 to 60 in. On most larger micrometers, especially in the size range above 6 in., the anvil can be replaced with interchangeable anvils in 1-in. increments, as shown in Fig. 5.50.

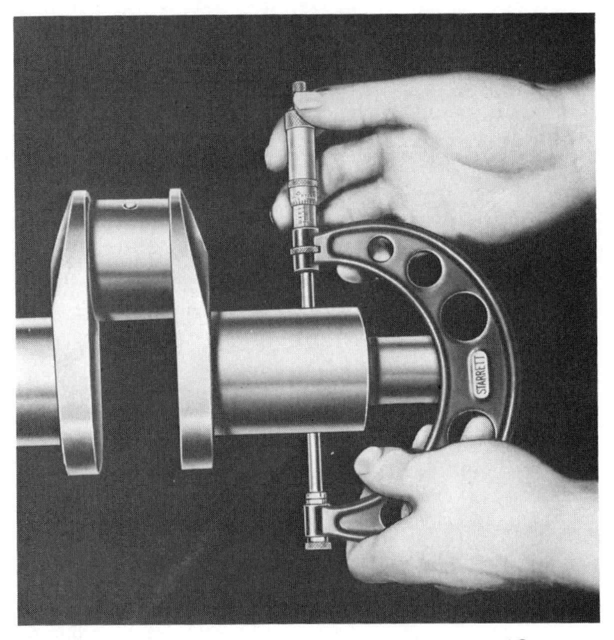

FIGURE 5.49 Using an outside micrometer. (*Courtesy L.S. Starrett Co.*)

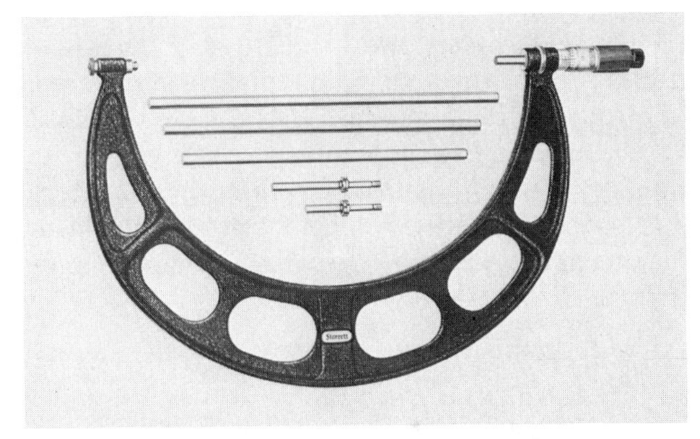

FIGURE 5.50 Micrometer with interchangeable anvils and standards. (*Courtesy L.S. Starrett Co.*)

Special Outside Micrometers. A number of special types of outside micrometers are available for use in machining and inspection applications. The *multianvil* micrometer shown in Fig. 5.51 can be fitted with several types of anvils

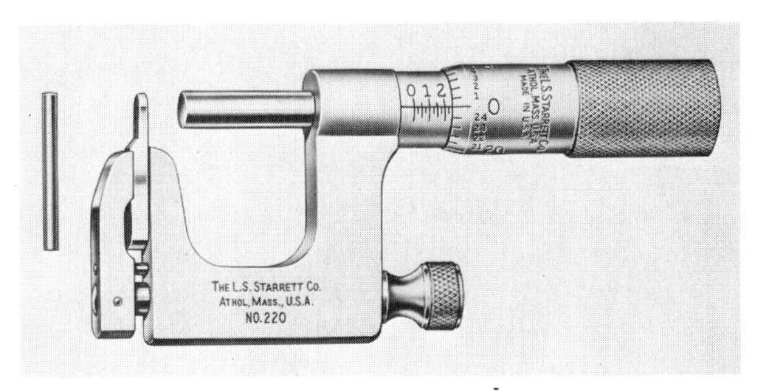

FIGURE 5.51 Multi-anvil micrometer. (*Courtesy L.S. Starrett Co.*)

and used as a height gage with the anvil and clamp removed.

The *V-anvil* micrometer is unique because the work being measured makes contact with the micrometer at three points. This is a particularly useful feature when three-fluted cutting tools such as counterbores and core drills must be measured (see Fig. 5.52). The V-anvil micrometer is also used for detecting out-of-roundness in round work finished by centerless grinding.

Screw thread micrometers measure the pitch diameter of screws (see Fig. 5.53) with sharp V, American National, or Unified thread forms. There are four separate micrometers in each 1-in. size range because each micrometer can only be used to measure threads within a limited pitch range. The pitch range is stamped on the thimble of the micrometer. The screw thread micrometer can be used to duplicate a thread by setting it with a thread plug gage or another thread of known accuracy.

Tube micrometers and *rounded anvil* micrometers measure the wall thickness of bushings, tubing, and other items that have curved surfaces. The tube micrometer (Fig. 5.54) can also be used for measuring from a hole to an edge within its 1-in. operating range. Both tubing micrometers and rounded anvil micrometers are available in metric form, with an operating range of 0 to 25 mm.

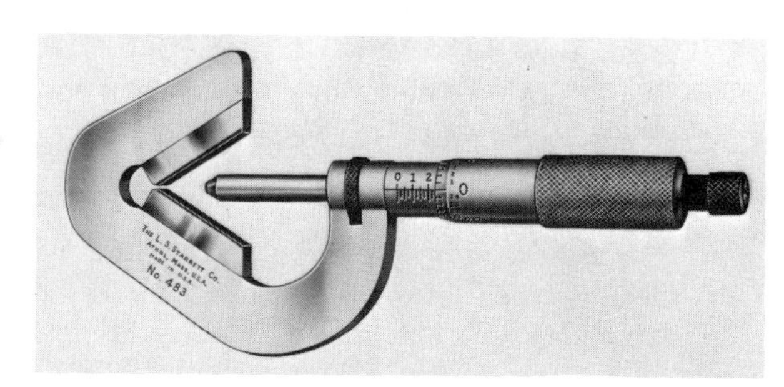

FIGURE 5.52 V-anvil micrometer. (*Courtesy L.S. Starrett Co.*)

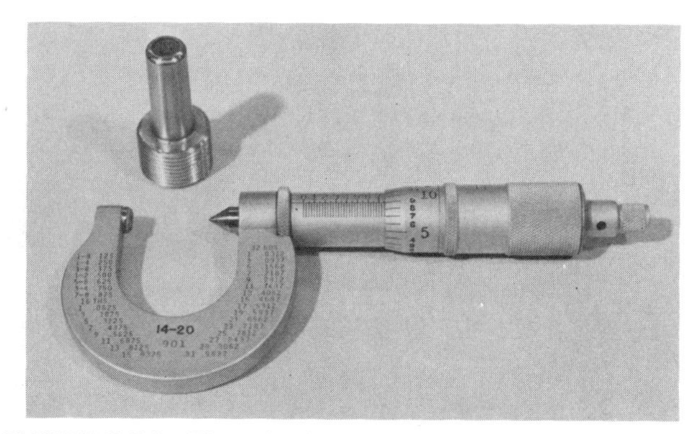

FIGURE 5.53 Thread micrometer.

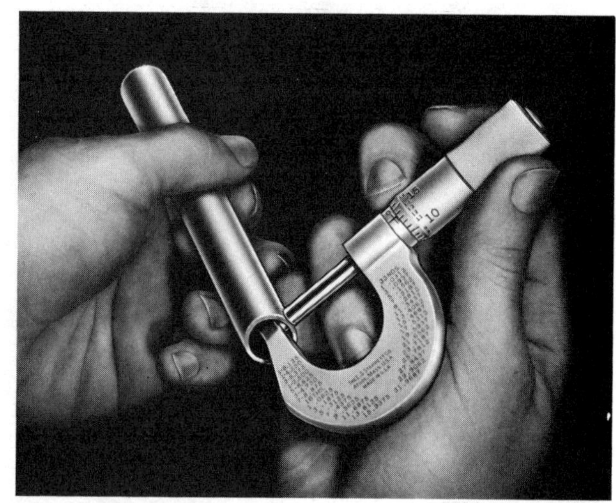

FIGURE 5.54 Tubing micrometer. (*Courtesy L.S. Starrett Co.*)

FIGURE 5.55 Bench micrometer. (*Courtesy L.S. Starrett Co.*)

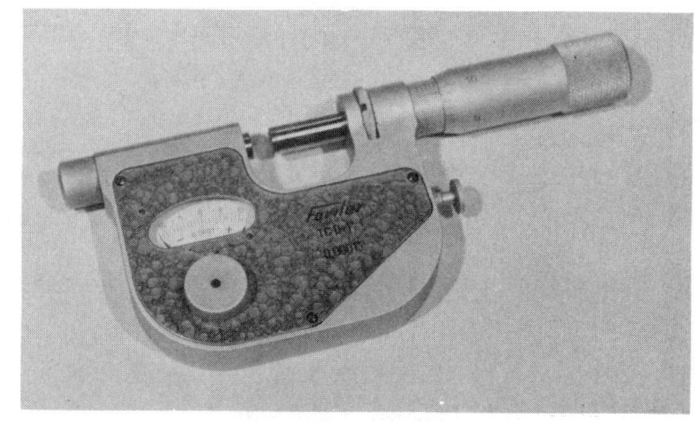

FIGURE 5.56 Direct-reading micrometer.

The *bench* micrometer (Fig. 5.55) in its various forms is used primarily in inspection work, but it can also be used in the shop. Readings can be taken directly to an accuracy of 0.0001 or 0.000050 in. when using the dial indicator. The bench micrometer can also be used as a comparator.

Direct-reading micrometers and *indicating* micrometers are becoming more popular. The direct-reading micrometer (see Fig. 5.56) has a digital readout in thousandths or ten-thousandths of an inch. Indicating micrometers are very useful when comparing the diameter of parts in inspection work. The micrometer is usually checked and set with gage blocks. The readout dial is graduated in 1/10,000-in. increments.

5.3.3 Inside Micrometers

Many inside micrometers are used in machine shop work and for inspection. In most cases the inside micrometer set includes a number of rods of different lengths and other attachments. The set shown in Fig. 5.57 has a range of 2 to 8 in., and sets with operating ranges of up to 32 in. are available. Metric inside micrometers are available with an operating range or 50 to 800 mm and can be read in increments of 0.01 mm. Some inside micrometers have tubular measuring rods to provide greater rigidity when large objects are measured (see Fig. 5.58).

The *three-point contact inside micrometer* shown in Fig. 5.59 eliminates several problems involved in accurately measuring holes. It is self-centering and self-aligning. Because it is read directly, it eliminates errors introduced by the use of telescoping gages, calipers, small-hole gages, and other transfer tools. Many different measuring arms are available to allow accurate measurement of the diameters of blind holes and to extend the operating range of the tool.

The three arms of the micrometer are moved outward simultaneously by a cone when the thimble is screwed in. A spring retracts the arms when the thimble is screwed out. The micrometer can be checked or set by accurate setting rings, which are included as part of each set. The operating range varies from 0.275 to 0.5000 in. for the smallest set of three micrometers and from 4.000 to 8.000 in. for the largest set of four micrometers.

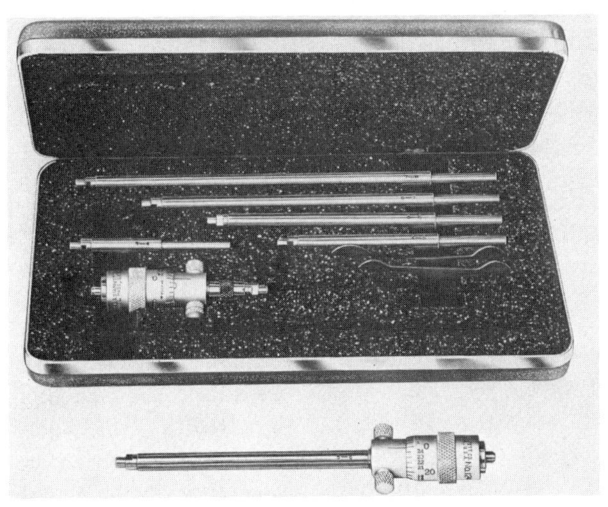

FIGURE 5.57 Inside micrometer set. (*Courtesy L.S. Starrett Co.*)

FIGURE 5.58 Inside micrometer with tubular extension rods. (*Courtesy L.S. Starrett Co.*)

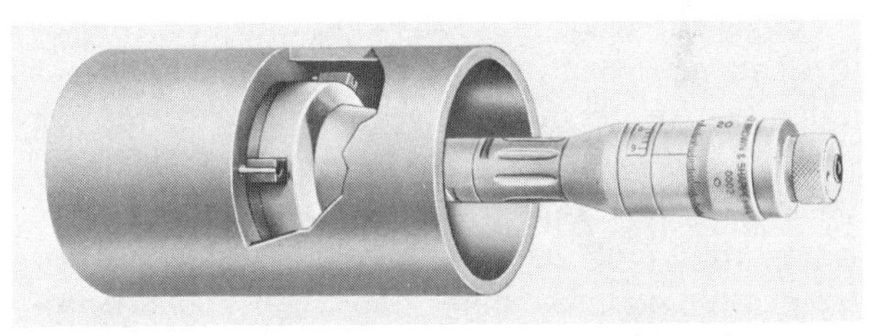

FIGURE 5.59 Three-point contact inside micrometer. (*Courtesy Browne & Sharpe Manufacturing Co.*)

5.3.4 Micrometer Depth Gages

Micrometer depth gages can be used to accurately measure the depth of slots, keyways, recesses, and blind holes. They also help determine the relative position of parts in setup and layout work. One type of micrometer depth gage has round measuring rods that rotate with the thimble and are held in place by a knurled screw cap on top of the thimble (see Fig. 5.60). The range of thimble movement is usually 1 in., and the graduations read from right to left on the sleeve.

The thimble is also graduated in the opposite direction, and the micrometer depth gage is read by adding the thimble reading to the value of the last graduation under the edge of the thimble. A micrometer depth gage is read in the following manner, for example:

The eighth mark on the sleeve representing 0.800 in. of measuring rod movement is covered by the lip of the thimble. $\quad 8 \times 0.100 = 0.800$

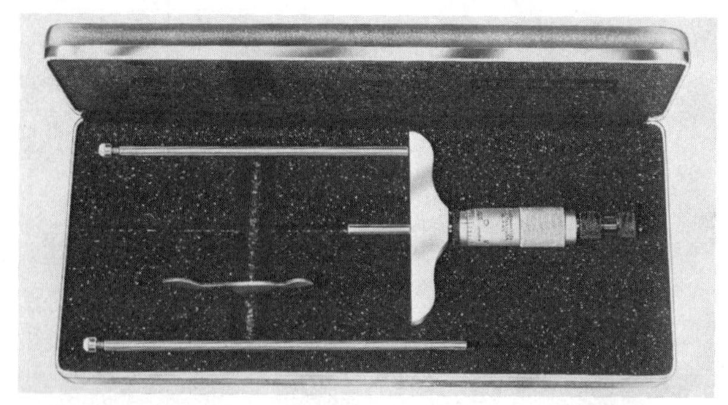

FIGURE 5.60 Micrometer depth gage set. (*Courtesy L.S. Starrett Co.*)

The first small mark (0.025 in.) on the sleeve is covered.	**1 × 0.025 = 0.025**
The thimble reading at the reference line on the sleeve is 0.014 in.	**14 × 0.001 = 0.014**
By adding this to the previous readings, a total reading of 0.839 in. is attained.	**0.839**

Several variations of the standard micrometer depth gage are particularly useful for certain types of depth measurement. The tool shown in use in Fig. 5.61, for example, has a blade type measuring rod that does not rotate and can be used to measure the depth of slots as narrow as 0.045 in. Another useful type of micrometer depth gage has a half base and can be used in confined areas, as shown in Fig. 5.62.

Check micrometer depth gage sets for accuracy by using gage blocks and a surface plate (see Fig. 5.63). Adjust the measuring rods by moving the nut at the end of the rod. Be certain that all parts of the tool are free of burrs, chips, and any other foreign matter before making adjustments.

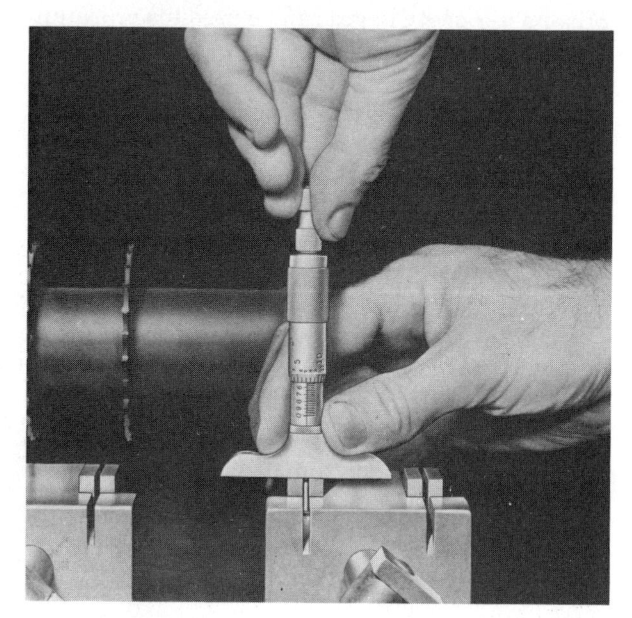

FIGURE 5.61 Blade-type micrometer depth gage. (*Courtesy L.S. Starrett Co.*)

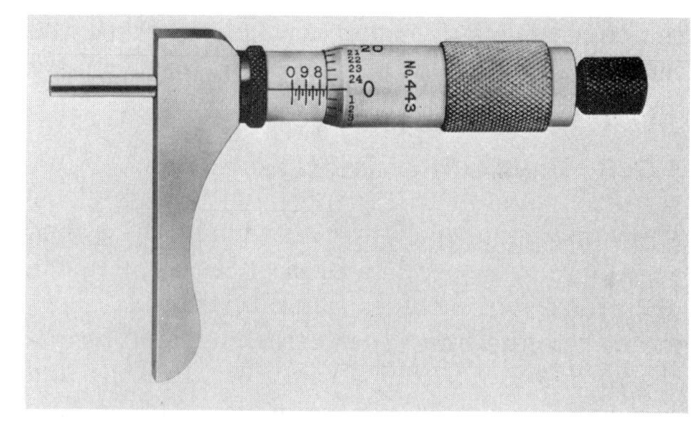

FIGURE 5.62 Half-base depth micrometer. (*Courtesy L.S. Starrett Co.*)

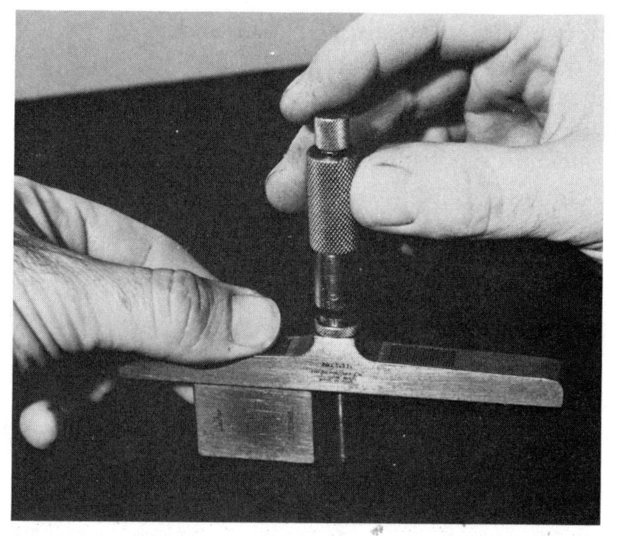

FIGURE 5.63 Checking a depth micrometer.

5.3.5 Vernier Calipers and Gages

The vernier principle, which was discussed with regard to micrometers, can also be applied to calipers, height gages, and protractors. The vernier caliper shown in Fig. 5.64 can be used to measure accurately in increments of 0.001 in. for internal, external, and depth measurements. Vernier calipers' flexibility and range of operation make them very useful tools.

The 25-Division Vernier Caliper. The vernier plate on this caliper has 25 divisions that are 0.600 in. in actual length, or equal to 24 divisions on the *graduated beam* (also called the *bar*). Each division on the vernier scale is 0.024 in. in length, and each division on the graduated beam, which is a true scale, is 0.025 in. in length. The vernier divisions are therefore 0.001 shorter than the divi-

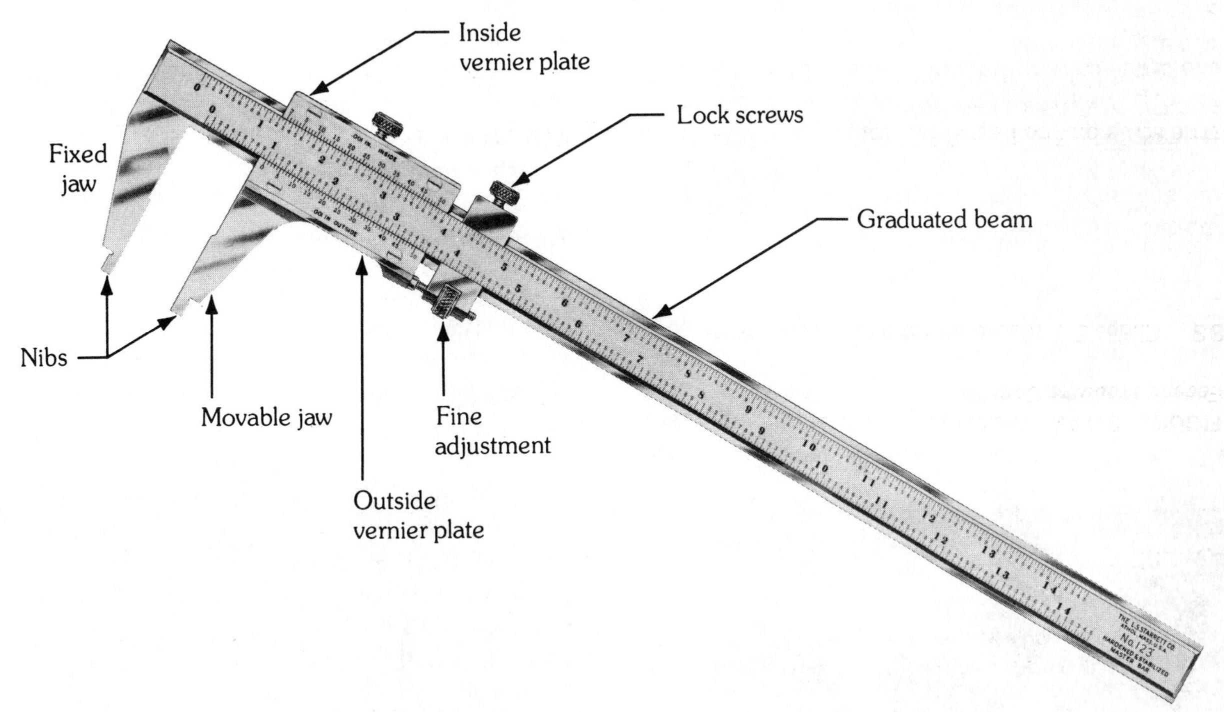

FIGURE 5.64 Vernier caliper. (*Courtesy L.S. Starrett Co.*)

sions on the beam, and thus only one line on the vernier scale lines up with a line on the graduated beam at any one time. (An exception is when the 0 on the vernier scale is exactly aligned with a line on the beam, in which case the line numbered 25 also lines up with a line on the beam.)

Reading a vernier caliper consists of determining the value of the various readings, starting with the largest, and then adding them. The setting shown in Fig. 5.65 is read in the following manner:

Note the largest reading in whole inches to the left of the 0 on the vernier scale.	**1.000**
There are four numbered *divisions of 0.100 in. each to the left of the 0 on the vernier scale.*	**4 × 0.100 = 0.400**
There is one whole division of 0.025 in. to the left of the 0.	**1 × 0.025 = 0.025**
The number 11 on the vernier scale is directly aligned with a line on the beam. Since each division on the vernier scale represents 0.001, multiply 0.001 by 11.	**11 × 0.001 = −0.011**
The total reading is therefore 1.436.	**1.436**

Note the relationship between the true scale and the vernier scale in Fig. 5.66.

FIGURE 5.65 Vernier caliper reading of 1.436 in. (*Courtesy L.S. Starrett Co.*)

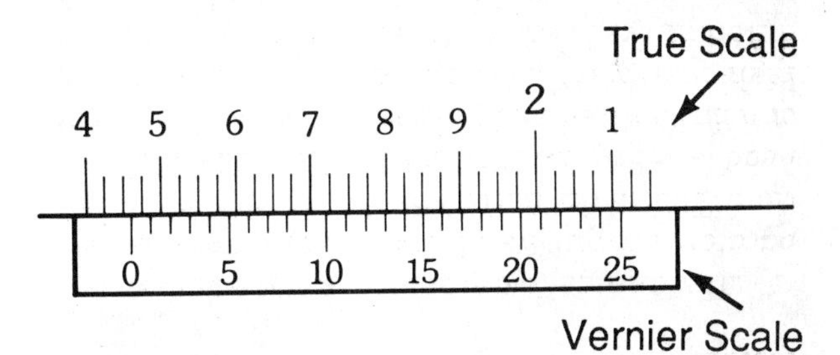

FIGURE 5.66 Vernier caliper reading of 1.464 in.

Metric Vernier Calipers. The graduated beam of the metric vernier caliper or height gage is divided into centimeters, millimeters, and 1/2

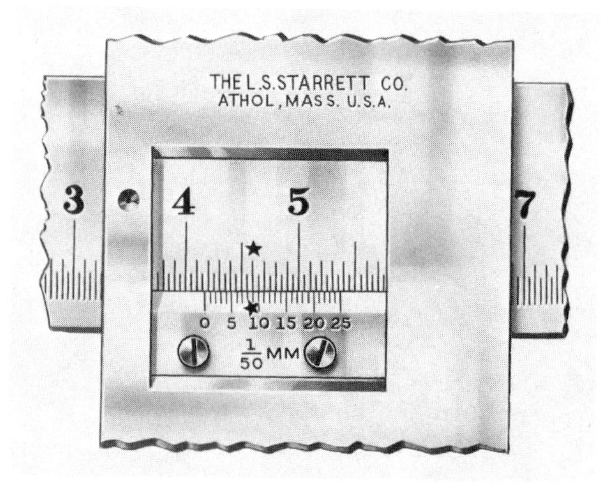

FIGURE 5.67 Metric vernier caliper reading of 41.68 mm. (*Courtesy L.S. Starrett Co.*)

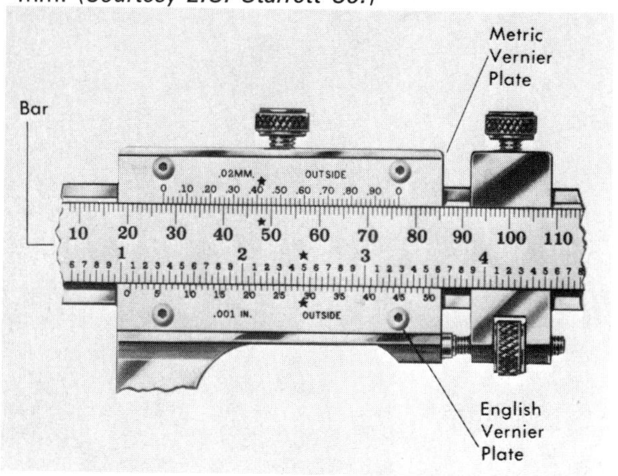

FIGURE 5.68 Vernier caliper with English and metric scales. (*Courtesy L.S. Starrett Co.*)

mm. The vernier, which has 25 divisions, provides the final reading in increments of 0.02 mm.

The 25 divisions of the vernier scale are equal to 24 divisions of 0.5 mm each on the beam. Because each division on the vernier scale is actually 0.48 mm long, the difference in length between the divisions on the true scale on the beam and the divisions on the vernier scale is 0.2 mm.

To read the metric vernier scale shown in Fig. 5.67, use the following procedure:

The largest reading in whole centimeters to the left of the 0 on the vernier scale is 4, or 40 mm.	**40.00**
There is one 1-mm increment to the left of the 0.	**1 × 1 = 1.00**
There is one 0.5-mm increment to the left of the 0.	**1 × 0.50 = 0.50**
The ninth line on the vernier scale is aligned with a line on the beam. Since each division of the vernier scale represents 0.02 mm, multiply 0.02 by 9.	**9 × 0.02 = 0.18**
The total reading is 41.68 mm.	**41.68**

Some metric vernier calipers and height gages have a 50-division vernier scale. In this case, the smallest division on the beam scale is 1 mm. The procedure for reading the tool is basically the same, and the smallest increment in the reading is 0.02 mm. The vernier caliper shown in Fig. 5.68 has a 50-division vernier for both the metric and inch scales.

Dial Calipers. Inside, outside, and depth measurements may be made with dial calipers. The dial is graduated in increments of 0.001 in. with each revolution of the needle indicating a jaw movement of 0.100 in. The jaws for outside measurements are longer than the inside measurement jaws and both are thinned to a knife edge for a part of their length to allow measurement in narrow slots and other confined spaces (see Fig. 5.69).

A depth gage in the form of a narrow strip with a thinned portion at the end is attached to the movable jaw. For depth measurements, the calipers read in the same manner as for inside or outside measurements. A knurled thumbwheel is used to move the movable jaw along the slide. A locking screw is provided at the top of the movable jaw. Since the fine-pitch rack that operates the gear train that moves the needle on the dial is exposed, the calipers should be kept as clean as possible to avoid damaging the delicate gears.

A variation of the dial caliper is the electronic

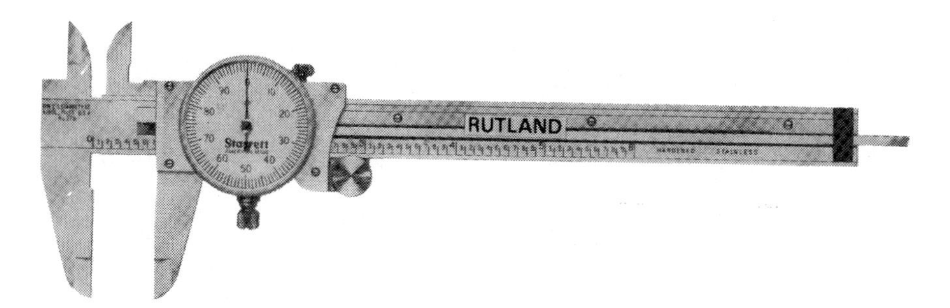

FIGURE 5.69 Dial caliper for internal, external, and depth measurement. (*Courtesy Rutland Tool and Supply Co.*)

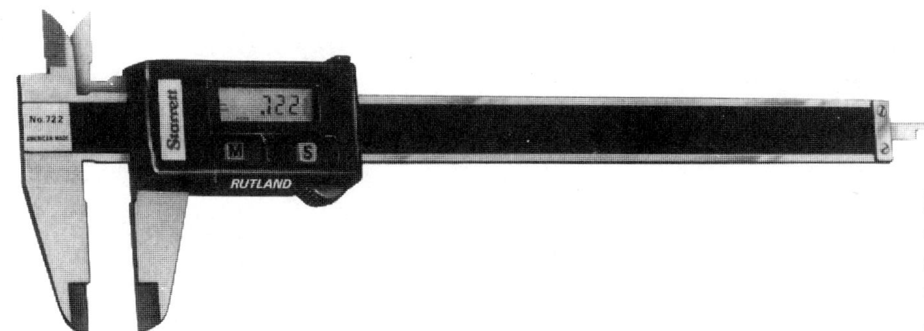

FIGURE 5.70 Digital readout caliper for internal, external, and depth measurement. (*Courtesy Rutland Tool and Supply Co.*)

caliper shown in Fig. 5.70. The position of the jaws is shown in a digital readout window that is part of the movable jaw. A knurled thumbwheel is used to move the movable jaw.

5.3.6 Vernier Height and Depth Gages

The *vernier height gage* is a precision tool used in inspection, layout, and setup jobs by toolmakers, inspectors, and machinists. It is usually used on a surface plate (see Fig. 5.71), along with gage blocks, angle plates, and a number of other accessories. It can be fitted with a scriber for layout work, with a depth attachment (see Fig 5.72), or with a dial indicator for inspection and setup operations.

Some vernier height gages are made with a slotted base so that measurements can be taken and lines can be scribed close to the surface plate (see Fig. 5.73). A depth attachment that can be substituted for the scriber is also shown.

Vernier depth gages can be used to measure the depth of slots, holes, and other recesses to an accuracy of 0.001 in., or 0.02 mm. This type of gage is read in the same manner as other vernier tools.

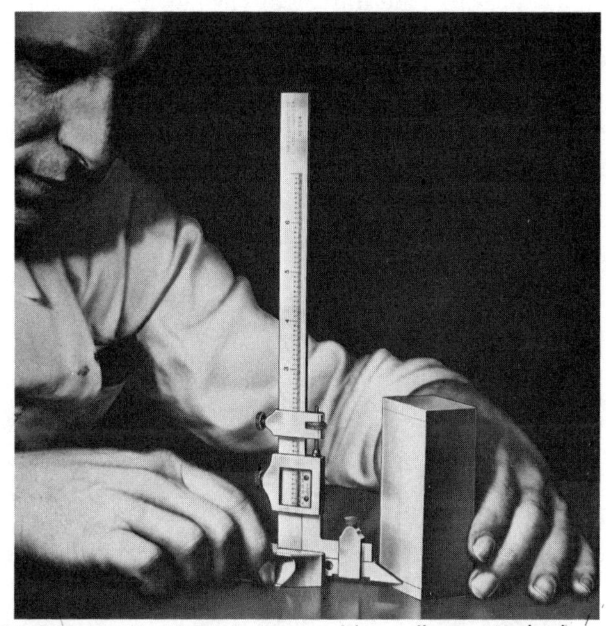

FIGURE 5.72 Height gage with scriber attached. (*Courtesy L.S. Starrett Co.*)

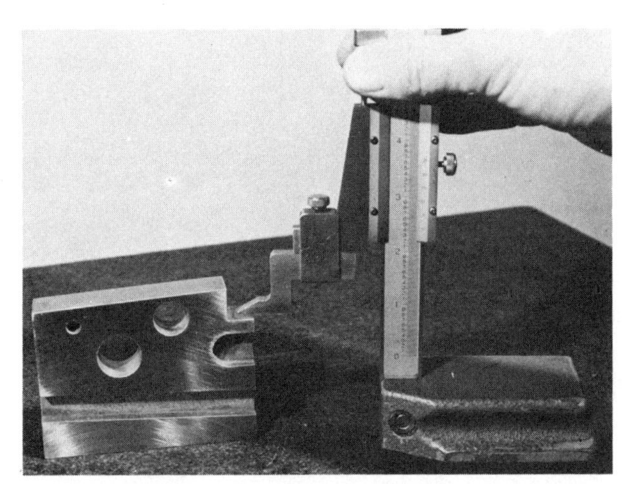

FIGURE 5.71 Vernier height gage in use.

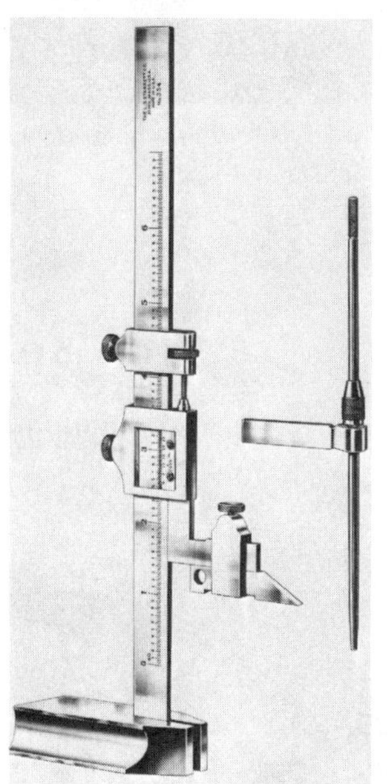

FIGURE 5.73 Height gage with slotted base and depth attachment. (*Courtesy L.S. Starrett Co.*)

5.3.7 Vernier Protractors

The universal bevel protractor (Fig. 5.74) measures angles in increments as small as one-twelfth of 1°, or 5′ (minutes of arc) (see Fig. 5.75). Each half of the dial is graduated in 1° increments, from 0 to 90° and back to 0. The vernier scale (see Fig. 5.76) is graduated from 0 to 60 minutes in 5′ divisions in both directions. A number of attachments, including blades of different lengths, are available to make the tool easier to use on a surface plate or elsewhere.

Reading the universal vernier bevel protractor consists of noting the largest incremental reading and adding the smaller readings to it as shown (refer to Fig. 5.76):

Since the 0 on the vernier plate lies to the left of 50 and the numbers on the true scale are in ascending order to the left, the largest increment is 50°.	**50° 00′**
Reading to the left *on the vernier plate, the fourth line on the vernier is aligned with a line on the true scale. Since each division on the vernier plate represents 5′, four divisions equal 20′.*	**4 × 5′ = <u>20′</u>**
The total reading is therefore 50° 20′.	**50° 20′**

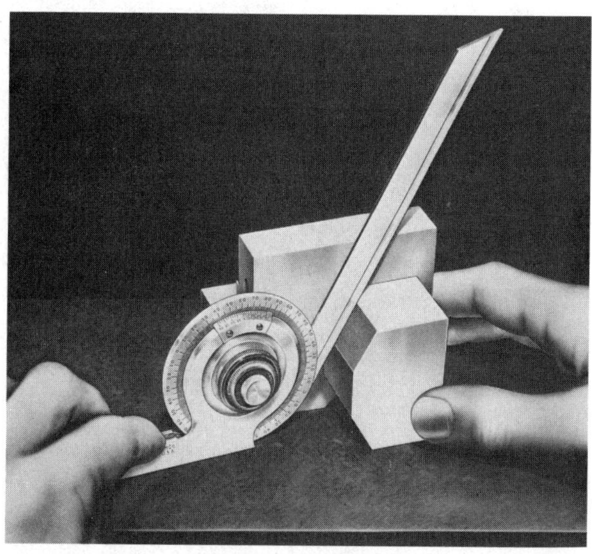

FIGURE 5.74 Using the universal bevel protractor. (*Courtesy L.S. Starrett Co.*)

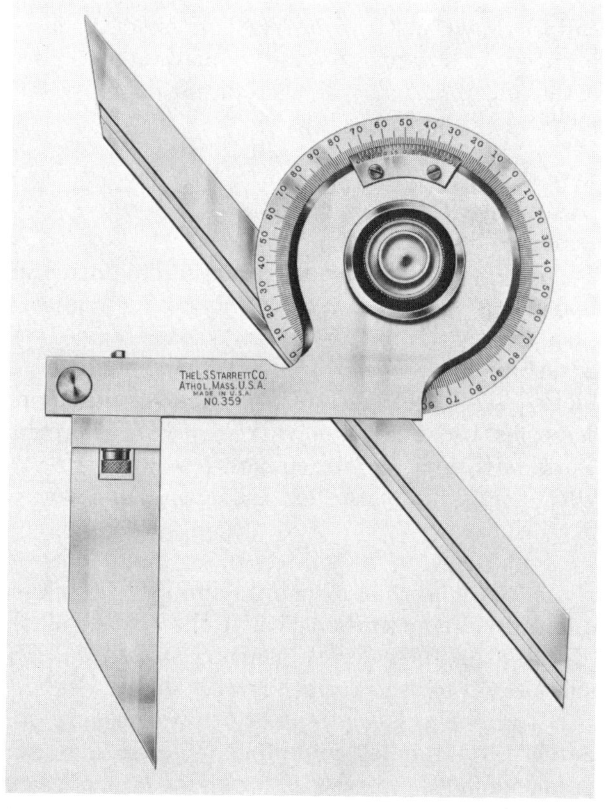

FIGURE 5.75 Universal bevel protractor. (*Courtesy L.S. Starrett Co.*)

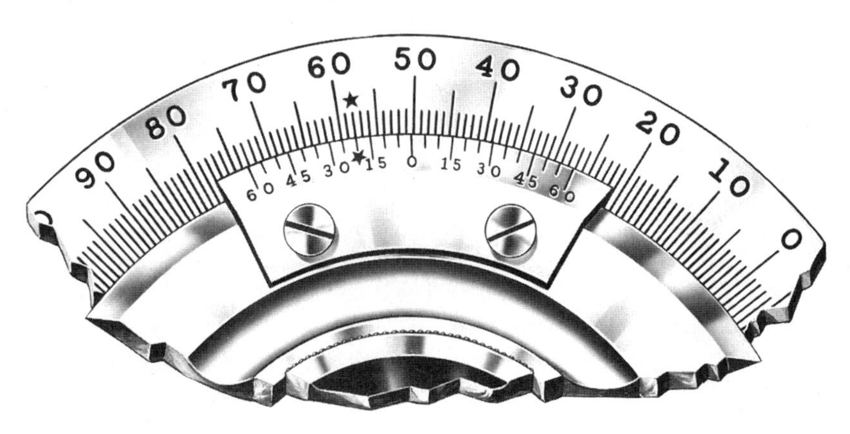

FIGURE 5.76 Protractor reading of 50°20′. (*Courtesy L.S. Starrett Co.*)

When both the 0 and the 60 on the vernier plate coincide with a line on the dial, the reading at the 0 line is the protractor reading in degrees.

5.4 DIAL INDICATORS AND RELATED TOOLS

Various dial indicators are extensively used in machine shop work, setup and layout operations, and in inspection of finished parts. Basically, the dial indicator is a mechanical means of multiplying motion. A movement of 0.001 in. at the contact point causes a movement of as much as 0.060 in. at the perimeter of the dial on a dial indicator with 0.001-in. divisions. Dial indicators with 0.0001-in. divisions are 10 times as sensitive. The dial indicator in Fig. 5.77 has 0.0001-in. graduations and an operating range of 0.025 in., or 2 1/2 revolutions of the needle. Metric dial indicators are available with 0.01- or 0.002-mm graduations for general shop and layout work.

As the sensitivity of an indicator increases, its operating range generally decreases. Some sensitive indicators with a large operating range have an additional dial to count the needle revolutions. The indicator shown in Fig. 5.78, for example, has 0.01-mm graduations, and one revolution of the needle represents 1 mm of contact point movement.

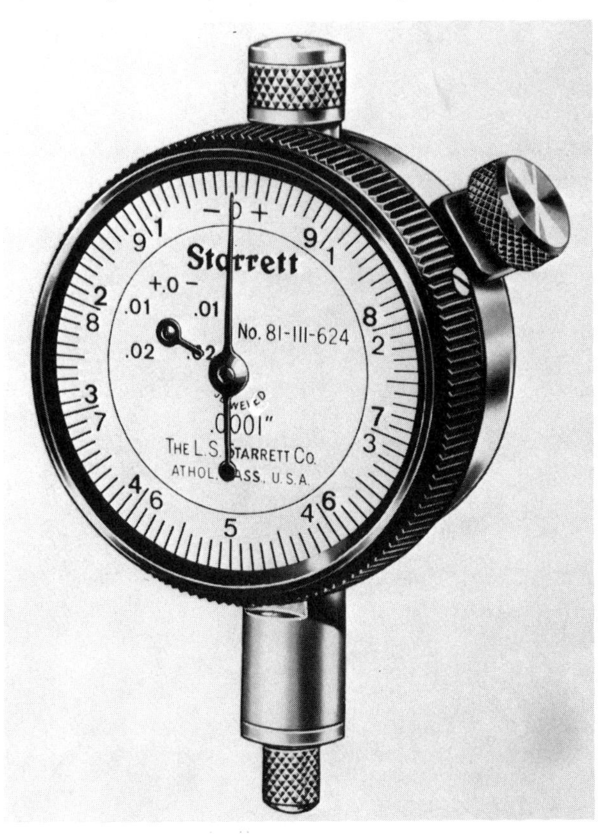

FIGURE 5.77 Dial indicator reads clockwise and counterclockwise in 0.0001-in. increments. (*Courtesy L.S. Starrett Co.*)

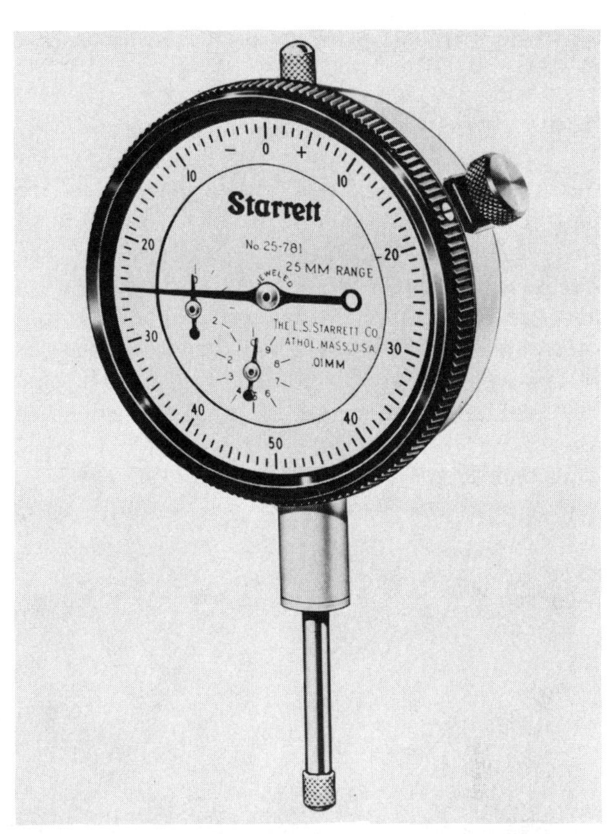

FIGURE 5.78 Metric dial indicator with 25-mm operating range. (*Courtesy L.S. Starrett Co.*)

FIGURE 5.79 Working parts of a dial indicator. (*Courtesy Federal Products Corp.*)

The operating range is 25 mm, or 25 needle revolutions.

Some dial indicators have the contact point shaft either perpendicular or parallel to the face. Generally, indicators with the shaft perpendicular to the face have a short operating range. The mechanism of a typical dial indicator is shown in Fig. 5.79. The dial graduations are *balanced*, as shown in Fig. 5.77, or *continuous*, with the numbers increasing clockwise all the way around the dial.

5.4.1 Dial Indicator Applications

Several examples of dial indicators in use are shown in Figs. 5.80 to 5.82. When these precision tools are used, a variety of accessories are necessary, as shown in Figs. 5.80 and 5.81. For inspection of parts, dial indicators can be mounted on stands, as shown in Fig. 5.82. Dial bench gages of this type can be set or checked with gage blocks.

Dial indicators can also be used as part of dial bore gages, as shown in Fig. 5.83. The diameter of holes can be read directly in increments as small as 0.0001 in. if the gage has been preset and irregularities such as taper and out-of-roundness can be detected.

FIGURE 5.80 Checking the runout of a hole. *(Courtesy L.S. Starrett Co.)*

FIGURE 5.82 Lever-type dial indicator mounted on a granite inspection fixture. *(Courtesy L.S. Starrett Co.)*

FIGURE 5.81 Checking a planer or milling machine setup. *(Courtesy L.S. Starrett Co.)*

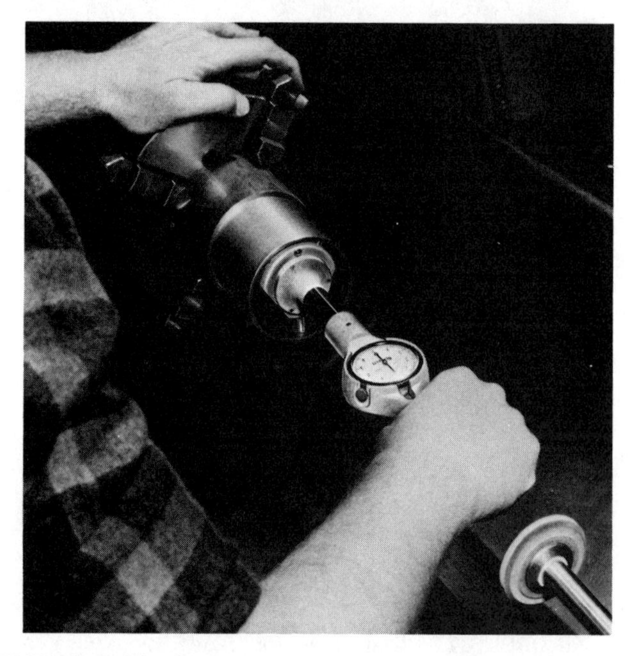

FIGURE 5.83 Dial bore gage used to check a hole diameter. *(Courtesy L.S. Starrett Co.)*

5.4.2 Dial Test Indicators

Dial test indicators have a limited operating range and are widely used in setting-up work in machines and on surface plates. They are usually clamped to a reference surface on the machine or held in a magnetic base fitted with adjustable swivels and clamps (see Fig. 5.84).

Dial indicators of the type shown in Fig. 5.85 are used for specialized applications such as measuring the diameter of an internal O-ring groove. The tool must be preset using a micrometer or a caliper since it gives comparative rather than absolute measurements.

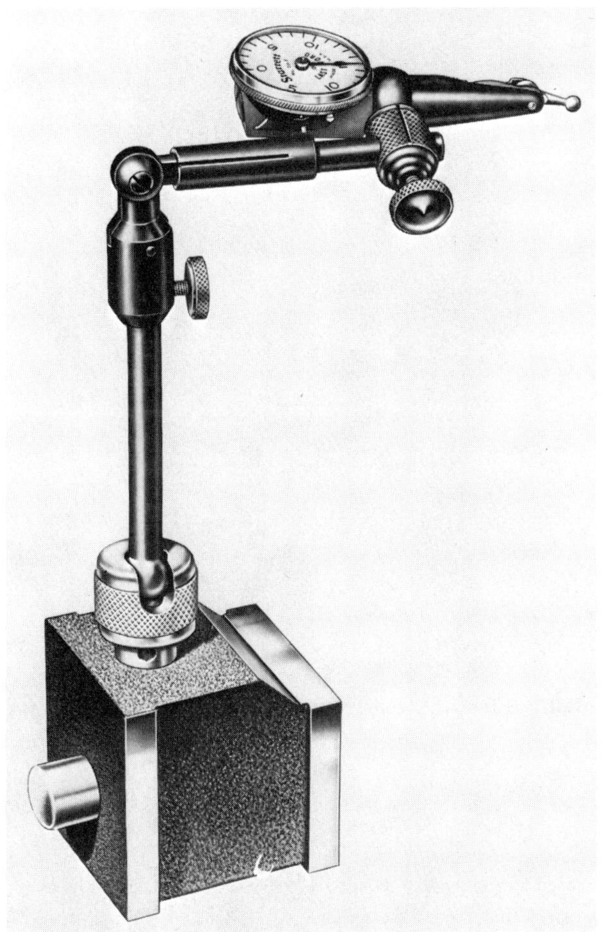

FIGURE 5.84 Dial test indicator. (*Courtesy L.S. Starrett Co.*)

5.5 GAGE BLOCKS, SINE BARS, AND RELATED TOOLS

As the need for precision in manufacturing, toolmaking, and inspection increased, tools that could be relied upon as standards became a necessity. Gage blocks, sine bars, snap gages, and other tools were developed to meet this need. They are now used extensively by machinists, toolmakers, and inspectors. These tools are used in conjunction with surface plates, dial indicators, micrometers, and other precision tools with which the machinist must be familiar.

5.5.1 Gage Blocks

Gage blocks are very precise tools that are used mainly for setting other tools, such as snap gages, height gages, and sine bars, and for checking the accuracy of measuring tools like micrometers (see Fig. 5.86). They can be made of alloy tool steel, chrome carbide, or tungsten carbide and are heat-treated, stabilized, and ground. They can be lapped to tolerances of as little as ±0.000001 in. (one-millionth of an inch) in length for laboratory-type blocks.

The federal standards for length, flatness, and parallelism for class AA, A and B blocks are shown in Table 5.1

TABLE 5-1

Federal standards for blocks

Class	*Length (in.)*	*Flatness (in.)*	*Parallelism (in.)*
AA	±0.000002	0.000004	0.000003
A	+0.000006 −0.000002	0.000004	0.000004
B	+0.000010 −0.000006	0.000006	0.000005

The specifications are for blocks 1 in. or less in length at a temperature of 68°F (20°C).

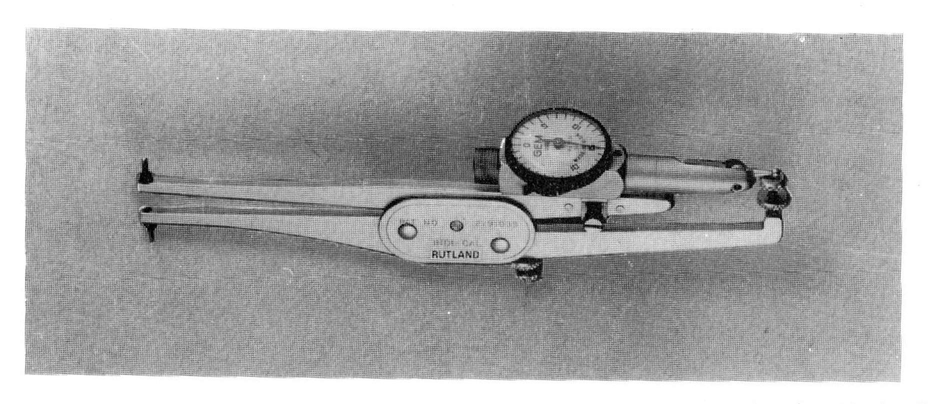

FIGURE 5.85 Dial indicator with attachment for internal groove measurement. (*Courtesy Rutland Tool and Supply Co.*)

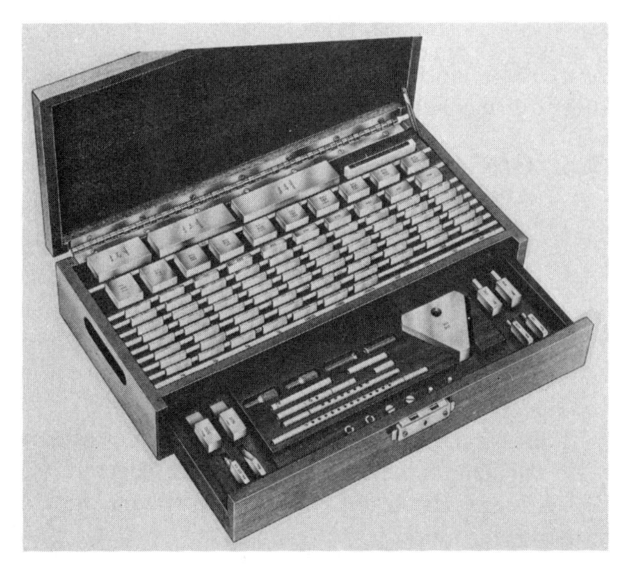

FIGURE 5.86 Gage block set. (*Courtesy L.S. Starrett Co.*)

Class AA gage blocks are often referred to as *master* gage blocks and are used only in temperature-controlled rooms. Class A blocks are used for inspection of machined parts, setting gages used in inspection work, and other activities not requiring the accuracy of master blocks. Class B gage blocks are generally referred to as *working blocks* and are used in shop activities requiring accurate measurement. Setting a comparator, as shown in Fig. 5.87, is an example of a gage block application.

Gage Block Sets. Gage block sets range from 3 to 111 blocks and are available in English and metric sizes. The most widely used set has 81 blocks in the following sizes:

9 blocks: 0.1001–0.1009 in.	0.0001-in. steps
49 blocks: 0.101–0.149 in.	0.001-in. steps
19 blocks: 0.050–0.950 in.	0.050-in. steps
4 blocks: 1.000–4.000 in.	1.000-in. steps

In some cases *wear blocks* are added to the set so that for most combinations only these blocks make contact with gages, tools, or the surface plate. When the wear blocks are no longer in tolerance, they are replaced.

Gage Block Combinations. If the proper combination of blocks is selected, well over 100,000 different dimensions can be achieved, with all but a few in 0.0001-in. increments. For example, if a dimension of 1.7384 in. is desired, the following procedure can be used. The object is to make up the stack with the smallest possible number of blocks.

Select a block that will eliminate the right-hand digit. Use block 0.1004 from the 0.0001 series.	**1.7384** **−0.1004** **1.6380**

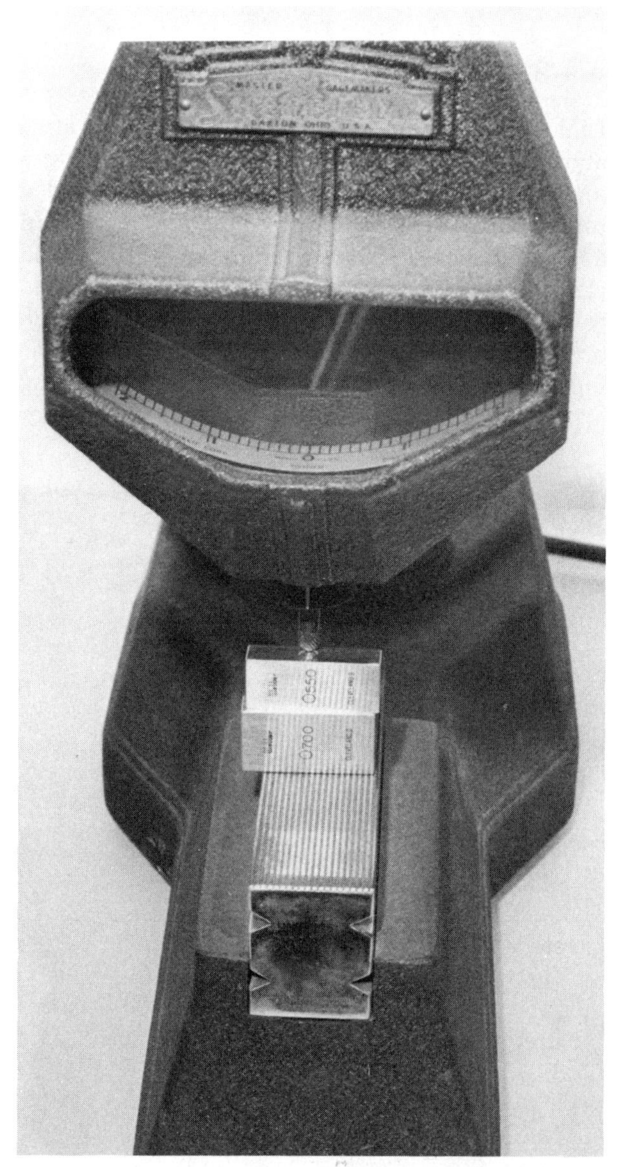

FIGURE 5.87 Setting a comparator with gage blocks.

Select a block that will eliminate as many digits as possible from the right of the decimal. Use block 0.138 from the 0.001 series.	**1.6380** **−0.1380** **1.5000**
Select the 0.500 block from the 0.050 series.	**1.500** **−0.500** **1.000**
Select the 1.000 block from the 1.000 series.	**1.000** **−1.000** **0.000**
Add the dimensions of the blocks selected to check for errors.	**0.1004** **0.1380** **0.500** **1.000** **1.7384 in.**

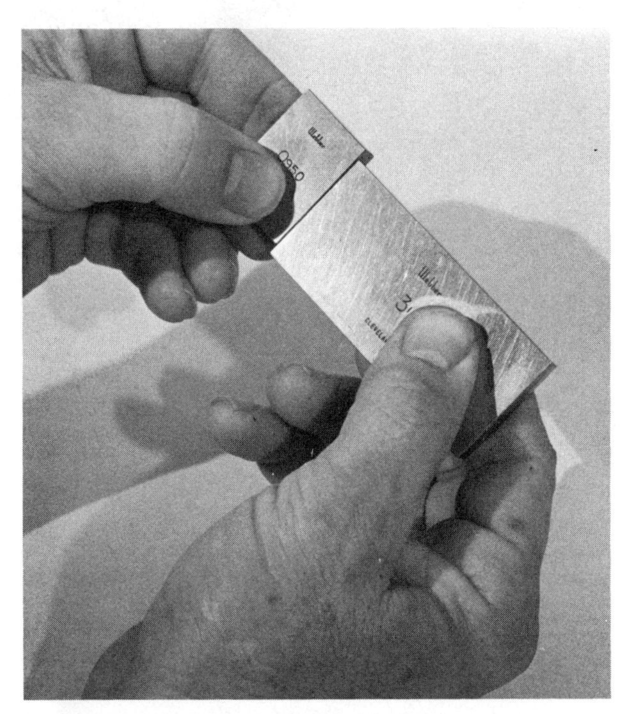

FIGURE 5.88 Wringing gage blocks together.

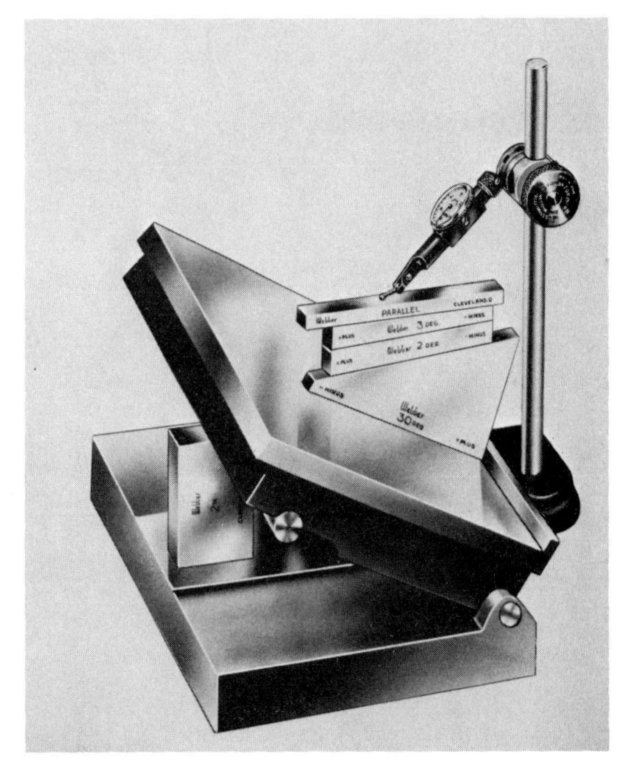

FIGURE 5.89 Checking an angle plate setting with angle gage blocks. (*Courtesy L.S. Starrett Co.*)

After the blocks have been selected, they must be wrung together. The surfaces of gage blocks in good condition are so flat that when they are wrung together they adhere very well. The blocks must be cleaned with a soft, *clean* cloth, and the mating surfaces should not be touched if possible. The procedure for wringing gage blocks is shown in Fig. 5.88.

Angle Gage Blocks. In machine shop or toolroom work it is often necessary to establish accurate angles for either setup or inspection operations. Angle gage blocks may be used for this operation. Angle gage blocks are made in the same manner as rectangular gage blocks and are available in *laboratory master* and *toolroom* accuracy classifications. With a set of 16 angle gage blocks, any angle between 0 and 99° can be set up in steps of 1″ (seconds of arc).

Angle gage blocks can be used instead of a sine bar for surface plate setups and also used for setting vises and fixtures for milling and grinding operations (see Fig. 5.89). The procedure for establishing an angle relative to a reference surface with angle gage blocks is quite simple and does not require the use of trigonometric tables. For example, to set up an angle of 28°30′, select a 30°, a 1°, and a 30′ block. Wring together the *minus* ends of the 1° and 30′ blocks and then wring the ends onto the *plus* end of the 30° block. The resulting angle between the surface of the top block and the surface plate is 28°30′ (see Fig. 5.90). Use

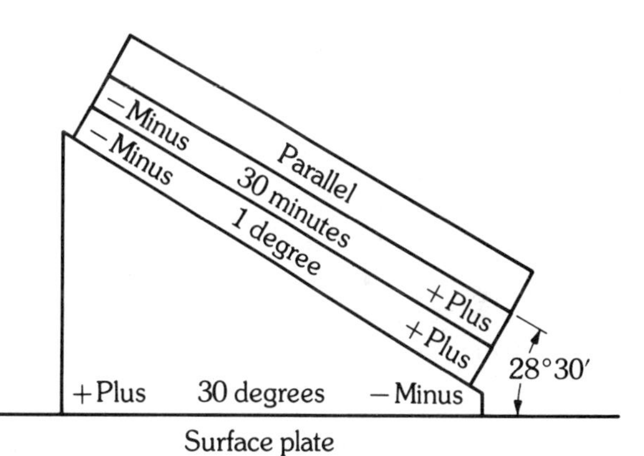

FIGURE 5.90 Angle gage block combination for 28°30′.

a parallel between the top angle block and the indicator or part.

Care and Storage of Gage Blocks. Because gage blocks are precision tools, they remain accurate only if the following five procedures are closely observed.

1. Keep the blocks clean and lightly oiled *at all times.*
2. Always store blocks in their wooden case when not using them.
3. Never leave blocks wrung together longer than necessary.

FIGURE 5.91 Plug gage. (*Courtesy Size Control Co.*)

FIGURE 5.92 Plug gage. (*Courtesy Size Control Co.*)

4. Do not touch the mating surfaces of the blocks with your bare fingers.
5. Disassemble stacks of blocks by sliding them apart gently.

5.5.2 Gages

Fixed and adjustable types of gages are widely used in machine shops for inspection of finished work. They are comparison-type measuring tools manufactured to precise specifications and usually made of hardened alloy steel. In some cases small gages are made of tungsten carbide. Larger gages are often fitted with tungsten carbide working surfaces.

Fixed gages. Fixed gages cannot be adjusted. They are hardened, ground, and lapped to the desired dimensions and must be inspected periodically for wear and damage. Fixed gages are used to determine whether a particular dimension on a machined part falls within a given tolerance.

Cylindrical *plug gages* are used to determine whether a machined hole is within tolerances. A typical cylindrical plug gage (see Figs. 5.91 and 5.92) has two diameters, usually labeled *go* and *no go.* The small-diameter plug, the go part of the gage, controls the minimum diameter of the hole, and it can be inserted into any hole larger than the minimum size allowed. The larger plug, the no-go part, determines the maximum dimension and will not enter an acceptable hole. For example, a cylindrical plug gage made for checking a 1.250-in.-diameter hole with a ±0.001-in. tolerance has a 1.249-in. go end and a 1.251-in. no-go end.

Taper plug gages and *thread plug gages* are two other common types of fixed gages. The taper plug gage is used to compare the diameter and angle of a tapered hole. If the angle is incorrect, the gage makes contact with the hole only at the large or small end. Most taper plug gages have two lines (see Fig. 5.93) scribed on the body, usually at the large end. The taper being measured is acceptable in terms of diameter if the end falls between the two lines on the plug gage.

Thread plug gages have a *go* end that is longer than the *no-go* end. The thread on the go end is of normal shape, but the threads on the no-go end have cutoff crests and relieved roots (see Fig. 5.94). The pitch diameter is at the large end of the tolerance. Thread plug gages for tapered pipe threads have a single end, and the depth to which the plug gage can be screwed into the thread determines if it is of acceptable quality. Thread plug gages are subject to severe wear and damage unless the threads to be gaged are cleaned and the plug gage is cleaned and oiled prior to its use.

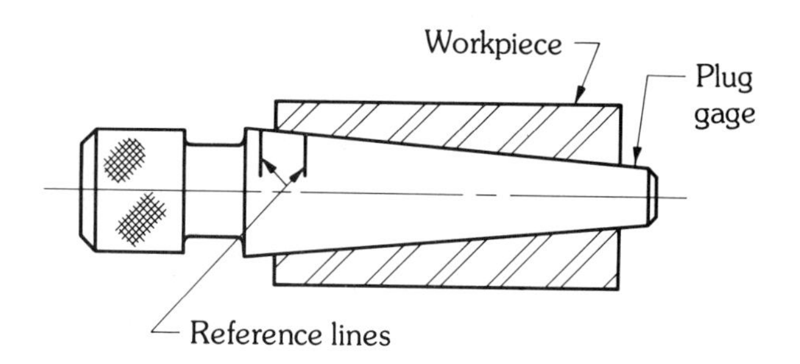

FIGURE 5.93 Taper plug gage hole.

FIGURE 5.94 Thread plug gage.

Ring gages are used to check the diameters of cylindrical and tapered parts and the pitch diameter of threads. *Plain* ring gages for checking cylindrical parts are made in pairs. The go gage has a plain knurled surface, and the no-go gage has a groove in its outside diameter.

Taper ring gages are used to check the angle of taper and the outside diameter of tapers. A step is usually machined on the small end, and lines indicating whether or not the taper is acceptable are scribed on the step. Both the part being measured and the gage must be clean and free of burrs.

Thread ring gages are used to compare the pitch diameter of a threaded part to a given accuracy standard. The body of the gage (see Fig. 5.95) has three radial slots, to allow for minor adjustment of the pitch diameter. A screw adjusts the gage. The thread specifications and the nominal pitch diameter are etched or stamped on the body of the gage. Both go and no-go thread ring gages are set by using *thread-setting* plug gages. Thread-setting plug gages have no chip clearance grooves and must never be used to check internal threads on workpieces.

Form gages, such as the radius gage in Fig. 5.96 and the thread pitch gage in Fig. 5.97, are used to check the profile of a part in terms of internal or external radii or the spacing (pitch) of threads. These gages are made of hardened alloy steel and are available in several size ranges.

Thickness gages, also known as *feeler gages,* are used for checking clearances between machine parts and for some setup operations. The 26 leaves in the typical thickness gage range from 0.0015 to 0.025 in. in thickness. The leaves of the gage can be used singly or in many combinations in 0.001-in. increments (see Fig. 5.98). The leaves of the thickness gage are made of heat-treated alloy steel and are individually marked.

FIGURE 5.95 Thread ring gages.

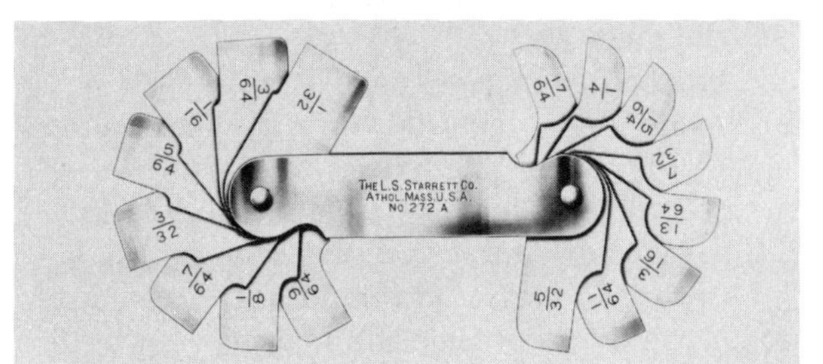

FIGURE 5.96 Radius gage. (*Courtesy L.S. Starrett Co.*)

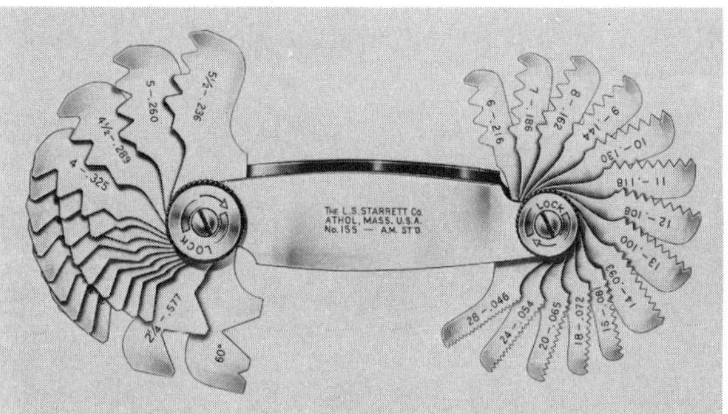

FIGURE 5.97 Thread pitch gage. (*Courtesy L.S. Starrett Co.*)

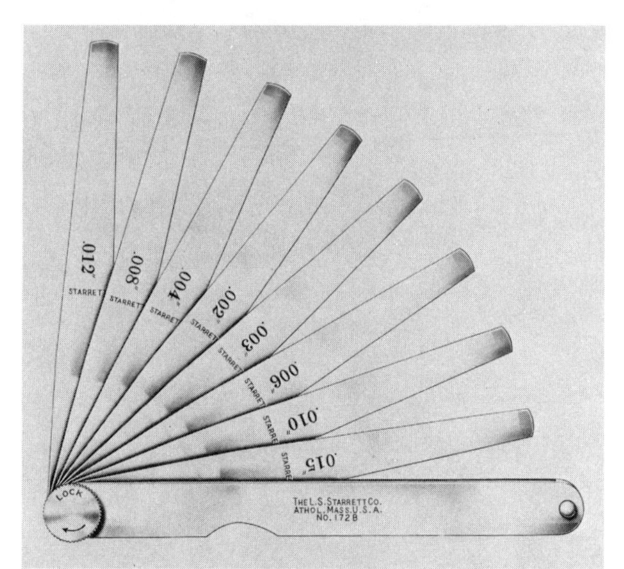

FIGURE 5.98 Thickness gage. (*Courtesy L.S. Starrett Co.*)

Adjustable Gages. A variety of gages that have a limited range of adjustments are used in production machine work. Such gages are usually preset in the toolroom or metrology lab to an accuracy level of one-tenth of the tolerance of the finished part. For example, if an adjustable snap gage is set to measure a part of nominal 1.000-in. size with a tolerance of ± 0.002 in., the go opening is set to 1.002 ± 0.0002 in. and the no-go opening is set to 0.0998 ± 0.0002 in. *Gage blocks* are used to set the gage.

Adjustable snap gages have a heavy C-shaped frame made from forged steel or high-quality cast iron (see Fig. 5.99). The fixed jaw is attached by screws, and the adjustable jaws are locked into position by screws after they are set.

Thread roll gages and *thread snap gages* are used to check the outside diameters of straight threads (see Fig. 5.100). In either case, the go and no-go jaws or rollers are set with a *thread-setting plug gage* or some other external thread of known accuracy.

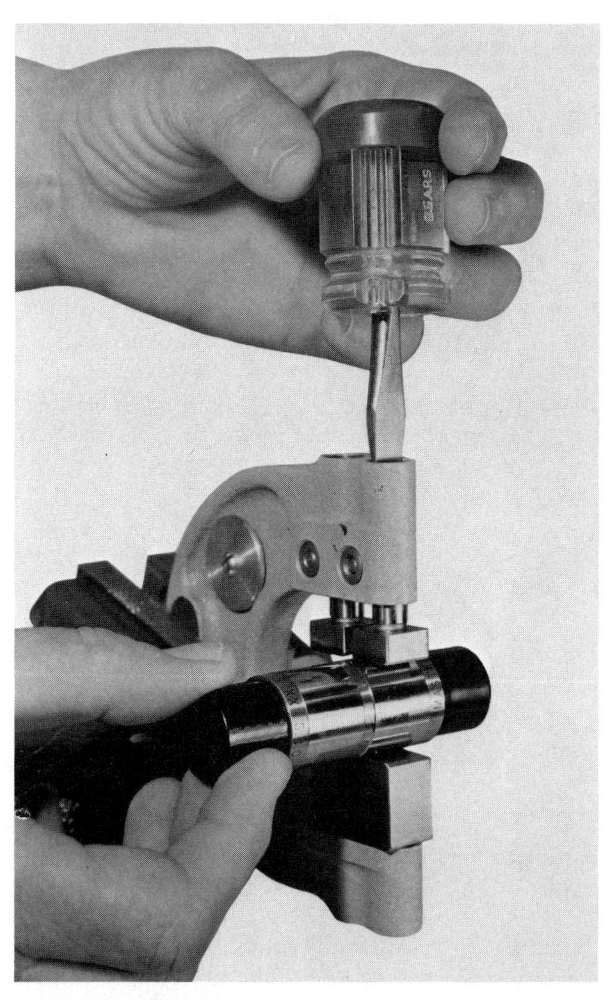

FIGURE 5.99 Adjustable snap gage. (*Courtesy Size Control Co.*)

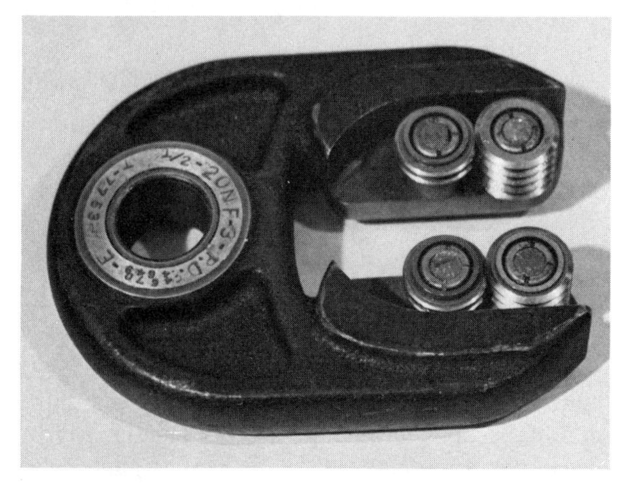

FIGURE 5.100 Thread roll gage. (*Courtesy Orange Coast College*)

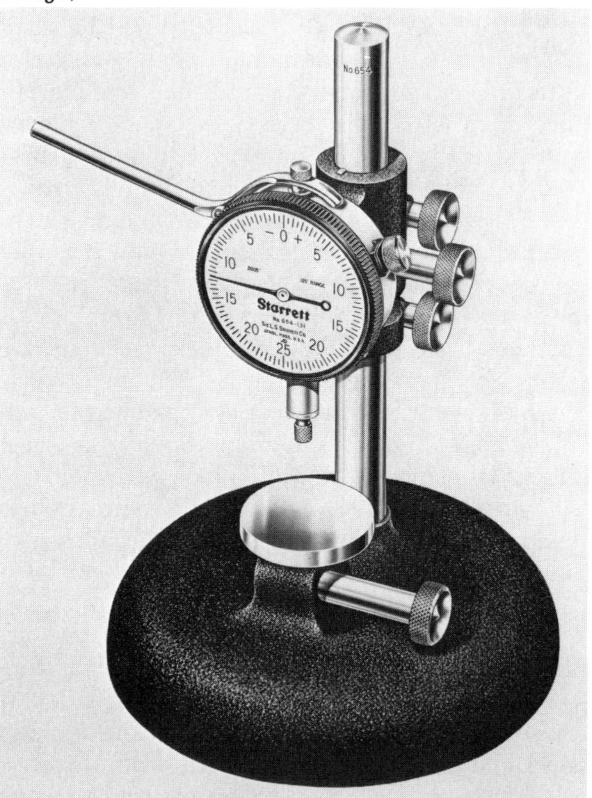

FIGURE 5.101 Dial-type comparator. (*Courtesy L.S. Starrett Co.*)

Both the portable- and bench-types *dial gages* can be used for *comparative* or *direct* readings. A simple comparator such as the one shown in Fig. 5.101 can be set with gage blocks. These comparators are usually fitted with high-quality dial indicators that read in 0.0001-in. increments. The dial indicator is set to read zero with the correct stack of gage blocks in place. Deviations in the workpiece are read on the dial indicator in 0.0001-in. increments. Marks can be put on the cover of the dial indicator to show allowable deviations.

The *dial sheet gage* in Fig. 5.102 is an example of a *direct-reading* dial gage. This gage gives

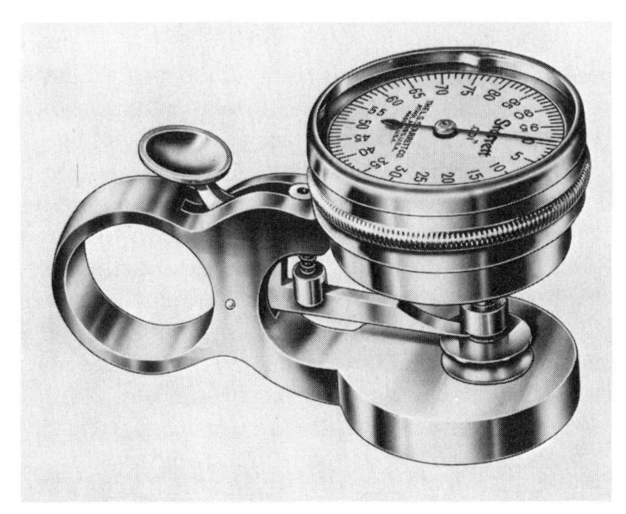

FIGURE 5.102 Dial sheet gage. (*Courtesy L.S. Starrett Co.*)

a direct reading in thousandths of an inch within the range of the dial indicator. The dial is set at zero by turning the knurled ring while the fixed and movable anvils are in contact.

Universal Precision Gage. This tool resembles the *planer and shaper gage* (see Fig. 5.103) which is used primarily for setting cutting tools, transferring measurements, and layout work. It can also be used for inspection operations when it is fitted with a dial indicator.

When a hardened scriber is attached, the tool can scribe accurate layout lines on workpieces (see Fig. 5.104). The height of the scriber point can be set with gage blocks. A level is built into the body of the gage, and a thumbscrew helps make fine adjustments. Extension rods, usually 1 and 3 in. in length, and an offset attachment that permits

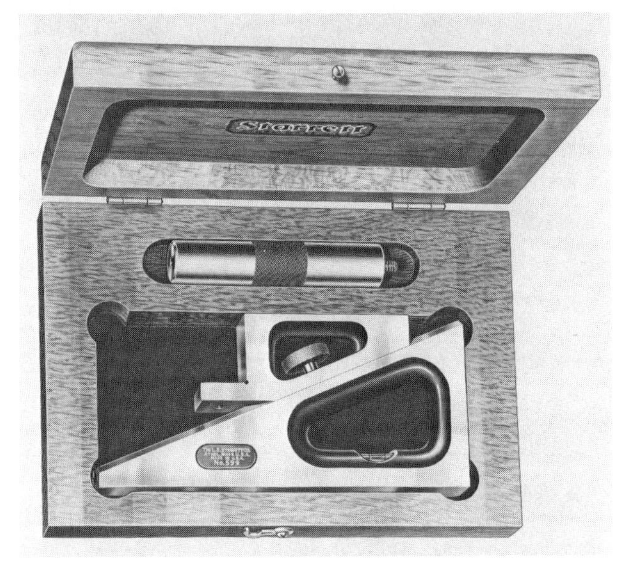

FIGURE 5.103 Planer and shaper gage. (*Courtesy L.S. Starrett Co.*)

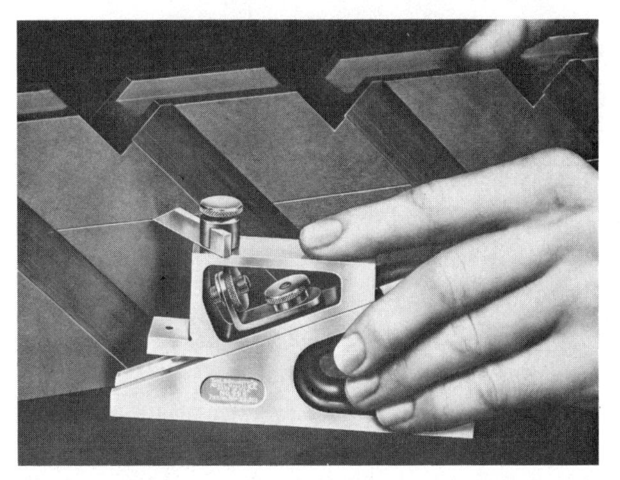

FIGURE 5.104 Universal precision gage fitted with scriber. (*Courtesy L.S. Starrett Co.*)

measurement to be taken below the baseline of the gage, are available. They extend the tool's operating range.

5.5.3 Sine Bars, Sine Plates, and Sine Tables

In the process of inspecting workpieces, making machine setups, or doing layout work, the machinist often needs to establish angles with a tolerance of $\pm 5'$ of arc or less. The angle is generally measured in relation to an accurate plane surface, such as the table of a machine or a surface plate. With any sine tool the desired angle is established by raising one end of the tool a predetermined amount. Gage blocks are usually used for this purpose, as shown in Fig. 5.105. The sine bar is always the *hypotenuse* of any angle that is formed relative to the surface plate, and the stack of gage blocks is the *side opposite.*

The *sine* of an angle is determined by dividing the *hypotenuse* into the *side opposite,* as shown in the following diagram and formula (see Fig. 5.106):

$$\text{sine of angle } A = \frac{\text{side opposite}}{\text{hypotenuse}}$$

Because sine bars and other sine tools are generally available in 5- or 10-in. lengths, the machinist

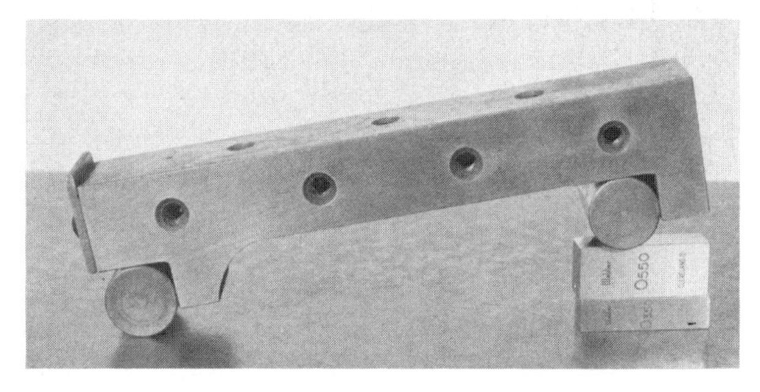

FIGURE 5.105 Setting a sine bar with gage blocks.

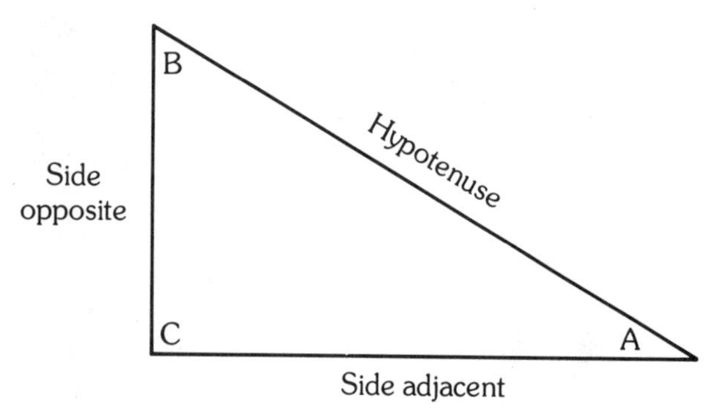

FIGURE 5.106 Right triangle.

can measure angles and make precise angular setups by using only simple calculations and trigonometric tables. The following examples illustrate typical sine bar applications.

To measure an unknown angle A using a 5-in. sine bar when the stack of gage blocks (the *side opposite* of the triangle) is 1.9381 in. high:

$$\text{sine of angle } A = \frac{\text{height of gage block stack}}{\text{length of sine bar}}$$

$$= \frac{1.9381}{5}$$

$$= 0.38762$$

angle A = 22°48′ to nearest whole minute, using a five-place table

To set up a 5-in. sine bar at a 31°55′ angle, because the angle and the length of the sine bar (5 in.) are known, the height of the *side opposite* must be found. (*Note:* Either a table of natural trigonometric functions or a table of sine bar constants may be used.)

$$\text{Sine of } 31°55' = 0.52869$$

$$\text{Sine of } 31°55' \times \text{length of sine bar} = \text{height of side opposite}$$

$$0.52869 \times 5 \text{ in.} = 2.64345 \text{ in.}$$

A stack of gage blocks as close as possible to this dimension is wrung together and forms the side opposite of the triangle when the sine bar is placed in position.

The body of the *sine bar* is made of ground and lapped heat-treated tool steel. The cylinders at each end of the sine bar are also made of hardened alloy steel and finished by lapping to an exact diameter. The center-to-center distance is critical, generally held to a tolerance of ±0.0002 in. for a 5-in. sine bar at 68°F (20°C). Sine bars are available in 5- or 10-in. lengths and can be fitted with a retaining edge at the end that rests on the surface plate.

Sine plates are used for both layout and inspection operations and are of *simple* or *compound* type. Simple sine plates (see Fig. 5.107) are made in 5- and 10-in. sizes and set up with gage blocks using the same procedure as with sine bars. The movable part of the sine plate has a series of drilled and tapped holes that are used to attach fixtures or clamps for holding the workpiece. A retaining edge is usually attached to the hinged edge of the plate to help hold the work or fixtures in position.

Compound sine plates are used for positioning a workpiece at a precise compound angle to the surface of the surface plate or machine table (see Fig. 5.108). They are available in 5- and 10-in. sizes, and the top table is drilled and tapped to allow easy attachment of the workpiece or fixtures. The base has raised lugs at one end; these lugs are accurately bored parallel to the bottom surface to accept the pivot shaft upon which the lower plate tilts. The lower plate also has lugs into which the pivot shaft for the upper plate is fitted.

Simple or compound magnetic sine plates can be used to hold parts for inspection, layout, or grinding operations (see Fig. 5.109). Ferrous metal parts can be held directly on the top plate, or a steel vise can be used to hold nonferrous parts. The lever on the side of the top plate moves the permanent magnets into alignment to activate the magnetic top plate.

FIGURE 5.107 Hinged sine plate. (*Courtesy Quality Grinding Co.*)

FIGURE 5.108 Compound sine plate. (*Courtesy Quality Grinding Co.*)

FIGURE 5.109 Magnetic sine plate. (*Courtesy Quality Grinding Co.*)

5.5.4 Height Gages

Various gages are used in layout work, setting up machines, and inspection processes. For layout and inspection work, the height gage is usually placed on a surface plate. Depending on the accuracy required, the machinist uses the graduations on the height gage, or sets the desired measurement with a stack of gage blocks.

Several accessories can be mounted on the head of height gages. For layout work, a scriber can be attached. A dial indicator can also be attached to the slide assembly (see Fig. 5.110). Depending on the type of dial indicator used, the height gage can be used to check hole locations, the height of objects, or perform other setup and inspection tasks.

A precision-type height gage can be used to set transfer tools, as shown in Fig. 5.111. This type of height gage eliminates the need for making up stacks of gage blocks. The movable vertical bar has ground and polished steps 1 in. apart. The bar is

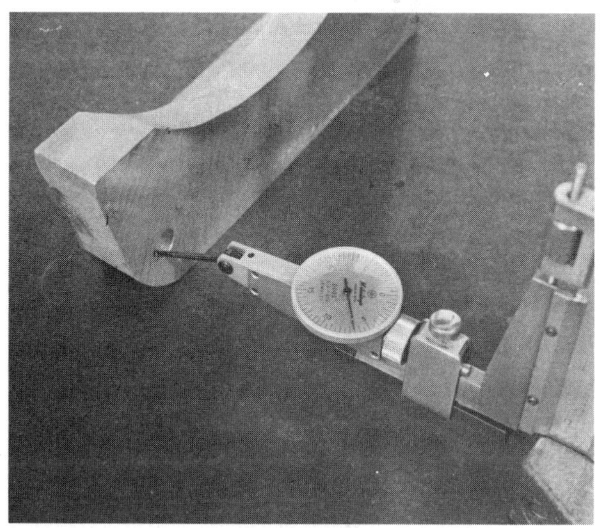

FIGURE 5.110 Height gage used with dial indicator.

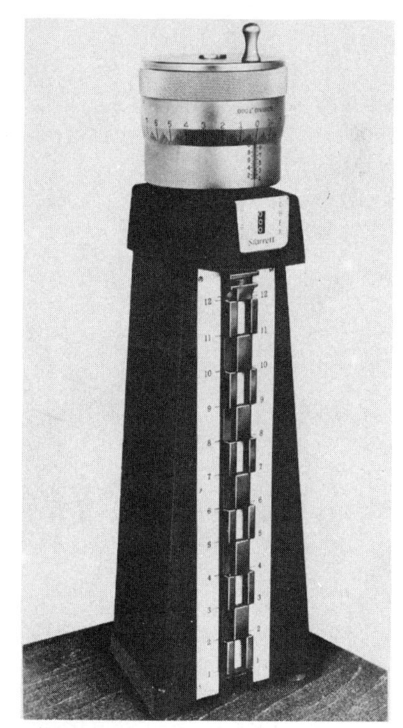

FIGURE 5.111 Precision-type height gage. (*Courtesy L.S. Starrett Co.*)

moved vertically by a 10-pitch screw, and the thimble at the top of the gage has 100 divisions, each representing 0.001 in. of vertical movement. The true scale on the barrel has major divisions 0.100 in. long. Measurements are read directly by adding the true scale and thimble readings.

5.5.5 Comparators

Different comparators are used in machine shops, primarily in inspection work. Comparators are used to determine the variation between a dimension on a workpiece and a predetermined standard. The comparator reading is usually in increments of 0.001 in. or less. The variation from the selected standard is amplified to make it more easily detectable.

Mechanical Comparators. *Mechanical comparators* are quite simple in concept and construction. The main parts are the *base, column,* and *head* (see Fig. 5.112). The base is made of granite or a normalized and precisely machined casting of high-quality iron. Hardened steel anvils of several different types can be fitted to it to accommodate many workpieces. The column is made of heavy-wall steel tubing or bar stock and usually has a rack attached to it or cut into its surface. A manually operated pinion engages with the rack for moving the head up or down.

The head consists of an accurate dial indicator that can be read in 0.0001-in. increments or a more complex device graduated in increments of ten millionths in. The scale is usually graduated with the zero at the center.

The comparator is usually set with gage blocks (see Fig. 5.113) or by using a workpiece of known accuracy as a master. When other parts are measured, an indication to the left of zero shows that the part is smaller than the nominal size. The tolerance (the deviation above or below a specified size) can be set manually on the dial by using two needles, or markers; one is to the right of zero and the other to the left. When a part is measured and the center needle, which is attached to the mechanism of the dial indicator, goes past either marker, the part is out of tolerance.

Air Gage Comparators. *Air gage comparators* are used to compare dimensions and to detect irregularities such as taper, waviness, out-of-roundness, or other workpiece characteristics. The basic operating principle of air gaging is that a particular clearance between the gaging head and the workpiece results in either a given *rate of airflow* reading or a *pressure* reading which is indicated

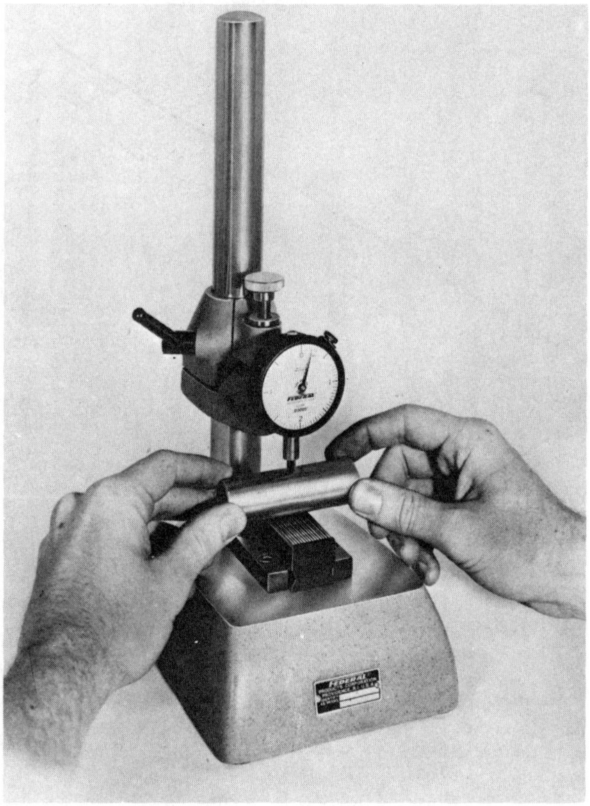

FIGURE 5.112 Mechanical comparator. (*Courtesy Federal Products Corp.*)

FIGURE 5.113 Setting a mechanical comparator.

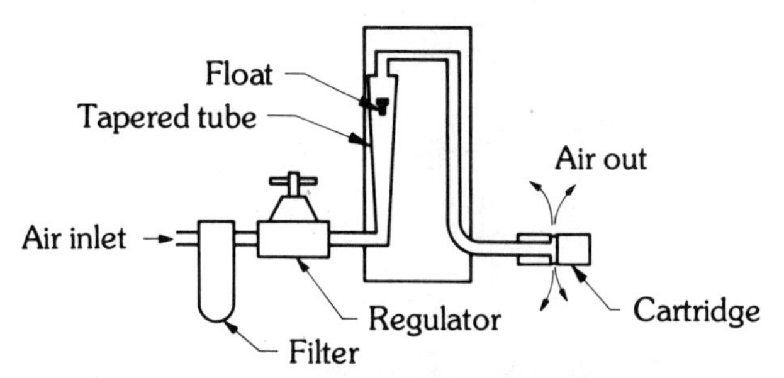

FIGURE 5.114 Schematic of column-type air gage.

on a precise differential air pressure meter. All air gages require a source of compressed air for their operation.

In the case of the *airflow rate,* or *column*-type gage, the air flows through the instrument as shown in Fig. 5.114. When the air reaches the *gaging head* or *cartridge,* its only means of flowing to the atmosphere is through the clearance between the outside diameter of the gaging head and the inside diameter of the part being evaluated. It is necessary to filter the air that enters the system and to have a means of accurately regulating air pressure.

As the air flows upward through a transparent tapered tube, a *float* of closely controlled size and weight is supported at a particular position by the moving column of air. The position of the float is set while the gaging head is inserted into a gage of the correct size, and air is flowing at a given pressure. The amount of deviation from the specified size can then be established in terms of how far the float rises or falls when parts at the large or small ends of the tolerance range are checked. For example, when a part is larger than the specified size, more air flows between the gaging head and the part. The rate of airflow through the tapered tube increases, and the float rises. If the part is above tolerance, the float rises above the upper limit mark.

The *back-pressure system* is quite simple. The air flows through a filter, and a regulator sets the pressure at the desired level. The air then flows through an adjustable restrictor to the gaging head. A sensitive air pressure gage is placed in the line between the adjustable restriction and the gaging head. Depending on the work to be checked, a master gage is put in position on the gaging head, and the indicator (the sensitive air pressure gage) is set to zero by the adjustable restriction. Parts larger than normal let more air escape through the orifices in the gaging plug. The needle on the indicator moves to the *plus* side, showing the amount of variation. An undersize part will restrict the airflow and cause a *minus* reading.

Differential pressure comparators (see Fig. 5.115) are designed with two air lines. The air that flows through the *reference* line goes through a nonadjustable orifice and leaves the system through the zero-setting valve. The reference line

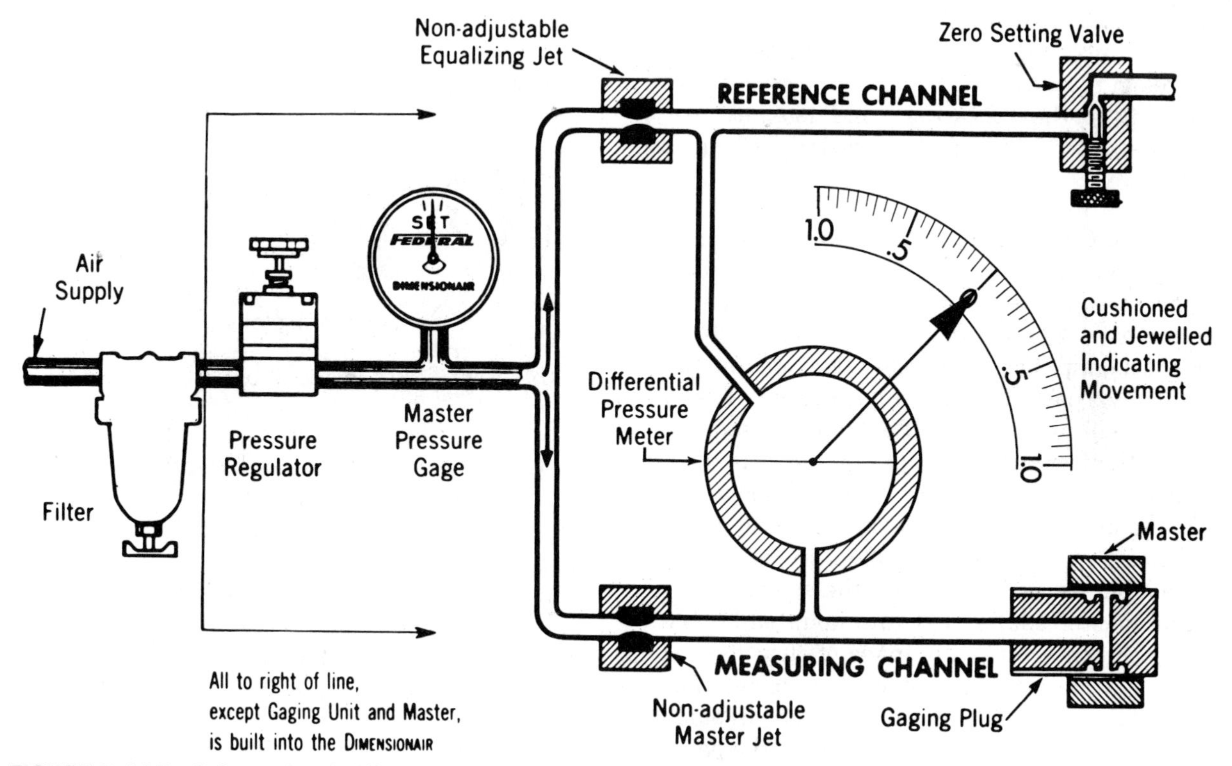

FIGURE 5.115 Schematic of differential pressure air gage. *(Courtesy Federal Products Corp.)*

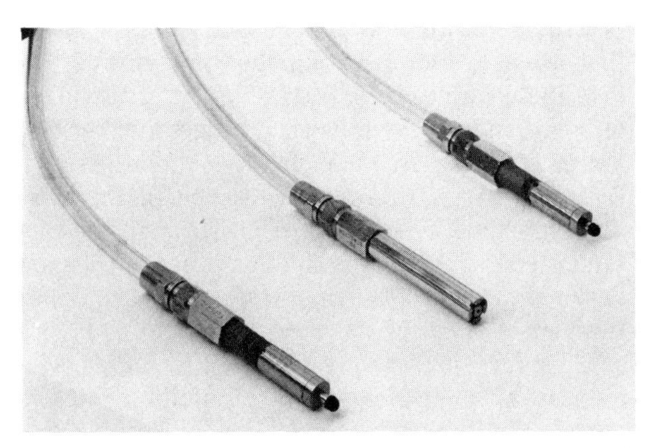

FIGURE 5.116 Pneumatic cartridge. (*Courtesy Federal Products Corp.*)

is also connected to the chamber that houses the diaphragm (or bellows) which actuates the sensitive indicator. The *measuring* line flows through a nonadjustable master orifice to the gaging head. The measuring line is also connected to the other side of the diaphragm attached to the sensitive indicator. The gage is set to zero by using the procedures described for other air gages.

All types of air gage comparators can be fitted with a large variety of attachments. Workpiece characteristics such as height, out-of-roundness, squareness, depth, diameter, or flatness can be checked. Pneumatic cartridges (see Fig. 5.116) can be used with float- or pressure-type gages for inspection and machine control applications. The cartridge consists of a spring-loaded plunger that precisely controls the flow of air when it is moved. Changes in plunger position cause corresponding movements on the indicating dial. Pneumatic cartridges can be fitted to a large variety of inspection fixtures.

Optical Comparators. *Optical comparators* (see Fig. 5.117) project an enlarged image of an object on a screen so that the object can be compared with a master. As shown in Fig. 5.118, the light source is a lamp. The light passes through the *condensing lens,* past the object to be evaluated, and through the *projection lens.* If a mirror is used, the image is then reflected onto the screen. The magnification of the object varies from 5 to about 100 times the original size. For example, with 100× magnification, an error of 0.002 in. in the length of a part shows as 0.200 in. on the screen.

The master charts that are mounted on the viewing screen are usually made of clear frosted plastic or some other translucent material. The outline of the object is drawn on the chart in black ink. The chart and the lens must be of the same magnification.

FIGURE 5.117 Optical comparator. (*Courtesy El Camino College*)

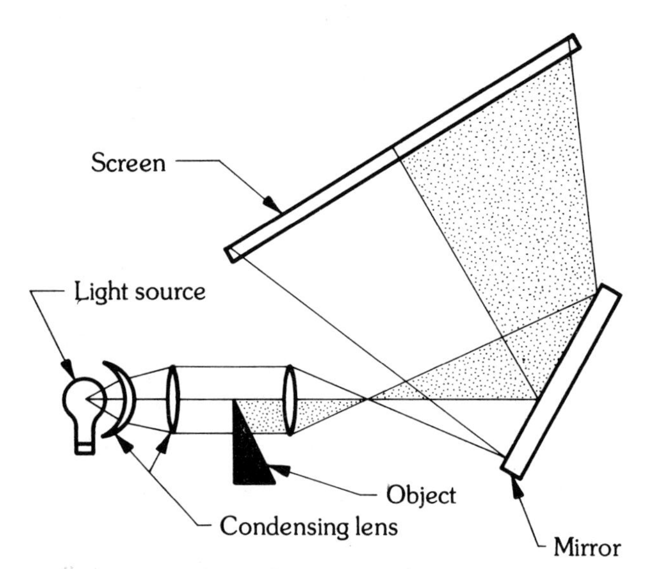

FIGURE 5.118 Schematic of optical comparator.

Optical comparators can be used to check the profile of gears, threads, and many other objects (Fig. 5.119). Other characteristics, such as internal and external radii, angles, length, and the pitch of threads, can also be evaluated. A number of accessories, ranging from simple V blocks to complex fixtures, are available for mounting workpieces on the table of the comparator.

Electronic Comparators. *Electronic comparators* are capable of detecting size variations as small as 1/1,000,000 in. The variation is shown on an electrical meter in greatly magnified form. The three main components of the system are the *gaging head, amplifier,* and *meter.* The movement of the probe on the gaging head above or below a given setting causes an increase or decrease in the resistance within an electrical circuit. The amplifier senses this change. It amplifies the change in accordance with the scale selected by the operator. This causes the meter needle to move in the proper direction. The amount of needle deflection indicates the deviation in size.

The gage shown in Fig. 5.120 can be used for many inspection procedures, including checking height, diameter, flatness, taper, and out-of-roundness of parts. The gage can be set with gage blocks.

In the continuing search for greater accuracy in machined components and better quality assurance procedures, new machines such as the one shown in Fig. 5.121 have been developed. These highly accurate and versatile machines are most often used in quality assurance laboratories and in the inspection departments of machine shops doing precision work. The machine provides for lateral, fore and aft, and vertical movement of the

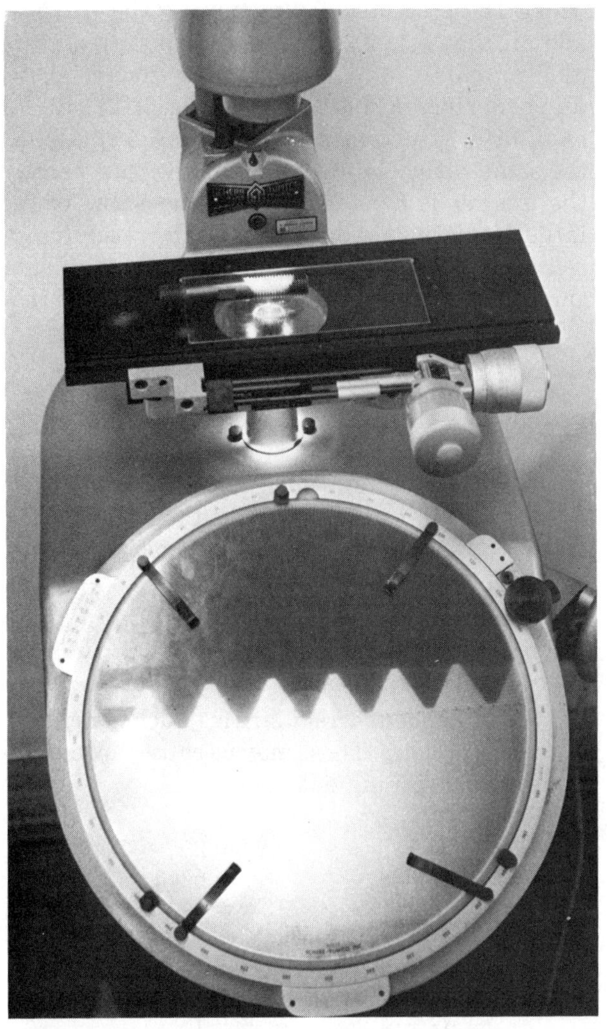

FIGURE 5.119 Close-up of optical comparator in use.

FIGURE 5.120 Electronic comparator. (*Courtesy L.S. Starrett Co.*)

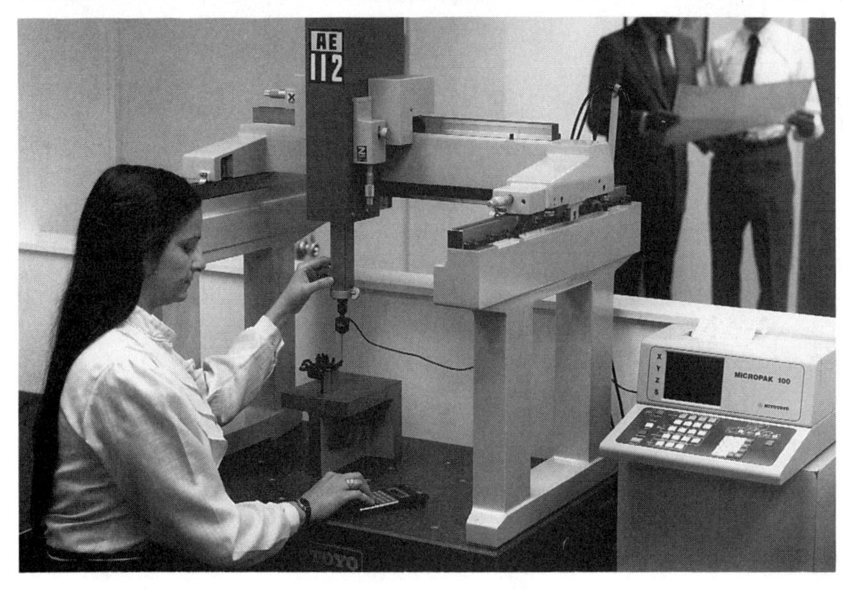

FIGURE 5.121 Special measuring machines are used to check dimensions and geometric relationships on complex machine parts. (*Courtesy Dimensional Inspection Laboratories*)

measuring head. Micrometer-type fine adjustment devices are provided for all axes of movement. The ways are precisely aligned with the surface of the black granite surface plate on which it is mounted. Depending on the probe or measuring head used, the readout can be on a gage read by the operator, or more commonly, in the form of a digital readout. On some machines of this general type a printed record of the dimensions checked can be produced.

5.5.6 Surface Roughness Measurement

Surface roughness indicators are used to determine the presence and extent of several types of surface irregularities on machined parts. The quality of a machined surface is important because rough, wavy, or irregular surfaces can cause shorter life or failure of a part or machine. On the other hand, because producing fine surface finishes is expensive, it is not desirable economically to use a better finish than needed.

In English measure, the unit of surface finish measurement is the *microinch* (μin.) (0.000001 μin.); in metric measure it is the *micrometer* (μm) (0.000001 μm). On some drawings no surface finish specifications are stated. It is assumed then that the finish produced by the specified operation, such as drilling or turning, will be satisfactory. If a particular type of surface finish is necessary it is designated in the title block of the drawing and on the drawing of the part.

The roughness of a surface can be expressed as an *arithmetic average* (AA) or *root mean square* (rms). The procedure for computing the AA is to add all the vertical variations without regard to plus or minus signs and to then divide by the number of increments taken (see Fig. 5.122). In this case, with 18 increments, the AA is 5.8 μin. When the same surface is evaluated by the rms method, the answer is 6.6 μin. because the value of the larger deviations is emphasized. The *mean line* or *reference line* is the imaginary line that runs through the center of the vertical variations that are referred to as roughness.

Besides roughness, there are other surface characteristics that are produced by various machining processes (see Fig. 5.123). These include:

Lay: the direction of the irregularities caused by machining. For example, the lay of the finish on a surface ground part is parallel to the table travel.

Waviness: vertical deviations from the mean line that are of much greater width than the roughness variations.

Flaw: random irregularities such as cracks, tears, or holes. These may be caused by material or machining defects.

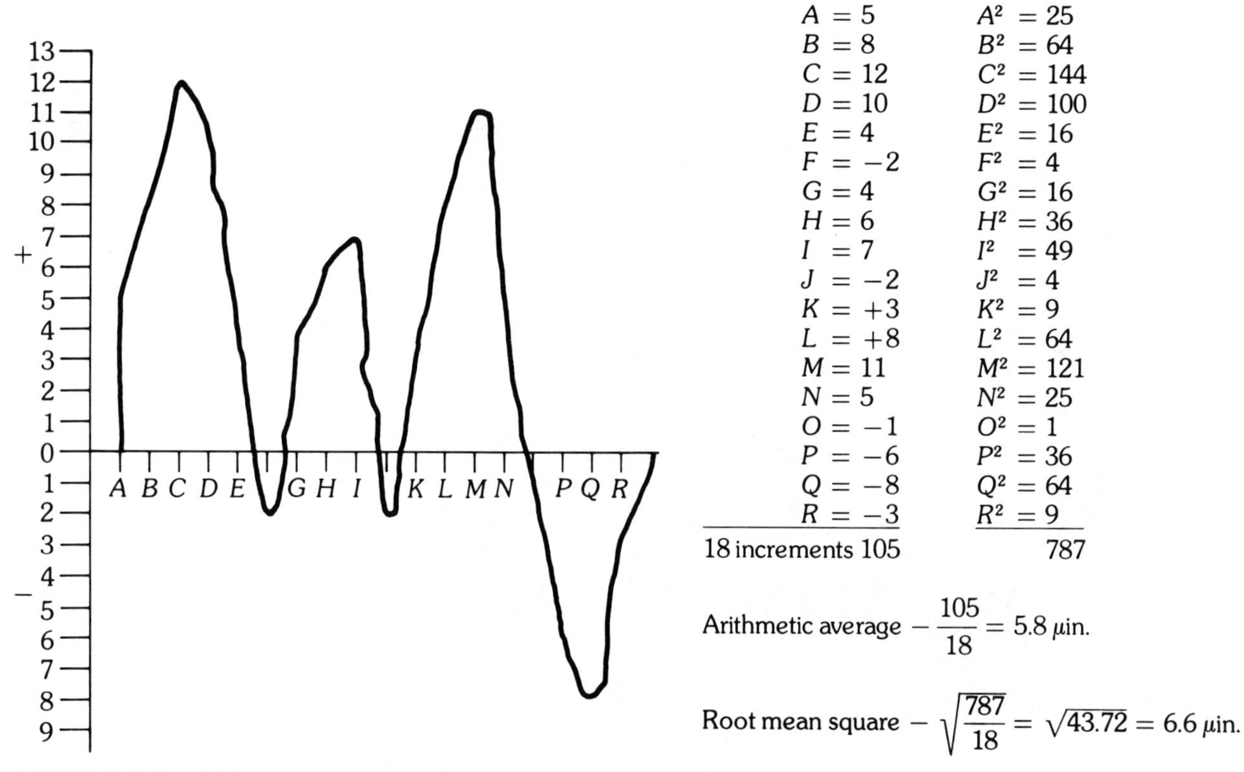

FIGURE 5.122 Arithmetic average and root-mean-square computations.

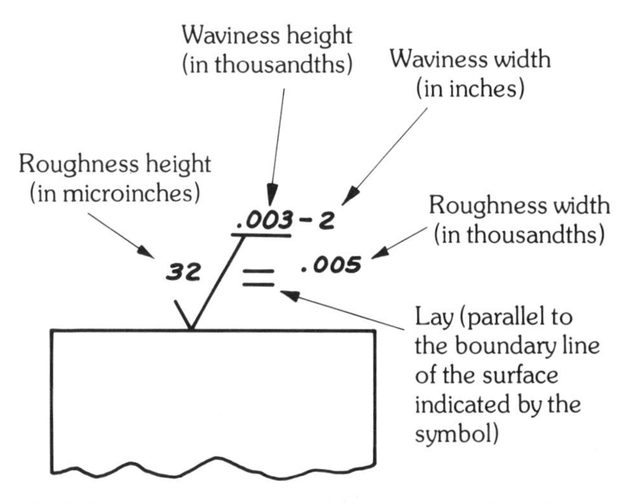

FIGURE 5.123 Symbols used to denote surface finish.

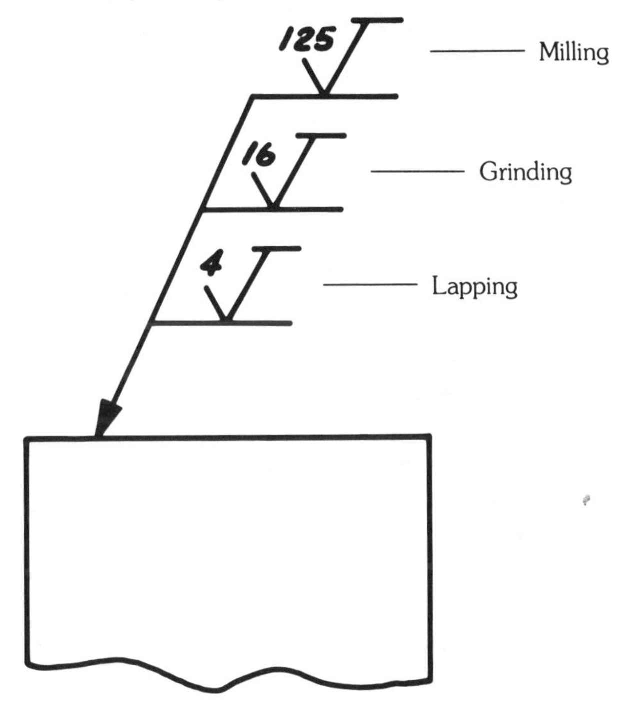

FIGURE 5.124 Finishes produced by various machining operations.

A typical example of the symbols used to denote the desired surface condition on a part is shown in Fig. 5.124. Multiple symbols can also be used to denote a sequence of operations. The surface roughness of finishes produced by common machining processes is listed in Table 5.2.

TABLE 5.2

Surface finishes in common machining operations

Operation	*Surface Finish (rms)*
Sawing	
Coarse pitch blade	325–400
Fine Pitch blade	250–350
Turning	175–250
Milling	150–250
Drilling	100–200
Reaming	50–125
Filing	50–150
Grinding	10–20
Honing	5–15
Lapping	1–4

Surface Indicators. Several types of *surface indicators* or *profilometers* are used to measure the surface roughness of a part. The main components of the instrument shown in Fig. 5.125 are the *stylus head, amplifier, meter, range selector switches,* and an *actuator* to move the stylus head across the workpiece. As the stylus point moves horizontally, it is also forced to move vertically by the roughness of the surface. Its vertical movement causes variations in the flow of electrical current through the circuit. This variation is amplified and read on the meter as the AA or rms of the roughness of the surface.

A less accurate but much quicker method of determining surface roughness is to use a surface finish comparator. The comparison can be made by

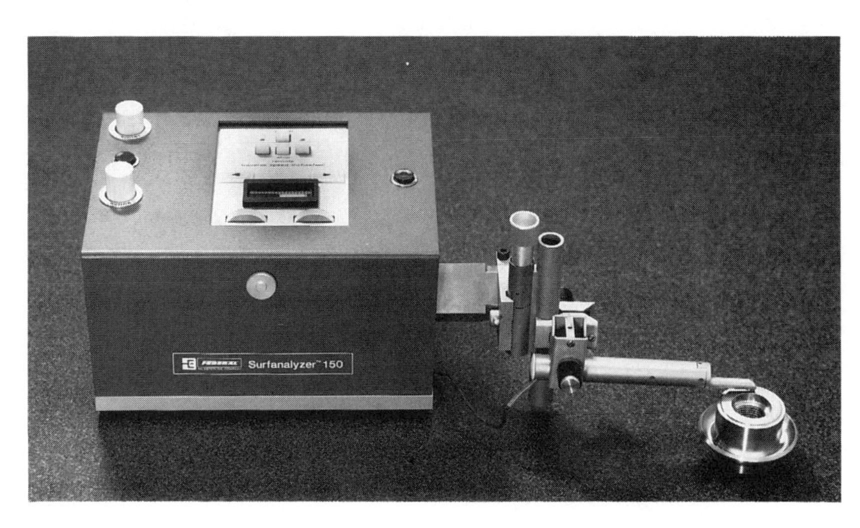

FIGURE 5.125 Surface finish measurement is an important part of quality assurance. (*Courtesy Dimensional Inspection Laboratories*)

sight and by using a fingernail to check the relative roughness of the machined part to the standard sample. This process is adequate when close fits are not required.

5.6 LAYOUT OPERATIONS

A thorough understanding of layout procedures and tools is an important part of the expert machinist's abilities. Layout work on complex and precise parts is a critical first step in the total machining process and can be the deciding factor in the success or failure of the machining operation. To be able to lay out work properly, the machinist must know how to interpret drawings and sketches and must have an understanding of tolerances and allowances.

5.6.1 Layout Tools and Materials

Many tools used in layout work are also used in other machine shop operations. The tools required range from a simple steel rule and scriber to precision gages, surface plates, gage blocks, and other complex equipment.

The tools and equipment shown in Fig. 5.126 are used in most layout operations. *Scribers* draw lines on surfaces that have been coated with *layout dye,* a fast-drying blue fluid that makes scribed lines easily visible. *Dividers* scribe circles and arcs and are available in several sizes. *Trammels* are used for scribing large circles and arcs. *Hermaphrodite* calipers are used to scribe lines parallel to an edge or a slot.

A *surface gage* can be used, as shown in Fig. 5.127 to scribe a line at a given distance above the surface plate. Because the position of the scriber is usually set visually with a steel rule, the location of the scribed line is not necessarily precise. A much more accurate way to scribe precisely located lines involves using a vernier height gage fitted with a scriber attachment (see Fig. 5.128).

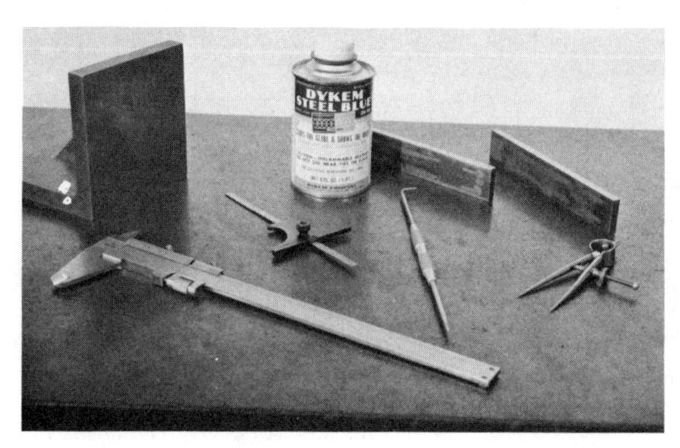

FIGURE 5.126 Layout tools.

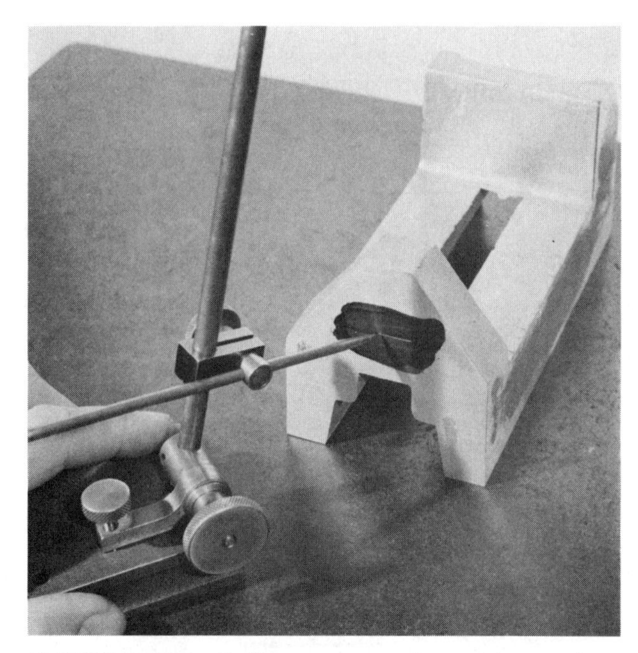

FIGURE 5.127 Scribing a layout line with a surface gage.

(a)

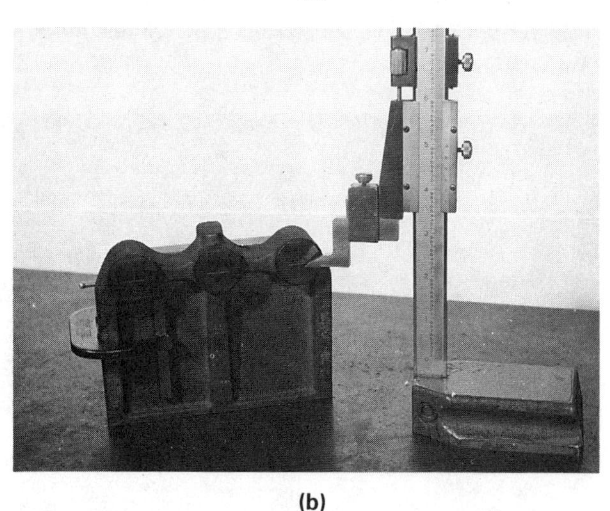

(b)

FIGURE 5.128 Scribing a layout line with a vernier height gage.

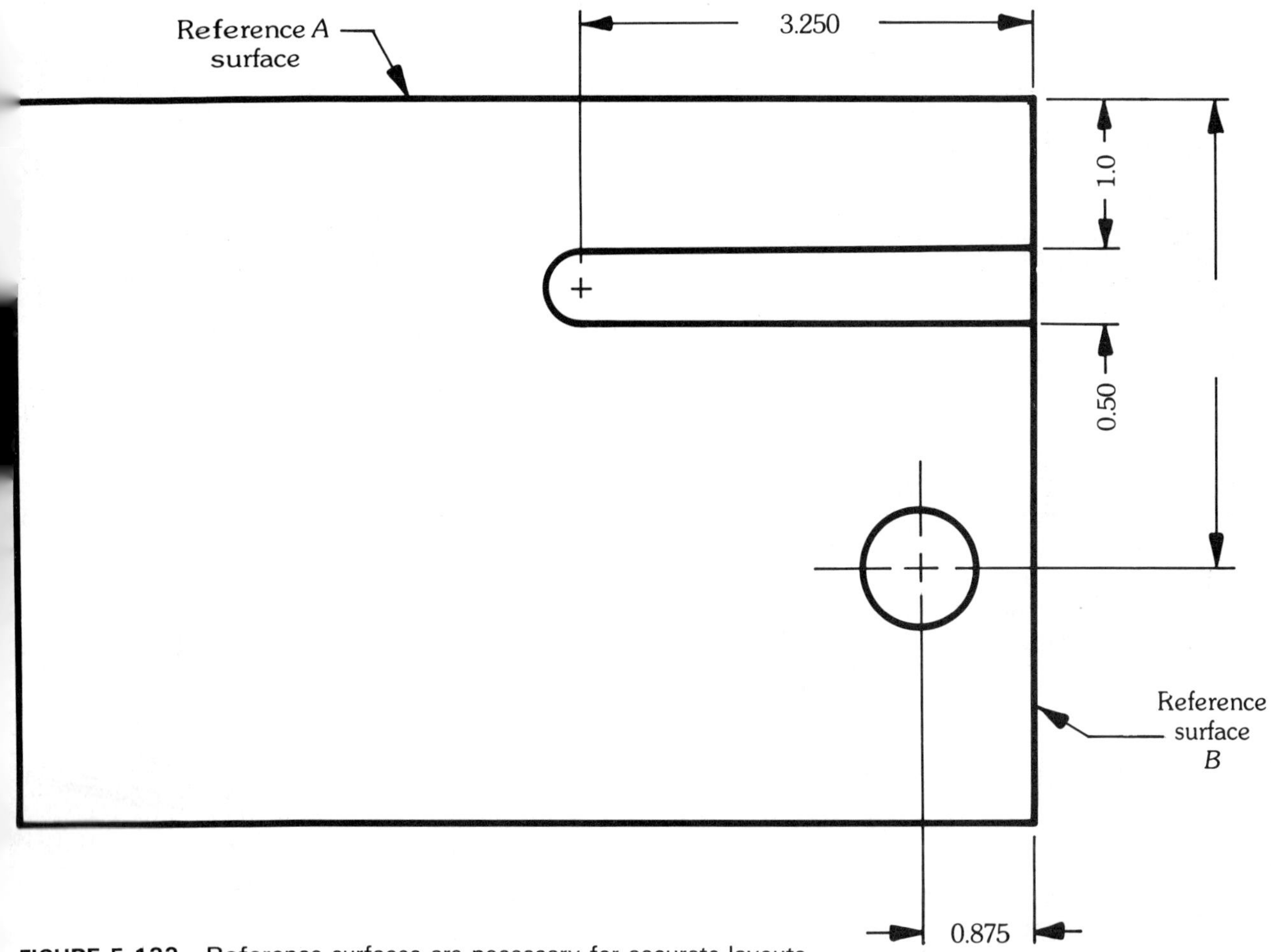

FIGURE 5.133 Reference surfaces are necessary for accurate layouts.

5.134). The intersections of lines that will form the center of arcs and circles to be drawn with dividers should be punched with a *prick punch* ground with a 60° included angle point. Dividers can then be used to scribe the necessary arcs and circles or to step off distances.

Angles can be scribed on a part with a protractor, as shown in Fig. 5.135. If the layout is being

FIGURE 5.134 Layout operation using height gage and scriber.

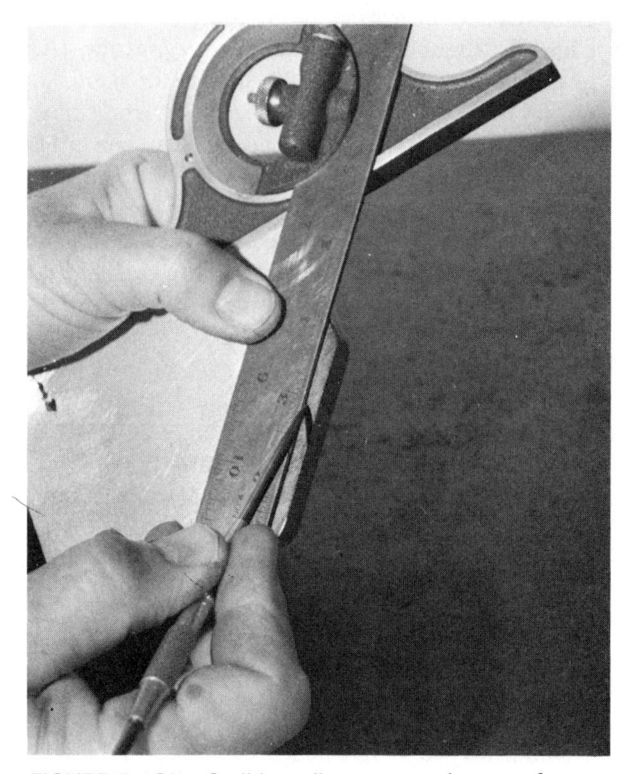

FIGURE 5.135 Scribing a line at an angle to a reference surface.

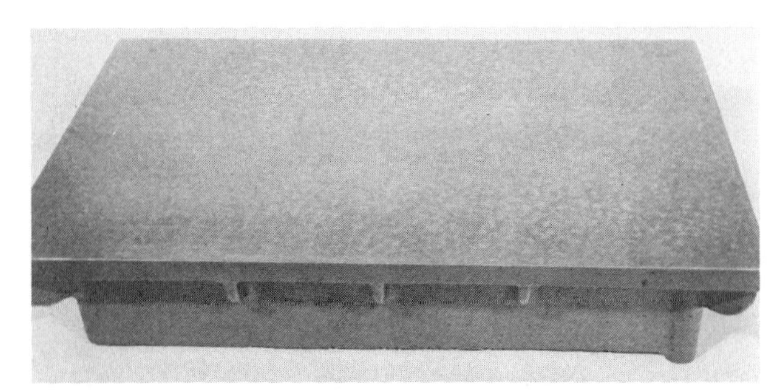

FIGURE 5.129 Cast iron surface plate.

FIGURE 5.130 Granite surface plate. (*Courtesy L.S. Starrett Co.*)

Surface plates are the reference surface in practically all layout operations and range in size from 8 by 12 in. to 48 by 144 in. They may be made of granite or a fine-grained high-quality cast iron (see Figs. 5.129 and 5.130). Laboratory grade (AA) granite surface plates are finished to an overall accuracy of ± 0.000025 in. A typical cast iron surface plate is finished by hand scraping to an overall accuracy of ± 0.0001 in. and is fitted with handles.

Keep surface plates covered when they are not in use. Clean angle plates, parallels, and measuring tools and gently place them on the surface plate to avoid nicking and scratching. Both cast iron and granite surface plates can be cleaned with solvents and clean rags. Keep cast iron surface plates very lightly oiled to prevent rust.

Angle plates, straightedges, and *parallels* made of granite are used in inspection and layout work. Straightedges and parallels can also be made of high-quality cast iron or hardened steel. Angle plates, particularly larger ones, are heavily ribbed iron castings (see Fig. 5.131). *V blocks* and *toolmaker's clamps* are also extensively used in holding work during the layout process (see Fig. 5.132).

Remember that the accuracy of layout work depends on your skill and the condition of the precision tools you use. The good machinist uses and maintain tools carefully.

FIGURE 5.131 Cast iron angle plate.

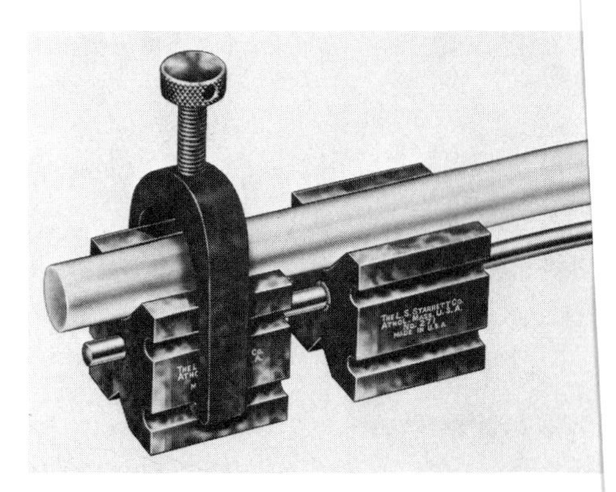

FIGURE 5.132 V block set and clamp. (*Co Starrett Co.*)

5.6.2 Basic Layout Procedures

The first step in doing accurate layout wor ing sure that the information on the dra thoroughly understood. This includes tol allowances, and the position of baselines ar lines that form the reference surfaces of th The relationship of angular surfaces, part compound angles, to the baseline of the obj be understood.

The first step in actually making the l to establish the centerline or the baseline stock or casting. When working with bar sto ticularly flat stock, it is helpful to establish curate surfaces at 90° to each other, as sh Fig. 5.133. Protractors, squares, hook rul other tools can then be used to measure ac ly from either surface. If a surface and an plate are being used, the part may be turn for more convenient use of the height gage cating hole centers and other positions (se

FIGURE 5.136 Scribing a line at a precise angle.

FIGURE 5.137 Laying out a wood vise casting.

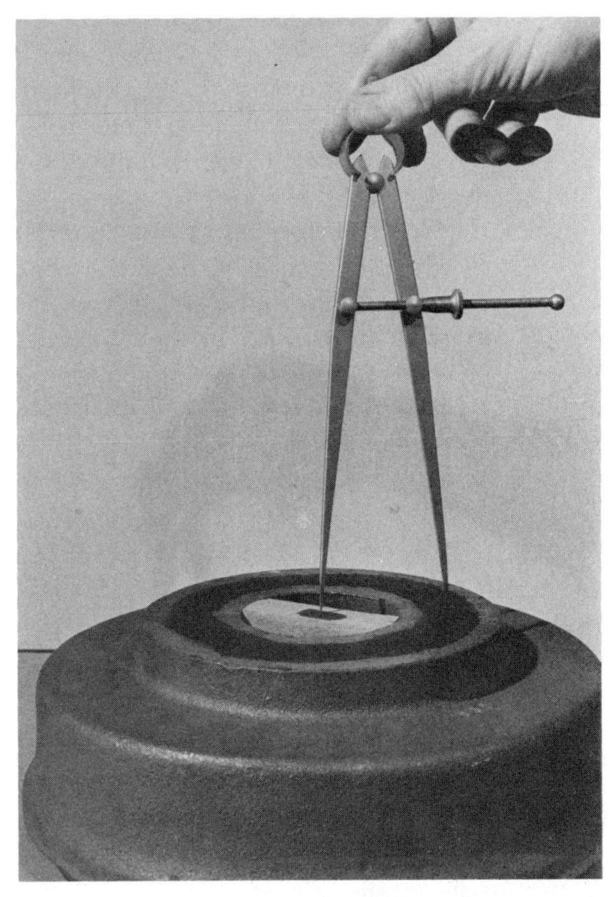

FIGURE 5.138 Laying out a casting using a wood bridge.

made on a surface plate, a sine bar or a universal vernier protractor can be used to set the part at the correct angle, and the scribe mark can be made with a scriber attached to a vernier height gage (see Fig. 5.136).

When laying out castings, finding a usable baseline or reference surface may be difficult, depending on the shape of the casting. In some cases, a reference surface or edge can be ground or rough-machined so that the edge of the casting can be placed on the surface plate or parallels (see Fig. 5.137). A centerline can then be established, and other parts of the layout can be completed.

In some cases the center of a cored or previously machined hole has to be accurately located. A piece of wood, preferably hardwood, can be cut and sanded so that it is a drive fit in the hole. After the center of the hole has been approximately located, a large-headed nail is driven into wood at this spot, and the final center location is made. The head of the nail can then be prick-punched and the layout completed (see Fig. 5.138).

The effective use of more complex measuring tools, an understanding of the metric system, and competency in planning and layout procedures are necessary parts of the machinist's preparation for advancement in the trade. Industrial products are becoming more sophisticated and complex, challenging the skill and imagination of those who work in the machine trades.

REVIEW QUESTIONS

5.2 NONPRECISION MEASURING TOOLS

1. Describe the two basic systems in which English (inch) measure rules are graduated.
2. How are angular settings made on the combination rule and depth gage?
3. How is an inside measurement made with a caliper rule?

4. What are the four major components of the combination set?
5. Explain the function of the center head.
6. Briefly describe a diemaker's square and explain how it is adjusted.
7. How is the cylindrical square read?
8. Briefly describe the operating mechanism of a spring-joint caliper.
9. For what purpose is a lock-joint transfer caliper used?
10. Describe a trammel set and briefly explain its purpose.
11. Describe the process of setting an inside caliper with a micrometer.
12. Briefly describe the telescoping gage and explain how it is used to transfer measurements to an outside micrometer.

5.3 PRECISION MEASURING TOOLS

1. Explain the relationship between the 40-pitch screw on a micrometer and the divisions on the sleeve.
2. Explain the relationship between the vernier scale on a micrometer and the divisions on the thimble.
3. List the sequence of steps in reading a vernier micrometer.
4. Why is the true scale reversed on depth micrometers?
5. How are the different-length rods installed in depth micrometers?
6. What is the pitch of the spindle screw on metric micrometers?
7. How is the sleeve of a metric micrometer graduated?
8. What two devices can be used on the thimble of a micrometer to provide uniform contact pressure when taking a measurement?
9. Describe the procedure for checking an outside micrometer for accuracy.
10. Briefly describe the V-anvil micrometer.
11. What is the main function of the bench micrometer?
12. What is the main operational advantage of the three-point contact inside micrometer?
13. List three functions of depth micrometers.
14. Briefly explain the operating principle of the 25-division vernier caliper.
15. List the sequence of steps in reading a metric vernier caliper.
16. List three accessories that can be used with a vernier height gage.
17. Briefly explain the operating principle of the universal bevel protractor.

5.4 DIAL INDICATORS AND RELATED TOOLS

1. How are dial bore gages set for use in inspection work?
2. Identify and describe one type of operating mechanism used in dial indicators.
3. Explain the relationship between the sensitivity of a dial indicator and its operating range.
4. What is the difference between balanced and continuous dial indicators?
5. List two uses for dial test indicators.

5.5 GAGE BLOCKS, SINE BARS, AND RELATED TOOLS

1. What are the three classes of gage blocks?
2. Describe the process of wringing together gage blocks.
3. Explain how, using the smallest number of blocks, a stack of gage blocks is assembled to achieve a particular dimension.
4. Why should the temperature of gage blocks be kept as near as possible to 68 °F (20 °C)?
5. How should gage blocks be cleaned?
6. Explain the operating principle of a go/no go plug gage.
7. On a thread plug gage, how are the threads on the no-go end different? Why?
8. Explain how a taper ring gage would be used to check the diameter and taper of a machined part.
9. How are adjustable snap gages set?
10. List three uses for sine bars.
11. What means are used to establish the height of the "side opposite" when a sine bar is set to a particular angle on a surface plate?
12. Describe the process of setting a height gage with gage blocks.
13. Briefly describe a precision height gage and explain how it is read.

14. What would you use to set a mechanical comparator on which a part with a tolerance of ± 0.0001 in. was to be measured?
15. Explain the basic operating principle of air gage comparators.
16. What type of gage or standard would be used to set an air gage comparator for checking the inside diameter of a part?
17. Name four characteristics of a workpiece that can be evaluated by air gaging.
18. Explain how the image of a part is amplified on an optical comparator.
19. Name the three main components of a typical electronic comparator and describe the function of each.
20. What are the two methods of expressing the roughness of a surface?
21. Define waviness and indicate how it is expressed in a surface finish designation.

5.6 LAYOUT OPERATIONS

1. What is the main function of a surface plate in layout work?
2. Of what two materials may surface plates be made?
3. Why is it important to establish a baseline, centerline, or both on a part as the first step in making a layout?
4. List three functions of dividers in layout work.
5. What tool or tools could you use to lay out an angle accurate to within $\pm 1/2°$?

STUDENT ACTIVITY

Sooner or later every machinist must face the problems involved in producing machined parts that meet tolerances and specifications set by the purchaser. In some cases, particularly in small shops, the machinist must not only produce the parts but must establish an inspection system to verify that parts are satisfactory.

The part shown below is the shaft on which the pulley of a belt sander rotates. The tolerances and allowances shown make it possible to assemble the mating parts nonselectively. That is, any pulley will fit on any shaft with a working clearance within the design tolerances. After examining the drawing, specify the equipment to be used in inspecting the part and specify the sequence of operations.

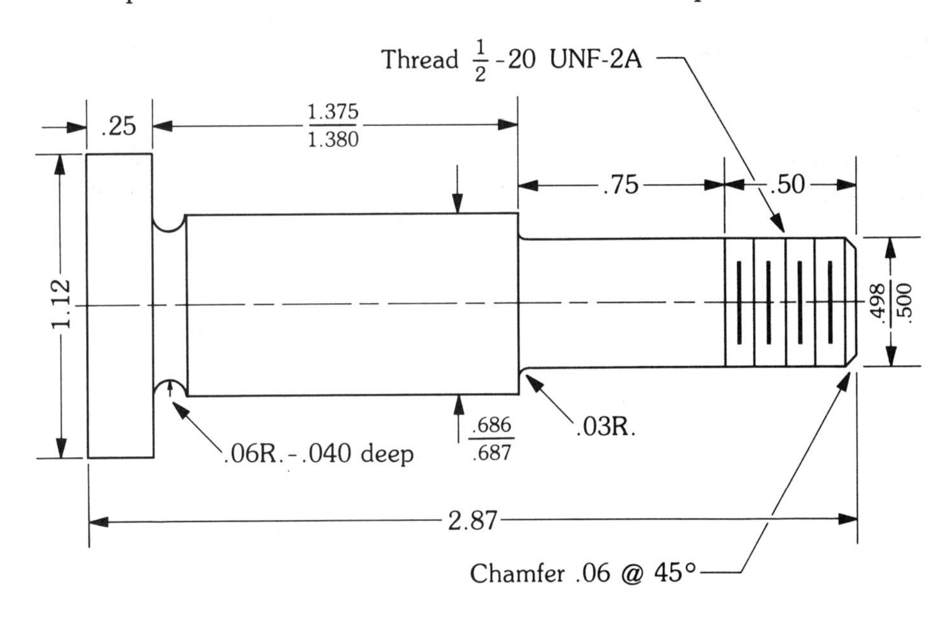

Notes:

1. Break all sharp edges.
2. $\frac{.498}{.500}$ and $\frac{.686}{.687}$ diameters must be concentric within .003″ TIR (total indicator reading).
3. Finish on $\frac{.686}{.687}$ diameter must be 60 rms or better.

6

Screw Threads and Threading Processes

6.1 INTRODUCTION

Screw threads are used for a variety of purposes and applications in the machine tool industry:

1. To hold or fasten parts together: as screws, bolts, and nuts
2. To transmit motion: as the lead screw moves the carriage on an engine lathe
3. To control or provide accurate movement: as the spindle on a micrometer
4. To provide a mechanical advantage: as a screw jack raises heavy loads.

When defining a screw thread, one must consider separate definitions for an *external thread* (screw or bolt) and an *internal thread* (nut).

An external thread is a cylindrical piece of material that has a uniform helical groove cut or formed around it. An internal thread is defined as a piece of material that has a helical groove around the interior of a cylindrical hole.

6.2 SCREW THREAD NOMENCLATURE

Screw threads have many dimensions. It is important in modern manufacturing to have a working

Opening photo courtesy of Teledyne Landis Machine Tool Co.

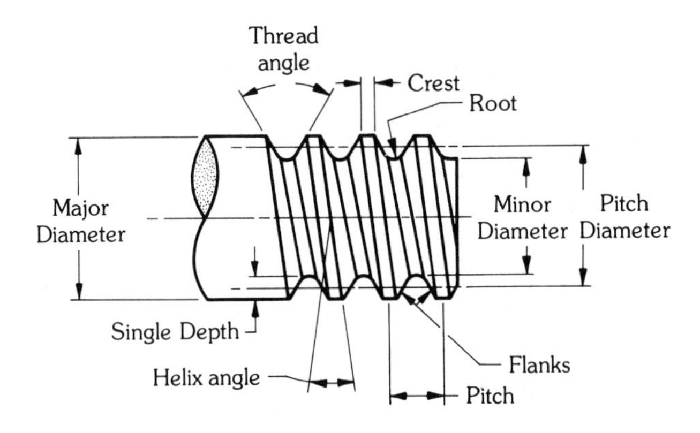

FIGURE 6.1 Screw thread nomenclature.

knowledge of screw thread terminology to identify and calculate the dimensions correctly (see Fig. 6.1).

The *major diameter* is the largest diameter of a screw thread. On an external thread it is the outside diameter; on an internal thread it is the diameter at the bottom or root of thread.

The *minor diameter* is the smallest diameter of a screw thread. On an external thread, the minor diameter is at the bottom of the thread; on an internal thread the minor diameter is the diameter located at the crest.

The *pitch diameter* is an imaginary diameter that passes through the threads at the point where the widths of the groove and the thread are equal. The pitch diameter is the most important dimension on a screw thread; it is the basis from which all thread measurements are taken.

The *root* is the bottom surface connecting two sides of a thread. The *crest* is the top surface connecting two sides of a thread. *Pitch* is the linear distance from corresponding points on adjacent threads. The pitch is equal to 1 divided by the total number of threads per inch [$P = 1/(\text{no. threads/in.})$]. A screw having a single lead with 16 threads per inch has a pitch equal to 1/16 in., commonly referred to as a "16-pitch thread."

The *lead* is the axial distance a threaded part advances in one complete rotation. On a *single lead* threaded part, the lead is equal to the pitch. Multiple-lead threads are discussed later in the chapter.

The *depth* is the distance, measured radially, between the crest and the root of a thread. This distance is often called the *depth of thread.*

The *flank* is the side of the thread. *Thread angle* is the angle between the flanks of the thread. For example, Unified and Metric screw threads have a thread angle of 60°. *Helix* is the curved groove formed around a cylinder or inside a hole.

A *right-hand thread* is a screw thread that requires right-hand or clockwise rotation to tighten it. A *left-hand thread* is a screw thread that requires left-hand or counterclockwise rotation to tighten it. *Thread fit* is the range of tightness or looseness between external and internal mating threads. *Thread series* are groups of diameter and pitch combinations that are distinguished from each other by the number of threads per inch applied to a specific diameter. The two common thread series used in industry are the *coarse* and *fine* series, specified as UNC and UNF.

6.3 SCREW THREAD FORMS AND THREAD STANDARDS

The screw thread dates back to 250 B.C., when it was invented by Archimedes. For centuries wooden screws, handmade by skilled craftsmen, were used for wine presses and carpenters' clamps throughout Europe and Asia. Precision in screw thread manufacture did not come into being until the screw-cutting lathe was invented by Henry Maudslay in 1797.

In the early 1800s, Maudslay also began to study the production of uniform and accurate screw threads. Until then no two screws were alike; manufacturers made as many threads per inch on bolts and nuts as suited their own needs. For example, one manufacturer made 10 threads per inch in 1/2-in.-diameter threaded parts, whereas another made 12 threads, and so forth. During this period the need for thread standards (a model or measure of comparison) became acute.

Despite many attempts at standardization, it was not until World War I that thread standards were developed. The thread profile was designated the *American National thread form* and was the principal type of thread manufactured in the United States until World War II.

During World War II the United States manufactured military equipment that used the American National thread form, which presented interchangeability problems with machinery made in Canada and Great Britain. Not until after World War II, in 1948, did these countries agree upon a *Unified* thread form to provide interchangeability of threaded parts. The Unified thread form is essentially the same as the old American National, except that it has a rounded root and either a rounded or flat crest. The Unified thread form is mechanically interchangeable with the former American National threads of the same diameter and pitch. Today it is the principal thread form manufactured and used by the United States.

6.3.1 The Unified Thread Form

The Unified screw thread has a 60° thread angle with a rounded root and a crest that is *flat* or *rounded* (see Fig. 6.2). As mentioned, this is the principal thread form used for screw thread fasteners used in the United States.

The Unified screw thread system includes six main thread series:

1. Unified Coarse (UNC)
2. Unified Fine (UNF)
3. Unified Extra-Fine (UNEF)
4. Unified 8-Pitch (8 UN)
5. Unified 12-Pitch (12 UN)
6. Unified 16-Pitch (16 UN)

The *coarse-thread series* (UNC) is one of the more commonly used series on nuts, bolts, and screws. It is used when lower-tensile-strength materials (aluminum, cast iron, brass, plastics, etc.) require threaded parts. Coarse threads have a greater depth of thread and are required on these types of materials to prevent stripping the internal threads.

The *fine-thread series* (UNF) is used on higher-tensile-strength materials where coarse threads are not required. Because they have more threads per inch, they are also used where maximum length of engagement between the external and internal threads is needed.

The *extra-fine thread series* (UNEF) is used when even greater lengths of engagement are required in thinner materials. *Eight-, 12-, and 16*-pitch threads are used on larger-diameter threads for special applications. The 8-pitch is generally regarded as a coarse thread for larger diameters, 12-pitch is the fine series, and 16 is the extra-fine thread used on the larger-diameter threads.

The relationship between the *pitch diameter* or *major diameter* determines the helix angle of that thread. For example, a 12-pitch (12 UN) thread with a 1.250-in. major diameter will have a greater helix angle than that of a 12-pitch thread with a 2.0-in. major diameter. Generally speaking, the lower the helix angle, the greater the tensile stress applied to the bolt for a given torque applied to the nut. The fastener with a lower helix angle will also resist vibration and loosening more effectively.

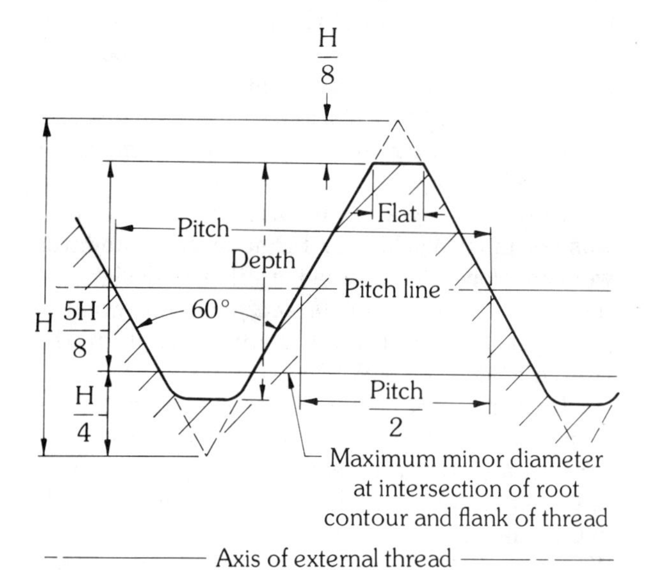

$$\text{Flat—}\frac{\text{external}}{\text{thread}} = 0.125 \times P$$

$$\text{Flat—}\frac{\text{internal}}{\text{thread}} = 0.25 = P$$

$$\text{Depth—}\frac{\text{external}}{\text{thread}} = 0.6134 \times P$$

$$\text{Depth—}\frac{\text{internal}}{\text{thread}} = 0.5413 \times P$$

FIGURE 6.2 Unified screw thread.

Unified Screw Thread Designation. Thread designation refers to identification of the major elements involved in manufacturing a particular screw thread. For example, a typical thread designation on a shop drawing might be

1/4–28 UNF–3B–LH

Examining each major element, we find:

1/4 inch is the nominal major diameter of the thread.

28 is the number of threads per inch.

UNF refers to the thread series, in this case, Unified Fine.

3B 3 indicates the fit; B indicates an internal thread. An A designates an external thread.

LH indicates that the thread to be manufactured will be left-handed. (*Note:* when there is no designation, the thread is assumed to be right-handed.)

Unified Screw Thread Classes (Fits). Before attempting to machine a screw thread, it is important that you understand the classes of thread fits. A thread *fit* is the relationship between an external and internal thread when assembled. Specifically, fit refers to the range of looseness or tightness between two mating parts. A particular class of fit is determined by the different amounts of allowance and tolerances specified.

The Unified Thread system uses three classes of fits. The external thread fits are 1A, 2A, and 3A;

internal thread fits are designated 1B, 2B, and 3B. Briefly these classes of fits are defined as

Class 1A, 1B: loose fit
Class 2A, 2B, medium fit
Class 3A, 3B, close fit

Loose fit, class 1 fit, is used when rapid assembly of thread parts is required. It is the loosest fit, with the greatest range of tolerances and allowance specified. Class 1A and 1B thread fits are used where there is a possibility that the threaded parts may become dirty, slightly damaged, or burred.

Medium fit, class 2A and 2B fits, are used for the majority of nuts, bolts, screws, and common fasteners. This class of fit provides minimum looseness between the external and internal threaded assembly.

Close fit, class 3A and 3B fits, are intended for applications that require closer tolerances than those specified by classes 2A and 2B. Automobile engines, aircraft assemblies, and environments where vibration is present require this fit. Classes 3A and 3B may require a wrench on initial assembly.

Special thread fits are sometimes required for threaded assemblies. The Unified system of thread fits accommodates these special fits by providing that any combination of external-thread class be used with any internal-thread class. For example, a class 2A external thread can be used with a class 1B, 2B, or 3B internal thread.

As mentioned, class of fit refers to the amount of allowance and tolerances specified for a particular class of fit. Complete information can be found in any machine handbook for specific dimensions, allowances, and tolerances.

6.3.2 Acme Screw Threads

Acme screw threads are manufactured for assemblies that require the carrying of heavy loads. They are used for transmitting motion in all types of machine tools, jacks, large C clamps, and vises. The Acme thread form has a 29° thread angle and a large flat at the crest and root (see Fig. 6.3).

Acme screw threads were designed to replace the Square thread, which is difficult to manufacture.

There are three classes of Acme threads (2G, 3G, and 4G), each having clearance on all diameters to provide for free movement. Class 2G threads are used on most assemblies. Classes 3G and 4G are used when less backlash or looseness is permissible, such as on the lead screw of a lathe or the table screw of a milling machine.

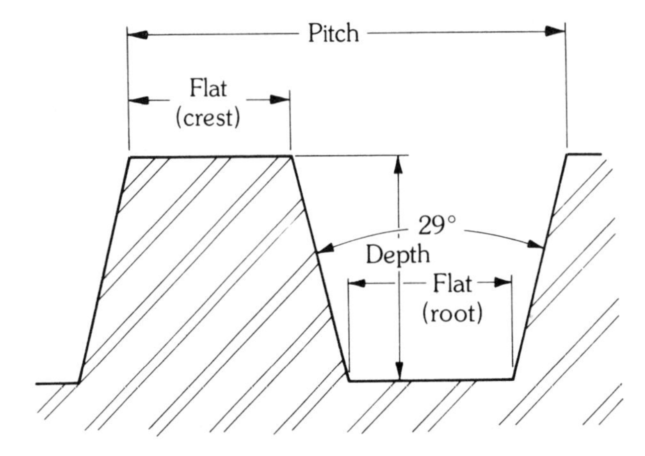

Depth = $0.5 \times P$
Flat (root) = $0.3707 \times P - .005$
Flat (crest) = $0.3707 \times P$

FIGURE 6.3 General-purpose Acme screw thread.

6.3.3 ISO Metric Threads

Metric screw thread standards (size and pitches) were established by the International Organization for Standardization (ISO). The ISO metric thread is used throughout Europe, Asia, and most of the rest of the world. Metric threads have a 60° thread angle and are similar to the Unified thread form (see Fig. 6.4).

Metric threads are designated by the capital letter M preceding the diameter and the pitch, both of which are specified in millimeters. This is somewhat different from the Unified screw thread, which is specified by diameter and the number of threads per inch. Most ISO metric thread sizes come in coarse, medium, and fine pitches. If a coarse thread is required for a part, it is listed without the pitch specified. For example, a coarse

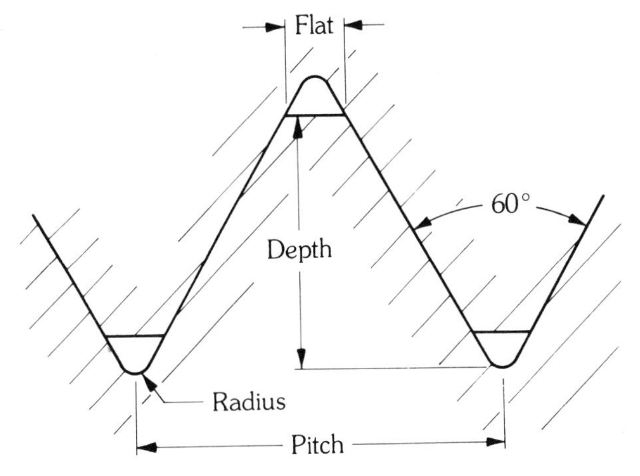

Depth = $0.6134 \times P$
Flat = $0.125 \times P$
Radius = $0.1443 \times P$

FIGURE 6.4 ISO Metric screw thread.

10-mm-OD thread is listed as M10. This thread has a pitch of 1.5 mm, and the pitch is usually omitted from the listing. A fine 10-mm thread has a pitch of 1.25 mm and is designated M10 × 1.25. The extra-fine series for a 10-mm thread is designated M10 × 075.

The ISO metric thread system has several classes of fit that are similar to Unified thread classes. Complete information on thread sizes and classes of fits can be found in any machine handbook.

6.3.4 Tapered Pipe Threads

Pipe threads, usually designated NPT (National Pipe Taper), are tapered threads (see Fig. 6.5) used for sealing threaded joints such as water and air pipes. Most pipe threads have a slight taper (3/4in./ft) and are cut using special pipe taps and dies. Pipe threads can also be machined using the taper attachment on an engine lathe.

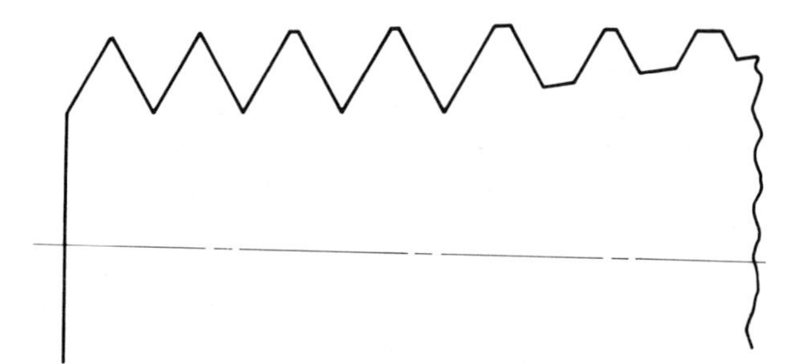

FIGURE 6.5 Tapered pipe threads.

6.4 MULTIPLE (LEAD) SCREW THREADS

Most screw threads have single leads, where the lead is equal to the pitch. A single lead thread has a single ridge and groove. A multiple thread has two or more separate ridges and grooves produced around the circumference (Fig. 6.6). This means that a multiple thread having two ridges and grooves around its circumference is known as a *double* thread; one having three grooves is a *triple* thread, and so forth. The lead of a double thread is twice that of a single thread having the same pitch; the triple thread has a lead three times the pitch of the single lead screw.

Multiple threads are used to obtain an increase in lead without weakening the threaded part on smaller diameter. For example, a 1/4-in.-diameter threaded part is required with a lead of 0.100 in. per revolution. A double lead 20-pitch thread could be cut. A double lead thread produces motion in one revolution equal to twice the pitch (or 2 × 1/20 = 0.100 in.). The equivalent single lead thread to produce a 0.100 in. per revolution requires 10 threads per inch. A 10-pitch thread on the 1/4-in. diameter is cut twice the depth and weakens the 1/4-in.-diameter screw.

Multiple threads can be cut using any of the common thread forms. However, the most often used thread forms for multiple threads are the Unified, the ISO Metric, and the Acme.

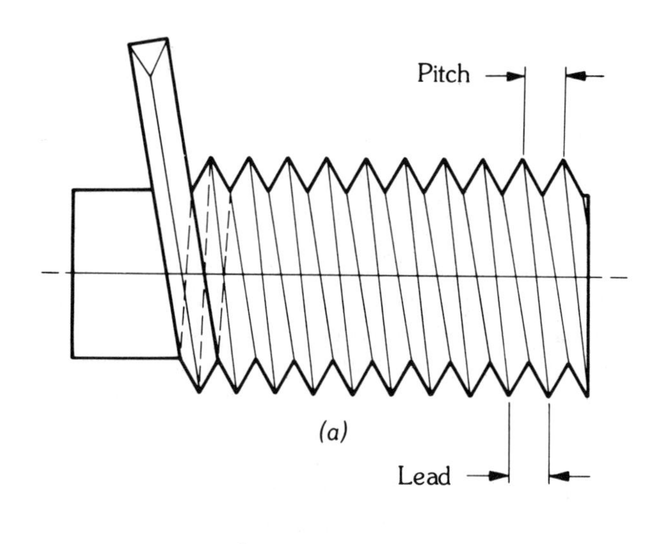

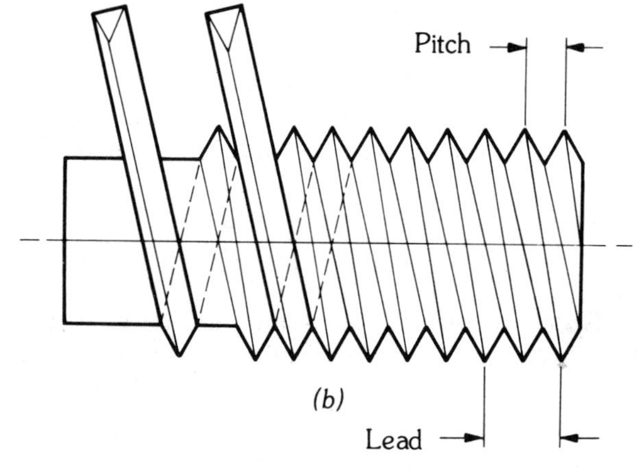

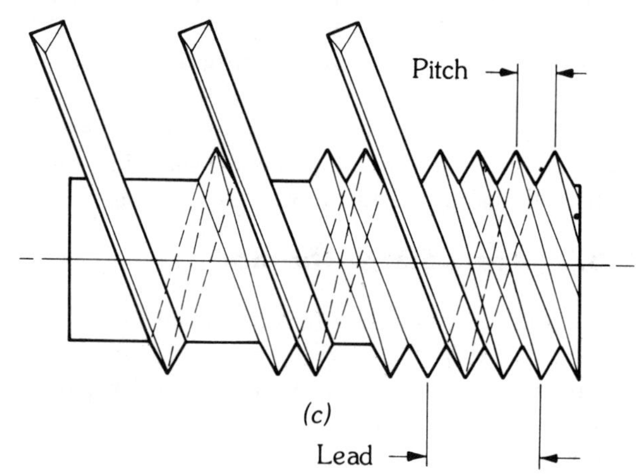

FIGURE 6.6 Pitch and lead relationship for (*a*) single, (*b*) double, and (*c*) triple threads.

6.5 METHODS OF MANUFACTURING SCREW THREADS

Screw threads are manufactured by a number of processes and types of equipment. The manufacturing method generally depends on workpiece size, production rate, accuracy required, and equipment available. The methods listed are those more commonly used in industry for producing screw threads.

6.5.1 Taps and Dies

Tapping and threading with dies (Fig. 6.7) are perhaps the fastest methods of producing threads on a part. Taps are used for producing internal threads, and dies are used for external threads. Taps or dies can be used to cut threads by turning them by hand or by using machine power (see Fig. 6.8). (*Note:* For a complete description of *hand* taps and dies and their use, see Sec. 4.4.)

In production work, self-opening die heads (Fig. 6.9) and collapsible taps (Fig. 6.10) are used to produce larger threads. Self-opening die heads and collapsible taps are tools that have cutting edges called *chasers* inserted into them. The chasers are not part of the tool body. As a result, when a thread is completed in production work, the individual chasers automatically pull clear of the part. The tap or die head can be withdrawn, resulting in time savings by not having to reverse the direction of the tap or die to withdraw it from the workpiece. Another advantage of using self-opening or collapsing mechanisms is that the individual chasers can be adjusted for a particular pitch diameter of a given thread size, allowing the choice of a variety of thread fits. The chasers also can be resharpened.

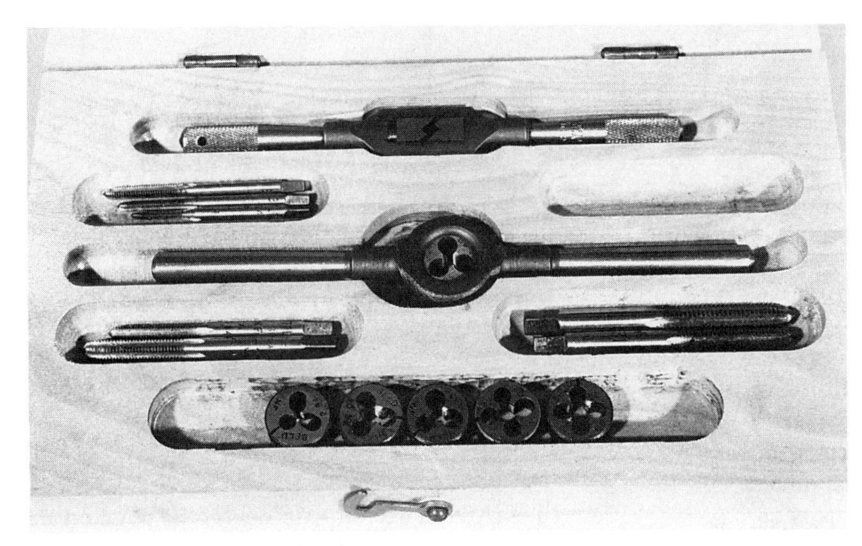

FIGURE 6.7 Thread taps and dies.

FIGURE 6.8 Using a tapping head to tap threads. (*Courtesy Tapmatic Corp.*)

Self-opening dies and collapsible taps are used on threading machines, turret and automatic lathes, and multiple-spindle drilling machines.

6.5.2 Thread Grinding

Grinding a screw thread generally is done when the hardness of the material makes cutting a thread with a die or a single point tool impractical. Grinding threads also results in greater accuracy and in superior surface finishes compared to what can be achieved with other thread-cutting operations. Taps, thread chasers, thread gages, and micrometer spindles all use ground threads.

Ground threads are produced by thread-grinding machines. A thread-grinding machine closely resembles a cylindrical grinder in appearance. It incorporates a precision lead screw to produce the correct pitch or lead on the threaded

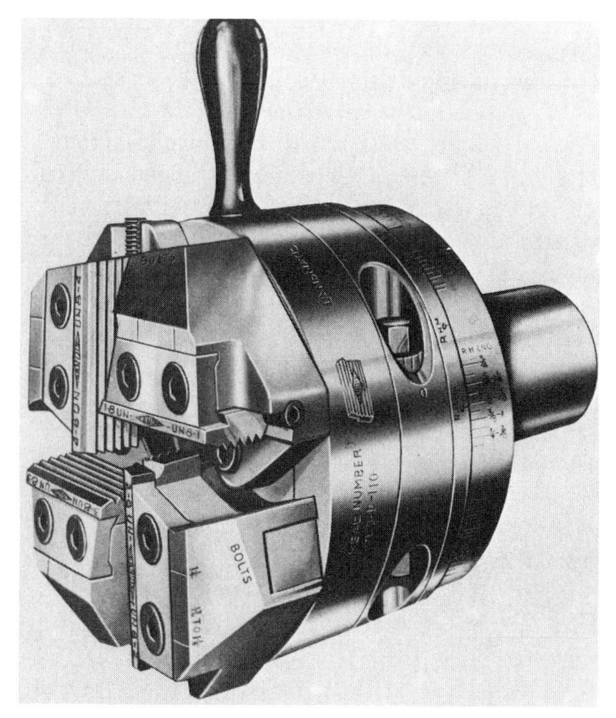

FIGURE 6.9 Self-opening die head. (*Courtesy Teledyne Landis Machine Tool Co.*)

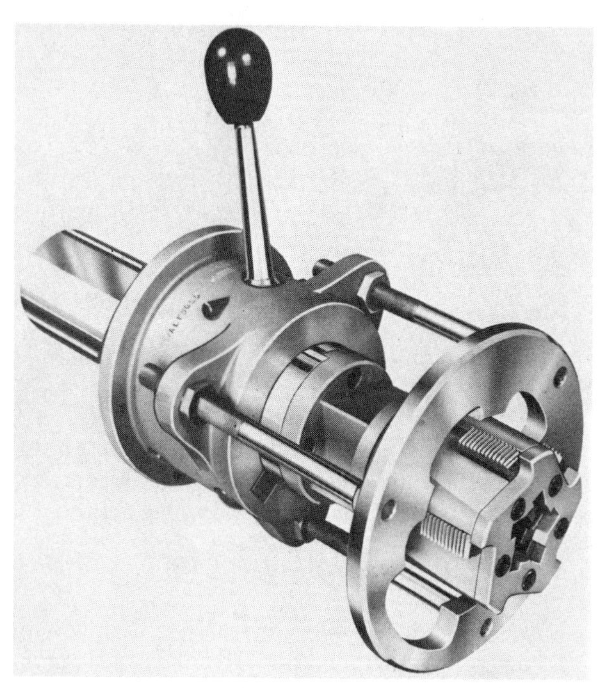

FIGURE 6.10 Collapsible tap. (*Courtesy Teledyne Landis Machine Tool Co.*)

part. Thread-grinding machines also have a means of dressing or truing the cutting periphery of the grinding wheel so it will produce a precise thread form on the part. Grinding wheels used in producing ground threads are single- or multiple-rib (Fig. 6.11). Single-rib types are used for grinding longer threads and feed longitudinally for the required

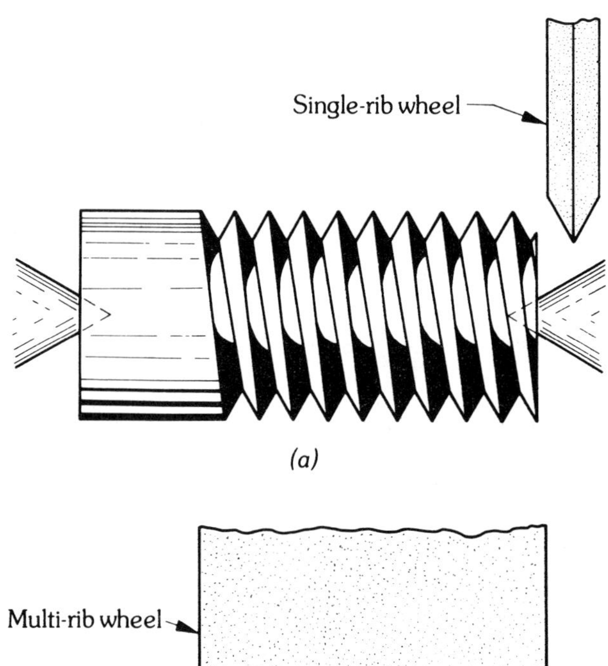

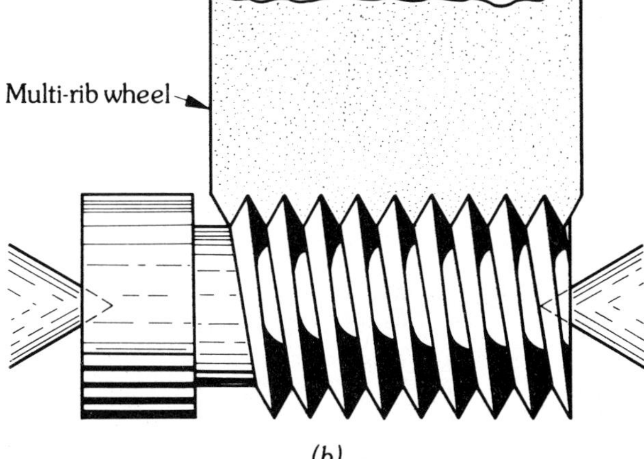

FIGURE 6.11 (*a*) Single- and (*b*) multi-rib thread grinding wheel.

length of thread. The multiple-rib type of grinding wheel is generally used for forming short threads. This type of wheel is "plunged" into the workpiece to produce the thread.

6.5.3 Thread Rolling

Thread rolling is the process of producing threads by displacing material rather than by removing it with a cutting tool. Thread rolling produces thread forms by rotating the work between hardened steel dies until the reverse of the form on the die face is reproduced on the blank. Because thread rolling does not remove any material, the size of the blank is very important. In general, the blank size is slightly less than the pitch diameter of the thread to be produced.

Thread rolling is usually accomplished by *flat-die* or *cylindrical-die rolling.* Flat-die rolling is done in a reciprocating die machine. The flat dies operate by rolling the blank across the face of a stationary die with a moving die. Flat-die rolling is

FIGURE 6.12 Cylindrical-die thread rolling machine. (*Courtesy Teledyne Landis Machine Tool Co.*)

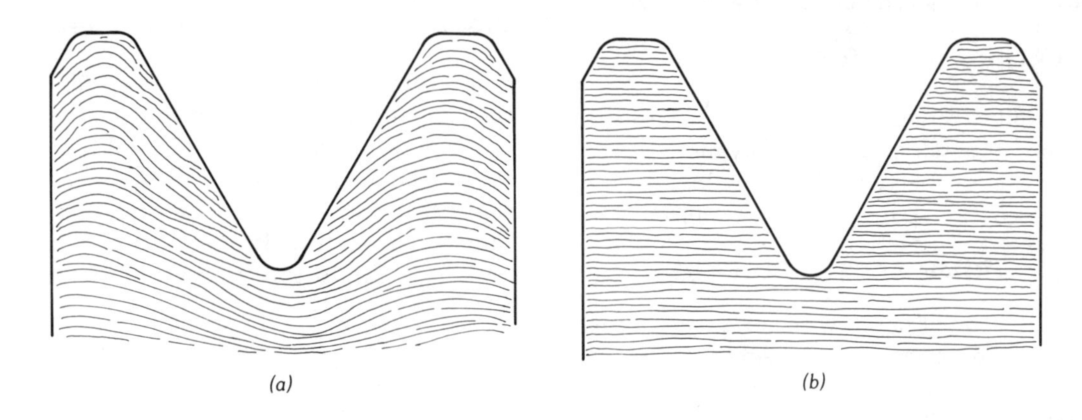

FIGURE 6.13 Comparison of (*a*) rolled and (*b*) machined threads and their grain structure.

used in the production of bolts. Cylindrical-die rolling machines have three or more cylindrical dies. The work blank rotates with the revolving dies as they move in to form the threads. Bolts with shoulders, pipe threads, and parts with large diameters are generally produced on the cylinderical-die rolling machines (see Fig. 6.12).

Thread rolling is considered one of the fastest methods of producing threads. Rolled threads have good surface finishes and physical properties. They provide greater shear and tensile strength than machined threads because the grain structure of the material is re-formed into lines following the thread contour (Fig. 6.13).

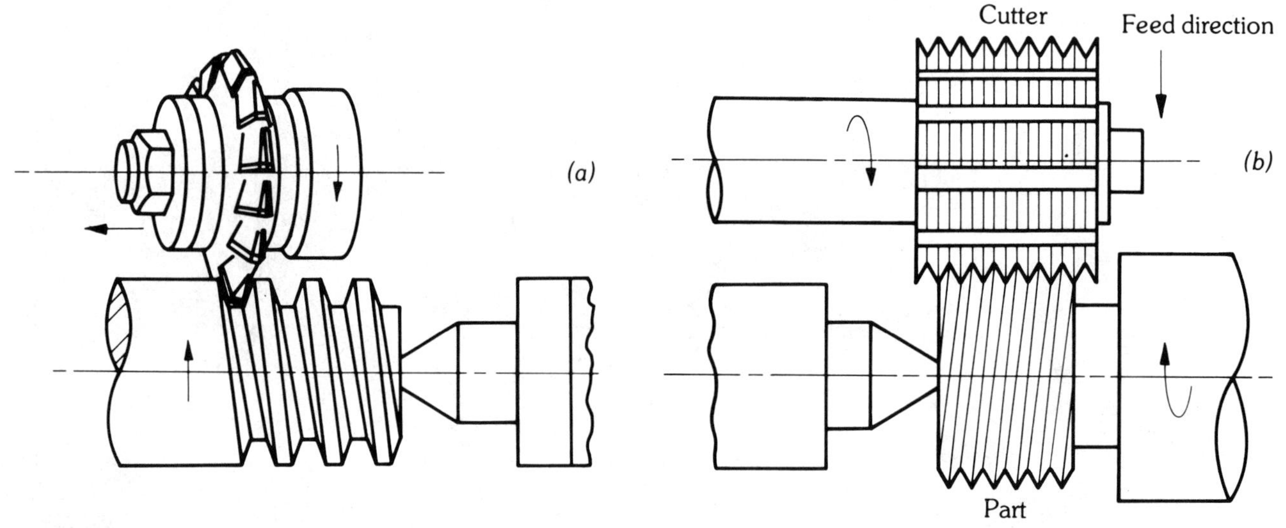

FIGURE 6.14 External thread milling with (*a*) single-row and (*b*) multiple-row tooth cutter.

6.5.4. Thread Milling

Thread milling has been an established method of manufacturing accurate screw threads for many years. Long screws, such as lead screws on lathes and multiple start threads, are often manufactured by milling.

Milling a screw thread is done with either a single- or multiple-lead milling cutter. The rotating cutter is fed into the work to the required depth. The work is then rotated and fed longitudinally at a rate that will produce the proper lead on the part (Fig. 6.14). Any class of fit or thread form can be manufactured by the thread milling process.

6.5.5 Cutting Threads On a Lathe

There are many advantages in using a lathe to manufacture screw threads. These advantages, and complete procedures for cutting screw threads, are explained in detail in Sec. 9.8.9 of Chap. 9.

6.6 MEASURING SCREW THREADS

Proper measurement of screw threads is very important to ensure proper fit between threaded assemblies. The methods used to measure threads depend on the accuracy required and equipment available. In simple repair work involving screw threads, mating the components parts may be satisfactory. However, in the manufacture of screw threads, there are many measuring methods that can be employed. Regardless of which method is actually used, the pitch diameter is always measured—directly or indirectly. Because the pitch diameter is the area where most of the mating threads contact, it is the most important dimension to be concerned with. Six of the common methods of measuring the pitch diameter are:

1. Screw thread micrometers
2. Screw thread comparator micrometer
3. Three-wire system
4. Go/no go thread gage
   (*a*) Plug gage
   (*b*) Ring gage
   (*c*) Snap gage
5. Optical comparator
6. One-wire method for Acme screw threads

6.6.1 Screw Thread Micrometers

Screw thread micrometers are used for measuring the pitch diameter of V-shaped threads such as the Unified, American National, and Metric thread forms. The thread micrometer has a V shaped anvil that fits over a thread and a cone-shaped spindle that is moved directly into a thread opposite the anvil (see Fig. 6.15). By having the spindle and

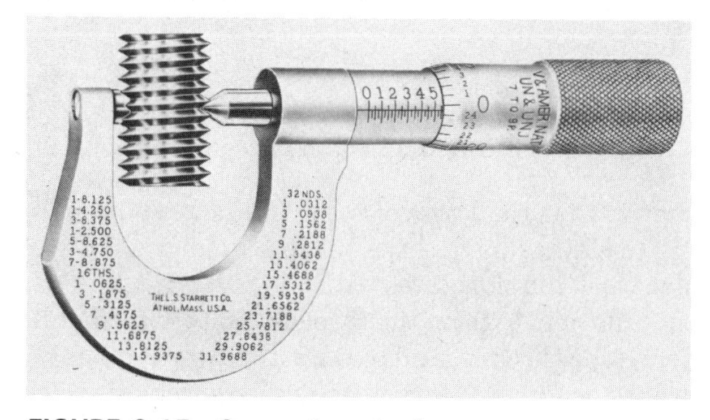

FIGURE 6.15 Screw thread micrometer.

anvil both contacting the sides of the thread, the pitch diameter is measured directly and can be read on the barrel and thimble. The movable spindle is slightly rounded so as not to contact the bottom of the thread. The anvil provides clearance to prevent it from touching the top of a thread.

Each thread micrometer will measure only a certain range of thread pitches, and the range of pitches each micrometer is designed to measure is marked on the frame. The common 1-in. thread mikes for measuring screw threads under 1 in. in diameter are 8 to 13 tpi (threads per inch), 14 to 20 tpi, 22 to 30 tpi, and 32 to 40 tpi.

When using a thread mike, always make sure that contact surfaces of the spindle and anvil are clean. Check the accuracy of the mike by bringing the spindle and anvil into contact. The reading should be zero. Greater accuracy can be obtained from thread micrometers by using a plug gage or standard for a comparison reading with the threaded part to be measured. See Chap. 5 for additional information on thread micrometers.

6.6.2 Screw Thread Comparator Micrometers

Screw thread comparator mikes can be used for accurately measuring threads. They differ from thread micrometers by having a conical anvil that is the same as the spindle (see Fig. 6.16). This measuring instrument does not read the pitch diameter directly as does the thread mike. Therefore, a plug gage or existing screw thread of the same size and pitch must be used in conjunction with the comparator mike. A reading is first obtained from a plug gage or existing screw thread and then compared with the threaded workpiece being measured.

6.6.3 Three-Wire System of Measuring Screw Threads

Threaded parts that require extreme accuracy are usually measured by the three-wire system. Three wires of equal diameter and an ordinary micrometer are used with this method. Two wires are placed in contact with adjacent threads on one side of the thread, and the third wire is in a position opposite the first two (see Fig. 6.17). A micrometer is then used to measure the distance over the three wires. It is important with this method that the wires have the following three characteristics:

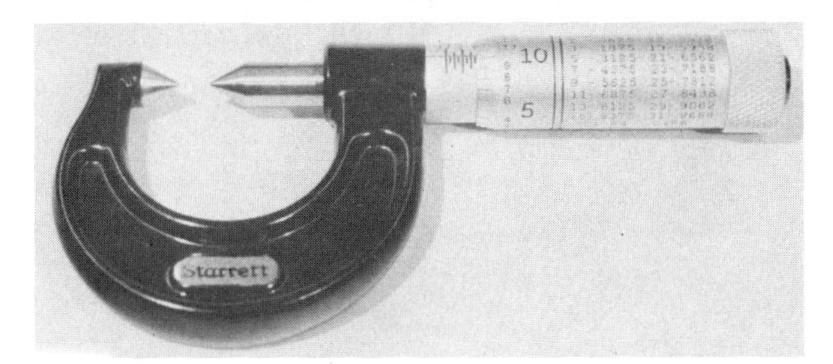

FIGURE 6.16 Screw thread comparator micrometer. *(Courtesy El Camino College)*

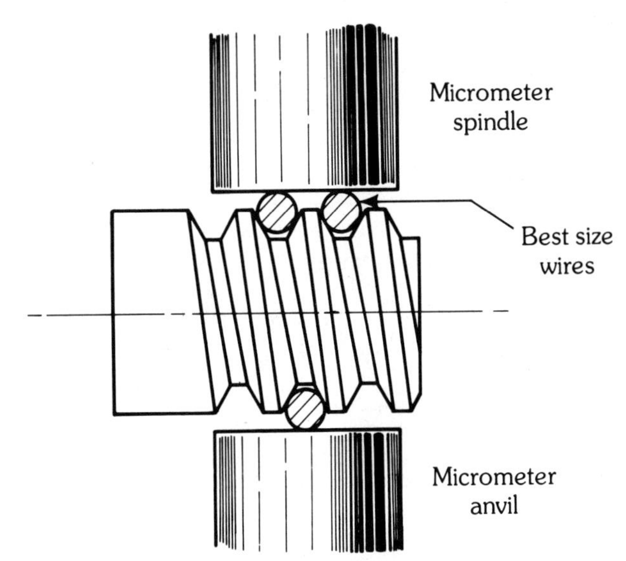

FIGURE 6.17 Three-wire method of measuring screw threads.

1. Be of equal diameter within $\pm$ 0.00002 in. (0.0005 mm)
2. Have hardened and lapped surfaces
3. Be the *best* (correct) size wire for the pitch line contact

The *best* size wire for pitch line contact is found using the following formula:

$$\text{best size wire} = 0.57735 \times \text{pitch}$$

Example: To find the best size wire for measuring a 1/2–13 UNC screw:

$$\begin{aligned}\text{best size wire} &= 0.57735 \times \text{pitch}\\ &= 0.57735 \times \frac{1}{13}\\ &= \frac{0.57735}{13}\\ &= 0.04441 \text{ in.}\end{aligned}$$

After the best size wire is determined, the following formula is used to find the correct pitch diameter when measuring over the three wires:

$$M = D + 3S - \frac{1.5155}{N}$$

where M = measurement over wires
D = diameter of thread
S = size of wire used
N = number of threads per inch

Example: To find the measurement over the wires for a 1/2–13 UNC screw thread, using a *best* size wire of 0.04441, we find

$$
\begin{aligned}
M &= D + 3S - \frac{1.5155}{N} \\
&= 0.500 + 3(0.04441) - \frac{1.5155}{13} \\
&= 0.500 + 0.13323 - 0.11658 \\
&= 0.5167 \text{ in.}
\end{aligned}
$$

Therefore, the measurement over the wires for a 1/2–13 UNC screw thread should read 0.5167 in. When using the micrometer to measure over the wires, apply moderate pressure to ensure greatest accuracy.

6.6.4 Go/No Go Thread Gages

Go/no go gages are manufactured in many sizes and styles to measure different sizes of shafts and holes. Thread go/no go gages may be of the plug gage, ring gage, and roll snap gage type.

Thread Plug Gages. Thread plug gages are used to check the pitch diameter of internal threads (see Fig. 6.18). They have two hardened, ground, and lapped threaded ends connected by a single handle. The longer threaded end is the go gage, and the shorter threaded end the no-go gage. The handle of the thread gage specifies the thread size, the number of threads per inch, and the class of fit for which it should be used. When using thread plug gages, be sure the go end enters the internal thread and turns freely for the entire length. The no-go end should just barely enter the internal thread and become snug if the thread is the correct size.

Thread plug gages are expensive measuring devices and so must be used carefully. To obtain accurate readings and prolong the life of the tool, remove loose chips and burrs from the thread before using this gage.

FIGURE 6.18 Thread plug gage. (*Courtesy Orange Coast College*)

FIGURE 6.19 Thread ring gages. (*Courtesy Orange Coast College*)

Thread Ring Gages. Thread ring gages are of the go/no-go variety and usually come in pairs for checking external threads (see Fig. 6.19). A small adjusting screw is incorporated into the thread ring gage and accurately sets the go and no-go limits from a standard or thread plug gage. The no-go ring gage is distinguished from the go ring gage by a small groove around its periphery. When using thread ring gages, be sure the go ring gage turns freely around the threaded part; the no-go gage will just start and become snug after about one turn. As with thread plug gages, extreme care should be taken to prevent any damage to these precision measuring tools.

Thread Snap Gages. Many sizes and styles of thread snap gages are used to accurately measure external threads. Snap gages generally are made with three or four threaded and adjustable rollers or gages (see Fig. 6.20). Accurate adjustments for different thread sizes and classes of fits are made from existing standards or thread plug gages. The unit is then used by holding the C-shaped frame and carefully applying pressure on the threaded part. The first set of rollers should pass over the threads being checked; the second set

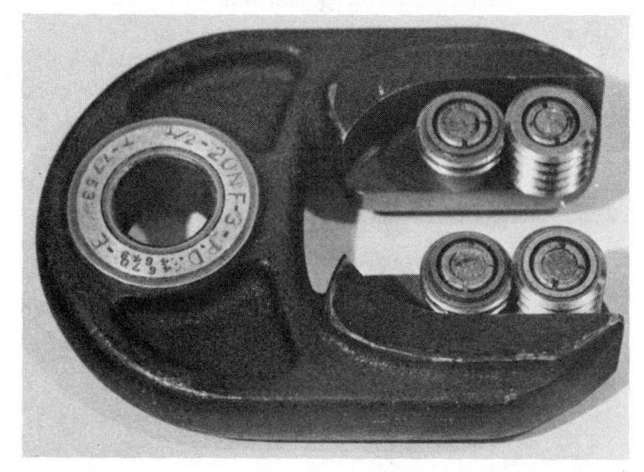

FIGURE 6.20 Thread snap gage. (*Courtesy Orange Coast College*)

of rollers should not. The first set of rollers are the go gages, and the second set are the no-go gages.

Thread snap gages have an advantage over ring gages in that they are faster to use because the gage does not have to be screwed on and off when measuring threads.

6.6.5 Optical Comparators

Optical comparators can be used for measuring screw threads. As explained in Chap. 5, they are instruments that project a magnified profile of a part onto a large screen. This profile can then be compared to a master form or template to gage its accuracy (Fig. 6.21). Optical comparators often have a range of magnification from true size to 125 times original size of the part. Very accurate comparisons can be made on threaded parts by comparing them to enlarged thread templates. The advantages of the optical comparator in measuring screw threads are that only templates of a thread series are needed, and any size (OD) threaded part can be measured using the correct template for the series (pitch) to be measured.

6.6.6 One-Wire Method for Acme Screw Threads

A single wire can be used to measure Acme threaded parts. If the proper wire size is placed in the thread groove of an Acme thread, the wire will be flush with the top of the thread (Fig. 6.22). To use this method, the outside diameter of the thread must be accurate and free from any burrs. The correct diameter of the wire is found as follows:

$$\text{wire size} = 0.4872 \times \text{pitch}$$

Example: The wire size for an 1–8 TPI Acme is

$$\begin{aligned}\text{wire size} &= 0.4872 \times \text{pitch}\\ &= 0.4872 \times \frac{1}{8}\\ &= \frac{0.4872}{8}\\ &= 0.0609 \text{ in.}\end{aligned}$$

The correct size of wire for measuring a 1–8 TPI Acme screw thread is therefore 0.0609 in.

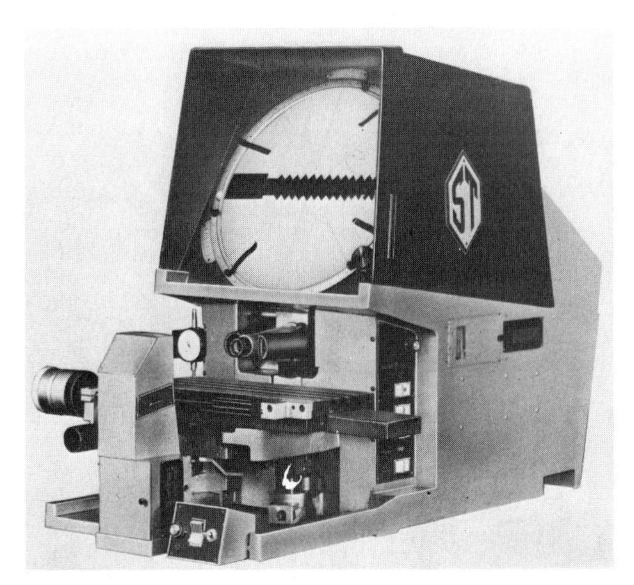

FIGURE 6.21 Measuring screw threads with an optical comparator. (*Courtesy Scherr Tumico, Inc.*)

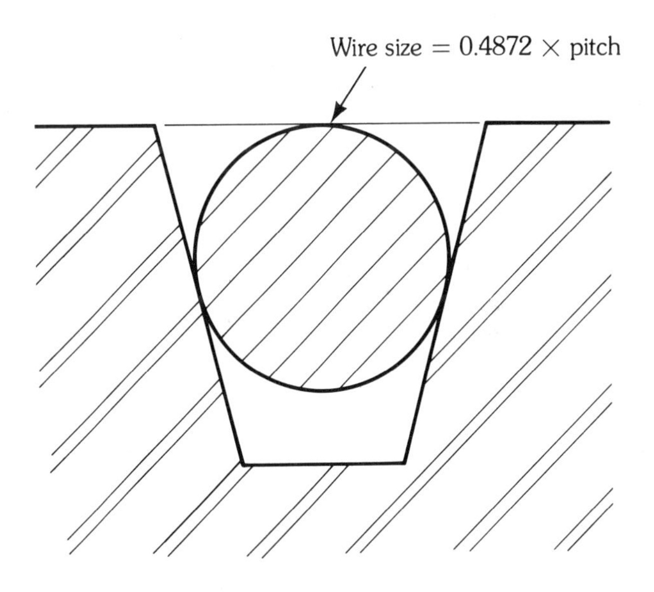

FIGURE 6.22 One-wire thread measurement method for Acme screw threads.

REVIEW QUESTIONS

6.1 SCREW THREADS

1. Explain the differences between an internal and external thread.
2. List four uses of screw threads, and give an example (application) of each.

6.2 SCREW THREAD NOMENCLATURE

1. Briefly define the following thread terms:
   (*a*) Major and minor diameters
   (*b*) Pitch diameter
   (*c*) Root

(*d*) Crest
(*e*) Pitch
(*f*) Lead
(*g*) Flank
(*h*) Thread angle
(*i*) Helix
(*j*) Right- and left-hand threads
(*k*) Thread fit
(*l*) Thread series

2. Sketch an internal and external thread. Label the thread terms (*a*) through (*h*) in question 1.

6.3 SCREW THREAD FORMS AND THREAD STANDARDS

1. Briefly describe the historical developments of the screw thread from 250 B.C. to the present Unified threads.
2. Explain the differences between coarse, fine, and extra-fine thread series for Unified threads.
3. Explain the meaning of the following Unified thread designation: 1/2–13 UNC–3A–LH.
4. Briefly explain the three classes of fits used for the Unified thread form.
5. Using a reference handbook, research and list the allowances and tolerances for the major, minor, and pitch diameters of the following Unified threads:
   (*a*) 1/2–13 UNC–2A
   (*b*) 1/4–28 UNF–3B
   (*c*) 3–16 UN–1A
6. List three applications of the Acme screw thread.
7. Compare and contrast the Unified thread with the ISO metric thread.

6.4 MULTIPLE LEAD SCREW THREADS

1. Why are multiple-lead screw threads manufactured?
2. What is the lead of each of the following screw threads?
   (*a*) Double lead 1/4–20 thread
   (*b*) Triple lead 1/2–thread
   (*c*) Single lead 1–8 thread
3. What thread forms may be used for multiple-lead screw threads?

6.5 METHODS OF MANUFACTURING SCREW THREADS

1. Explain the operating principle of the self-opening die and collapsible tap.
2. On what type of machine tools may self-opening dies and collapsible taps be used?
3. When are thread-grinding machines used for manufacturing screw threads?
4. What two types of grinding wheels are used in producing screw threads?
5. Briefly describe the thread-rolling process in the manufacture of screw threads.
6. What are the advantages of thread rolling compared to other manufacturing methods for producing screw threads.
7. Explain the process of thread milling. What type of screw threads is usually manufactured by thread milling?

6.6 MEASURING SCREW THREADS

1. What is the most important dimension to be concerned with on a screw thread when measuring a threaded part?
2. List six measuring devices that are commonly used when measuring a threaded part, and briefly describe the operating principle of each.
3. Calculate the best size wire for measuring a 3/4–10 UNC screw.
4. What are some additional applications, other than measuring screw threads, of an optical comparator.
5. Calculate the correct wire size for measuring a 2 1/4–6 Acme screw thread.

7

Metal Sawing Operations and Equipment

7.1 INTRODUCTION

Several types of power saws are extensively used in machine shops. Once sawing was considered a secondary machining process and saws were used mostly for cutting bar stock in preparation for other machining operations. In recent years, the development of new types of saws and better blade materials have made metal sawing a much more effective, versatile, and economical process. In many cases bandsaws are now being used as the primary means of shaping certain types of metal parts.

When the proper sawing machines and blades are used, sawing is one of the most economical means of cutting metal. The saw cut (kerf) is narrow, and relatively few chips are produced in making a cut. When a bandsaw is used for cutting the contours of complex shapes, only a small portion of the metal removed is in the form of chips. Therefore, the power used in removing large amounts of waste metal is at a minimum.

Sawing is a basically simple process. As the blade moves past the work, each tooth takes a cut. Depending on the thickness or diameter of the work, the number of teeth cutting at one time varies from 2 to 10 or more. Saws may be of the

---

Opening photo courtesy of Morey Machinery Co.

continuous cutting (band or rotary) or *reciprocating* type.

During their work, machinists will use many types of power saws and cut a number of different metallic and nonmetallic materials. The cutting speeds and characteristics of the materials must be understood before the proper blades and operating conditions can be selected. Saws are an effective and efficient category of machine tools found in almost every type of machine shop. The good machinist must know how to use them well.

7.2 BANDSAWS

The metal-cutting bandsaw has been extensively used in metalworking for only about 50 or 60 years. It is probably the most versatile of all machine tools because with it, cuts of unrestricted length and of almost any shape can be made. It can also be used for operations other than sawing, including such *band-machining* operations as band filing, polishing, and friction cutting.

7.2.1 Vertical Bandsaws

All vertical bandsaws, regardless of whether they are light-, medium-, or heavy-duty machines, are made up of certain basic components. Although these major parts of the machine may be made by different methods, depending on the manufacturer, their function in the operation of the machine is essentially the same.

Vertical bandsaws are available in sizes and configurations ranging from light-duty hand-fed machines to heavy-duty machines with power-fed tables. The light-duty machines usually have two wheels and are driven through a variable speed belt drive, V belts and step pulleys, or some other type of speed-change mechanism. Blades ranging from 3/16 in. (4.7 mm) to 5/8 in. (15.9 mm) in width can be used on light-duty machines (see Fig. 7.1).

FIGURE 7.1 Light-duty vertical bandsaw.

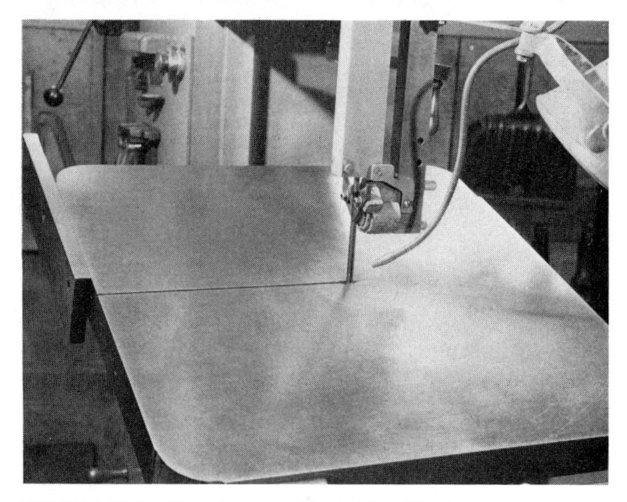
FIGURE 7.2 Bandsaw with table tilted.

Table Types. The *table* of the vertical metal-cutting bandsaw is usually made of cast iron and fitted with a tilting mechanism so that simple or compound angle cuts can be made. On *fixed-table* machines, the table does not move with the work, but can be tilted, as shown in Fig. 7.2, 45° to the right and 10° to the left on most machines. The work can be fed and guided manually, or a weight-operated feed mechanism can be used to supply the feed pressure (see Fig. 7.3).

Vertical bandsaws with *power tables* are generally heavy-duty machines. The feed pressure is provided by the mechanism that moves the table; the feed rate can be varied by the operator.

There is usually enough power available to make effective use of high-speed steel or tungsten carbide saw blades rather than the high-carbon steel blades used on light-duty machines. Coolant systems are also widely used on power-table machines, thus allowing higher cutting speeds and higher feed rates along with longer blade life. Many types of fixtures can be used on power-table

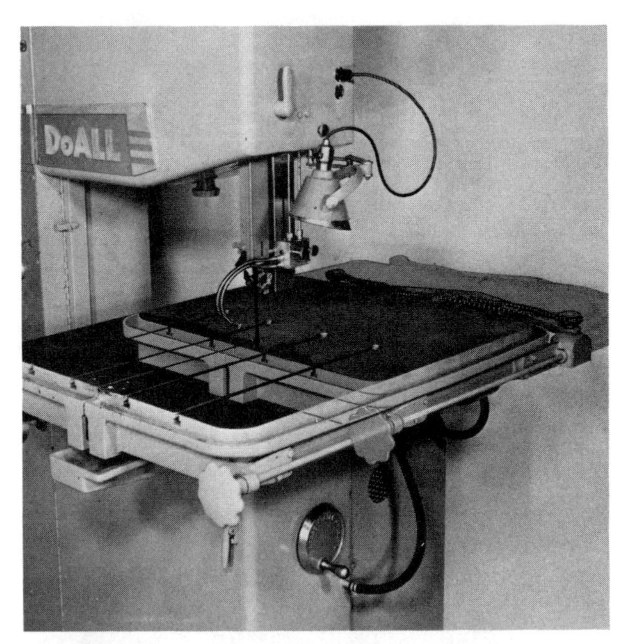

FIGURE 7.3 Feed mechanism operated by weights. (*Courtesy DoALL Co.*)

FIGURE 7.4 Bandsaw with power-feed table. (*Courtesy DoALL Co.*)

machines, particularly when they are used for repetitive operations. A typical power-feed table machine is shown in Fig. 7.4.

Base. The *base* of the machine may be of welded steel or cast iron and usually houses the electric motor, belts, speed-reducing mechanisms, and wheel-speed control. It is a rigid structure that has doors for access to the lower wheel and the other components housed within it. The base usually has lugs or feet that allow the machine to be bolted to the floor.

Column. The column may be a steel weldment or an iron casting, and in some machines it is an integral part of the base. It is almost always at the operator's left side. If the machine has three wheels, the column houses one of the wheels (see Fig. 7.5).

The controls for the machine, including the switch, band tension indicator, and cutting-speed indicator, are usually placed in the column. If the machine has a blade-welding and grinding attachment, it is generally attached to the column also. Because the column carries the loads imposed by blade tension and feed pressure, it must be rigid.

FIGURE 7.5 Three-wheel bandsaw. (*Courtesy DoALL Co.*)

Wheels. Vertical bandsaws may have two, three, or four wheels. Two- and three-wheel bandsaws are the most common types. As seen in Fig. 7.5, the three-wheel arrangement offers the advantage of large throat capacity with relatively small-diameter wheels. The wheels are usually iron or aluminum castings that are fitted with hard rubber tires set into a groove on the rim. The upper wheel (or wheels, in the case of a three-wheel saw) rotates on ball or roller bearings and has adjustments for controlling the track of the blade. The lower wheel is keyed and set screwed to the drive shaft. All the wheels are *crowned* (higher in the center), to help keep the blade track on the center of the wheel.

Head. The head is cast or fabricated as an integral part of the column and provides support for the upper wheel, the blade tensioner, and a number of other parts of the machine. The blade tensioner is usually a screw and nut arrangement that moves the entire upper wheel up or down in a slide or set of ways. The head must be rigid because it has to withstand the twisting and bending loads caused by blade tension and feed pressure. The blade guide, which can be adjusted up or down to provide blade support immediately above the work, is attached to the head by a slide and locking device.

Accessories. Most bandsaws that do not have a coolant system have an *air pump* that directs a stream of air at the point where the blade is cutting the workpiece. This removes the chips, letting the operator see the layout lines clearly, and provides some cooling.

If the machine has a *fluid coolant system,* the tank and pump are usually located in the base. The pump is controlled by a separate switch. Coolant systems are usually found on medium- and heavy-duty vertical bandsaws.

Blade-welding attachments, which are a specialized form of electric butt-welding machines, are a standard accessory on almost all bandsaws. The blade welder, as shown in Fig. 7.6, usually consists of cast copper or bronze blade clamps, a grinder, a saw thickness gage, and the necessary switches and operating levers.

Weight-operated feed devices can be used on bandsaws not fitted with power feed attachments. This reduces operator fatigue and generally results in more uniform feed rates and longer blade life.

Other attachments such as fixtures for cutting arcs and circles, ripping fences, and miters are used extensively on bandsaws. Special fixtures for holding specific types of workpieces are often designed for use in mass-production applications.

7.2.2 Horizontal Bandsaws

Because horizontal bandsaws are used primarily for cutting bar stock and structural shapes, they are also known as *cutoff* saws. The band-type cutoff saw is widely used because it is easy to set up and makes a narrow saw cut, thus requiring less power to operate and wasting less material. The cutting action is continuous and rapid. The blade is supported close to either side of the material being cut, so the cut is accurate if the machine is properly adjusted and the blade is in good condition.

FIGURE 7.6 Blade welding attachment. *(Courtesy DoALL Co.)*

Horizontal bandsaws range in capacity from small, fractional horsepower machines, which can also be used in the vertical position (see Fig. 7.7), to large, heavy-duty industrial saws (see Fig. 7.8). They all have certain parts and characteristics in common, however.

Base. This part of the machine is usually of welded steel construction, although some castings may be used. The base houses the motor and drive mechanisms, the coolant pump, and the hydraulic pump and serves as the coolant reservoir. The top of the base carries the vise or fixture for holding the material to be cut and the support upon which the saw head pivots. If a hydraulically operated vise is used, the hydraulic cylinder is generally attached to the base.

Saw Head. The saw head assembly usually consists of a welded steel or cast frame to which

FIGURE 7.7 Small horizontal bandsaw. (*Courtesy W.F. Wells and Sons*)

FIGURE 7.8 Heavy-duty horizontal bandsaw. (*Courtesy DoALL Co.*)

other parts, such as the wheels and blade guides, are attached. A pivot casting and shaft allow the saw head assembly to move up and down. The two wheels on which the band runs are usually made of hardened steel and are mounted on double ball bearings on the larger saws. On some light-duty saws, the drive wheel is keyed directly to the output shaft of the drive mechanism. The *idler* wheel is adjusted manually for blade tension and track on light-duty saws. On heavy-duty machines an initial tension adjustment is made manually, and full tension is applied hydraulically when the machine is in operation.

The *saw guides* are an important factor in accurate cutoff operations. The saw blade has to twist as it leaves the idler pulley, and the guides make the blade travel perpendicular to the material being cut (see Fig. 7.9). Tungsten carbide inserts are used to minimize wear (Fig. 7.10).

Controls and Accessories. On light-duty saws, the controls are simple, consisting mainly of an off–on switch, a means for changing blade speed,

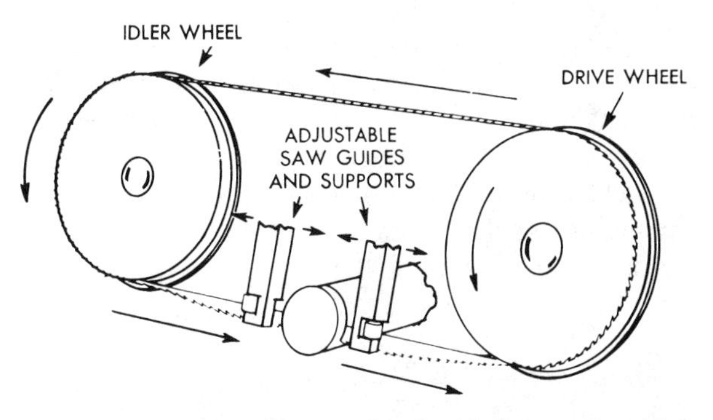

FIGURE 7.9 Cutoff saw wheels, blades, and guides. (*Courtesy DoALL Co.*)

FIGURE 7.10 Tungsten carbide blade guides. (*Courtesy DoALL Co.*)

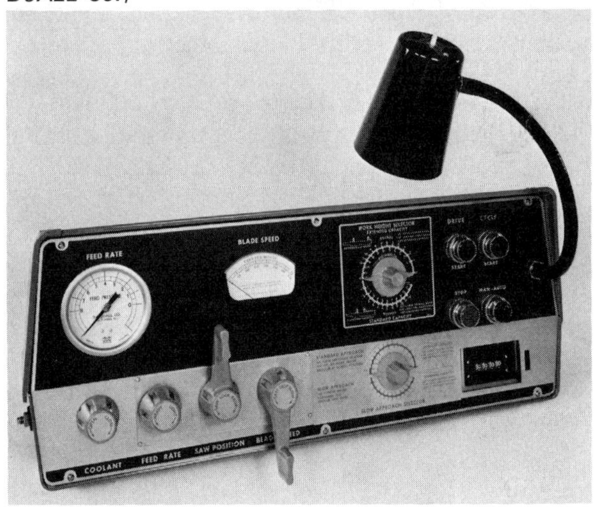

FIGURE 7.11 Control panel. (*Courtesy DoALL Co.*)

and possibly a control for feed pressure. On the larger machines a control panel (see Fig. 7.11) is usually mounted on the saw head. It consists of the necessary switches, valves, and instruments that indicate blade speed in feet per minute, feed rate in inches per minute, and other factors, such as blade tension. Some machines used for production work are capable of fully automatic operation and can be preset to cut a given number of pieces of work. A counter is usually part of the instrumentation on semiautomatic and automatic machines.

There are *coolant systems* on almost all medium- and heavy-duty horizontal bandsaws. The coolant extends blade life and allows higher cutting speeds and metal removal rates. The rate of coolant flow is controlled by the operator. Solid lubricants such as wax or grease can also be used. Wax in stick form is usually applied manually to the blade on light-duty machines.

7.3 BANDSAW BLADES

The sawtooth-type blade is most commonly associated with band machining operations in the machine shop. A wide variety of other bands are available for different applications. Some, like the knife-edge band, are used only for cutting textiles and paper and are not discussed in this text. The special types of bands, such as *diamond edge, filing,* and *polishing,* are covered because they are used in machining operations.

7.3.1 Bandsaw Blade Characteristics

Efficient bandsawing of metal requires a basic knowledge of the types of bandsaw blades and their operating characteristics and limitations. Machinists must also have an understanding of the machinability of the material they will be cutting, the speed at which the material should be cut, and other factors affecting blade selection.

Blade Materials. Bandsaw blades are available in three categories of materials: carbon alloy steel, high-speed steel, and tungsten carbide. *Carbon steel* blades are generally used with light- and medium-duty machines for maintenance, light manufacturing, and toolroom operations. They can be used dry or with coolant. The cutting speed, which is always specified in feet per minute, is lower for carbon steel blades than for the other types because the teeth on carbon steel blades soften at about 450°F. The blade has either a hardened cutting edge and a flexible, relatively soft back or a hard edge and a spring-tempered, medium-hard back. Carbon steel blades can be used on vertical or horizontal machines.

High-speed-steel blades are usually capable of cutting speeds almost double the cutting speed of carbon alloy steel blades because the teeth can withstand temperatures of about 1000°F. They are intended for use with coolant on medium- and heavy-duty machines, although they can be used dry for light-duty applications. High-speed-steel blades are commonly used on horizontal cut-off machines.

There are two methods by which high-speed-steel blades are manufactured. One method is to make the entire blade of high speed steel. The other method involves welding high-speed steel to the edge of a flexible carbon band (see Fig. 7.12). In either case the teeth are hardened to over 60 on the Rockwell C scale. Blades with high-speed-steel teeth and carbon steel bands can be welded on the blade welders found on most bandsaws or can be joined by scarfing and silver brazing.

Tungsten carbide blades consist of a flexible, fatigue-resistant band to which individual tungsten carbide teeth are bonded. Because tungsten carbide is more heat resistant than either carbon alloy steel or high-speed steel, tungsten carbide blades can be run at high speeds. They are generally used in both vertical and horizontal heavy-duty saws with power feed mechanisms and are the most expensive type of blade.

Spiral-Edge Sawbands. The spiral-toothed sawband (see Fig. 7.13) has a 360° cutting edge and can make a minimum radius of one-half of the blade diameter. Spring-tempered blades are used for wood and plastics, and hard-edged blades are used for metals. A major advantage of the spiral-edge sawband is that intricate contours can be cut without turning the workpiece. Special saw guides

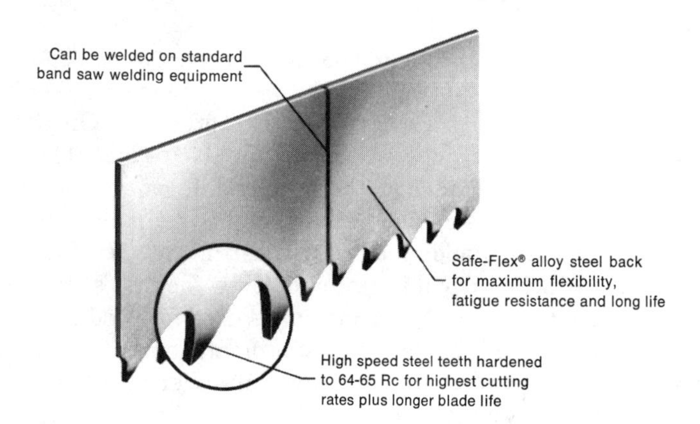

FIGURE 7.12 High-speed steel teeth welded to carbon steel band. (*Courtesy L.S. Starrett Co.*)

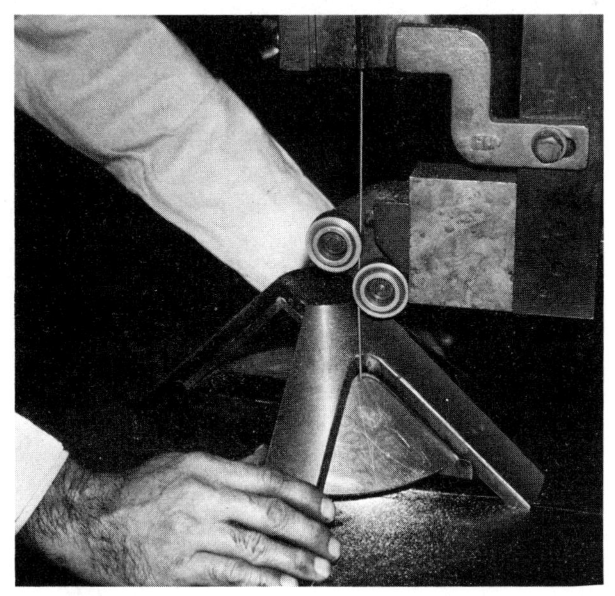

FIGURE 7.13 Spiral-toothed blade. (*Courtesy DoALL Co.*)

with rubber-coated rollers are used with spiral-edge bands. Note the position of the grooved rollers relative to the blade.

Diamond-Edge Sawbands. When extremely hard metallic or nonmetallic materials must be cut, diamond-edged sawbands can be used. The sawbands are available in widths ranging from 1/4 to 1 in. (6.35 to 25.4 mm) and are usually operated at 2000 to 3000 ft. (609 to 913 m) per minute. Because a coolant is required for effective use of diamond-edge sawbands, they are generally used on power-table machines. Such bands are especially effective for cutting materials such as glass, quartz, granite, and fiberglass laminate.

Tooth Types and Shapes. The basic bandsaw blade tooth forms are shown in Fig. 7.14. Carbon alloy and high-speed-steel blades are available

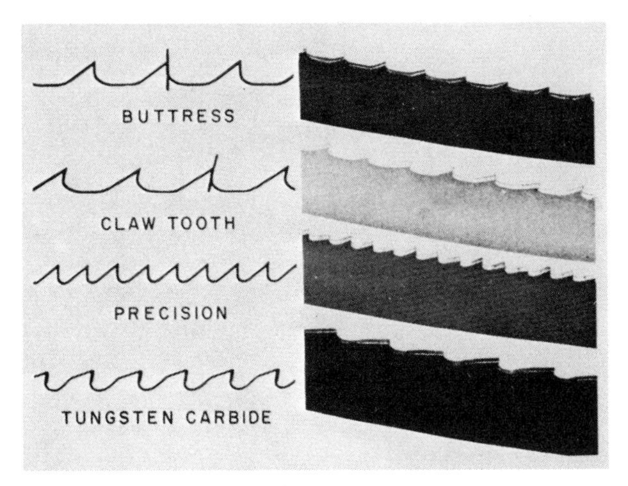

FIGURE 7.14 Basic bandsaw blade tooth forms. (*Courtesy DoALL Co.*)

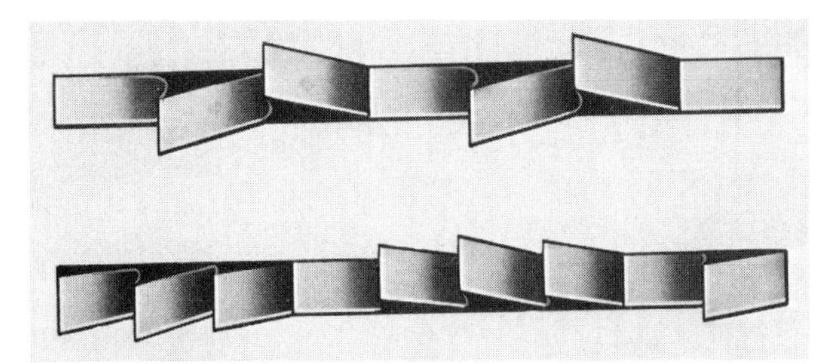

FIGURE 7.15 Wavy set (top) and raker set (bottom) blades. (*Courtesy L.S. Starrett Co.*)

in the first three shapes shown, and some manufacturers offer variations of these basic tooth shapes. The *precision* tooth form is the most widely used, especially for sawing ferrous metals and some of the harder nonferrous metals. The rake angle is 0°, and the blades are available in wavy or raker set (see Fig. 7.15).

Pitch. The *pitch* of the blade (number of teeth per inch) varies from 32 teeth per inch for narrow, wavy set blades used in cutting tubing and thin materials to 2 teeth per inch for blades 1 in. (25.4 mm) or wider used for cutting heavy sections of relatively soft materials. The wider and coarser blades almost always have raker set, and the *hook* or *claw* tooth shape is used with softer materials. If fine-pitch blades are used for cutting soft materials, the *gullet* (cavity between the teeth) can become clogged and prevent the blade from cutting, especially if no coolants or lubricants are used.

Width and Gage. Bandsaw blades are available in widths ranging from 3/16 to 2 in. (4.7 to 50.8 mm). Narrow blades are used for contouring and cutting small radii. Wider blades are more stable, and it is recommended that the widest blade usable for a particular radius be selected. For straight cuts, the widest blade that the machine will accept should be used. On production-type cutoff saws, wide blades are generally used because they withstand high feed pressures and make straight cuts.

The *gage* of the blade is the thickness of the blade stock. It ranges from 0.025 in. (0.63 mm) for blades up to 1/2 in. (12.7 mm) in width to 0.050 in. (1.27 mm) for 2-in. (50.8-mm) blades. Remember that the width of the saw cut, or kerf, is always wider than the gage of the blade because the teeth have *set*.

Although heavy gage blades are stronger, the wider kerf can be a disadvantage when cutting expensive materials on a production basis.

Set. The teeth on bandsaw blades are offset from the centerline of the blade, as shown in Fig.

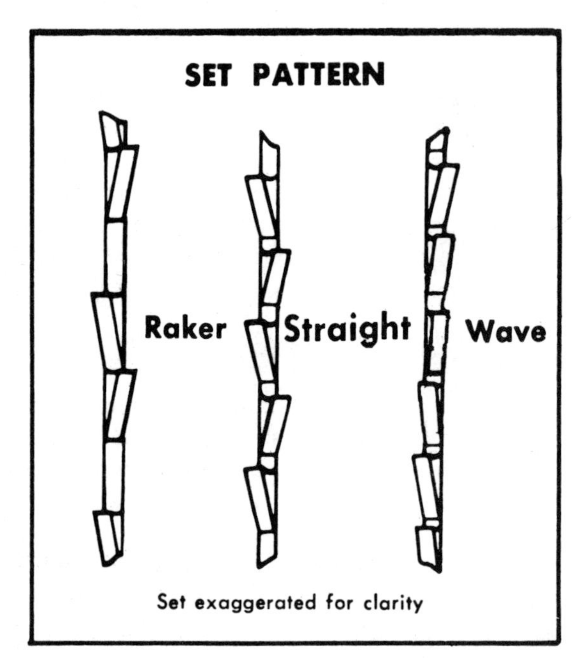

FIGURE 7.16 Three common set patterns. (*Courtesy DoALL Co.*)

7.16. The amount of side-clearance angle on the teeth determines the width of the kerf, thus providing clearance for the band.

A saw blade must have set to cut contours because the clearance provided by the wide kerf allows the work to be turned as the cut progresses. A narrow blade with a given amount of set can be used to cut a small radius because the work can be turned at a relatively large angle to the path of the blade without binding (see Fig. 7.17).

The two types of set most often used on bandsaw blades are *raker* and *wave*. As shown in Fig. 7.15, the raker set blade has one tooth offset to the left, one to the right, and one with no offset. This arrangement is repeated for every group of three teeth.

Wave set blades have groups of three teeth alternately set to the right and left. In some cases, one tooth is left straight (no offset) between the offset groups. Blades of this type are generally used for cutting tubing and other materials with thin or varying cross-sectional shape.

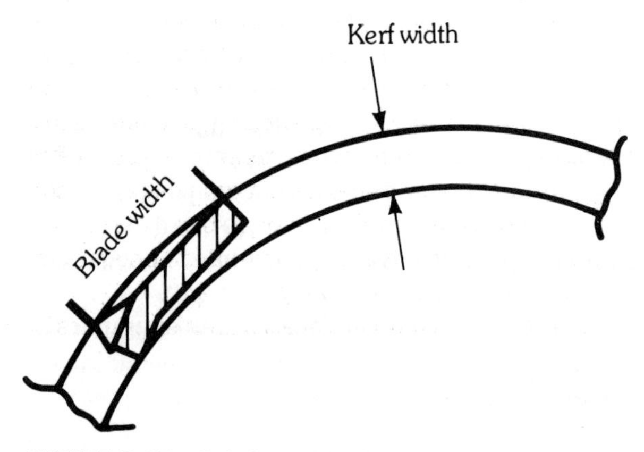

FIGURE 7.17 Relation of kerf width and blade width to radius of curved cut.

7.3.2 Blade Selection

The process of choosing the best bandsaw blade for a particular job must start with an evaluation of the material to be cut. Such factors as hardness, machinability, cross-sectional shape and area, minimum radius of curves, and minimum thickness of irregular shapes must be considered.

After the material to be cut has been properly identified, use the selector on the machine (see Fig. 7.18) to help select the proper blade and cutting speed. Tables and selectors are helpful, but the operator often must make choices that affect the three variables present in every sawing operation: *cutting rate, tool life,* and *accuracy.* Generally, increasing any one variable results in a decrease in one or both of the others. For example, an increase in cutting rate always reduces tool life and may affect accuracy.

Follow these nine general recommendations for best results:

1. When making straight cuts, use the widest blade possible.
2. In contour sawing, use the widest blade allowable for the smallest radius to be cut.
3. Use blade guides of the proper width for the blades.

FIGURE 7.18 Selector for computing blade speeds. (*Courtesy DoALL Co.*)

4. Place the upper blade guide assembly as close to the work as possible.
5. Use coarse-pitch blades for materials such as aluminum to avoid clogging the teeth.
6. Use wave-set blades on tubing and thin materials.
7. When in doubt about the cutting speed to use on a particular material, use a low cutting speed to avoid ruining the blade.
8. Avoid excessive feed rates, especially when using narrow blades, to prevent distortion and blade damage.
9. Use a cutting lubricant whenever possible to extend blade life.

7.3.3 Blade Welding

Practically all vertical metal-cutting bandsaws have an attachment for electrically butt-welding blades. It is usually set on the column of the machine at the operator's left and consists of a blade cutter, a small grinding wheel, and the butt-welding machine (see Fig. 7.19). The blade-welding attachment can be used for making sawbands from bulk saw-blade stock or for welding bands that have been cut and inserted into a hole in a workpiece that is to be band-sawed internally.

The importance of making good welds in sawbands cannot be overemphasized. Breakage caused by poor welding, improper joint finishing, or improper heat treatment is time consuming and potentially dangerous.

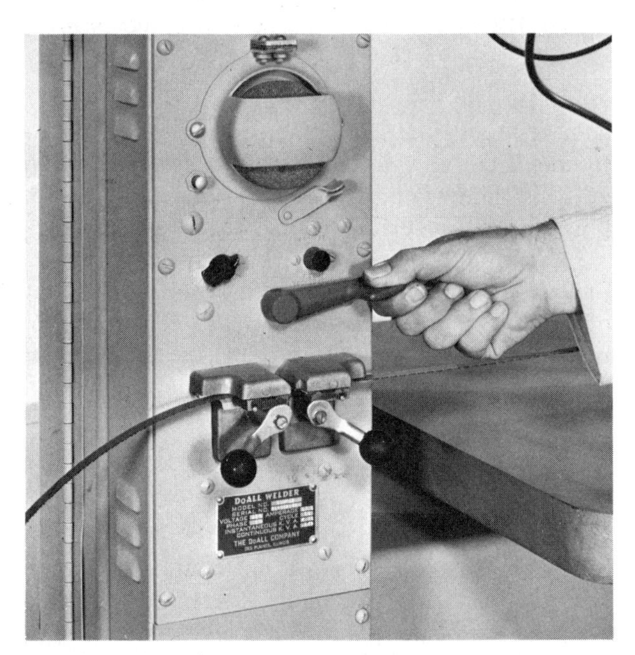

FIGURE 7.19 Bandsaw blade welder in use. (*Courtesy DoALL Co.*)

Butt Welder. The resistance-type butt welders found on almost all vertical bandsaws operate by causing electrical current to flow through the ends of the bandsaw blade while pressure is being applied. The high resistance where the blade ends meet causes the metal to become white-hot momentarily, and the blade ends fuse. Provision is made for annealing (softening) the welded joint. As the operator presses the anneal button for a very short time, current flows through the completed joint until the joint heats to a dull red. The joint then anneals as it cools slowly.

Blade Preparation and Welding. To make uniformly strong and durable joints in bandsaw blades, follow these eight procedures:

1. Cut the blade stock to length, using the shear on the machine.
2. Prepare the ends of the blade by holding the two ends together with the teeth facing in opposite directions and grinding the ends of the blade. The ends will match even though the ground end is not at exactly 90° to the centerline of the blade (see Fig. 7.20).

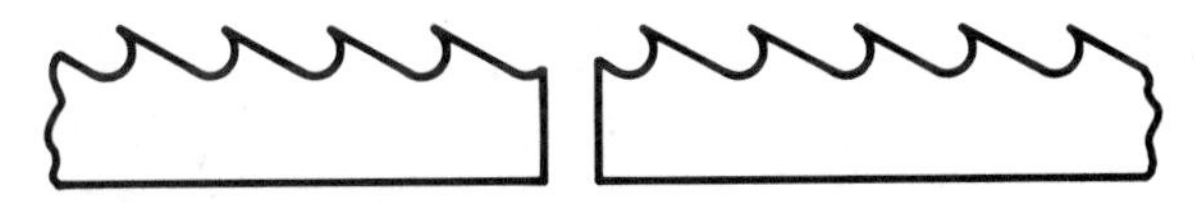

FIGURE 7.20 Blade ends prepared for welding.

3. Make sure that the jaws of the butt welder are *clean.*
4. Clamp the blade ends in the jaws, leaving a 0.010-in. gap between the blade ends.
5. Set the controls to the proper blade width and firmly depress the lever.
6. After the joint cools, release the lever, move the blade out to the edge of the jaws, reclamp it, and anneal it.
7. Remove the blade and grind off the flash (thickening of the welded area) to the *exact* blade thickness—no less.
8. Remove the burr at the back of the blade. It is now ready for use.

In cases where a butt welder is not available, bandsaw blades can be joined by silver brazing. When this process is used, the blade is *scarfed* as shown in Fig. 7.21 and held in a fixture. After the blade ends are aligned, the silver brazing operation is completed and the excess brazing material

FIGURE 7.21 Scarfed blades prepared for silver brazing.

is carefully filed off. As a rule, no annealing is necessary.

7.3.4 Using Bandsaws Safely

To use bandsaws safely and effectively, several basic factors about the nature of the operation must be understood. Since the bandsaw blade is usually quite thin and narrow in relation to the size of the object being cut, it must be well supported. On upright bandsaws the upper blade guide is vertically adjustable to accommodate different workpieces and adjustments are provided for different blade thicknesses and widths. There is also an adjustment for blade tension which will vary for different blade widths.

The following precautions and procedures should be observed when using bandsaws.

1. Set the upper guide 1/8 to 1/4 in. above the work.
2. Follow the manufacturer's blade tension recommendations.
3. Keep fingers and hands away from the blade and use push sticks or holding devices for small or unusually shaped parts. Use an inverted drill press vise to hold round stock when it is being bandsawed.

7.4 BAND MACHINING OPERATIONS

The categories of work briefly described here account for most of the band machining operations used in metalworking.

7.4.1 Cutoff Sawing

Although cutoff sawing can be done on any type of vertical or horizontal bandsaw, the majority of it is done on powerful horizontal machines. A variety of work-holding devices and fixtures can be used to hold tubing, angle iron, and other shapes (see Fig. 7.22).

Blade selection is important in terms of economy and the finish on the material that is cut. The precision tooth type of blade (see Fig. 7.14) is used extensively with the recommended pitch ranging from 10 teeth per inch for sections up to 3/8 in. thickness to 4 teeth per inch for material over 3 in. (9.52 mm) thick. Manufacturers' manuals should be consulted when heavy cuts are being attempted. The claw tooth type of blade is used when cutting some tough steels because the tooth penetrates the surface of the work more easily.

FIGURE 7.22 Multiple cutoff operation. (*Courtesy DoALL Co.*)

FIGURE 7.23 Indexing mechanism for production sawing. (*Courtesy DoALL Co.*)

Stock feeders are often used on cutoff machines, along with an indexing mechanism that allows the operator to automatically repeat cuts of preselected lengths (see Fig. 7.23). Almost all cutoff operations are done with a liquid coolant delivered to the saw cut by a pump.

7.4.2 Contour Sawing

Contour sawing, both internal and external, is one of the most versatile operations that can be done with the bandsaw. It may range from simple shapes cut on a fractional horsepower machine (see Fig. 7.24) to complex internal cuts made with

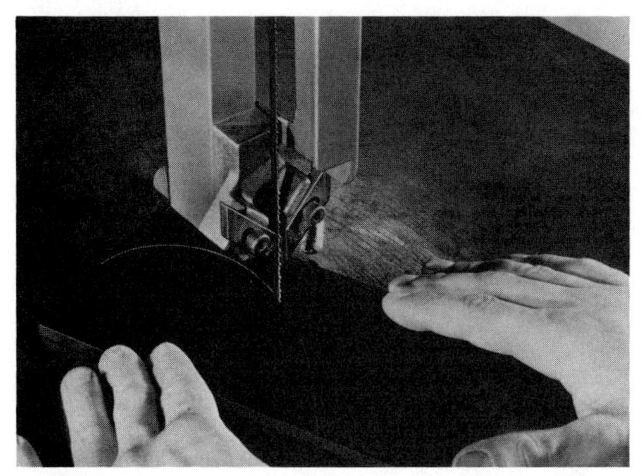

FIGURE 7.24 Cutting a curved surface. (*Courtesy L.S. Starrett Co.*)

FIGURE 7.25 Internal sawing operation.

FIGURE 7.26 Internal work with radiused corners.

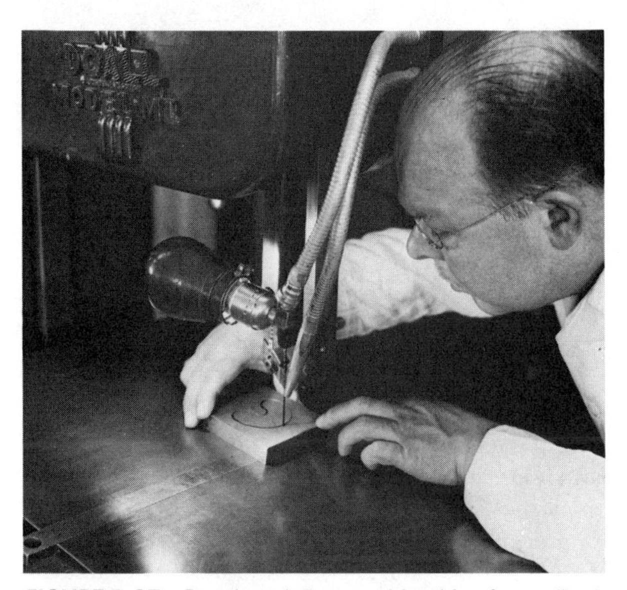

FIGURE 7.27 Punch and die cut with table of saw tilted. (*Courtesy DoALL Co.*)

tilting table machines (see Fig. 7.25). Blade selection is important when cutting complex contours, especially when small radii or corners are involved. Select the widest blade that will allow turns of the proper radius.

For internal work, a hole must be drilled so that the blade can be passed through it and rewelded. For plain contouring, the hole is drilled perpendicular to the face of the workpiece. When the internal shape has corners, as shown in Fig. 7.26, holes must be drilled at the corners so that the blade can be turned and the cut started in another direction.

Some types of punch-and-die sets can be cut on the bandsaw, saving much time and money, by using the technique shown in Figs. 7.27 and 7.28. The hole for the blade is drilled at an angle that is determined by the thickness of the die. The blade is threaded through the hole and welded, and then the table of the machine is tilted to the proper angle and the cut is made. The blade is then cut

FIGURE 7.28 Punch and die set. (*Courtesy DoALL Co.*)

to remove it from the outer part of the die, and the die is finished by hand filing or other methods.

Sawing tubing at an angle, as shown in Fig. 7.29, requires the use of a holding fixture if the job

FIGURE 7.29 Cutting round tubing at an angle.

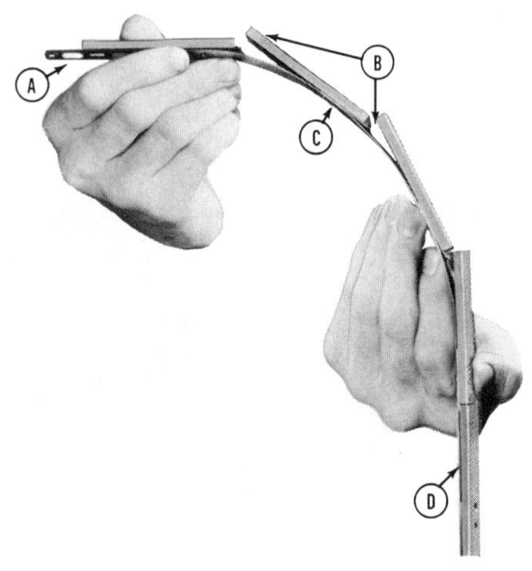

FIGURE 7.30 File band with inserted segments. (*Courtesy DoALL Co.*)

is to be done safely and accurately. The fixture shown is a combination miter gage and clamp that moves in a groove in the table of the bandsaw.

7.4.3 Band Filing and Related Processes

File bands are used extensively for both interior and exterior finishing operations on the band machine. The file band consists of file segments riveted near one end to a flexible steel band, as shown in Fig. 7.30. As the band moves around the wheels of the machine, the loose end of the file segments lifts off the band. When the band comes off the wheel, the segments lie flat against the band and lock to form a smooth filing surface.

Types of File Bands and Applications. File bands are available in flat, oval, and half-round shapes, in a variety of tooth shapes and coarseness. The most widely used file bands range from 1/4 to 1/2 in. (6.35 to 12.7 mm) in width. A representative group of file types is shown in Fig. 7.31.

In terms of applications, band filing can be used to finish both curved and straight surfaces in diemaking and in finishing other intricate shapes. Flat surfaces and the outside of curves are machined with flat-band files; oval or half-round files are used for internal contours.

Band Polishing and Applications. Abrasive bands can be used instead of a saw blade on vertical band saws with the appropriate accessories. A rigid backup support is used in place of the saw guides, as shown in Fig. 7.32.

The abrasive bands use either silicon carbide or aluminum oxide abrasive and are available in 50, 80, and 150 grits (abrasive grain size). The 50-grit band is used for heavy stock removal and on soft materials, with the machine running at about 500 to 700 surface feet per minute. The finer grit bands are used at higher speeds.

Many materials, ranging from very soft to very hard, may be processed on band polishing machines. For example, tungsten carbide tools and other very hard materials can be machined effectively with silicon carbide belts, as shown in Fig. 7.33.

7.4.4 Friction Sawing

Friction sawing is a unique process. A bandsaw blade with dull teeth traveling at very high speed

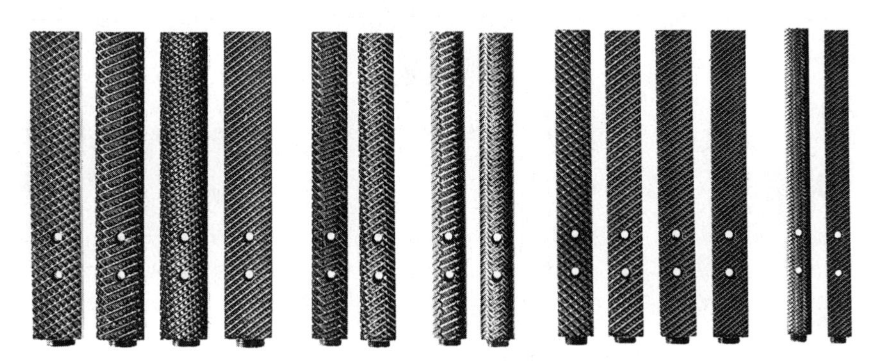

FIGURE 7.31 Types of file bands. (*Courtesy DoALL Co.*)

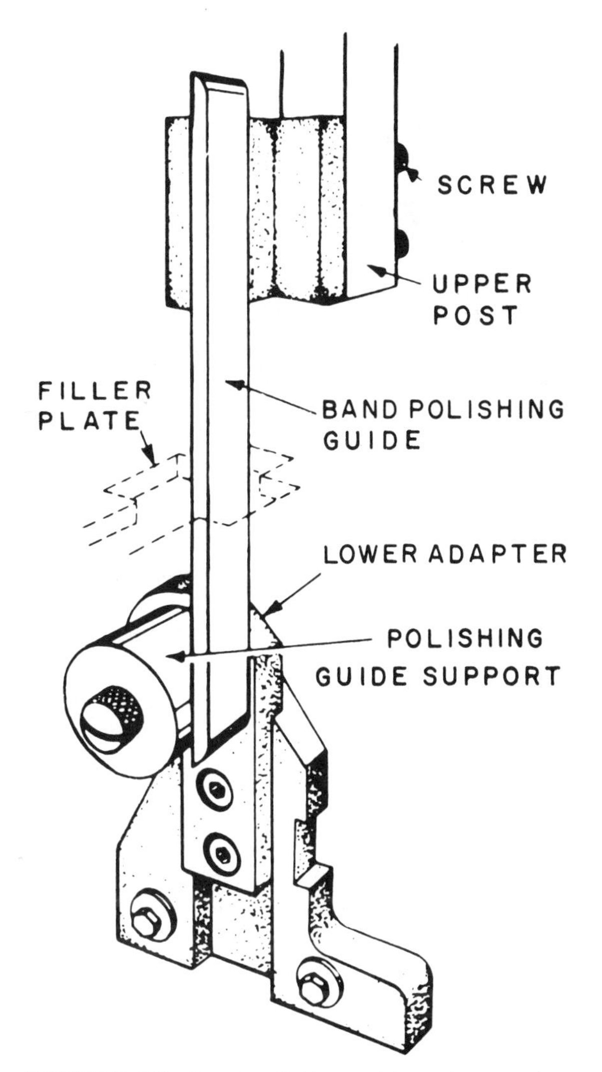

FIGURE 7.32 Band polishing guide and support. *(Courtesy DoALL Co.)*

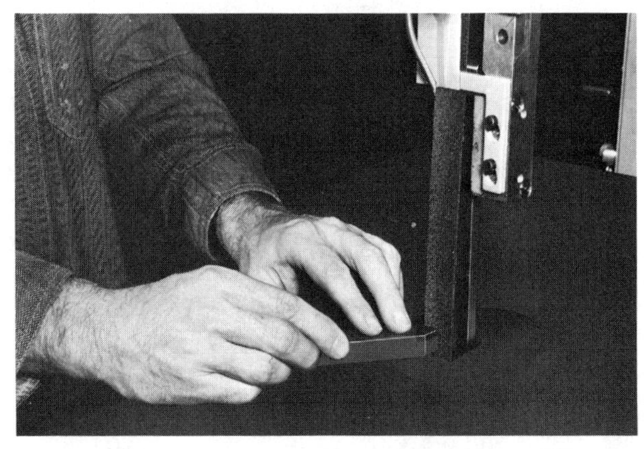

FIGURE 7.33 Polishing a tungsten carbide tool bit with a silicon carbide belt. *(Courtesy DoALL Co.)*

—6000 to 15,000 surface feet (1830 to 4575 m) per minute—is used to cut both hard and soft ferrous metals. It works particularly well on metals that have poor heat conductivity because the heat-affected zone remains very small. It is the fastest method of cutting ferrous metals less than 1 in. thick.

FIGURE 7.34 Friction sawing blades are run at high speed. *(Courtesy DoALL Co.)*

As the blade contacts the work, the metal at the point of contact immediately becomes white hot and is carried out by the teeth. The blade itself remains relatively cool because during its operating cycle it is in contact with hot metal for only a short time (see Fig. 7.34).

Applications. Thin materials such as steel tubing and sheet can be cut very rapidly without distortion. When cutting sheet metal, friction sawing is often faster than shearing. For materials under 5/8 in. (15.87 mm) thick, friction sawing is faster than flame cutting and in most cases makes a cleaner cut. A major advantage of friction sawing is its ability to cut materials regardless of their hardness. High-speed-steel cutters, as shown in Fig. 7.35, can be cut readily if the machine has

FIGURE 7.35 Hard materials can be friction-sawed. *(Courtesy DoALL Co.)*

enough power to operate the blade at the correct speed.

7.5 RECIPROCATING SAWS

Power hacksaws have been used for many years for cutting bar stock and structural steel shapes such as angles and channels. The saws range in size and complexity from the simple, low-power machines used for light work to heavy-duty saws capable of cutting many bars at one time. Reciprocating-type power hacksaws cut for only about half of the total operating time. For this reason they are not as popular at the present time as are bandsaws, which cut continuously.

7.5.1 Characteristics and Construction Features

Almost all power hacksaws share certain construction features, although they may vary in capacity.

Base. On dry machines, the bed of the machine is supported by legs because there is no need for a coolant tank. Wet machines usually have a boxlike base that serves as a coolant tank and houses the pump for the coolant (see Fig. 7.36). The base is cast iron or fabricated from steel by welding.

Bed. The beds of all power hacksaws, regardless of size, are usually heavy iron castings. The operating mechanisms, including the vise, crank, and pivots for the overarm, are mounted to the bed. The vise can be swiveled, usually to a 45° angle, and can be moved fore and aft. The hydraulic or mechanical mechanism for feeding the blade into the work on the cutting stroke and lifting it on the return stroke is mounted on the bed. The feed pressure is usually adjustable.

Overarm and Blade Holder. The overarm is pivoted at the rear of the machine (see Fig. 7.37),

FIGURE 7.36 Power backsaw with coolant system.

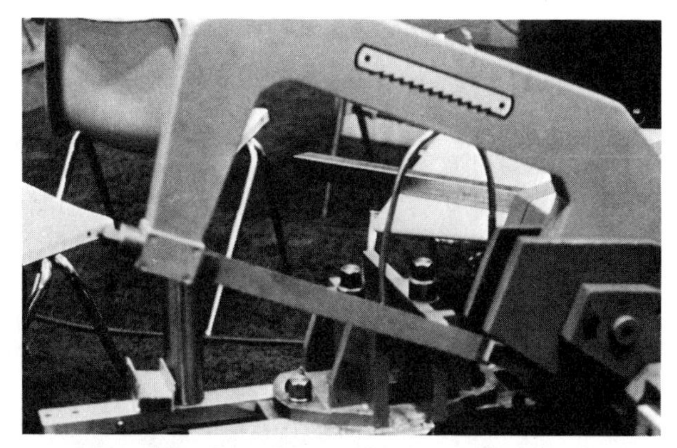

FIGURE 7.37 Overarm and blade holder on heavy duty reciprocating saw.

or the machine has two vertical columns or ways upon which the reciprocating blade holder moves vertically. In either case, the overarm brings the blade in contact with the work and applies feed pressure on the cutting stroke. The blade holder is a rigid casting to which the blade is attached by pins and a tensioning screw.

7.5.2 Power Hacksaw Blades

Power hacksaw blades are available in many sizes, pitches, and materials. Selecting a blade depends on such factors as the (1) power and operating speed of the machine, (2) hardness of the material to be cut, (3) thickness of the material, and (4) blade life desired.

Blade Characteristics. Power hacksaw blades are available in lengths of 14 to 30 in. (355 to 762 mm). The larger blades are usually thicker and wider because they must be capable of withstanding more feed pressure and tension in powerful machines. The thickness ranges from 0.050 to about 0.100 in. (1.27 to 2.54 mm), and the width ranges from 1 to 2 1/2 in. (25.4 to 63.5 mm). Because power hacksaw blades are thicker than bandsaw blades, more material is consumed per cut. Using wide blades helps in making straight cuts when heavy feed pressures are used.

Power hacksaw blades usually have *raker* set (see Fig. 7.15). As the blade wears, the outer edges of the teeth wear away, resulting in a narrower kerf. If a blade breaks or must be replaced before a cut is completed, the new blade should not be fed into the existing cut because it will bind and probably break. The workpiece should be rotated in the vise so that the new blade starts a fresh cut.

The *pitch* of power hacksaw blades varies from 14 to 3 teeth per inch. Small blades used for cutting thin materials or tubing usually have 14 teeth per inch, and large blades used for cutting large, relatively soft bar stock have 3, 4, or 6 teeth per inch. The larger space between the teeth provides space for larger chips and allows more rapid cutting.

Blade Materials. High-speed-steel blades with either flexible or all-hard backs are widely used in heavy-duty power hacksaws. For lighter duty operations, molybdenum and tungsten alloy steel blades are generally used. *Composite* blades have high-speed-steel teeth welded to a flexible, fracture-resistant alloy steel back.

7.5.3 Using the Power Hacksaw

Effective use of the power hacksaw requires an understanding of the characteristics of the materials being cut, machine capabilities, cutting speeds, and blade factors. The following six procedures should be used for best results:

1. Determine the recommended cutting speed for the material to be cut. (If the machine you are using is single-speed, it will be operating at the proper speed for the harder materials, such as alloy steels and cast iron.)
2. Select the proper blade. Use fine-pitch blades for tubing and thin materials, and coarse-pitch blades for large bars of soft materials.
3. Make sure the blade is rigidly installed and properly tensioned. A loose blade wears rapidly or flexes and breaks.
4. Use light feed pressure on tubing and thin materials to avoid breaking teeth out of the blade. Adjust feed pressure as necessary for thicker materials.
5. When replacing a blade that broke before completing the cut, do not feed the new blade into the old cut. Rotate the workpiece in the vise and start a new cut because the new blade will bind in the narrow kerf made by the worn blade.
6. When possible, use cutting fluids on all materials except cast iron.

7.6 CIRCULAR SAWS

Circular metal cutting saws, also known as *rotary saws,* can have blades as large as 72 in. (1.83 m) in diameter. The cutting action of the teeth is similar to that of a milling cutter (see Fig. 7.38), and the finish on the material is generally smooth and geometrically accurate. The term *cold sawing* is applied to circular saw operations that use toothed blades turning at relatively slow speeds. *Friction sawing,* which is done with a circular disk running at very high speeds [up to 15,000 surface feet (4575 m) per minute] is used only on metals that have a low rate of heat conductivity, such as stainless steel, alloy steels, and cast iron.

FIGURE 7.38 Large rotary saw.

7.6.1 Characteristics and Construction Features

Cold Saws. Most cold saws, regardless of size, consist of a base, drive mechanism, blade arbor, vise, feed mechanism, and necessary guards and switches. On some small saws the blade is fed into the work by hand (see Fig. 7.39). On larger machines, the feed mechanism is pneumatically or

FIGURE 7.39 Hand-fed cold saw.

hydraulically operated. The rate of feed is controlled by the operator.

The base of the machine or the vise can be swiveled to make angular cuts. In some cases two machines can be set up on a single work stand for production operations.

Cold Saw Blades. Blades smaller than 18 in. (457 mm) in diameter usually have the teeth cut directly in the rim of the saw disk. For cutting soft materials, the teeth are spaced farther apart, as in the case of bandsaw and power hacksaw blades, so that the *gullet* (the space between the teeth) will be large enough to accommodate large chips. When cutting thin tubing or other thin materials, use saw blades with closely spaced teeth to avoid chattering and tooth breakage. Cold saw blades with the teeth cut directly on the periphery of the disk may be made of high-carbon or high-speed steel.

Larger blades usually have segmented teeth, as shown in Fig. 7.40. The body of the blade is made of a tough, resilient alloy steel, and the inserted teeth are made of high-speed steel or tungsten carbide. The individual teeth or segments of three or four teeth are wedged or riveted to the blade and can be easily replaced if a tooth is damaged or broken. Larger cold saw blades can cut a kerf as wide as 1/4 in. (6.35 mm) and remove metal rapidly.

FIGURE 7.40 Segmented saw blade. (*Courtesy Nu-Tool Saw Co.*)

7.6.2 Using Circular Saws

Because cold-type circular saws with blade diameter sizes of 10 to 18 in. (254 to 457 mm) are widely used in machine shops and other metal fabrication shops, these six operating suggestions are limited to these machines:

1. Determine the recommended cutting speed (in surface feet per minute) for the material to be cut. Adjust the speed if the machine is of the multiple-speed type.
2. *Do not* attempt to cut hardened steel on this type of machine.
3. Start the cut slowly, with light feed pressure, especially when cutting thin material.
4. Use fine-pitch blades when cutting thin-walled tubing or other thin materials.
5. When practical, use cutting lubricants to increase blade life. Wax-type lubricants in stick form are suggested.
6. Securely clamp materials to be cut.

7.7 ABRASIVE CUTOFF MACHINES

Abrasive cutoff machines are used in many shops to cut metallic and nonmetallic materials. Because an abrasive—usually aluminum oxide—is used as the cutting tool, hardened steel can be cut without being annealed. The cutting action here is faster than on other types of cutoff machines.

7.7.1 Characteristics and Construction Features

Abrasive cutoff machines range in size from the simple bench-mounted machine shown in Fig. 7.41 to much larger and complex floor-mounted machines. Because of the speed of the cutting action, machines of this type require more power than saws of comparable blade diameter. For example, the machine shown in Fig. 7.41 has a 10-in. (25.4-cm) abrasive wheel that is driven by a 3-hp motor. In general, the wheel size ranges from 8 to 36 in., and the motor size from 2 to 50 hp.

The basic frames of abrasive cutoff machines are of either cast iron or welded steel. The structure must be rigid so that unwanted side loads are not applied to the blade as it is cutting. The guard for the wheel is fabricated from steel sheet or plate or is cast.

Abrasive cutoff machines may be of the *wet* or *dry* type. A wet-type machine that has a 20-in. (50.8-cm)-diameter wheel and a 20-hp motor is shown in Fig. 7.42. The flow of coolant, usually

FIGURE 7.41 Hand-fed abrasive cutoff machine. (*Courtesy Everett Industries, Inc.*)

FIGURE 7.42 Wet abrasive cutoff machine. (*Courtesy Everett Industries, Inc.*)

water and an antirust chemical of some type, is controlled by the operator. The coolant tank is separate, as shown, or built into the base of the machine.

Some larger cutoff machines have power feed mechanisms and oscillators. The oscillator moves the abrasive disk back and forth in the cut as feed pressure is applied. This reduces the amount of blade in contact with the work at any given time and reduces the power input required to cut solid bar stock of a given cross-sectional area.

The *spindle* of the machine runs in ball bearings and drives the abrasive disk. There is usually a multiple V-belt drive between the motor and the spindle.

The abrasive disks usually have a resinoid bonding agent, although rubber can be used on smaller wheels. Glass fiber is sometimes impregnated in the disk to increase its strength. Abrasive disks work efficiently at surface speeds of 12,000 to 15,000 surface feet (3660 to 4575 m) per minute.

7.7.2 Using Abrasive Cutoff Machines

Abrasive cutoff machines (Fig. 7.43) are used in many types of metalworking shops because of their versatility and rapid cutting characteristics. The machinist should know how to use this type of machine safely and how to determine which materials can be cut with abrasive disks. The following seven operating suggestions are basic; also study the manufacturer's operating recommendations.

1. When installing a new disk, be sure that the spindle flange and washer are clean and smooth. Tighten the nut to specifications.

FIGURE 7.43 Abrasive cutoff machine in use. (*Courtesy El Camino College*)

2. If the machine is a *mitering* type, check the vise or head location.
3. Check the condition of the abrasive disk. Look for cracks and nicks in the outer edge.
4. Clamp the work securely.
5. Make *sure* that the guards are securely mounted. *Never* operate an abrasive cutoff machine without the guard in place.
6. Feed the disk into the work smoothly.
7. Always wear eye and face protection.

Many types of saws are being used extensively in machine shops because of their versatility and economy of operation. When used with imagination, they can become the prime means of doing certain machining operations and serve as auxiliary machines. Proper use of saws of all types depends largely on the skill of the machinist, an understanding of the characteristics of various types of saws and the materials to be cut.

REVIEW QUESTIONS

7.2 BANDSAWS

1. Name three operations other than sawing for which vertical bandsaws can be used.
2. By what two methods may the work be fed on a fixed-table machine?
3. Why are some bandsaws made with three wheels?
4. Explain why coolant is needed in some bandsawing operations.
5. Of what material are blade guides on some bandsaws made? Why?

7.3 BANDSAW BLADES

1. Name three materials from which bandsaw blades are made.
2. Briefly describe the two methods by which high-speed steel blades are made.
3. Describe a spiral-edge saw blade.
4. List three materials that may be cut with a diamond-edge sawband.
5. Explain why coarse-pitch blades should be used when cutting soft materials.
6. What is meant by *wave set*?
7. Why should the widest blade possible be used in any bandsawing operation?
8. Briefly explain the operation of a butt welder for bandsaw blades.
9. What other means can be used to join bandsaw blades?

7.4 BAND MACHINING OPERATIONS

1. Briefly explain the procedure of making an internal cut with a vertical bandsaw.
2. What would be the effect of using a fine-pitch blade to saw aluminum plate?
3. On what type of bandsaw is band filing usually done? Why?
4. Briefly explain the process of friction sawing.
5. On what types of materials is friction sawing most effective? Why?
6. Explain why a fine-pitch precision-type blade should be used when cutting thin-walled tubing?

7.5 RECIPROCATING SAWS

1. Why should wide blades be used on a reciprocating saw when making cuts that require heavy feed pressure?
2. Why should a new cut be started when a blade is replaced even though the old cut was not completed?
3. Describe a *composite* power hacksaw blade.
4. What are the effects of installing a power hacksaw blade with insufficient tension?
5. Why must light feed pressure be used when cutting tubing or thin-walled structural shapes?

7.6 CIRCULAR SAWS

1. Describe two basically different metal cutting processes that can be done with circular saws.
2. What two materials can be used for the teeth of segmented-tooth saw blades?
3. How is the cutting speed specified for circular saws?

7.7 ABRASIVE CUTOFF MACHINES

1. What is the abrasive that is generally used in abrasive cutoff disks?
2. What bonding agent and reinforcing material are generally used in abrasive cutoff disks?

3. What cutting fluid is used on wet-type abrasive cutoff machines?
4. At what speeds (in *surface feet per minute*) are abrasive cutoff disks usually operated?
5. What personal protective equipment must be worn when using an abrasive cutoff machine?

8

Drilling Machines and Operations

8.1 INTRODUCTION

One of the most important and essential tools in any metalworking shop is the *drilling machine* or *drill press.* Although the drilling machine is used primarily for drilling holes, it is often used for reaming, boring, tapping, counterboring, countersinking, and spotfacing.

All drilling machines operate on the same basic principle. The spindle turns the cutting tool, which is advanced either by hand or automatically into a workpiece that is mounted on the table or held in a drill press vise. Successful operation of any drilling machine requires a good knowledge of the machine, proper setup of the work, correct speed and feed, and proper use of cutting fluids applied to the cutting tool and work.

8.2 TYPES OF DRILLING MACHINES

Many types and sizes of drilling machines are used in manufacturing. They range in size from a simple bench-mounted sensitive drill press to the large multiple-spindle machines able to drive many drills at the same time.

Opening photo courtesy of Summit Machine Tool Co.

8.2.1 The Sensitive Drilling Machine

The most common type of drill press and the simplest to operate is the *sensitive* type [Fig. 8.1(*a*)], so-called because the feed lever is operated by hand and allows the operator to "feel" the operating action of the cutting tool. Stepped V-belt pulleys [Fig. 8.1(*b*)] or variable-speed drives [Fig. 8.1(*c*)] are used to change the spindle speeds on the most sensitive drill presses. Sensitive drill presses generally have a maximum capacity for drills up to 1/2 in. Some larger floor models have tapered spindles. On this type the drill chuck can be removed, and larger drills with tapered shanks can be used. The table can be lowered or raised by hand or by a crank through a rack and pinion gear.

8.2.2 The Upright Drilling Machine

The *upright* drilling machine (Fig. 8.2), sometimes called the floor type, is similar to the sensitive type. It is somewhat larger and more rugged. It is equipped with a power feed device for automatically feeding the cutting tool, and it usually has a speed change gearbox that allows a variety of spindle speeds for different size cutting tools.

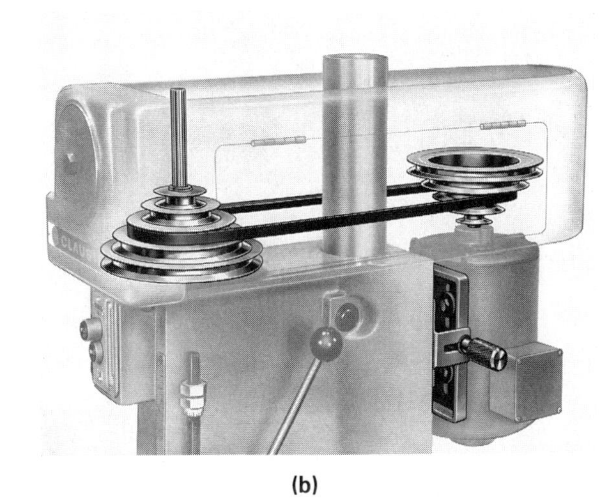

(b)

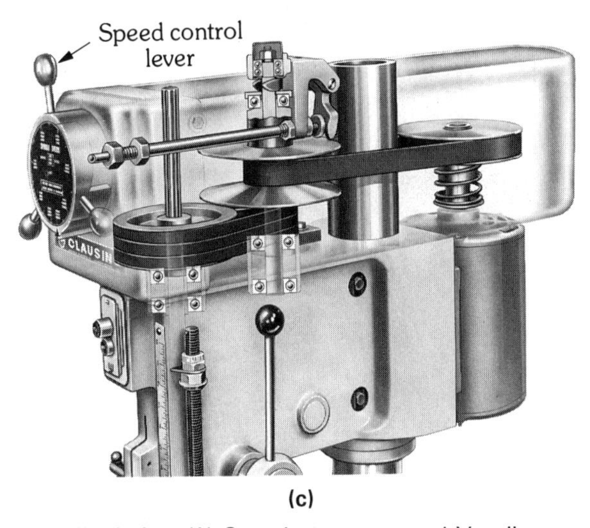

(c)

(a)

FIGURE 8.1 (*a*) A sensitive drill press is used for drilling smaller holes. (*b*) Speeds on a stepped V pulley drive are changed by changing the position of the V belt. (*c*) Speeds on a variable-speed drive mechanism are changed by the handwheel on the head. The spindle must be revolving when this is done. (*Courtesy Clausing Industrial, Inc.*)

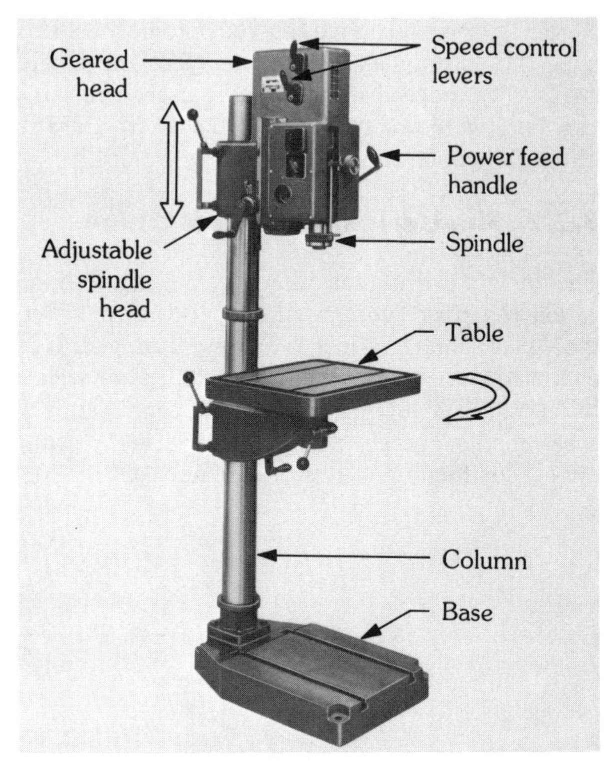

FIGURE 8.2 The upright drill press is more rugged than the sensitive type and has power feed. (*Courtesy Arboga Machine Co.*)

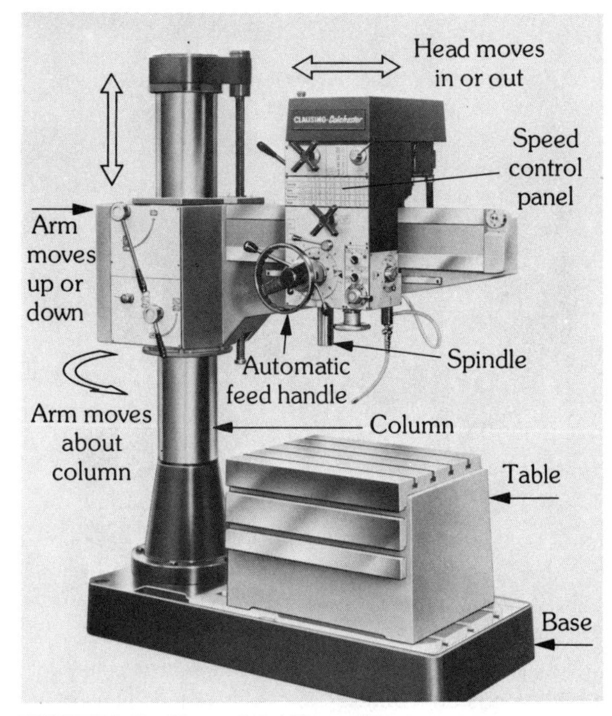

FIGURE 8.3 The radial drill machine is designed for drilling larger work. (*Courtesy Clausing Industrial, Inc.*)

8.2.3 The Radial Drilling Machine

The *radial* drill (Fig. 8.3) is more adaptable to a larger variety of jobs. It can easily handle very large work that cannot be mounted on a worktable. The large arm that extends from the column can be rapidly raised, lowered, and swung about the column by power to any desired location. The drilling head moves back and forth on this arm and, on universal models, can be swiveled to drill holes at an angle. These features permit drilling holes in many different locations without moving the mounted part.

Radial drilling machines have more horsepower than other types, permitting a variety of different-size cutting tools to be used.

8.2.4 Special Types of Drilling Machines

The *gang* drill press (Fig. 8.4) is made up of many drilling heads placed side by side on a long common table and base. With gang drilling machines, drills, reamers, and other cutting tools can be used on successive spindles. The part is moved from spindle to spindle where the different operations are performed without changing tools and adjusting spindle speeds.

Multiple-spindle drilling machines (Fig. 8.5) permit drilling several holes at one time in the part. Generally, the spindles—from 2 to more than 30—are driven by one drive gear in the head through universal-joint linkage. Figure 8.6 illustrates a multiple-spindle drill head attachment used on a conventional upright drilling machine.

Turret drill presses (Fig. 8.7) are used when several different sizes of drills or other cutting tools

FIGURE 8.4 Gang drilling machine. (*Courtesy Clausing Industrial, Inc.*)

FIGURE 8.5 A multiple-spindle drilling machine drills many holes at the same time. *(Courtesy Miles Standard Corp.)*

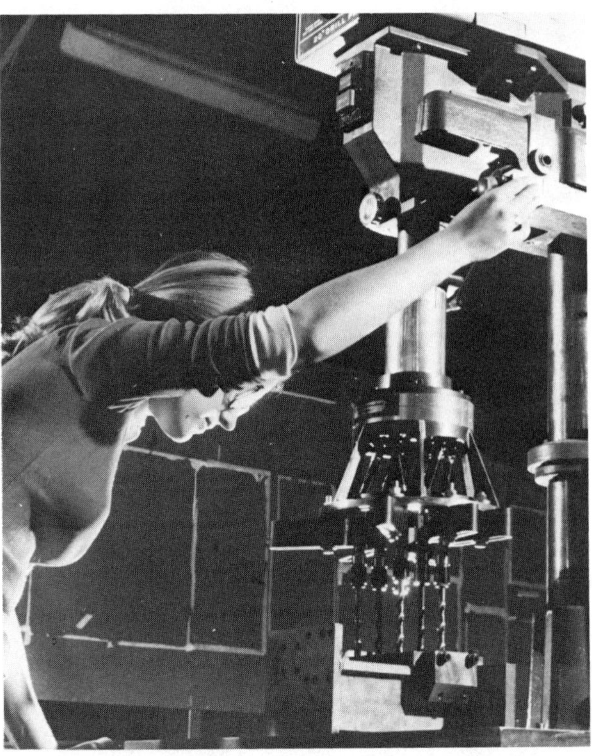

FIGURE 8.6 A multiple-spindle attachment is often used on a drill press. *(Courtesy Delta International Machine Corp.)*

FIGURE 8.7 Turret drilling machines have many different cutting tools on a turret. *(Courtesy Strippit-Houdaille, Inc.)*

are employed. The turret is indexed automatically to locate the cutting tool that is needed. Applications of the turret drill press are similar to those of the gang drilling machine.

Numerically controlled (NC) drilling machines (Fig. 8.8) are frequently used in industry to accurately locate the position of the work to be drilled. NC drilling machines are similar to those previously discussed, except that the worktable is automatically positioned to the desired location. NC machine tools are discussed in greater detail in Chap. 16.

FIGURE 8.8 Numerically controlled drilling machines automatically position the worktable. (*Courtesy Kanematsu-Gosho, Inc.*)

8.3 CONSTRUCTION FEATURES OF DRILLING MACHINES

Rigid and accurate construction of drilling machines is important to obtain proper results with the various cutting tools used. The sensitive drilling machine construction features are discussed in this section because its features are common to most other drilling machines (see Fig. 8.9).

8.3.1 Base

The *base* is the main supporting member of the machine. It is a heavy gray iron or ductile iron casting with slots to support and hold work that is too large for the table.

8.3.2 Column

The round *column* may be made of gray cast iron or ductile iron for larger machines or steel tubing for smaller bench drill presses. It supports the table and the head of the drilling machine. The outer surface is machined to function as a precision way of aligning the spindle with the table.

8.3.3 Table

The *table* can be adjusted up or down the column to the proper height. It can also be swiveled around the column to the desired working position. Most worktables have slots and holes for mounting vises and other work-holding accessories. Some tables are semiuniversal, meaning that they can be swiveled about the horizontal axis.

8.3.4 Head

The *head* houses the spindle, quill, pulleys, motor, and feed mechanism. The V belt from the motor drives a pulley in the front part of the head, which in turn drives the spindle. The spindle turns the drill.

8.3.5 Spindle and Quill Assembly

The *spindle* rotates within the *quill* (Fig. 8.10) on bearings. The quill moves vertically by means of a rack and pinion. The quill assembly makes it pos-

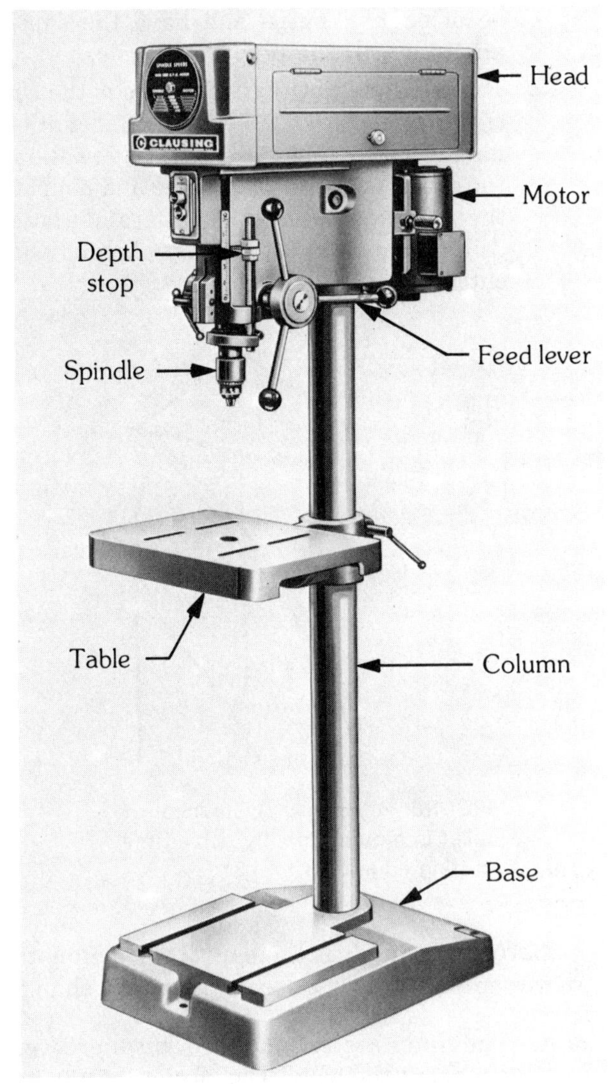

FIGURE 8.9 Principal parts of a sensitive drill press. (*Courtesy Clausing Industrial, Inc.*)

sible to feed or withdraw the cutting tool from the work. Located on the lower end of the spindle is either a Morse tapered hole or a threaded stub where the drill chuck is mounted. For drilling larger holes, the drill chuck is removed and Morse tapered cutting tools are mounted.

8.3.6 Size Classification of Drilling Machines

The size (capacity) of a drilling machine is determined by all the following features:

1. Twice the distance from the center of the spindle to the inner face of the column
2. The maximum length of quill travel
3. The size of the Morse taper in the spindle
4. The horsepower of the motor

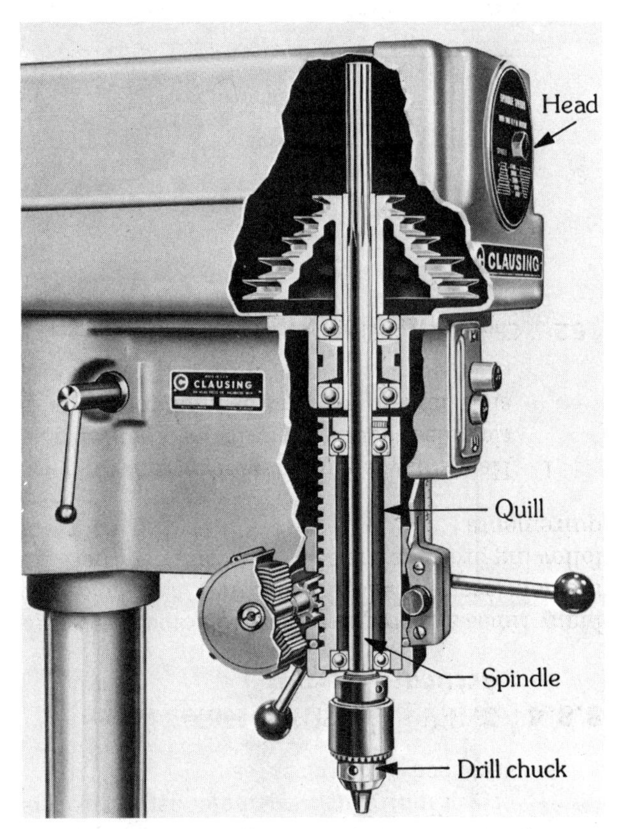

FIGURE 8.10 Spindle and quill assembly. (*Courtesy Clausing Industrial, Inc.*)

8.4 CUTTING TOOLS FOR DRILLING MACHINES

Drilling machines can be used for operations other than drilling a hole. The standard cutting tools used to perform these operations are now described.

8.4.1 Twist Drills

The tool generally used to drill a hole in metal is called a *twist drill.* The twist drill is formed by forging and twisting grooves in a strip of steel or by milling a cylindrical piece of steel.

Twist drills may be made from carbon steel, high-speed steel, or tungsten carbide. High-speed steel is the most generally used material and is designated by HSS on the shank. High-speed drills can be run twice as fast as carbon steel drills. Tungsten carbide drills are often used in production work and can be run two to three times faster than high-speed-steel drills. Figure 8.11 shows the three principal parts of a twist drill: the point, body, and shank.

Point. The *point* of a twist drill is the entire cone-shaped surface at the cutting end (Fig. 8.12).

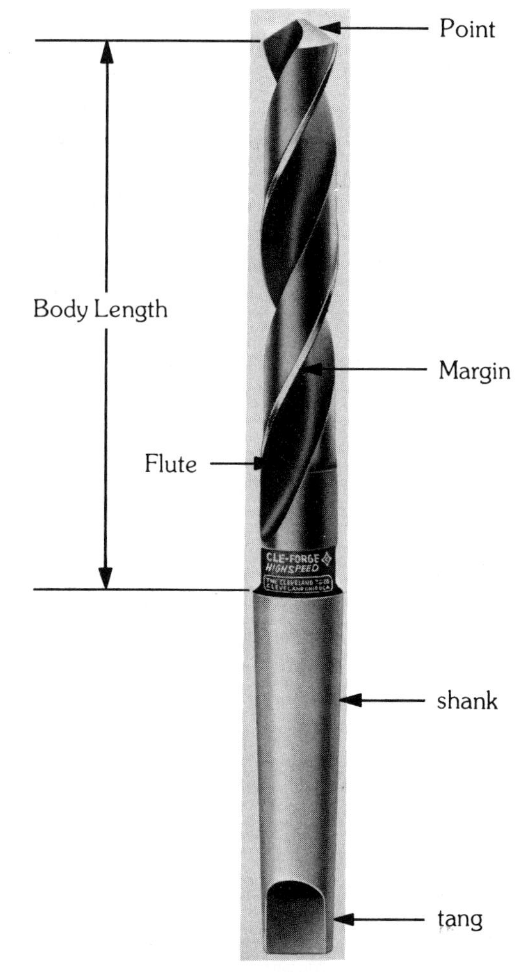

FIGURE 8.11 Parts of a twist drill. (*Courtesy Cleveland Twist Drill Co.*)

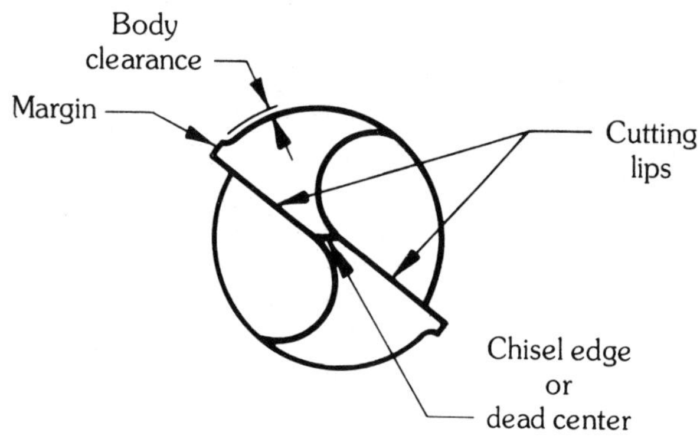

FIGURE 8.12 Parts of a drill point.

1. Located at the tip of the cone is the *dead center* or *chisel edge.* The dead center does not do any of the cutting; it pushes the material around until it becomes part of the chip.
2. Extending outward from the dead center is the actual *cutting edge* or *lip.* The lips must be of equal lengths and have the same angle to drill the proper size hole.
3. Directly behind the cutting lips is the *lip clearance* or *relief,* which is the amount of metal that has been ground away to allow the cutting edge to drill a hole and not rub the workpiece (Fig. 8.13). General-purpose drills have 8 to 12° of relief behind the cutting edge.

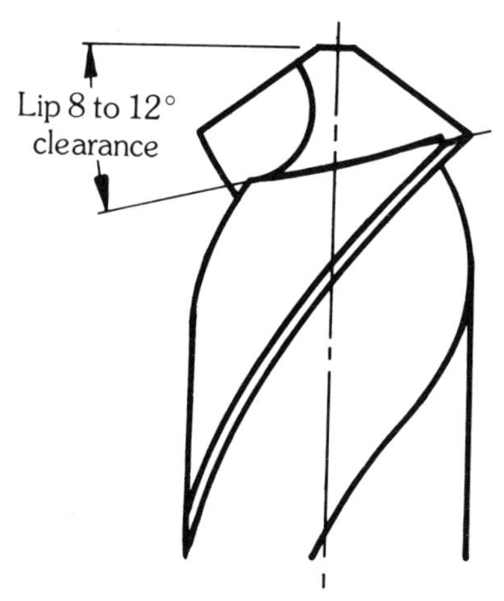

FIGURE 8.13 Lip clearance or relief is ground directly behind the cutting edge at 8 to 12°.

Body. The *body* is the part of the drill with grooves and lies between the point and the shank.

1. The *flutes* are two or more spiral grooves cut about the body. They form the cutting edges on the point, allow cutting fluid to reach the cutting edges, and cause the chips to curl and escape from the hole.
2. The *margin* is the narrow surface along the flute. It is relieved to provide body clearance. This clearance is called the *land* and reduces friction between the drill body and the hole. The margins determine the size of the hole to be drilled and keep the drill aligned during the drilling operation.
3. The *web* (Fig. 8.14) is the narrow section between the flutes. It runs the entire length of the body and gradually increases in thickness toward the shank, giving additional strength to the twist drill.

Shank. The *shank* is the part that fits into the spindle or chuck of the drill press. Most drills that are 1/2 in. (12.7 mm) and under in diameter have straight shanks. Larger drills have Morse tapered shanks, although some drills larger than

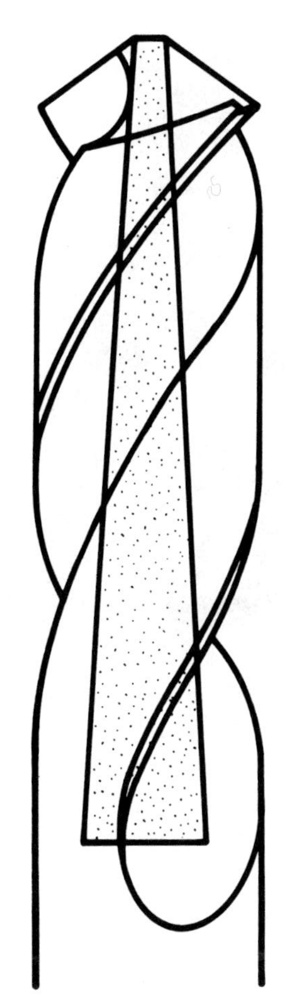

FIGURE 8.14 The web is the tapered section between the flutes.

1/2 in. may have 1/2 in. straight shanks. *Morse tapered shanks* include eight different sizes, ranging from No. 0 to No. 7, No. 7 being the largest. Tapered shank drills have a *tang* at the extreme end of the drill that helps drive the drill and provides a means of removing it from the spindle. For complete specifications and sizes of Morse tapers, see Appendix F.

8.4.2 Twist Drill Size Classification

There are four systems of designating the size of twist drills: fraction, number, letter, and metric. A complete table of sizes is provided in Appendix B.

The *fractional* size drills range in size from 1/64 in. (0.40 mm) to more than 4 in. (101.6 mm), in increments of 1/64 in.

Number sizes range from No. 80 (0.0135 in. diameter), the smallest, to No. 1 (0.228 in. diameter), the largest. They are used to fill in between the sizes of smaller fractional size drills.

The *letter* size drills range from A (0.234 in. diameter) to Z (0.413 in. diameter). Letter-size drills also dovetail in between fractional size drills.

Metric (*millimeter*) sizes are manufactured according to the metric system of measurement and range in size from 0.35 to 50.5 mm.

8.4.3 Sharpening the Twist Drill

Drill Grinding Machines. Where there is a considerable amount of drilling done, many larger shops use an automatic drill grinding machine to sharpen their drills (Fig. 8.15). Drill grinding machines quickly sharpen a large variety of drill sizes. They also sharpen the drill cutting lips to the correct point angle, clearance angle, and to equal length.

Some smaller shops use the drill grinding attachment, which fastens onto an ordinary bench or pedestal grinder (see Fig. 8.16). This attachment gives the same results as the drill grinding machine. However, the drill must be sharpened manually once the machine has been set up.

Sharpening a Drill by Hand. Often it is necessary for the machinist to grind (sharpen) a drill by hand. To grind a drill properly, three important factors must be considered:

FIGURE 8.15 Automatic drill grinding machine. (*Courtesy Cherry Textron*)

FIGURE 8.16 A drill grinding attachment is attached to a pedestal grinder for sharpening drills. (*Courtesy Orange Coast College*)

1. The *length* of the cutting lips must be equal. If they are not, the drill will not drill the proper size hole for which it was intended. Figure 8.17 shows the results of improperly ground drills not having equal cutting lips. The drill point gage (Fig. 8.18) shows the proper technique of measuring the angle and length of the cutting lip of a drill. If a drill point gage is not available, any suitable rule or protractor can be used to check the cutting lip.

2. Proper cutting lip *clearance* is essential to successful drilling. Lip clearance is the relief ground directly behind the cutting lip so only the cutting lips will enter the metal. If there is not enough clearance behind the lip, the point of the drill rubs and cannot cut the metal. However, if too much clearance angle is ground, the cutting lips dull rapidly or the outer corners break off. Table 8.1 indicates the correct clearance angle for various types of materials.

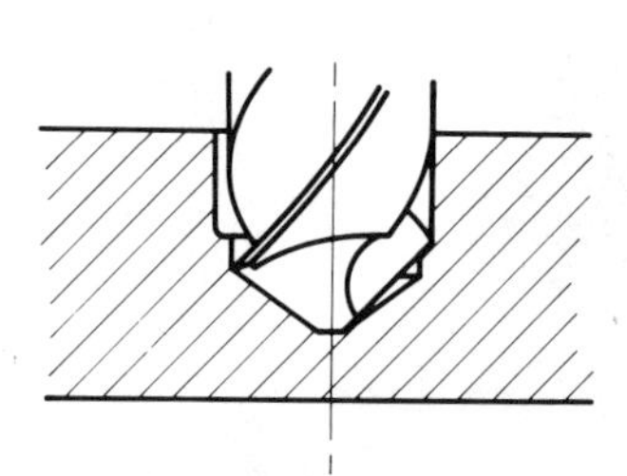

Cutting lips of unequal angles cause one cutting edge to work harder.

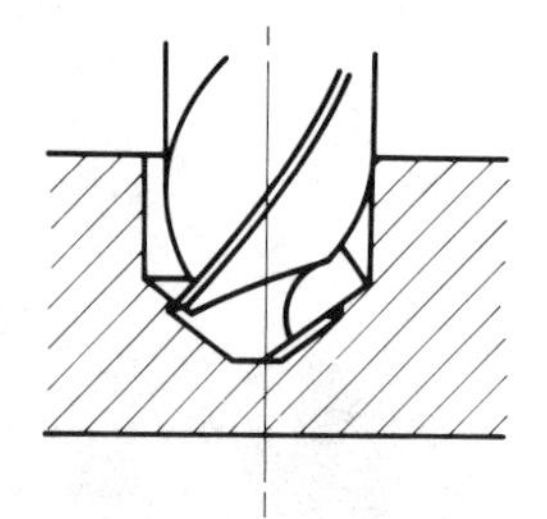

Oversize holes are the result when the cutting lips are ground unequal in length.

FIGURE 8.17 Effects of improperly ground cutting lips.

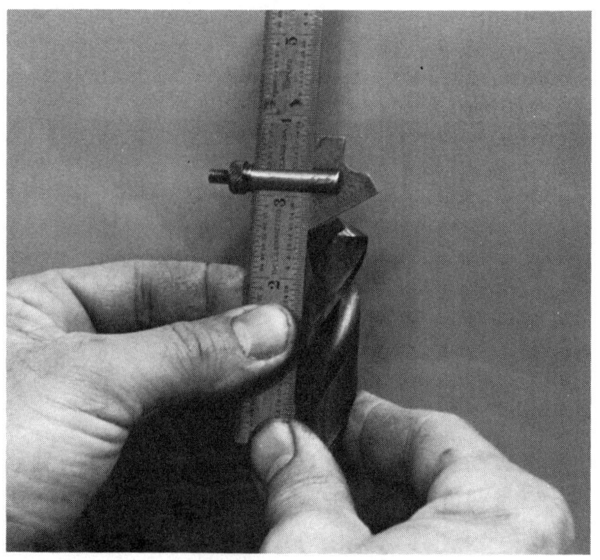

FIGURE 8.18 The drill point gage is used to check the length of cutting lips and point angle.

TABLE 8.1
Clearance angles

Material to Be Drilled	*Degrees of Clearance*
Mild steel	8–12
Alloy steels	6–10
Aluminum	12–15
Brass	10–12
Cast iron	10–14
Plastics	12–15

3. The *included angle* of the drill must be correct. The general-purpose drill uses an angle of 118 or 59° on each side of the centerline (see Fig. 8.19). For tougher and harder metals, an included angle of 130 to

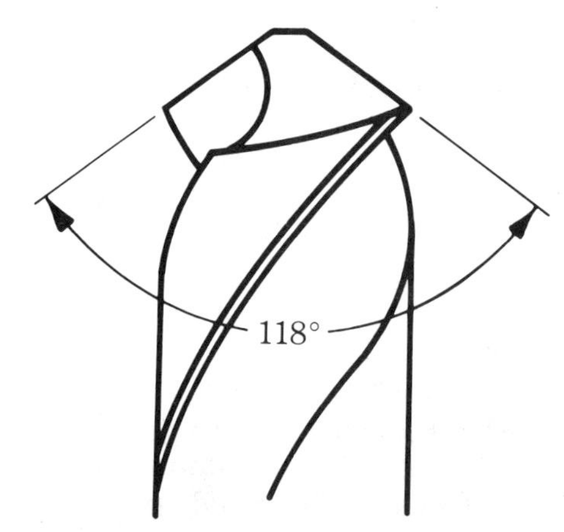

FIGURE 8.19 For general-purpose drilling, the drill point angle is generally ground to 118°.

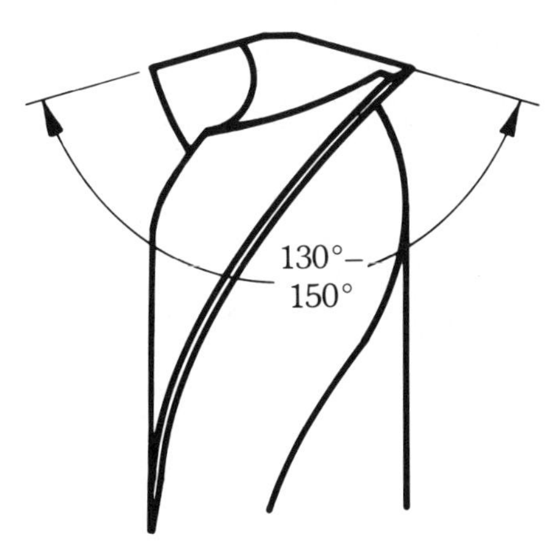

FIGURE 8.20 For harder materials a point angle of 130 to 150° is used.

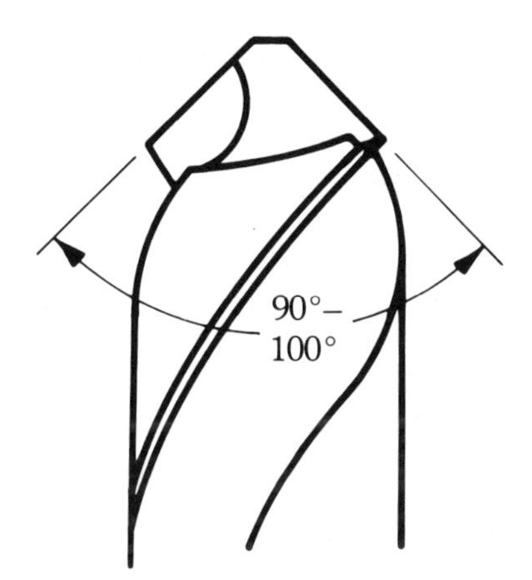

FIGURE 8.21 For soft materials use a point angle of 90 to 100° for best results.

150° is generally ground on the drill for better performance (see Fig. 8.20). For drilling softer materials, such as aluminum or plastic, an angle of 90 to 100° is preferred (Fig. 8.21).

To hand sharpen a drill, follow these six steps:

1. Hold the drill about 60° with the face of the grinding wheel (see Fig. 8.22).
2. Hold the shank a little lower than the hand holding the body, and slowly press the drill against the grinding wheel. (*Note:* Make sure that the cutting lip touches the grinding wheel first.)
3. Slowly lower the shank of the drill while applying a slight pressure against the grinding wheel.
4. Rotate the drill one-half turn, without moving the position of hands, and grind the other cutting edge.

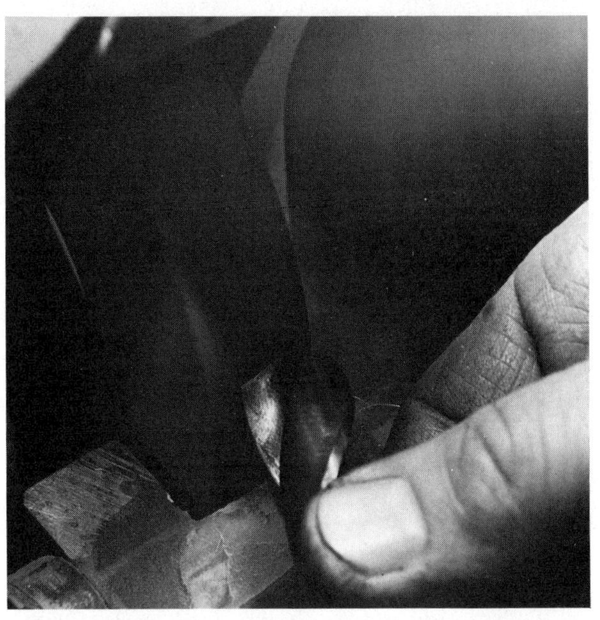

FIGURE 8.22 Using a pedestal grinder to resharpen a drill.

5. Check the point angle and length of cutting lips with a drill point gage (see Fig. 8.18). Also, check the clearance angle with a protractor.
6. Continue to sharpen the cutting edges until the cutting lips are sharp and the margins are free from wear.

Hints on drill sharpening

1. Never twist the drill while sharpening.
2. Do not grind too much off one cutting edge before grinding the other.
3. Grind from the cutting edge to the heel.
4. To avoid overheating the drill, cool it in water frequently during the sharpening procedure.
5. After the drill has been sharpened many times, the *web* becomes thicker and drilling a hole becomes more difficult. Figure 8.23 shows how to *thin* the web of a drill. Thin the web by holding the cutting edge parallel to the side of the grinding wheel. Grind back into the flutes on both sides until the web is the desired thickness.

8.4.4 Drill Cutting Speeds and Feed Speeds

Cutting speed may be referred to as the rate that a point on a circumference of a drill will travel in 1 minute. It is expressed in feet per minute or

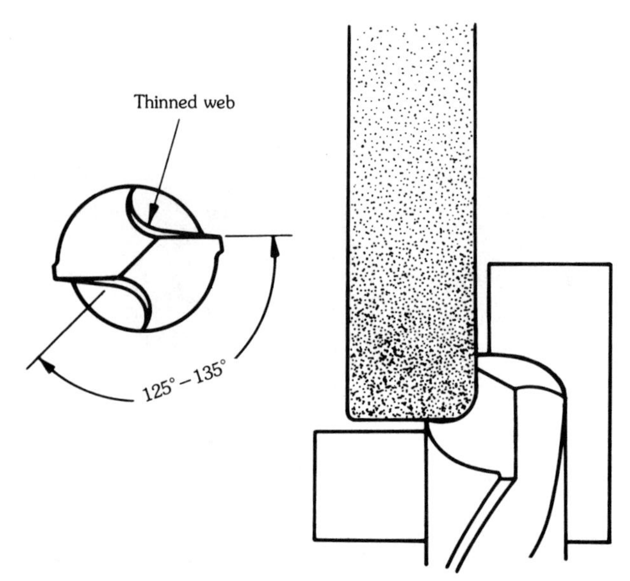

FIGURE 8.23 Correct method of thinning the web of a twist drill.

TABLE 8.2

Cutting speeds for high-speed drills[a]

Material	*Cutting Speed* ft/min	*Cutting Speed* m/min
Steel		
Mild	100–120	30–36
Tool and alloy	50–80	15–24
Cast Iron		
Gray	100–120	30–36
Malleable	80–100	24–30
Aluminum	200–300	60–90
Brass, machining	100–200	30–60
Copper	50–80	15–24

[a]Carbon-type drills use one-half of these cutting speeds.

meters per minute. Cutting speed is one of the most important factors that determines the life of a drill. If the cutting speed is too slow, the drill might chip or break. A cutting speed that is too fast rapidly dulls the cutting lips.

Cutting speeds depend on the following seven variables:

1. The type of material being drilled. The harder the material, the slower the cutting speed.
2. The cutting tool material and diameter. The harder the cutting tool material, the faster it can machine the material. The larger the drill, the slower the drill must revolve.
3. The types and use of cutting fluids allow an increase in cutting speed.
4. The rigidity of the drill press.
5. The rigidity of the drill (the shorter the better).
6. The rigidity of the work setup.
7. The quality of the hole to be drilled.

Each variable should be considered prior to drilling a hole. Each variable is important, but the work material and its cutting speed are the most important factors. Table 8.2 shows the various cutting speeds for the common types of materials. To calculate the revolutions per minute (rpm) rate of a drill, the diameter of the drill and the cutting speed of the material must be considered. Use the following formula when calculating the rpm rate of an inch-based drill:

$$\text{rpm} = \frac{12 \times \text{cutting speed (CS)}}{D \times \pi\,(\text{in.})} \qquad \text{ft/min}$$

Explanation: 12 is used to multiply the material cutting speed to change feet into inches. D equals the diameter of the drill and is multiplied by π (3.1416) to obtain the circumference. A shorter version of this formula is generally used because most drill presses cannot be set on an exact calculated rpm. The shorter version is found by using 3 in place of π (3.1416), as shown:

$$\text{rpm} = \frac{12 \times \text{CS}}{3 \times D} = \frac{4 \times \text{CS}}{1 \times D} \quad \text{or simply} \quad \frac{4 \times \text{CS}}{D}$$

Example: A 1/4-in. hole is to be drilled in mild steel. Using Table 8.2, the cutting speed selected is 110 (the middle). Using the formula $4 \times CS/D$ gives

$$\text{rpm} = \frac{4 \times 110}{\frac{1}{4}} = \frac{440}{\frac{1}{4}} = 1760$$

Formula for the metric system:

$$\text{rpm} = \frac{\text{CS (m/min)}}{D \times \pi\ (\text{m})}$$

Explanation: There is no shorter version for the metric system for calculating the rpm. Also, because the diameter of the drill is given in millimeters, it must be converted to meters. To do this, divide the diameter by 1000. For example:

$$12\ \text{mm} \div 1000 = 0.012\ \text{m}$$

Example: A 12-mm drill is to be used to drill a hole in gray cast iron at a speed of 33 m/min. Using the formula, we obtain

$$\text{rpm} = \frac{CS}{D \times \pi} = \frac{33}{0.012 \times 3.1416}$$
$$= \frac{33}{0.0377} = 875$$

Drilling Feeds. The feed of a drill can be defined as the distance the drill advances into the work for each revolution of the drill. Drilling feeds are generally expressed in thousandths of an inch per revolution of the drill. Table 8.3 shows the correct drilling feeds for various sizes of drills. As noted in the table, the larger the drill, the greater the drilling feed. Some smaller drilling machines may not have automatic feeding devices; therefore, the drill must be fed manually. When feeding by hand, try to feel the drill cutting, and watch the chips coming from the hole. Correct feeding pressures result in uniform spiral chips forming outward from the hole. Too heavy a feed rate may chip the cutting lips or break the drill. Feed rates that are too light cause a squeaking or chattering noise and quickly dull the cutting lips.

TABLE 8.3
Drilling feed rates

	Feed rate per Revolution	
Drill Size	*in.*	*mm*
1/8 and smaller	0.001–0.002 or feed by hand	0.025–0.05
1/8 to 3/8	0.002–0.005	0.05–0.13
3/8 to 5/8	0.005–0.008	0.13–0.2
5/8 to 1	0.008–0.015	0.2–0.40
1 and larger	0.015–0.025	0.40–0.65

Cutting Fluids for Drilling Operations. Cutting fluids are essential for best results in drilling operations. Cutting fluids help cool and lubricate both the cutting tool and workpiece. For a complete listing of the various types of cutting fluids and their purposes, refer to Chap. 15.

8.5 OTHER TYPES OF CUTTING TOOLS USED IN DRILLING MACHINES

As mentioned, the twist drill is the primary cutting tool used in the drill press. However, other types of drills and cutting tools are sometimes needed for special setups and operations.

8.5.1 Core Drill

Figure 8.24 shows the multiflute *core drill.* It generally has three or four flutes and ranges in size from 1/4 to over 2 in. (6.35 to 51 mm) in diameter. Core drills are generally used to increase the size and finish of cored holes in castings. Because the core drill has three or four flutes, it also has the same number of margins and cutting lips, thereby permitting greater feed rates and more accurately drilled holes. The core drill requires a preexisting hole before it can be used as the web is blunt and cannot penetrate solid metal.

8.5.2 Countersinks

Countersinking is the process of machining a conical-shaped recess in a previously drilled hole (see Fig. 8.25). The most common countersinks available are the 60, 82, and 100° types. Center holes used in lathe work require 60° countersinks.

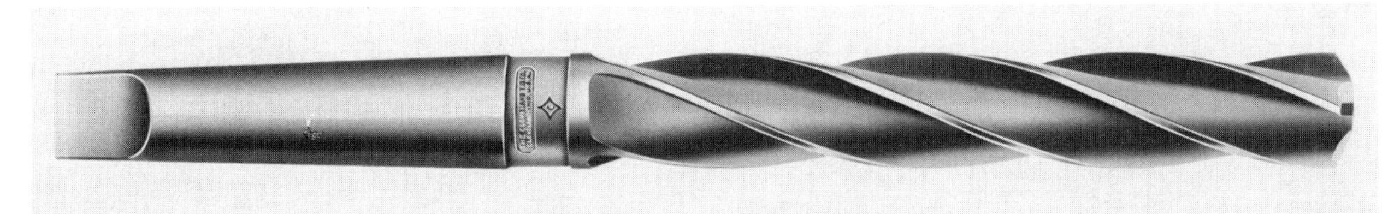

FIGURE 8.24 The core drill is used to increase the size of cored holes in castings. (*Courtesy Cleveland Twist Drill Co.*)

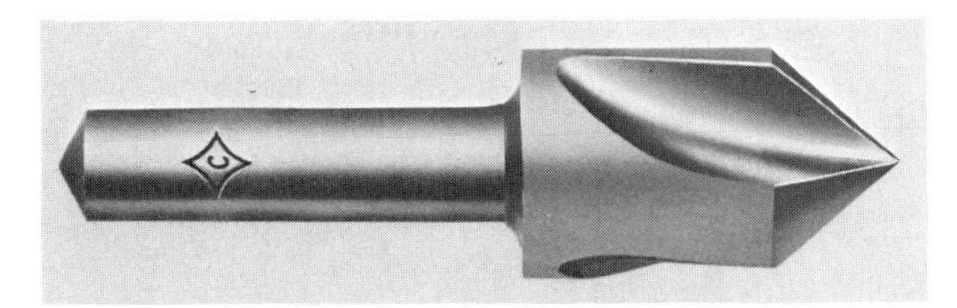

FIGURE 8.25 A countersink is used to make a conical recess at the end of a drilled hole. (*Courtesy Cleveland Twist Drill Co.*)

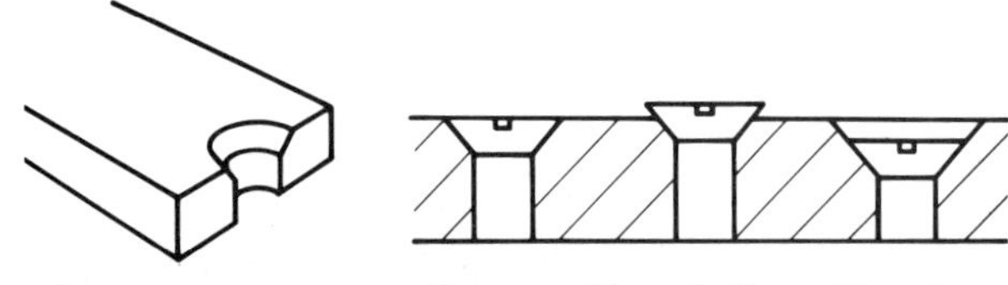

FIGURE 8.26 Correct and incorrect depths of a countersunk hole.

Flat-head machine screws and some rivets have an angle of 82°. Aircraft rivets generally have an angle of 100° and require a 100° countersink. Countersinking is easily accomplished in the drill press, but the spindle must revolve *slowly,* about one-fourth the rpm of drilling the equivalent-size hole. Care must be taken to countersink a hole the proper depth. It should be deep enough so that the head of the flat-head screw or rivet fits into a hole flush with the top of the material (see Fig. 8.26).

8.5.3 Counterbores and Spotfacers

Counterboring is the process of machining a cylindrically shaped hole in a previously drilled hole. The counterbore (Fig. 8.27) provides a recess for cap screws, bolt heads, and fillister-head screws. Counterboring is similar to countersinking; that is, a slower rpm is used, and the proper depth is important to ensure the correct fit of the fastener in the hole. Counterbores are available in many sizes, with straight or taper shanks. The pilot, located at the end, is used with the counterbore to help locate the hole quickly and guide the cutting lips in the hole (Fig. 8.28). Spotfacers are similar to counterbores and are used to machine surfaces perpendicular to a hole axis to provide a level seat for the heads or nuts of threaded fasteners.

8.5.4 Reamers

Reamers are used to produce smooth and accurate holes. There are a variety of reamers. Some are turned by hand, and others use machine power.

Machine Reamers. Machine reamers are used on both drilling machines and lathes for roughing and finishing operations. Machine reamers are available with tapered or straight shanks, and with straight or helical flutes (Fig. 8.29). Tapered shank reamers fit directly into the spindle, and the straight shank reamer, generally called the *chucking reamer,* fits into a drill chuck.

Rose reamers (Fig. 8.30) are machine reamers that cut only on a 45° chamfer (bevel) located on the end. The body of the rose reamer tapers slightly (about 0.001 in. per inch of length) to prevent binding during operation. This reamer does not cut a smooth hole and is generally used to bring a hole to a few thousandths undersize. Because the rose reamer machines a hole 0.001 to 0.005 in. (0.025 to 0.127 mm) under a nominal size, a hand reamer is used to finish the hole to size.

Fluted reamers (Fig. 8.31) are machine reamers used to finish drilled holes. This type of reamer removes smaller portions of metal compared to the rose reamer. Fluted reamers have more cutting edges than rose reamers and therefore cut a smoother hole. Fluted reamers cut on the chamfered end as well as the sides. They are also available in solid carbide or have carbide inserts for cutting teeth.

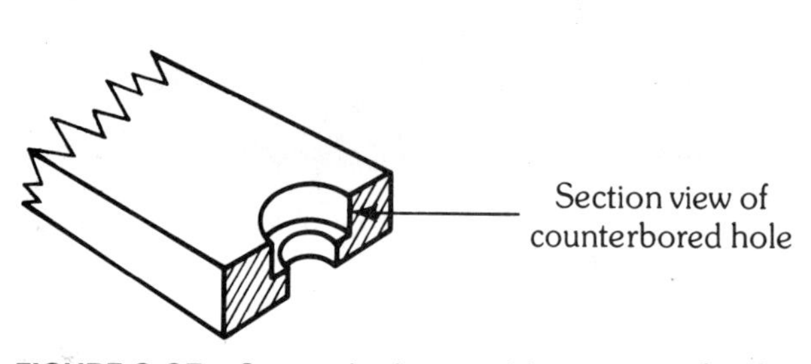

FIGURE 8.27 Counterboring provides a recess for the head of a capscrew or nut.

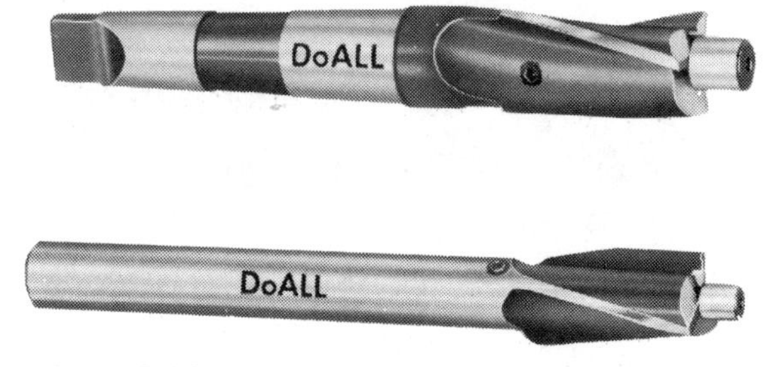

FIGURE 8.28 Counterbores and pilots. (*Courtesy DoALL Co.*)

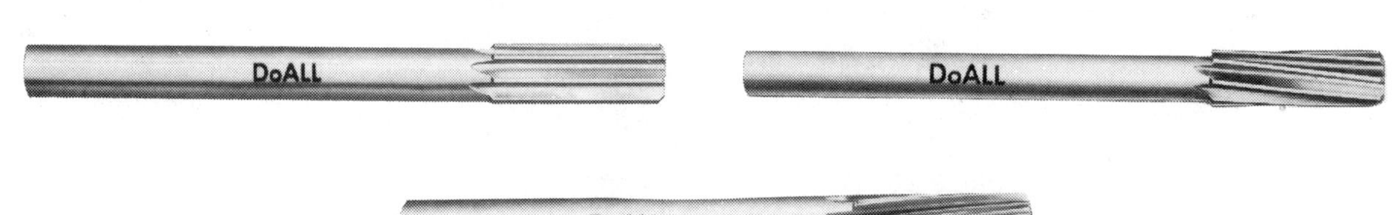

FIGURE 8.29 Reamers are used to finish a drilled hole. (*Courtesy DoALL Co.*)

FIGURE 8.30 A rose reamer cuts on the end. (*Courtesy DoALL Co.*)

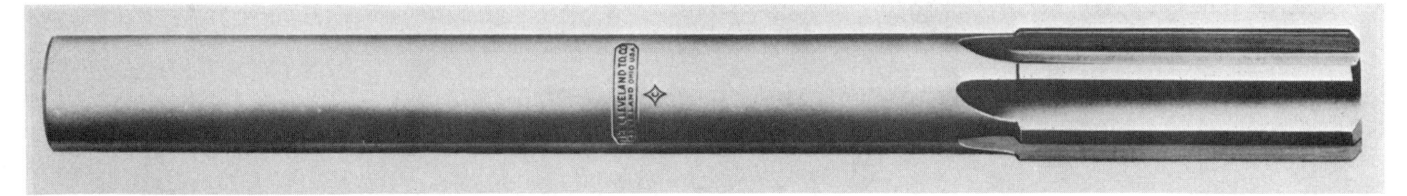

FIGURE 8.31 Fluted reamers cut on both the sides and the end. (*Courtesy Cleveland Twist Drill Co.*)

Shell reamers (Fig. 8.32) are made in two parts: the reamer head and the arbor. In use, the reamer head is mounted on the arbor. The reamer head is available in either a rose or flute type, with straight or helical flutes. The arbor is available with either a straight or tapered shank. The shell reamer is considered economical because only the reamer is replaced when it becomes worn or damaged.

Hand Reamers. *Hand reamers* (Fig. 8.33) are finishing reamers distinguished by the square on their shanks. They are turned by hand with a tap wrench that fits over this square. This type of reamer cuts only on the outer cutting edges. The end of the hand reamer is tapered slightly to permit easy alignment in the drilled hole. The *length* of taper is usually equal to the reamer's diameter. *Never* turn hand reamers *by machine power,* and you *must* start them true and straight. Hand reamers should never remove more than 0.001 to 0.005 in. (0.025 to 0.127 mm) of material. They are available from 1/8 to over 2 in. (2.5 to 50 mm) in diameter and are generally made of carbon steel or high-speed steel.

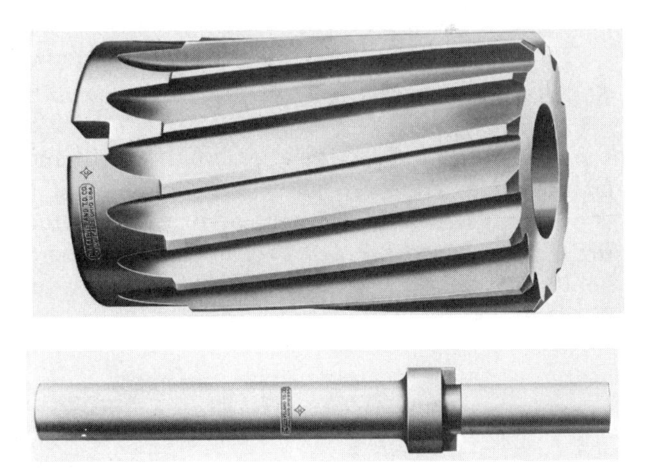

FIGURE 8.32 Shell reamers come in two parts, the reamer and the arbor. (*Courtesy Cleveland Twist Drill Co.*)

Taper hand reamers are hand reamers made to ream all standard size tapers. They are made for both roughing (Fig. 8.34) and finishing tapered holes (Fig. 8.35). Similar to the straight hand reamer, this taper should be used carefully, and never with machine power.

Adjustable reamers (Fig. 8.36) are used to produce any size hole within the range of the reamer. Their size is adjusted by sliding the cutting blades to and from the shank. These blades are moved by the two adjusting nuts located at each end of the blades. Adjustable hand reamers are available in sizes from 1/4 to over 3 in. diameters (6.35 to 75 mm). Each reamer has approximately 1/64-in. adjustment (0.4 mm) above and below its nominal diameter.

Expansion hand reamers (Fig. 8.37) are like the adjustable reamers but have a limited range of

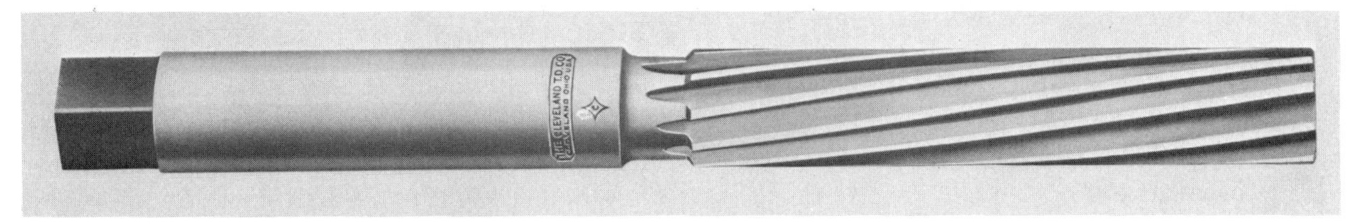

FIGURE 8.33 Hand reamers are finishing reamers and are distinguished by the square on the end of the shank. (*Courtesy Cleveland Twist Drill Co.*)

FIGURE 8.34 Roughing taper hand reamer. (*Courtesy DoALL Co.*)

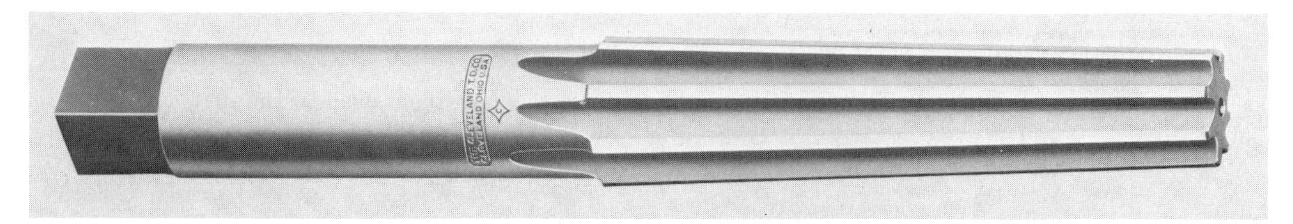

FIGURE 8.35 Finishing taper hand reamer. (*Courtesy Cleveland Twist Drill Co.*)

FIGURE 8.36 Adjustable hand reamer. (*Courtesy Cleveland Twist Drill Co.*)

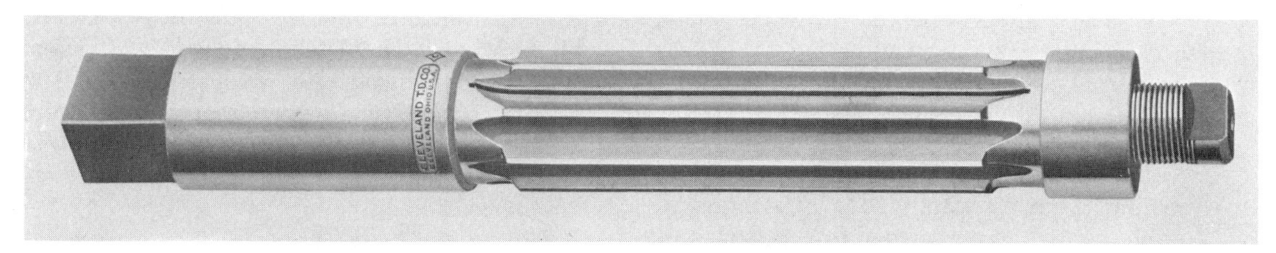

FIGURE 8.37 Expansion hand reamer. (*Courtesy Cleveland Twist Drill Co.*)

approximately 0.010-in. adjustment. Expansion reamers have an adjusting screw at the end of the reamer. When turned, this adjusting screw forces a tapered plug inside the body of the reamer, expanding its diameter. Expansion reamers are also available as machine reamers.

Care of Reamers. Because reamers are precision finishing tools, they should be used with care:

1. Reamers should be stored in separate containers or spaced in the tooling cabinet to prevent damage to the cutting edges.
2. Always use cutting fluids during reaming operations, except with cast iron.
3. *Never* turn any reamer backward or you will dull the cutting edges.
4. Remove any burrs or nicks on the cutting edges with an oilstone to prevent cutting oversize holes.
5. Always use a slower speed with machine reamers. Use a speed equal to one-half of the cutting speed for a drill of the same diameter.

8.5.5 Center Drills

Center drills (Fig. 8.38) are a combination drill and countersink. They are often used in place of a center-punched hole for more accurately located drilled holes (see Sec. 8.8.4). The point acts as a small pilot hole, and the tapered portion accurately positions the drill prior to drilling.

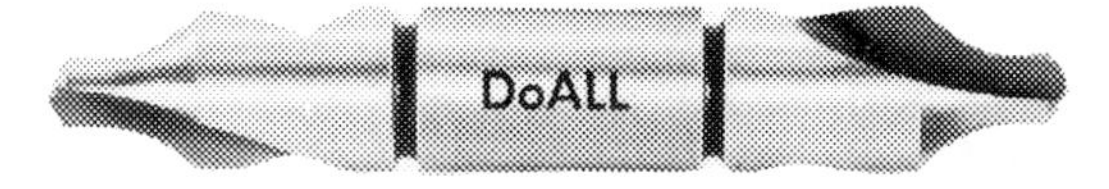

FIGURE 8.38 A center drill is a combination drill and countersink. (*Courtesy DoALL Co.*)

8.5.6 Boring Bars

Boring is the operation of producing a very straight and accurate size hole. When a drill size is not available, a hole is generally bored to a finish size (see Sec. 8.8.9). Figure 8.39 shows an adjustable *boring head* and *boring bars.* Boring heads and bars are available in a variety of sizes. The boring head can be adjusted to bore holes within 0.001 in. (0.025 mm) or closer.

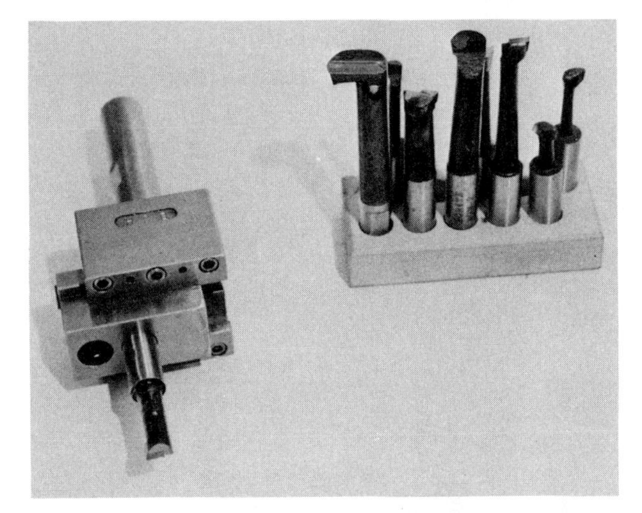

FIGURE 8.39 Adjustable boring head and boring bars.

8.6 CUTTING TOOL HOLDING DEVICES

Some cutting tools used in the drill press can be held directly in the spindle hole of the machine; others must be held with a *drill chuck, sleeve,* or *socket.*

8.6.1 Drill Chucks

Cutting tools with straight shanks are generally held in the *drill chuck.* Two common types of drill chucks often used are the type that uses a *chuck key* (Fig. 8.40) to lock the cutting tool and the keyless type (Fig. 8.41). When using the type requiring a chuck key, *always be sure to remove it before turning on the machine power.*

8.6.2 Sleeves

Cutting tools with tapered shanks are available in many different sizes. When a cutting tool that has a smaller taper than the spindle taper is used, a sleeve must be fitted to the shank of the cutting tool (Fig. 8.42).

FIGURE 8.40 Key type of drill chuck.

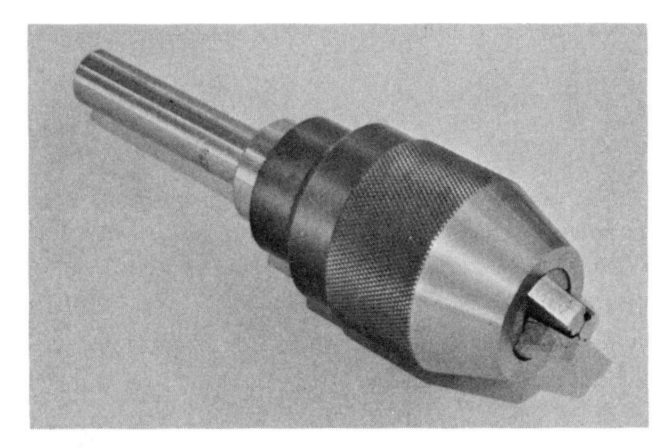

FIGURE 8.41 Keyless drill chuck.

FIGURE 8.42 A drill sleeve is fitted to a tapered drill shank that is too small for the drill press spindle.

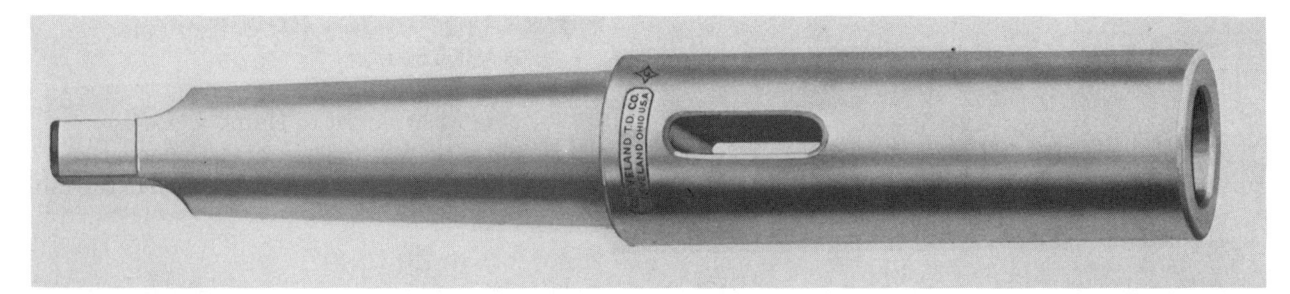

FIGURE 8.43 The drill socket is used to adapt a drill shank that is too large for the spindle.

FIGURE 8.44 Correct way to remove a drill from a spindle using a drill drift.

FIGURE 8.45 A drill press vise is used to hold work of regular shapes. (*Courtesy Universal Vise & Tool Co.*)

8.6.3 Sockets

If the cutting tool has a tapered shank larger than the spindle taper, the *socket* (Fig. 8.43) is used to reduce it to the correct size. Figure 8.44 shows the correct way to remove taper shank cutting tools from the spindle of the drill press using a drill drift.

8.7 WORK-HOLDING DEVICES

Work must be clamped rigidly to the table or base of the drilling machine. When improperly clamped, the work moves during an operation and the cutting tool generally breaks. Serious accidents may occur from setups that move or spin during an operation.

8.7.1 Vises

Vises (Fig. 8.45) are widely used for holding work of regular size and shape, such as flat, square, and rectangular pieces. *Parallels* are generally used to support the work and protect the vise from being drilled. Vises should be clamped to the table of the drill press to prevent them from spinning during operation.

FIGURE 8.46 An angle vise is used when drilling an angle hole in the work. (*Courtesy Universal Vise & Tool Co.*)

The *angle vise* (Fig. 8.46) tilts the workpiece and provides a means of drilling a hole at an angle without tilting the table.

8.7.2 Angle Plates

An *angle plate* (Fig. 8.47) supports work on its edge. Angle plates accurately align the work perpendic-

FIGURE 8.47 The angle plate and C clamp are used to secure work for drilling operations.

ular (90°) to the table surface, and they generally have holes and slots to clamp them to the table and hold the workpiece.

8.7.3 V Blocks

V blocks are used to support round work. They are generally used in pairs (Fig. 8.48) and are fastened to the drill press table with clamps and T bolts.

FIGURE 8.48 V blocks are used to support round workpieces.

8.7.4 Strap Clamps, T Bolts, and Step Blocks

Strap clamps are used to clamp work directly to the table (Fig. 8.49). They are supported on the outer end with *step blocks* and fastened to the table with *T bolts.* Strap clamps are available in a variety of sizes and shapes (Fig. 8.50) to fulfill most clamping arrangements. Always place the T bolt as close as possible to the work to ensure maximum clamping pressure (Fig. 8.51).

8.7.5 C Clamps

A *C clamp* is useful for clamping work directly to the table of the drill press (Fig. 8.52). It may also be used when the table is not provided with slots for T bolts.

8.7.6 Drill Jigs

A *drill jig* is a production tool used when a hole, or several holes, must be drilled in a large number of *identical parts* (Fig. 8.53). The drill jig has

FIGURE 8.49 Strap clamps secure irregularly shaped parts directly to the table.

FIGURE 8.50 Strap clamps come in a variety of sizes.

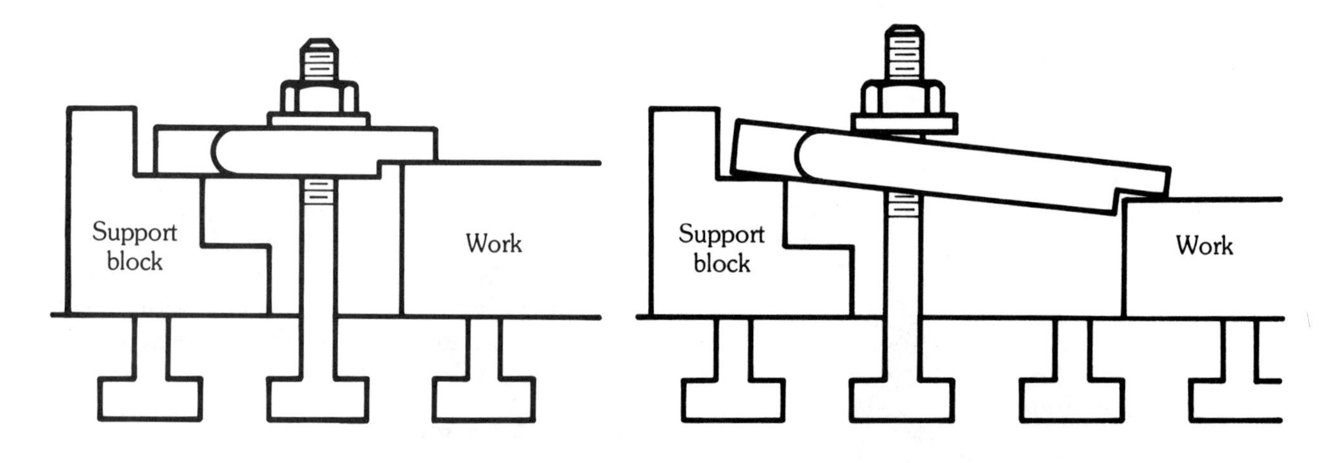

Correct way to clamp work.

Improperly clamped work; the bolt is too far from the work.

FIGURE 8.51 Proper way to use strap clamps.

FIGURE 8.52 The C clamp is used to hold the work directly to the table when work cannot be mounted in a vise.

FIGURE 8.53 Drill jigs are used when a large number of identical parts are to be drilled.

several functions. First, it is a work-holding device, clamping the work firmly. Second, it locates work in the correct position for drilling. The third function of the drill jig is to guide the drill straight into the work. This is accomplished by use of drill bushings. Drill jigs are designed and manufactured for a part when quantity production is necessary.

8.8 OPERATIONS PERFORMED ON THE DRILL PRESS

Many different setups and operations may be performed on a drill press. In Sec. 8.8.1 we explain those operations performed most frequently.

8.8.1 Drill Press Safety

1. Always wear eye protection when using the drill press.
2. Make sure that the work being drilled is held securely.
3. Do not touch or attempt to stop a revolving spindle with your hands.
4. Always use a chipbrush or pliers to remove chips, and only after first stopping the spindle.
5. Keep your head away from a turning spindle.
6. Never adjust the work while the machine is running.
7. Always remove the chuck key after installing or removing a cutting tool.

8.8.2 Drilling Smaller Holes

Holes smaller than 1/2 in. (12.7 mm) are generally drilled with straight-shank twist drills held in a drill chuck. Here is the procedure:

1. Examine the shop drawing carefully to determine the exact hole location. Lay out the hole location and mark the intersecting lines with a center punch.
2. Select the proper-size drill. Check the drill by examining the stamped size on the shank. If the size is worn off the shank, check the drill with a drill gage, or measure across the margins with a micrometer (Fig. 8.54).

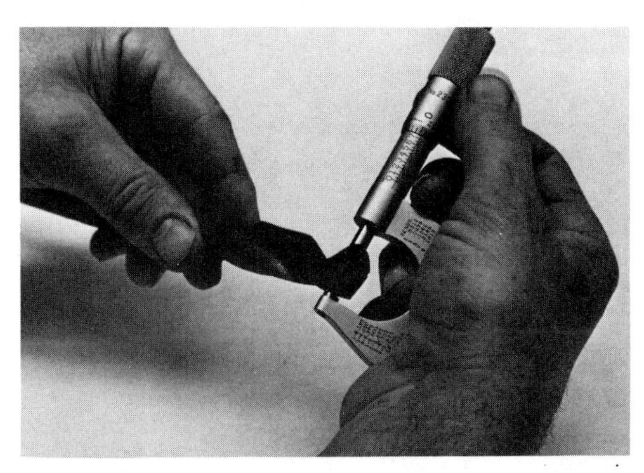

FIGURE 8.54 Measuring the size of a drill with a micrometer.

3. Secure the drill in the chuck.
4. Calculate the correct rpm and adjust the machine to operate at or below the calculated rpm.
5. Start the drill press and check to see if the drill is running true. If it is wobbling, the shank could be bent, or the drill was placed off-center into the chuck.
6. Secure the workpiece in a vise or other suitable work-holding device (see Sec. 8.7). Fasten the work-holding device to the table of the machine, loosely, for now.
7. Move the drill down with the feed handle and move the workpiece, or table, until the point of the drill is properly aligned in the center-punched hole. Raise the table if necessary.
8. After the drill and workpiece are properly aligned, tighten the work-holding device securely to the table.
9. Turn on the power, bring the drill to the workpiece, and carefully feed it into the workpiece a small distance. Raise the drill and examine the work to see if the drill started in the proper location. If not, use a round nose or cape chisel (Fig. 8.55) to cut a groove to *draw* the drill point back to center.

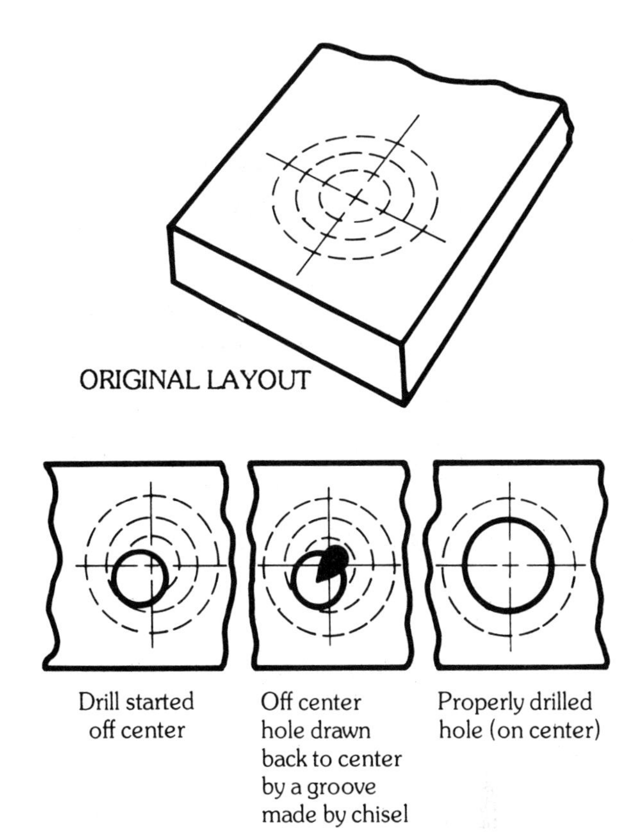

FIGURE 8.55 Drawing the drill point to center.

10. If the drill and workpiece are correctly aligned, apply a cutting fluid and continue the drilling. Apply even pressure on the feed to keep the drill cutting freely. This can be accomplished by using the automatic feed if available.
11. Raise the drill frequently and apply additional cutting fluid. When the drill begins to break through the work, ease up on the feed pressure to prevent the drill from digging in. (*Note:* Thin material may grab the drill as it breaks through. To prevent this, back the thin material with a wooden block and clamp both to the table.)
12. When the hole is completely drilled, back the drill out of the hole and turn off the power.

8.8.3 Drilling Larger Holes

Holes larger than 1/2 in. (12.7 mm) are drilled with Morse tapered shank drills that fit directly into the machine spindle. As a drill increases in size, the web becomes thicker, making a pilot hole necessary. On larger drills, the web should also be thinned (see Sec. 8.4.3).

1. The procedure for drilling larger holes begins as the procedure for drilling smaller holes (Sec. 8.8.2). That is, the layout is made, the workpiece is secured, and the hole is located. The next procedure to be considered is that of drilling a pilot or lead hole prior to drilling the larger hole (Fig. 8.56). The pilot hole reduces the feed pressure required because the chisel point is free to revolve in the pilot hole. A general rule is to select a pilot drill about equal to or slightly smaller than the web thickness of the larger drill to be used.

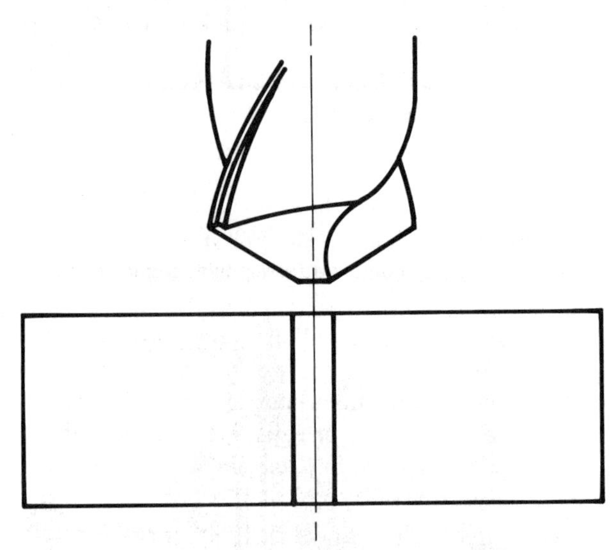

FIGURE 8.56 A pilot hole reduces feed pressure required when using a larger drill.

2. After the pilot hole has been drilled, remove the chuck and clean the tapered hole in the spindle. Place the drill into the spindle, and tap the end of the drill with a lead or soft-faced hammer to seat it properly. Finally, drill the hole, using the correct feed and rpm rates.

8.8.4 Drilling Precision Holes to an Accurate Layout

Many times a job requires close tolerances between the hole locations and an accurately sized hole. The following procedure is used to accomplish these requirements:

1. Holes that require greater accuracy in location also require accuracy in layout. Precision layout procedures and tools are a must. For further details on layout accuracy, refer to Chap. 5.
2. After completing the layout, *lightly* prick punch each intersecting line where a hole is to be drilled. Carefully examine each prick punch mark with a magnifying lens, and correct any mark that is not on the center of the intersecting lines.
3. Mount the workpiece in a suitable work-holding device and place it on the table of the drill press. (*Note:* Secure work as described in Sec. 8.8.2.)
4. Insert a *wiggler* or *center finder* into the drill chuck. Turn the machine power on, and center the wiggler point with a piece of stock (Fig. 8.57). (*Note:* Keep fingers away from the wiggler point when centering.)

FIGURE 8.57 Correct way to center a center-finder or wiggler.

5. Position the work until the wiggler point is correctly aligned in the center of the prick punched mark. Be careful not to insert the wiggler too far into the location mark or the wiggler will go off center.
6. Secure the workpiece and table and recheck the alignment.
7. The *center drill* (Fig. 8.58) is used next as a starter hole for the drill. Center drill the location carefully. Apply a suitable cutting fluid. Drill just deep enough so that the tapered portion of the center drill penetrates the work.

FIGURE 8.58 A center drill is used as a starter hole for a drill.

8. Replace the center drill with a drill approximately 1/8 in. *under* the finish size drill. The preliminary drilled hole acts as a pilot hole for the finish drill. This procedure helps the finish drill obtain a more accurate size hole and smoother finish.

8.8.5 Drilling Blind Holes

A blind hole is a hole that has not been drilled completely through the workpiece. The depth of a hole is measured by the distance the hole goes in the work at its full diameter (Fig. 8.59). To drill a hole to a required depth, set the *depth stop* on the drill press to the depth specified. This can be accomplished by first drilling the top of the workpiece with the point of the drill to margin depth. Next, using a steel rule, adjust the nut on the stop to the proper depth (Fig. 8.60).

If the drill press does not have a depth stop or other provisions for gaging the depth of a hole, mark a line with a pencil to show the depth of the hole. Sometimes blind holes are required to have a flat bottom. This is accomplished by first drilling to the required depth with a standard drill and then finishing with a flat-bottomed drill (Fig. 8.61).

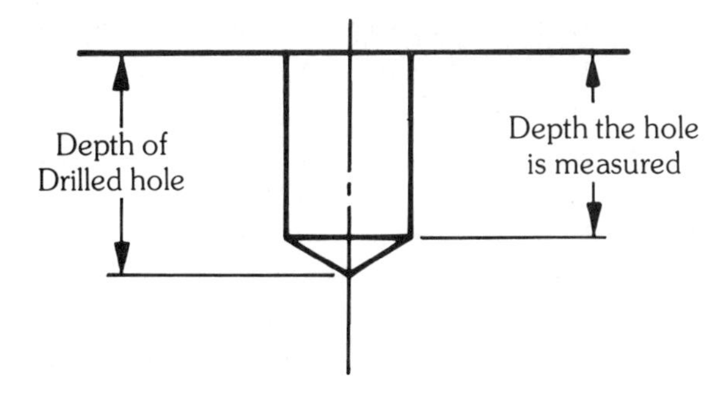

FIGURE 8.59 The depth of a hole is measured by its depth at full diameter.

FIGURE 8.60 Adjusting the depth stop on a drill press.

FIGURE 8.61 A flat bottom drill is used to produce a flat bottom in a blind hole.

8.8.6 Reaming Holes in a Drill Press

Selecting the Proper Reamer. Selecting the proper reamer to fulfill the job depends on the following five factors:

1. Degree of accuracy required
2. Finish desired
3. Type of material to be reamed
4. Size of the hole to be reamed
5. Amount of material to be removed by the reamer

When a shop drawing requires a hole with a high degree of accuracy and smooth finish, a hand reamer is generally used. The hand reamer produces a more accurate hole compared to that of a machine reamer. Because the hand reamer removes a smaller amount of metal (about 0.001 to 0.005 in.) (0.025 to 0.127 mm), it also leaves a smoother finish than a machine reamer. Often the two are used in conjunction. That is, a hole might first be machine-reamed a few thousandths undersize, then finished with a hand reamer.

Hand reamers are very fragile cutting tools. The cutting edges act as a scraping tool more than a cutting tool. When holes in harder and tougher metals are being reamed, the cutting edges of a hand reamer often crack or break. Machine reamers have cutting edges that are thicker and more durable and should be used when reaming tougher and/or harder metals.

Sometimes a shop print requires a reamed hole that is not standard in size. For example, it might ask for a hole size of 0.511. This size hole is between the standard size range for most machine reamers, that is, nominal sizes in increments of 1/64 in. Therefore, a hole 0.511 in. in diameter could be hand-reamed with either an adjustable or expansion reamer.

One operational characteristic of machine reamers is that of *chattering,* which is the vibration of either the tool or the workpiece. It is caused by the action of the cutting tool. Chatter affects the accuracy and finish of a hole. To avoid chatter, follow these procedures. First, remember that most reamers are made with unevenly spaced flutes and/or helical flutes. Helical fluted reamers are the best to use to avoid chatter. A second precaution is to select the shortest reamer available. Longer reamers chatter more often than shorter ones. Finally, always make sure that the work setup is rigid and the cutting tool is revolving at less than one-half the recommended drilling speed.

Machine Reaming (*See* Fig. 8.62)

1. Locate and mount the workpiece on the drill press table.
2. For holes up in 1 in. (25. 4 mm), select a drill size 1/64 in. (0.4 mm) *smaller* than the reamer. Use a drill size of 1/32 in. (0.8 mm) for holes larger than 1 in. (25.4 mm). Drill the hole.
3. Replace the drill with the proper reamer without changing the position of the workpiece.

FIGURE 8.62 Reaming operations are often performed on a drill press.

4. Set the machine for the proper rpm rate. *Remember,* this will be approximately one-half the drilling rpm rate.
5. Start the machine and feed the reamer into the work. Feed rates for machine reamers will be two to three times more than when drilling a hole. Use the proper cutting fluid.
6. Withdraw the reamer from the hole before stopping the machine.

Hand Reaming

1. Locate and mount the workpiece.
2. Select a drill that will allow the hand reamer to remove only 0.001 to 0.005 in. (0.025 to 0.127 mm) of metal. Sometimes a drill this size is not available, so the hole must be brought to the proper size either by boring (see Sec. 8.8.9) or by using a rose reamer. Machine the hole to proper size.
3. Replace the cutting tool with a 60° pointed center in a drill chuck.
4. Select the proper hand reamer and put it in an adjustable tap wrench.
5. Place the end of the reamer into the hole. Position the 60° point into the center drilled hole of the reamer (Fig. 8.63). This procedure helps steady the reamer and keep it straight in the hole during the reaming operation.
6. Turn the reamer with the tap wrench clock-

FIGURE 8.63 A 60° center is used to guide the hand reamer when used in a drill press.

wise, by hand, while applying pressure to the feed handle of the drill press.

7. Continue to turn the reamer until the hole is reamed all the way through. Be sure to use plenty of cutting fluid.
8. Never force the reamer during the operation. Always turn the reamer clockwise, even when withdrawing it from the hole.

8.8.7 Countersinking on a Drill Press

1. Mount the work and drill the proper size hole. (*Note:* This would be the body size of the fastener.)
2. Place the countersink in the drill chuck. Adjust the machine to one-fourth the rpm rate of the drill size.
3. Start the machine and slowly feed the countersink into the hole (Fig. 8.64).
4. Continue to feed until the tapered portion is large enough for the flat-head screw or rivet to fit flush with the work (Fig. 8.26).

8.8.8 Counterboring on a Drill Press

1. Mount the work and drill the proper-size hole. (*Note:* This would be the body size of the bolt.)
2. Select the proper-size counterbore with a pilot equal in diameter to the drilled hole.
3. Secure the counterbore in the drill chuck and adjust the rpm rate to one-fourth of that used for drilling.
4. Feed the counterbore close to the drilled hole, and check that it fits the hole without binding (Fig. 8.65).

FIGURE 8.64 Countersinking in a drill press requires lower rpm.

FIGURE 8.65 Counterboring a part in a drill press.

5. Start the machine and feed the counterbore to the proper depth. Use plenty of cutting fluid.

8.8.9 Boring on a Drill Press

1. Mount the workpiece to the drill press table in a suitable work-holding device. [*Note:* Upright (or larger) drilling machines give better results when boring because they are more rigid than sensitive drill presses.]

2. Locate and drill the workpiece. For boring most jobs, drilling the hole to 1/16 to 1/8 in. (1.5 to 3.2 mm) *undersize* is sufficient.
3. Mount the boring head in the machine spindle.
4. Select the largest boring bar available that will fit into the drilled hole, and fasten in the boring head.
5. Feed the boring head down to the drilled hole, and adjust it so that the cutting edge of the boring bar just touches the inside of the hole.
6. Next, withdraw the boring bar from the hole and adjust the boring head for a cut of 0.025 to 0.030 in. (0.64 to 0.76 mm).
7. Calculate the cutting speed and adjust the machine to the proper rpm rate.
8. Start the machine and carefully lower the boring head so that the cutting edge of the bar is just above the top of the hole (Fig. 8.66).

FIGURE 8.66 Boring operations require rigid machines and setups.

9. Engage the power feed and apply cutting fluid during the boring operation.
10. After the first cut is completed, measure the hole.
11. Continue the boring operation, taking successive cuts until the hole has been bored to the proper size. [*Note:* The last cut should probably be more shallow (approximately 0.005 to 0.010 in. or 0.127 to 0.254 mm) to ensure a smooth finish. Finer feeds are also used for the finish cut.]

8.8.10 Tapping on a Drill Press

Tapping on the drill press may be done in two ways. First, tapping may be done with a *tapping attachment.* The tapping attachment holds the tap and is rotated by the drill press spindle (Fig. 8.67). Tapping attachments have a friction clutch that slips and stops the rotation of the tap if it starts to bind. They also have a reversing mechanism that reverses the rotation of the tap. This is accomplished by reversing the feed direction on the drill press when the hole has been tapped to the finish length.

FIGURE 8.67 A tapping attachment permits the tapping operation on a drill press to be performed using machine power.

The second method of tapping on the drill press is turning the tap by hand:

1. Locate and mount the workpiece on the drill press table.
2. Drill the hole to the correct *tap drill size* for the tap to be used. (*Note:* This information is provided in Appendixes L and M.)
3. Chamfer the hole slightly with a countersink. This helps start the tap straight into the hole.
4. Without moving the table or shifting the workpiece, mount a 60° center point in the drill chuck.
5. Install the tap in a tap wrench, and place the tap into the drilled hole. Next, lower the 60° point into the center hole in the tap shank (Fig. 8.68).
6. Always apply a cutting fluid or a suitable tapping fluid when tapping by hand or with a tapping attachment.
7. Rotate the tap wrench by hand in a clockwise direction into the hole, keeping a light pressure on the tap with the 60° point. This procedure ensures that the tapped hole will be straight.

8. Continue to tap the hole until it is finished. If the tap starts to bind in the hole, remove the 60° point and rotate the tap a couple of turns in the *opposite direction.* Turning the tap backward helps break the chips that are causing the tap to bind.
9. For additional information on selecting taps and tapping procedure, refer to Chap. 6.

FIGURE 8.68 Hand tapping a part on a drill press.

REVIEW QUESTIONS

8.2 TYPES OF DRILLING MACHINES

1. Why is the word *sensitive* used to name the sensitive drill press?
2. How is the table on the sensitive drill press moved?
3. How does the upright drill press differ from the sensitive type?
4. How does the radial drilling machine operate for locating the workpiece in drilling operations?
5. Compare the gang drilling machine with the multiple-spindle drilling machine.

8.3 CONSTRUCTION FEATURES OF DRILLING MACHINES

1. Briefly describe the function of the following parts of a drilling machine:
   (a) Base
   (b) Column
   (c) Table
   (d) Head
   (e) Spindle
   (f) Quill
2. Name four features that are used to specify the size of a drilling machine.

8.4 CUTTING TOOLS FOR DRILLING MACHINES

1. What three cutting tool materials may be used in the manufacture of twist drills?
2. Name the three *major* parts of a twist drill, and describe their various functions.
3. List the various systems used to classify twist drills by size.
4. How are the cutting lips accurately measured when resharpening a twist drill?
5. What is the result if too much clearance is ground on the cutting lips of a twist drill?
6. A 3/4-in hole is to be drilled in a piece of aluminum with a cutting speed of 250 ft/min. Calculate the correct rpm rate.
7. Calculate the correct rpm rate for a 14-mm drill that is to be used for drilling a piece of tool steel at 15 m/min.
8. Describe seven variables that should be considered prior to selecting a cutting speed for a drill.

8.5 OTHER TYPES OF CUTTING TOOLS

1. Briefly describe the following types of cutting tools and their functions:
   (a) Core drills
   (b) Countersinks
   (c) Counterbores
2. Explain two differences between hand and machine reamers.
3. What is the difference between shell and fluted reamers?
4. Define the purposes of reaming, boring, and spotfacing.
5. List five ways to care properly for reamers.
6. What is the purpose of a center drill?
7. What three features are unique to the boring head versus other cutting tools?

8.6 CUTTING TOOL HOLDING DEVICES

1. How are cutting tools with straight shanks generally held in a drill press?

2. How is a taper shank drill held securely if the shank is too small for the spindle? Too big?
3. How are taper shank tools removed from the spindle of the drill press?

8.7 WORK-HOLDING DEVICES

1. What happens if the work moves while a hole is being drilled?
2. List two common uses for parallels in a drill press setup.
3. How is round work generally held for drilling operations?
4. What are three purposes of drill jigs?

8.8 OPERATIONS PERFORMED ON THE DRILL PRESS

1. List five important safety rules for drilling operations.
2. How is a drill measured for its correct size?
3. Why is cutting fluid used when drilling a hole?
4. Why should you ease up on the feed pressure of a drill when it breaks through the material?
5. When is the web "thinned" on a twist drill? How much should a web be thinned?
6. What device is used to locate layout lines accurately on the work?
7. What feature on a drill press is used to gage the depth for drilling a blind hole?
8. How much metal should a *hand* reamer remove in inches? In millimeters?
9. How can chattering be eliminated during machine reaming operations?
10. Briefly describe the proper setup procedure for hand reaming in a drill press.
11. What is the major difference between countersinking and counterboring a drilled hole?
12. How much material should be left prior to the finish cut on a boring operation?
13. Why is the tap frequently turned backward during a hand-tapping operation?

9

Engine Lathes and Operations

9.1 INTRODUCTION

The *engine lathe* is one of the oldest and most important machine tools. It is the most common type of metalworking lathe used in industry and is generally classified as a nonproduction machine tool. The name engine lathe dates back to when the lathe was powered by steam engines. The engine lathe is the most versatile machine tool; with proper setups and accessories it can perform such machining operations as facing, straight and taper turning, drilling, threading, milling, grinding,

FIGURE 9.1 Typical engine lathe operation.

boring, forming, and polishing. Each operation is discussed in detail in this chapter. All operations performed on the engine lathe remove metal in essentially the same manner; that is, the cutting tool is fed against the work and shapes the workpiece, which is securely mounted and rotating (see Fig. 9.1).

9.2 TYPES OF ENGINE LATHES

Many types of metalworking lathes are used today in industry; special types of lathes are fully covered in Chap. 10. Engine lathes are classified into two types: the bench model and the floor model.

9.2.1 Bench Model

Figure 9.2 illustrates the bench model engine lathe. It is used primarily for light-duty work and training. It is a small lathe and, as shown, mounted on some type of bench or cabinet, with shelves, drawers, and racks for tools and accessories.

FIGURE 9.2 Bench model engine lathe. (*Courtesy South Bend Lathe, Inc.*)

FIGURE 9.3 Floor model engine lathe for school shop or light industrial use. (*Courtesy Republic-Lagun Machine Tools Co.*)

FIGURE 9.4 Larger floor model engine lathe used in industry. (*Courtesy Republic-Lagun Machine Tools Co.*)

9.2.2 Floor Model

Floor model engine lathes come in assorted sizes. Figure 9.3 illustrates the lighter-duty floor model used in school shops and industry. Larger floor models as shown in Fig. 9.4 require a much greater area in a shop and are used primarily for machining long shafts of large diameter. Lathes of this type require considerable horsepower to remove the amount of metal needed to complete the job.

9.3 LATHE CLASSIFICATION BY SIZE

Lathes are classified to size by the largest diameter (in inches) that may be revolved; this is commonly called the swing of a lathe. The second factor is the total length between the centers (Fig. 9.5), with the tailstock at the end of the ways.

Lathes vary in size from a 6-in. (152.4-mm)-diameter swing to well over a 100-in. (2.54-m)-diameter swing. However, the lathes most commonly used in industry have a 10- to 30-in. (254- to 762-mm)-diameter swing.

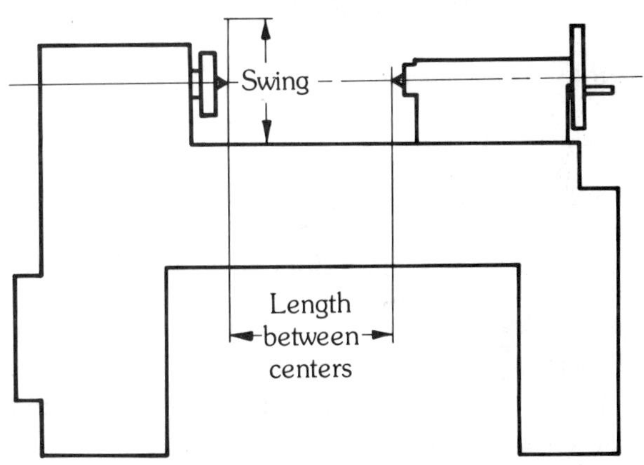

FIGURE 9.5 Lathe classification by swing and length between centers.

9.4 LATHE CONSTRUCTION FEATURES AND FUNCTIONS

Before attempting to operate an engine lathe, the machinist must become familiar with the basic lathe construction features (Fig. 9.6) and the functions of the various components. The manufacturers' operator's manual should be consulted for additional information as well as for periodic maintenance of a particular lathe. The five main parts of a lathe are the *bed, headstock, tailstock, carriage assembly,* and *quick-change gearbox.*

9.4.1 Bed

The bed is the main foundation of any lathe and supports all the other parts. It is generally made of gray or ductile cast iron or fabricated from steel by welding. On the top of the bed are precision-machined *ways.* Most lathes have two sets of ways: outer ways and inner ways. The headstock and tail-

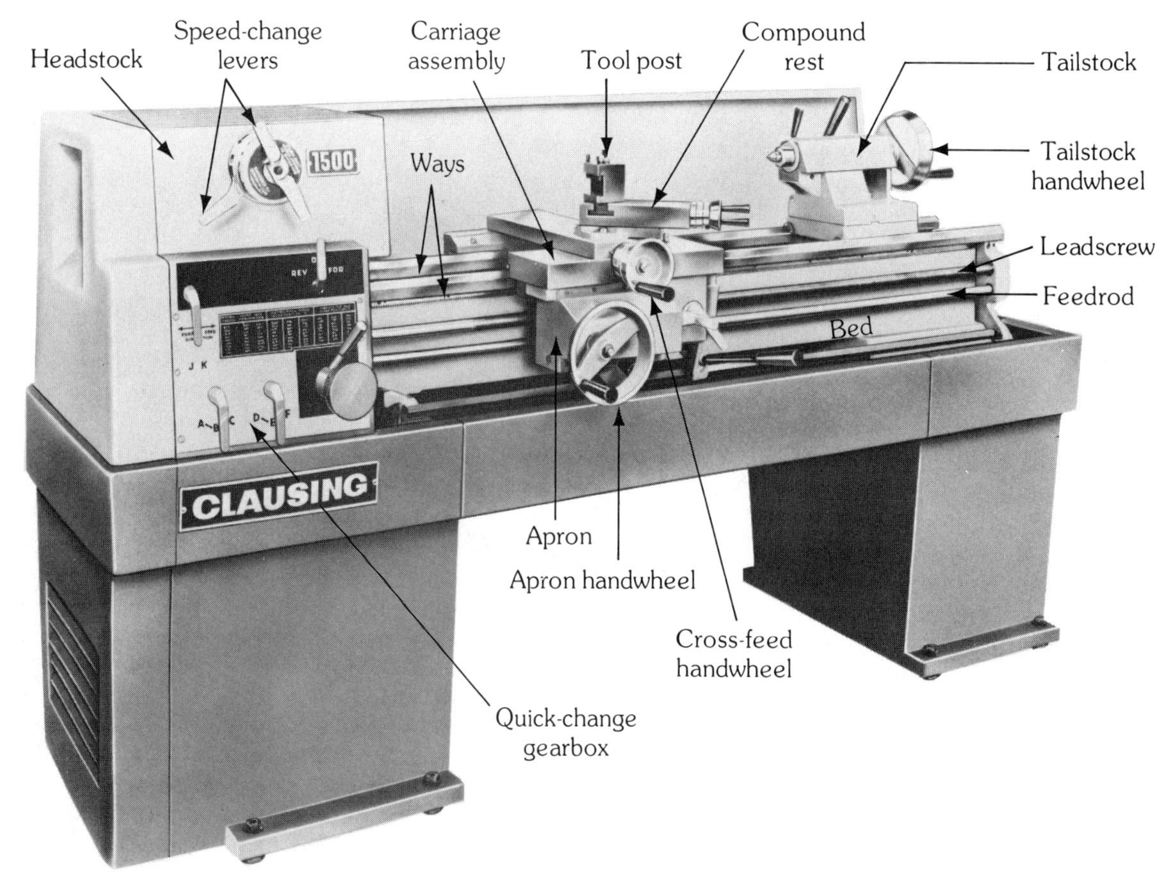

FIGURE 9.6 Engine lathe construction features. (*Courtesy Clausing Industrial, Inc.*)

stock are mounted on the inner ways, which keep them perfectly aligned with each other. The outer ways guides the longitudinal movement of the carriage assembly and aligns it with the centerline of the lathe.

The *ways* are a key factor in the accuracy of a lathe. They are made of cast iron, induction-hardened ductile iron, or hardened steel. They are precision machined, ground, and hand scraped to provide the exactness needed for the other mating parts. Tools and other equipment should not be carelessly placed on the ways because they may get damaged. The ways should be covered when machining materials such as wood, phenolic, rubber, and other abrasive materials. Oil the ways for further protection and to enable the carriage assembly to move about easily.

9.4.2 Headstock

The headstock is mounted on the left end of the lathe (Fig. 9.7). It contains the necessary driving mechanisms—either pulleys or gears—to drive the spindle. The *spindle* is the unit in the headstock that turns the workpiece and is where work-holding attachments mount. The spindle must be mechanically true, so it is mounted in special bearings such as preloaded double-row ball bearings or tapered rollers. Spindles are hollow to accommodate longer stock and have internal machined tapers to hold centers and other tooling with tapered shanks. The outside of the spindle is machined to accept various work-holding attachments like chucks and faceplates. This is called the *spindle nose.* There are three common types of spindle noses: threaded, tapered key, and camlock. Figures 9.8 to 9.10 illustrate each type of spindle nose.

Power is transmitted to the spindle from an electric motor mounted behind or inside the machine. Various spindle speeds are selected by either changing the position of the belt on different-sized pulleys for belt-driven lathes or changing levers that shift gears for the gearhead lathes (Fig. 9.11). Newer lathes sometimes have variable-speed spindle drives; simply by turning a wheel or dial, an infinite range of speeds between minimum and maximum speeds can be selected.

The *feed reverse* lever is also located on the headstock. It reverses the rotation of the feed rod and lead screw, which, in turn, changes the direction of movement of the carriage.

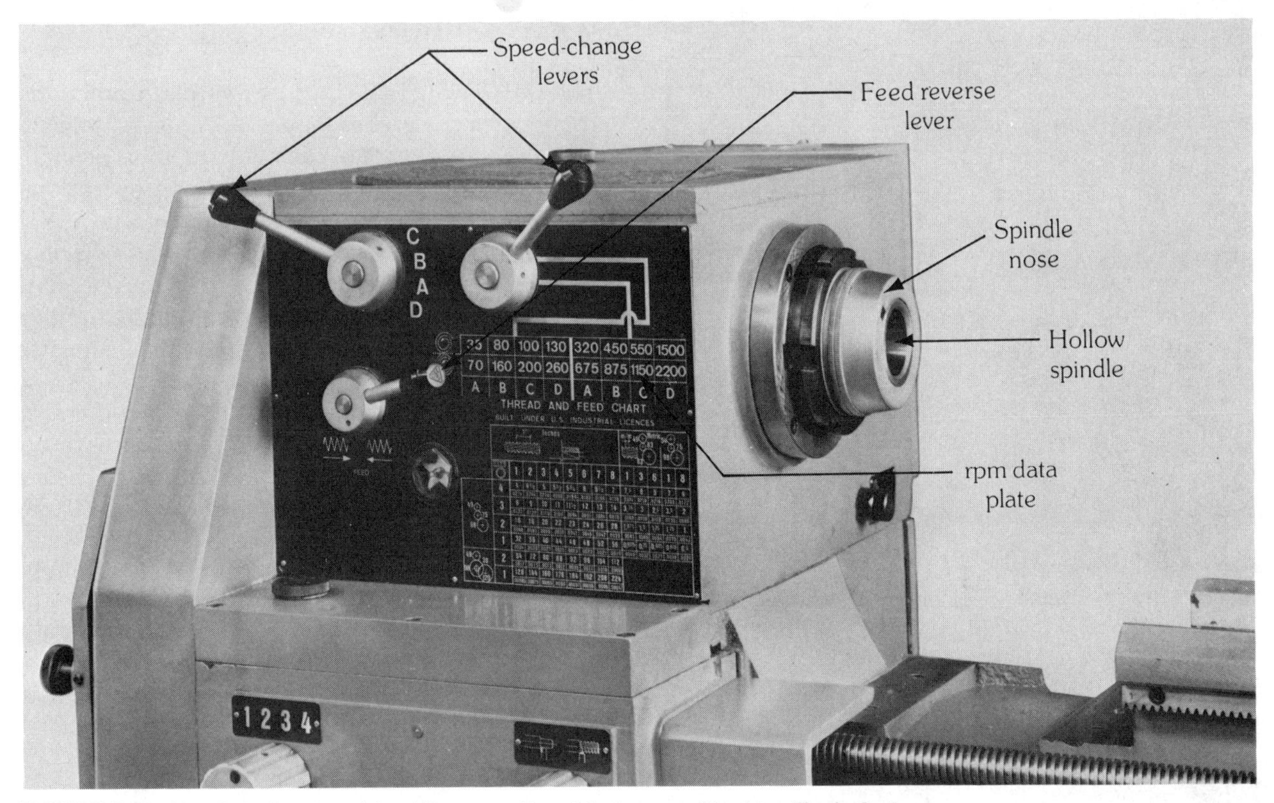

FIGURE 9.7 Headstock assembly. (*Courtesy Republic-Langun Machine Tools Co.*)

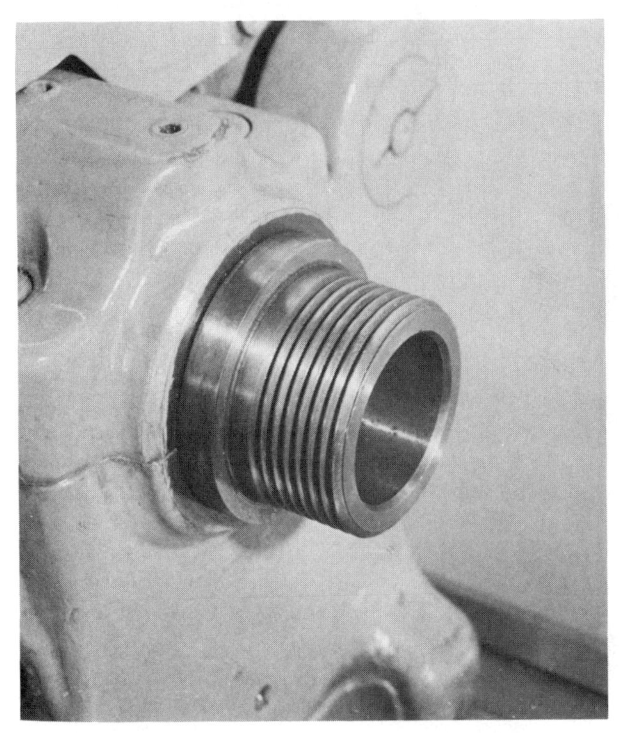

FIGURE 9.8 Threaded spindle nose.

FIGURE 9.9 Tapered key spindle nose.

FIGURE 9.10 Camlock spindle nose. (*Courtesy South Bend Lathe, Inc.*)

9.4.3 Tailstock

The tailstock (Fig. 9.12) has two main functions. It supports the end of longer workpieces, and the *tailstock spindle* holds such cutting tools as drills and reamers for internal machining operations. The spindle is graduated to control the depth of drilling operations. To accommodate various lengths of work, the tailstock can be moved along the ways and is fastened into position by the *tailstock clamp.* The tailstock spindle can be adjusted longitudinally by rotating the *handwheel* and locked into position by the *tailstock spindle lock.*

The tailstock is made in two parts, with adjusting screws holding the parts together. The adjusting screws allow the top part of the tailstock to move toward or away from the operator, for turning tapered parts or aligning the tailstock spindle true with the headstock spindle (Fig. 9.13). The

FIGURE 9.11 Geared head lathe. (*Courtesy Clausing Industrial, Inc.*)

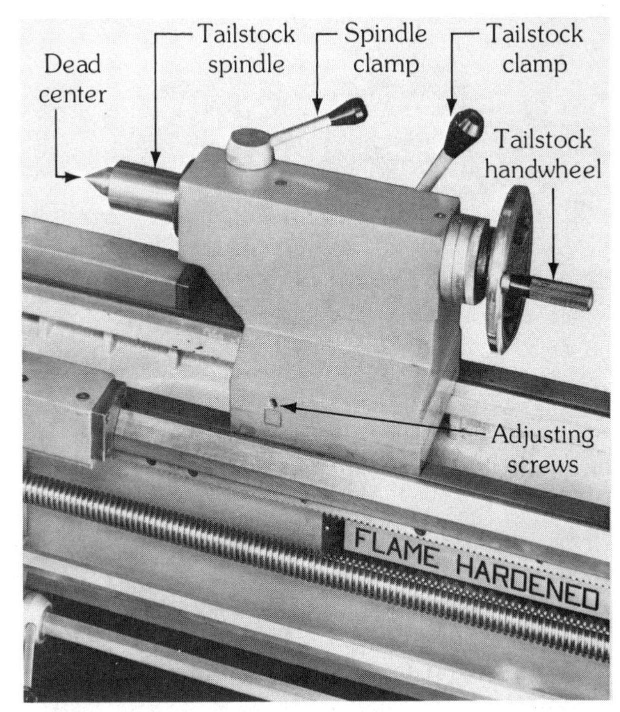

FIGURE 9.12 Tailstock. *(Courtesy Republic-Lagun Machine Tools Co.)*

FIGURE 9.13 Adjusting screw on the tailstock allows the tailstock to be moved off-center for turning tapers. *(Courtesy Le Blond-Makino Machine Tool Co.)*

tailstock parts must be realigned exactly on center when turning a cylindrical part.

9.4.4 Carriage Assembly

The carriage assembly carries the tool along the ways longitudinally. It has five major parts (Fig. 9.14):

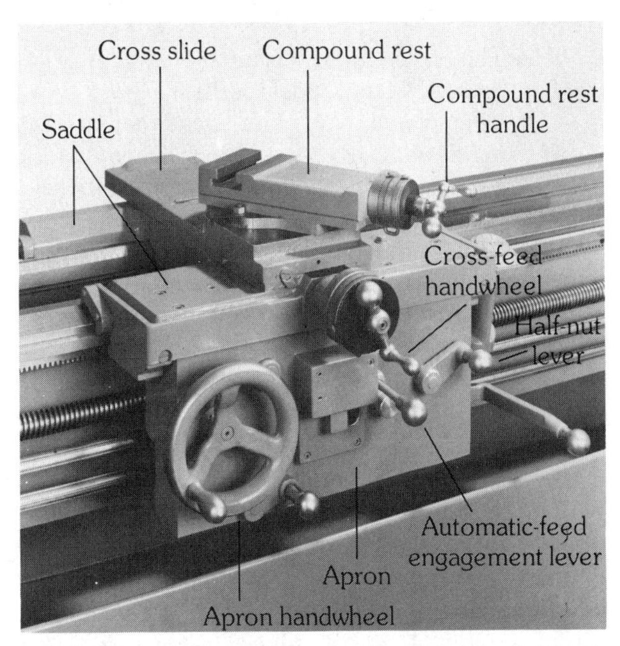

FIGURE 9.14 Carriage assembly. *(Courtesy LeBlond-Makino Machine Tool Co.)*

1. The *saddle* is an H-shaped casting fitted to the outer set of ways.
2. The *cross slide* is mounted on top of the saddle and moves the cutting tool laterally across the bed by means of a *cross-feed handwheel.* The cross-feed handwheel has a micrometer collar that allows the cutting tool to remove metal in thousandths of an inch (Fig. 9.15).

FIGURE 9.15 The cross-fed micrometer collar allows the cutting tool to be moved accurately in either thousandths of an inch or millimeters.

3. The *compound rest* is mounted on top of the cross slide and supports the *tool post.* It can be swiveled to any angle for taper turning or threading operations. The compound rest is moved *manually* by the *compound-rest feed handle and screw.* Like the cross-feed handle, it has a micrometer collar graduated in thousandths of an inch for accurate cutting tool settings.
4. The *apron* is mounted beneath the front of the saddle and houses the carriage and cross-slide control mechanisms. The *apron handwheel* is used to move the carriage assembly manually along the ways by means of rack and pinion gears. On some lathes the apron handwheel has a graduated sleeve. The importance of this sleeve is explained in our discussion of machining shoulders in Sec. 9.8.2. The apron also contains the *automatic feed lever* used for engaging power feeds of either the carriage assembly or cross slide. When using power feeds, the *feed change lever* is set to either the cross feed or carriage feed. The thread chasing dial and split-nut lever are also mounted on the apron and are used for threading operations. These two parts are discussed in detail later.
5. The *toolpost assembly* (Fig. 9.16) is mounted in the T slot of the compound rest. The toolpost clamps the toolholder in the proper position for machining operations.

9.4.5 Quick-Change Gearbox

The quick-change gearbox is mounted on the left side of the bed and below the headstock (Fig. 9.17). It houses the necessary gears and other mechanisms that transmit various feed rates from the headstock spindle to either the *lead screw* or *feed rod.* The lead screw advances the carriage during threading operations; the feed rod moves the carriage during turning, boring, and facing operations. Most feed rods and lead screws have shear pins or slip clutches to prevent damage to the gear trains. It is good practice to disengage the lead screw during straight turning operations to eliminate the possibility of rags and other objects being tangled in the revolving screw. This also helps preserve the accuracy of the lead screw.

FIGURE 9.16 Toolpost assembly.

FIGURE 9.17 Inside view of a quick-change gear box. (*Courtesy Republic-Lagun Machine Tools Co.*)

9.5 LATHE ACCESSORIES AND ATTACHMENTS

There are many accessories and attachments that are used on a lathe to increase the variety of operations that can be performed. *Work-holding devices* hold the work securely in the proper position for machining. *Work-supporting devices* are used along with work-holding devices to secure longer workpieces during machining operations.

9.5.1 Work-Holding Devices

Three-Jaw Universal Chuck. The three-jaw universal chuck, as its name implies, has three jaws that move in unison toward or away from the center of the chuck by a scroll mechanism (Fig. 9.18). This simultaneous motion causes the jaws to locate the work on center to within a few thousandths of an inch. The universal chuck is used for holding round or hexagonally shaped workpieces. On some chucks each jaw is made in two pieces and the top piece can be reversed for holding larger workpieces. The universal chuck is not considered highly accurate over a wide range of diameters.

Four-Jaw Independent Chucks. The four-jaw chuck (Fig. 9.19) has four jaws, each of which can be adjusted *independently* by a screw and half-

FIGURE 9.18 Three-jaw universal chuck. (*Courtesy Pratt Burnerd America, Division of Clausing Industrial, Inc.*)

FIGURE 9.19 Four-jaw independent chuck. (*Courtesy Pratt Burnerd America, Division of Clausing Industrial, Inc.*)

nut when turned by the chuck wrench. Round, square, or irregularly shaped workpieces can be held either on or off center. The jaws on the independent chuck can be reversed for holding larger diameter work or for gripping on the inside of hollow workpieces. The four-jaw independent chuck holds a workpiece more rigidly than the universal because it has one additional jaw. The independent chuck can be used to hold parts to a high degree of accuracy and concentricity because each jaw can be adjusted individually.

Collet Chucks. The collet chuck is one of the most accurate work-holding devices. The *draw-in collet chuck* uses spring collets that are available with round, square, rectangular, or hexagonal openings. Each spring collet has been *accurately made to hold smooth workpieces* that vary only a few thousandths of an inch from a nominal size; therefore hot-rolled steel and castings should not be used with collets.

Figure 9.20 shows how the draw-in collet attachment is mounted in the headstock of a lathe. As the handwheel is turned in a clockwise motion, the hollow draw bar turns. Located near the spindle nose is the collet, which fastens to the draw bar with threads. As the draw bar is tightened, the tapered part of the collet is forced against a tapered adapter, causing the collet to tighten on the workpiece. Collets give a high degree of repeated accuracy and concentricity.

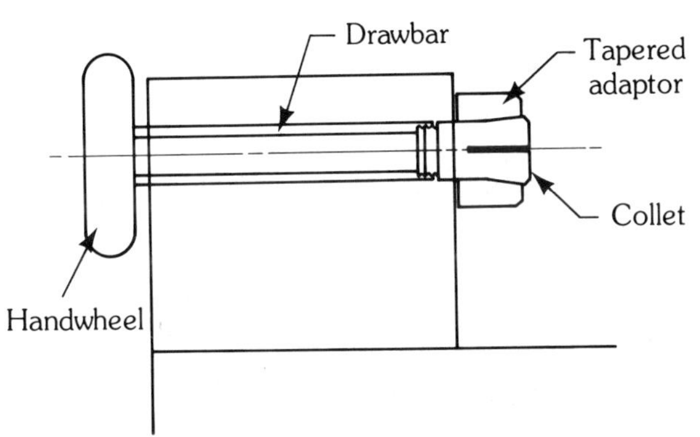

FIGURE 9.20 Phantom view of the draw-in collet assembly.

The *rubber flex* or so-called *Jacobs collet chuck* is another type of collet chuck used on lathes. Its collet will hold only round workpieces. However, each has a range approximately 1/8 in. larger than its nominal size (Fig. 9.21). The rubber flex collet chuck, unlike the spring collet, does not utilize a draw bar to tighten the collet. When the handwheel on the chuck is turned clockwise, it forces the collet against a taper, causing it to tighten around the workpiece. To further ensure that the workpiece is secure, a couple of impact rotations are necessary in the clockwise direction. This procedure is reversed to release a workpiece.

Faceplates. When a workpiece is too large, has an odd shape, or cannot be mounted in a chuck, a faceplate (Fig. 9.22) is generally used. Faceplates have slots so that bolts and clamps can be used to secure the workpiece.

FIGURE 9.21 Rubber-flex collet assembly.

FIGURE 9.22 Types of faceplates used on engine lathes. (*Courtesy Clausing Industrial, Inc.*)

9.5.2 Work-Supporting Devices

Lathe Centers. Lathe centers are commonly used to support a workpiece between the headstock and tailstock (Fig. 9.23). The center that is mounted in the headstock spindle is called a *live center* and revolves with the workpiece. The tailstock center is generally called a *dead center* and does not usually rotate. Some tailstock centers are manufactured with antifriction bearings, and the tip rotates along with the work while the body and taper that fit into the tailstock remain stationary. This type of center is generally considered a live center, as is the headstock center. Tailstock live centers permit heavier machining operations to be performed and are used where higher rpms are required. All centers have a 60° included angle at the tapered point, and therefore each workpiece must have a 60° tapered hole drilled prior to

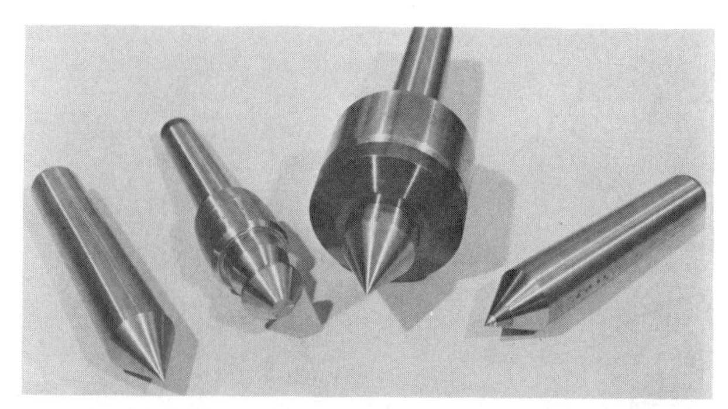

FIGURE 9.23 Lathe centers. (*Courtesy Orange Coast College*)

mounting it between centers. Centers are used in conjunction with *driving plates* and *lathe dogs,* as shown in Fig. 9.24.

Another type of tailstock center commonly used is the *bell* or *pipe* center. It also revolves with the workpiece and is used to support the end of longer, hollow cylindrical shape such as tubing or pipe.

Steady Rest. The steady rest is mounted directly to the ways on a lathe and supports the outer end of longer work. This setup permits machining operations to be performed on or near the end of the workpiece (Fig. 9.25). Steady rests can be positioned anywhere along the ways as needed. They also can be used to support the center of very long workpieces that have been mounted between centers. On very large lathes more than one steady rest may be used. Steady rests require that the surface of the material be round and smooth where the jaws are in contact.

Follow Rest. Follow rests are bolted to the carriage assembly and travel with the cutting tool. They prevent long slender workpieces from springing up and away from the cutting tool during turning and external threading operations (Fig. 9.26). Generally, the jaws of the follow rest contact the finished part of the workpiece.

Mandrel. Mandrels are hardened and ground pieces of cylindrical steel that have a slight taper from one end to the other. Workpieces with a smooth hole are placed on the small diameter of the mandrel and moved by an arbor press to the larger (marked) end until tight. Next, the workpiece and mandrel are mounted between centers on the lathe (Fig. 9.27). Work is generally supported on a mandrel when external operations must be concentric with the hole of the part. Gear blanks and pulleys are examples of such parts that are generally machined on a mandrel.

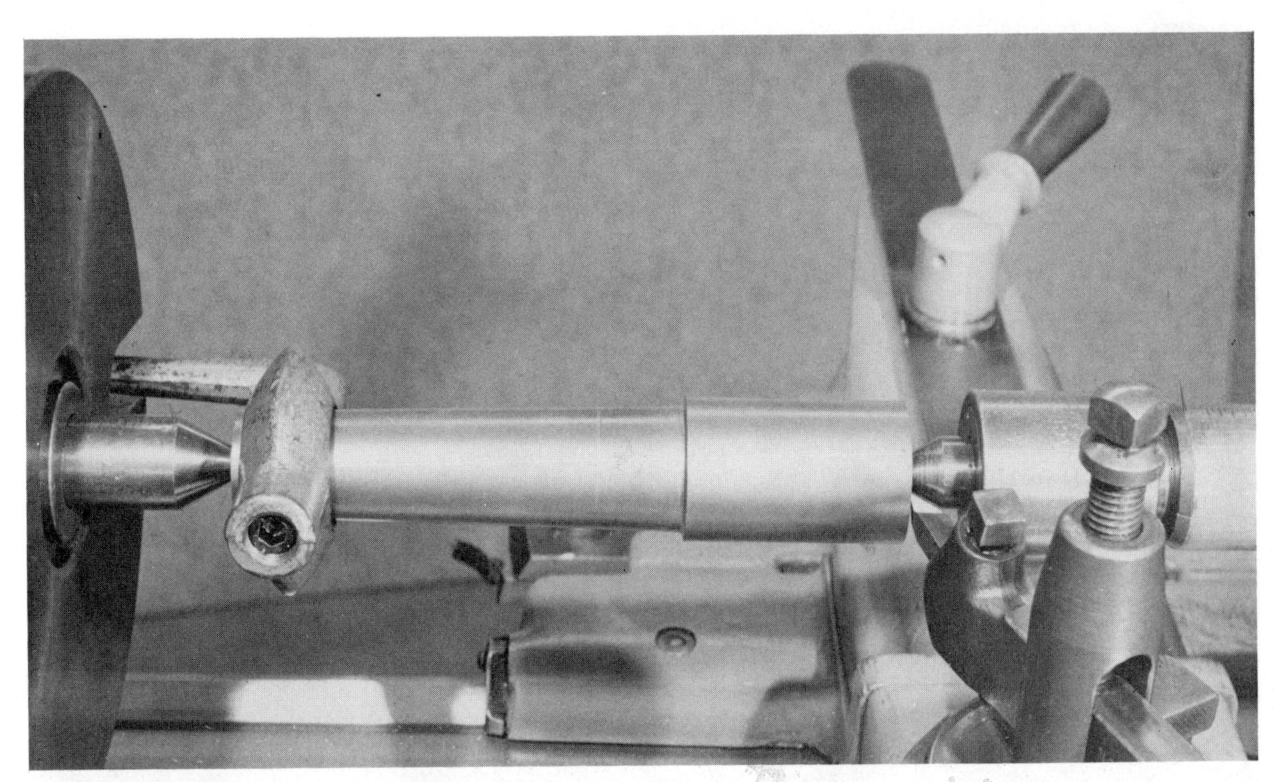

FIGURE 9.24 Part being turned between centers.

FIGURE 9.25 Steady rest supporting the end of a workpiece during a facing operation.

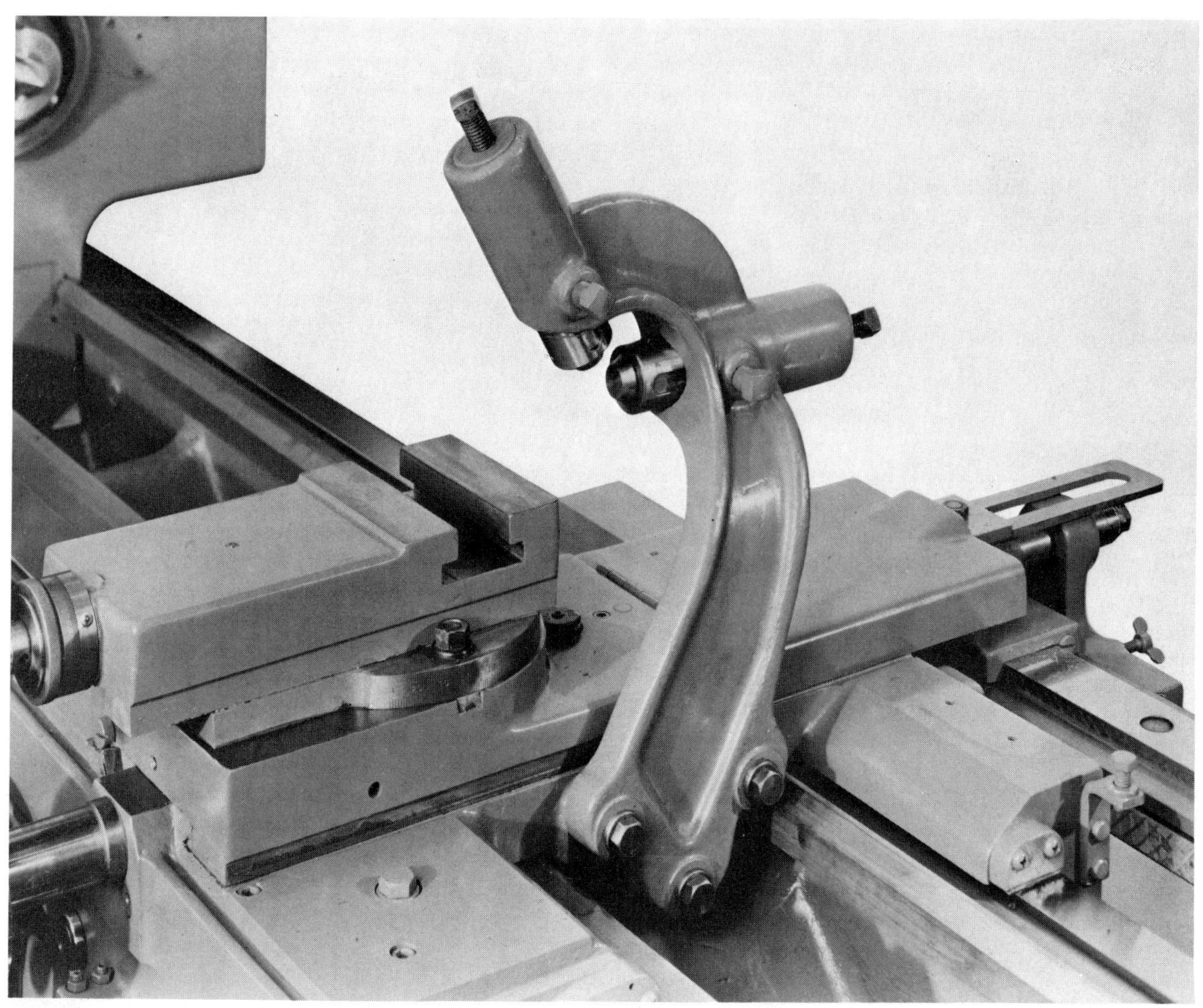

FIGURE 9.26 A follow rest is bolted to the saddle. (*Courtesy Le Blond-Makino Machine Tool Co.*)

Arbors. For special setups, or when a mandrel of the correct size is not available, the machinist often machines an arbor (Fig. 9.28). This arbor, sometimes called a spud, mounts the work for additional machining operations on the outer diameter. The shank of the arbor, diameter A, is generally held with a three- or four-jaw chuck. The smaller diameter B is machined to a size that will give a close fit with the hole in the workpiece. The length of diameter B is slightly shorter than the total length of the work, thus allowing the parts to be tightened against the shoulder with a capscrew and washer.

9.5.3 Toolposts and Toolholder

Cutting tools are held in a variety of different types of toolholders.

Rocker-Arm Toolpost. The rocker-arm toolpost is generally standard equipment and comes with most lathes (Fig. 9.29). It mounts on the compound rest and has a slot where three different kinds of toolholders are fitted: the left-hand offset, the right-hand offset, and the straight toolholder.

The *left-hand offset* toolholder (Fig. 9.30) is used when operations are required near the chuck and for cutting from the right to the *left* or from the tailstock to thc headstock.

The *right-hand offset* toolholder (Fig. 9.31) is used when machining the workpiece from the left to the *right* or from the headstock to the tailstock.

The *straight* toolholder (Fig. 9.32) has no offset and is used for machining the workpiece in either direction. This toolholder is generally used for most operations.

Quick-Change Tool-Holding Assemblies. The quick-change toolholder is used when multiple machining operations on many workpieces are needed. Quick-change toolholders are the most popular type used in industry and are more rigid than the rocker-arm toolposts. Preset cutting tools

FIGURE 9.27 Mandrel used to hold a workpiece during a facing operation.

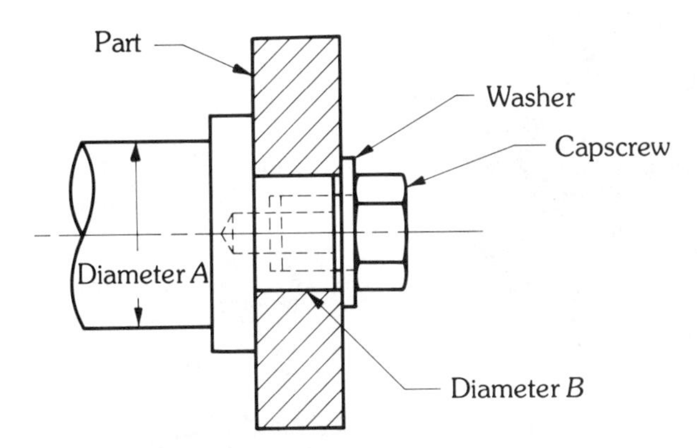

FIGURE 9.28 An arbor is often made for special jobs.

(tool bits) are mounted in blocks that can be quickly changed and accurately located (Fig. 9.33). The clamping lever located at the top secures the cutting tool rigidly in place.

Turret Toolholders. Another toolholder commonly used for multiple operations is the turret

FIGURE 9.29 Rocker-arm toolpost.

FIGURE 9.30 Left-handed offset toolholder.

FIGURE 9.31 Right-hand offset toolholder.

FIGURE 9.32 Straight toolholder.

FIGURE 9.33 Quick-change tool-holding assembly with various types of cutting tools. (*Courtesy of Aloris Tool Co.*)

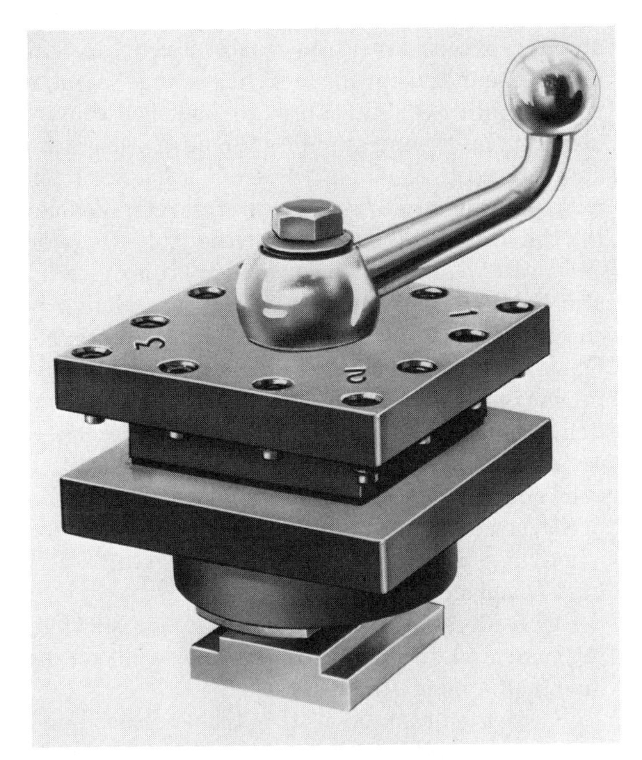

FIGURE 9.34 A turret toolholder holds four types of cutting tools. (Courtesy Clausing Industrial, Inc.)

FIGURE 9.35 Open-side toolholder.

type. It has slots for mounting four different types of toolbits. Each toolbit is rotated to the proper position and locked into place by the locking handle on top of the holder (Fig. 9.34).

Open-Side Toolholder. The open-side toolholder is used for single cutting tools and is more rigid than the toolpost type. It is generally used in conjunction with carbide toolbits for heavy cuts (Fig. 9.35).

9.5.4 Taper Attachment

A taper attachment (Fig. 9.36) is a convenient way to machine tapers. It can be used to cut both internal and external tapers and is mounted directly behind the carriage assembly. The taper attachment allows for long tapers but is limited to tapers up to 20° included angle or approximately 4 in. taper per foot. Use of the taper attachment is discussed in Sec. 9.8.3.

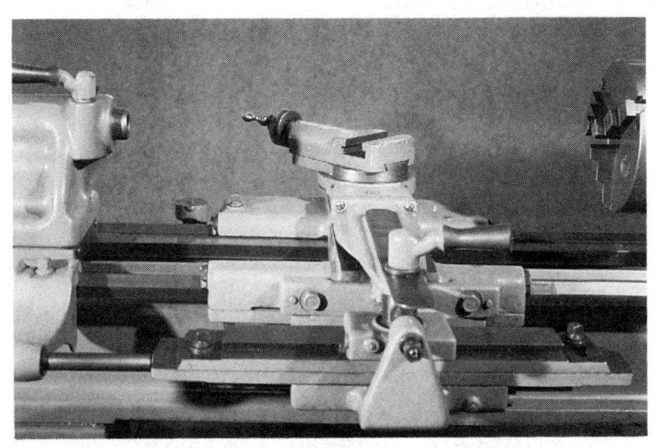

FIGURE 9.36 Telescopic taper attachment.

9.6 CUTTING TOOL MATERIALS

Lathe cutting tools are single-point cutting tools, generally called *toolbits.* They are usually manufactured from one of four materials: high-speed steel, cast alloys, cemented carbides, or aluminum oxide, each having its unique advantages. All have the same basic characteristic: the cutting tool material is harder than the material to be cut. So that the most effective cutting material may be selected for a particular machining operation, knowledge of the capabilities of cutting tools is a necessity.

9.6.1 High-Speed Steel

High-speed steel (HSS) is probably the most common cutting tool material used on an engine lathe. Toolbits made of HSS are inexpensive (compared to other materials), capable of withstanding heavy shocks or interrupted cuts, and retain their sharp cutting edge at temperatures as high as 1100°F. High-speed-steel toolbits are easier to grind than those made of other types of material and are available in a variety of sizes and shapes.

High-speed-steel toolbits are made from many materials; molybdenum, vanadium, tungsten, chromium, and cobalt are the principal elements used to make high-speed cutting tools.

General-purpose HSS toolbits are classified as 18–4–1, that is, 18 percent tungsten, 4 percent chromium, and about 1 percent vanadium, with the remainder being iron. The carbon content of HSS is generally 0.75 to 0.9 percent. When higher cutting speeds are needed, the cobalt or high vanadium types are used.

Cobalt high-speed steels contain cobalt in amounts varying from 5 to 12 percent. The presence of cobalt increases the toolbit's "red hardness," the ability to cut at elevated temperatures. Abrasion resistance is improved somewhat with cobalt, but there is a decrease in the ability of the material to withstand shocks.

High *vanadium* high-speed steels contain up to 5 percent vanadium and somewhat more carbon. The increase of vanadium in high-speed steels provides improved resistance to abrasion.

9.6.2 Cast Alloys

Cast alloy toolbit materials are composed of varying amounts of tungsten, cobalt, and chromium. The combination of these materials allows cast alloy cutting tools to retain their hardness up to 1300°F. Above 1500°F the tools lose their hardness, but when cooled they reharden without heat treatment. Cast alloy toolbits are nonmachinable. They are cast to shape and finished by grinding and polishing operations.

As a rule, cast alloy toolbits are more brittle than tools made from high-speed steels, but they will withstand 20 to 50 percent higher cutting speeds and more severe abrasion. This material is more difficult to grind than high-speed steels, and this material should *never* be cooled in water during the grinding process. Cast alloys are recognized by such trade names as Stellite, Armide, and Tantung.

9.6.3 Cemented Carbides

Cemented carbides are composed primarily of carbon mixed with tungsten, tantalum, or titanium powders and bonded by cobalt in a sintering process (powdered metallurgy). For example, a typical tungsten carbide toolbit is composed of 94 percent tungsten and 6 percent pure carbon by weight. Varying amounts of cobalt (the binder) are used to regulate the toughness of the cutting tool. Higher amounts of cobalt result in greater toughness, with a decrease in the hardness of the cutting tool material. Titanium and tantalum are added as alloying elements in varying amounts to change the cutting characteristics of cemented carbides. These factors make it necessary to be very careful when considering the use of carbides as a cutting tool. Many factors must be analyzed prior to using this material, and the manufacturers' specifications should be consulted prior to selection (see also Chap. 15). However, the disadvantage of having to select the proper type of carbide is far outweighed by the advantages carbides give in machining. Cemented carbides have excellent red hardness capabilities, can remove large amounts of materials in a short period of time, have high hardness, and are capable of cutting speeds three to four times greater than high-speed-steel cutting tools.

Considerable care is required when sharpening cemented carbides. Special wheels—silicon carbide or diamond—must be used.

Cemented carbide cutting tools are used on a lathe in two ways: as tips brazed on special shanks and as throwaway inserts.

Brazed carbides are available in a variety of shapes and sizes (Fig. 9.37). The carbide tip is brazed to a steel shank, and the shank is held in the toolholder of the lathe.

Throwaway inserts are generally used more often in industry than brazed tips. They are held in special holders by some type of clamping device (Fig. 9.38). When a cutting edge on a carbide insert becomes dull, it is simply indexed to the next side, clamped, and ready for continued use. This means a square-shaped insert that is held in a negative toolholder can be indexed four times, turned over, and indexed four more times, for a total of *eight* usable cutting edges. The carbide insert is then thrown away. This feature saves time in pro-

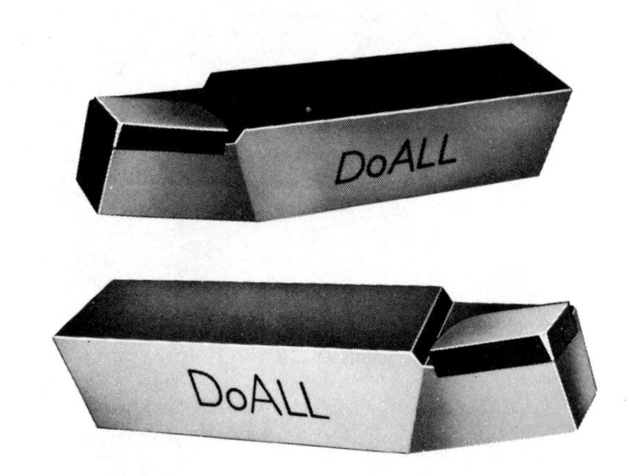

FIGURE 9.37 Brazed-type carbide cutting tools. (*Courtesy DoALL Co.*)

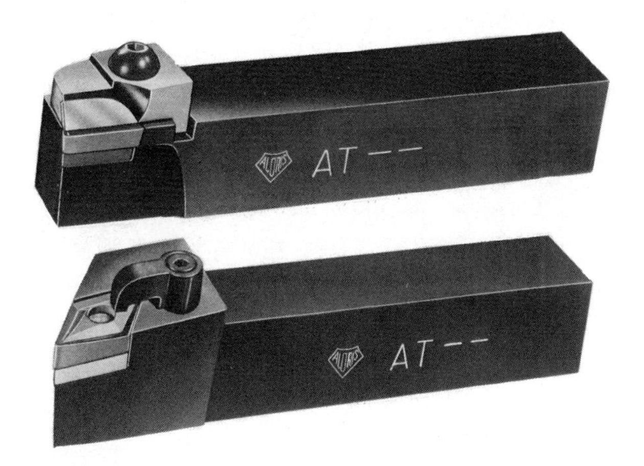

FIGURE 9.38 The throw-away insert carbide is most often used in industry. (*Courtesy Aloris Tool Co.*)

duction operations by eliminating the necessity of resharpening a dull tool.

9.6.4 Oxides

One of the latest developments in cutting tool materials is the use of aluminum oxide. Toolbits made with this material are generally referred to as ceramic cutting tools (Fig. 9.39). The manufacturing process of ceramic tools is similar to that of carbide cutting tools. Aluminum oxide is used as the principal material, with nickel as the binder. Pressures exceeding 4000 lb and temperatures as high as 3000 °F are used in the sintering process. Like carbides, ceramics are generally used as throwaway inserts.

Ceramic cutting tools have very low heat conductivity and high compressive strength but are quite brittle. Extreme care must be taken while using ceramics, and setups must be rigid to avoid breaking the cutting tool. Ceramic tools can outperform carbides in rigid and powerful machines. They are not generally used where interrupted cuts are required. They are very resistant to abrasive wear and can withstand cutting speeds two to three times faster than carbide tools.

FIGURE 9.39 Ceramic throw-away inserts.

9.7 CUTTING TOOL NOMENCLATURE AND GEOMETRY

A machine tool can perform efficiently only when correct cutting tools are used. The tool must have a keen, well-supported cutting edge and be ground for the particular metal being machined. When operating a lathe, it is important to understand the basic terms used to describe a toolbit and the purpose of each angle ground onto the cutting tool.

The lathe toolbit consists of a face, cutting edge, flank, nose, and base (Fig. 9.40). The *face* is the top of a toolbit and is the surface where the chip passes as it is cut away from the workpiece. The *cutting edge* is that part of the toolbit that does the actual cutting of the workpiece.

The *flank* is the surface directly below the cutting edge. The *nose* is the tip of the toolbit and is formed by the side and end edges. The *base* is the bottom surface of the tool.

9.7.1 Cutting Tool Angles

Relief or clearance angles are ground on the end and side of a lathe toolbit to keep it from rubbing on the workpiece. *Side relief* (Fig. 9.41) is the angle ground directly below the cutting edge on the flank of a tool. *End relief* (Fig. 9.42) is the angle ground from the nose of the tool. Relief angles are necessary to permit only the cutting edge to touch the workpiece.

Rake angles are ground on a tool to provide a smooth flow of the chip over the toolbit and away from the workpiece. A lathe tool generally uses two rake angles: side and back.

Side rake (Fig. 9.43) is ground on the tool face away from the cutting edge. Side rake influences the angle at which the chip leaves a workpiece. Tougher materials, such as tool steels, require smaller angles of 5 to 8 ° to prolong tool life. Softer materials such as aluminum need 20 to 30 ° of side rake for faster removal of metal. Most *general-purpose* lathe toolbits use a side rake of 14 °.

Back rake (Fig. 9.44) angles influence the angle at which the chip leaves the nose of the *toolbit.* Back rake angles are established in two ways: by the position of toolholder, or by grinding it onto the

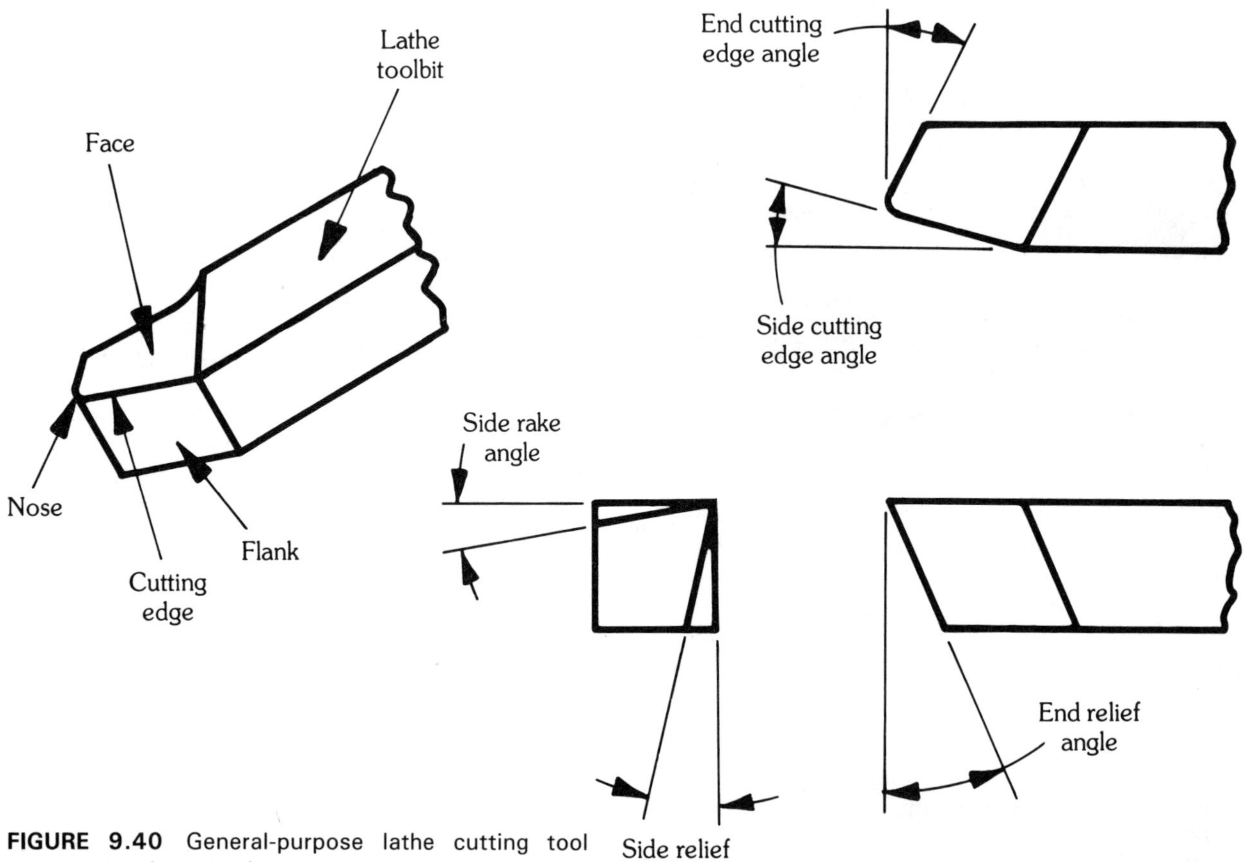

FIGURE 9.40 General-purpose lathe cutting tool geometry and nomenclature.

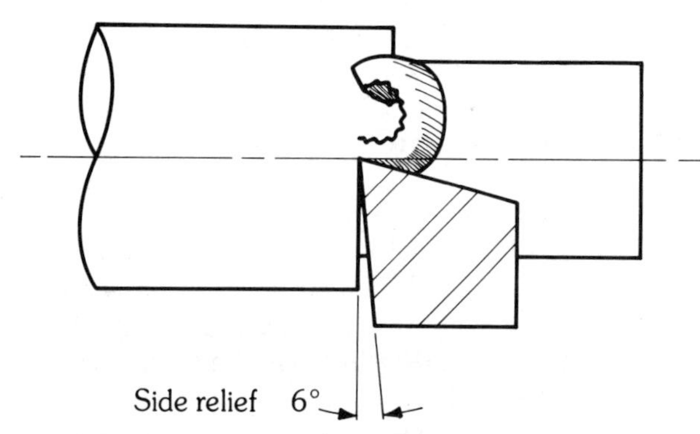

FIGURE 9.41 Side relief angles are ground directly below the cutting edge.

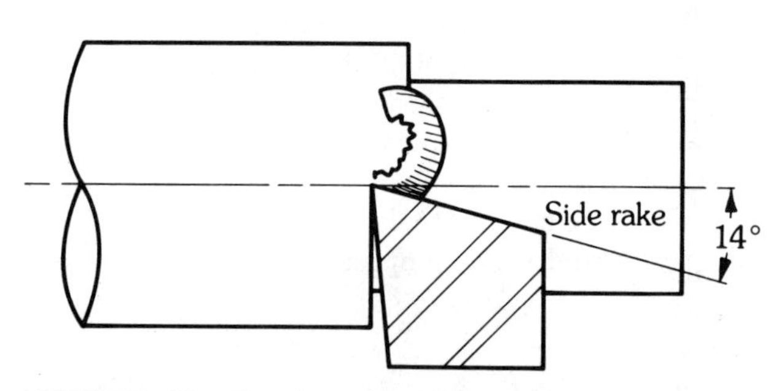

FIGURE 9.43 The side-rake angle is ground on the tool face away from the cutting edge.

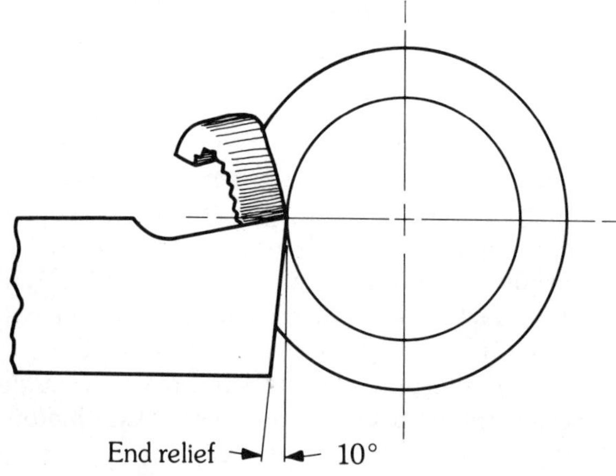

FIGURE 9.42 End relief angles are ground directly below the nose of the cutting tool.

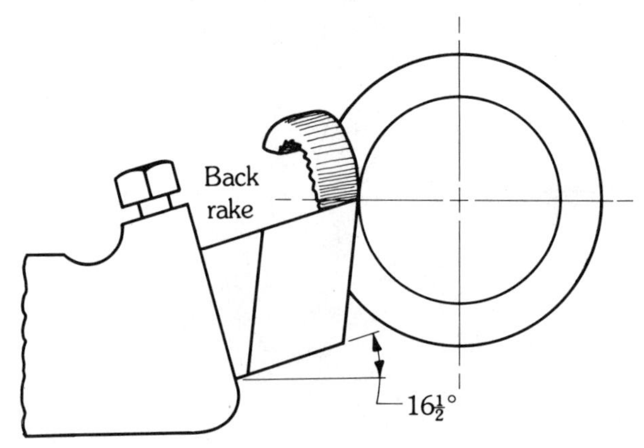

FIGURE 9.44 This toolholder positions the cutting tool with the back rake.

face of the tool. Some toolholders position the cutting tool with a back rake angle of 16 1/2°. If the toolholder does not have a back rake angle incorporated into it, the toolbit should be ground with 8 to 10° of back rake, as shown in Fig. 9.45.

End- and side-cutting edge angles are ground on a toolbit so that the tool can be mounted in the correct position for various machining operations (Fig. 9.46). The end-cutting edge angle is usually 20 to 30° and allows the cutting tool to machine close to the chuck during turning operations. Side-cutting edge angles are approximately 15° and allow the flank of the toolbit to approach the workpiece first, thus reducing the initial shock of the cut from the point. This angle also spreads the material over a greater distance on the cutting edge, thereby thinning out the chip (Fig. 9.47).

The *nose radius* is the rounded tip on the point of a toolbit (Fig. 9.48). The nose radius has two functions: to prevent the sharp, fragile tip from breaking during use, and to provide a smoother finish on the workpiece during machining operations. The amount of radius varies, depending on the type and depth of cut. For example, a tool with a large nose radius is generally a stronger tool and gives a smoother finish. Smaller-diameter workpieces require a smaller nose radius to minimize excessive tool pressure and reduce the possibility of chatter. Generally, a smaller nose radius is also used when deep cuts at heavier feeds are required. A nose radius of approximately 1/32 in. (0.77 mm) works well with most operations.

9.7.2 Types of Lathe Toolbits

Many types of toolbits are used on the engine lathe. Most toolbits are designed to cut in one direction only. They are usually classified into two categories: left-cut toolbits and right-cut toolbits (Fig.

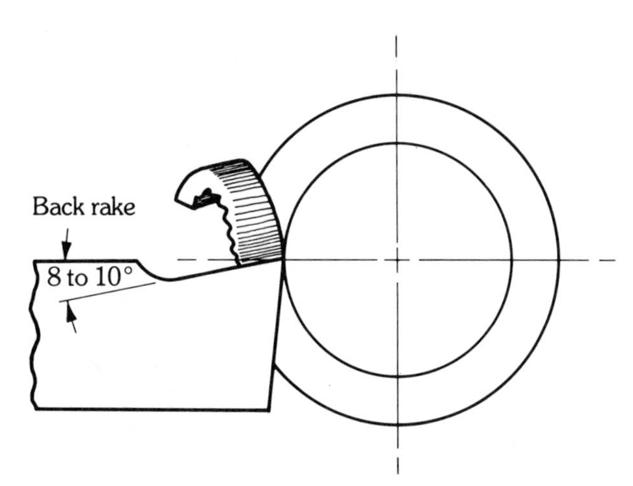

FIGURE 9.45 This tool requires the back rake to be ground onto the face of the toolbit.

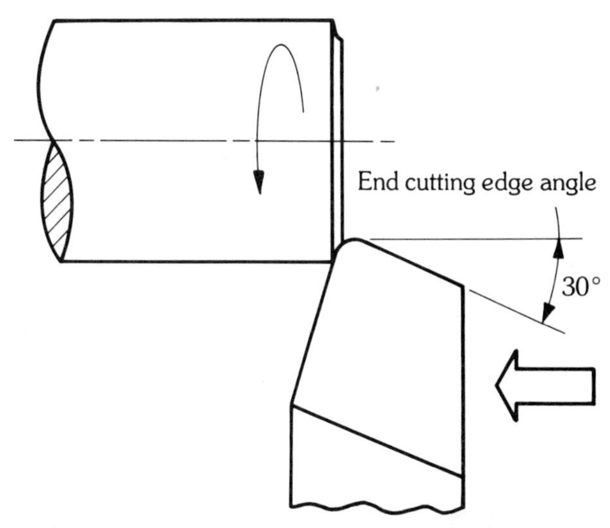

FIGURE 9.46 The end cutting edge angle allows the toolbit to machine close to the chuck.

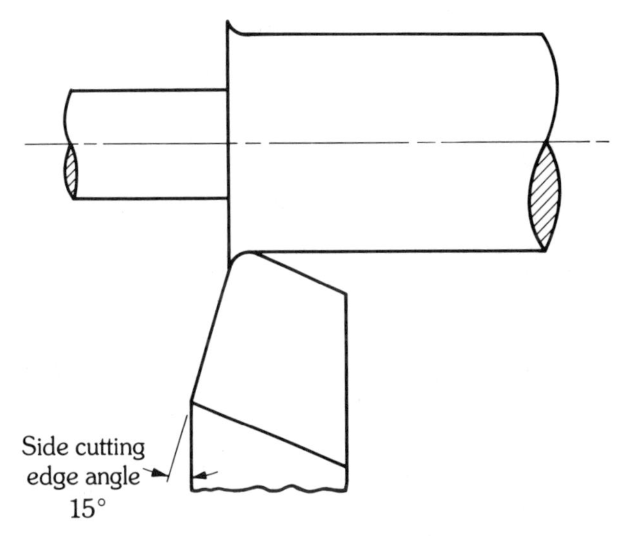

FIGURE 9.47 The side cutting edge angle allows the flank of the toolbit to approach the work first.

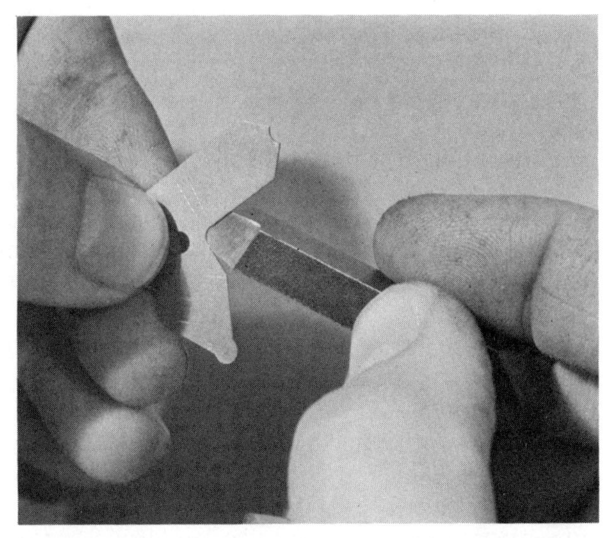

FIGURE 9.48 The nose radius on a cutting tool prevents the tip from breaking and produces a smoother finish.

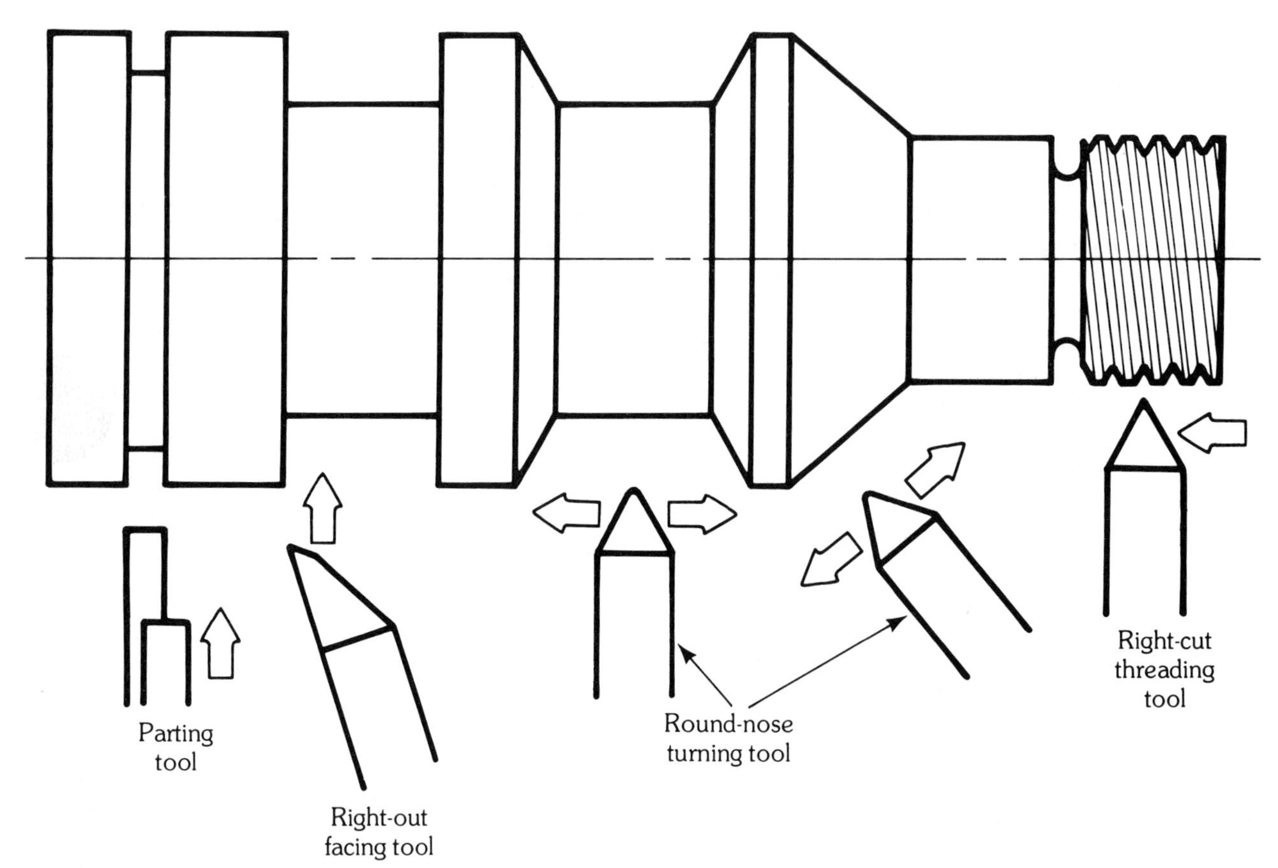

FIGURE 9.49 Various lathe cutting tools and lathe operations.

9.49). *Left-cut* toolbits cut *from* the *left* to the right or toward the tailstock of a lathe. *Right-cut* toolbits cut *from* the *right* to the left or toward the headstock.

Finishing Tools. Finishing tools have larger nose radii to produce finer finishes on surfaces of workpieces. A light cut and a fine feed are used with these tools.

Roughing Tools. Roughing tools have a smaller rounded nose, which permits deep cuts at heavier feeds. Roughing tools are used to machine the majority of the metal from a workpiece, leaving just enough material for a finish cut.

Facing Tools. Facing tools are used to square the end of a piece of stock with a smooth finish. Right-cut facing tools are ground to permit facing operations close to the tailstock center. Left-cut facing tools are used for facing the inside of a shoulder.

The *cutoff or parting tool* cuts only on the end. It is used for grooving or cutting off stock from the workpieces, which are held in the lathe.

The *round-nose tool* is a tool that cuts in both directions. It is sometimes used to form a radius at the corner of a shoulder.

9.7.3 Grinding a General-Purpose Lathe Toolbit

High-speed-steel toolbits are best ground on an ordinary pedestal, or bench-mounted, grinding machine equipped with aluminum oxide wheels. Generally, the grinder should have a coarse grit wheel for "roughing out" the toolbit and a finer grit wheel for finish grinding. The face of each wheel must be dressed properly.

The general-purpose toolbit is an example of proper toolbit grinding procedures. It is a right-cut tool generally used for facing, straight turning, and taper turning operations. It can be used for roughing as well as finishing operations by changing the nose radius and oilstoning the tool (Fig. 9.50):

1. Hold the tool firmly while resting the hands on the grinder toolrest.
2. Do not apply excessive pressure against the grinding wheel. Too much pressure could overheat and burn the tool and fingers, causing the toolbit to be ripped from your grip and possibly injuring you. Cool the tool in water often while grinding high-speed-steel tools to prevent overheating and burning.

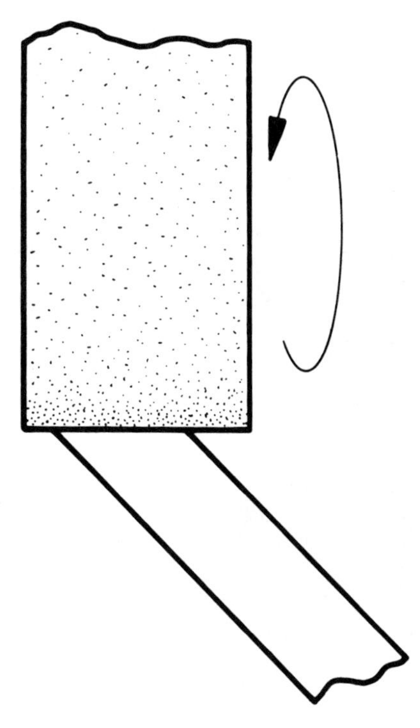

FIGURE 9.50 Grinding the end relief and end-cutting edge angles.

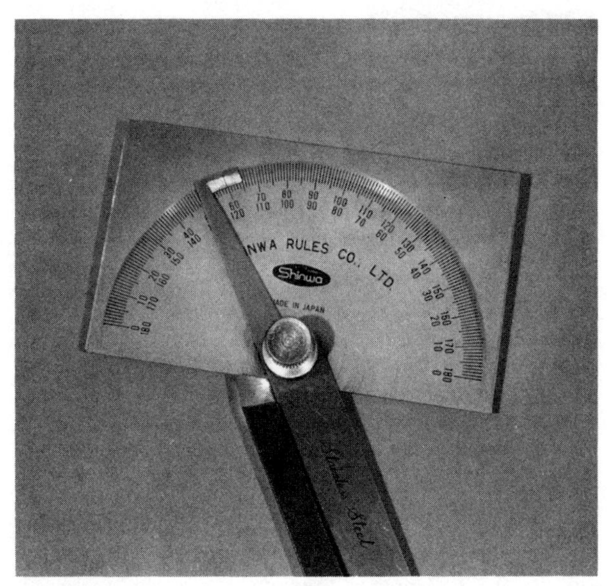

FIGURE 9.51 Measuring the end-cutting angle with a protractor.

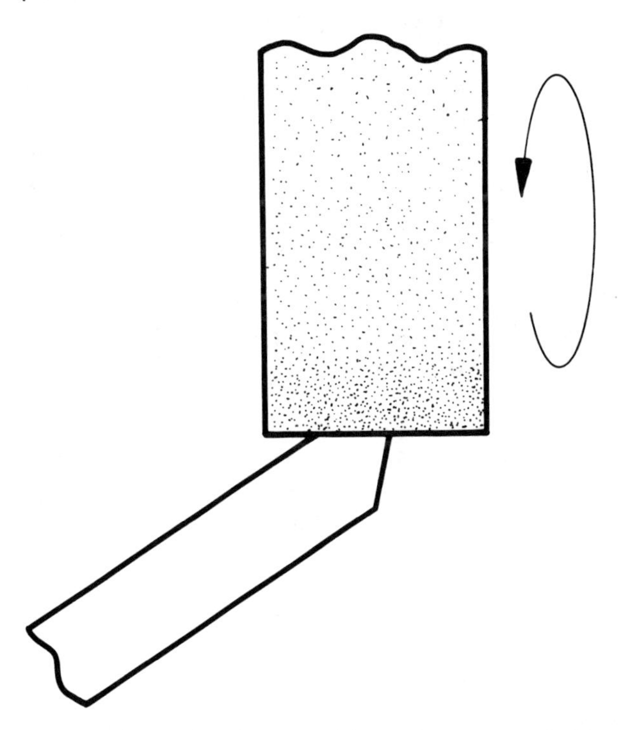

FIGURE 9.52 Grinding the side-cutting edge and side relief angles.

3. While grinding, move the toolbit back and forth across the face of the wheel without changing its position.
4. The first angles ground are the end relief and end-cutting edge angles. Hold the toolbit against the rotating wheel and tilt the bottom inward 26° (only 10° is required with straight toolholders described in Sec. 9.5.3) for the end relief angle and to the right 30° for the end-cutting edge angle. [*Note:* Beginners should use a suitable gage or protractor to measure each angle during grinding (Fig. 9.51). Laying out the angles of the toolbit on a piece of cold-rolled steel and practicing grinding may help the beginner.]
5. Grind the side relief and side-cutting edge angles. Tilt the bottom of the side inward approximately 6° for the relief angle and to the right 15° for the side-cutting edge angle (Fig. 9.52). When ground, these two angles form the cutting edge. The cutting edge should be the same length as the size of the toolbit; that is, a 3/8-in. square toolbit has a 3/8-in. cutting edge.
6. The last angle ground is the side rake angle. To grind the side rake, hold the toolbit shank approximately 45° downward to the axis of the wheel and the bottom inward 14° (Fig. 9.53). When grinding the side rake, be sure that the top of the cutting edge is not ground below the top of the toolbit. Doing so reduces the efficiency of the tool and wastes expensive material.
7. Grind a small nose radius of a 1/64- to 1/32-in. point on the toolbit. The nose radius should extend from the tip of the cutting edge to the base of the toolbit. When grinding the nose radius, only a slight amount of pressure is needed—too much causes a larger radius than desired. (*Note:*

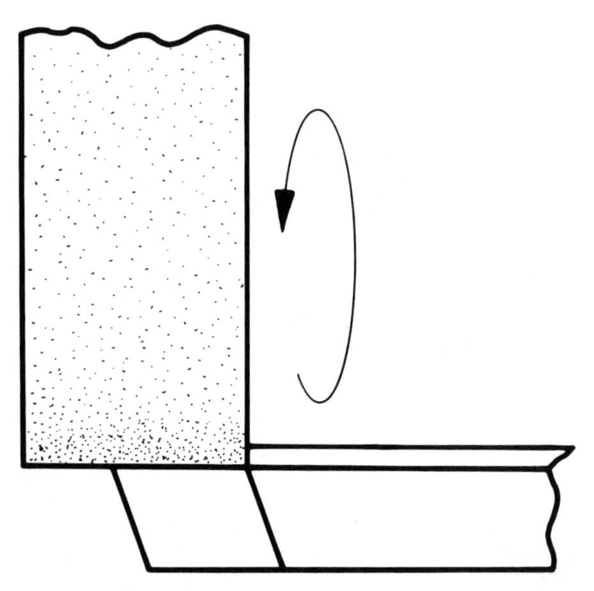

FIGURE 9.53 Grinding the side-rake angle.

Some machinists prefer to *hone* the radius with an oilstone.)

8. Hone the cutting edge slightly with a whetstone or oilstone to prolong the life of the toolbit and improve surface finish. Be sure and hold the toolbit firmly against the stone to avoid changing the angles or inadvertently rounding the cutting edge (Fig. 9.54).

FIGURE 9.54 When honing the toolbit, hold it firmly against the oilstone.

9.7.4 Cutting Speed, Depth of Cut, and Feed

Cutting speeds, the depth of cut, and the feed rate of the toolbit are important factors in the machining time of a part. Each of these important factors is discussed in this section.

Cutting Speed and Lathe RPM. *Cutting speed* is the distance (in feet or meters) that the circumference of the work moves past the cutting tool in 1 minute. Cutting speed is expressed in *surface feet per minute or meters per minute.* For example, if it were possible to measure the length of a continuous chip removed from the workpiece in 1 minute, the result would be a measurement of the cutting speed of that particular material.

Cutting speeds are established by two factors: the type of material to be machined and the cutting tool material to be used in the machining operation. Table 9.1 shows the optimum cutting

TABLE 9.1

Lathe cutting speeds for high-speed-steel cutting tools

	Surface Feet per Minute		*Meters per Minute*	
Type of Material	*Roughing Cut*	*Finishing Cut*	*Roughing Cut*	*Finishing Cut*
Aluminum	200–400	300–500	60–120	90–150
Brass—free-turning	100–200	200–300	30–60	60–90
Bronze (hard)	60–90	100–125	18–27	30–37
Cast Iron				
Hard	60–70	80–100	18–21	24–30
Soft	80–90	90–100	24–27	27–30
Steel				
Mild or free-machining	90–150	150–200	27–45	45–60
Medium-carbon	60–80	80–120	18–24	24–36
High-carbon or alloy	50–60	60–90	15–18	18–27

speeds of the more common metals using HSS cutting tools. These cutting speeds are based on the best results achieved through careful research and testing in industry.

Notice that each category has a range for both roughing and finishing cuts. A range of cutting speeds is used because each lathe machining operation is unique. In other words, there are seven other factors to consider:

1. Depth of cut
2. Size and shape of the work
3. Design of the toolbit
4. Rigidity of the work setup
5. Rigidity of the lathe
6. Use of cutting fluids
7. Power available

The proper *lathe rpm* rate is achieved by the formulas given below. When working with inch-based systems, use

$$\text{lathe rpm} = \frac{12 \times \text{cutting speed (ft./min.)}}{\pi \times D \text{ (in.)}}$$

Explanation: Multiplying the material cutting speed by 12 changes feet into inches. D equals the diameter of the work in inches and is multiplied by π (3.1416) to obtain the circumference. A shorter version of this formula is generally used because most lathes cannot be set on an exact calculated rpm. The shorter version is found by using 3 in place of π (3.1416), as shown.

Example: A 3-in.-diameter piece of aluminum is to be machined on a lathe at a speed of 300 ft/min (mean speed for roughing cut). What should the rpm rate be?

$$\text{rpm} = \frac{4 \times \text{CS}}{D} = \frac{4 \times 300}{3} = \frac{1200}{3} = 400$$

The correct rpm rate would be 400.

When working with the metric system, the following formula is used:

$$\text{rpm} = \frac{\text{CS (m/min)}}{\pi \times D \text{ (m)}}$$

Explanation: There is no shorter version for the metric system. Also, because the diameter of the work is usually given in millimeters, it is necessary to change the millimeters into meters. This must be done so that the units (meters) are the same above and below the line. To do this, divide the work diameter by 1000.

Example: Calculate the proper lathe rpm for a piece of high-carbon steel 25 mm in diameter. Use a speed of 16 m/min.

$$\text{rpm} = \frac{\text{CS}}{\pi D} = \frac{16}{3.1416 \times 0.025}$$
$$= \frac{16}{0.07854} = 204$$

The correct rpm rate would be 204.

Depth of Cut. The depth of cut can be defined as the amount of metal being machined. The depth is measured at a right angle to the surface of the work and is the difference between the original surface and the surface left by the toolbit (machined surface). There are two kinds of cuts used on lathe operations: the *roughing cut* and the *finishing cut.* The roughing cut removes the majority of material from the workpiece, leaving enough metal for the final or finishing cut. Sometimes it is necessary to use several roughing cuts prior to the finishing cut. Figure 9.55 shows a roughing cut of 1/16 in. (1.588 mm). Here, the depth of cut is 1/16 in.; however, note that a 1/16-in. cut would reduce the diameter twice that amount, or in this case, 1/8 in. (3.175 mm).

Finishing cuts are used to make the work surface smooth and accurate. Fine feeds, sharp tools, and small depth of cuts (0.010 to 0.025 in.) are used for finishing cuts.

Feed. Feed is the distance the tool advances along the work surface for each revolution of the spindle. It is usually expressed in thousandths of an inch. For example, a feed of 0.010 in. requires

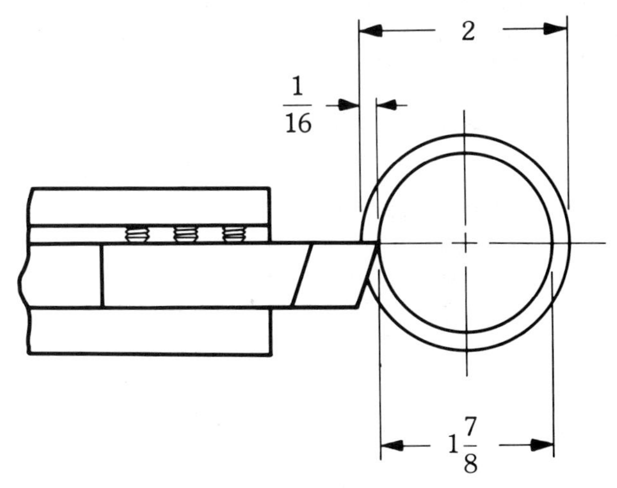

FIGURE 9.55 A 1/16-in depth cut will reduce the diameter by 1/8 in.

TABLE 9.2
Lathe feed rates for high-speed-steel cutting tools

Type of Material	Inches		Millimeters	
	Roughing Cut	Finishing Cut	Roughing Cut	Finishing Cut
Aluminum	0.015–0.030	0.002–0.005	0.40–0.75	0.051–0.13
Brass—free-turning	0.010–0.020	0.002–0.005	0.25–0.50	0.051–0.13
Bronze, hard	0.010–0.020	0.002–0.005	0.25–0.50	0.051–0.13
Cast Iron				
Hard	0.010–0.020	0.003–0.010	0.25–0.50	0.075–0.25
Soft	0.015–0.025	0.003–0.010	0.40–0.65	0.075–0.25
Steel				
Mild or free-turning	0.010–0.020	0.002–0.005	0.25–0.50	0.051–0.13
Medium-carbon	0.010–0.020	0.002–0.005	0.25–0.50	0.051–0.13
High-carbon or alloy	0.005–0.015	0.002–0.005	0.13–0.40	0.051–0.13

100 revolutions of the spindle to move the tool 1 in. (there are 100 increments of 0.010 in. in 1 in.). There are no specific rules to use when selecting feeds because the feed varies according to the type of metal being machined, the depth of the cut, and the surface finish required. Table 9.2 lists suggested feed rates for various types of metals and the type of cut being used for HSS cutting tools. Generally, coarser feeds are used on softer metals and rough turning, and finer feeds are used for harder metals and finishing cuts.

9.8 EXTERNAL LATHE OPERATIONS

There are many different types of machining operations performed on the engine lathe. These operations generally are of two types: *external* and *internal*. External machining operations include facing, straight and taper turning, parting, knurling, forming, threading, and grinding. Internal operations include drilling, reaming, boring, counterboring, threading, and grinding. Each external lathe operation is discussed in this section. We discuss internal lathe operations in Sec. 9.9.

9.8.1 Facing Operations

Facing is done on a lathe to square the ends of a workpiece (Fig. 9.56). When a piece of material is selected for a job, it generally is cut a little longer than the finished size required. Facing brings the part to the required length, leaving the ends square to the axis of work.

Facing is most often done in some type of chuck. Three- and four-jaw chucks are generally used for larger workpieces; collet chucks can be used for smaller work. Sometimes facing is done while the work is held between centers. The procedure for facing in a chuck begins by mounting the chuck on the lathe.

FIGURE 9.56 Facing operations are done to square the end of a part.

Mounting a Lathe Chuck or Other Work-Holding Devices:

1. Place a wooden cradle on the ways of the lathe next to the spindle.
2. Place a *three-jaw universal chuck* or *four-jaw independent* chuck on top of the cradle.
3. Clean the spindle nose and the inside of the chuck, where it is to be mounted, thoroughly with a clean rag (Fig. 9.57).
4. Slide the cradle and chuck on the spindle nose and tighten the chuck securely.

The remainder of the procedure for installing the chuck will vary depending on the type of spindle nose. The three somewhat different procedures are outlined below.

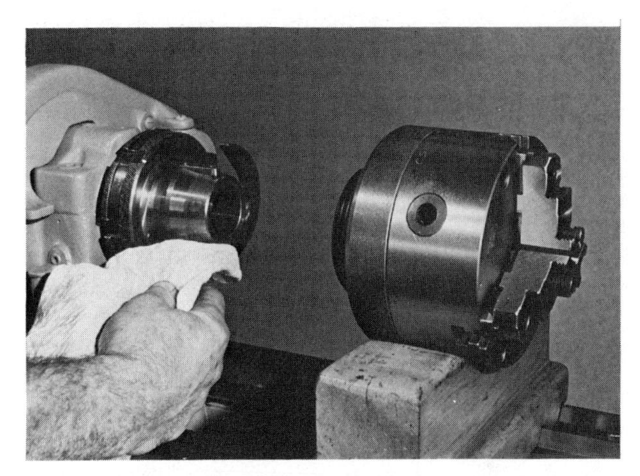

FIGURE 9.57 It is important when mounting a chuck that both the spindle nose and chuck are clean.

For tapered spindle noses:

1. Rotate the chuck in the cradle until the keyway is aligned with the keyway on the spindle nose.
2. Tighten the collar with a spanner wrench in a counterclockwise direction.
3. Set the lathe gearbox to the lowest rpm position, or engage the back gears to keep the spindle from rotating during the tightening operation.
4. To remove a chuck from a tapered spindle nose, reverse the procedure.

For threaded spindle noses:

1. Thoroughly clean the threads on the spindle nose and chuck.
2. Rotate the spindle clockwise by hand after the chuck and cradle have been moved directly onto the threaded spindle.
3. After the threads have been engaged, rotate the chuck until it is seated firmly against the chuck shoulder. If the chuck will not go all the way on by hand, remove it and check for chips in the threads or damaged threads.
4. *Note:* Never use machine power when installing or removing a chuck.
5. To remove a chuck from a threaded spindle nose, rotate the chuck until a square wrench hole is at a top position and place the lathe in back gear to lock the spindle.
6. Place a chuck wrench in the chuck and pull toward you sharply.
7. Continue to rotate the chuck counterclockwise, by hand, until the chuck is free from the spindle and resting on the cradle. In some cases it may be necessary to place a square bar between the chuck jaws to remove a chuck.

For cam lock spindle noses:

1. Using a chuck wrench, turn each cam lock until the registration line on the spindle nose is aligned with the line on the cam lock (Fig. 9.58).

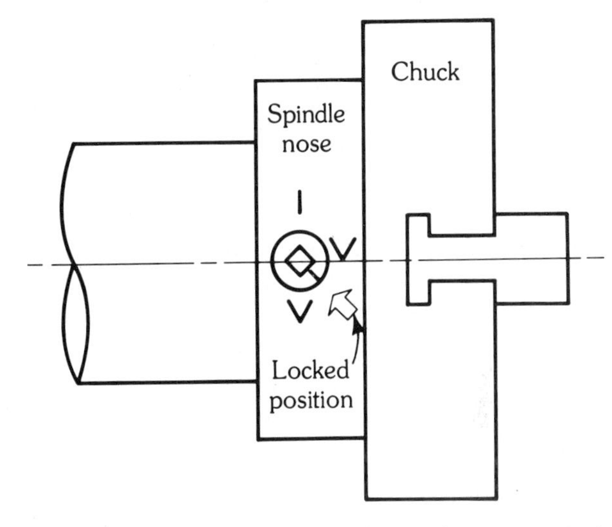

FIGURE 9.58 Aligning registration marks on a camlock spindle nose.

2. Rotate the lathe spindle until the holes in the spindle nose are aligned with the studs on the chuck. Clean the studs and taper on the chuck with a rag.
3. Slide the chuck into place, and with a chuck wrench tighten each cam lock in a clockwise direction.
4. To remove the chuck from a cam lock spindle nose, reverse the procedure. (*Note:* It may be necessary to tap the chuck lightly with a lead or plastic hammer to break the chuck loose from the spindle nose.)

Facing Work Held in a Three-Jaw Universal Chuck:

1. Position the workpiece in the universal chuck so that slightly more material projects from the chuck than is needed for the facing operation (see Fig. 9.59).
2. Using a chuck wrench, tighten the workpiece securely. (*Note:* After tightening always remove the chuck wrench *immediately.*)

FIGURE 9.59 When facing, try not to extend the end of the part from the chuck any more than necessary.

3. Swivel the compound to the right until the feed handle is just past the cross-feed handle (approximately 30°). Tighten the compound rest in this position.
4. Place the toolpost assembly in the left side of the *T* slot on the compound rest. Hand-tighten the assembly.
5. Put a right-cut facing tool or general-purpose toolbit in the toolholder and hand-tighten the setscrew.
6. Set the height of the facing tool to the centerline of the workpiece by moving the toolholder until the tip of the toolbit is the same height as the point of a tailstock lathe center (Fig. 9.60).

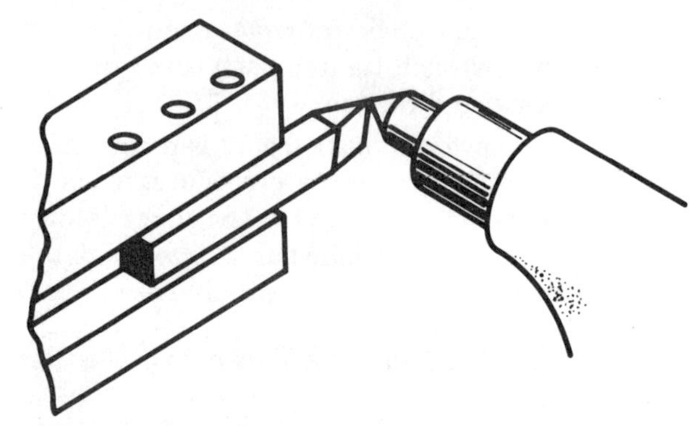

FIGURE 9.60 Setting the toolbit to the correct height using the tailstock center.

7. Next, swivel the toolholder carefully until the cutting edge of the toolbit is in the correct position (Fig. 9.61). (*Note:* Be sure not to disturb the position of the toolbit height during this procedure.)

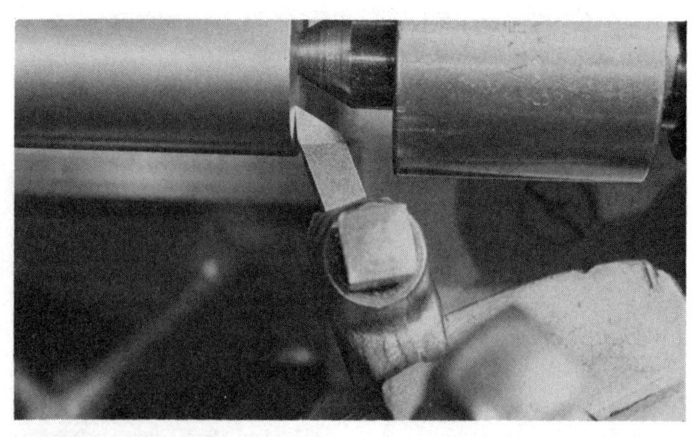

FIGURE 9.61 The correct position of the toolbit cutting edge for facing operations.

8. Using the proper wrench, tighten the toolholder.
9. Using the carriage hand wheel, move the carriage assembly so that the cutting edge is approximately 1/4 in. (6.5 mm) away from the center of the workpiece.
10. Set the lathe rpm rate for the diameter and type of material to be faced.
11. Start the lathe and slowly move the carriage until a chip begins to form from the cutting edge. Lock the carriage with the carriage lock screw to prevent the tool from pushing away from the work.
12. Turn the cross-feed handle slowly until the end of the workpiece has been completely faced. [*Note:* On diameters larger than 1 in. (25.4 mm), the power cross feed is generally used.]
13. Continue to face the workpiece until the end is square and to the length desired. (*Note:* With this setup the tool may be fed in either direction.)

Hints on facing:

1. If the toolbit is not on exact center, a small projection (fin) will be left on the end of the work. Reset the toolbit height accordingly if this occurs and continue the facing operation.
2. If accurate lengths are required, the compound rest should be set to 30° to the right of the cross-feed handle (Fig. 9.62). At this position the longitudinal movement of the cutting tool is one-half the movement of the compound rest. For example, if the compound feed is moved 0.020 in., the tool moves longitudinally (toward the headstock) 0.010 in.

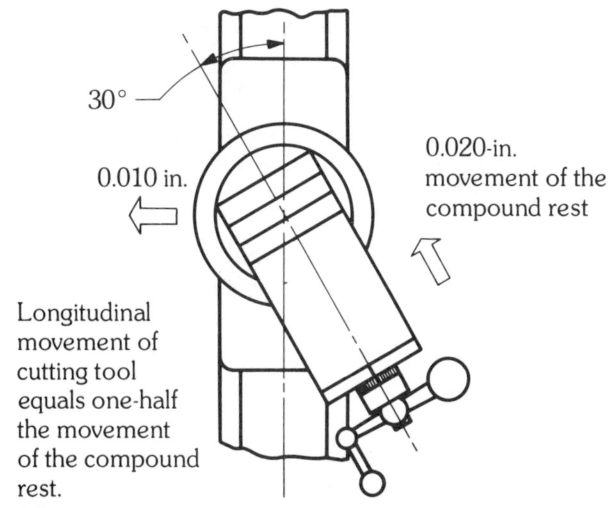

FIGURE 9.62 Setting the compound rest at 30° for accurate longitudinal movement.

3. Another method of accurately facing a part to length is to mount an indicator, as shown in Fig. 9.63. With this setup the button of the dial indicator rests against the saddle; therefore, the carriage assembly can be moved an accurate amount for the facing operation.

FIGURE 9.63 Using a dial indicator for accurate movement of the cutting tool for facing operations. (*Courtesy South Bend Lathe, Inc.*)

4. When a large amount of material is to be removed by facing, it may be advisable to rough-machine part by feeding the tool sideways in a series of steps. This "stepping off" procedure is quicker than conventional radial facing. The finish cut is generally taken using the conventional procedures.

Facing Work Held in a Four-Jaw Independent Chuck. The procedure for facing in an independent chuck is the same as in a universal chuck except for mounting the workpiece. The independent chuck is often used when the work must be mounted exactly on center. Another feature of the independent chuck is the ability to mount the work off center for machining eccentrics. The procedure for mounting a workpiece on center in an independent chuck is as follows:

1. Using the proper chuck wrench, open each jaw so that the total opening is slightly larger than the diameter of the workpiece. Use the concentric rings that are on the face of the chuck as a guide so that each jaw is approximately the same distance from the center of the chuck (Fig. 9.64).

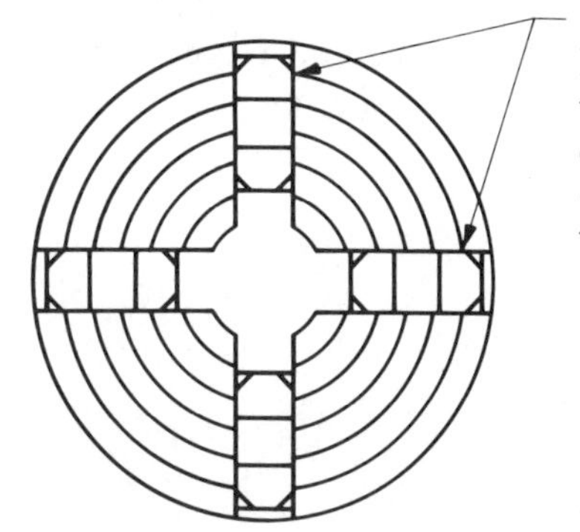

FIGURE 9.64 Using the concentric rings on the face of the four-jaw chuck to align the chuck jaws.

2. Tighten each jaw against the workpiece with even pressure, being careful not to lose the approximate workpiece center location.
3. Install a lathe toolholder and toolpost assembly in the compound rest. Position the toolholder backward so that the butt end is close to the work, and tighten the toolpost screw fingertight (Fig. 9.65).

FIGURE 9.65 Butt end of a toolholder being used to help true a workpiece mounted in a four-jaw independent chuck.

4. Rotate the chuck slowly by hand and observe the clearance between the toolholder and workpiece. If the toolholder touches the work on one side (called the high side) and there is a space on the opposite side, the work is not centered.
5. To center the workpiece, loosen the jaw on the side opposite the high side. Tighten the jaw next to the high side. This procedure may need to be done several times to center the workpiece properly. (*Note:* Do not loosen more than one jaw at a time. Otherwise, the work may fall out of the chuck.)
6. For more accurate centering, use a dial indicator. Mount the dial indicator in the toolpost of the lathe, as shown in Fig. 9.66.

FIGURE 9.66 The dial indicator is a more accurate method of centering the work in a four-jaw independent chuck.

7. Move the carriage and cross feed until the dial indicator is in contact with the work. Set the indicator to zero.
8. Rotate the chuck by hand and note the high and low sides.
9. As before, loosen the low-side jaw and tighten the high-side jaw. Using the indicator as reference, adjust the high-side jaw only one-half the error. For example, if the dial indicator indicates 0.020 in. (0.5 mm) difference between the high and low sides, adjust the high-side jaw half of the difference, or simply 0.010 in. (0.25 mm).
10. Continue to adjust the *first* two opposing jaws until the dial indicator reads zero.
11. Adjust the other set of opposing jaws until they read zero on the indicator.
12. Rotate the chuck and recheck each jaw. If the workpiece is centered accurately, the dial hand on the indicator will stand still.

Facing Work Held between Centers. Sometimes it is necessary to face the ends of a workpiece that is being held between centers (Fig. 9.67). Before this can be done, it is necessary to prepare both the workpiece and lathe.

To prepare the workpiece for facing between centers, you have to drill a 60° center hole on each end of the workpiece before the workpiece can be mounted between centers on a lathe. The 60° center hole provides a space for the sharp point of the centers located in the headstock and tailstock. The tailstock center hole also provides a reservoir for lubricating oil.

Center holes are usually drilled with a combination drill and countersink, commonly called a center drill (Fig. 9.68). Center holes should be drilled to a depth that will provide a good bearing surface for the center. Figure 9.69*(a)* and *(b)* show the effects of center holes that have been drilled either too shallow or too deep. Figure 9.69*(c)* shows the proper depth the center hole should be drilled.

To *drill center holes on a lathe,* accurately mount the workpiece in either a three- or four-jaw chuck. Next, install a drill chuck in the tailstock. Fasten the center drill in the drill chuck. Position the tailstock near the workpiece and clamp it in place. Then, start the lathe and crank the tailstock handwheel in and out slowly until the center drill

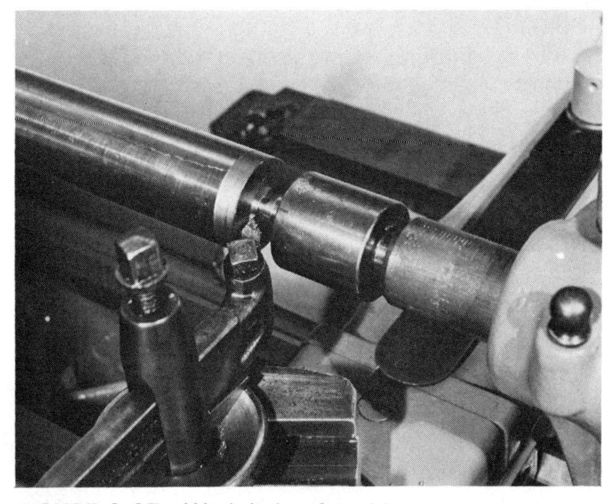

FIGURE 9.67 Work being faced between centers.

FIGURE 9.68 The center drill is used for drilling center holes for mounting a part between centers.

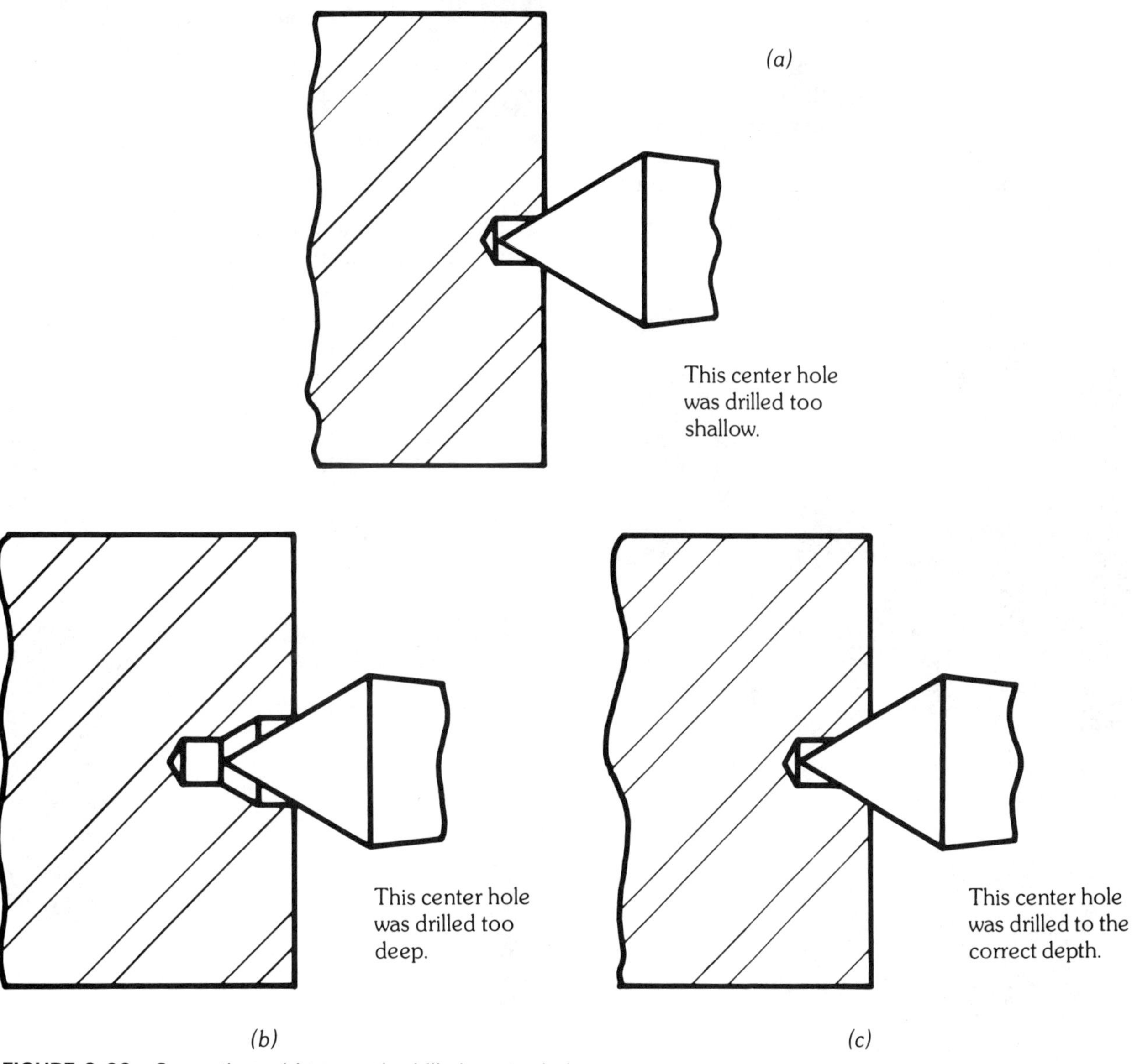

FIGURE 9.69 Correctly and incorrectly drilled center holes.

has reached the proper depth. Use a cutting fluid on steel, brass, bronze, and aluminum.

To reduce the possibility of center drill breakage, the spindle speed should be computed for the size of the tip on the center drill. Usually, this will result in running the lathe at relatively high spindle speed and using a light, intermittent infeed to avoid overloading the center drill tip.

If the workpiece is too large to fit through the spindle of the lathe, use a steady rest to support the outer end during the center drilling operation (Fig. 9.70).

To *drill center holes on a drill press,* make sure that the center of each end is first located. Locate the center of each end with a center head (Fig. 9.71). Next, position the work so that the layout lines on the end of the workpiece and the point of the center drill are aligned (Fig. 9.72). Drill both ends using this procedure. Use a cutting fluid while center-drilling steel.

To prepare the lathe for facing between centers, you have to *mount the lathe centers.* To do this:

1. Thoroughly clean the headstock and tailstock lathe centers. Before installing the centers, thoroughly clean both the headstock and tailstock spindles. (*Note:* Never attempt to clean the headstock spindle while it is revolving.)

FIGURE 9.70 Steady rest being used to support a long workpiece during a center drilling operation.

FIGURE 9.71 Center head being used to find the center of a piece of round stock.

FIGURE 9.72 Drilling a center hole in a drill press.

2. Install both centers by using a quick forcing motion into each spindle.
3. To remove a center from a tailstock, turn the handwheel so that the quill moves back into the tailstock. The end of the tailstock screw forces the center out of the spindle. To remove the headstock center, put a *knock-out* rod through the outer end of the spindle. Grasp the center with the right hand and gently tap the center with the knock-out rod held in the left hand (Fig. 9.73).

FIGURE 9.73 Removing a center from the headstock of a lathe.

Checking the Alignment of Lathe Centers. For accurate work the tailstock and headstock centers must be in perfect alignment. Figure 9.74 shows the results of facing the end of a workpiece when the centers are not aligned.

Three methods are used to align lathe centers. *Visual alignment* is used when great accuracy is not required. Visual alignment is accomplished by

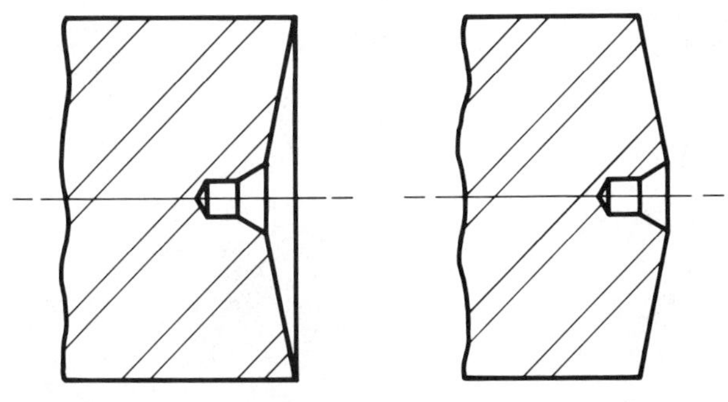

FIGURE 9.74 Effects of the tailstock being off center during a facing operation.

checking the registration lines at the back of the tailstock or the tip of the head and tailstock centers.

In *the trial-cut method,* a sample workpiece is turned, each end is measured, and the sizes are compared. *The test bar method* requires mounting a precision parallel bar between centers. A dial indicator is used to check for accuracy.

To use the *visual alignment method:*

1. Loosen the tailstock from the bed with the tailstock clamp lever.
2. The tailstock is moved in the proper direction by the adjusting screws or by bolts located at the base of the tailstock. Loosen one adjusting screw and tighten the opposite one. Selection of the screw to be tightened depends on the direction the tailstock needs to be moved.
3. Continue this procedure until the tailstock is properly aligned by:
   (a) Checking the position of the two lines on the back of the tailstock (Fig. 9.75)
   (b) Checking the points of each center (Fig. 9.76).
4. Tighten the loosened screw to secure the tailstock in place.

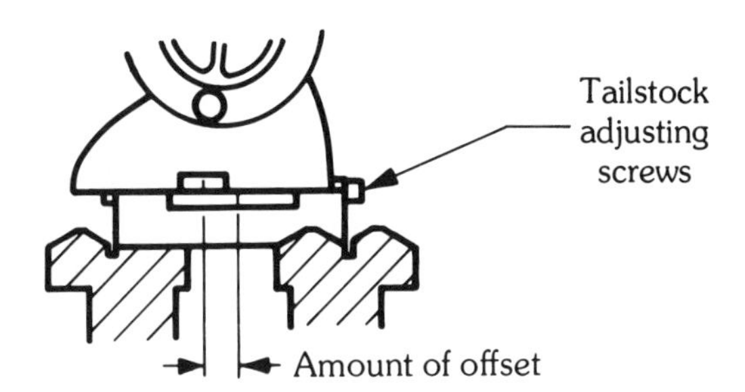

FIGURE 9.75 Checking the position of registration lines on a tailstock.

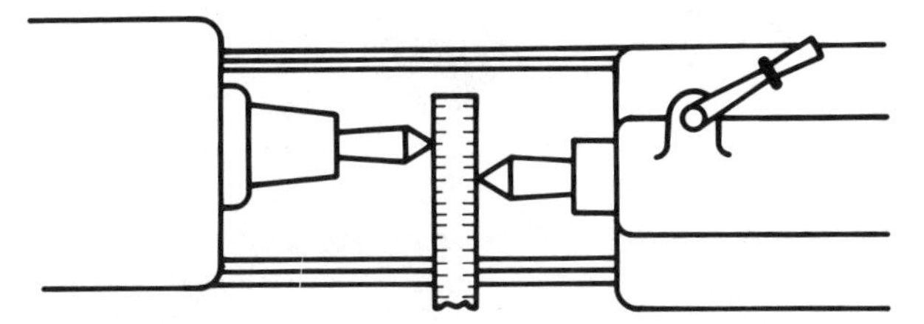

FIGURE 9.76 Aligning head and tailstock centers.

To use the *trial-cut method:*

1. Mount the workpiece between centers.
2. Take a trial cut along the entire length of the workpiece.
3. With a micrometer, measure the diameters of the work at both ends (Fig. 9.77).

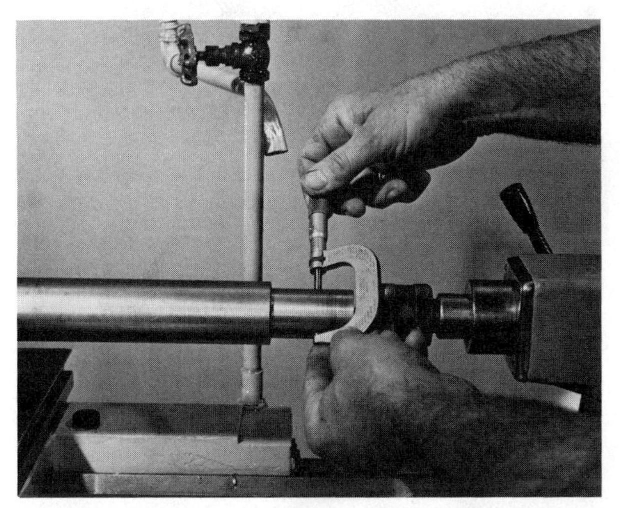

FIGURE 9.77 A micrometer is used to check the size of a part during the trial-cut method of checking tailstock alignment.

4. If the two diameters are not the same, move the tailstock one-half the distance between the two micrometer readings. Move the tailstock toward you if the measurement at the tailstock is larger, away from you if it is smaller.
5. Accurate tailstock movement is accomplished by the following sequence:
   (a) Move the butt end of a lathe toolholder slowly toward the tailstock spindle. Using a paper shim, stop the toolholder when the paper shim is snug between the spindle and the toolholder (Fig. 9.78).
   (b) Set the reading on the micrometer collar on the cross-feed handle to zero.
   (c) Back the toolholder away from the tail-

FIGURE 9.78 Using a paper shim to accurately check motion of the tailstock.

stock spindle *one full revolution* and then move it slowly back. (*Note:* Moving the toolholder one full turn away from the spindle eliminates all the backlash from the cross-feed mechanism.)

(*d*) Adjust the tailstock until the spindle touches the paper shim held between the toolholder and the spindle.

(*e*) Recheck the alignment by taking another trial cut.

To use the *test bar method:*

1. Select a suitable test bar (see your instructor) and place it between centers.
2. Adjust the test bar snugly and tighten the tailstock in place.
3. Place a dial indicator in a toolpost and adjust the button so that it is on the centerline of the test bar (see Fig. 9.79).
4. Starting from the headstock, set the indicator to zero and move the carriage and dial indicator along the test bar until the indicator is at the tailstock end.
5. Note the reading and adjust the tailstock until it has the same reading as the headstock (Fig. 9.79).
6. Recheck the alignment and adjust accordingly.

FIGURE 9.79 Checking tailstock alignment using a test bar and dial indicator. The indicator should read zero at both the head and tailstock.

Facing the End between Centers. Figure 9.80 shows the proper setup for end facing between centers:

1. Clean the headstock and tailstock spindles and install a lathe center in each spindle.
2. Clean the center holes in the workpiece and the points on the lathe centers.
3. Mount a driving plate on the spindle nose of the lathe.
4. Place a lathe dog on one end of the workpiece and hand-tighten the setscrew. Apply center point lube to the center hole on the opposite end.
5. Place the end of the workpiece with the dog on the live center and position the workpiece in line with the dead center.
6. Slide the tailstock to the work and clamp in place so that the dead center is about 2 in. from the work.
7. Turn the tailstock handwheel until the dead center supports the work. Adjust the dead center so that the workpiece is held snugly and tighten the spindle clamp.
8. Position the tail of the dog in one of the slots on the drive plate and tighten the dog on the workpiece. [*Note:* The tail of the dog should fit freely in the slot of the driving plate (Fig. 9.81). Also, the work should move freely on centers and not be too tight.]
9. Place a right-cut facing tool in a toolholder and hand-tighten it. Place the toolpost and toolholder on the left side of the

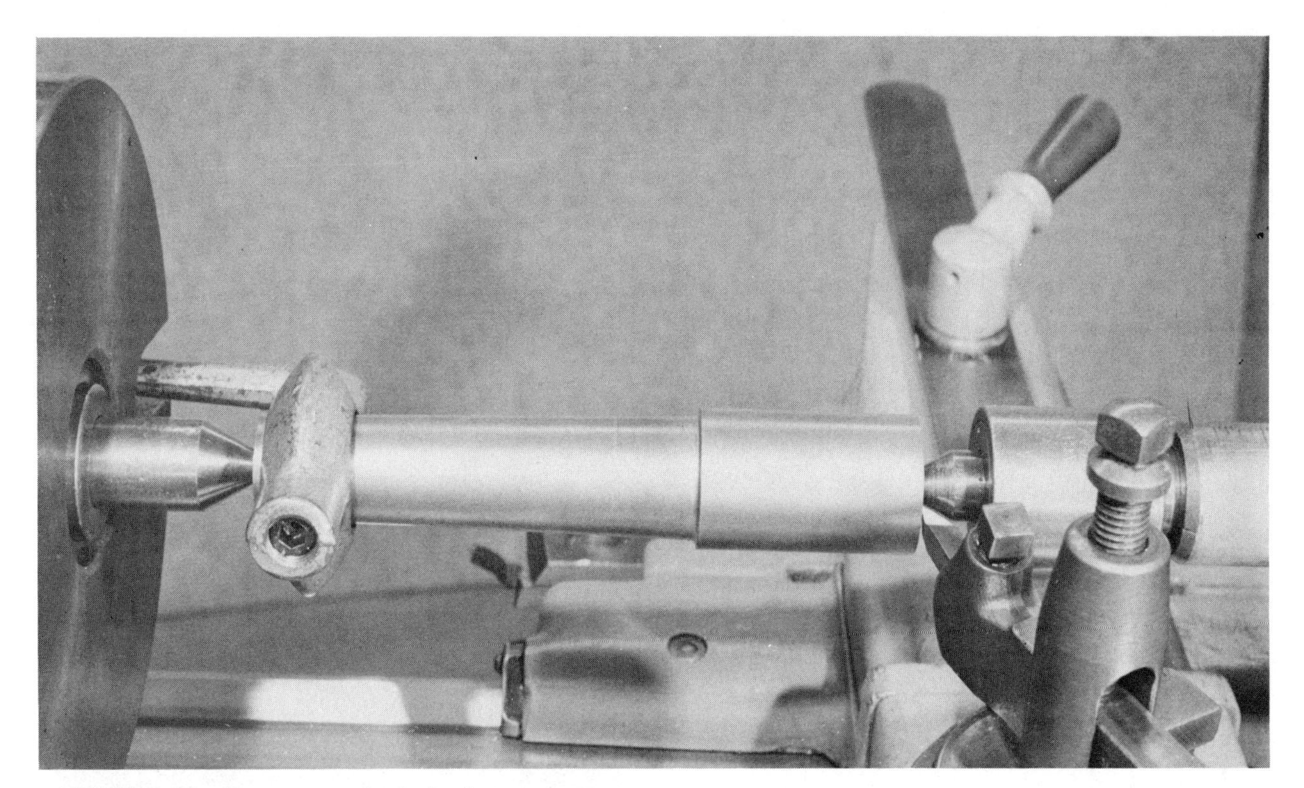

FIGURE 9.80 Proper setup for facing between centers.

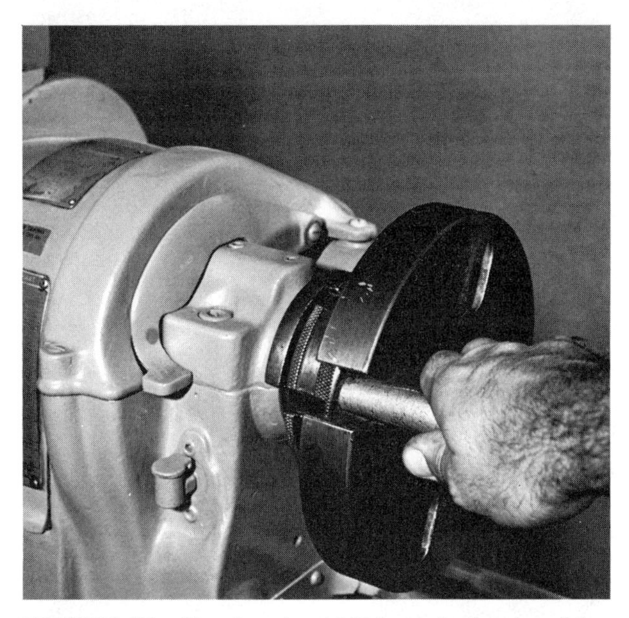

FIGURE 9.81 The dog should fit freely in the slot of the driving plate.

FIGURE 9.82 Proper position of the cutting tool for facing between centers.

compound rest and hand-tighten the assembly.

10. Set the height of the tool to the centerline of the workpiece by removing the workpiece and using the tailstock center as a reference.
11. Position the cutting edge of the toolbit so that the *point* is slightly to the left, as shown in Fig. 9.82. Tighten the toolholder and toolbit in place.
12. Set the lathe rpm rate for the diameter and type of material to be faced.
13. Start the lathe; using the carriage handwheel, move the carriage until a small cut is taken from the work. Lock the carriage in place.
14. Feed the cutting tool using the cross-feed handle. Generally, feed the tool toward the center of the work for roughing cuts and from the center toward the outside for

finishing cuts. (*Caution:* When facing between centers, avoid bumping the point of the toolbit against the hardened center.)

15. Continue to face the end of the workpiece until the end is square and to the required length.

Hints on facing between centers:

1. When facing is done between centers, a small amount of material (burr) is left next to the center hole because of the position of the cutting tool and tailstock center. A bell-type center drill eliminates this problem (Fig. 9.83).

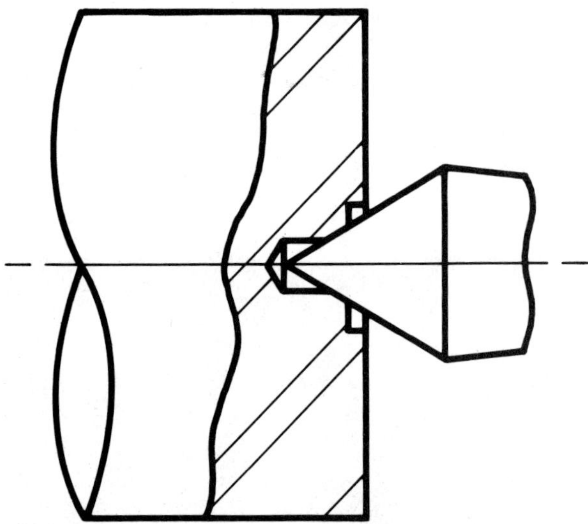

FIGURE 9.83 The bell-type center hole eliminates the problem of a burr being produced.

2. The half center may also be used in the tailstock spindle to avoid the problem of removing the burr (Fig. 9.84). The half center is used only for facing operations.
3. Another method of mounting the work between centers is shown in Fig. 9.85. The center in the headstock spindle and driving plate have been eliminated. A universal chuck is used; a scrap piece of material is mounted and a 60° point is machined by using the compound rest (see Sec. 9.8.3). The dog is then driven by placing the end between the chuck jaws.

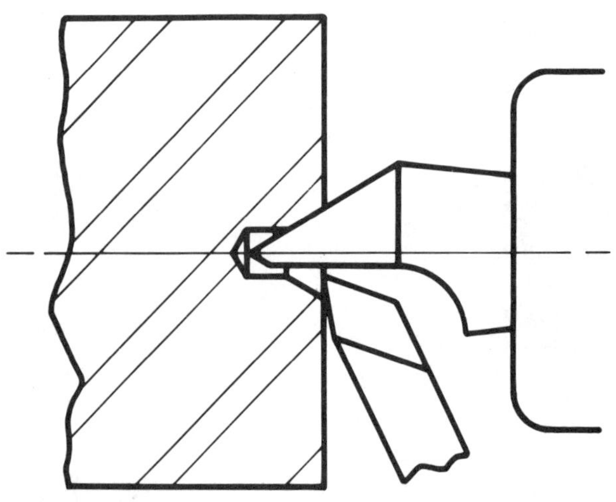

FIGURE 9.84 The half center is used when facing between centers to eliminate the burr around the hole.

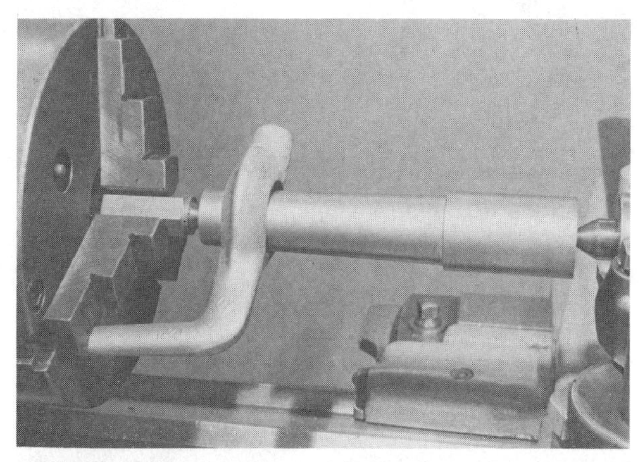

FIGURE 9.85 Another method of turning between centers is to machine a 60° point in a universal chuck and drive the dog between chuck jaws.

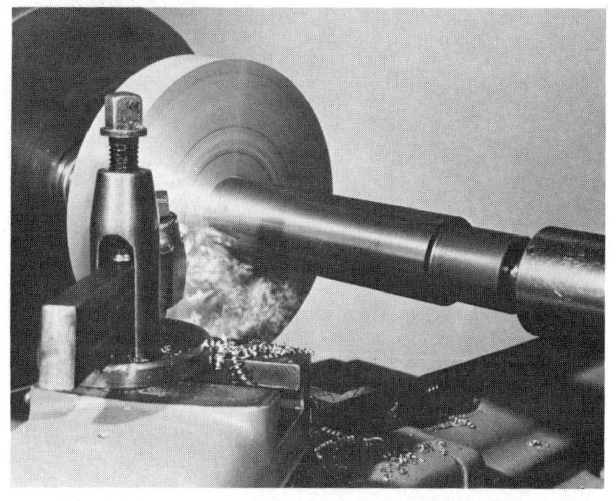

FIGURE 9.86 Facing work that has been mounted on a mandrel.

Facing a Workpiece Mounted on a Mandrel. Figure 9.86 illustrates facing a workpiece mounted on a mandrel. Mandrels (discussed in Sec. 9.5.2) are work-supporting devices that are mounted between centers. The procedure for facing is similar to that described in facing work held between centers. Since the mandrel is also turning, caution must be exercised that the point of the facing tool does not come in contact with the hardened mandrel. Damage to both the cutting tool and mandrel may result if this occurs.

9.8.2 Straight Turning Operations

Turning on a lathe is the operation of reducing the diameter of a workpiece to a specified size. *Straight turning* operations (Fig. 9.87) require that the diameter be the same at each end of the workpiece. Turning is accomplished by revolving the work while the tool is fed *longitudinally* (parallel to the centerline of the lathe). The principles, methods, and cutting action of the tool are essentially the same during turning operations, whether the work is being held between centers or in a chuck. With this thought in mind, the following straight turning procedure will have the workpiece held between centers.

FIGURE 9.87 Turning a part on a lathe.

Straight Turning a Workpiece Held between Centers. There are two procedures for straight turning a workpiece held between centers: *rough turning* and *finish turning.*

Rough turning procedure:

1. Mount the workpiece between centers as discussed in Sec. 9.8.1. (*Note:* Make sure that the live and dead centers are aligned and that the live center is running true.)
2. Set the compound rest slightly to the right of the cross-feed handle, far enough so that the cross-feed micrometer collar can be seen easily.
3. Mount the toolpost assembly on the left-hand side of the compound rest. Place a suitable right-hand-cut roughing tool in the toolholder and hand-tighten it. Adjust the toolbit on center, and position the cutting edge as shown in Fig. 9.88. Tighten the toolpost and toolbit.
4. Move the carriage so that it will be in the position where the cut will end. Check to see if the dog clears the compound rest and toolpost assembly by rotating the workpiece by hand. If it does not, readjust

FIGURE 9.88 Correct position of the cutting tool for turning operations.

the toolpost assembly for the necessary clearance.

5. Set the lathe rpm rate for the diameter and type of material to be machined. Set the quick-change gearbox for the feed rate (see Table 9.2).
6. Move the carriage back to the tailstock end.
7. Start the lathe and take a trial cut. The trial cut is hand fed and should be approximately 0.015 to 0.020 in. (0.5 mm) deep and 1/4 in. long (6.5 mm). After the trial cut, move the carriage so that the toolbit is past the right end of the work.
8. Stop the lathe and measure the diameter of the work. The first cut is generally a *roughing cut.* Roughing cuts, as discussed, remove most of the excess material from the workpiece. The roughing cut should be as heavy (deep) as the lathe and cutting tool will stand. If the workpiece is very oversized, several roughing cuts may be necessary. With the micrometer collar on the cross-feed handle, adjust the tool for the proper depth of roughing cut.
9. Start the lathe and hand-feed the carriage so that a 1/4-in. (6.5-mm)-long cut is taken. Stop the lathe and measure the diameter for size. Readjust the depth of cut if necessary. (*Note:* This procedure eliminates the possibility of going too deep and causing the workpiece diameter to be undersized.)
10. Restart the lathe and engage the longitudinal feed lever. Apply a suitable cutting fluid to the cutting point of the tool (see Chap. 15).
11. As the work is being machined, observe the cutting action of the toolbit. Note the following two conditions:
    (a) Chips should be coming off the work freely in small sections. Long coils of chips are dangerous and can be broken if the power feed is momentarily disengaged. (*Never* touch the chips with your hands because they are razor sharp and hot.)
    (b) Heavy roughing cuts may cause the workpiece to overheat and expand. The expansion of the workpiece results in scoring or binding of the dead center. Applying a liberal amount of the proper cutting fluid may help prevent the workpiece from overheating. However, if overheating does occur, loosen the tailstock slightly, add some lubricant to the center hole, or use a revolving tailstock center.
12. Continue the turning operation. When the cutting tool reaches the required length, disengage the feed and stop the lathe.
13. If additional roughing cuts are necessary, return the carriage to the starting position and continue the turning procedure until the workpiece is to the required diameter.
14. To rough-turn the part of work held by the lathe dog:
    (a) Remove the workpiece from the lathe.
    (b) Secure the dog on the machined end.
    (c) Apply a lubricant to the center hole, and replace the work between centers.
    (d) Rough-turn to the same diameter.

Finish-turning procedure:

1. Mount a finishing toolbit in the toolholder and adjust it as described in the roughing procedure.
2. Set the correct lathe rpm rate and the quick-change gearbox for a slow feed rate (see Table 9.2).
3. Start the lathe and slowly turn the cross-feed handle until the point of the toolbit just scratches the surface of the revolving workpiece.
4. Observe the reading on the cross-feed micrometer collar, or carefully set the collar to zero.
5. Position the carriage so that the cutting tool is beyond the right end of the workpiece.
6. Using the micrometer collar as an indicator, move the cutting tool the exact amount required for the finishing cut.
7. Take a trial cut of about 1/4 in. (6.5 mm).
8. Stop the lathe and measure the diameter to see if it is within limits. If necessary, readjust the depth of cut.
9. Start the lathe and engage the longitudinal feed lever.
10. Continue the turning operation to the required length, disengage the feed, and stop the lathe. Note the setting on the mi-

crometer dial. Move the toolbit away from the work, and position the carriage until the tool clears the right end of the work.

11. To finish turning the other end, remove the workpiece from the lathe and mount the lathe dog on the opposite end. (*Note:* To prevent marring the newly machined surface, place a piece of soft copper or aluminum around the diameter of the workpiece before mounting the lathe dog.)
12. Install the workpiece, reposition the tool to the previous setting on the collar, start the lathe, and again take a trial cut.
13. Stop the lathe and measure the workpiece diameter. If the tool setting is the same, the diameters should be the same. If they are not, reset the tool to the correct depth.
14. Start the lathe and engage the power feed.
15. Continue the turning operation until the toolbit reaches the previous finish cut.

Hints on straight turning operations:

1. The micrometer collar is very important to successful turning operations. Machinists should become acquainted with the micrometer collar for each lathe they will be using because the collars differ somewhat. For example, if the micrometer collar indicates 0.025 in. of tool movement, 0.025 in. will be removed from the *radius* of the workpiece. This type of collar (or lathe) is called *indirect reading.* However, some lathes have *direct-reading* collars. That is, the micrometer collar reads the amount removed from the *diameter* of the workpiece. Therefore, a collar movement of 0.025 in. will remove 0.025 in. on the diameter. If in doubt, take a small trial and check with a micrometer.
2. Another important consideration during turning operations is the *backlash* in the cross-feed unit. Backlash is the play or clearance between the cross-feed screw and nut assembly. The amount of backlash varies with different lathes. Older lathes have more backlash than newer lathes because the screw and nut assembly have more wear. Accurate tool settings are accomplished by always feeding the tool toward the workpiece on successive cuts. If this is not possible, or if the tool is fed too far, turn the feed handle at least a half turn in the opposite direction (out) and then back (in) to the correct reading to compensate for backlash.
3. Caution must be used when rough-turning long, slender workpieces. If a heavy cut is used on long, slender work, the workpiece may spring up or away from the cutting tool. If this occurs, reduce the amount of the depth of cut or use the follow rest (see Sec. 9.5.2).
4. When greater accuracy is required during turning operations, set the compound rest at 84° to the cross slide (Fig. 9.89). At this position a 0.001-in. movement of the compound rest will move the cutting tool 0.0001 in.

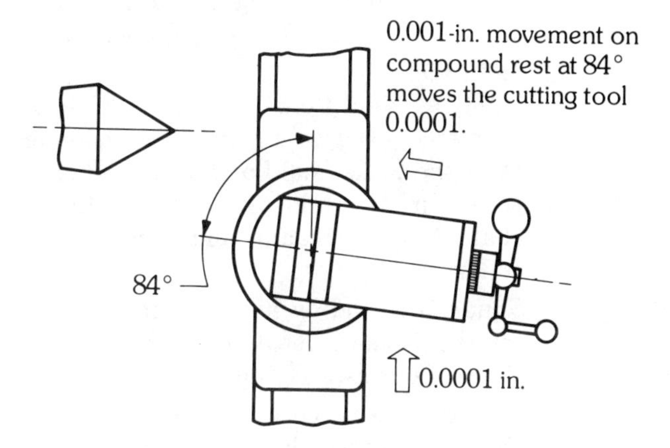

FIGURE 9.89 The compound rest set at 84° for accurate settings of the cutting tool.

Turning to a Shoulder. When the workpiece is turned to two or more diameters, the step between the diameters is known as a *shoulder.* Three types of commonly machined shoulders are shown in Fig. 9.90. The filleted corner gives the greatest strength at the shoulder of the workpiece, particularly if it is to be hardened.

The setup for turning to a shoulder is similar to that of straight turning. Additional procedures are as follows:

1. Mount the work in the lathe and face the end. The faced end will be the reference for the length of the various turned diameters.
2. With a rule or hermaphrodite caliper, lay out the shoulder distance as specified on the shop drawing. Position the point of the toolbit at this mark, and cut a small groove to make the mark easier to see.
3. Rough- and finish-turn the small diameter to within 1/32 in. of the required length. (*Note:* If a filleted shoulder is to be turned,

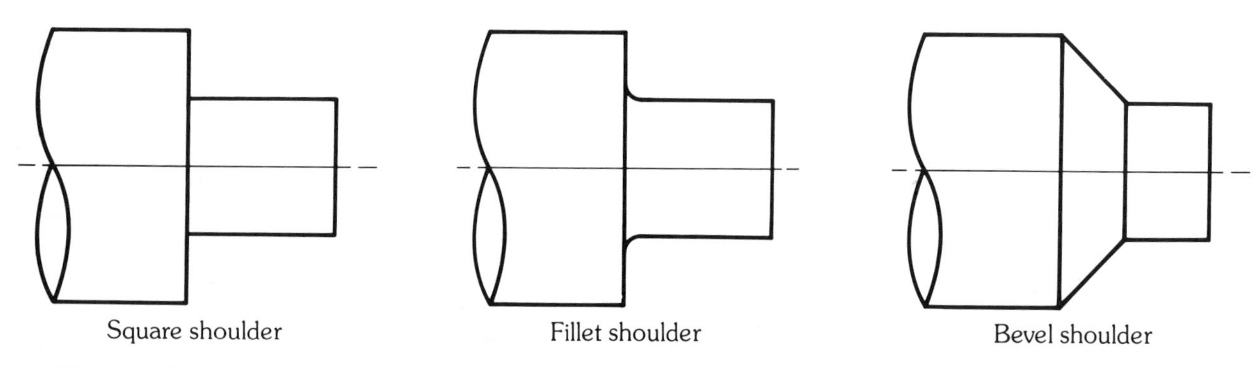

FIGURE 9.90 Three common kinds of shoulders.

leave sufficient material to allow the proper size radius to be formed.)

4. Mount a suitable facing toolbit in the toolholder, and position the tool for a facing operation.
5. Apply a layout fluid or chalk to the small diameter of the work next to the shoulder, start the lathe, and hand-feed the toolbit so that it just removes the layout fluid or chalk. Note the reading on the micrometer collar.
6. Square the end by hand-feeding the tool toward the shoulder line and then feeding the tool out as a facing operation.
7. If successive cuts are needed to finish the shoulder, return the toolbit to the small diameter, using the same reading on the micrometer dial, and continue as before.

Hints on shoulder turning:

1. For filleted shoulders, use a radius gage when grinding the nose of the toolbit as a guide for the proper radius (Fig. 9.91).

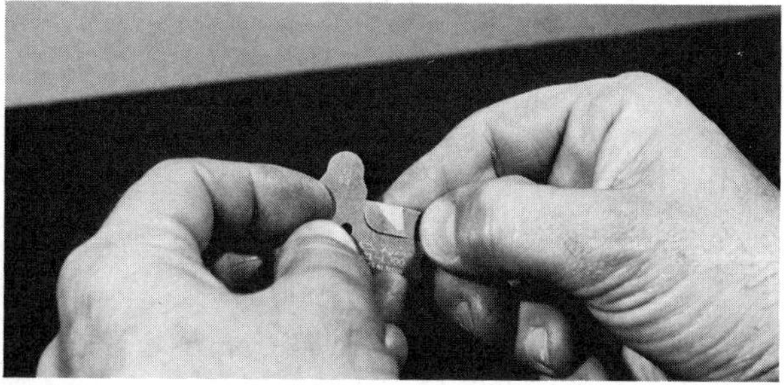

FIGURE 9.91 A radius gage is used to accurately measure the nose radius of a toolbit.

2. For chamfered shoulders, position the cutting edge of the tool to the required angle of chamfer and feed against the shoulder.
3. When more than one shoulder is required on the workpiece, use the same procedure as described, and rough-turn each diameter, the largest first to the smallest last. After rough turning each diameter, finish-turn each in the manner described, and then finish-machine each shoulder.
4. Some machinists mount a dial indicator as shown in Fig. 9.92. By using a dial indicator, exact lengths may be machined from the end of the part to the shoulder. If the lathe is equipped with an apron handwheel micrometer collar, it may be used to give exact lengths.

FIGURE 9.92 A dial indicator is used to accurately machine lengths. (*Courtesy South Bend Lathe, Inc.*)

9.8.3 Tapers and Methods of Turning Tapers

Often a workpiece requires that a tapered diameter be machined. A *taper* is the uniform difference in diameters measured along the axis of the work. Tapers can be specified as taper per foot, taper per inch, or in degrees.

Tapered holes are used on the spindles of drilling machines, lathes, mills, and other machine tools for accurately aligning tapered shank tools such as drills, reamers, and centers. Another use for tapers is to align assembled parts accurately.

Common Types of Tapers. The *Morse taper* is used for lathe and drilling machine spindles and has an approximate taper of 5/8 in./ft. Morse tapers range in size from No. 0 to No. 7, with No. 7 being the largest. Specifications and dimensions for the various Morse tapers are given in Appendix F.

The *Jarno taper* is used mainly on cylindrical grinding machine spindles and some lathe and drill press spindles. The Jarno taper has 0.600 in./ft of taper and ranges in size from No. 2 to No. 20.

The *Brown and Sharp taper* is used in older milling machine spindles. It has a taper of approximately 1/2 in./ft and ranges in size from No. 4 to No. 12. The Brown and Sharp tapers have been mainly replaced by the *standard milling machines taper*. Brown and Sharp, Morse, and Jarno tapers are known as self-holding tapers because they require no other device to hold them in spindles.

The *standard milling machine taper* has a taper of 3 1/2 in./ft. It is a steeper taper than most others and self-releasing. This taper is generally used on modern milling machine spindles. It has a key drive to prevent slipping and a drawn-in bolt to hold it securely in the spindle. This taper is available in four sizes, 30 (the smallest) to 60 (the largest). The standard milling machine taper is known as an easy-release taper because it requires a drawbar to hold it firmly in place.

Tapered pins used in assembly work have a taper of 1/4 in./ft. They range in size from No. 6/0 to No. 10.

Measuring Tapers. It is important to measure the amount of taper before turning the work to finish size. Tapers can be measured by four methods:

1. Measuring the difference between the small and large diameters
2. Using the sine bar
3. Comparing with gages or existing parts
4. Using the taper micrometer

Tapers can be measured by carefully marking two lines exactly 1 in. apart on the taper parallel to the end of the work. Then measure each diameter at the line with a micrometer. The difference in readings is the amount of *taper per inch of the work*. Multiplying this by 12 will give the taper in inches per foot.

The *sine bar* is used when accurate taper measurements are necessary. The basic principles of using the sine bar are explained in Chap. 5; Fig. 9.93 shows the application.

Taper plug and ring gages accurately check standard or special tapers. To use a plug or ring gage, draw three light chalk lines equally spaced along the length of the taper on the workpiece. Insert the gage and rotate it in a counterclockwise direction for approximately one-third turn. (*Note:* Rotating the gage counterclockwise keeps the gage from sticking. If the chalk lines do not rub off evenly, the taper is incorrect.)

FIGURE 9.93 A taper being measured with a sine bar.

Prussian blue can be used instead of chalk on fine finishes or ground surfaces for even greater accuracy. Existing parts, such as spindles, shanks, or sleeves can be used in place of manufactured gages to check tapers.

The *taper micrometer* is a fast and accurate way to measure tapers. It can be used for checking external and internal tapers, and it eliminates the need for a costly inventory of plug and ring gages.

Turning Tapers. The three methods commonly used in the machining of tapers are:

1. Offsetting the tailstock
2. Using the taper attachment
3. Positioning the compound rest

The *tailstock offset method* (Fig. 9.94) is probably the oldest method of turning tapers. It is generally used when a taper attachment is not available. To turn tapers by this method, move the tailstock off center (offset) the amount necessary to produce the taper required. Offsetting the tailstock has three disadvantages:

1. It is a slow process. The tailstock center must be moved from the axis of the lathe for turning tapers, and therefore it must be realigned for straight turning operations.
2. By having the tailstock offset, the center holes will not have a true bearing surface

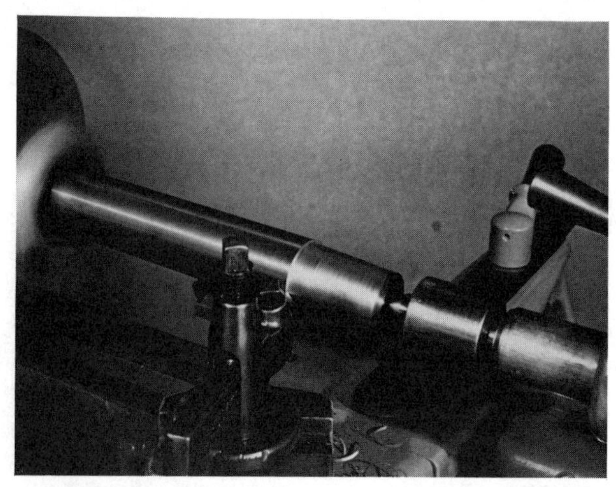

FIGURE 9.94 The tailstock offset method is sometimes used to machine external tapers.

for the lathe centers. This bearing surface becomes less and less satisfactory as the amount of offset is increased. Therefore, center holes will generally wear out of their true locations.

3. The tailstock method can be used only on external taper turning operations.

The first step in *calculating the amount of offset* is to find the taper, in inches per foot. To calculate the amount of taper per foot (tpf), use the following formula:

$$\text{tpf} = \frac{\text{large diameter} - \text{small diameter}}{\text{length of taper (in.)}} \times 12$$

or

$$\text{tpf} = \frac{D - d}{L} \times 12$$

When the tpf is known, use the following formula for calculating the amount of offset:

$$\text{offset} = \frac{\text{tpf} \times \text{total length of workpiece (in.)}}{24}$$

There are two methods of measuring the offset: the *paper shim technique* and the *dial indicator method.* You can use the graduations on the end of the tailstock, but this quick and easy method is not accurate. You can also use a rule and measure the setover distance between the headstock and tailstock centers. For more accurate settings, the *micrometer collar* on the cross-feed handle or a dial indicator may be used. To use the *micrometer collar,* follow these five steps:

1. Mount a toolholder so that the butt end is pointing toward the tailstock.
2. Adjust the tailstock spindle out a few inches and lock it in place.
3. With the cross-feed handle, move the toolholder toward the spindle until a piece of paper is lightly pinched between the spindle and toolholder (Fig. 9.95). Note the reading on the collar.

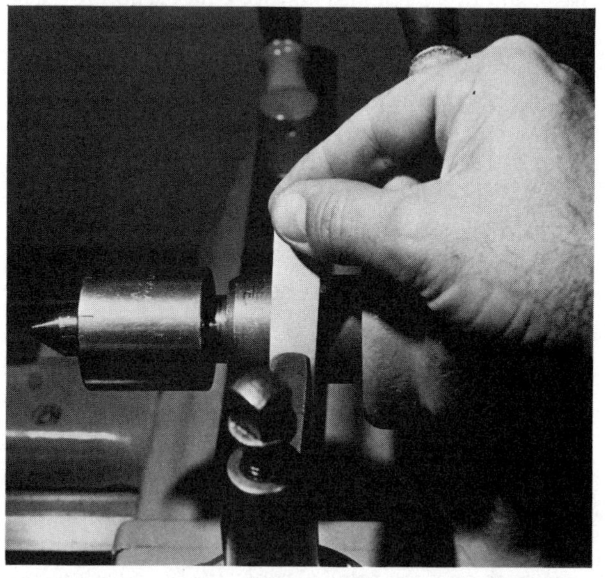

FIGURE 9.95 Using a paper feeler and the cross-feed micrometer collar to measure the tailstock offset.

4. Using the graduations of the collar, run the cross slide away from the tailstock spindle until the distance between the toolholder and the spindle is equal to the amount of offset required. [*Note:* Be sure to take up the lost motion in the cross-feed screw by running the cross slide out farther than the desired reading (approximately one-half turn) and then returning it to the correct position.]
5. Adjust the tailstock toward the toolholder until the paper is pinched between the spindle and toolholder, as before.

The *dial indicator* method is similar to the paper shim technique (Fig. 9.96). This method involves four steps:

1. Mount a dial indicator in the toolpost of the lathe.
2. Adjust the spindle of the tailstock out a few inches and lock it into place.
3. With the cross-feed handle, move the dial indicator toward the tailstock spindle until the dial reads approximately 0.010 in.

FIGURE 9.96 A dial indicator is used to accurately measure the tailstock offset.

(0.025 mm). Set the dial indicator to zero and note the reading on the micrometer collar.

4. Finally, adjust the tailstock toward the dial indicator until the dial reads the correct offset amount.

Turning a taper by offsetting the tailstock involves eight steps:

1. Adjust the tailstock for the required amount of offset. (*Note:* Offset the tailstock toward the operator for a taper that is small at the tailstock end, away for a taper that is large at the tailstock end.)
2. Mount the workpiece between centers, the same as for straight turning.
3. The cutting tool is set *exactly* on center for taper turning. If not, the work will not have true conical shape, resulting in an inaccurate taper.
4. Start the cut at the tailstock end and take successive roughing cuts until the workpiece taper is approximately 0.060 to 0.080 in. (1.5 to 2.0 mm) oversized.
5. Measure the taper using one (or more) of the methods mentioned previously.
6. If necessary, readjust the tailstock offset until the taper is correct.
7. Finish turning the workpiece until the taper is to the required size.
8. After the taper turning operation is complete, realign the tailstock to center.

Using the *taper attachment* (Fig. 9.97) is the most accurate method of turning both external and internal tapers. It offers these six advantages:

1. It is quick and easy to set up and use.
2. The alignment of the lathe centers is not disturbed.
3. Internal tapers may be machined.
4. A much greater range of tapers may be produced.
5. A variety of work-holding devices can be used.
6. Duplicate tapers can be turned on any length of work.

All taper attachments operate on the same general principle; however, they may differ in *construction.* Each type of taper attachment has a *guide bar* that swivels in a horizontal plane in either direction. The angle to which the guide bar is set determines the angle duplicated between the centerline of the lathe and one side of the workpiece.

Graduations at each end of the bar indicate the amount required for obtaining taper settings in degrees or taper per foot.

The *guide shoe* or *sliding block* moves along the guide bar, which is held stationary during the taper operation. The guide shoe, in turn, is connected in some manner with the cross slide. (This feature varies among different manufacturers.) This assembly lets the cross slide move as the carriage feds along the ways of the lathe, thus producing a taper.

Two types of taper attachments are generally used on the lathe: the plain and the telescopic. With the *plain taper attachment,* the bolt or nut connecting the main cross-feed screw with the cross slide must be disconnected. This permits the cross slide to move laterally as the guide shoe moves along the guide bar. With this type of attachment the compound rest is set at 90° (perpendicular) to the longitudinal axis of the lathe. At this setting the compound rest is used to control the depth of cut during taper operations.

With the *telescopic taper attachment* it is not necessary to disengage the cross-feed screw and nut. The depth of cut can also be controlled with the cross-feed handle. The axial movement of the tool is made possible by a splined, or telescopic, assembly between the cross-feed handle and cross-feed screw.

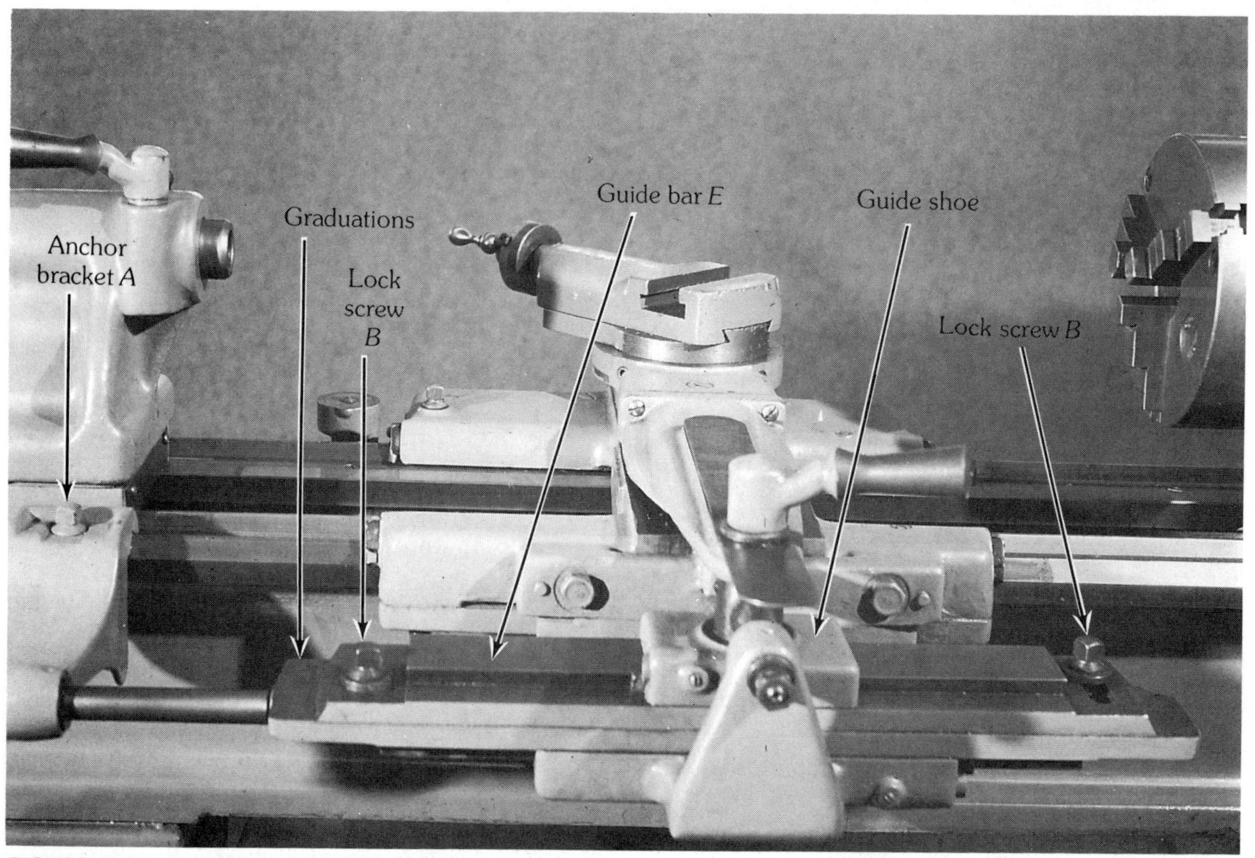

FIGURE 9.97 The most convenient way to turn a taper is to use the taper attachment.

To turn tapers with a taper attachment, follow the following nine-step procedure:

1. Loosen lock screws *B* at each end of the guide bar, set the guide bar to the angle or taper per foot desired, and tighten the lock screws.
2. Mount the workpiece in the lathe.
3. Set the cutting tool exactly on center.
4. Position the center of the taper attachment so that it is opposite the center of the workpiece; then tighten anchor bracket *A* to the ways of the lathe. (*Note:* If a plain taper attachment is used, disengage the cross-feed screw from the cross slide and use it to connect the guide shoe and cross slide.)
5. Move the carriage to make sure that there is enough travel to complete the taper cut on the workpiece. If not, readjust as necessary.
6. Position the cutting tool approximately 1 in. (25.4 mm) beyond the end of the workpiece before starting the cut. This procedure removes any backlash or play in the taper attachment.
7. Take a trial cut on the workpiece, stop the lathe, and check the taper. Readjust the taper attachment accordingly.
8. Continue to rough-turn the taper on the workpiece.
9. Finish turning the taper until the required size is obtained.

The *compound rest* is generally used for turning and boring short and steep tapers. It is the easiest method of producing a taper. However, the length of taper is limited to the amount of movement of the compound-rest feed screw unless the feed screw is retracted and the carriage is moved and locked in a new location. Figure 9.98 shows the compound rest being used to machine a taper.

To turn a taper using the compound rest, follow this procedure:

1. Loosen the compound-rest lock screws.
2. Swivel the compond rest counterclockwise to one-half the included angle required. (*Note:* Fig. 9.99 shows the correct position of the compound rest for various angles required. Tighten the lock screws.)
3. Mount the workpiece and set the cutting tool exactly on center.

FIGURE 9.98 The compound rest is used to machine short and steep tapers.

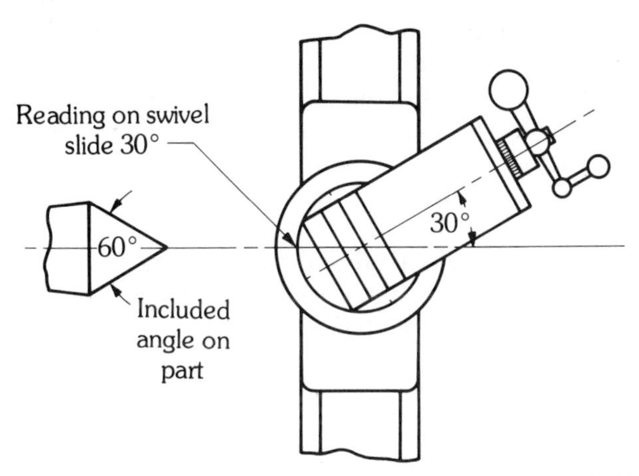

FIGURE 9.99 The compound rest is swiveled one-half the included angle to be machined on the part.

4. Move the carriage so that the cutting edge of the toolbit is close to the end of the workpiece. Lock the carriage in place.
5. Start the lathe; using the compound rest feed handle, feed the compound rest by hand to cut the taper.
6. After the first roughing cut, stop the lathe and check the taper. Readjust the compound rest if necessary.
7. Continue the rough turning procedure and finish turning the workpiece. (*Note:* Handfeed the compound rest slowly to obtain a fine finish on the workpiece during the finish-turning operation.)

9.8.4 Parting or Cutting Off Stock in a Lathe

When the workpiece has material to be removed from the end, it is common practice to cut off (or part off) this excess while the work is being held

FIGURE 9.100 Cutting off stock in a lathe requires an ample application of cutting fluid.

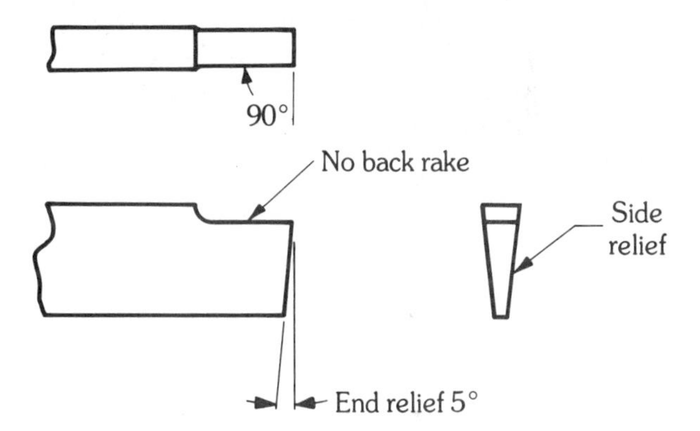

FIGURE 9.101 Proper tool geometry for a cutoff tool.

in the lathe (Fig. 9.100). In addition, often several parts are finish machined on one single piece of material and later cut off.

Cutting off stock is generally done while the workpiece is being held in a chuck or collet but *never* when it is mounted between centers. Figure 9.101 shows a typical cutoff tool with the proper clearance angles. The procedure for cutting off is as follows:

1. Mount the workpiece in a chuck or collet.
2. Set the cutoff tool exactly on center and position it at 90° to the workpiece. (*Note:* Make sure that the tool does not have excessive overhang from the toolpost. Hold the toolholder as rigidly as possible.)
3. Set the lathe rpm rate to about one-fourth the speed required for turning.
4. Move the carriage until the tool is at the

correct location on the workpiece for cutting off.

5. Lock the carriage to the bed of the lathe.
6. Start the lathe and slowly feed the cutting tool into the workpiece. If chatter occurs, reduce the rpm rate.
7. Apply ample cutting fluid directly on the groove of the part produced by the cutting tool.
8. Continue to feed the tool until the excess piece is parted completely from the work.

9.8.5 Grooving or Necking in a Lathe

Grooving operations are often done on workpiece shoulders to ensure the correct fit of mating parts [Fig. 9.102*(a)*]. When a thread is required to run the full length of the part to a shoulder, a groove is usually machined to allow full travel of the nut [Fig. 9.102*(b)*].

Grooving the workpiece prior to cylindrical grinding operations allows the grinding wheel to completely grind the workpiece without touching the shoulder [Fig. 9.102*(c)*]. Another use for grooving is the machining of a recess for O-rings or snap rings.

Whenever possible, rounded-bottomed grooves are used because they reduce the possibility of cracks in the part, especially if the part is to be heat treated.

Grooving tools are usually ground to the dimensions and shape required for a particular job. Most grooving tools are similar in appearance to the cutoff tool mentioned previously, except that the corners are carefully rounded.

9.8.6 Knurling in a Lathe

Knurling is somewhat different from most external lathe operations. It is a *forming* or embossing operation rather than a cutting operation; that is, no metal is removed. Knurling is generally done to provide a gripping surface to handles of tools, nuts, screws, and other hand tools.

Knurling the surface of a workpiece actually increases the diameter slightly because some of the metal is raised during the operation. Because of this feature, knurling is sometimes done to increase the workpiece diameter when press or interference fits are necessary.

The knurling toolholder contains small hardened wheels (called knurls). These knurls are available in diamond and straight-line patterns and in coarse, medium, and fine pitches (spacing of pattern) (Fig. 9.103). Knurling tools are also

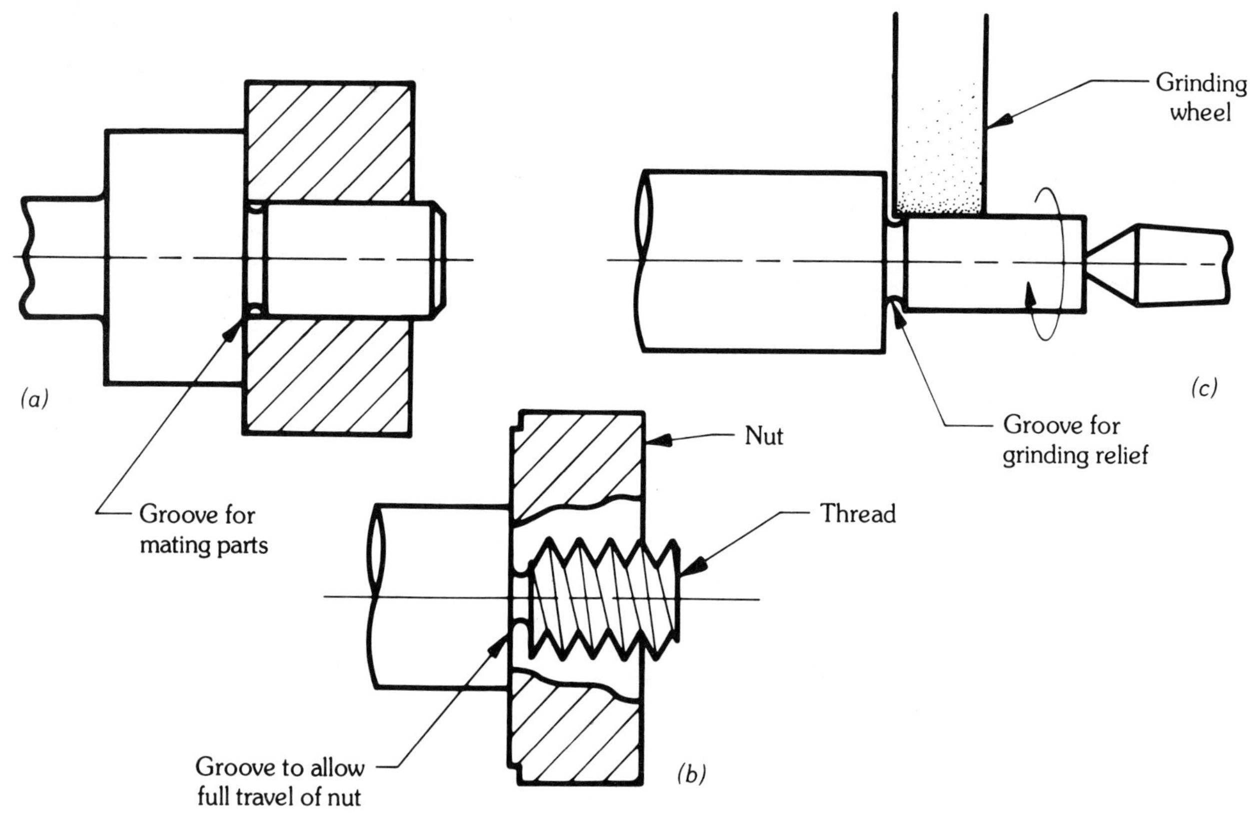

FIGURE 9.102 Grooving operations are often performed on a lathe.

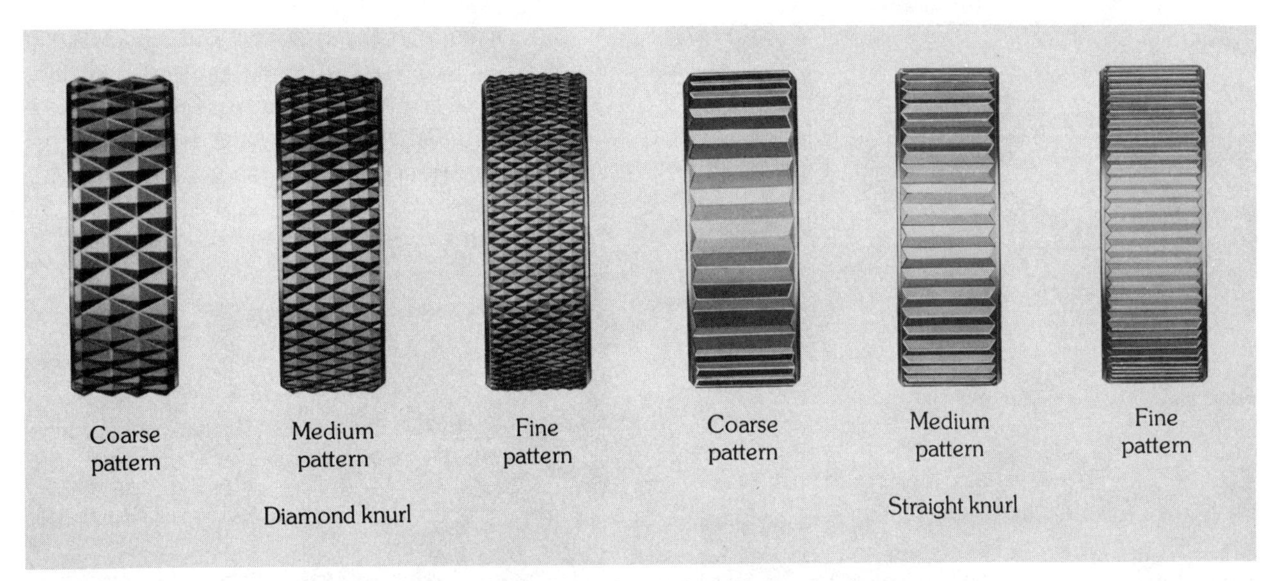

FIGURE 9.103 Various knurling patterns. (*Courtesy South Bend Lathe, Inc.*)

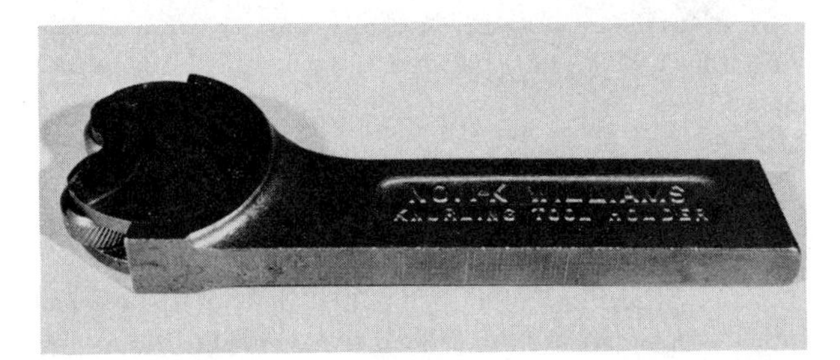

FIGURE 9.104 Self-centering knurling tool.

available with three sets of rollers mounted in the toolholder (Fig. 9.104).

The procedure for knurling is similar to that of straight turning. However, knurling requires rigid setups because of the extreme pressure used. It is best to mount the workpiece in a chuck, with the outer end supported by a tailstock center. Knurling the workpiece that is held in a collet could distort the collet due to the pressures involved.

Here is the nine-step *procedure for knurling:*

1. Mount the workpiece and mark the length to be knurled.
2. Set the knurling toolholder well back in the toolpost. Position the toolholder so that the top and bottom wheels are an equal distance above and below the center of the work (Fig. 9.105) and the face of the wheels are at a slight angle (1 to 2°) to the work. This slight angle helps set the pattern more easily.
3. Set the lathe rpm rate to approximately one-fourth the speed required for turning.
4. Set the quick-change gearbox for a coarse feed rate of 0.015 to 0.020 in.

FIGURE 9.105 Knurling in a lathe requires rigid setups.

5. Position the carriage so that the knurling tool is at the right edge of the section to be knurled.
6. Start the lathe and force the knurling tool into the work, abruptly, to approximately 0.020 to 0.025 in. (0.6 mm). Stop the lathe and check the knurl for proper form—a perfect diamond pattern should appear (Fig. 9.106). If not, readjust the height of the tool slightly and proceed at a new location on the workpiece. (*Note:* When the tool does produce the proper knurl, going over the bad section should correct it.)
7. When the diamond shape is obtained, engage the power feed and apply a liberal amount of cutting fluid.
8. When the knurling tool has reached the re-

FIGURE 9.106 Make sure that the correct knurling pattern is obtained before engaging the power feed.

quired length, *without* disengaging the power feed, stop the lathe spindle.

9. Check the knurling pattern for proper depth. If necessary, reverse the feed direction, start the lathe, and take another pass, increasing the depth of the knurling tool. Usually, two passes bring the knurl to the proper depth.

9.8.7 Filing and Polishing in a Lathe

Filing and polishing are sometimes used to finish a workpiece. *Filing* is usually done to remove sharp burrs that occur either on the end of the workpiece or on the shoulders. The file slightly rounds off these sharp corners.

Sometimes filing is done to improve the surface finish. However, it is not a substitute for a properly ground toolbit. Careless filing results in out-of-round and inaccurate work. The nine steps for filing are as follows:

1. Mount the workpiece in a lathe. Remove any rings or watches and tuck in any loose clothing that might get caught in the lathe.
2. Set the lathe rpm rate to twice the speed used for normal turning operations.
3. Disengage the feed rod and lead screw.
4. Use a single-cut smooth, mill, or long angle lathe file. Make sure that the file is equipped with the proper handle.
5. Start the lathe, hold the file handle firmly in the left hand, and support the outer end with the right (Fig. 9.107).

FIGURE 9.107 Left-handed filing in the lathe is the safest.

6. Stroke the file slowly, using light pressure, for its full length.
7. Keep a slight pressure on the file on the return stroke. Each stroke should overlap the previous by one-half the width of the file.
8. Leave only 0.002 to 0.003 in. (0.05 to 0.07 mm) on the work for filing.
9. Rub some chalk on the file to help prevent metal filings from clogging the file teeth. Clean the file frequently with a file card.

Polishing is used to produce a fine finish on the work. It is usually done on decorative pieces to give them a mirrorlike finish. The procedure for polishing involves six steps:

1. Mount the workpiece in the lathe. Remove any jewelry and tuck in loose clothing.
2. Place paper towels or shop rags on the ways of the lathe directly beneath the workpiece. This protects the bearings, ways, and other working mechanisms from abrasive grit during the polishing operation.
3. Set the lathe for the highest rpm rate possible. Disengage the lathe feed rod and lead screw.
4. Select a suitable abrasive cloth. Generally, aluminum oxide is used for steels and cast iron; silicon carbide is used with nonferrous metals. Use a piece 8 to 10 in. long (20 cm) by 1 in. wide (25 mm). Normally, 100 to 125

grit gives suitable finishes. Use finer grits for super finishes.

5. Start the lathe, place the abrasive cloth around the work and hold the ends (Fig. 9.108). Move the cloth back and forth along the workpiece. Applying cutting oil helps improve the finish and extends the life of the abrasive.

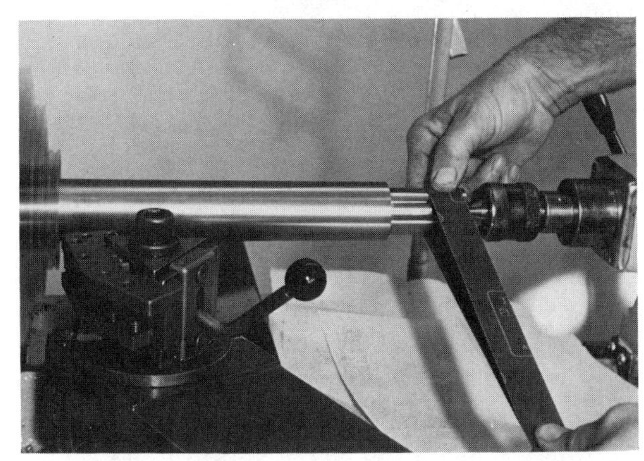

FIGURE 9.108 One method of polishing in a lathe: the left hand holds the outer end of the abrasive cloth, and the right hand secures it against the work.

6. Alternatively, place the strip of abrasive cloth on a file as shown in Fig. 9.109 and move it across the work as for filing.

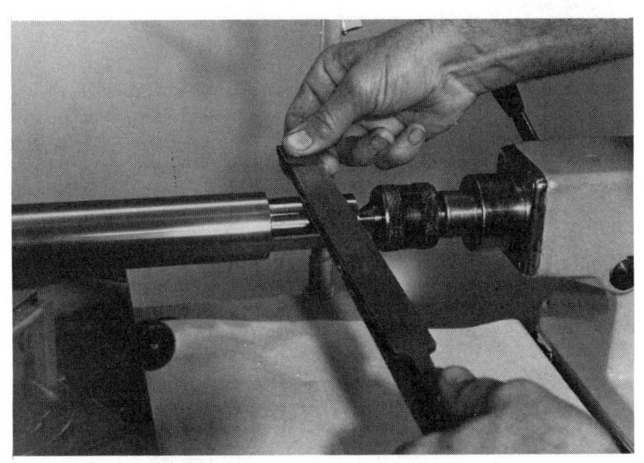

FIGURE 9.109 Another way of polishing in a lathe is to place the abrasive cloth on a file.

Hints on polishing:

1. Polishing heats the work. For accurate measurements, always cool the workpiece first.
2. Do not hold the cloth in one place because this will cause rings to be cut on the work surface. Always move the cloth from side to side.
3. A very fine (600 grit) abrasive cloth or crocus cloth gives fine finishes. Use successively finer cloths to remove any scratches. Then use the 600-grit or crocus cloth to finish.

9.8.8 Grinding in a Lathe

Grinding operations can be performed in a lathe with a toolpost grinder (Fig. 9.110). Straight or tapered surfaces can be ground on workpieces by using this attachment. Grinding in the lathe is usually performed when a cylindrical grinder is not available or when it is not convenient to move the workpiece from the lathe to a grinder. To grind in the lathe:

1. Place heavy cloth or canvas over the ways of the lathe for protection.
2. Mount the toolpost grinder on the compound rest and adjust the spindle on center.
3. Select and mount the proper grinding wheel (see Chap. 13) on the grinder.
4. Dress the grinding wheel true.
5. Mount the workpiece in the lathe.
6. Set the lathe for a slow spindle speed. (Smaller work may need higher speeds.)
7. Set the quick-change gearbox for a feed rate of approximately 0.015 to 0.020 in.
8. Start the lathe so that the workpiece is revolving in the reverse direction. (*Note:* The best practice is to have the workpiece and grinding wheel revolving in the same direction.)
9. Start the grinder and slowly feed the cross

FIGURE 9.110 Grinding a shaft in a lathe with a toolpost grinder. (*Courtesy Clausing Industrial, Inc.*)

slide until the grinding wheel just touches the revolving work.

10. Hand-feed the carriage slowly along the entire length of workpiece. This removes any high spots along the workpiece surface.
11. Feed the grinding wheel so that only 0.001 to 0.002 in. (0.02 to 0.05 mm) is removed from the diameter per pass. Use the power feed and reverse the direction at the end of each pass.
12. Take successive cuts until the work is ground to the required diameter.

Hints on grinding in the lathe:

1. Avoid heavy cuts because they tend to overheat and possibly warp the workpiece. Also, overheating a hardened part may cause the temper to be drawn.
2. For smooth and accurate finishes, grind the work to within a few thousandths of an inch and redress the grinding wheel for the finish cut.
3. When grinding precision parts that require close dimensions, always let the grinding wheel take several passes without additional in-feed. This is called sparking out the part and ensures a rounder and more accurate diameter.

9.8.9 Cutting External Screw Threads on a Lathe

The engine lathe is indispensable for machining various sizes and types of screw threads. Screw threads are machined in a lathe by using a single-point tool ground to the shape of the thread form desired (Fig. 9.111). The threading operation is accomplished on the lathe by having the carriage, which moves the threading tool, connected by gearing to the spindle. The ratio of this gearing determines the number of threads per inch that will be cut. For example, to cut 10 threads per inch on a part, the quick-change gearbox is adjusted so that the carriage moves exactly 1 in. while the spindle rotates 10 times.

The engine lathe offers many advantages over other methods of manufacturing screw threads. However, the chief disadvantage of the lathe is that it is a slower method than most. The advantages of the lathe include the following:

1. Any thread series (pitch) can be cut on the engine lathe (the quick-change gearbox determines this).

FIGURE 9.111 Threads of all sizes and forms may be cut in a lathe.

2. Threads of any diameter or length can be machined. This, of course, is determined by the capacity of the lathe.
3. Both external and internal threads can be produced.
4. By grinding the proper tool, any thread form (for example, Unified, metric, Acme) can be machined.
5. Left- or right-hand threads can be cut by reversing the lead screw.
6. Multiple-lead threads can be produced on the lathe by indexing the work, using the thread chasing dial, or manipulating the gears on the end of the quick-change gearbox.

Machining threads on the engine lathe is a very precise and exacting operation and one of the most challenging. It requires a thorough knowledge of screw thread principles and setup procedures. Chapter 6 gives detailed explanations of the different types of screw thread forms, measurement, classes of fit, and so on, that should be reviewed prior to threading on the lathe. In this chapter we deal only with the essential operations in cutting threads on the lathe.

The Threading Tool. The threading tool used on the lathe is a single-point tool ground to the *form* of the particular thread that will be machined. For example, Unified and Metric threads have a form with a 60° included angle and a specified nose radius. Acme threads have a 29° included angle and a different nose width on the tool for each pitch. The Square thread uses a form tool with square, or 90° sides. Therefore, since each type of screw thread requires a toolbit ground to that form, an accurately ground threading tool is very important in producing an accurate, clean thread.

Figure 9.112 shows the correct relief and rake angles for the Unified threading tool. The point of the tool has a 60° included angle and is checked with the center gage. The *side relief* is generally 3 to 5° on both sides of the tool. A greater amount of side clearance may be needed for coarser threads (2- or 4-pitch) to ensure that the toolbit does not rub on the side of the thread groove. The *front relief* is generally 8 to 10°. The *backrake* angle on the threading tool must be 0°. The type of toolholder used is the determining factor as to how the rake angle is ground. If a *turret* or *quick-change toolholder* is used, the top of the tool is not ground because the toolholder automatically provides the necessary 0° back rake angle (Fig. 9.113). If the lathe is equipped with the standard toolpost and a toolholder that incorporates a 16 1/2° angle, the top of the toolbit must be ground that same 16 1/2° to provide 0° back rake (Fig. 9.114).

The procedure for grinding the unified threading tool is as follows:

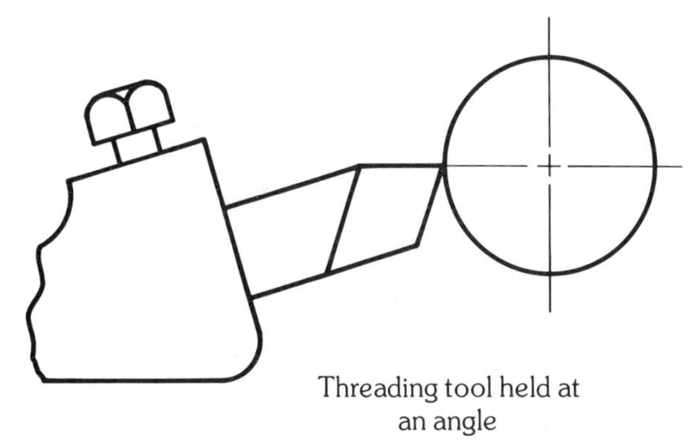

FIGURE 9.114 Threading tools that are held at an angle must have the top ground to provide zero rake angle.

1. Grind the front relief angle (8 to 10°).
2. Grind the side relief angles (3 to 5°) to form a 60° included angle.
3. Grind the top rake angle (16 1/2° downward toward the point) if necessary.
4. Carefully grind the proper radius or flat on the tip of the toolbit; this is determined by the pitch of the thread and measured by the screw thread gage (Fig. 9.115).

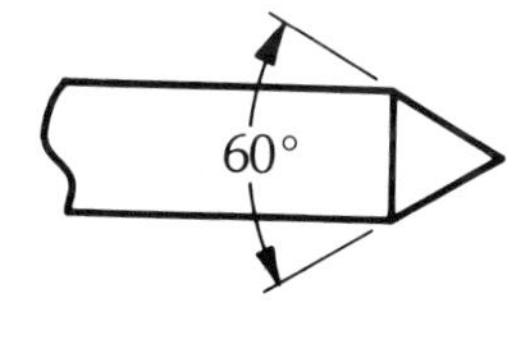

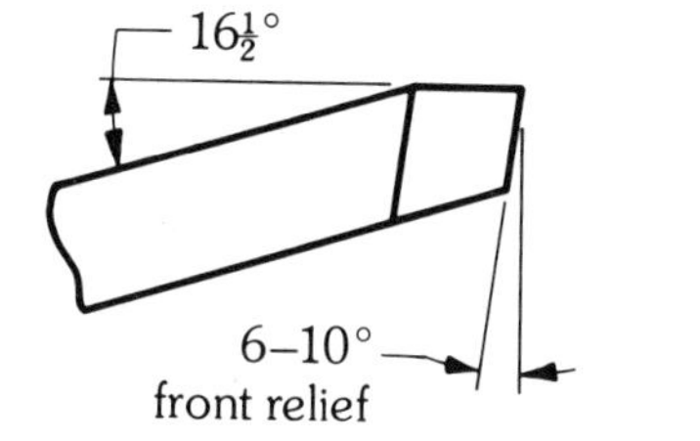

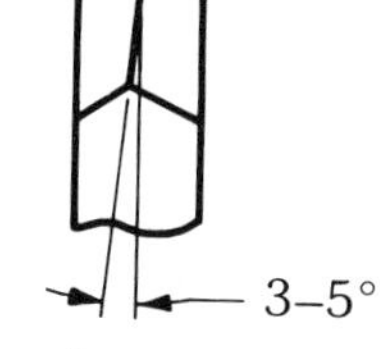

FIGURE 9.112 Unified thread tool geometry.

FIGURE 9.115 The screw-thread gage is used to measure the proper nose radius on threading tools.

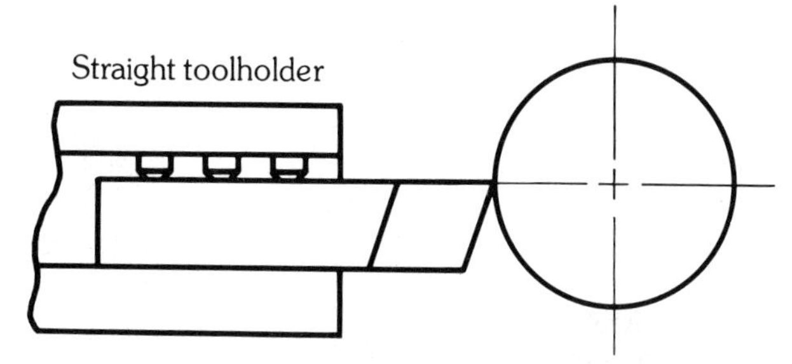

FIGURE 9.113 With straight toolholders the top of a threading tool is not ground. Thus, the rake angle is zero degrees.

Generally, on thread pitches of 20 and over, an oil stone rather than a grinding wheel is used to produce the radius.

Preparing the Lathe. Before cutting a thread on the lathe it is important to prepare the lathe properly. First the compound rest and threading tool have to be set. To set the *compound rest,* use the following procedure:

1. To cut a right-hand external Unified thread, set the compound rest at 29° to the

FIGURE 9.116 When cutting right-hand screw threads, set the compound rest at 29° to the right of the cross slide.

right of the cross slide (Fig. 9.116). For left-hand threads, set the compound rest 29° to the left. The compound rest is set at an angle for two reasons:

(a) The depth of cut is controlled with the compound rest feed screw during the "roughing-out" stages of the threading operation. This allows the left side of the threading tool to do most of the cutting. Finish cuts are taken with the cross slide, allowing the tool to feed straight in and finish both sides of the thread.

(b) With the compound rest at 29° it is possible to reset the threading tool if it has been removed for regrinding, or to finish a partially completed thread at another time. (*Note:* This procedure is discussed later.)

2. Mount the threading tool in the toolholder and place it in the compound rest. Adjust the tool exactly on center. If the tool is set above center, the included angle will be less than the angle ground on the toolbit; if set below center, the included angle will be greater. Place the V notch of the center gage over the tool, and align the straight backside parallel to the workpiece (Fig. 9.117). This procedure aligns the threading tool perpendicular to the work. Place a piece of white paper on the cross slide to help make this alignment procedure easier to see.

Next, *adjust the quick-change gearbox* for the proper number of threads per inch. The quick-change gearbox, when set correctly, gives the proper relationship between the rotation of the spindle and travel of the carriage to cut a thread of the correct pitch. Each square of the index plate on the quick-change gearbox contains two numbers. One number represents the feed of the carriage in thousandths of an inch per spindle revolution and is used when selecting feed rates for turning and facing operations. The second number, a *whole* number, indicates the number of threads per inch that will be machined during a threading operation. This index plate also gives directions as to where to place the various levers to achieve the ratio desired.

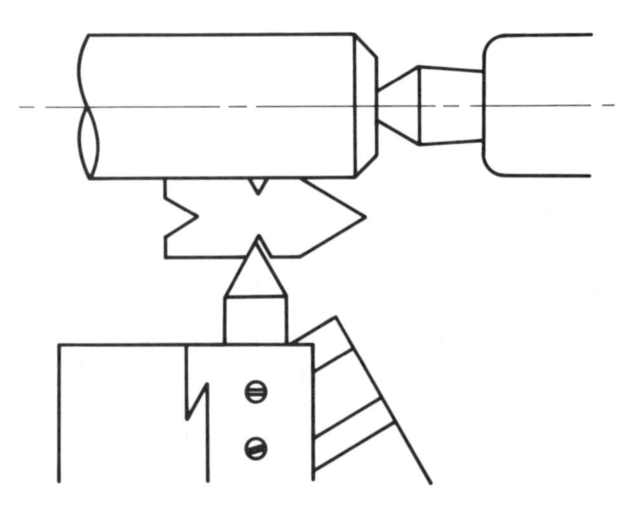

FIGURE 9.117 The center gage is used to align the threading tool square to the work.

To adjust the quick-change gearbox, follow these two steps:

1. From the blueprint or working drawing, find the number of threads per inch required on the part. For example, the drawing might read 28 UNF; the *28* indicates that 28 threads per inch are required.
2. Locate the box on the index plate that contains the *whole* number, 28 (Fig. 9.118).

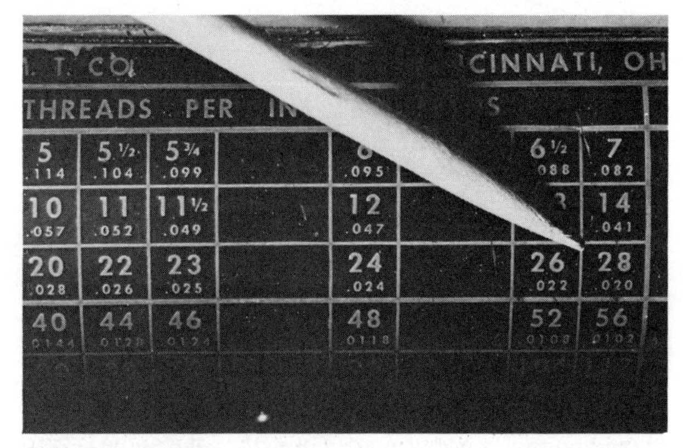

FIGURE 9.118 Locating the correct thread pitch number on the quick-change gearbox.

Place the levers (usually two or three) accordingly, as the index plate indicates. (*Note:* Lathe quick-change gearbox levers differ slightly, but most are similar. Follow the directions on the plate.)

Most modern lathes have a *thread chasing dial.* It is generally mounted to the right of the carriage (Fig. 9.119). On the end of the thread chasing dial is a worm gear that meshes with the lead screw of the lathe and causes the dial to revolve.

FIGURE 9.119 The thread chasing dial and half-nut lever are mounted to the right of the carriage.

The thread chasing dial is generally marked with eight equally spaced lines. Four of these lines are numbered; four lines are unnumbered and are called half-lines. The thread chasing dial indicates to the operator exactly when to engage the half-nut lever during the threading operation (Fig. 9.120). The thread chasing dial is used as follows:

1. Engage the half-nut lever at any point where it will mesh to produce threads that are the same or multiples of the number of threads per inch on the leadscrew. *Example:* If the lathe is equipped with an 8-pitch lead screw, 8-, 16-, 32- and so on, pitch threads may be cut by engaging the half-nut lever wherever it will enter.
2. To cut an even number of threads (6, 8, or 10), engage the half-nut lever at *any* of the eight lines on the chasing dial.
3. To cut an odd number of threads (3, 5, 7, etc.), engage the half-nut lever at *any of the numbered* lines on the dial.
4. For "half" threads (3 1/2, 6 1/2, etc.), engage the half-nut lever only at each half-revolution of the dial. Therefore, use the *numbered lines* 1 and 3 *or* 2 and 4.
5. For "quarter" threads (2 1/4, 4 1/4, etc.), engage the half-nut lever only at each full revolution of the chasing dial. In other words, select a numbered line and use *only that numbered line* during the threading operation.

FIGURE 9.120 The thread chasing dial indicates when to engage the half-nut lever for taking successive cuts during threading operations.

External Thread Cutting Procedures

1. Mount the workpiece so that the outside diameter runs true, either between centers or in a chuck with a tailstock center supporting the outer end.
2. Adjust the lathe for the proper rpm rate, about one-fourth the normal cutting speed for turning. (The beginner might first try a speed under 100 rpm.)
3. Set the compound rest at 29°, swinging the handwheel end toward the tailstock for right-hand threads.
4. Mount the toolbit in the toolholder and adjust exactly on center. With a center gage align the threading tool perpendicular to the workpiece. Tighten the toolpost toolbit and assembly.

5. Set the quick-change gearbox for the proper threads per inch.
6. Start the lathe and chamfer the end of the work with the side of the threading toolbit. The depth of chamfer should approximately equal the minor diameter of the thread.
7. With the lathe still running, *slowly* turn the cross-feed handle until the tool just scratches the work. Zero the cross-feed and compound-rest micrometer collars.
8. It is good practice, especially for the beginner, to machine an undercut at the end of the thread before threading on the lathe. Figure 9.121 shows the proper undercut on the work. This undercut makes it easier to pull the threading tool out at the end of each successive pass. This particular undercut was made with the threading tool to the exact minor diameter of the thread. Other types of undercuts might have round or square bottoms. Table 9.3 lists the amount of infeed of the compound rest when set at 29° for various thread series. Using the figure from this table, feed the tool to the proper depth when machining an undercut.

FIGURE 9.121 A small groove or undercut is machined before the threading operation begins so that the threading tool is easier to pull out at the end of the cut.

TABLE 9.3

Depth settings for cutting Unified and ISO Metric threads with compound rest at 29°

Unified Threads per Inch	*Depth of Feed (in.)*	*Metric Pitch*	*Depth of Feed (mm)*
40	0.017	0.45	0.26
36	0.019	0.50	0.30
32	0.022	0.60	0.36
28	0.026	0.70	0.43
24	0.029	0.80	0.48
20	0.036	1.00	0.61
18	0.040	1.25	0.77
16	0.045	1.50	0.92
14	0.051	1.75	1.07
13	0.056	2.00	1.23
11	0.066	2.50	1.55
10	0.073	3.00	1.84
9	0.080	3.50	2.15
8	0.090	4.00	2.47
7	0.105	4.50	2.76
6	0.122	5.00	3.08
4	0.183		

9. Position the tool so that it clears the end of the workpiece.
10. Using the *compound-rest feed handle,* feed the tool 0.002 in. for a *trial cut.*
11. Place your left hand on the cross-feed handle; with the right hand, engage the half-nut lever at the correct line on the thread chasing dial.
12. At the end of the cut, disengage the half-nut lever and back out the tool by turning the *cross-feed handle* counterclockwise.
13. Stop the lathe and check for the proper number of threads per inch with a thread pitch gage or rule (Fig. 9.122).
14. Return the carriage to the beginning of the thread and turn the *cross-feed handle until the micrometer collar reads zero.* (*Note:* Repeat this procedure on each successive pass.)

FIGURE 9.122 Checking the proper number of threads per inch using a thread pitch gage.

15. Continue to take roughing cuts by feeding the tool with the *compound rest.* The amount of feed per cut must decrease on successive passes because the thread becomes deeper and a broader face is contacted by the threading tool. Table 9.3 gives the total amount of in-feed with the compound rest set at 29°. Use this table as a guide for roughing cuts.
16. Two or three finish cuts of 0.001 in. per pass with the *cross slide* should be sufficient to bring the thread to the proper fit. (See Chap. 6 for classes of fit and methods of measuring threads.) If you use the cross slide for finish cuts, the threading tool will machine both sides of the thread, thus giving each the proper finish.

Hints on threading:

1. Apply a suitable cutting fluid during the threading operation.
2. Just prior to the finishing cuts, *lightly* file the tops of the thread to remove any burrs that may have been formed during the roughing cuts and that might cause the thread to be oversized and have a rough finish.
3. Avoid engaging the half-nut lever at the wrong line on the thread chasing dial because this usually ruins a thread.
4. To cut a *left-hand thread,* observe the following five differences:
   - *(a)* Set the compound rest to 29° to the *left* of the cross-feed handle.
   - *(b)* Machine an undercut at the end of the length of the thread. This is necessary because the threading tool starts its cut at this groove.
   - *(c)* Reverse the direction of the lead screw so that the carriage moves toward the tailstock.
   - *(d)* Start the threading operation at the undercut and machine the thread by feeding toward the outer end of the workpiece.
   - *(e)* Always use a ball bearing tailstock center when cutting left-hand threads since the thrust forces are toward the tailstock.

During the threading operation it sometimes becomes necessary to *reset the threading tool* (realign the tool to follow the original thread groove) for any of the following four reasons:

1. The tool becomes dull and needs resharpening.
2. For various reasons, the tool gets out of alignment.
3. To machine threads on a previously threaded part.
4. To machine partially completed threads on the work at a later time.

Resetting the tool is best accomplished by using the cross-feed screw and the compound-rest feed screw to move the tool to the correct position. Here is the six-step procedure:

1. Mount the threading tool so that it is on center and properly aligned to the workpiece. Make sure the tool is clear of the work.
2. Start the lathe and engage the half-nut lever at the proper line on the thread chasing dial.
3. Let the carriage move a short distance, and then stop the lathe. (*Note:* Do not disengage the half-nuts.)
4. Position the threading tool into the thread groove by moving both the compound-rest and cross-feed screws (Fig. 9.123).
5. Set the cross-feed micrometer collar to zero.
6. Back out the tool by using the cross-feed handle, disengage the half-nuts, and proceed with the threading operation.

Cutting Metric (ISO) Threads in a Lathe. Many nations have now adopted metric (ISO) screw

FIGURE 9.123 To reset a threading tool, turn the cross-feed and compound-rest handles.

threads as their standard thread. The metric thread form is identical to that of the Unified: both having an included angle of 60°. The basic difference between the two is the designation of the pitch. Unified threads are specified in terms of the number of threads per inch, whereas on metric threads the pitch is stated as the distance between the crests of the adjoining threads.

Metric threads can be cut on inch-based lathes by using two change gears of 50 and 127 teeth. The number of teeth on these two gears represents the relationship between the English (inch) and the metric (millimeter) systems of measurement. Mathematically, this relationship is expressed as follows: 1 in. is equivalent to 2.54 cm. Thus the ratio is

$$\frac{1}{2.54}$$

Therefore, to find the change gears, we multiply the top and bottom of this fraction by 50:

$$\frac{1 \times 50}{2.54 \times 50} = \frac{50}{127}$$

When we place these two gears in the gear train of the lathe—the 50-tooth gear on the spindle and the 127-tooth gear on the lead screw—the lathe is geared to cut a given number of threads per centimeter. For example, if a metric thread is to be cut with a pitch of 1.25 mm we must next convert this to the number of threads per centimeter:

$$\frac{10 \text{ mm}}{1.25 \text{ mm}} = 8 \qquad (\textit{Note: } 10 \text{ mm} = 1 \text{ cm})$$

The next step is to set the quick-change gearbox to 8 threads per inch. With the 50- and 127-tooth gears mounted in the gear train, the lathe will cut 8 threads per centimeter, which is the same as a 1.25-mm pitch thread.

After the quick-change gearbox has been set for the proper threads per centimeter, the setup procedures are essentially the same as for cutting a Unified thread, with one important difference: When cutting metric threads on an inch-based lathe, the thread chasing dial is ignored. That is, once the half-nut lever is engaged, it is *never* disengaged during the entire threading operation. The tool is returned to the starting position for successive passes, first, by pulling the cutting tool out of the thread at the end of the cut, and second, by reversing the spindle rotation to return the tool to the starting position.

Cutting Acme Threads in a Lathe. The Acme screw thread was developed to carry heavy loads. It is gradually replacing the square thread because it is much easier to manufacture. The Acme thread has an included angle of 29°. The cutting tool is ground to form an included angle of 29° and is measured with an Acme thread gage. The Acme thread gage is used as follows:

1. Grind the side-cutting edge angles to form an included angle of 29°. Check this angle with the large V notch on the gage. Be sure to grind enough clearance on the sides to prevent them from rubbing during the threading operation. On large-diameter Acme threads the helix angle of the thread must be taken into consideration when grinding the tool, or the side of the tool will rub on the work.
2. Carefully grind a flat on the end of the Acme thread tool to fit the proper notch along the side of the gage (Fig. 9.124). Selecting the proper notch depends on the number of threads per inch to be cut. For example, if 4 threads per inch are to be cut, grind the flat to fit the notch marked 4.
3. The Acme thread gage is also used to correctly align the threading tool (as a center gage is used for a V thread) during the setup procedure.

When an Acme thread (Fig. 9.125) is being cut, there should be a clearance of 0.010 in. between the top of the thread of the screw and the bottom of the thread of the nut it fits. To get this clearance, the screw is cut 0.010 in. deeper on the minor diameter on each side of the thread (therefore, a

FIGURE 9.124 Measuring the correct flat on the nose of an Acme threading tool.

FIGURE 9.125 Machining an Acme thread on a lathe. (*Courtesy Aloris Tool Co.*)

total of 0.020 in. on the diameter) and 0.010 in. deeper on the major diameter of the internal thread or nut. This clearance of 0.010 in. is necessary to prevent the screw and nut from binding during assembly.

The half-angle of an Acme thread is 14 1/2°; therefore, when cutting right-hand external threads, position the compound rest at 14° to the right of the cross-feed handle and 14° to the left for left-hand threads. An angle of 14° rather than 14 1/2° is used because it is slightly less than half of the included angle of 29°.

The remainder of the procedure for cutting an Acme screw thread is essentially the same as for the Unified thread, except for the depth of the cut. Because the Acme threading tool has a larger flat on the nose, smaller depth cuts are necessary to prevent the tool from chattering or digging in during the operation.

Hints on cutting Acme threads:

1. It is a good procedure when cutting an Acme with a larger pitch, such as 2, 3, 4, or 6, *first* to grind a threading tool one size smaller. The undersized tool is used to remove most of the material. In other words, to cut an Acme having two threads per inch, use a tool ground for three threads; rough-out the thread and then use the proper tool (two threads per inch) to finish the thread. The procedure generally provides a better finish on the thread.
2. When cutting an Acme thread, turn a short section of 1/8 to 1/16 in. on the end of the work to the minor diameter plus 0.010 in. clearance. This helps indicate when the thread is to the full depth during the thread-cutting procedure.

9.9 INTERNAL MACHINING OPERATIONS

Many internal machining operations are performed on the engine lathe. Internal operations require that the workpiece be held in a chuck rather than between centers so that the end of the workpiece is accessible. When internal operations are being performed on a workpiece and the cutting tool is held in the tailstock, such as a drill, it is *essential* that the tailstock be aligned accurately with the headstock. Otherwise, the hole may not be concentric to the workpiece, and it could break the cutting tool.

9.9.1 Drilling a Hole in a Lathe

Perhaps the most common internal operation performed on the lathe is drilling a hole (Fig. 9.126). Drilling is also performed prior to reaming, boring, turning internal tapers, and internal threading operations. The procedure for drilling a hole is as follows:

1. Mount the workpiece in a suitable chuck. Make sure that the work is running true. [*Note:* If the hole must be concentric with the outer diameter, an independent chuck or collet chuck (for smaller diameters) may be best as the work-holding device.]
2. Face the end of the workpiece square to ensure that the center drill will start properly. A rough end may cause the

FIGURE 9.126 Drilling a hole using a straight-shank drill bit and a drill chuck. Use ample cutting fluid during drilling operations.

FIGURE 9.127 Center drilling a part before a drilling operation helps the larger drill start on center.

center drill to walk off center, causing the drill hole to be eccentric or crooked.

3. Install a drill chuck into the tailstock spindle. Make sure that the spindle and drill chuck shank are clean and free of burrs.
4. Tighten a center drill in the drill chuck. (*Note:* A center drill is used prior to the twist drill to ensure that the twist drill starts and drills straight.)
5. Position the tailstock so that the center drill is approximately 1 in. (25.4 mm) away from the end of the workpiece. Tighten the tailstock in place.
6. Set the lathe for a spindle speed of 900 to 1000 rpm. Higher spindle speeds prevent the small point of the center drill from breaking.
7. Start the lathe, and with the tailstock handwheel slowly feed the center drill. The center drill should be fed into the end of the work approximately half of the length of the tapered portion. Use a cutting fluid several times during the center drilling operation to clear the center drill of chips (Fig. 9.127).
8. Stop the lathe, remove the center drill, and place the correct size of twist drill into the drill chuck and tighten it. (*Note:* Twist drills over 1/2 in. in diameter have tapered shanks and are mounted directly into the tailstock spindle in place of the drill chuck. However, when you are drilling holes over 1/2 in. in diameter, it is first desirable to drill a pilot hole using a drill equal to or slightly smaller than the web thickness of the larger drill. See step 15 for mounting tapered shank drills.)
9. Position the tailstock so that the point of the drill just clears the end of the work, and tighten it.
10. Set the lathe spindle to the proper speed for the type of material and size.
11. Start the lathe and slowly feed the drill into the work. If the hole is deep, back out the drill frequently to help clear the chips and prevent the drill from binding.
12. Apply a suitable cutting liquid during the drilling operation.
13. Reduce the feed pressure as the drill breaks through the work.
14. Disengage the tailstock binder nut and pull back the tailstock to remove the drill from the work.
15. When using drills over 1/2 in. in diameter with tapered shanks, remove the drill chuck and install the drill directly into the tailstock spindle. To prevent larger drills from slipping in the tailstock, install a dog on the shank, and position the tail of the dog on the compound rest, as shown in Fig. 9.128.

FIGURE 9.128 When using taper shank drills in the tailstock, use a lathe dog to prevent the drill from twisting in the tailstock.

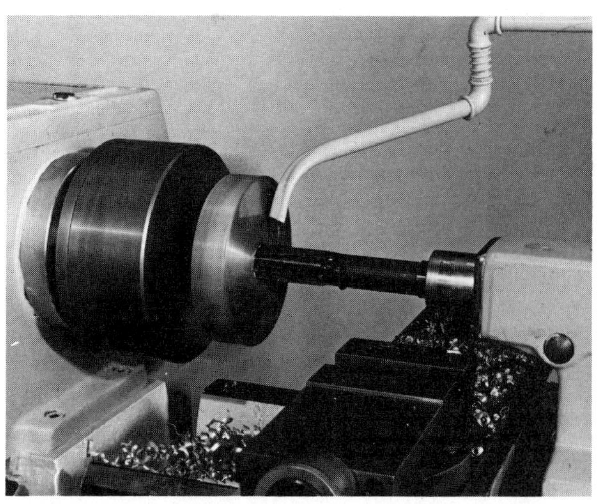

FIGURE 9.129 Reaming in a lathe requires ample quantities of cutting fluid and one-half the rpm rate of an equal-sized drill.

Hints on drilling on a lathe:

1. To gage the depth of a blind hole, use the graduations on the tailstock spindle.
2. To drill a hole on longer workpieces that have large diameters, use the steady rest to support the outer end during the drilling operation.
3. When you are drilling holes over 1/2 in. in diameter, it is often desirable first to *pilot* drill the hole with a smaller drill. The diameter of the pilot drill should be equal to or slightly smaller than the web thickness of the larger drill.

9.9.2 Reaming a Hole in a Lathe

Reaming improves the accuracy of diameter and the finish of a drilled hole. Chapter 8, in which we describe and explain the operation of various types of reamers, should be examined carefully prior to using these tools. Here is the procedure for reaming in a lathe:

1. Mount the workpiece in the lathe and face the end.
2. Drill the proper-sized hole prior to reaming. For holes under 1 in. (25 mm), drill a hole 1/64 in. (0.4 mm) undersized; for holes over 1 in. (25 mm), drill the hole 1/32 in. (0.8 mm) undersized.
3. Mount the proper-sized reamer in a drill chuck or the tailstock spindle. Remember that tapered-shank cutting tools require a lathe dog to prevent them from slipping in the spindle.
4. Adjust the spindle speed to aproximately one-half the drilling rpm rate, or slower.
5. Start the lathe and slowly feed the reamer into the drilled hole. Apply a suitable cutting fluid during the reaming operation.
6. Back the reamer out of the hole occasionally to allow the chips to clear from the reamer and hole. Continue to apply a cutting fluid (Fig. 9.129).
7. When the reamer reaches the required length, stop the lathe and back the reamer out of the hole.

Hints on reaming:

1. Never reverse the spindle direction during reaming because this may damage the reamer's cutting edges.
2. As mentioned, reaming produces a hole with a smooth finish and of an accurate size. However, it is important to remember that the reamer does not correct an improperly drilled hole. That is, if the hole is drilled off center or crooked, the reamed hole also will be off center or crooked. Therefore, when extreme accuracy is required, first *bore* the drilled hole to produce a straight hole, then ream it to size. (Boring is explained in the next section.)
3. Hand-reaming operations are sometimes done on the lathe. The procedure for hand reaming is similar to that for machine reaming, except that the reamer is turned by hand (Fig. 9.130). The outer end of the reamer is supported by a tailstock center,

FIGURE 9.130 When hand reaming in a lathe, the work does not revolve.

and an adjustable wrench or a small tap handle turns the reamer.

9.9.3 Boring a Straight Hole in a Lathe

Boring is the operation of producing very straight and accurate holes of any diameter. Boring is often done to straighten a drilled hole prior to the reaming operations. Boring in a lathe involves 13 steps:

1. Mount the workpiece in the lathe facing the end.
2. Drill a hole approximately 1/16 to 1/8 in. (1.5 to 3 mm) undersized.
3. Select the largest-diameter boring bar available that will fit the drilled hole and mount it in the boring bar toolholder (Fig. 9.131). (*Note:* Do not extend the boring bar any farther than necessary to reach the proper depth of hole to be bored. This reduces the possibility of chatter.)
4. Adjust the boring bar on center. Tighten the boring toolholder first and then the boring bar in the toolholder.
5. Adjust the lathe for the proper spindle speed according to the diameter and type of material to be machined.
6. Start the lathe and move the carriage and cross slide so that the toolbit will just touch the surface inside the drilled hole.
7. Back the tool out of the hole and set the depth of cut for the roughing cut. [*Note:* Leave approximately 0.015 in. (0.3 mm) for the finish cut.]
8. Engage the power feed and take the roughing cut. Apply a suitable cutting fluid during the boring operation.
9. Disengage the feed at the end of the cut, stop the lathe, and back the boring tool out of the hole.
10. Carefully measure the diameter of the bored hole with a telescopic gage, inside micrometers, or vernier caliper (Fig. 9.132).

FIGURE 9.131 Select the largest boring bar that will fit into the hole during a boring operation in a lathe.

FIGURE 9.132 Checking a bored hole size using a vernier caliper.

11. Adjust the tool for the proper depth of cut for the finish cut, start the lathe, and take a trial cut of about 1/4 in. (6.3 mm).
12. Stop the lathe spindle and measure the finish cut to make sure that it is the correct size. Readjust the depth of cut as necessary.
13. Start the lathe and continue the finish cut, applying a cutting fluid.

Hints on boring in a lathe:

1. Toolbits used in boring bars are similar to turning toolbits. They must have end and side relief and side rake. End relief on a boring toolbit must be increased as the diameter of the hole to be bored decreases. Always use a sharp tool when boring a hole.
2. If chattering develops during the boring operation, reduce the lathe spindle speed, or try a smaller depth of cut. Sometimes chatter occurs if too small a boring bar is being used or the bar has excessive *overhang.* An excessively large nose radius on the toolbit may also cause chatter.
3. Tapered holes can be bored in the lathe by using either the compound rest or the taper attachment (Fig. 9.133).

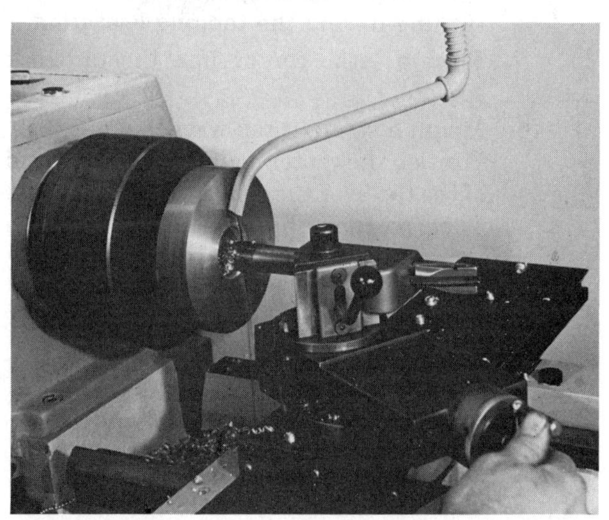

FIGURE 9.133 A tapered hole being bored using the compound rest.

9.9.4 Cutting an Internal Thread in a Lathe

Internal threads are cut in the lathe with a boring bar. Internal threading is similar to both external threading and boring operations. Generally, internal screw threads under 1/2 in. (13 mm) in a diameter are machined on the lathe by using a tap (Sec. 9.9.5). To cut internal threads:

1. Mount the workpiece and drill or bore a hole of the correct size for the thread to be machined (see Appendices L and M for tap drill sizes). Determine this correct size hole by dividing the number of threads per inch to be machined into 1 and subtracting the result from the major diameter of the thread, or simply:

$$\text{sz of hole} = \text{major dia} - \frac{1}{\text{no. of thds/in.}}$$

For example, if a 1–8 UNC thread is required, the size of the hole would be:

$$\text{size of hole} = 1 - \frac{1}{8} = \frac{7}{8}$$

Therefore, the hole size required would be 7/8 in. This formula provides 75 percent of the thread depth for the internal thread. For all practical purposes, this is sufficient.

2. Set the compound rest 29° to the *left* of the cross slide.
3. Set the quick-change gearbox for the required proper number of threads per inch and adjust the lathe spindle for one-fourth the speed of boring.
4. Install an internal threading tool that has been ground to the proper thread form into a suitable boring bar.
5. Place the boring bar into the boring bar toolholder, and adjust the toolbit on center.
6. With a center gage, set the threading tool at right angles to the work, as shown in Fig. 9.134. After the tool has been set on center at right angles to the work, carefully tighten the toolholder and boring bar in place.
7. Start the lathe and slowly bring the threading tool toward the workpiece until it just scratches the surface inside the hole.

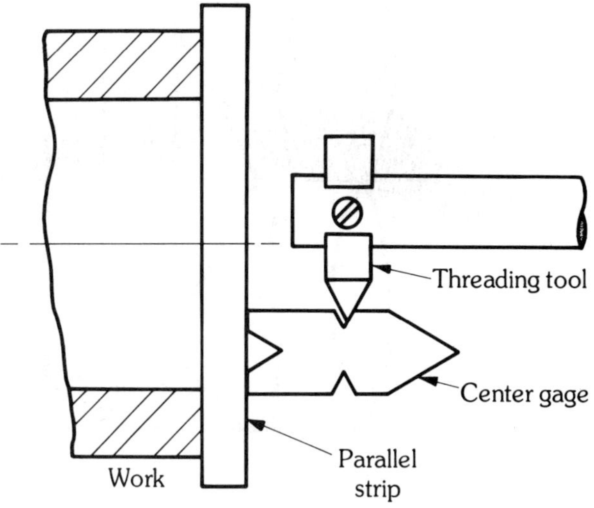

FIGURE 9.134 Using a center gage to properly set an internal threading tool square to the work.

8. Stop the lathe and set the micrometer collars on the cross-feed and compound rest to zero.
9. Back the threading tool out of the hole, start the lathe, and proceed by taking successive roughing cuts, using the compound rest to feed the tool. The roughing cuts should be small for internal threading (0.003 to 0.005 in.) to prevent unnecessary chatter during the operation.
10. As in external threading, several finish cuts of 0.001 in. should be taken with the *cross feed* until the thread is at the proper depth.
11. Deburr the crests of the threads with a scraper.

Hints on internal threading:

1. Remember to feed the tool *in* after each successive pass before returning it to the beginning of the thread. This procedure is just the opposite of that for external threading.
2. When cutting an internal thread only partway in a hole, machine a recess equal to the major diameter of the thread at the end of thread (Fig. 9.135). This recess provides a space for the threading tool at the end of each cut.
3. Boring a slight recess at the beginning of the thread and equal to the major diameter helps determine when the proper depth has been reached (Fig. 9.135).
4. A mark of some kind on the boring bar helps indicate the proper place to disengage the half-nut when the thread length is reached.

To cut internal threads in a lathe *with a tap:*

1. Mount the workpiece in the lathe.
2. Face the end and drill the correct-size hole for the thread to be tapped. The correct hole size may be found in Appendix F or by using the formula discussed in the procedure for cutting internal threads.
3. Slightly countersink the end of the hole to help start the tap straight.
4. Set the lathe for the lowest rpm rate available. This keeps the spindle from turning during the tapping operation.
5. Place a lathe center into the tailstock spindle.
6. Mount a straight tap wrench on the tap. Position the center drilled hole on the end of the tap over the center in the tailstock. Then position the tailstock so that the tapered portion of the tap thread is located in the drilled hole (Fig. 9.136).
7. Apply a suitable cutting fluid.
8. With the tailstock handwheel, apply slight pressure to the end of the tap; at the same time, turn the tap clockwise (for right-hand threads) into the hole.

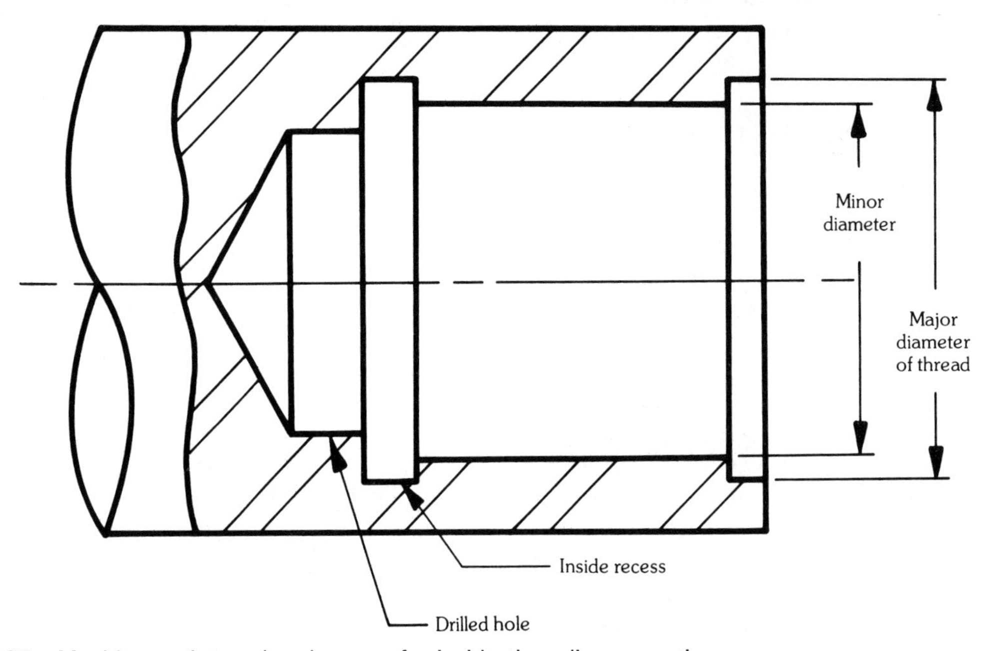

FIGURE 9.135 Machine an internal end recess for inside threading operations.

FIGURE 9.136 Hand tapping in a lathe requires a tailstock center to support the outer end of the tap.

9. Continue to turn the tap while supporting the end with the center. If the tap becomes difficult to turn, back the center away from the tap and turn the tap counterclockwise to break the chips.
10. Proceed with the tapping operation until the threaded hole is the proper length.

FIGURE 9.137 Large capacity lathe with follower rest and steady rest. Note that the bed rests directly on the floor. (*Courtesy Summit Machine Tool Manufacturing Co.*)

While the lathe is one of the oldest machine tools, it is also one of the most versatile. Continuing development by machine tool manufacturers has helped to produce lathes that are more powerful, more precise, and capable of a greater variety of operations. The large industrial lathe shown in Fig. 9.137 is an example of this trend toward greater sophistication and power in machines.

REVIEW QUESTIONS

9.2 TYPES OF ENGINE LATHES

1. Compare and contrast bench and floor model engine lathes.

9.3 LATHE CLASSIFICATION BY SIZE

1. What two measurements are used to indicate the size of a lathe?
2. In what common size range are most of the lathes used in industry?

9.4 LATHE CONSTRUCTION FEATURES AND FUNCTIONS

1. Briefly describe the functions of the five major components of a lathe.
2. Why is it important to protect the ways of a lathe? How should they be protected?
3. Explain the major differences between the feed rod and lead screw.

9.5 LATHE ACCESSORIES AND ATTACHMENTS

1. What is the difference between workholding and work-supporting devices?
2. Explain the major differences between the universal and independent chucks.
3. List two types of collet chucks and describe their differences.
4. When is a faceplate used on a lathe?
5. List and explain the application of the two types of lathe centers.
6. Describe the function of the following four devices:
   (*a*) Steady rest
   (*b*) Follow rest
   (*c*) Mandrel
   (*d*) Arbor
7. Describe the differences between the following toolholders:
   (*a*) Rocker
   (*b*) Quick-change
   (*c*) Turret
   (*d*) Open side
8. What is the purpose of the taper attachment?

9.6 CUTTING TOOL MATERIALS

1. List the advantages and disadvantages of the following cutting tools:

(a) High-speed steel
(b) Cast alloys
(c) Cemented carbides
(d) Oxides

9.7 CUTTING TOOL NOMENCLATURE AND GEOMETRY

1. Explain the following terms:
   (a) Face
   (b) Cutting edge
   (c) Flank
   (d) Nose
   (e) Base
2. What is the purpose of the relief angle on a lathe cutting tool?
3. What are the two rake angles used on toolbits? Why are they necessary?
4. Why are cutting edge angles ground into a toolbit?
5. What are two purposes of the nose radius?
6. Explain why the general-purpose toolbit is used for most lathe operations.
7. Briefly describe the steps necessary for grinding a general-purpose toolbit.
8. What is cutting speed?
9. Calculate the proper rpm rate for the following situations:
   (a) Machining a 1 3/4-in. piece of aluminum at 200 ft/min
   (b) Turning a 6 1/8-in. piece of soft cast iron at 84 ft/min
   (c) Turning a 30-mm piece of brass at 40 m/min
   (d) Machining a 12.7-mm piece of alloy steel at 19 m/min
10. What feed rates are generally used for roughing and finishing cuts?

9.8 EXTERNAL LATHE OPERATIONS

1. Why is facing done on a workpiece?
2. State the three different types of spindle noses, and briefly describe the setup procedure for changing work-holding devices.
3. Why is it important to make sure that the spindle nose and mating surface are clean before mounting a work-holding device?
4. How can accurate lengths be machined on the work when facing?
5. Briefly describe how to center a workpiece accurately in an independent chuck.
6. Explain the procedure for drilling center holes in the lathe.
7. Briefly describe the three methods of aligning the tailstock center in line with the headstock center.
8. How can the burr be eliminated from the center hole while facing the work being held between centers?
9. Define straight and taper turning operations.
10. Explain the difference between roughing and finishing cuts.
11. Describe the differences between direct and indirect micrometer collars.
12. What is backlash? How can backlash influence tool settings?
13. Sketch the various types of shoulders, and explain the differences between each type.
14. List five types of tapers commonly found on machine tools and equipment.
15. Briefly explain four methods of measuring tapers.
16. How can tapers be machined on a part using the lathe? What are the advantages of each method?
17. Calculate the amount of offset required for the tailstock when the taper is 4 1/2 in./ft.
18. Briefly describe the procedure for cutting off work in the lathe.
19. State three reasons for grooving a part.
20. What is the purpose of knurling?
21. What are the feed rate and rpm rate used during a knurling operation?
22. Explain the differences between filing and polishing in the lathe.
23. Why is it safer to file left-handed in the lathe?
24. What happens to the work when the operator presses too hard on the file during the filing operation?
25. What rpm rate should be used for polishing in the lathe? Fast? Slow? Why?
26. Why must the ways of the lathe be protected during the lathe grinding operations?
27. List seven reasons why the lathe is used to machine screw threads.
28. Briefly describe the procedure for grinding a unified threading tool.

29. Why is the compound rest set to 29° during a lathe threading operation?
30. Explain why it is necessary to have the cutting tool exactly on center while machining a thread on a lathe.
31. When is the half-nut lever engaged while threading odd number pitches?
32. What role does the quick-change gearbox have during threading operations?
33. How does the procedure differ when machining a left-hand thread versus a right-hand thread?
34. Briefly describe how to reset the threading tool during a threading operation.
35. How can metric threads be machined on an inch-based lathe?
36. What are the differences in the setup procedures for machining an Acme thread versus a Unified thread?

9.9 INTERNAL MACHINING OPERATIONS

1. Briefly describe the setup procedure for drilling a hole using the lathe.
2. What can be done to prevent a tapered-shank drill from spinning or slipping in the tailstock spindle?
3. What rpm rate is used while reaming in a lathe?
4. Explain how a crooked drilled hole can be machined straight prior to reaming in a lathe.
5. What size boring bar should be used when boring a hole using a lathe?
6. How can chattering be eliminated during a boring operation?
7. Calculate the correct-size hole for machining a 1 1/4–16 UNF Unified internal thread.

10

Special Lathes

10.1 INTRODUCTION

The basic engine lathe, which is one of the most widely used machine tools, is very versatile when used by a skilled machinist. However, it is not particularly efficient when many identical parts must be machined as rapidly as possible. As far back as 1850 there were efforts to develop variations of an engine lathe that could be operated by a relatively unskilled person for mass producing machined parts. The cutting tools were preset, or "set up" by a skilled machinist, and usually several cutting tools were in operation at the same time, reducing the time spent in machining each part. This is still the basic concept on which mass-production type lathes are based.

The turret lathe and automatic screw machine, in their various forms, have been developed and improved with the objectives of producing machined parts more rapidly and accurately at lower cost. On most machines of this type, the power available at the spindle has been greatly increased to take advantage of better cutting tool materials. Mechanical power, in electrical, hydraulic, or pneumatic form, has replaced human muscle power for such functions as feeding tools, operating chucks or collets, and feeding bar stock in the machine.

Because of the complexity of such machines, planning the sequence of operations, tool selection, and quality assurance have become highly special-

Opening photo courtesy of Hardinge Brothers, Inc.

ized operations done by production planners or, in some cases, by experienced *set-up people.* The actual tooling of the machine and the trial runs are done by setup experts who usually were experienced machinists before they learned this specialty.

10.2 TURRET LATHES

Turret lathes are *horizontal* or *vertical* (see Figs. 10.1 and 10.2), depending on the location of the axis of the spindle of the machine. Regardless of type, all turret lathes share some common characteristics. All turrets are mounted on a slide or on the ways; most vertical turrets have provisions for mounting five tools, and horizontal turrets have provisions for mounting six or more tools. Each tool does one or more machining operations on the workpiece.

Ram-type turret lathes have a cross slide that accommodates four or more tools and that can be fed laterally and longitudinally. The versatility of the turret lathe makes it a valuable machine tool in almost all types of manufacturing.

FIGURE 10.1 Horizontal turret lathe. (*Courtesy Bardons and Oliver Inc.*)

10.2.1 Horizontal Turret Lathes

Horizontal turret lathes are of the *ram* (see Fig. 10.3) or *saddle* type (see Fig. 10.4) and can be set up for either *chucking* or *bar* work. These lathes are manufactured in a wide range of sizes and can

FIGURE 10.2 Vertical turret lathe. (*Courtesy Summit Machine Tool Manufacturing Corp.*)

FIGURE 10.3 Ram-type turret in operation. (*Courtesy Gamet, Inc.*)

be manually operated, semiautomatic, fully automatic, or numerically controlled. In recent years new and different turret arrangements, such as that shown in Fig. 10.5, have been introduced.

Major Construction Features. All horizontal turret lathes, regardless of size, share certain major construction characteristics. The *bed* of the machine is usually a high-quality alloy iron or ductile iron casting that has been normalized prior to machining. The ways and other surfaces on which parts of the machine slide are surface-hardened and ground for wear resistance. A typical bed casting for a modern turret lathe is heavily ribbed and very rigid.

The *headstock* and *spindle* assemblies are heavy and rigid so that metal can be removed at high rates while maintaining accuracy. The spindle is usually driven by a gear train that is also used to select different spindle speeds. On some light- and medium-duty turret lathes the spindle is driven by multiple V belts, and speed changes are made in a remotely located gearbox. The spindle is usually a hollow alloy steel forging that is carried in ball or roller bearings which can take a combination of axial (thrust) and radial (side) loads. The size of the hole in the spindle determines the bar stock capacity of the machine. The nose of the spindle may be fitted with collets, chucks, or other fixtures to hold the workpiece.

Ram-Type Turret Lathes. On the ram-type turret lathe (see Fig. 10.3), the turret is mounted on a ram, or slide, that can be moved fore and aft by power or manually. The saddle in which the slide moves is clamped in position on the ways of the machine at the proper distance from the spindle. The turret indexes one position each time the slide is moved to the end of its travel farthest away from the spindle. On a typical machine with a six-position turret there are six adjustable stops on the end of the slide at the operator's right. Each stop

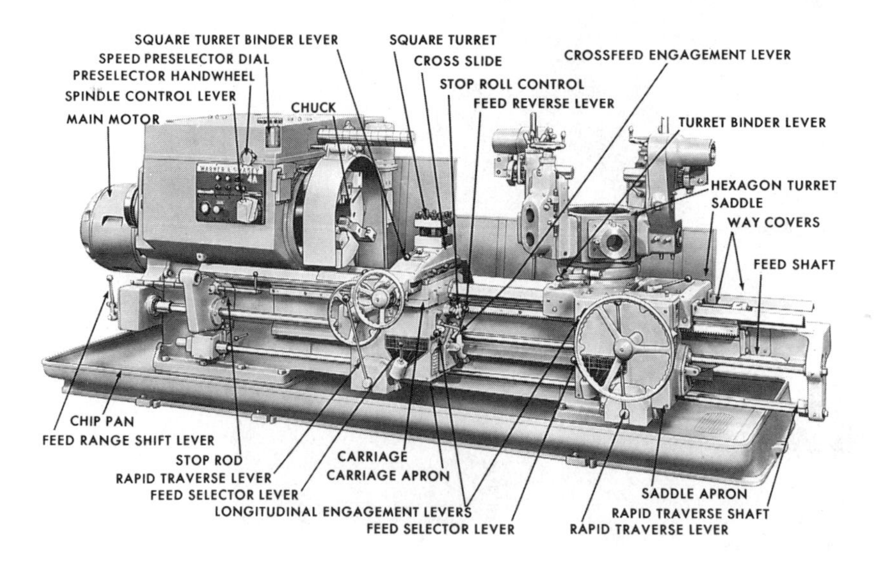

FIGURE 10.4 Saddle-type turret lathe. (*Courtesy Warner & Swasey Co.*)

FIGURE 10.5 Turret that rotates on horizontal axis. (*Courtesy Kanematsu-Gosho, Inc.*)

FIGURE 10.6 Inclined axis-type of turret. (*Courtesy Gamet, Inc.*)

can be adjusted to provide for a specific depth of cut for the tool mounted on the corresponding turret face.

The turret is usually a high-quality iron casting that rotates around a vertical shaft set into the end of the slide nearest the chuck. The indexing mechanism locks the turret into position. The thrust loads produced by the cutting action are taken by the vertical shaft and the locking mechanism.

Each face of the turret is perpendicular to the longitudinal axis of the lathe and is machined to accept many different tools. On some smaller machines the vertical shaft about which the turret rotates is inclined toward the headstock (see Fig. 10.6), and the turret faces are machined so that they are at right angles to the machine centerline.

Ram-type turret lathes have a carriage on which the cross-slide turret is mounted. This adds to the versatility of the machine because an additional four or five tools can be brought in contact with the work. Depending on the size of the lathe, the cross slide on the carriage can be operated with either a hand lever of a conventional handwheel and cross-feed screw. Handlever cross slides most often are found on smaller turret lathes. In either case, adjustable stops that limit cross slide travel are provided.

The *carriage* on a ram-type turret lathe can be power-fed or moved manually. Depending on the job, it can also be locked in position on the ways when only the cross slide movement is needed. The components on a ram-type turret lathe are shown in Fig. 10.7.

Saddle-Type Turret Lathes. On the saddle-type turret lathe the turret is mounted on the sad-

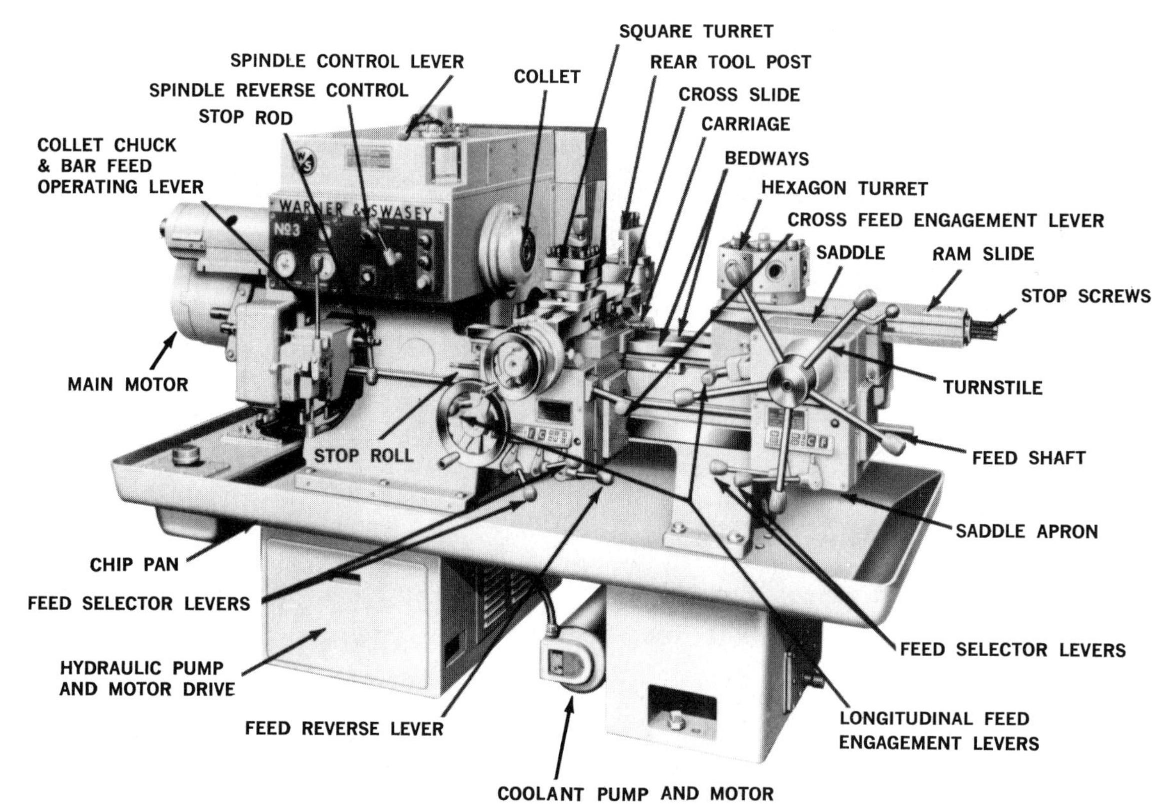

FIGURE 10.7 Ram-type turret lathe components. (*Courtesy Warner & Swasey Co.*)

FIGURE 10.8 Small saddle-type turret lathe for precise work. (*Courtesy Hardinge Brothers, Inc.*)

dle, and the entire assembly moves along the ways (see Fig. 10.4). This arrangement is used mostly on larger lathes because of the strength and rigidity it provides, but it can also be used on small-capacity lathes (see Fig. 10.8). On some lathes of this type the saddle has a cross slide so that the cutting tools can be moved perpendicular to the longitudinal axis of the machine for facing operations. The saddle has power feed, and on some machines there is a rapid traverse mechanism for moving the heavy assembly quickly.

There may also be a separate carriage on saddle-type turret lathes. This carriage can be of the side-hung type, which rides on the front ways and has a supplementary slide near the bottom of the bed casting, thus allowing very heavy cuts to be taken. The type of carriage that rests on both ways of the lathe can also be used, although it sometimes restricts the overall diameter of the work that can be machined.

Innovations. Major variations in turret lathe design are shown in Figs. 10.9, 10.10, and 10.11. The lathe shown in Fig. 10.9 has longitudinal ways on which a heavy cross slide is mounted. The cross slide carries two turrets, one rotating about the horizontal axis and one about the vertical axis. Internal and external operations may be done simultaneously.

The machine shown in Fig. 10.10 has no horizontal bed. The axis about which the eight-sided turret rotates is horizontal, and the turret moves on a set of angled ways at the rear of the machine.

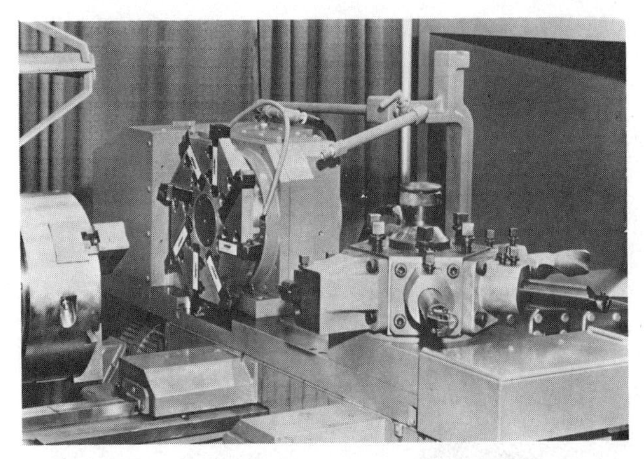

FIGURE 10.9 Turret axis is parallel to the spindle axis. (*Courtesy American Machine Tool Co.*)

FIGURE 10.10 The turret moves fore and aft on ways at the rear of the machine.

FIGURE 10.11 Slant bed chucker and bar machine. (*Courtesy Hardinge Brothers, Inc.*)

Bar and chucking work can be done on this type of machine. Spindle speeds can be programmed to change automatically, and the rate of feed is usually infinitely variable.

10.2.2 Vertical Turret Lathes

Vertical turret lathes share several characteristics with the larger horizontal turret lathes and vertical boring mills. These lathes are used to machine large and bulky objects.

General Construction Features. As shown in Fig. 10.12, the *spindle* of the machine is mounted vertically in the base and supports the *table*, which can range in diameter from 24 in. to about 54 or 60 in. The table is a heavy iron casting that is slotted to accept the T bolts or other work-holding devices such as the individually mounted chuck jaws shown. The spindle and table are capable of supporting heavy loads.

The vertical ways of the machine allow movement of the entire cross-slide assembly either up or down to accommodate work of different heights. One or two vertical slides are usually mounted on the cross slide; when two are used they can be moved independently of each other. One vertical slide carries the turret, which usually has five or six faces for mounting tools. A horizontal slide for turning operations may also be provided.

Although the controls are different on vertical turret lathes, the machining operations performed on the workpiece are similar to those done on horizontal turret lathes. Turning, boring, facing, grooving, chamfering, and other operations can be performed. Machines of this type have power feeds and rapid traverse mechanisms for moving the cross and vertical slides.

Vertical turret lathes can be operator-controlled, semiautomatic, or numerically controlled. The various functions and operations on an operator-controlled machine require almost constant attention from the machinist. Semiautomatic machines can be controlled by a series of plate or drum cams and adjustable stops and are generally used for production work. Numerically con-

FIGURE 10.12 Vertical turret lathe. (*Courtesy WCI Machine Tools and Systems, formerly Bullard*)

trolled machines are controlled by a program on punched tape. For a description of numerical control, refer to Chap. 16.

10.2.3 Cutting Tools and Accessories

Many different types of cutting tools are available for turret lathes. The tools can be used alone or in combinations to perform the typical turret lathe operations.

Turning Tools. The basic turning tool that is mounted on the turret is the *box tool* (see Fig. 10.13). It is used to reduce the diameter of bar stock, and it may have one or more cutting tools. The bar stock is supported by two rollers opposite the tool, and the tool and rollers are adjustable so that any desired diameter can be cut. Most box tools have hollow shanks to accommodate longer workpieces. When two cutting tools are used in tandem on a box tool, the leading tool is for roughing, and the second tool usually is for the finishing cut.

Knee tools (see Fig. 10.14) can be used for reducing the outside diameter of forgings, castings, or bar stock. One or more cutting tools can be used; the tool position is adjustable. On some knee tools a drill or boring bar can be placed in line with the spindle of the machine so that both internal and external operations can be done at the same time.

Boring and Drilling Tools. Slide tools (see Fig. 10.15) are used for boring, backfacing, and internal grooving operations. For grooving and backfacing, the slide and tool are moved vertically by the hand lever shown at the top. A micrometer stop

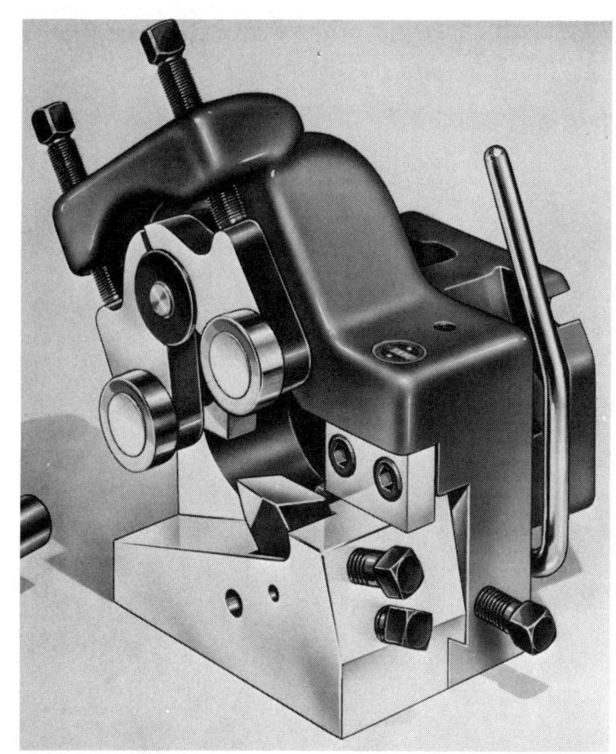

FIGURE 10.13 Box tool. (*Courtesy Warner & Swasey Co.*)

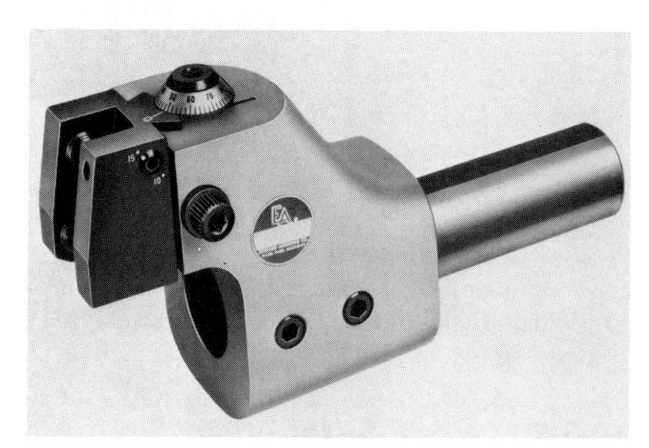

FIGURE 10.14 Knee tool. (*Courtesy Edward Andrews Co.*)

is provided for cutting grooves and recesses to specified depths.

A widely used tool that combines two operations in one turret position is the *combined stock stop* and *starting drill* shown in Fig. 10.16. When the drill is retracted, the end of the tool positions bar stock in the chuck or collet. The turret is then retracted, and the drill is advanced and locked in position. The starting drill is then fed into the work only far enough to provide an accurate starting hole for the subsequent drilling operations.

Many different types of drills and reamers can be used in turret lathe operations. For large-

FIGURE 10.15 Slide tool. (*Courtesy Warner & Swasey Co.*)

FIGURE 10.16 Combined stock stop and centering drill. (*Courtesy Edward Andrews Co.*)

FIGURE 10.17 Spade drill. (*Courtesy DoALL Co.*)

FIGURE 10.18 Rose chucking reamer. (*Courtesy DoALL Co.*)

diameter shallow holes, spade drills (see Fig. 10.17) are very effective. Conventional twist drills in either jobbers' lengths or stub lengths for small shallow holes are also used extensively. Step drills can produce holes with two or more diameters.

Several different types of reamers can be used on turret lathes. Rose chucking reamers (see Fig. 10.18) are used primarily for roughing operations. Stub screw machine reamers, which are short and rigid, are also widely used in turret lathe and automatic screw machine operations.

Threading Tools. External screw threads can be cut with a single-point tool or a die. They can also be rolled. A self-opening die head (see Fig. 10.19) may have four or more chasers and is

FIGURE 10.19 Self-opening die head. (*Courtesy Teledyne Landis Machine Tool Co.*)

mounted on the turret. It is fed manually to engage it with the part to be threaded, and light pressure is maintained as the die head advances. When the turret slide stop is reached, the die head advances slightly and opens so that it may be withdrawn over the completed threads.

A single-point tool is usually mounted on the cross-slide carriage of a ram-type turret lathe. Its movement is controlled by a lead screw and half-nuts. On some machines the single-point threading system uses threaded sleeves and followers and is mounted behind the headstock of the machine.

Thread rolling die heads produce threads by displacing rather than cutting the metal (see Fig. 10.20). The die head is mounted on the turret and fed onto the chamfered end of the workpiece. After the thread has been rolled to the proper length, the head disengages and is withdrawn.

FIGURE 10.20 Thread rolling die head. *(Courtesy Teledyne Landis Machine Tool Co.)*

Internal threads are cut with taps or a single-point tool. The taps can be straight or tapered; when a solid tap is used, the machine spindle must be reversible. Collapsible taps (see Fig. 10.21) can also be used, especially for tapping larger-diameter holes. Most larger taps of either the collapsible or solid type have inserted chasers that can be adjusted to cut the pitch diameter desired and can be removed from the tap body for reconditioning.

10.2.4 Basic Setups and Operations

Because the function of any turret lathe is to produce accurate parts rapidly, the *setup time* for a particular job and other time-related factors must be carefully considered. On very short production runs the machinist or production planner must decide whether the time and cost required to set up the turret lathe are justified because the cost can be spread only over a small number of parts.

The next factor to consider is the *cycle time.* This is the time it takes to machine a part, remove it from the machine, and place another part to be machined in position. The time required to handle the workpiece is very short in bar work, where a bar feeder and manual or power collet closers are used. In chucking operations, the operator must remove the finished workpiece and place an unmachined part in the chuck or fixture. This may be a significant part of the total time per part. Therefore, the use of power-operated chucks and

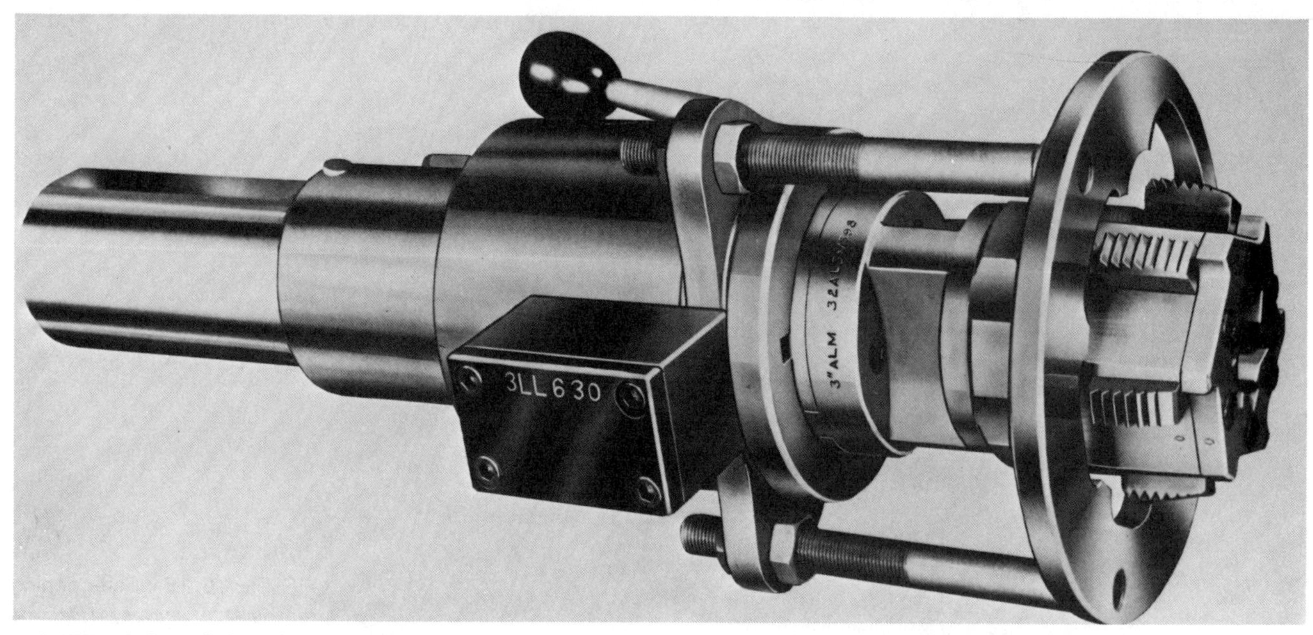

FIGURE 10.21 Collapsible tap. *(Courtesy Teledyne Landis Machine Tool Co.)*

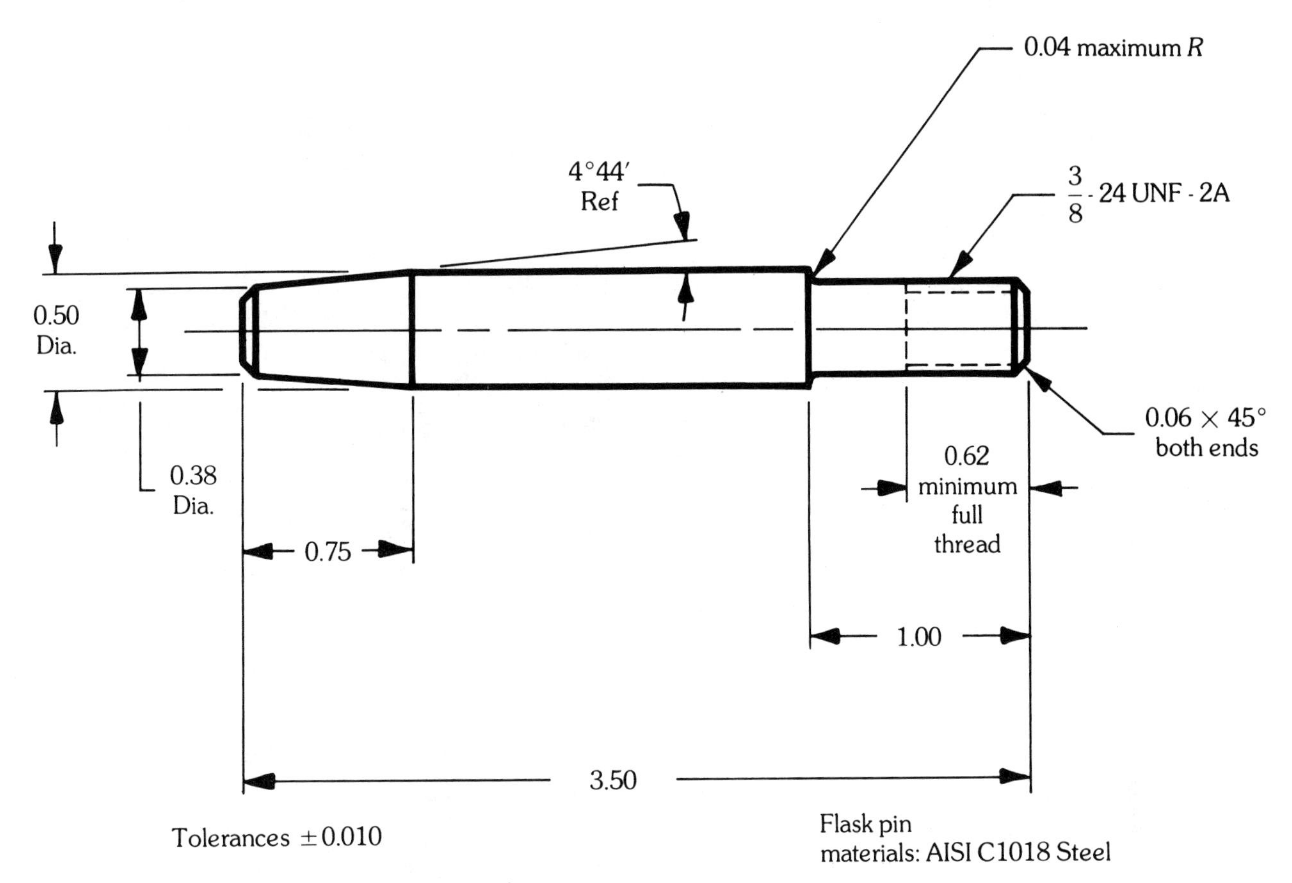

FIGURE 10.22 Drawing of simple bar stock part.

special fixtures for holding castings and forgings is common practice since both time and the chance of errors is reduced.

The *cutting time* is the time in which one or more cutting tools are in operation. It can be shortened by careful planning of the sequence of operations and the use of multiple tools whenever possible. For example, turning and boring operations can be done simultaneously with some knee-type tools mounted on the turret. Multiple-tool external operations can also be done with tools mounted on the cross slide.

The proper selection of cutting tools and accessories for turret lathe operations is based on the experience of the setup person and skill in interpreting the drawings of the part to be made. The accuracy requirements are a major factor in selecting the tools used and the sequence of operations. The geometric relationship of the various surfaces must also be considered, especially for parts that will require no further machining.

Figure 10.22 is a drawing of a typical bar-type job for a turret lathe. The sequence of operations in Fig. 10.23 traces the positioning and machining steps from the initial placement of the bar stock to the cutting off of the finished part. When large numbers of such parts are required, an automatic screw machine may be used.

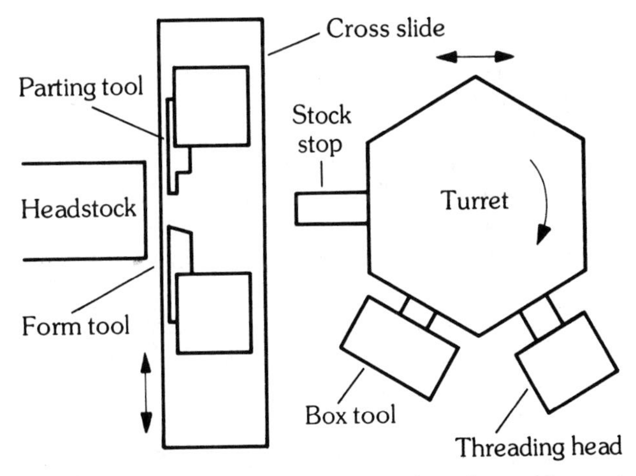

FIGURE 10.23 Sequence of operations for making part shown in Fig. 10.22.

10.3 AUTOMATIC LATHES AND SCREW MACHINES

The automatic screw machine, which was first patented in 1873, has been used extensively in the metal machining industry. It was developed to meet the need for very large numbers of small, precise parts. As the screw machine evolved into various forms over the years, it continued to serve basically the same purpose. Many changes and improvements have made screw machines more efficient, including the development of multiple-spindle machines, introduction of better cutting tool materials, and the capability for holding closer tolerances.

10.3.1 Automatic Lathes

Automatic lathes are made in numerous sizes and shapes. In many ways these lathes resemble turret lathes and are capable of doing the same types of work. The cutting tools used on both the turret and cross slides of automatic lathes are similar to the ones used on manually controlled turret lathes. The main difference lies in the control systems. Automatic lathes are controlled by plate or drum cams, punched cards or tape, or other electric or hydraulic means.

A typical example of a small single-spindle automatic lathe is shown in Fig. 10.24. This lathe has a bar capacity of about 1 in. and can also be set up for chucking operations. There are no controls for manual operations of the machine, and all the slides are air-operated. The control system is electrical. This type of machine is programmed by the board shown in Fig. 10.25 and does not use cams, tape, or punched cards.

The inclined turret has six tool positions and two cross slides, one vertical and one horizontal. Accessories such as air-operated drills can be added for auxiliary operations.

On some larger automatic lathes (also known as single-spindle automatic chucking machines), the turret is mounted on a heavy tubular member above the chuck (see Fig. 10.26). Machines of this type usually have two cross slides that operate independently of each other; these machines are used for external contours and facing operations. The tools on the turret can be used for both external and internal operations, including threading. The motions of all the tools and spindle-speed changes are controlled by stops on the selector drum switch (see Fig. 10.27).

FIGURE 10.25 Control board for automatic lathe shown in Fig. 10.24. (*Courtesy REM Sales*)

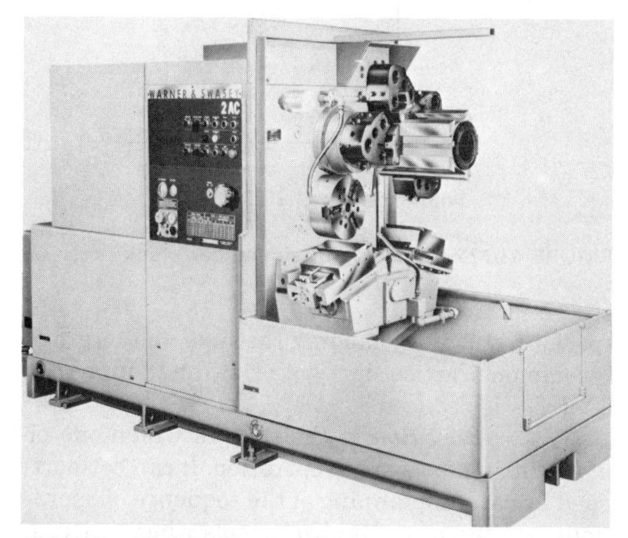

FIGURE 10.26 Single-spindle automatic chucking machine. (*Courtesy Warner & Swasey Co.*)

FIGURE 10.24 Small single-spindle automatic lathe. (*Courtesy REM Sales*)

10.3.2 Automatic Screw Machines

In many ways, single-spindle automatic screw machines resemble small automatic turret lathes. Both machines are used to make small, accurate parts in very large numbers.

Single-Spindle Automatic Screw Ma-hines. The machine shown in Figure 10.28 is typical of most of the American-type automatic screw machines. All the functions are controlled by drum or plate cams operating through a series of linkages and rocker arms. Because the machines are set up for bar work only, an automatic bar stock

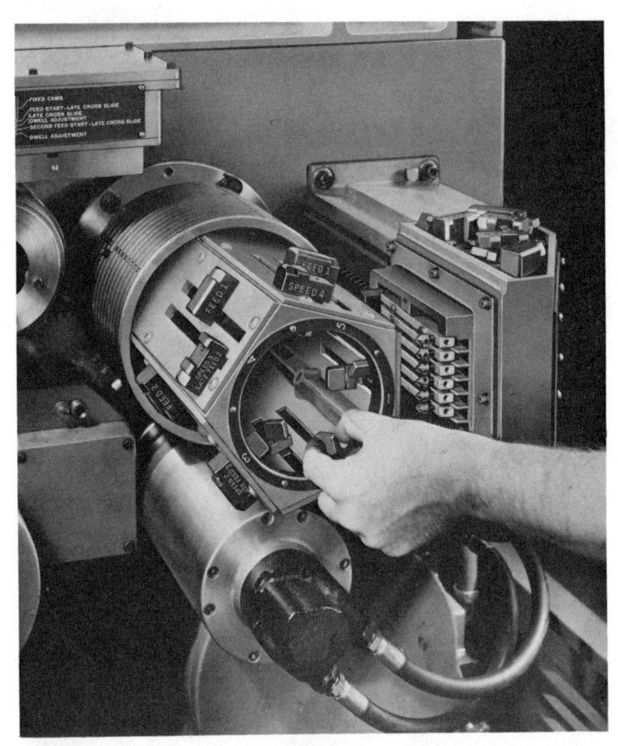

FIGURE 10.27 Machine functions are controlled by the drum selection switch. (*Courtesy Warner & Swasey Co.*)

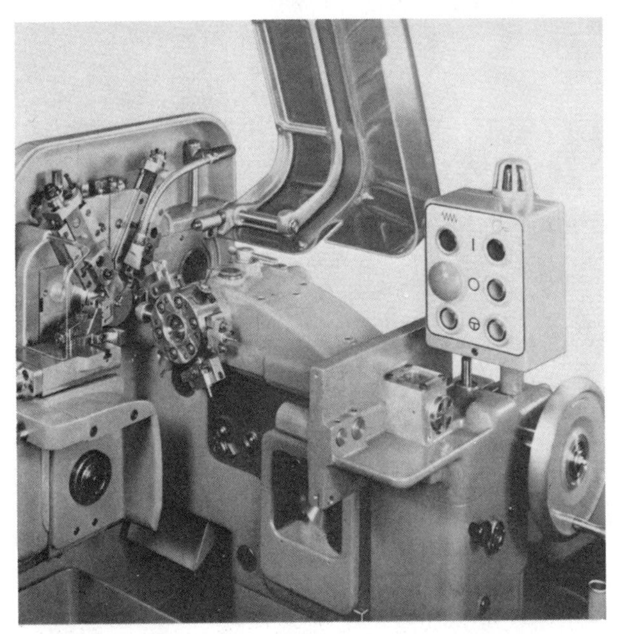

FIGURE 10.28 Small single-spindle automatic screw machine. (*Courtesy B.S.A.*)

feeder is provided. The spindle is usually reversible on these machines so that tapping operations with small solid taps can be performed.

Single-spindle screw machines are generally used for making small parts from bar stock or tubing (see Fig. 10.29). The bar capacity of such machines ranges from under 1/2 in. (12.7 mm) to 2 in. (51 mm). Collets, generally of the push type, hold the bar stock or tubing while it is being machined.

The plate cams that are used on single-spindle machines are mounted on a shaft that runs parallel to the spindle (see Fig. 10.30). On some machines, additional cams can also be mounted on a cross shaft. One turn of the shaft that carries the cams is required to complete one operating cycle of the machine. The operating cycle generally involves six turret movements and several cross-slide movements, including the cut-off operation. Spindle-speed and feed-rate changes can also be programmed to occur during the operating cycle, and controlled by cam movement.

The cutting tools and accessories used on single-spindle automatic screw machines are generally similar to those used on small turret lathes and automatic lathes. Drills, reamers, counterbores, and similar tools are usually of *stub* length for rigidity and generally carbide tipped (see Fig. 10.31). Collapsible taps and self-opening die heads are also extensively used for threading operations.

Swiss-Type Automatic Screw Machines. These machines (see Fig. 10.32) have some unique features. They were developed for the machining of very precise small parts. Machines of this type can have a maximum bar diameter capacity ranging from 1/8 (3 mm) to about 1 in. (26 mm). Tolerances of 0.0002 in. or less can be maintained on small internal and external diameters. Machines designed for making watch and instrument parts are capable of even closer tolerances. The finish on the parts is usually very good, partly because of the high spindle speeds at which this type of machine operates.

One unique feature of the Swiss-type automatic screw machine is that the headstock moves fore and aft, carrying the rotating bar stock with it. The headstock can move as much as 4 in. The bar stock is gripped in a collet and moves through a bushing made of tungsten carbide or some other hard, wear-resistant material. The side loads imposed by the cutting tools are carried by the bushing rather than by the spindle.

The tools for external operations are mounted in the five holders shown in Fig. 10.33. The movement of the rotating bar stock past the tools at a predetermined feed rate produces the desired outside diameters. The tool slides are mounted so that the cutting edge of the tool contacts the workpiece very close to the carbide bushing through which the rotating bar stock is fed. This allows long,

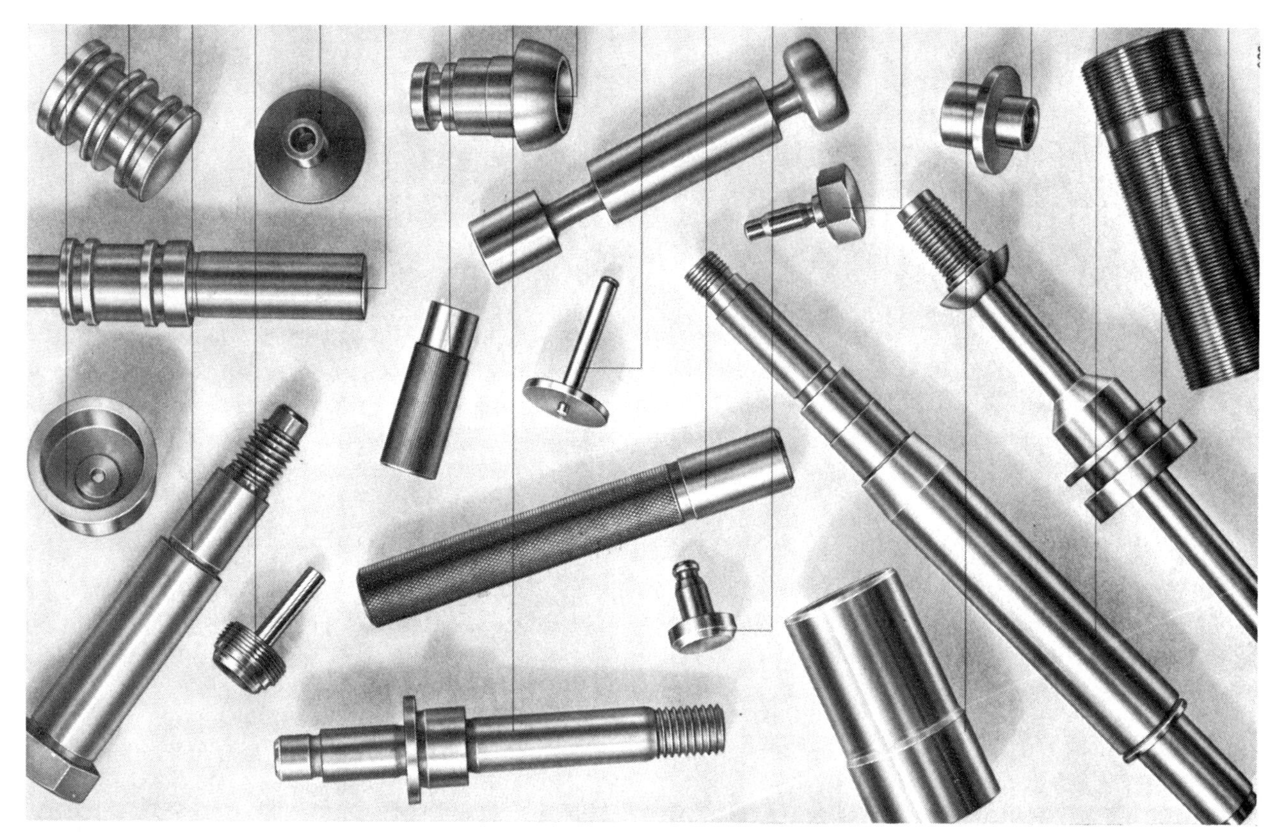

FIGURE 10.29 Typical parts made on automatic screw machines. (*Courtesy Tornos Bechler U.S. Corp.*)

FIGURE 10.30 Cams that control machine functions. (*Courtesy Tornos Bechler U.S. Corp.*)

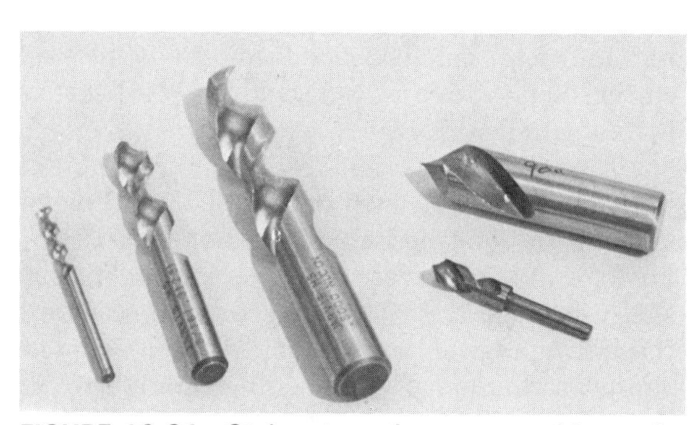

FIGURE 10.31 Stub automatic screw machine tools.

FIGURE 10.32 Swiss-type automatic screw machine. (*Courtesy Tornos Bechler U.S. Corp.*)

small-diameter workpieces to be machined to close diametric and concentricity tolerances.

Single-point tools of either carbide or high-speed steel are used for turning operations. Form and cutoff tools that are fed radially into the workpiece can also be mounted on the radial slides. They are fed inward while the spindle is rotating but not moving axially unless a taper is desired.

Internal operations can be performed by mounting a turret that can bring as many as six

FIGURE 10.33 Tooling and tool slides on Swiss-type automatic screw machine. (*Courtesy Tornos Bechler U.S. Corp.*)

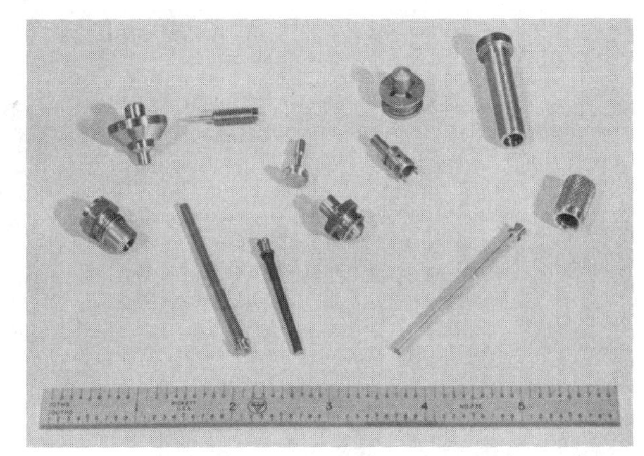

FIGURE 10.34 Typical parts machined on Swiss-type automatics.

tools in contact with the end of the workpiece in the desired sequence. In some cases more than one tool is mounted in a given turret position so more than six different internal operations can be performed. The axis around which the turret rotates is horizontal and parallel to the spindle axis. It is offset enough to bring the center line of each turret position in line with the spindle axis. Tools such as stub drills, boring bars, reamers, and form tools can be mounted in the turret.

Almost all tool- and spindle-movement functions are controlled by a series of plate or drum cams that are mounted on a camshaft in the base of the machine. The process of designing the cams and setting up the machine is time-consuming and expensive, but the low cost per part after the machine is operating properly justifies the expense. Some typical parts made on Swiss-type automatic screw machines are shown in Fig. 10.34.

The demand for very accurate small machined parts has increased greatly in recent years and this has caused machine tool builders to develop new machines as well as new concepts in precision turning. Introduction of electronic controls and gaging devices has resulted in machines that require less operator attention. In-process gaging as well as monitoring of tool condition are now relatively common features on the Swiss-type automatic screw machine shown in Fig. 10.35.

Another interesting innovation is the use of two sliding headstocks in tandem on the same slide, thus reducing unproductive time in the operating cycle. Spindle speeds are very high in machines of this type, with small-diameter parts being machined at speeds up to 20,000 rpm. This reduces machining time and results in very good surface finishes. Effective cutting fluids are very important in automatic screw machine operations.

Multiple-Spindle Automatic Screw Machines. The typical multiple-spindle machine is very complex and used only where very large numbers of parts are to be made. Depending on the work-holding devices used, multiple-spindle automatics can be set up for either bar or chucking work. When bar work is done, the bar or tubing is fed to each spindle by bar feeders in the stock reel (see Fig. 10.36). For chucking operations, the parts to be machined are loaded by hand or from a magazine loader. Finished parts are generally unloaded automatically and carried away by a chute or conveyor. Machines of this type are classified by the largest bar diameter that can be accommodated or by the outside diameter and capacity of the chucks used.

One main advantage of the multiple spindle machine is that as many parts can be machined simultaneously as there are spindles. This reduces the cycle time for each part to the time required for the longest operation, plus the time to move the spindle position. When simple parts are being made, two identical sets of tooling may be used, each set using one half of the available spindles. With this arrangement, two parts are completed each time the spindle indexes.

This type of machine may have four, five, six, or eight spindles. The spindles and the mechanisms for rotating them are housed in the *spindle carrier* (see Fig. 10.37). The *end tool slide*, which does not rotate but moves in or out on the spindle carrier centerline, carries one or more tools for each spindle position. Therefore, all the tools on the end tool slide are fed simultaneously. When the tool slide operations are completed, the tool slide retracts rapidly and the spindle carrier indexes to the next position.

Tools for internal and external operations can be mounted on the tool slide. The motion of the slide is usually provided by a drum cam that rotates once for each cycle. Disk cams are also used

FIGURE 10.35 Double headstock Swiss-type automatic screw machine. (*Courtesy Tornos Bechler U.S. Corp.*)

FIGURE 10.36 Multiple-spindle automatic screw machine. (*Courtesy Cosa Corp., Machine Tool Division*)

FIGURE 10.37 Spindle carrier and cutting tools. (*Courtesy Cosa Corp., Machine Tool Division*)

on some machines to provide tool motion and other functions.

A cross slide is provided for each spindle position (see Fig. 10.37). Because the cross-slide movement is perpendicular to the spindle center line, the major operations done with cross-slide tools are facing, external contouring, and grooving with form tools, and cutting off the finished part. Operations such as cross drilling, slitting, and tapping can also be done with cross-slide mounted tools that are individually powered.

FIGURE 10.38 Tracer unit mounted on an engine lathe.

10.4 TRACER LATHES

Tracer and duplicating lathes are used in the production of parts that have multiple diameters, complex contours, or both. Lathes of this type are used in manufacturing rather than job shop operations. They are generally of the single-spindle type, but multiple-spindle machines are also used where production runs are long or the part is complex.

Tracer attachments are available for engine lathes and are often used in low-volume production work (see Fig. 10.38). Tracer units can also be mounted on single-spindle automatic lathes or turret lathes.

A master template is required for each different part to be made with a tracer. The template can be accurately cut from flat plate, or it can be a turned part identical to the parts to be made. A stylus follows the contours of the template and sends commands to the actuator in the tracing unit. The actuator generally uses hydraulic pressure to move the cutting tool.

Larger tracer lathes may be manually controlled or automatic. In most cases the tool head and actuator are mounted behind and above the spindle center line of the lathe (see Fig. 10.39). This makes it easier for the operator to load and unload the machine if a magazine loading system is not used.

Conventional preset tools and cam-operated tool slides can be used on the carriage of some tracer lathes. These tools can be used to do some roughing as well as finishing operations. The template-controlled tracer is used for finishing and contouring operations. External threading can also be incorporated into the operation cycle for work done between centers.

FIGURE 10.39 Single-spindle tracer lathe. *(Courtesy Schaefer Sales Co.)*

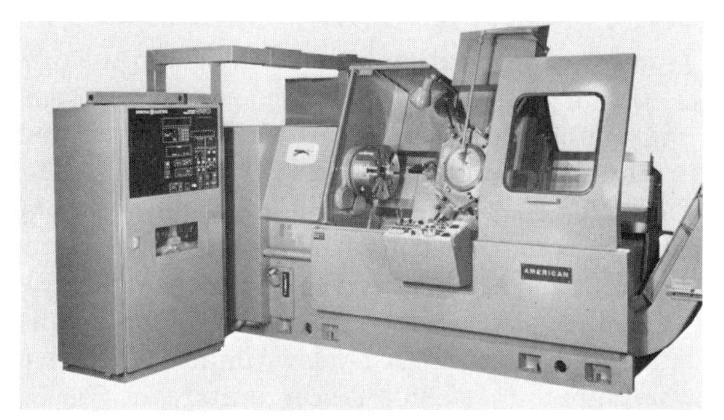

FIGURE 10.40 Numerically controlled chucker. (*Courtesy American Machine Tool Co.*)

10.5 NUMERICALLY CONTROLLED LATHES

Numerical control has been applied to practically all types of machine tools, including lathes. A major feature of numerically controlled (NC) lathes is their flexibility. The number and complexity of commands that can be delivered to the spindle and tooling of an NC lathe is far greater than with cam-controlled machines. Therefore, it becomes possible to machine very complex parts with a minimum of tooling and setup time.

The chucker shown in Fig. 10.40 is typical of smaller NC lathes. The spindle nose accepts either

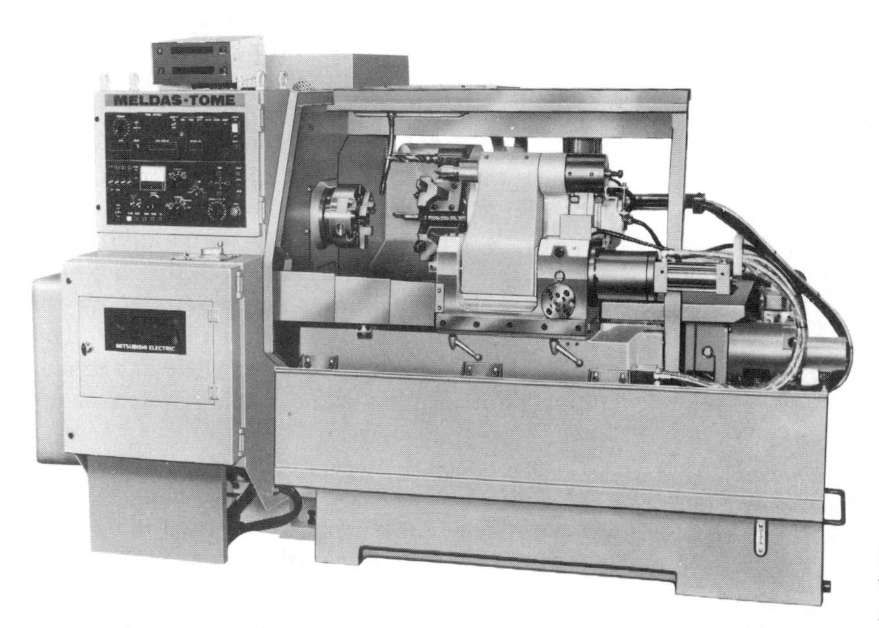

FIGURE 10.41 Numerically controlled turning center. (*Courtesy Kanematsu-Gosho, Inc.*)

FIGURE 10.42 Slant bed chucker and bar machine. Note computerized control panel. (*Courtesy Hardinge Brothers, Inc.*)

collets or chucks, and the eight-position turret is mounted on a cross slide. By simultaneously moving the two slides, tapers and contours can be produced. Threading can be done with dies, taps, or single-point tools.

A larger type of NC lathe is shown in Fig. 10.41. The axis of the turret is parallel to the axis of the spindle. The turret has 10 to 14 tool positions, depending on the model of the machine. Machines of this type are usually very powerful and may have electric motors of 30 to 50 hp. This results in a high rate of metal removal and high productivity.

The machine shown in Fig. 10.42 is an example of one of the more modern types of production lathes being used where multiple, precise turning operations are necessary.

REVIEW QUESTIONS

10.2 TURRET LATHES

1. What are the two basic types of turret lathes?
2. What are the two major types of work that can be done in turret and other special lathes?
3. Briefly describe the spindle and bearing assembly of a typical horizontal turret lathe.
4. How is the turret indexed after each operation is completed?
5. How does the turret mounting differ between saddle-and ram-type turret lathes?
6. By what means are workpieces usually mounted on the table of a vertical turret lathe?
7. Briefly describe a box tool.
8. Why are stub-length drills often used on turret lathes?
9. Briefly describe a self-opening die head.
10. By what two methods may internal threads be cut on a turret lathe?
11. What is meant by *cycle time* in turret lathe operation?

10.3 AUTOMATIC LATHES AND SCREW MACHINES

1. Why were automatic screw machines invented and developed?
2. What three types of control systems can be used on automatic lathes?
3. What work-holding devices can be used on automatic chucking machines?
4. How is the workpiece usually held in small automatic lathes set up for bar work?
5. Briefly explain how motion is transmitted from the plate cam to the tool slide on an automatic screw machine.
6. List two methods of producing external threads on an automatic screw machine.
7. What is the major operating difference between Swiss and other types of small automatic screw machines?
8. What two cutting tool materials are usually used on automatic screw machines?
9. What is the main operational advantage of the multiple-spindle automatic screw machine?
10. List two reasons for machining small parts at high spindle speeds.

10.4 TRACER LATHES

1. Explain how the motion of the cutting tool is controlled on a typical single-spindle tracer lathe.
2. What types of parts can be produced on tracer lathes?
3. How can templates for tracer lathes be made?

10.5 NUMERICALLY CONTROLLED LATHES

1. What is one main advantage of numerically controlled lathes?
2. Is a template required to machine contours with an NC lathe? Why?

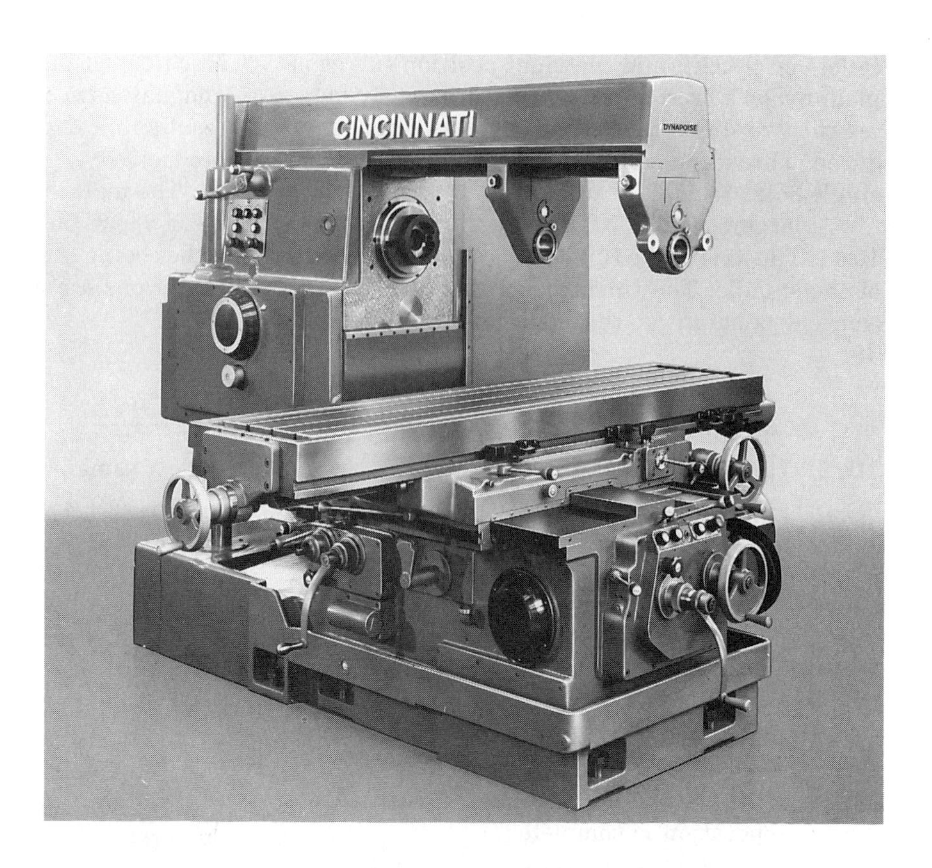

11

Milling Machines and Operations

11.1 INTRODUCTION

Milling is the process of machining metal by means of a rotating cutter with one or more teeth. Milling dates back about 200 years, and the process appears to have first been used in France, mostly for machining gears for clocks. In the United States, Eli Whitney used early types of milling machines for manufacturing parts for firearms and similar items. The introduction of Whitney's plain milling machine in 1818 was a major step forward in milling as a machining process.

By about 1850, milling machines were fairly widely used. The demands for mass-produced firearms and other complex products during the Civil War caused rapid development of machine tools, including the milling machine. A major development was the introduction of the universal milling machine and accessories such as the dividing head. When fitted with the proper formed cutters, the universal milling machine was used to produce gears, tools, and many complex parts.

Since the U.S. Civil War, the development of milling machines, cutters, and accessories has been rapid. Presently a wide variety of milling machines are in use, along with many accessories, cutters, and fixtures. A good machinist must be able to at

---

Opening photo courtesy of Cincinnati Milacron Inc.

least select the proper machine, cutters, and accessories for a job and perform basic milling operations. These are very necessary skills because milling machines are found in almost every type of machine shop. If the basic milling operations are well learned, the machinist will be able to perform more difficult operations and use more complex machines with relative ease.

11.2 TYPES AND CONSTRUCTION FEATURES

The many types of milling machines used in machine shops have been grouped into three general classes. The common subtypes are also identified and discussed. This portion of the chapter describes the construction features and physical characteristics of the machines.

11.2.1 Column and Knee Milling Machines

Column and knee milling machines are made in both vertical and horizontal types, in sizes ranging from the small bench mill (see Fig. 11.1) to the massive and powerful machine shown in Fig. 11.2. Versatility is a major feature of knee and column milling machines. On a basic machine of this type, the table, saddle, and knee can be moved as shown in Fig. 11.3. Many accessories, such as universal vises, rotary tables, and dividing heads, further increase the versatility of this type of machine (discussed in Sec. 11.3).

FIGURE 11.1 Small horizontal milling machine of the knee and column type.

FIGURE 11.2 Large horizontal milling machine. (*Courtesy Cincinnati Milacron Inc.*)

FIGURE 11.3 Large knee and column vertical milling machine. (*Courtesy Cincinnati Milacron Inc.*)

Major Components. Regardless of whether the machine is of the vertical or horizontal type, several components on all knee and column milling machines are similar, except for size and minor variations because of manufacturer's preference. These similarities are described in terms of general shape, geometric relationship to the rest of the machine, function, and the material from which the components are made.

Column. The *column*, which is usually combined with the *base* as a single casting (see Fig. 11.4), is cast gray iron or ductile iron. In the last 25 or 30 years, ductile iron has been used almost exclusively for castings of this type because it is stronger than gray iron, more stable, and capable of being surface hardened. The column houses the spindle, bearings, and the necessary gears, clutches, shafts, pumps, and shifting mechanisms for transmitting power from the electric motor to the spindle at the selected speed. The gears usually run in oil and are made of carburized alloy steel for long life. Some of the necessary controls are usually mounted on the side of the column, as shown in Fig. 11.5.

FIGURE 11.4 Base and column casting for horizontal milling machine. (*Courtesy Republic-Lagun Machine Tools Co.*)

The *base* is usually hollow, as shown in Fig. 11.4, and in many cases serves as a sump for the cutting fluid. A pump and filtration system can be installed in the base. The hole in the center of the base is the support for the screw that raises and lowers the knee.

The machined vertical slide on the front of the column may be of the square or dovetail type, as shown in the cross-sectional view in Fig. 11.6. The *knee* moves up and down on this slide. The slide must be machined at a 90° angle to the face of the column in both the lateral and vertical planes. The tolerances are very close and are usually expressed in *minutes* (1/60 of 1°) or *seconds* (1/3600 of 1°) of arc. The large hole in the face of the column casting is for the *spindle*. The hole is very accurately bored perpendicular to the front slide in two planes and parallel to the upper slide.

On some milling machines, the overarm consists of one or two heavy steel bars that can be moved forward to carry the outboard support for the arbor on which milling cutters are mounted (see Fig. 11.7). On milling machines with a single

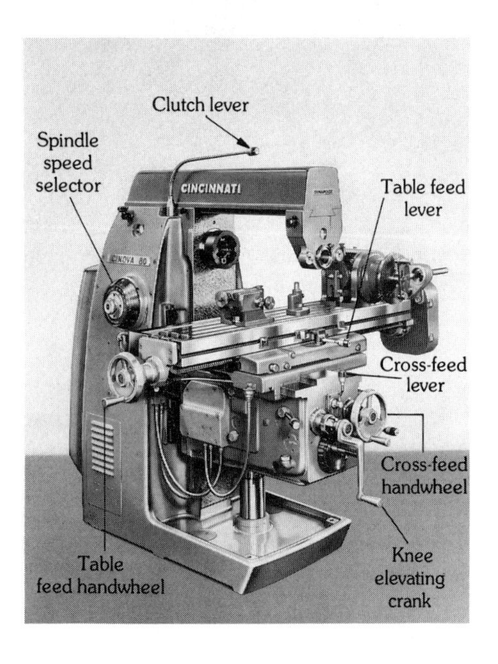

FIGURE 11.5 Controls for a typical horizontal milling machine. (*Courtesy Cincinnati Milacron Inc.*)

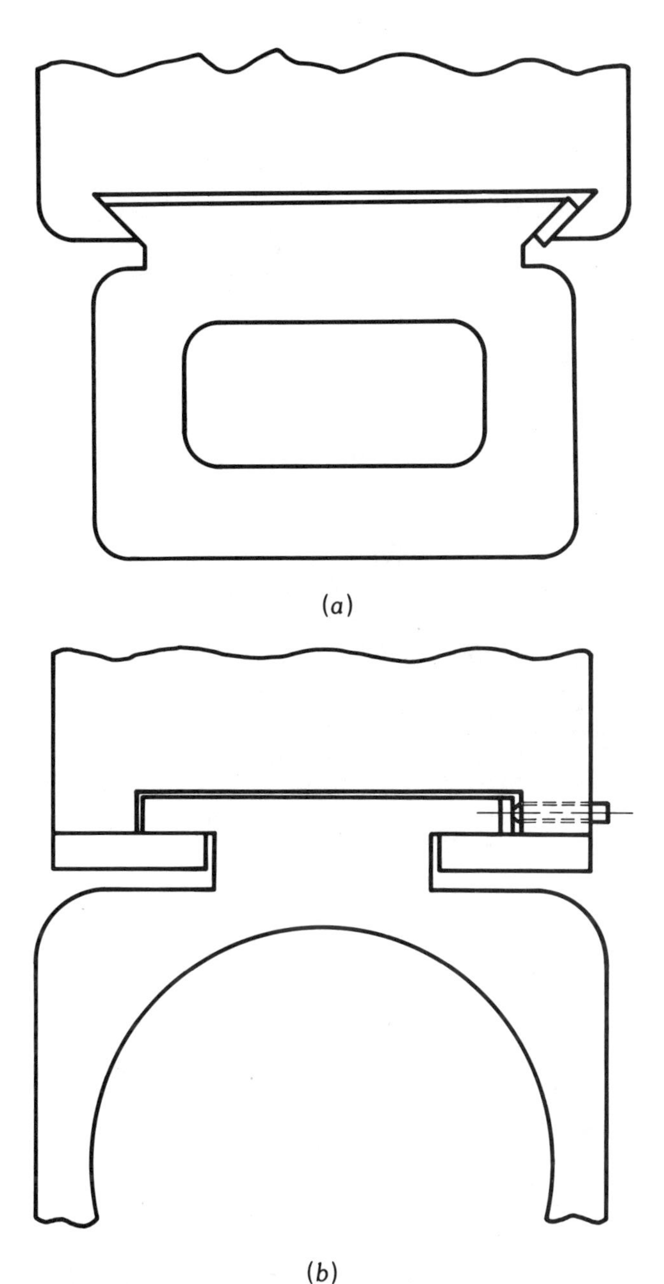

FIGURE 11.6 (*a*) Dovetail or (*b*) square slides may be used on the face of the column.

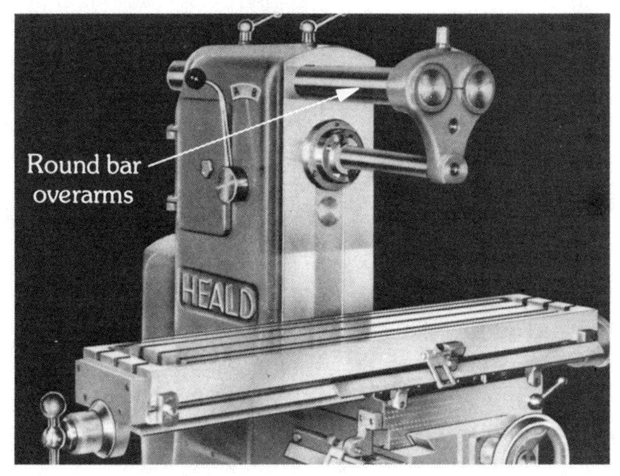

FIGURE 11.7 Round bar overarms on a horizontal milling machine.

round overarm, a vertical head can be attached, as shown in Fig. 11.8.

In some cases a cast overarm is used, as shown in Figs. 11.5 and 11.9. The casting is made of high-

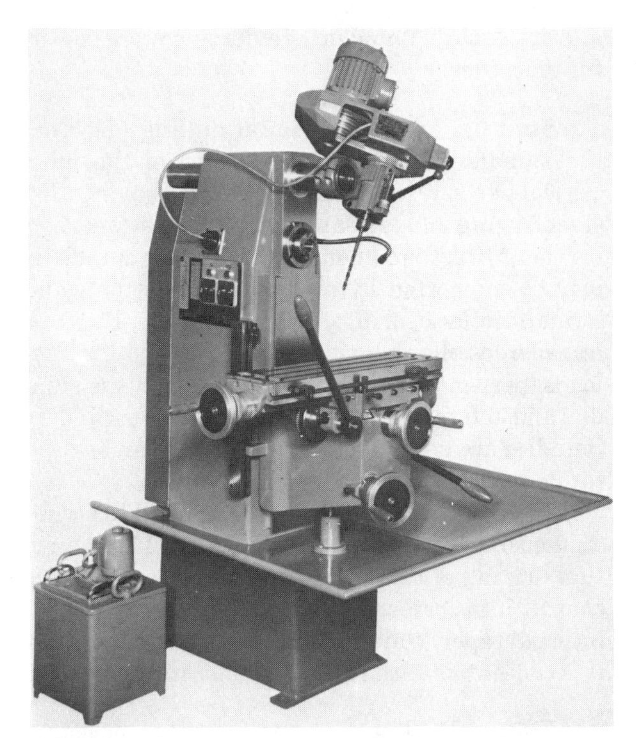

FIGURE 11.8 Vertical head attached to round overarm on horizontal mill. (*Courtesy Republic-Lagun Machine Tools Co.*)

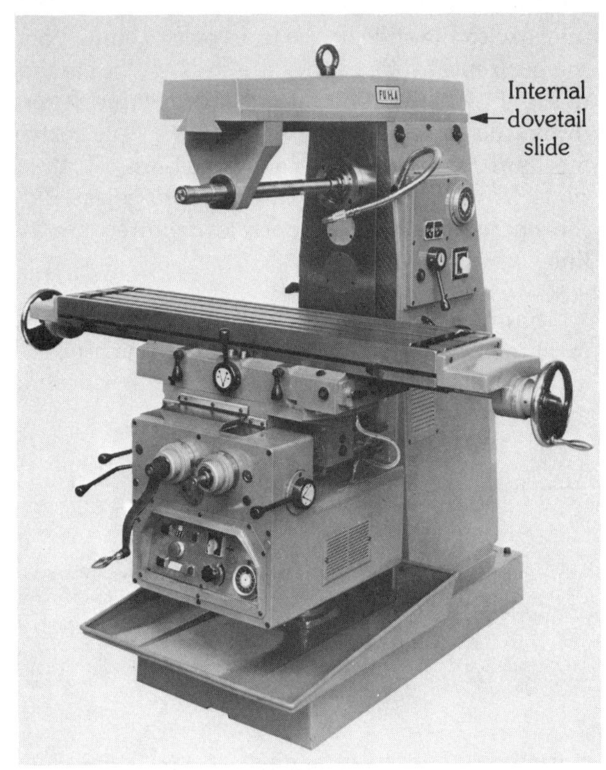

FIGURE 11.9 Cast overarm on a horizontal milling machine. (*Courtesy Republic-Lagun Machine Tools Co.*)

quality gray cast iron or ductile iron and is usually partly hollow. A dovetail slide is machined on its lower face and this mates with corresponding dovetails on the top of the column and on the overarm supports. Cast overarms are very rigid and are usually found on heavier and more powerful machines.

Spindle. On a horizontal milling machine, the spindle (see Fig. 11.10) is one of the most critical parts. It is usually machined from an alloy steel forging and is heat-treated to resist wear, vibration, thrust, and bending loads. The spindle is usually supported by a combination of ball and straight roller bearings, as shown in Fig. 11.10, or by tapered roller bearings that absorb both radial loads (perpendicular to the centerline of the spindle) and end thrust loads (in line with the spindle). Spindles are hollow so that a drawbar can be used to hold arbors securely in place.

The front of the spindle is machined to accept standard arbors, as shown in Fig. 11.11. The actual driving of the arbor is done by the two keys that fit into corresponding slots in the arbor. The internal taper, which is accurately ground so that it is concentric with the spindle, locates the arbor.

FIGURE 11.11 Spindle nose. Note taper and driving lugs. (*Courtesy El Camino College*)

Knee. The knee is a casting that is moved up or down the slide on the front of the column by the elevating screw. Two dovetail or square slides are machined at 90° to each other. The vertical slide mates with the slide on the front of the column, and the horizontal slide carries the *saddle.* The casting in Fig. 11.12 is for a larger milling machine. It contains the necessary gears, screws, and other mechanisms to provide power feeds in all directions. Various feed rates can be power selected by the operator with the controls mounted on the knee.

FIGURE 11.12 Partly machined knee casting for knee and column milling machine. (*Courtesy Republic-Lagun Machine Tools Co.*)

Saddle. The saddle for a plain milling machine is a casting with two slides machined at an exact 90° angle to each other. The lower slide fits the slide on the top of the knee, and the upper slide accepts the slide on the bottom of the table. The surfaces of the slides that make contact with the knee and the table are parallel to each other. Locks for both the cross slide and table are fitted to the saddle, along with the nuts that engage with the cross-feed and table feed screws.

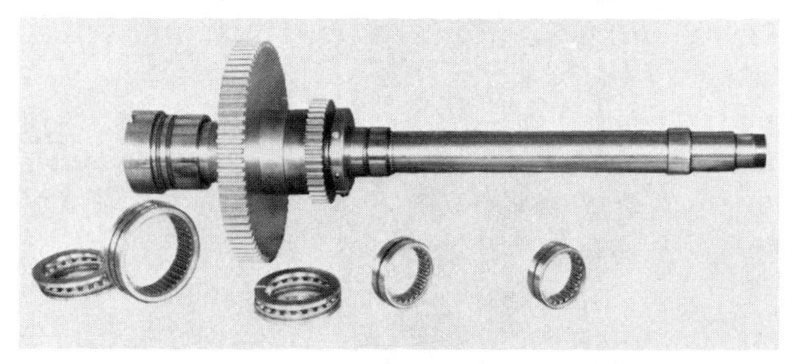

FIGURE 11.10 Spindle and bearings for a horizontal milling machine. (*Courtesy Republic-Lagun Machine Tools Co.*)

On a *universal* milling machine, the saddle is made in two pieces and is more complex because it must allow the table to swivel through a limited arc (see Fig. 11.13). The lower part has a dovetail slide that fits the top of the knee, and a circular

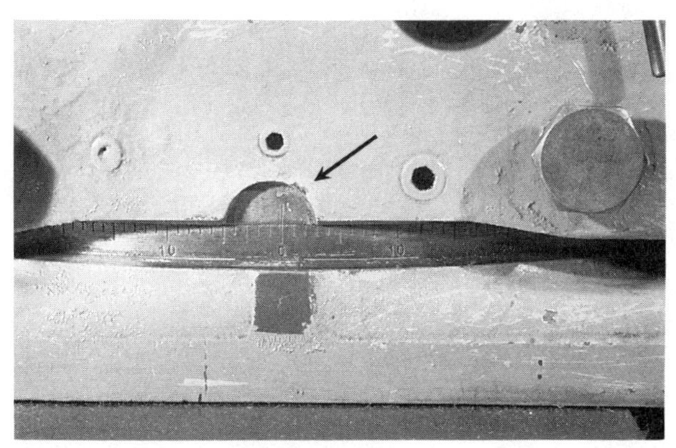

FIGURE 11.13 Saddle on universal milling machine. Note swivel graduations. (*Courtesy El Camino College*).

slide above it is graduated in degrees for a small portion of its periphery. The upper portion of saddle consists of a circular face that fits against the lower circular slide, a central pivot point, and a dovetail slide that accepts the table. Locking bolts moving in a circular T slot are provided so that the two parts of the saddle can be locked in any position.

Table. Milling machine tables vary greatly in size, but generally they have the same physical characteristics. The bottom of the table has a dovetail slide that fits in the slide on top of the saddle. It also has bearings at each end to carry the table feed screw. The top of the table is machined parallel with the slide on the bottom and has several full-length T slots for mounting vises or other work-holding fixtures.

A dial graduated in thousandths of an inch is provided to allow for accurate table movement and placement. The table feed screw usually has an Acme thread.

Vertical Milling Machines. Milling machines with vertical spindles are available in a large variety of types and sizes. The vertical mill in Fig. 11.14 has a plain table. The head, which houses the spindle, motor, and feed controls, is fully *universal* and can be placed at a compound angle to the surface of the table. The *ram*, to which the head is attached, can be moved forward and back and locked in any position. A turret on top of the column allows the head and ram assembly to swing laterally, thus increasing the reach of the head of the machine.

Some ram-type vertical milling machines of the type shown in Fig. 11.15 can be used for both vertical and horizontal milling. On ram-type vertical mills that have the motor in the column, power is transmitted to the spindle by gears and splined

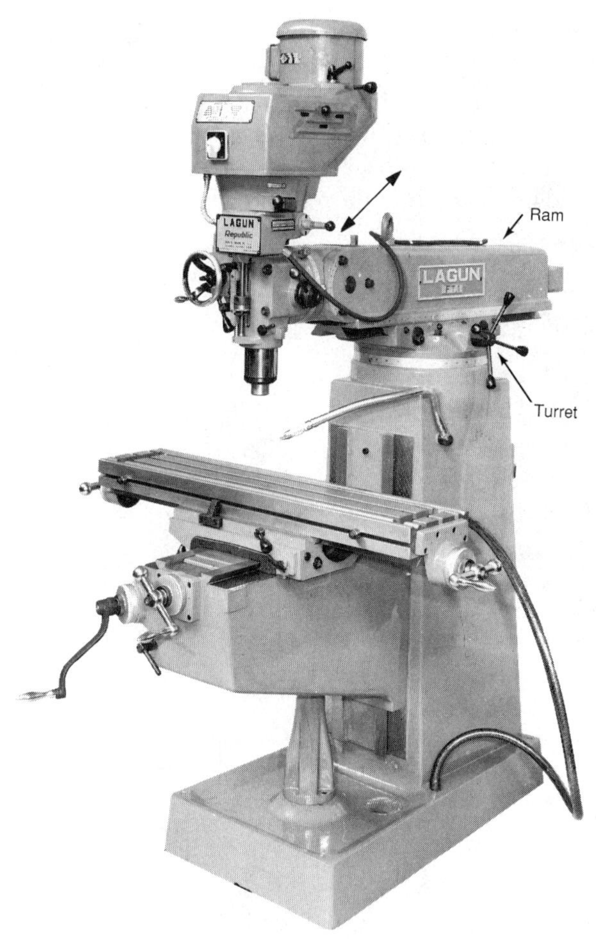

FIGURE 11.14 Vertical milling machine with universal head. (*Courtesy Republic-Lagun Machine Tools Co.*)

shafts. Some heavy-duty vertical mills (see Fig. 11.16) have a spindle and head assembly that can be moved only vertically by a power feed mechanism or manually. These are generally known as *overarm*-type vertical milling machines.

11.2.2 Production-Type Milling Machines

Generally, production-type milling machines are less versatile than the typical knee and column horizontal or vertical machine. They usually have greater rigidity, strength, and power and are ideal for heavy manufacturing applications. The skill requirements for operators are lower once the machine has been properly set up. The work is usually held in fixtures, and in some cases a number of pieces can be machined at one time.

Most production-type milling machines are semiautomatic or fully automatic in operation, with the cycle controlled mechanically, electrically, or by a combination of both means. The operator loads and unloads workpieces and starts the machine. The operator may also be responsible for

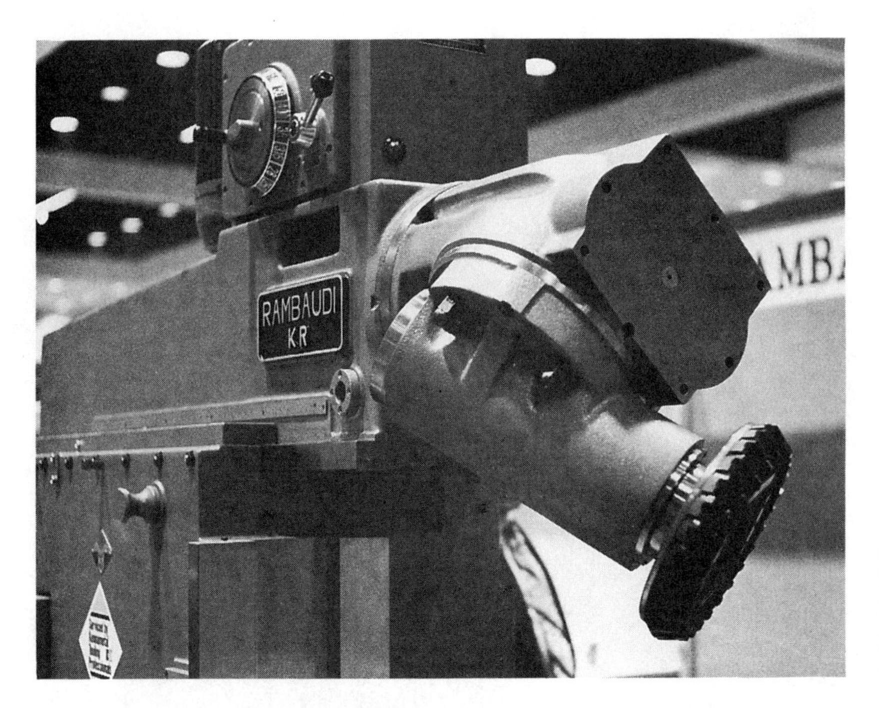

FIGURE 11.15 Ram-type vertical milling machine with universal head. Note that the cutter head is at a compound angle to the table. (*Courtesy Schaefer Sales Co.*)

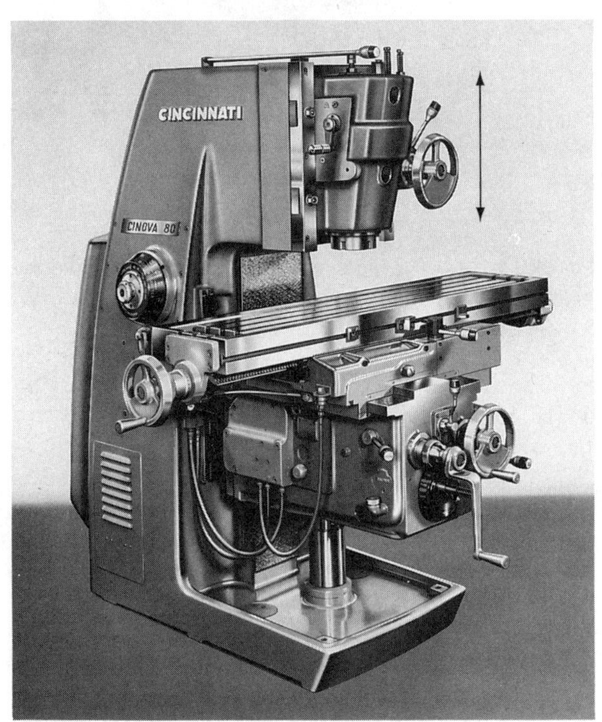

FIGURE 11.16 Overarm-type vertical mill with plain head. (*Courtesy Cincinnati Milacron Inc.*)

FIGURE 11.17 Single-spindle bed-type milling machine. (*Courtesy Cincinnati Milacron Inc.*)

checking the machined part with fixed gages or other measuring tools.

Fixed-Bed Machines. On machines such as the *simplex* or single-spindle mill shown in Fig. 11.17 the table moves only longitudinally. The feed is hydraulic or run by a screw and nut arrangement. Usually, there is no provision for hand feeding the table.

The position of the spindle can be adjusted vertically while the machine is being set up, and then it is locked in position. On some machines, however, there is a mechanically operated arrangement for causing the spindle to rise or drop as the table moves. This type of machine can be used for simple profiling operations.

A typical *duplex* machine is shown in Fig. 11.18. The table can move only longitudinally between the heads and spindles. Each head can be adjusted for height independently and can be fitted with a variety of cutters. Machines of this type are very effective in doing facing operations on

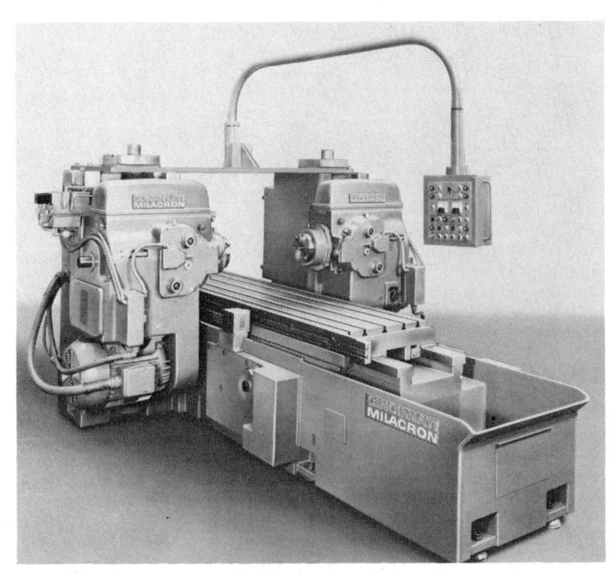

FIGURE 11.18 Two-spindle bed-type milling machine. (*Courtesy Cincinnati Milacron Inc.*)

parts that are held in fixtures and fed past the cutters. Usually, two sets of fixtures for holding work are used so that unmachined parts can be loaded while the other parts are being machined.

A more complex version of the duplex mill is the *triplex* machine in Fig. 11.19. It is used for machining three surfaces at once on the part that is being moved past the cutters by the table. The two side cutter heads can be set at an angle, raised, or lowered to accommodate workpieces of different shapes and sizes. They can also be run at different spindle speeds. The upper head can be moved both vertically and sideways. The various functions on machines of this type are usually controlled mechanically or hydraulically. The operating cycle is usually fully automatic and may include changes in feed rate and spindle speed.

Rotating-Table Milling Machines. Machines of this type are used mostly in factories in-

FIGURE 11.19 Three-spindle bed-type milling machine. (*Courtesy Cincinnati Milacron Inc.*)

FIGURE 11.20 Rotating table-type vertical milling machine. (*Courtesy Ingersoll Milling Machine Co.*)

volved in high-volume production. The machine can have one or more spindles, usually of the vertical type, as shown in Fig. 11.20, and the workpieces are held in a series of fixtures mounted on the table of the machine. The movement of the table is usually continuous, and the main duty of the operator is to place unmachined parts in the fixtures and remove machined pieces. Inserted-tooth face milling-type cutters are commonly used on machines of this type.

11.2.3 Special-Purpose Milling Machines

As industrial products have become more complex, new and unusual variations of the more common milling machines have been developed. The objectives are to accommodate larger work, make many duplicate parts, locate holes and surfaces precisely, or to do other unusual machining jobs. The special milling machines generally require skilled operators, particularly when complex setup work is required.

Planer-Type Milling Machines. The general arrangement of these types of machines are similar to that for planers, except that in place of individual tool bits, milling heads are installed (see Fig. 11.21). The table of the machine carries the work past the rotating cutter heads, which are individually powered and can be run at different speeds if necessary. As many as four cutter heads can be used, with two mounted on the cross rail and two on the vertical pillars.

Planer-type milling machines are used mostly

FIGURE 11.21 Planer-type milling machine. (*Courtesy Ingersoll Milling Machine Co.*)

for machining parts like the bedways for large machine tools and other long workpieces that require accurate flat and angular surfaces or grooves, such as wing spars for high-performance aircraft.

Profile Milling Machines. Two-dimensional profiling can be done by using a template, as shown in Fig. 11.22, or with a numerically controlled vertical milling machine. Some profilers have several spindles, and a number of duplicate parts can be produced in each cycle. Hydraulic-type profilers have a stylus that is brought into contact with the template to start the operation. The operator then moves the stylus along the template, causing hydraulic fluid under pressure to flow to the proper actuating cylinders. The table moves the work past the cutter, thus duplicating the shape of the template.

Diesinking and other processes involving the machining of cavities can be done on three-dimensional profilers. An accurate pattern of the cavity is made of wood, plaster, or soft metal. The stylus follows the contour of the pattern, guiding the cutter as it machines out the cavity. Numerically controlled milling machines can also be used for this type of work (see Chap. 16).

Jig Borers. Some jig boring machines resemble vertical milling machines in several respects, (see Fig. 11.23), but jig borers are much more accurately constructed. They are intended primarily for doing boring operations on hole locations held to very close tolerances. Digital readout systems

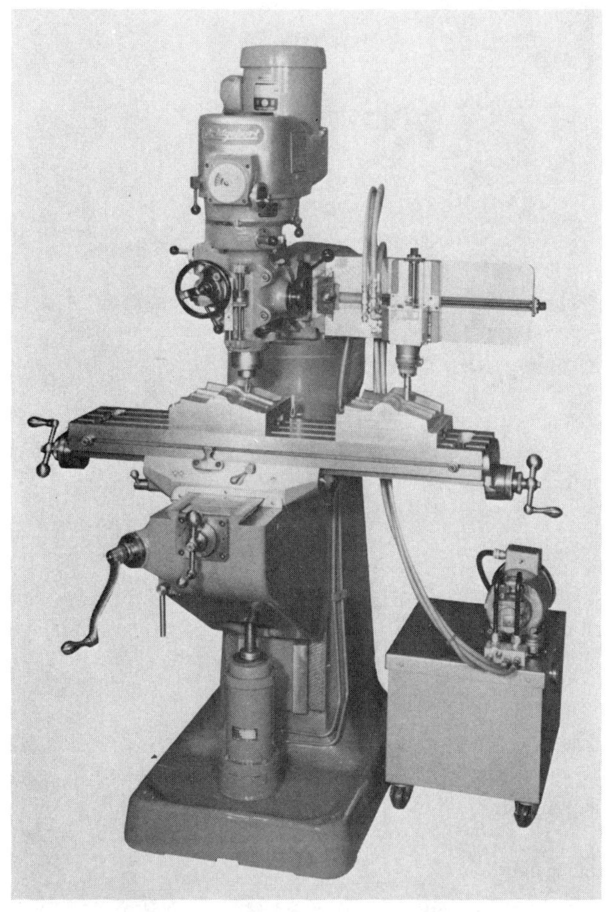

FIGURE 11.22 Profiling operation using a hydraulic tracer. (*Courtesy Republic-Lagun Machine Tools Co.*)

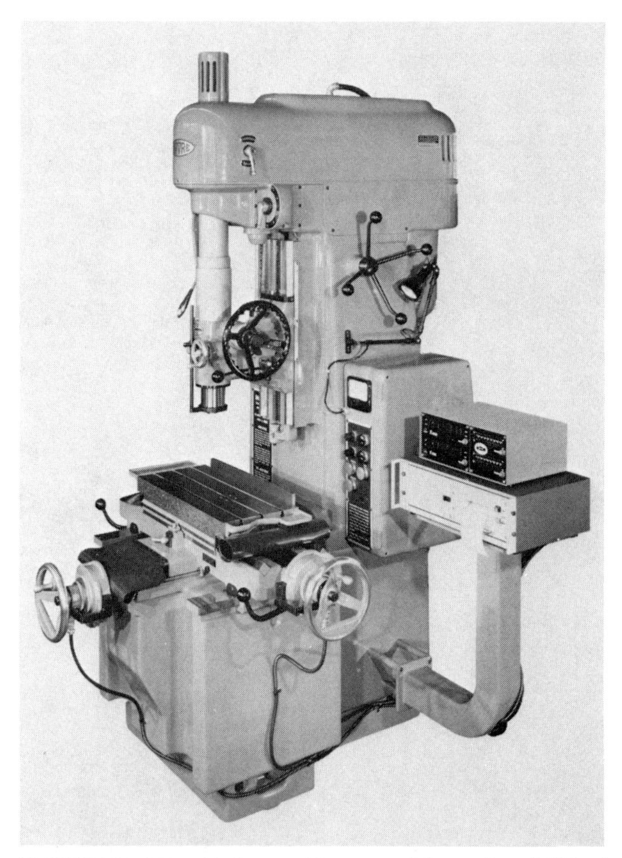

FIGURE 11.23 Jig borer. (*Courtesy Moore Special Tool Co., Inc.*)

are used on some jig borers. In some cases, end measuring rods of precise length are used to establish table location. Several elementary jig boring operations can be done on a good vertical milling machine if the spindle is in good condition and the necessary measuring tools are available.

11.3 MILLING MACHINE ATTACHMENTS AND ACCESSORIES

Many accessories have been developed for milling machines. Some are specialized and can be used for only a few operations; others, such as vises, arbors, and collets, are used in almost all milling operations.

11.3.1 Special Heads

Several types of special heads have been developed for use on horizontal or vertical milling machines. The function of such accessories is to increase the versatility of the machine. For example, a vertical head can be attached to a conventional horizontal column and knee milling machine, thus greatly increasing its usefulness, especially in small shops with a limited number of machines.

Vertical Heads. Vertical heads are generally attached to the face of the column or to the overarm of a horizontal milling machine. The head is a *semiuniversal* type, as shown in Fig. 11.24, which pivots only on the axis parallel to the center line of the spindle, or it is *fully universal.* Fully universal heads (see Fig. 11.15) can be set to cut compound angles. Both types of heads are powered by the spindle of the milling machine and accept standard arbors and collets.

Rack-Milling Attachment. The rack-milling attachment in Fig. 11.25 bolts to the spindle housing of the milling machine. Its spindle is at a right angle to the main spindle of the machine. Both spur and helical racks can be milled with this attachment, and it can also be used to mill worms, as shown. Some rack-milling attachments have an outboard support for the spindle, which makes it possible to take heavier cuts.

Slotting Attachment. This attachment, which is bolted to the column of a horizontal mill-

FIGURE 11.24 Semiuniversal head on horizontal milling machine. (*Courtesy Cincinnati Milacron Inc.*)

FIGURE 11.25 Rack milling attachment milling a worm. (*Courtesy Republic-Lagun Machine Tools Co.*)

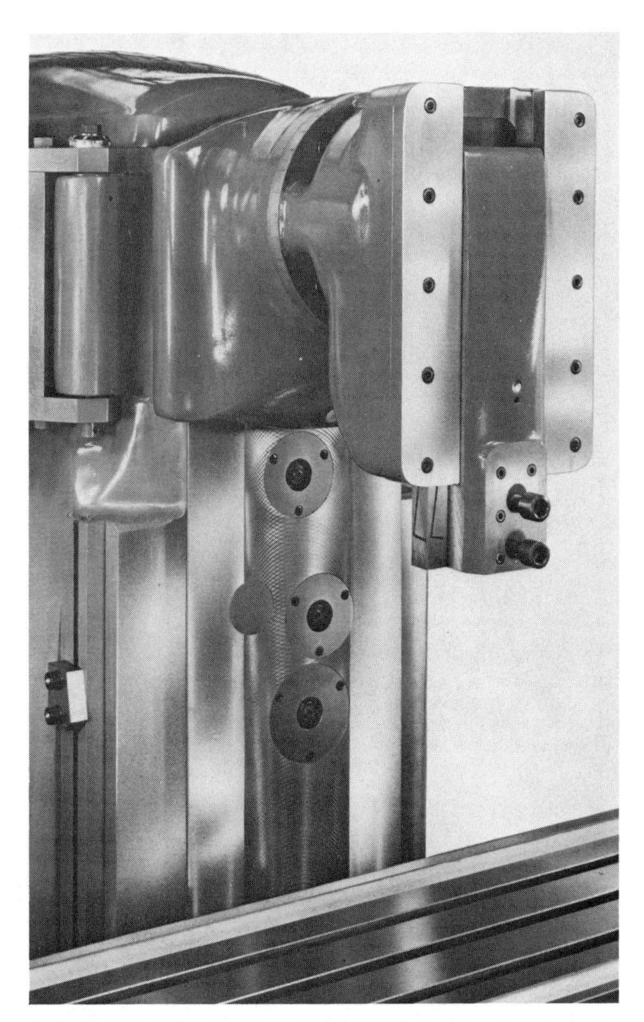

FIGURE 11.26 Slotting attachment on a horizontal milling machine. (*Courtesy South Bend Lathe, Inc.*)

ing machine, can be swiveled 90° in either direction from the vertical position (see Fig. 11.26). It is used primarily in toolmaking and prototype work for cutting keyways, internal splines, and square or rectangular cavities. The crank that actuates the reciprocating slide is driven directly by the spindle, and the stroke is adjustable.

High-Speed Milling Attachments. When spindle speeds beyond the operating range of the machine are necessary, high-speed attachments can be placed on both horizontal and vertical milling machines. A gear train is generally used to step up the speed as much as 6 : 1, which allows more efficient use of small cutters. A high-speed milling attachment that is fully universal is shown in Fig. 11.27.

FIGURE 11.27 High-speed universal milling head. (*Courtesy Cincinnati Milacron Inc.*)

11.3.2 Vises and Fixtures

In all milling operations, the work is held by fixtures, vises, or clamping arrangements. In most cases, the work is held stationary in relation to the table while it is being machined, but work held in indexing heads and rotary tables can be moved in two planes while machining operations are in progress.

Plain Vise. Plain milling vises, as shown in Fig. 11.28, are actuated by an Acme threaded screw, and the movable jaw moves on either a dovetail or rectangular slide. The vises are usually cast of high-grade gray cast iron or ductile iron and can be heat-treated. Steel keys are attached in slots machined into the bottom of the vise parallel with and perpendicular to the fixed jaw to allow accurate placement on the milling table. The jaw inserts are usually heat-treated alloy steel and are attached by cap screws. Vises of this type are classified by the jaw width and maximum opening.

Cam-operated plain milling vises (see Fig. 11.29) are widely used in production work because of the savings in time and effort and the uniform clamping pressures that can be achieved.

Swivel-Base Vise. A swivel-base vise is more convenient to use than the plain vise,

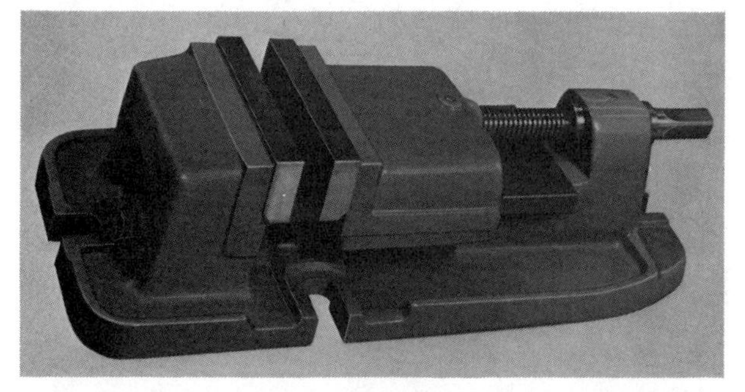

FIGURE 11.28 Plain milling vise. (*Courtesy Cincinnati Milacron Inc.*)

FIGURE 11.29 Cam-operated swivel base milling vise. (*Courtesy Chicago Tool Eng.*)

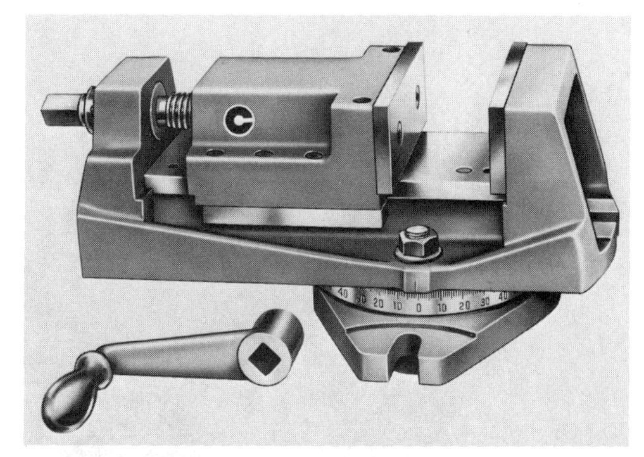

FIGURE 11.30 Swivel base vise. (*Courtesy Clausing Industrial, Inc.*)

although it is somewhat less rigid in construction. The base, which is graduated in degrees (see Fig. 11.30), is slotted for keys that align it with the T slots in the table. The upper part of the vise is held to the base by T bolts that engage a circular T slot.

The swivel-base vise, when used on a milling machine with a semiuniversal head, makes possible milling compound angles on a workpiece.

Universal Vise. A universal vise is used mostly in toolroom, diemaking, and prototype work (see Fig. 11.31). The base of the vise is graduated in degrees and held to the table by T bolts. The intermediate part of the vise has a horizontal pivot upon which the vise itself can rotate 90°. Because there are several joints and pivots in the vise assembly, the universal vise is usually the least rigid of the various types of milling machine vises.

Another type of universal vise is shown in Fig. 11.32. It allows a third plane of movement, providing easy setting of any compound angle. Vises of this type are used only for light milling operations.

Angle Plates. Several types of angle plates can be used to hold work or work-holding fixtures for milling. Plain angle plates are available in T-slotted or blank form and are usually strong iron castings. Adjustable angle plates may tilt in one direction only or have a swivel base (see Fig. 11.33). They are very useful for milling workpieces that are irregular in shape and cannot be held easily in a vise.

Holding fixtures that are a combination of a simple angle plate and a collet are sometimes used to hold round or hexagonal work for milling. The collet-holding fixtures may be manually or air operated. Both fixtures can be bolted to the milling table in the vertical or horizontal position or

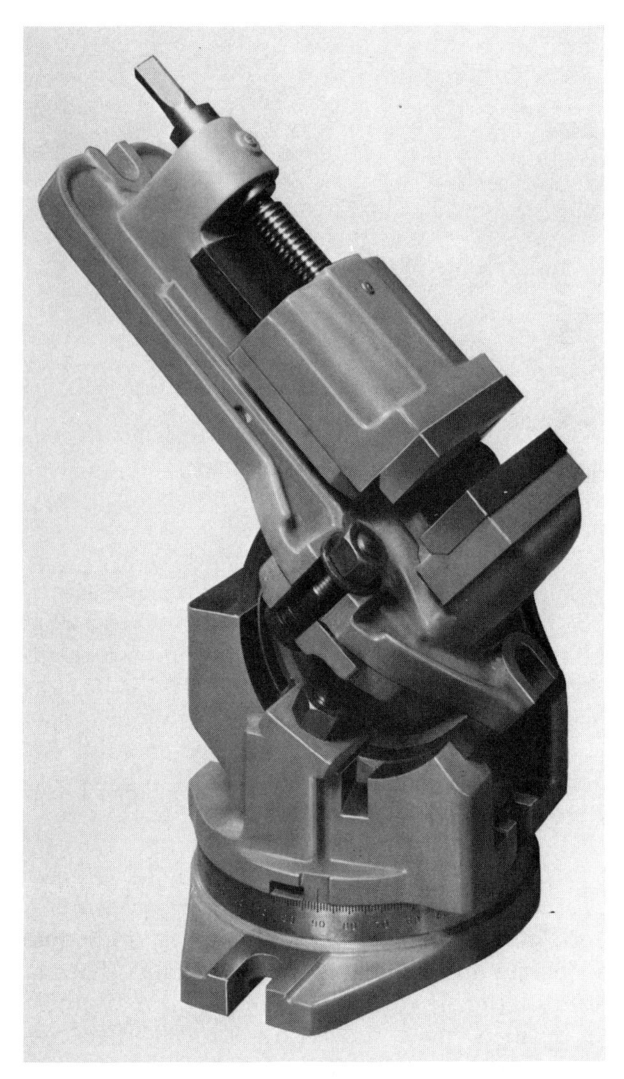

FIGURE 11.31 Universal milling vise. (*Courtesy Cincinnati Milacron Inc.*)

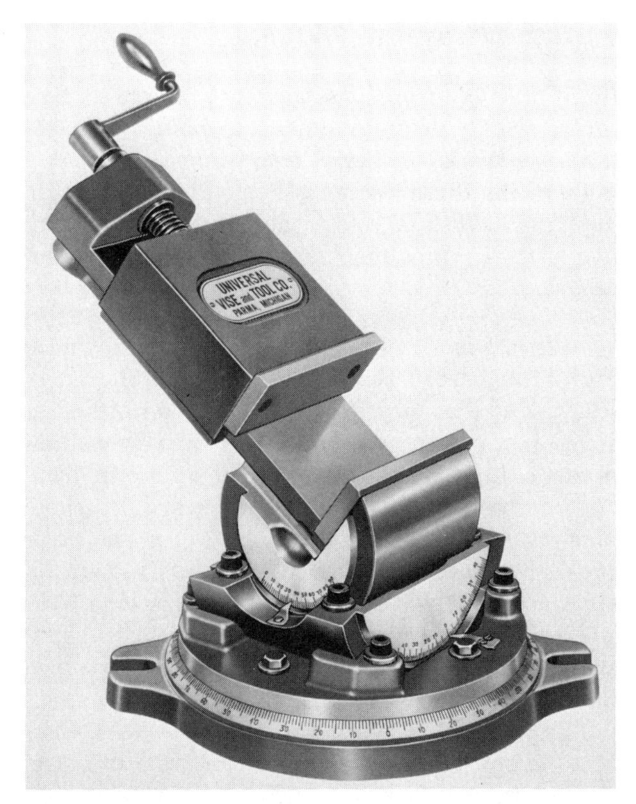

FIGURE 11.32 Universal three-way milling vise. (*Courtesy Universal Vise & Tool*)

attached to an adjustable angle plate for holding workpieces at simple or compound angles to the table or other reference surface.

Indexing Heads. The indexing head, also known as the *dividing* head, can be used on vertical and horizontal milling machines to space the cuts for such operations as making splines, gears, worm wheels, and many other parts requiring accurate division. It can also be geared to the table screw for helical milling operations such as cutting flutes in twist drills and making helical gears.

Indexing heads are of the plain or universal type. Plain heads cannot be tilted; universal heads can be tilted to the vertical or any intermediate position. The spindle of the indexing head can be fitted with a chuck, as shown in Fig. 11.34, or with other work-holding devices, including collets or a center.

FIGURE 11.33 Tilting angle plate with swivel base.

A complete indexing head set usually includes the *indexing head,* the necessary *index plates,* a *gear set* and *gear case* for connecting the gear to the table screw for helical milling, a *footstock,* and in some cases a *center rest* for supporting work held on centers (see Fig. 11.35).

FIGURE 11.34 Indexing head with chuck.

FIGURE 11.35 Indexing head with accessories. (*Courtesy South Bend Lathe, Inc.*)

Most indexing heads have a worm and wheel reduction ratio of 40 : 1, requiring 40 turns of the hand crank to make the spindle revolve once. When the necessary index plates are available, all divisions to and including 50 can be achieved by plain indexing. For some numbers above 50, differential indexing is necessary. A detailed discussion of the use of the indexing head follows later in this chapter.

In recent years, programmable precision indexers (see Fig. 11.36) have become fairly common in shops doing work that requires accurate spacing of complex hole patterns or surfaces. The indexer may be mounted with the axis of the chuck vertical or horizontal, and in some cases the chuck may be replaced with a specially made holding fixture or a faceplate. If necessary, a tailstock may be used to support the end of the workpiece. The controller is capable of storing a series of programs, each of which may incorporate as many as 100 operational steps or positions.

Rotary Table. Rotary tables are available in a wide range of sizes and can be used on both vertical and horizontal milling machines (Fig. 11.37). Most can also be clamped with the face at a 90° angle to the surface of the milling machine table. The face of the rotary table has four or more T slots and an accurately bored hole in the center, which is concentric with the axis about which the table rotates.

The base of the rotary table, which houses the worm drive mechanism, is graduated in degrees, and the handwheel can be graduated in increments as small as 5′, or 1/12 of 1°. On some rotary tables an index plate may be attached to the base, and a sector arm and indexing crank arrangement similar to one on an indexing head can be used (Fig. 11.38).

Rotary tables can also be geared to the table feed screw, as shown in Fig. 11.39. When set up in this manner, the rotary table can be used to make plate cams and to generate a number of other irregular shapes.

(a)

(b)

FIGURE 11.36 Programmable precision indexer (*a*) and controller (*b*). (*Courtesy Rutland Tool and Supply Co.*)

FIGURE 11.38 Indexing plate-mounted on worm shaft on rotary table. (*Courtesy South Bend Lathe, Inc.*)

FIGURE 11.39 Rotary table geared to table feed screw. (*Courtesy Cincinnati Milacron Inc.*)

FIGURE 11.37 Rotary table. (*Courtesy Universal Vise & Tool Co.*)

FIGURE 11.40 Tilting rotary table. (*Courtesy Rutland Tool and Supply Co.*)

The axis of the tilting rotary table shown in Fig. 11.40 can be positioned at any desired angle between 0 and 90° relative to the surface of the milling machine table. This is a particularly useful feature when the tilting rotary table is used on a plain or semiuniversal milling machine. Another variation of the standard rotary table is shown in Fig. 11.41. Two slides that move at a 90° angle to each other are built into the top of the rotating table. This feature makes it possible for the skilled machinist to do a large variety of jobs that require

FIGURE 10.41 Rotary table with cross slides. (*Courtesy Rutland Tool and Supply Co.*)

irregular hole patterns, slots, arcs, angular surfaces, or any combination of these operations.

11.3.3 Arbors and Collets

Several basic types of arbors and collets are used to hold milling cutters and to transmit power from the spindle to the cutter. Regardless of type, they are usually precisely made of alloy steel and heat-treated for wear resistance and strength.

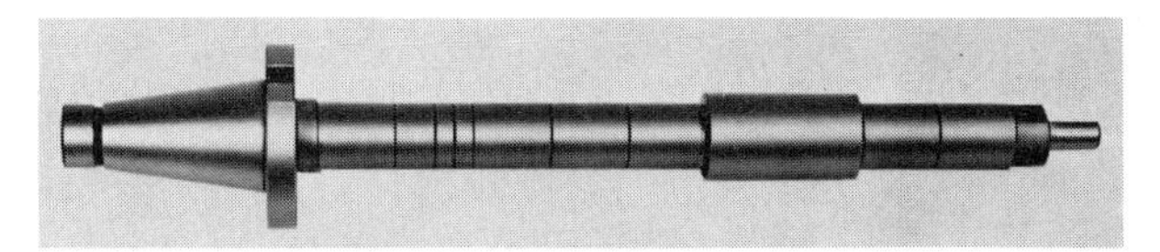

FIGURE 11.42 Style A arbor. (*Courtesy Cincinnati Milacron Inc.*)

Arbors. Arbors for horizontal milling machines are available in three basic types: style A (Fig. 11.42), style B (Fig. 11.43), and style C (Fig. 11.44). A draw bolt that goes through the spindle of the machine screws into the small end of the taper and draws the arbor tightly into the tapered hole in the milling machine spindle. Power is transmitted from the spindle to the arbor by two short keys that engage with the slots on the flange of the arbor.

Style A arbors consist of the tapered portion that fits the spindle, the shaft on which the cutter or cutters fit, the spacers, and the nut.

The shaft has a keyway along its entire length. The outboard end of the arbor has a *pilot* that fits into a bronze bushing in the outboard support of the milling machine overarm, as shown in Fig. 11.45. One or more cutters can be mounted on the arbor, either adjacent to each other or separated by spacers and shims. Style A arbors are used primarily for light- and medium-duty milling operations.

Style B arbors are used for heavy milling operations, especially where it is necessary to provide support close to a milling cutter, such as in

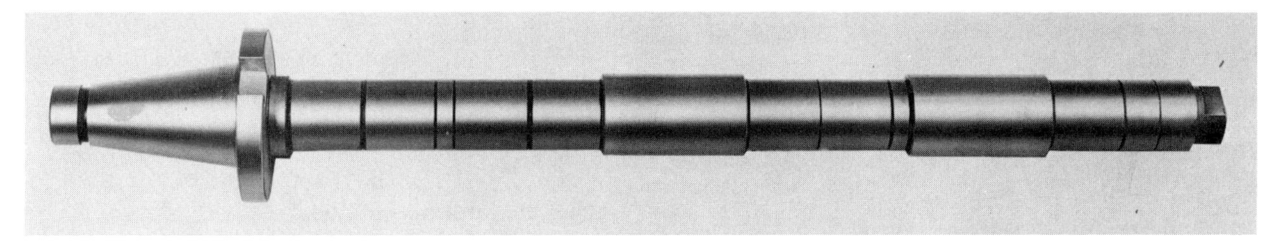

FIGURE 11.43 Style B arbor. (*Courtesy Cincinnati Milacron Inc.*)

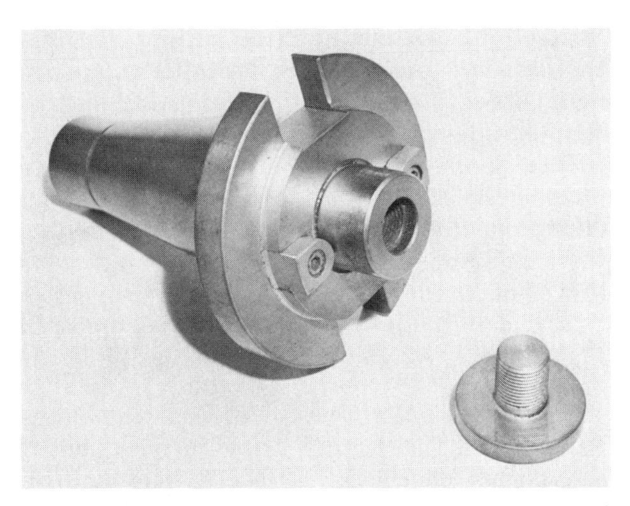

FIGURE 11.44 Style C arbor. (*Courtesy Cincinnati Milacron Inc.*)

FIGURE 11.45 Style A arbor in use. (*Courtesy El Camino College*)

FIGURE 11.46 Style B arbor used in a straddle milling operation.

the straddle milling operation shown in Fig. 11.46. One or more bearing sleeves may be placed on the arbor as near to the cutters as possible. An outboard bearing support is used for each bearing sleeve on the arbor.

Style C arbors are used to hold and drive shell end mills and some types of face milling cutters and require no outboard support. In some cases, they can also be fitted with adapters for mounting other types of cutters. A style C arbor with a shell end mill mounted on it is shown in Fig. 11.47.

Arbors are classified by type, size of the standard milling machine taper, shaft size, and length from shoulder to nut. They are available in different lengths. A 51B24 arbor, for example, has a No. 50 taper and a 1-in.-diameter shaft, is a style B arbor, and has a shoulder-to-nut length of 24 in.

Some adapters that resemble style C arbors in several respects are shown in Fig. 11.48. The rears

FIGURE 11.47 Style C arbor with shell end mill. (*Courtesy El Camino College*)

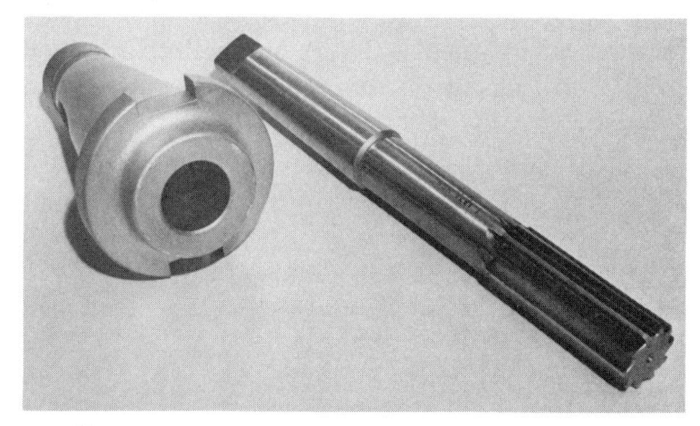

FIGURE 11.48 Adapters for end mills and Morse taper tools.

of the adapters fit the tapered opening in the spindle. The front accommodates straight shank end mills up to 1 in. or larger in size or tools with Morse taper shanks, such as twist drills or reamers.

Collets. On some vertical milling machines the spindle is bored to accept a collet that has a partly straight and partly tapered shank, such as the R-8 solid-type collet shown in Fig. 11.49. The collet is secured by a drawbar that is screwed into a tapped hole in the back of the collet and tightened from the top of the spindle. Some milling machine manufacturers offer collet arrangements that do not need a drawbar. Collets of this type can be closed with a lever-operated cam, as shown in Fig. 11.50, or with a large locking nut.

Special Toolholders. For some operations that require the use of tools with nonstandard shank sizes, chucks can be used to hold the tool. These chucks are available with Morse taper or straight shanks. Either type can be used in mill-

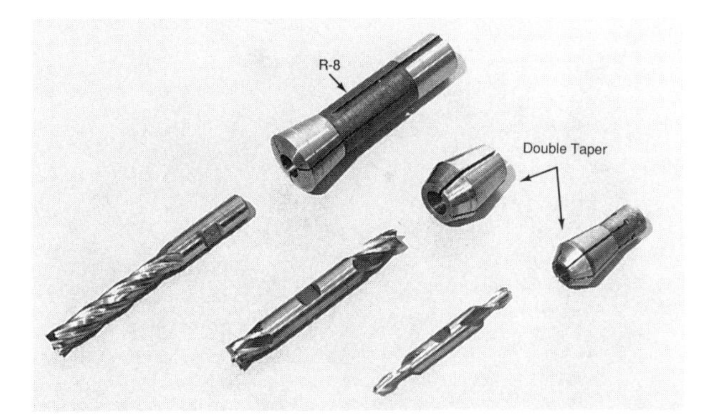

FIGURE 11.49 Collets for holding end mills.

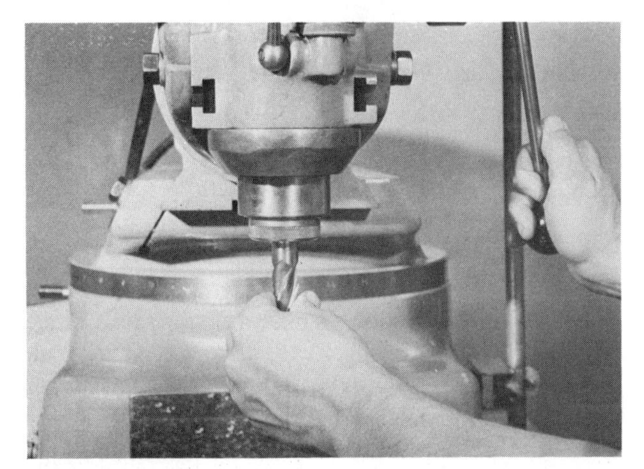

FIGURE 11.50 Closing a collet with a feed-lever-operated cam.

FIGURE 11.51 Chuck used to hold drill in spindle of vertical mill. (*Courtesy El Camino College*)

ing machines when the proper adapters or collets are available. A typical drilling operation in a vertical milling machine is shown in Fig. 11.51.

Offset boring heads are often used in vertical milling machines for boring, facing, chamfering, and outside diameter turning operations. They are available with straight, Morse taper, or standard milling machine taper shanks and usually have three mounting holes for boring bars. Two of the holes are usually parallel with the centerline of the tool, and one is perpendicular to the centerline. The head shown in Fig. 11.52 can be used for both boring and facing operations. Some boring heads have two adjusting mechanisms, and the movable slide can be adjusted accurately in increments of 0.0001 in.

FIGURE 11.52 Offset boring head. (*Courtesy Criterion Machine Works*)

Flycutters can be used for facing operations. The tool shown in Fig. 11.53 can be fitted with several different types of shanks. The tools in cutters of this type are adjusted as shown in Fig. 11.54 so that both a roughing and finishing cut are taken in one pass.

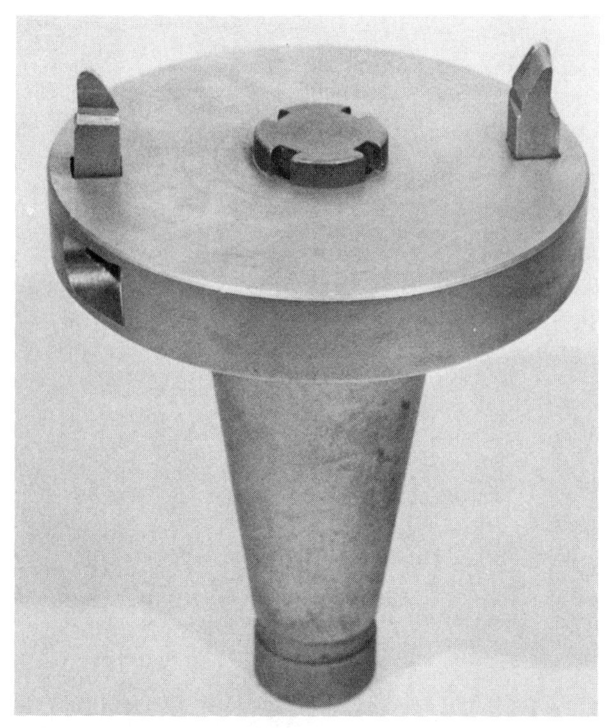

FIGURE 11.53 Fly cutter head on style C arbor.

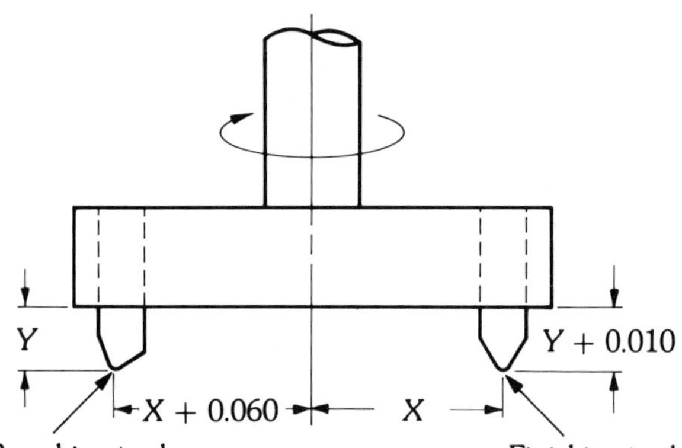

FIGURE 11.54 Tools adjusted for roughing and finishing cuts in one pass.

11.4 MILLING CUTTERS

The variety of milling cutters available for all types of milling machines helps make milling a very versatile machining process. Cutters are made in a large range of sizes and of several different cutting tool materials. The successful and efficient machinist or milling machine operator must be able to select cutters that are compatible with the machine and the material being machined.

11.4.1 Plain Milling Cutters

Plain milling cutters cut only on the periphery and are used to machine flat surfaces and slots. The surface that is produced by the teeth of the cutter is always parallel to the spindle of the machine. Plain milling cutters range in width from about 0.032 in. to as much as 8 in. and are almost always made of high-speed steel.

FIGURE 11.55 Plain milling saw.

Saws. Narrow cutters are generally known as *saws* (see Fig. 11.55) and vary in diameter from 2 1/2 to 8 in. Milling saws usually have fairly small teeth and are almost always ground with side clearance to reduce friction when making deep cuts. Some metal slitting saws are available with carbide-tipped teeth.

Light-Duty Plain Milling Cutters. These cutters (see Fig. 11.56) have fairly small teeth and are used with fairly low chip loads and feed rates. They range from 2 1/2 to 4 in. in diameter and have from 14 to 20 teeth. Cutters more than 3/4 in. in width have left- or right-hand helical teeth machined at an 18 or 20° helix angle, as shown in Fig. 11.56 (right). The helical teeth reduce the possibility of chatter because each tooth engages the work gradually and cuts with a mild shearing action.

Heavy-Duty Plain Milling Cutters. These cutters, shown in Fig. 11.57, are available in widths ranging from 2 to 6 in. and have 8 or 10 teeth. The *gullet* (the space between the teeth) is large, and heavy cuts can be taken at high feed rates without clogging the cutter with chips. The helix angle is about 45°, and the teeth have a 10° positive rake. The arbor hole size ranges from 1 in. for the 2 1/2-in.-diameter cutters to 2 in. for the 4 1/2-in.-diameter cutters.

Some plain milling cutters are made with 52 or 60° helix angle for milling soft steel or brass alloys (see Fig. 11.58). Cutters of this type may have as few as 4 teeth and are capable of taking heavy cuts at high speeds.

The larger plain milling cutters are usually mounted on style B arbors so that arbor support

FIGURE 11.56 (Top) plain straight-tooth and (bottom) plain helical milling cutters. (*Courtesy Cleveland Twist Drill Co.*)

bearings can be placed as near the spindle as possible.

11.4.2 Side-Milling Cutters

Side-milling cutters have teeth on the periphery of the cutter and on one or both sides. The sides of the teeth are relieved so that only the cutting edge contacts the work. Separate chips can be cut by the teeth on the periphery and sides, although this is not the case in simple slotting operations.

Straight-Tooth Side-Milling Cutters. These cutters (see Fig. 11.59) are widely used for

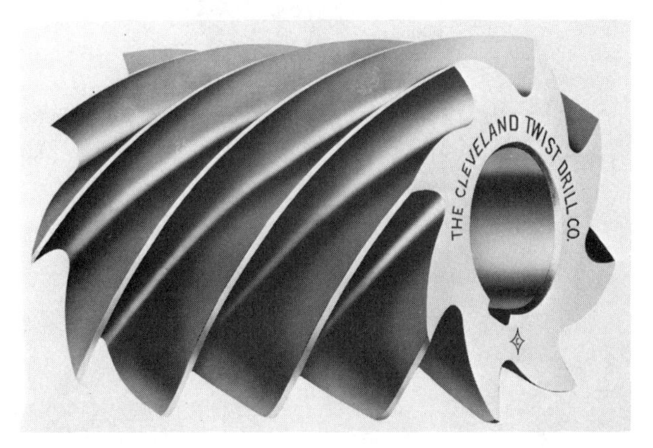

FIGURE 11.57 Heavy-duty plain milling cutter. (*Courtesy Cleveland Twist Drill Co.*)

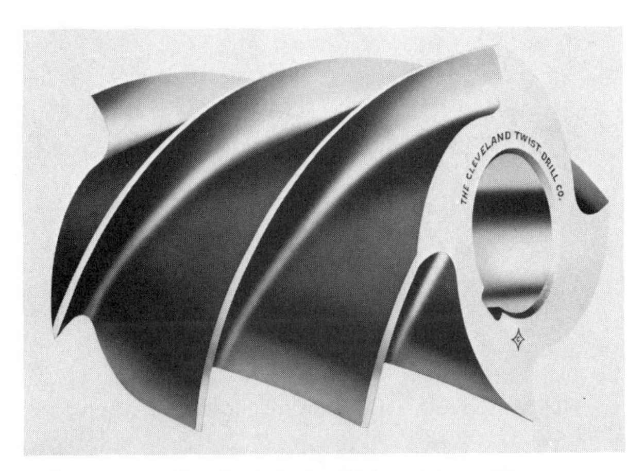

FIGURE 11.58 High-helix-angle plain milling cutter. (*Courtesy Cleveland Twist Drill Co.*)

FIGURE 11.59 Straight-tooth side milling cutter. (*Courtesy Cleveland Twist Drill Co.*)

FIGURE 11.60 Half-side milling cutter. (*Courtesy Cleveland Twist Drill Co.*)

slotting operations, gang milling, and straddle milling. They are usually available in 2- to 8-in. diameters and 3/16- to 1-in. widths. The teeth generally have a 10° positive rake angle. Remember that when cutters of this type are sharpened on both the periphery and sides they will no longer be suitable for cutting in one pass slots that are the nominal width of the cutter.

Half-side-milling cutters (see Fig. 11.60) are available in diameters up to 8 in. and face widths up to 1 in. in both right- and left-hand cuts. They are used extensively for straddle-milling operations and are capable of taking heavy cuts. The teeth are helical and usually have a rake angle of 10° or more.

Stagger-Tooth Side-Milling Cutters. These cutters (see Fig. 11.61) are used for deep slotting and heavy side-milling operations. The teeth on the periphery are helical, with alternate right- and left-helix angles. Side thrust is eliminated, and heavy cuts can be taken.

Stagger-tooth cutters are generally available in widths from 3/16 to 3/4 in. and diameters from

FIGURE 11.61 Stagger-tooth, side milling cutter. (*Courtesy Cleveland Twist Drill Co.*)

2 1/2 to 8 in. The cutters can be used for straddle and gang milling.

11.4.3 Inserted Tooth Cutters

Cutters of the type shown in operation in Fig. 11.62 are used for rapid metal removal on powerful milling machines. They are generally used for facing operations on both vertical and horizontal production milling machines. Variations of this type of cutter can be used for a combination of facing, boring, and chamfering operations.

Face Mills. Face mills vary in diameter from about 4 to 16 in. or more. The body of the cutter is made of alloy steel and is machined to accept inserts of various types. The angles at which the cutter insert pockets are machined determine whether the cutter will have negative, neutral, or positive rake. On some mills the cutter inserts can be indexed. With a square insert, and the inserts set at a negative rake angle, as many as eight cutting edges can be used. The cutter inserts are usually chamfered or radiused on the corner that enters the work. The insert is also tilted so that the part behind the cutting edge has a clearance angle of 5 to 10°. Smaller face mills are usually mounted on standard shell end mill adapters. Large cutters are mounted on special adapters as close as possible to the end of the machine spindle.

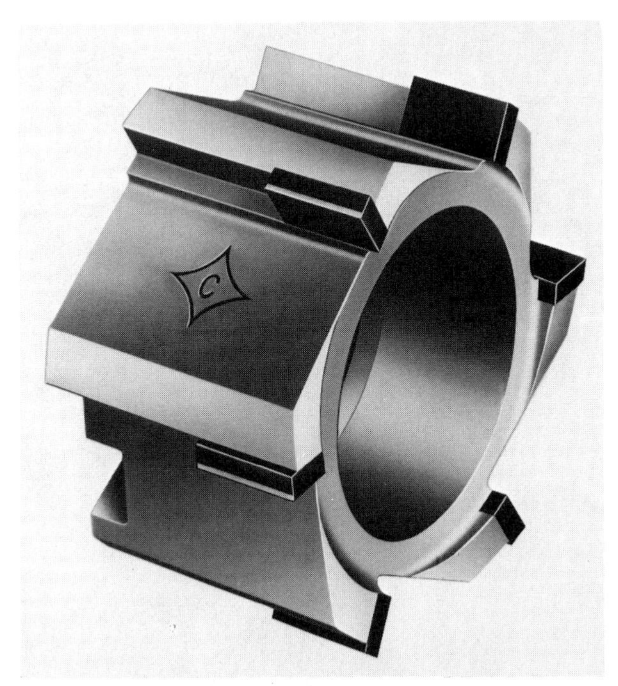

FIGURE 11.63 Carbide-tipped shell end mill.

FIGURE 11.62 Inserted-tooth facing cutter. (*Courtesy Cincinnati Milacron Inc.*)

Brazed-in Inserts. Some milling cutters are made with the carbide inserts nickel-silver-brazed in place. The shell end mill shown in Fig. 11.63 is typical of this type of cutter. These cutters are available in diameters ranging from 1 1/4 to 6 in. and mount on a standard shell end mill adapter.

11.4.4 End Mills

End mills can be used on vertical and horizontal milling machines for a variety of facing, slotting, and profiling operations. Solid end mills are made from high-speed steel or sintered carbide. Other types, such as shell end mills and fly cutters, consist of cutting tools that are bolted or otherwise fastened to adapters.

Solid End Mills. Solid end mills have two, three, four, or more flutes and cutting edges on the end and the periphery. Two-flute end mills can be fed directly along their longitudinal axis into solid material because the cutting faces on the end meet. Three- and four-fluted cutters with one end-cutting edge that extends past the center of the cutter can also be fed directly into solid material.

Solid end mills are double (see Fig. 11.64) or single ended, with straight or tapered shanks (see Fig. 11.65). The end mill can be of the stub type, with short cutting flutes, or of the extra-long type for reaching into deep cavities. On end mills designed for effective cutting of aluminum, the helix angle is increased for improved shearing action and

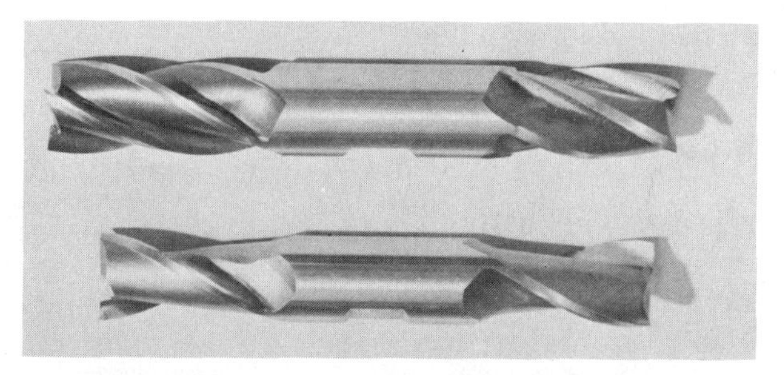

FIGURE 11.64 Two- and four-flute end mills.

FIGURE 11.65 (Top) Morse taper shank and (bottom) straight shank end mills. (*Courtesy Cleveland Twist Drill Co.*)

chip removal (see Fig. 11.66), and the flutes may be polished. An end mill designed for rapid metal removal is shown in Fig. 11.67.

Shell End Mills. Solid shell end mills are usually made of high-speed steel and mounted on adapters that fit into the milling machine spindle. They range in diameter from 1 1/4 to 6 in. and are available in right-hand helix–right-hand cut and left-hand helix–left-hand cut (see Fig. 11.68). Shell end mills with serrated teeth on the periphery as shown in Fig. 11.69 are used for roughing operations.

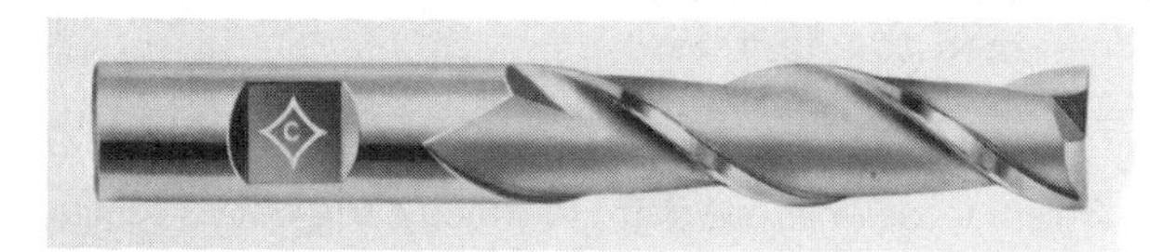

FIGURE 11.66 High-helix-angle end mill for aluminum. (*Courtesy Cleveland Twist Drill Co.*)

Special End Mills. Ball end mills are available in diameters ranging from 1/32 to 2 1/2 in. in single- and double-ended types. A typical single-ended two-flute ball-end mill is shown in Fig. 11.70.

Single-purpose end mills such as Woodruff keyseat cutters, corner rounding cutters, and dovetail cutters are used on both vertical and horizontal milling machines. They are usually made of high-

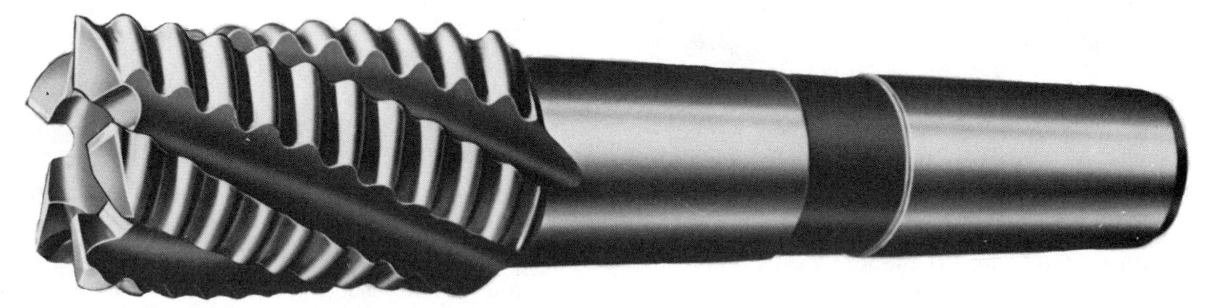

FIGURE 11.67 End milling for roughing cuts. (*Courtesy Bridgeport Machines, Inc.*)

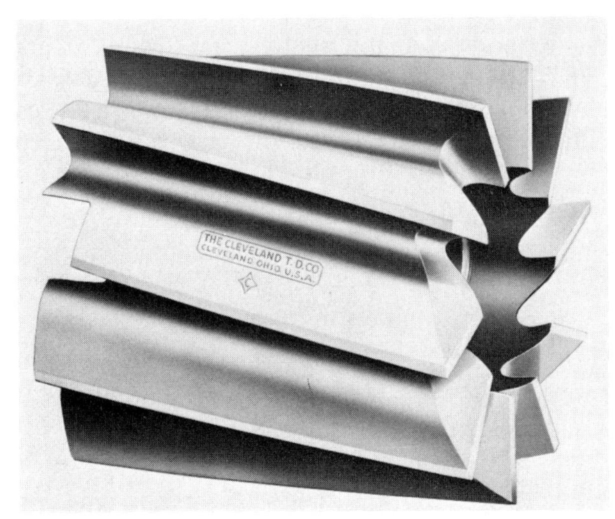

FIGURE 11.68 Shell end mill. (*Courtesy Cleveland Twist Drill Co.*)

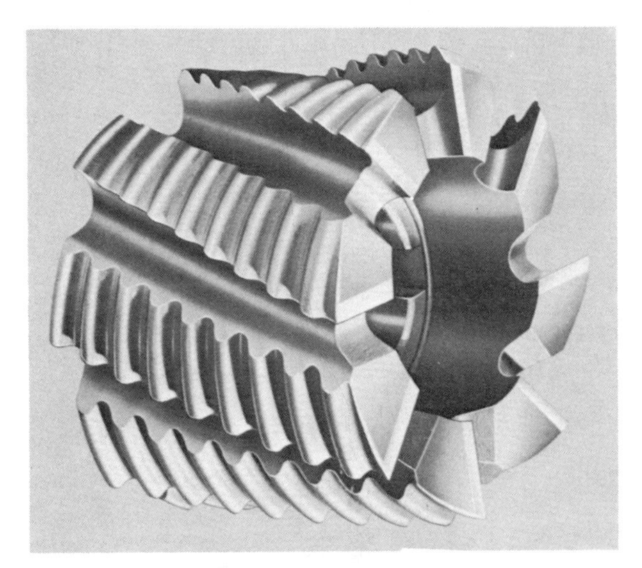

FIGURE 11.69 Shell end mill for roughing cuts.

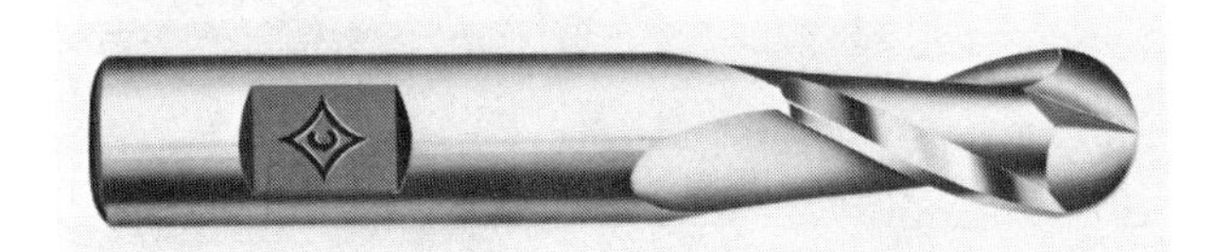

FIGURE 11.70 Ball end mill. (*Courtesy Cleveland Twist Drill Co.*)

speed steel and may have straight or tapered shanks (see Fig. 11.71).

Many more types of special end mills are used in milling operations; some are designed and made for one or a very limited number of operations.

11.4.5 Special Cutters

Almost all special milling cutters are used to produce certain forms and shapes on the finished part.

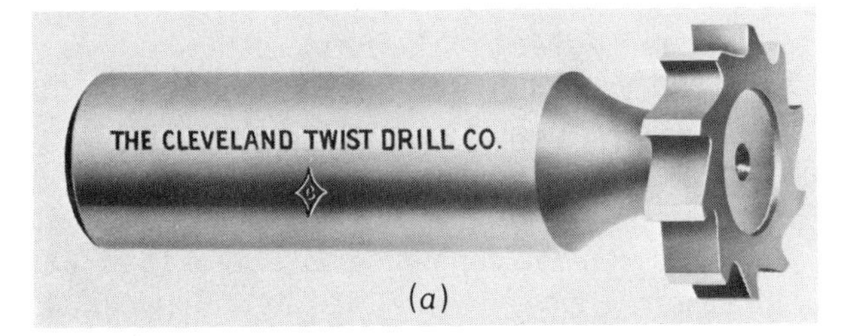

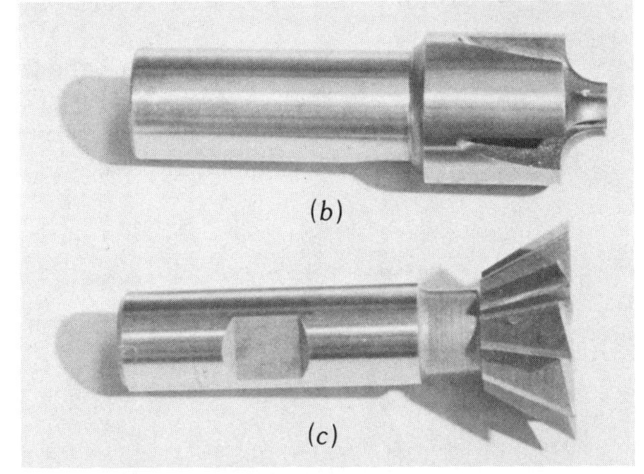

FIGURE 11.71 Special end mills: (*a*) Woodruff keyseat cutter; (*b*) corner rounding cutter; (*c*) dovetail cutter; (*d*) dovetail cutter in use. [*a*] (*Courtesy Cleveland Twist Drill Co.*) [*b*] (*Courtesy Cincinnati Milacron Inc.*)

This shape can be external or internal, and it can be produced partly by the cutter and partly by movement of the work or the machine.

Form Cutters. The form cutters shown in Figs. 11.72 and 11.73 are used for making angular grooves. The points of the teeth are sharp or rounded, as necessary. *Single-angle* cutters are *plain* or *side cutting.* Plain cutters have cutting edges only on the conical surface. In either case, the cutter is classified by the angle between the face perpendicular to the arbor and the conical cutting surface.

Double-angle cutters are classified by the included angle between the two cutting faces. For example, a 45° double-angle cutter produces a symmetrical groove with an included angle of 45°.

FIGURE 11.72 Single-angle cutter.

FIGURE 11.73 Double-angle cutter. (*Courtesy Cleveland Twist Drill Co.*)

Concave and *convex* cutters (see Figs. 11.74 and 11.75) for use on horizontal milling machines are available in various radii or diameters. They are ground with zero rake so that they can be sharp-

FIGURE 11.74 Concave cutter. (*Courtesy Cleveland Twist Drill Co.*)

FIGURE 11.75 Convex cutter. (*Courtesy Cleveland Twist Drill Co.*)

FIGURE 11.76 Gear-tooth cutter.

ened by grinding only the face. They are form-relieved, and the contour of the cutter is not affected by the sharpening procedure. Corner-rounding cutters, which are essentially one-half of a concave cutter, are also available.

Gear Cutters. Cutters for involute gears are made in sets of eight for each diametral pitch. Since gear tooth cutters are a type of form cutter, the face is radial, with neither positive nor negative rake. Like other form cutters, they are form-relieved and sharpened by grinding the face only. A typical gear tooth cutter is shown in Fig. 11.76.

11.5 ALIGNMENT AND SETUP PROCEDURES

The safe and efficient use of any milling machine is largely dependent on how well the machine and work-holding fixtures are aligned and how rigidly and securely the work is set up. Work that is not held securely may move during the machining process, causing the part to be ruined. There is also the possibility that improperly held work will be pulled out of the fixture or vise, resulting in damage to the machine and possible injury.

11.5.1 Machine Alignment

The machinist must be familiar with the procedures for aligning the worktable and head of those milling machines that require alignment. Plain milling machines, both vertical and horizontal, do not require table alignment in normal use because the table cannot be swiveled. Overarm-type vertical milling machines that do not have either a swiveling or fully universal head do not require head alignment because the head can move only up or down.

Table Alignment. The table on universal milling machines must be checked for alignment whenever it is being returned to the 0° position or when a job involving precise angular relationships is to be done. *Never* trust the graduations on the saddle or on any other part of the machine when really accurate work must be done.

The basic setup for aligning a universal milling table with the face of the column is shown in Fig. 11.77. A rigid mounting for the dial indicator is necessary so that flexing and slippage do not alter the readings. The dial indicator must be mounted to the *table* of the machine because the motion of the table relative to the face of the column is being checked.

The following seven-step procedure is suggested for table alignment:

1. Clean the table and column face, making sure there are no nicks that rise above the surface.
2. Attach the dial indicator.
3. Bring the dial indicator in contact with one edge of the column face.
4. Move in one-fourth of the indicator's operating range, and zero the dial by turning the bezel.
5. Move the table manually and note any changes in the dial indicator reading. (*Note:* The gibs on the knee, saddle, and table must be in good condition and properly adjusted to eliminate lost motion. The table *cannot* be adjusted accurately if there are any loose mating surfaces.)

FIGURE 11.77 Aligning the universal table.

6. If any error is noted, loosen the saddle clamps and move the table *one-half* of the error in the proper direction and retighten. (*Example:* If the dial reads 0 at the right-hand side of the column and 0.008 in. at the left-hand side, shift the table so that the reading becomes 0.004 in.)
7. Repeat the process until the dial indicator reading does not change as it traverses the column face.

Head Alignment. The head on vertical milling machines with semiuniversal or universal heads must be checked before doing jobs requiring accurate alignment between the head and table. The setup for aligning the spindle of a vertical milling machine perpendicular to the table is shown in Fig. 11.78. It is particularly important to align the spindle perpendicular to the table when drilling, boring, and flycutting operations are performed. For example, flycutting with the spindle slightly tilted generates a surface that is either beveled or concave, depending on the direction of feed.

The dial indicator must be rigidly attached to a fixture held in a collet in the spindle. The diameter of the circle that the dial indicator makes should be slightly smaller than the width of the table. The table must be clean and free of nicks before the aligning operation is started. If a steel or cast iron plate of *known flatness* and *parallelism* is available, it can be placed on the table so that the plunger on the dial indicator will not drop into the T slots.

FIGURE 11.78 Aligning the head on a vertical milling machine.

The following procedure is suggested for aligning the head of vertical milling machines with semiuniversal or universal heads:

1. Clean the table thoroughly and place a flat and parallel plate on it if one is available.
2. Attach the dial indicator to the spindle.
3. Feed the spindle down, with the dial plunger at the operator's right or left side until it registers about one-fourth of its operating range, and zero it.
4. Carefully rotate the spindle one revolution.
   (*a*) If the head is of the swiveling or *semiuniversal* type, the fore and aft readings in line with the cross feed will be identical; the right and left readings may vary if the head is not vertical.
   (*b*) If the head is of the universal type, the reading may vary on both axes.
5. Adjust a semiuniversal head by loosening the head and swiveling it so that the dial indicator reading is cut in half. (*Example:* If the highest dial indicator reading is 0.010 in., stop the rotation of the spindle at that point. The head should be moved so that the reading is reduced to 0.005 in.) Recheck by rotating the spindle and readjusting the head if necessary.
6. Universal heads should be adjusted in one plane at a time. (*Example:* Adjust the head to correct fore-and-aft tilt until both the fore and aft readings are zero. Then adjust the head so that the right and left readings are zero.)
7. Securely tighten all head locking bolts and recheck.

11.5.2 Vise and Fixture Alignment

Assuming that the head and table of the milling machine are properly aligned and all the gibs are adjusted, the next major procedure is to align vises and other fixtures used to hold workpieces. It is not advisable to trust the angular markings on most vises and fixtures if accurate work is to be done.

Vise Alignment. As shown in Fig. 11.79, the solid jaw of the vise is always the reference surface. The dial indicator is attached to the arbor of a horizontal milling machine or held in the collet of a vertical milling machine.

The following procedure is suggested for aligning plates and other fixture with major flat surfaces:

FIGURE 11.79 Aligning the vise.

1. Clean all parts thoroughly. Make sure that there are no burrs or nicks on mating surfaces.
2. Lightly clamp the vise or fixture in approximately the correct position.
3. Bring the dial indicator in contact with one end of the part to be aligned. Move in about one-forth of the indicator's operating range and zero the indicator.
4. Move the table or cross slide the full length of the jaw or fixture.
5. Note the variation in indicator reading, and move the vise or fixture in the appropriate direction, using a soft hammer.
6. When the indicator shows no deviation, retighten all bolts and recheck.

Angular Settings. In some cases vises or fixture have to be precisely set at an angle to the line of table or cross-feed travel. Several methods can be used, depending on the angle involved and the precision required. One method is to use the markings on the vise and base, as shown in Fig. 11.80.

A method of setting a swiveling vise at a precise angle is to use a vernier bevel protractor. The protractor is set to the correct angle and carefully and lightly clamped in a horizontal position in the vise. For example, if the solid jaw of the vise is to be set at an angle of 8°30′ to the line of table travel, the vernier bevel protractor is set to a complementary angle of 81°30′. With one leg of the protractor clamped in the vise, the other leg is roughly parallel to the table travel. The final position of the vise is then set by using a dial indicator attached to the arbor or held in a collet.

Similar procedures can be used to set universal vises, angle plates, and other holding fixtures.

FIGURE 11.80 Setting the vise at an angle.

Remember that dial indicators, protractors, and other precision measuring tools must be used carefully and not be subjected to excessive clamping force.

11.5.3 Basic Setup Procedures

The versatility of the milling machine allows workpieces of many different shapes and sizes to be held and machined. The skilled machinist must be able to select the correct setup tools and equipment from a large array of clamps, straps, bolts, jacks, blocks, vises, and other fixtures. Proper work setup results in more accurate machining done safely and with a minimum loss of time.

Mounting the Work. When the work is held in a vise, it should be placed and supported so that the loads imposed by the cutter are directed at the solid jaw of the vise (see Fig. 11.81). Parallel bars should be used under any work that is too thin to protrude above the jaws of the vise.

When castings are held in a vise, the part that

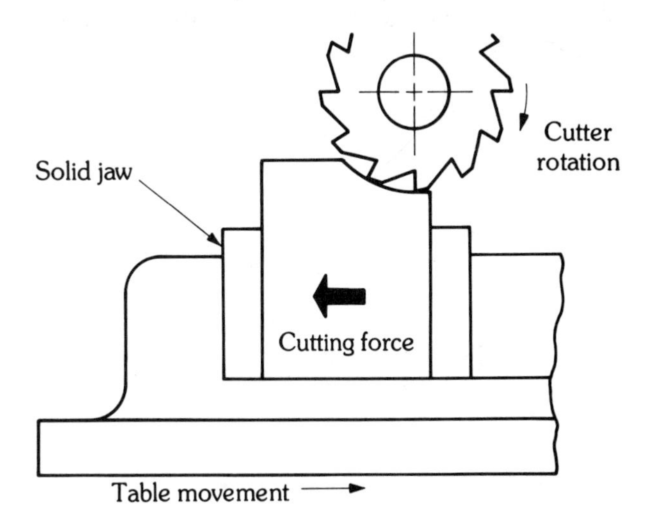

FIGURE 11.81 Direct cutting pressures against the solid jaw.

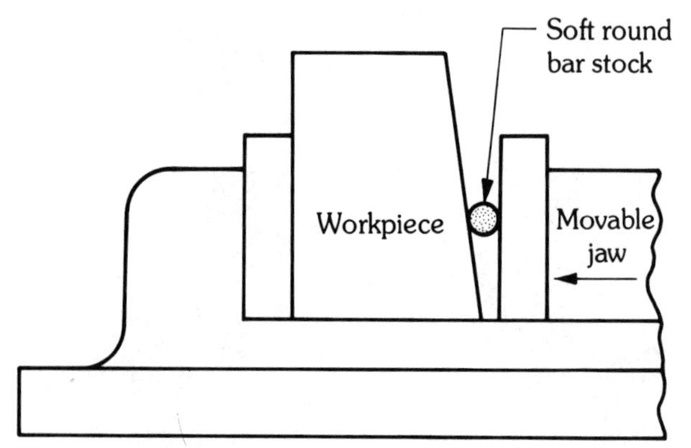

FIGURE 11.82 Round bar stock between the movable jaw and the workpiece.

FIGURE 11.84 Straps and T bolts used to hold work on the table.

contacts the fixed jaw should be ground as smooth as possible on a disk or belt sander. If this is not possible, place a sheet of soft aluminum or copper between the vise and the casting. Place a piece of round mild steel or aluminum between the movable jaw and the casting, as shown in Fig. 11.82. This protects the rear jaw and applies pressure more evenly to the workpiece. The round bar should be placed in line with the screw of the vise if possible.

A combination of straps, clamps, T bolts, step blocks, and other devices can be used to hold work on the milling table. The work must be set up so that it is held in the proper position and resists the forces of the cutting action. Stop blocks, as shown in Fig. 11.83, are used to withstand the cutting forces.

When straps, step blocks, and T bolts are used to hold work, as shown in Fig. 11.84, the straps must be level so that the T bolts will not tilt when they are tightened. Washers must always be used between the nut and the top of the strap. The T bolts must be placed as close to the workpiece as possible.

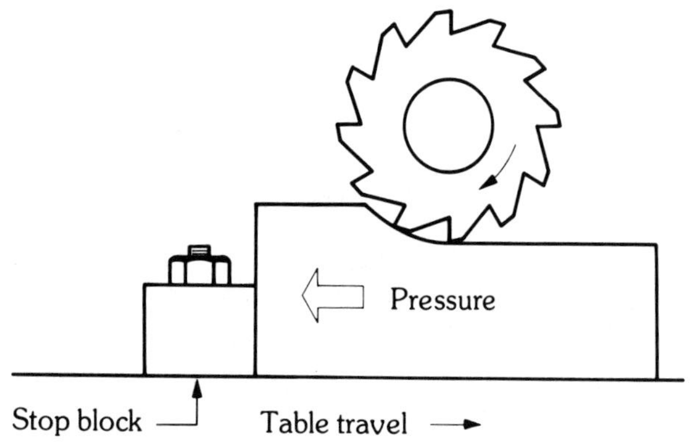

FIGURE 11.83 Stop block used to hold workpiece.

11.6 BASIC MILLING OPERATIONS

Before any milling job—no matter how simple—is attempted, the machinist has to make several decisions. In addition to selecting the best means of holding the work and the most appropriate cutters, the machinist must make an initial estimate of the cutting speed and feed rate that will provide good balance between rapid metal removal and long cutter life.

11.6.1 Cutting Speed and Feed Rate

Proper determination of cutting speed and feed rate can be done only when the following eight factors are known:

1. Type of material to be machined
2. Nature of heat treatment, if any
3. Rigidity of the setup
4. Physical strength of the cutter
5. Cutting tool material
6. Power available at the spindle
7. Type of finish desired
8. Cutting fluid to be used, if any

Several of these factors affect cutting speed only, and some affect both cutting speed and feed rate. The tables in machinists' handbooks provide approximate figures that can be used as starting points. After the cutting speed is chosen, the spindle speed must be computed and the machine adjusted.

Cutting Speeds. The cutting speed of a milling cutter is expressed in *surface feet per minute* (sfpm) or *meters per minute* (mpm) and is the distance that a point on the periphery of a cutter travels in 1 minute at a particular rotational speed.

Machinists, setup people, and those who estimate the time required for milling operations must know how to compute the rpm at which the cutter must run when the recommended cutting speed and the cutter diameter are known. Because the cutting speed (CS) is given in feet or meters per minute, the diameter of the cutter must be converted into feet or meters either before or during the computation. In the following examples, both methods are shown.

Example: A 4-in.-diameter cutter will be used to machine aluminum at 350 sfpm. Find the rpm rate at which it should be run.

Method A

Convert the cutter diameter into feet.

$$4 \div 12 = 0.333$$

$$\text{rpm} = \frac{\text{CS (fpm)}}{\pi \times D \text{ (ft)}} = \frac{350}{3.14 \times 0.333} = \frac{350}{1.05} = 333.3$$

Method B

The standard formula for finding rpm is

$$\text{rpm} = \frac{4 \times \text{CS}}{D}$$

The formula was derived as follows and involves rounding off 3.1416 to 3:

$$\text{rpm} = \frac{\text{CS}}{3.1416 \times (D/12)} = \frac{12 \times \text{CS}}{3 \times D} = \frac{4 \times \text{CS}}{D} = \frac{4 \times 350}{4} = 350 \text{ rpm}$$

The answers differ by less than 17 rpm, and since the speeds on most milling machines can be adjusted only in increments, the difference is not a problem.

In some cases, the machinist must compute the cutting speed when the machine rpm and the cutter diameter are the known factors. The formula for this computation is

$$\text{CS} = \frac{D \text{ (in.)} \times \pi \times \text{rpm}}{12}$$

The operation can be simplified by rounding off (3.1416) to 3. The formula then is

$$\text{CS} = \frac{D \text{ (in.)} \times 3 \times \text{rpm}}{12}$$

or

$$\text{CS} = \frac{D \text{ (in.)} \times \text{rpm}}{4}$$

If the diameter of the cutter is expressed in feet initially, the formula is

$$\text{CS} = D \text{ (ft)} \times \pi \times \text{rpm}$$

No division is necessary, and the computation can be done mentally in many cases if π (3.1416) is rounded off to 3.

The following sample problem is solved by each of the two methods outlined.

Example: Find the cutting speed of a 6-in.-diameter cutter being operated at 160 rpm (π is rounded off to 3).

Method A

The cutter diameter is expressed in inches.

$$\text{CS} = \frac{6 \times 3 \times 160}{12} = \frac{6 \times 160}{4} = 6 \times 40 = 240 \text{ sfpm}$$

Method B

The cutter diameter is expressed in feet.

$$\text{CS} = 0.5 \times 3 \times 160 = 240 \text{ sfpm}$$

Recommended cutting speeds for milling some of the more common materials with high-speed-steel cutters are shown in Table 11.1. Because each job is somewhat different, the machinist may alter the cutting speed according to personal judgment. For example, if sintered carbide is used as the cutting tool material, the cutting speed may be increased up to 500 percent of the basic speed for high-speed-steel cutters, depending on conditions. It is up to the machinist to decide if the machine has enough power and if the workpiece and setup are rigid enough to withstand the pressure and loads caused by high cutting speeds.

For finishing cuts, especially when cutting fluid is used, the cutting speed may be raised as much as 20 to 30 percent on some materials. A reduction in feed rate is generally recommended when taking finishing cuts.

TABLE 11.1

Recommended cutting speeds

Material to Be Cut	Cutting Tool Material: Sintered Carbide ft/min	Sintered Carbide m/min	High-Speed Steel ft/min	High-Speed Steel m/min
Aluminum	800–1500	240–450	300–800	90–240
Brass, leaded	300–450	90–135	100–250	30–75
Bronze				
Soft	200–350	60–105	80–200	24–60
Hard	150–300	45–90	70–150	21–45
Cast iron				
Gray iron	125–200	40–60	50–90	15–27
Ductile iron	150–250	45–75	70–120	21–36
Magnesium	900–1600	270–480	400–1200	120–360
Steel				
Free machining	250–350	75–105	125–175	40–55
Mild	150–250	45–75	75–150	25–50
Medium-carbon	150–200	40–60	60–120	18–36
High-carbon	100–175	30–55	50–90	15–27
Stainless	100–220	30–65	40–90	12–27
Titanium	125–175	40–55	25–70	8–21

Note: The figures are only recommendations. The final decision on the cutting speeds for a particular job must be made after evaluating the rigidity of the workpiece and setup, machine condition, power available, variations in material, and other pertinent factors.

TABLE 11.2

Recommended chip loads for high-speed-steel cutters

Type of Cutter	Aluminum in.	Aluminum mm	Bronze in.	Bronze mm	Cast Iron in.	Cast Iron mm	Free-Machining Steel in.	Free-Machining Steel mm	High-Carbon Steel in.	High-Carbon Steel mm
Slitting saw	0.004	0.10	0.003	0.08	0.003	0.08	.004	0.10	0.003	0.08
Form cutter	0.005	0.12	0.005	0.12	0.005	0.12	.006	0.15	0.003	0.08
End mill	0.010	0.25	0.008	0.20	0.007	0.18	.007	0.18	0.004	0.10
Plain helical mill	0.016	0.40	0.012	0.30	0.009	0.22	.010	0.25	0.008	0.20
Side mill	0.012	0.30	0.009	0.22	0.008	0.20	.008	0.20	0.007	0.18

Note: The figures are only recommendations. The final decision on the feed rate for a particular machining operation can be made only after the strength and size of the cutter, rigidity of the workpiece and setup, number of teeth in contact, and other pertinent factors have been evaluated.

Feed Rate. In choosing a feed rate for a milling operation, the machinist must consider several factors, such as the power available at the machine spindle, the rigidity of the setup, the strength of the workpiece, and the strength of the cutter. When conditions permit, the machinist must learn to make full use of the capabilities of cutter and machine, since productivity is generally a measure of the rate of metal removal (see Table 11.2).

On milling machines the rate of feed is expressed in *inches per minute* (ipm) or *millimeters per minute* (mmpm) of table travel and range from 0.5 ipm (12.5 mmpm) to 30 ipm (750 mmpm). The feed rate is adjusted independently by a gearbox (see Fig. 11.85) or by adjusting the speed of a variable speed electric motor.

Two major factors in determining feed rate for a particular job are the number of teeth on the cutter and its strength. A tooth on a thin slitting saw, for example, might be able to take a chip only one-fourth or one-fifth the thickness of a chip taken by a face mill. This is because the cutter is fragile and the spaces between the teeth (gullets) are small. The thickness of the chip taken by each tooth is referred to as the *chip load* or *feed per tooth.* The recommended chip load for most common milling cutters and materials is shown in Table 11.2. The chip load can be altered at the discretion of the

FIGURE 11.85 Feed-change mechanism.

machinist. The recommended procedure is to start with a relatively small chip load and gradually increase it as conditions permit.

The basic procedure for determining the feed rate for a particular job is as follows:

1. Consult the appropriate chart to find the recommended chip load.
2. Count the number of teeth on the cutter.
3. Determine the spindle speed at which the cutter will be operated.
4. Insert each factor into the following formula to compute feed rate:

$$\text{feed rate} = \text{chip load} \times \text{number of teeth} \times \text{rpm}$$

Example: A 5 in.-diameter slitting saw with 40 teeth is being used to cut a slot in soft gray cast iron at 80 sfpm (64 rpm). The recommended chip load is 0.004 in. per tooth, according to the chip load table in a reference handbook. Thus

$$\begin{aligned}\text{feed rate} &= 0.004 \text{ in.} \times 40 \times 64 \\ &= 10.24 \text{ ipm}\end{aligned}$$

The feed rate selector on the machine should be set to the nearest feed rate *below* the computed figure for the initial cuts.

Direction of Feed. For a given direction of rotation of the cutter, the work may be fed in one of two directions. If the work is fed in the opposite direction to the cutter tooth travel at the point of contact [see Fig. 11.86*(a)*], the process is called *conventional* milling. As the work advances into the rotating cutter, the chip that is taken gets progressively thicker, as shown in Fig. 11.87*(a)*. The action of the cutter forces the work and the table against the direction of table feed and eliminates

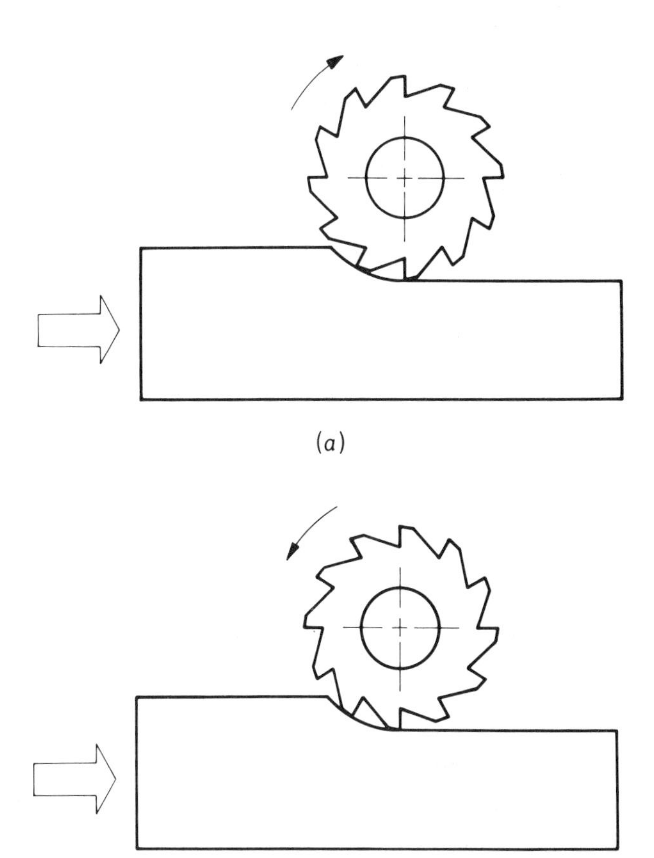

FIGURE 11.86 *(a)* Conventional (up) milling and *(b)* climb (down) milling.

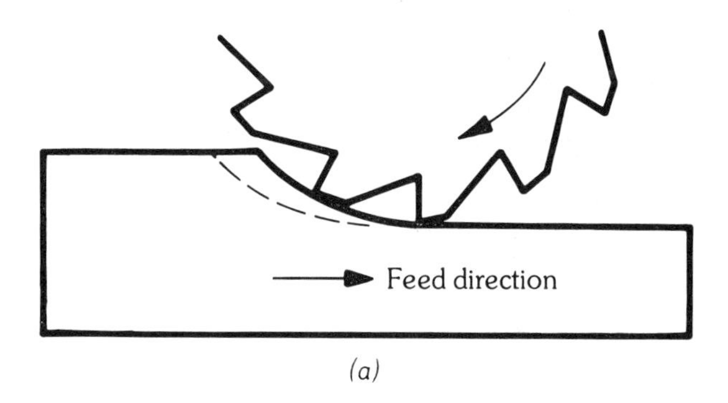

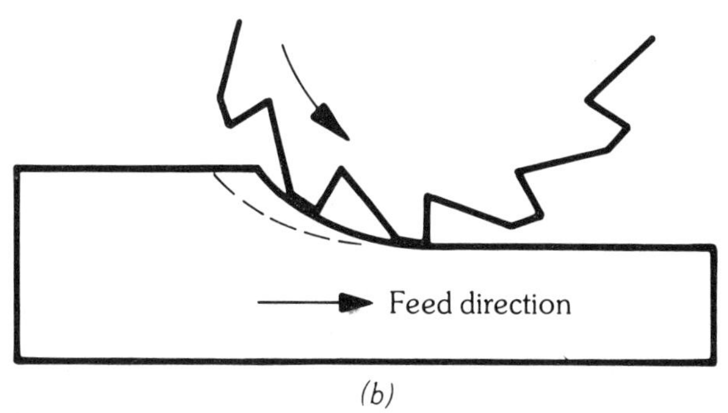

FIGURE 11.87 Change in chip thickness during cut in *(a)* conventional milling and *(b)* climb milling.

backlash due to a worn table feed screw and nut or lack of a backlash eliminator. In conventional milling, each cutter tooth enters clean metal gradually, and the shock loads on each tooth are minimized. Also, when machining castings and forgings that may have sand inclusions or a hard skin on the surface, the cutter does not make direct contact with these abrasive materials. Conventional milling (also known as *up* milling), is the most commonly used method since most older milling machines do not have backlash eliminators. When taking deep cuts, such as in heavy slotting operations, the cutter tends to pull the workpiece out of the vise or fixture since the cutting force is directed upward at an angle. Care must be taken to see that the workpiece is securely held in a vise or with strap clamps.

In climb milling (also known as *down* milling) the cutter teeth that are engaged with the work travel in the same direction as the workpiece [see Fig. 11.86*(b)*]. Generally, more metal can be removed for a given power input when climb milling is used. Therefore, it is widely used on large, rigid production machines that have backlash eliminators on the table screw. The cutter enters the top of the workpiece [see Fig. 11.87*(b)*] and takes a chip that gets progressively thinner as the cutter tooth rotates. Climb milling is used only on materials that are free of scale and other surface imperfections that would damage cutters.

11.6.2 Plain Milling

Plain milling is the process of milling a surface that is parallel to the axis of the cutter and basically flat. It is done on plain or universal horizontal milling machines with cutters of varying widths (see Fig. 11.88) that have teeth only on the periphery.

Cutter Selection and Machine Setup. The first step is to select a plain milling cutter of the proper diameter and width, making sure it is sharp. Then, determine whether you will need a style A or style B milling arbor. When you have selected the proper arbor and cutter, set up the machine in the following manner:

1. Make sure that the cutter, collars, arbor, and spindle are clean and free of nicks and burrs.
2. Mount the arbor in the spindle, locking it tightly in place with the drawbolt.
3. Place the collars, cutter, and key in position. If you use a style B arbor, place the bearing supports as close to the cutter as possible.
4. Secure the outboard bearing support to the overarm *before* tightening the arbor nut.
5. If you use a slabbing cutter with a high helix angle, choose the direction of rotation that will direct the side thrust *toward* the headstock. During the cutting operation, the chips should be moved *away* from the column by the helix angle of the cutter. (*Note:* Remember that *all* helical cutters produce side thrust.)
6. Compute the required spindle speed and feed rate and adjust the machine accordingly.
7. Plan the sequence of cuts to be taken: one or more roughing cuts and a finishing cut.
8. Choose the direction of feed. Do not climb mill unless the table has a backlash eliminator. Do not climb mill castings or forgings.
9. Adjust the table feed stop mechanism (trip dogs) to provide table travel slightly longer than the part to be machined.
10. Provide for flow of cutting fluid to the cutter if necessary.

FIGURE 11.88 Plain milling operation using a helical cutter. (*Courtesy El Camino College*)

You are now ready to make a trial cut. Remember that the cutting speeds and feed rates in Tables 11.1 and 11.2 are only general recommendations. The machinist must evaluate each job and make changes accordingly. It is best to start with fairly low cutting speeds and feed rates. Excessive cutting speeds quickly dull a cutter, causing lost time to change the cutter and expense to resharpen it. Excessive feed rates generally result in broken or quickly dulled cutters, or distortion of the workpiece if it is thin or fragile.

11.6.3 Side Milling

For side milling, a cutter that has teeth on the periphery and on one or both sides is used. When a single cutter is being used, as shown in Fig. 11.89, the teeth on both the periphery and sides may be cutting. The machined surfaces are usually either perpendicular or parallel to the spindle. Angle cutters can be used to produce surfaces that are at an angle to the spindle for such operations as making external dovetails or flutes in reamers.

Cutter Selection and Machine Setup. Cutter selection is a major factor in any side- or straddle-milling operation, particularly when close tolerances must be maintained. Another factor to consider is the depth of slots or vertical surfaces to be cut. When selecting the diameter cutter to be used, the size of the arbor and the outside diameter of the collars must also be considered. The smallest diameter cutter that will do the job should be used.

For side-milling operations involving a single cutter, the setup procedures are essentially the same as for plain milling (see Sec. 11.6.2). The following two additional precautions should be noted, however, particularly when accurate slots are being cut:

1. Closely inspect the hub of the cutter and the arbor collars for nicks and other irregularities. A wobbling cutter cuts an oversize slot and quickly becomes dull.
2. Measure the width of the cutter with a micrometer to determine if it is undersize because of previous sharpening of the side teeth. (For stagger-tooth cutters, use a thin parallel on either side of the cutter and measure. Then subtract the thickness of both parallels.)

FIGURE 11.89 Slotting operation using a stagger-tooth side milling cutter. (*Courtesy El Camino College*)

Straddle Milling. In a typical straddle-milling setup (see Fig. 11.46) two side-milling cutters are used. The cutters are half-side or plain side-milling cutters and have straight or helical teeth. Stagger-tooth side-milling cutters can also be used.

The cutters cut on the inner sides only or on the inner sides and the periphery. If the straddle-milling operation involves side and peripheral cuts, the diameter of the two cutters must be exactly the same. When cutters with helical teeth are used, the helix angles must be opposite.

Since straddle-milled surfaces must be parallel to each other and are usually held to close tolerances in terms of width, the condition and size of the collars and shims that separate the cutters is important. The arbor must also turn as true as possible to avoid cutting the workpiece undersize.

Usually, a combination of collars and steel shims can be assembled to provide the correct spacing between cutters. For some production operations, a special collar can be made from alloy or medium-carbon steel, heat-treated, and surface ground to length. The faces *must* be perpendicular to the bore and parallel to each other. The cutters should be keyed to the arbor, and the outboard bearing supports must be placed as close to the cutters as possible.

A variation of straddle milling is used when two inward facing surfaces must be cut parallel to each other. If half-side milling cutters are used, the blank sides face inward.

Gang Milling. In gang milling, three or more cutters are mounted on the arbor, and several horizontal, vertical, or angular surfaces are machined in one pass. A typical gang milling setup is shown in Fig. 11.90. When making a gang milling setup, several different types of cutters can be used, depending on the job to be done. Cutters used for producing vertical or angular surfaces must be of the side-cutting type; plain milling cutters of the proper width can be used for horizontal surfaces. In some cases face mills with the teeth facing inward can be used at one or both ends of the gang milling setup.

When only one wide plain helical milling cutter is used as part of a gang milling setup, the side thrust caused by that cutter should be directed *toward* the spindle of the machine. If possible,

FIGURE 11.90 Gang milling operation. (*Courtesy Cincinnati Milacron Inc.*)

interlocking cutters with opposite helix angles should be used to eliminate side thrust and reduce the possibility of chatter.

Because of the time and effort involved in setting up the milling machine for gang milling, the process is used mainly for production work. Since all or almost all of the workpiece is being machined at one time, power and rigidity are very desirable features in the machine being used. Every effort should be made to control vibration, including the use of support bars that are bolted to both the knee and the outboard bearing support.

11.6.4 Slitting and Related Operations

Milling *saws* of either the plain or side-cutting type are used for slitting operations. Slitting is usually done on horizontal milling machines, but it can also be done on vertical mills by using the proper adapters and accessories.

Cutoff Operations. Milling saws are often used for precision cutoff operations, as shown in Fig. 11.91. The workpiece can be cut exactly to length, thus eliminating another machining operation. Since milling saws are available in face widths of 0.032 in. or less, the savings when cutting off expensive materials are an important factor. Milling saws are relatively fragile, so great care must be taken during the setup and operating phases of the job. The following six basic procedures apply to cutoff and slotting operations:

1. Make sure that the arbor, collars, and cutter are free of nicks and chips.
2. Mount the cutter on the arbor, tightening the arbor nut only after the outboard bearing support is secure.

FIGURE 11.91 Cutoff operation using a milling saw. (*Courtesy El Camino College*)

3. Check to see that the cutter does not wobble or run out of round excessively. (*Note:* A cutter that runs out-of-round takes an excessive chip load for part of a revolution and may break.)
4. Calculate the cutting speed and chip load. Be conservative.
5. Make sure that the work to be cut is mounted rigidly in a vise or fixture.
6. Provide the appropriate cutting fluid, if needed. (*Note:* When deep slots in cast iron are being cut, carefully applied compressed air provides both a cooling and a cleaning effect.)

You are now ready to make a trial cut.

Cutting Slots and Fins. Metal-slitting saws of various diameters and widths are also used to cut slots. A number of identical saws can also be mounted on the same arbor for cutting fins. The parts shown in Fig. 11.92 were machined in this manner. When the thickness of the fins must be held to close tolerances, spacers are usually machined and surface ground to provide the necessary accuracy. The diameter of all cutters must be identical. A matched set of matched plain milling saws can be made by sharpening them while they are being held on a single arbor.

FIGURE 11.92 Finned heads machined by gang milling with slitting saws.

FIGURE 11.93 Face milling operation.

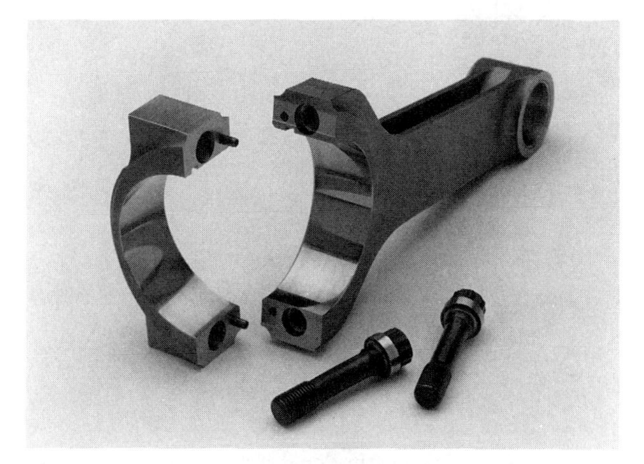

FIGURE 11.94 Racing engine connecting rod requiring extensive milling operations. (*Courtesy Cosworth Engineering, Inc.*)

FIGURE 11.95 Top contours and side recesses on racing pistons are formed by milling. (*Courtesy Cosworth Engineering, Inc.*)

11.6.5 Face Milling

Face milling can be done on vertical and horizontal milling machines. It produces a flat surface that is perpendicular to the spindle on which the cutter is mounted. The cutter ranges in size and complexity from a simple single-tool flycutter to an inserted-tooth cutter with many cutting edges (see Fig. 11.93). Large face mills are usually mounted rigidly to the nose of the spindle. They are very effective for removing large amounts of metal, and the workpiece must be securely held on the milling table.

11.6.6 Contour Milling

Objects that have unusual or irregular shapes, such as the connecting rod in Fig. 11.94 and the pistons in Fig. 11.95, usually require a combination of milling operations that are often a challenge to the machinist. When such items are made in large quantities special fixtures are usually made to position each part for the operation being performed. In some cases it is also necessary to design and make special cutters to produce the desired contours and recesses. Depending on the nature of the part being manufactured it may be necessary to use both vertical and horizontal milling machines as well as a variety of standard workholding devices such as angle plates and rotary tables.

Cutter Selection and Machine Setup. The cutter is held in a collet on light vertical mills (see Fig. 11.96) and mounted directly to the spindle nose on heavier machines. No arbors are required, except for shell end mills, which need a style C arbor. When face mills are used on horizontal milling machines, the overarm support is retracted.

When flycutters or face mills of any sort are used on vertical milling machines with tilting heads, the position of the head must be checked and adjusted if necessary. Unless the axis of the spindle is perpendicular to the table in both planes, the

FIGURE 11.96 Flycutting operation.

cutter cannot produce a flat surface that is parallel to the table.

When the machine has enough power and is rigid enough, face mills can withstand heavier chip loads than any other type of milling cutter. In many cases the chip load is double the amount allowed for other cutters. Nevertheless, the machinist should make the initial cuts with a conservative rate of feed, especially if an irregular surface on a forging or casting is being machined.

When a flycutter with two cutting tool inserts is being used, the roughing and finishing cuts can be taken simultaneously if the cutter is properly set up. On a cutter of this type (see Figs. 11.53 and 11.54), one tool is slightly farther from center than the other. The cutter with the wider sweep is the *roughing* tool and must be set to about 0.005 to 0.010 in. *less* depth than the finishing tool. The feed rate must be calculated as if a *single-point* finishing tool were being used.

11.6.7 End Milling

End milling is probably the most versatile milling operation. Many types of end mills can be used on both vertical and horizontal milling machines. End mills are available in sizes ranging from 1/32 to 6 in. (for shell end mills) and in almost any shape needed.

Cutter Selection and Machine Setup. All end mills with straight shanks are held in collets or end mill holders of some type. The collets are usually tightened by a draw bar or other locking device. Cutters that are held in end mill holders are usually secured by a screw that is tightened onto a flat spot on the shank of the cutter. Some larger end mills are also available with taper shanks. The taper is usually of the Morse, or Brown and Sharpe series, and the small end has a tang or is threaded internally for a draw screw.

End mills can be used for slotting, contouring, making cavities, and many other operations. A typical contouring operation on a numerically controlled vertical mill is shown in Fig. 11.97. The straight-shank end mill is held in an end mill holder, which in turn is held in the spindle.

In Fig. 11.98 a blind keyseat is being cut in a shaft with a two-lipped end mill. The cutter is held in a collet. Since the cutter must be fed directly into solid metal, a two-lipped cutter was used. A three- or four-lipped cutter could have been used as long

FIGURE 11.97 Contouring operation on numerically controlled vertical mill.

FIGURE 11.98 Cutting a blind keyseat with a two-lipped end mill. (*Courtesy El Camino College*)

as one of the end cutting lips extended to, or slightly past, the center of the cutter.

The procedure for cutting a keyseat with an end mill, which is similar to many other end-milling operations, is as follows:

1. Select a two-lipped end mill of the proper size. Inspect it to see that both flutes are sharp and that the outside diameter has not been ground undersize.
2. Install the end mill in the collet and check to see that it runs true. (*Note:* A milling cutter that does not run true cuts a slot wider than the diameter of the cutter.)
3. Center the workpiece under the cutter, and feed the work upward until a spot equal to the diameter of the cutter is machined. Apply cutting fluid if needed.
4. Zero the knee elevating screw and feed the work the desired amount up into the cutter. Make sure that the spindle, table, and saddle locks are tight.
5. Lock the knee, loosen the table lock, and feed the table in the desired direction.

Remember that end mills break easily. Use a moderate, even feed rate if you are feeding the table by hand. Very small end mills are particularly fragile, especially when chips tend to become caught in the flutes. Using a mist coolant system or carefully applied compressed air reduces clogging of the flutes and cutter breakage.

11.7 ADVANCED MILLING OPERATIONS

When you have learned the basic milling operations and the necessary related technical information, you are ready to master some of the more common advanced milling operations. This segment of the chapter is by no means a full treatment of advanced milling practice; you should become familiar with the handbooks and other literature dealing with specialized milling operations.

11.7.1 Indexing

A number of advanced milling operations involve the use of the dividing head and rotary table, and indexing procedures are required. Basically, indexing is the process of accurately spacing holes, gear teeth, or other machined areas on the perimeter or face of a workpiece. In some cases either the rotary table or indexing head can be geared to the lead screw of the table on a universal milling machine to produce helical gears and similar items.

As shown in Fig. 11.99, the crank and indexing pin are connected to the spindle by an accurately made compound gear train that consists of two spur gears, and a worm and wheel set. The gear ratio between the crank and the spindle is usually 40 to 1.

Direct Indexing. Direct indexing is also known as *quick indexing;* even though it is a simple procedure, it is also limited in scope. The indexing is done with a plate that has three-hole circles containing 24, 30, and 36 holes. Any number that divides evenly into 24, 30 or 36 is one of the numbers of divisions that can be indexed directly, as shown in Table 11.3.

The direct indexing plate is attached to the nose of the spindle of the dividing head, and the worm is disengaged from the worm wheel. The spindle can be turned until the proper hole aligns with the pin on the plunger. The pin then enters the hole, and the spindle is locked until the cut is completed. The workpiece can be held in a chuck

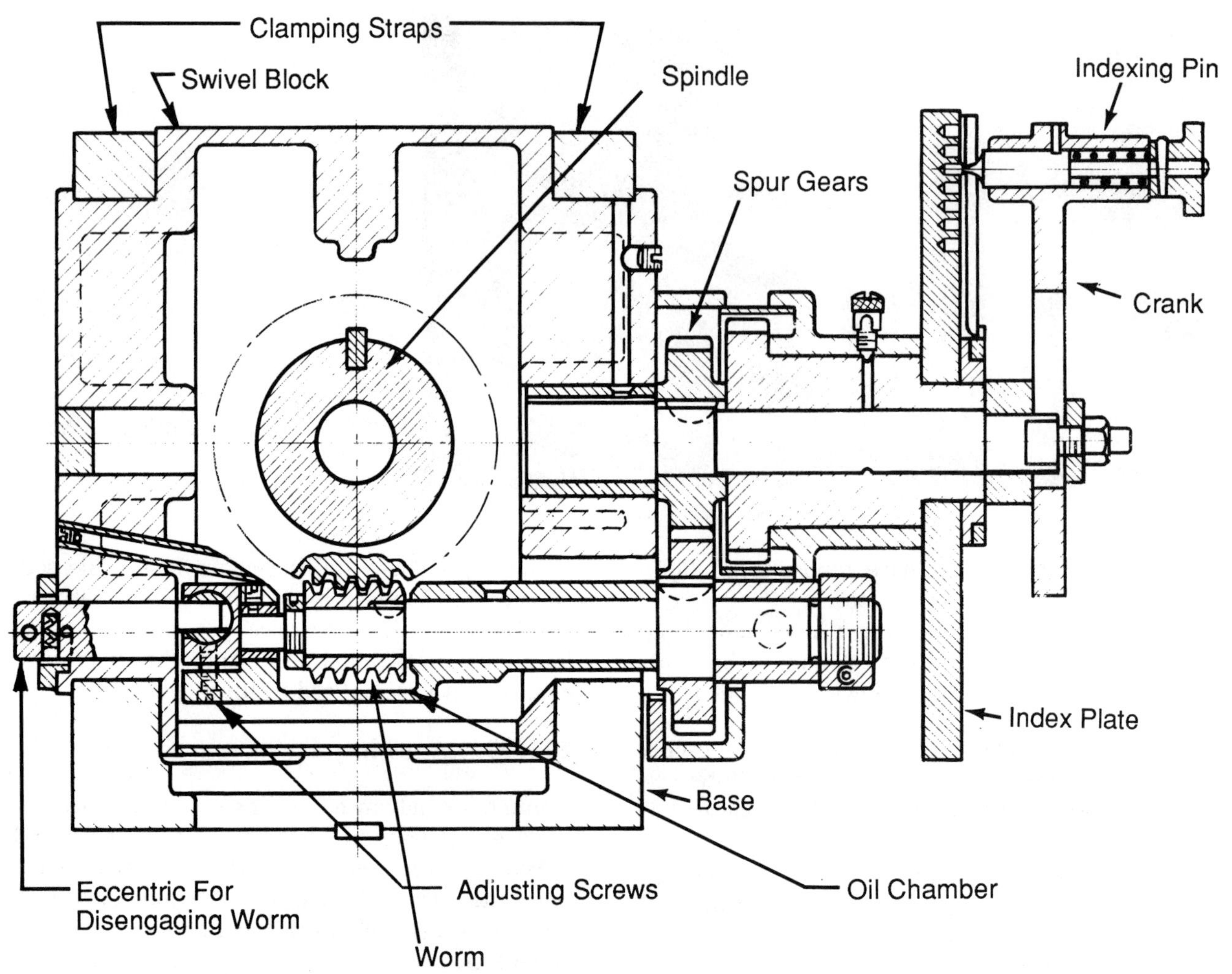

FIGURE 11.99 Cross-sectional view of indexing head. (*Courtesy Cincinnati Milacron Inc.*)

TABLE 11.3
Direct indexing

Possible Divisions	*Usable Plates* 24	30	36
2	×	×	×
3	×	×	×
4	×		×
5		×	
6	×	×	×
8	×		
9			×
10		×	
12	×		×
15		×	
18			×
24	×		
30		×	
36			×

or on centers, depending on the chuck's shape and size.

A typical direct indexing job might be to cut a square on the end of a shaft. Since the number 4 is divisible evenly into either 24 or 36, either hole circle can be used (refer to Table 11.3). If the 24-hole circle is used, the procedure is as follows:

1. Engage the plunger pin in one hole, lock the spindle, and machine one side of the square.
2. Unlock the spindle and move the index plate *six* holes (one-fourth of 24). Do *not* count the hole in which the pin was located.
3. Lock the spindle and machine the second side.
4. Repeat the process until all the sides are machined.

Simple Indexing. Simple indexing, also known as *plain indexing,* is a more versatile proc-

FIGURE 11.100 Indexing head. (*Courtesy Cincinnati Milacron Inc.*)

ess than direct indexing because a much wider range of divisions can be indexed. The worm is engaged with the worm wheel, which is attached to the spindle of the dividing head. The spindle is turned by turning the index crank, which is at a 90° angle to the spindle centerline. Forty turns of the index crank turns the spindle one revolution.

The index plate, as shown in Fig. 11.100, is attached to the housing in which the worm is turned by the index crank. Some indexing heads come with one plate that has different hole circles on each side of the plate. Other heads are equipped with two or more plates, each of which has several hole circles. The indexing head that uses only one plate has the following hole circles available:

Side 1: 24, 25, 28, 30, 34, 37, 38, 39, 41, 42, 43
Side 2: 46, 47, 49, 51, 53, 54, 57, 58, 59, 62, 66

Another manufacturer, who provides two plates with the dividing head, offers the following hole circles:

Plate 1: 15, 16, 19, 23, 31, 37, 41, 43, 47
Plate 2: 17, 18, 20, 21, 27, 29, 33, 39, 47

For an indexing head with 40 : 1 ratio between the crank and the spindle, the number of turns and/or fractional parts of a turn necessary to cut a particular number of divisions is computed by using the following rules:

$$\text{number of turns of index crank } (T) = \frac{40}{\sqrt{\text{number of divisions } (N)}}$$

Any number of divisions that divides *evenly* into 40 requires only complete turns of the index crank, when using the proper hole circle, with the sector arms locked in one position.

Any number of divisions that can be machined by simple indexing that does *not* divide evenly into 40 requires full and/or partial turns of the crank. The correct hole circle and proper placement of the sector arms are required.

If the number of divisions is *less* than 40, more than one turn of the index crank is required.

If the number of divisions is *more* than 40, less than one turn of the index crank is required.

The basic process of setting up the indexing head for simple indexing is illustrated in the following three examples:

Example A: Index for 8 divisions:

$$\text{number of turns} = \frac{40}{8} = 5$$

Any hole circle can be used. Place the sector arms so that one hole shows between them, and carefully bring the index plunger pin around to this hole on the fifth turn when indexing.

Example B: Index for 34 divisions:

$$\text{number of turns} = \frac{40}{34} = 1\frac{6}{34}$$

The whole number indicates that *one* full turn of the crank is required. The fractions indicate that the index plunger must be moved a further 6/34 of a turn. By reducing the fraction 6/34 to 3/17 and examining the available hole circles, it can be seen that 17-, 34-, and 51-hole circles are available, depending on the make of indexing head used.

Assume that a plate with a 51-hole circle is already mounted on the machine. Since 3/17 equals 9/51, we can set the sector arms so that 9 holes *plus the one occupied by the index plunger pin* are between the arms. Each time the index crank is turned one turn and 9 holes, the sector arms must be advanced in the same direction so that the rear arm contacts the plunger pin.

Example C: Index for 72 divisions:

$$\text{number of turns} = \frac{40}{72} = \frac{5}{9}$$

Since the number of divisions is more than 40, the index crank will be turned less than one turn. The fraction 40/72 is reduced to 5/9, and a hole circle into which 9 divides evenly must be found. Circles with 18, 27, or 54 holes can be used. If the 27-hole circle is selected, both the numerator and the denominator of the fraction 5/9 are multiplied by 3, which gives 15/27. The sector arms are set to include 15 holes *plus* the one in which the plunger pin is located.

Angular Indexing. In certain milling operations it is sometimes necessary to establish exact angular relationships between holes, surfaces, and other machined areas. The indexing head can be used to perform this operation by utilizing the index plates and sector arms. With some indexing heads, divisions of less than 1° can be accurately established.

Because one turn of the indexing crank rotates the spindle 1/40 of a turn, the angular movement of a point on the spindle is $360 \div 40$, or 9°. Therefore, holes or other machined areas spaced less than 9° apart require less than one turn of the crank, and those spaced more than 9° apart require more than one turn.

Angular indexing is a convenient way of locating machined areas that are not uniformly spaced around a circle because on drawings the location is given in angular terms from a reference point. On the part shown in Fig. 11.101, for example, hole 1 is considered the reference, or 0° point. Hole 2 is spaced 70° clockwise, and hole 3 is spaced 105° clockwise from hole 2, or 175° from the reference point.

The procedure for determining the indexing head setup for spacing these holes is as follows:

$$\text{number of turns} = \frac{\text{number of degrees to be indexed}}{9}$$

Therefore, for locating hole 2 relative to hole 1,

$$\text{number of turns} = \frac{70}{9} = 7\frac{7}{9}$$

Any hole circle into which 9 will divide evenly can be used. Assume that a 54-hole circle is available:

$$\frac{7}{9} \times 54 = \frac{42}{54} \quad \text{or} \quad \text{42 holes on 54-hole circle}$$

The sector arms are then set for 42 holes, *not* counting the hole occupied by the pin. After hole 1 is drilled, the crank is turned seven complete turns and 42 holes.

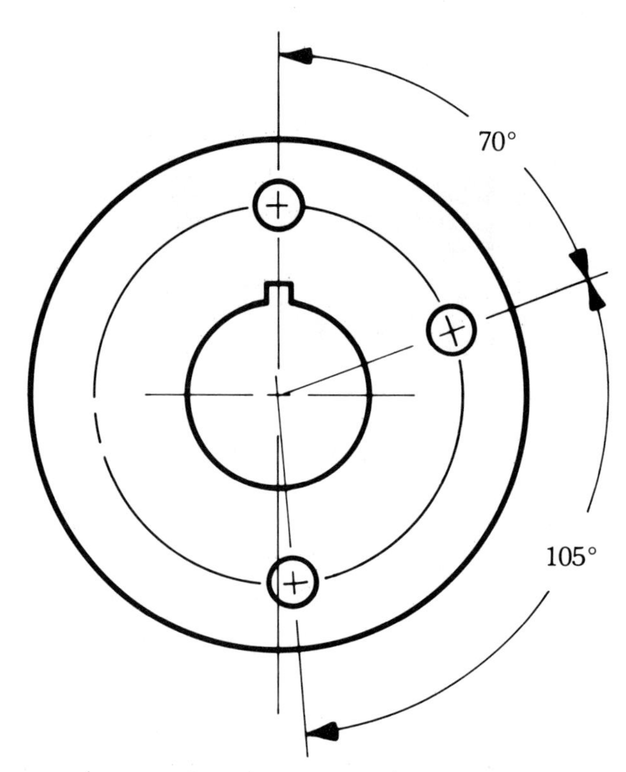

FIGURE 11.101 Angular indexing operation.

After hole 2 is drilled, the computation for hole 3, which is 105° from hole 2, is done as follows:

$$\text{number of turns} = \frac{105}{9} = 11\frac{6}{9}$$

Since $6/9 \times 54 = 36/54$, the sector arms are set for 36 holes as explained. The crank is turned 11 turns and 36 holes to advance to hole 3.

Differential Indexing. This procedure is used for indexing divisions that are beyond the range of simple or angular indexing. In recent years the availability of wide-range indexing heads (see Fig. 11.102) has reduced the need for differential indexing, which requires a gear train between the spindle and the hollow shaft on which the index plate is mounted.

In simple indexing, the index plate remains stationary while the index crank is turned because it is locked in position to the case of the index ring head. When differential indexing is being done, the indexing plate moves at a predetermined rate either clockwise or counterclockwise. The rate and direction of movement of the index plate are determined by the ratio and number of gears in the gear train. All manufacturers of indexing heads that are capable of differential indexing provide charts

FIGURE 11.102 Wide-range indexing head. (*Courtesy Cincinnati Milacron Inc.*)

FIGURE 11.103 Indexing head with chuck. (*Courtesy SMW Systems*)

showing the gears to be used for a particular number of divisions.

Precision indexers of the electro-mechanical type as shown in Fig. 11.103 are very versatile and may be used for angular as well as regular indexing.

Rotary Table Indexing. Rotary tables are used for many indexing operations. The lower part of the rotating table is graduated in degrees, and the dial near the handcrank is divided into smaller increments. Some rotary tables are fitted with indexing plates, sector arms, and indexing cranks similar to those used on indexing heads (see Fig. 11.33). This greatly increases the accuracy to which the rotary table can be set and allows it to be used for both simple and angular indexing.

The ratio of the worm and worm wheel must be known before any computations can be made. A rotary table with a 90 : 1 worm reduction, for example, rotates 4° for each turn of the crank (since $90 \times 4° = 360°$). To compute the indexing for a particular number of divisions, the following formula is used, assuming the worm is of the single lead type:

$$\text{number of turns} = \frac{\text{number of teeth on worm wheel}}{\text{number of divisions}}$$

Example: Set up the rotary table that has an index plate for a drilling operation requiring 52 equally spaced holes. The worm wheel ratio is 80 : 1.

$$\text{number of turns} = \frac{80}{52} = 1\frac{28}{52} = 1\frac{7}{13}$$

Since 13 will divide evenly three times into 39, the 39-hole circle on the index plate is selected, and 21 holes (3×7) are left between the sector arms, *not* counting the one occupied by the index pin.

Once the rotary table is set up, the procedure for doing the machining operation is as follows:

1. Center the rotary table under the vertical axis of the spindle.
2. Move the table (or cross slide) of the milling machine an amount equal to the *radius* of the hole circle to be drilled. Lock *all* slides.
3. Drill the first hole, making sure that the *left* sector arm is against the index pin.
4. Unlock the rotary table lock and turn the crank clockwise one full turn and 21 holes farther. Lock the rotary table lock and move the sector arms clockwise until the left sector arm touches the pin.
5. Drill the next hole and repeat the processes until all holes are drilled.

Rotary tables are very convenient accessories, especially when large worm wheels or gears are being machined. The rotary table can also be mounted so that its axis is horizontal, as shown in Fig. 11.104. Some rotary tables are geared to the table feed screw through a gearbox, as shown in Fig. 11.105. With this arrangement and the proper relationship between table feed rate and rotary motion, such items as scrolls and constant-rise

FIGURE 11.104 Rotary table mounted with table face vertical. (*Courtesy El Camino College*)

FIGURE 11.105 Rotary table geared to table feed screw. (*Courtesy Cincinnati Milacron Inc.*)

plate cams can be machined. The rotary table can also be fitted with fixtures for holding workpieces and rotated past the cutters without any table feed or cross feed.

11.7.2 Helical Milling

Helical milling operations are necessary when making helical gears, cams, reamers, taps, cutters, and other similar objects. A helix is a curve that moves around a cylindrical object and advances at a uniform rate, as shown in Fig. 11.106*(a)*. The *lead* is the distance that a point travels axially (along the axis) of the cylinder in one revolution. The helix angle is determined by the lead *and* the diameter of the cylinder. The diagram in Fig. 11.106*(b)* shows the development of a helix. The vertical line (side a or πD) is the circumference of the cylindrical object on which the helix is cut. The horizontal line (side b) represents the lead. The hypotenuse of the triangle (side c) is the developed length of the helix. For threads and gears the helix angle is computed by using the pitch diameter for establishing the length of side a.

Milling such a curve requires that the table feed mechanism be geared to the indexing head at the proper ratio. On most milling machines with provisions for gearing the dividing head to the table, helices with leads ranging from 0.670 to 60 in. can be machined. The tables in various machinist's handbooks generally provide the necessary information for choosing gears and universal table angles.

Computations. Before the process for determining the gears to be used can be started, the lead of the milling machine to be used must be found. The lead of the milling machine is the distance the table travels while the indexing head spindle turns one revolution. This assumes gears of the same size on the table screw and the worm gear stud on the indexing head. The following formula is used:

$$\text{lead of milling machine} = \text{lead of table feed screw} \times \text{number of index crank turns to turn indexing head spindle once}$$

For a milling machine that has a four-thread-per-inch table screw (0.250 in. lead), the lead is computed as follows, with a 40 : 1 indexing head:

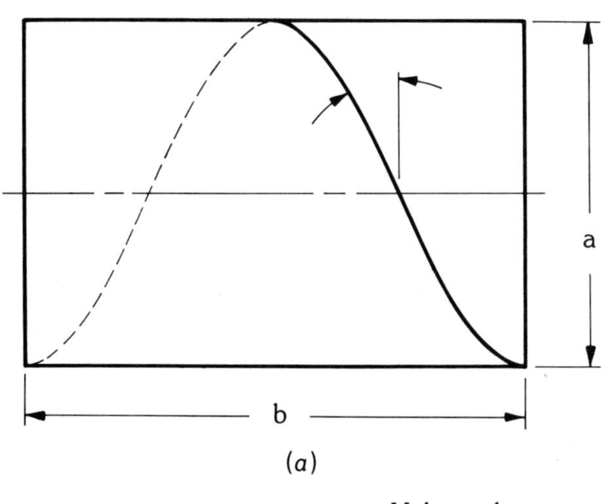

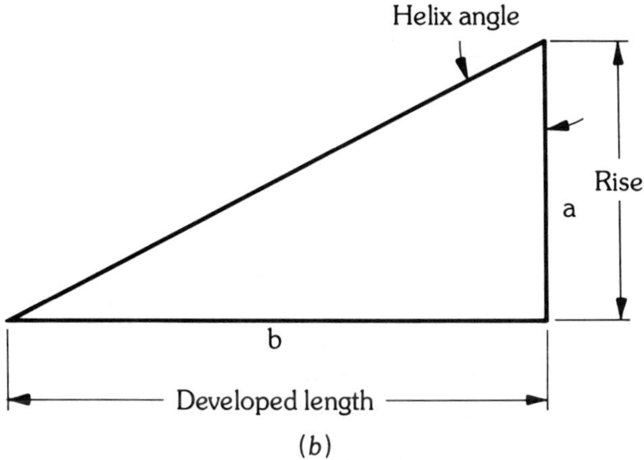

FIGURE 11.106 (*a*) Helix (*b*) developed helix.

$$\text{lead} = 0.250 \text{ in.} \times 40$$
$$= 10.0 \text{ in.}$$

Once the lead of the machine is known, the procedure shown in the following example can be used to find the gears required for a particular lead. Assume that gears with the following number of teeth are available: 24, 28, 32, 40, 44, 48, 56, 64, 72, 86, and 100.

Example: If the milling machine lead is 10 in., find the gears that must be used to cut a helix with a 48-in. lead. First, set up the fraction 48/10, which establishes the ratio between the driven and driving gears. Since a compound train of four gears must be used, the fraction is split as shown by factoring. The numerator and denominator of each fraction are then multiplied as shown to produce numbers that are available in the gear sets.

$$\frac{48}{10} = \frac{6 \times 8}{2 \times 5} = \frac{(6 \times 12) \times (8 \times 8)}{(2 \times 12) \times (5 \times 8)}$$
$$= \frac{72 \times 64 \text{ driven gears}}{24 \times 40 \text{ driving gears}}$$

FIGURE 11.107 Gearing for helical milling operation. (*Courtesy Cincinnati Milacron Inc.*)

It can be seen that the 72- and 24-tooth gears form one part of the *compound gear train.* The 72-tooth gear is mounted on the worm gear stud of the indexing head, and the 24-tooth gear is mounted on the intermediate stud (see Fig. 11.107). The 64-tooth gear is also mounted on the intermediate stud and keyed to the 24-tooth gear. The 40-tooth gear is mounted on the end of the lead screw. (**Note:** If the first driving gear and the last driven gear turn in the same direction, a right-hand helix will be cut. If a left-hand helix is desired, an idler gear must be inserted into the gear train.)

Another major step in preparing for a helical milling operation is to find the angle to which the universal table must be turned. If the lead of the helix and the diameter of the part are known, the following formula can be used to find the tangent of the angle of the helix [angle *A*, Fig. 11.106(*b*)] and then the angle itself. The first stop is to find the circumference by multiplying the diameter *(D)* by π (3.1416):

$$\tan A = \frac{\text{circumference}}{\text{lead}}$$

Example: Assume that the diameter of the part is 3.0 in. and the lead of the helix is 24.0 in.:

$$\tan A = \frac{3 \times 3.1416}{24} = \frac{9.4248}{24}$$
$$= 0.3927$$
$$\text{angle } A = 21°26'$$

The universal table must be set at an angle of 21°26′ in the proper direction. To cut a right-hand helix, the table must be moved counterclockwise when viewed from above. When certain helical milling operations are done on a vertical milling machine and an end mill is used, a milling machine with a plain table can be used. A universal milling attachment can also be used for milling steep helices, such as worms for worm and wheel sets.

11.7.3 Gear Cutting

Gears of various types are used in almost all kinds of machinery for the positive transmission of power, and they are mass produced in large numbers. Since the processes for mass producing gears are beyond the scope of this chapter, only the procedures for making racks, spur gears, and helical gears by use of the milling machine will be discussed. A significant number of gears are made singly or in small lots for repair purposes and for custom or prototype work using conventional milling machines and accessories.

Basic Gear Types. Gears are classified according to tooth shape, shaft arrangement, pressure angle, and other characteristics.

Spur gears (see Fig. 11.108) have teeth that are parallel to the axis of the gear. These gears can be machined by milling, hobbing, or with a gear shaper that uses a reciprocating cutter. They are the simplest form of gear and are relatively inexpensive to make. Because the teeth are parallel to the axis, no end thrust is produced as power is transmitted, but they tend to be noisy because each tooth comes into contact with the mating tooth along its entire length at one time.

Helical gears (see Fig. 11.109) may have parallel shafts or shafts that run at an angle to each other, and the teeth are machined so that they follow a curved path around the outside of the blank. There is always more than one tooth in contact at one time, so the flow of power from one gear to another is smooth, and the gears are quiet when properly made. Since the teeth are at an angle, side thrust is produced as power is transmitted, so thrust bearings must be provided. The side thrust increases as the helix angle of the teeth is increased.

FIGURE 11.109 Helical gear.

An *internal gear* set is composed of a *pinion* (the small external gear) and the ring gear (see Fig. 11.110), which has internal teeth that are machined on a gear shaper or by broaching in some cases. There is no side thrust, unless the gears are of the helical type. Because several teeth are in partial or full contact at all times, large amounts of power can be transmitted quietly and efficiently. The *driving* and *driven* gear both turn in the same direction. The driving and driven shafts are parallel but are not on the same center line.

FIGURE 11.108 Spur gears.

FIGURE 11.110 Internal gear.

FIGURE 11.111 Rack and pinion gear.

The *rack and pinion* set shown in Fig. 11.111 is of the spur, or straight-toothed, type, but helical racks and pinions are also available. A rack is essentially a gear of infinitely large diameter. It can be the driving or driven member. Spur rack and pinion sets are not as smooth in operation as helical sets, but they do not produce undesirable side thrust.

Herringbone gears are used when the smooth power-transmission qualities of helical gears must be achieved without side thrust. Basically, a herringbone gear set is composed of a left- and a right-hand helical gear machined on one gear blank. Gears of this type are used where large amounts of power must be transmitted through a compact and strong gear case.

Spur bevel gears have teeth whose axes meet at the apex of a cone when projected to the center line of the shaft. Gear sets of this type that have the same number of teeth, or a 1 : 1 ratio, are called *miter gears. Angular* bevel gears are machined so that the axis of the shafts on which they are mounted meet at an angle other than 90°. Spur bevel gears can be cut on the milling machine but are usually manufactured on gear shapers.

Spiral bevel gears (see Fig. 11.112) transmit power more smoothly than spur bevel gears because several teeth are always in contact. The axis of the shafts on which the two gears are mounted intersect at a 90° angle since they are on the same plane. The teeth on this type of gear are cut on special machines called gear generators, which use large inserted tooth cutters.

Hypoid gears resemble spiral bevel gears except that the shafts are at a 90° angle but their center lines are offset and do not intersect. They are used extensively in automotive rear-axle assemblies. The curved teeth are cut on gear-generating machines, and the gears can transmit large amounts of power smoothly. Since there is sliding motion between the teeth, special high-pressure lubricants are necessary.

Worm gear sets consists of a *worm* (or worm gear) and a *worm wheel* (see Fig. 11.113). The worm, which looks like a slightly deepened Acme thread, may be of the single entry or multiple entry. Single-entry worms provide the largest speed

FIGURE 11.112 Spiral bevel gear.

FIGURE 11.113 Worm gear.

reduction and are *irreversible* (the worm wheel cannot transmit power to the worm). This is also the case with multiple-entry worm and worm gear sets when the reduction ratio is large. Because of the sliding action by which worm gears operate, the worm and worm wheel are usually made of dissimilar materials. For example, a bronze worm wheel may be driven by a steel worm gear.

Gear Terms. The basic terms used in describing gears and gear teeth are now briefly defined (see Fig. 11.114). In almost all cases the terms apply to both metric and English measure, or inch gears. Most terms commonly used in describing gears are:

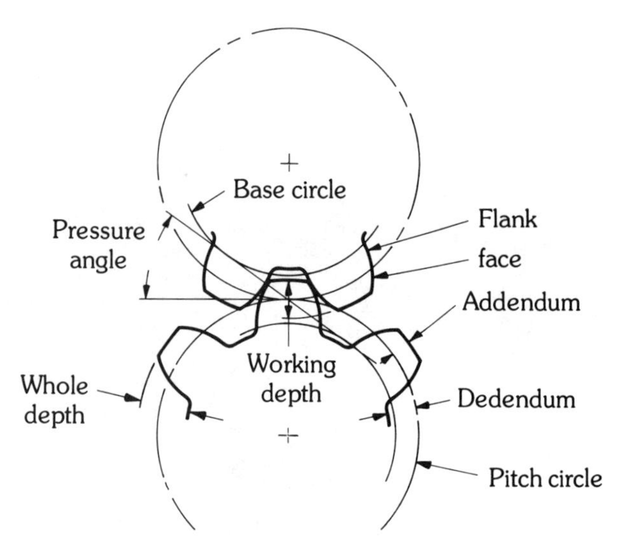

FIGURE 11.114 Gear terminology.

1. *Addendum:* height of the tooth above the pitch circle.
2. *Base circle:* circle from which an involute tooth curve is generated.
3. *Circular pitch:* length of the arc of the pitch circle from a point on one tooth to a corresponding point on the adjacent tooth.
4. *Circular thickness:* thickness of the tooth at the pitch circle.
5. *Clearance:* radial distance between the top of a tooth and the bottom of the mating tooth space.
6. *Dedendum:* distance between the pitch circle and the bottom of the tooth space.
7. *Diametral pitch:* ratio of the number of teeth to the number of inches of pitch diameter. Typical gear tooth sizes are shown in Fig. 11.115.
8. *Involute:* curved line described by the end of a taut string when unwound from the *base circle* of a gear (see Fig. 11.116).

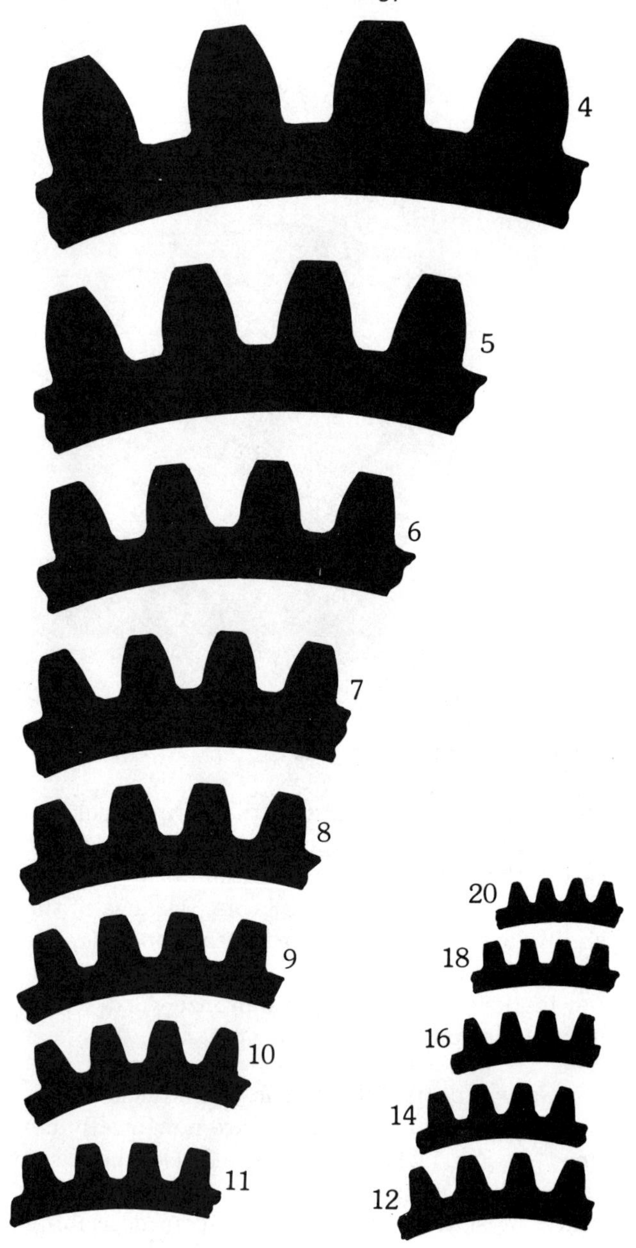

FIGURE 11.115 Gear tooth sizes.

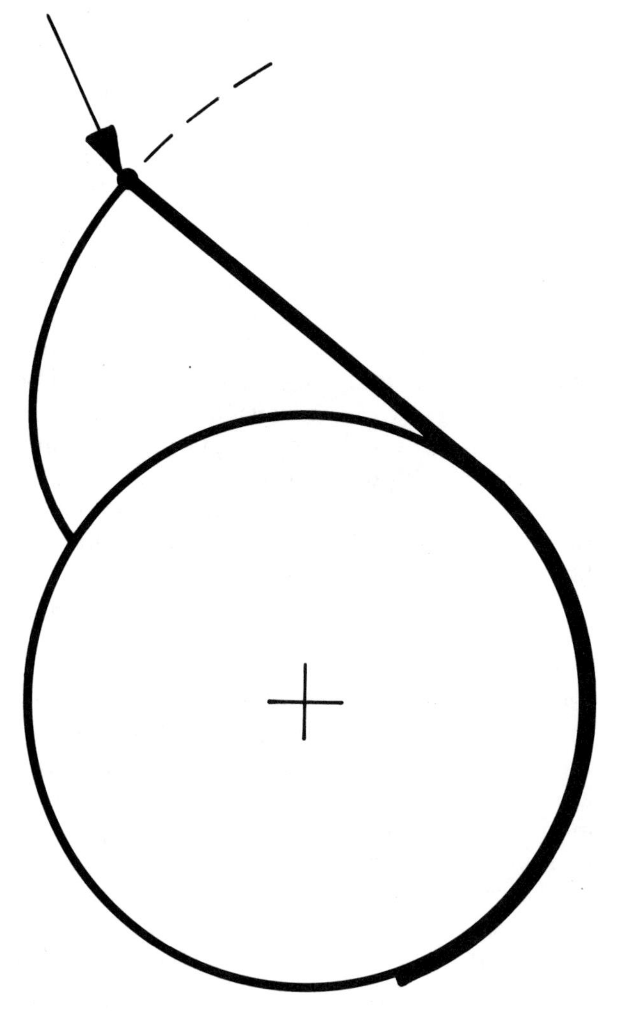

FIGURE 11.116 Development of an involute curve.

9. *Module:* pitch diameter of a gear divided by the number of teeth. The module is equal to the addendum.
10. *Outside diameter:* outside diameter of the blank on which the gear teeth are machined.
11. *Pitch diameter:* diameter of the pitch circle.
12. *Pressure angle:* angle between the line of action at the contacting tooth faces and a line tangent to both pitch circles.
13. *Root circle:* circle that coincides with the bottom of the tooth spaces.
14. *Whole depth:* entire depth of the tooth.
15. *Working depth:* distance the tooth extends into the space between the teeth of the mating gear.

Basic Spur Gear Computations. The following procedures that are discussed are those most often used in preparing to machine spur gears. For more complex gear cutting operations and for information on gear design procedures, consult a machinist's handbook. The basic procedures and formulas used in spur gear calculations are in Table 11.4. Four examples are worked out to illustrate the processes involved in preparing gear blanks and selecting the proper cutters.

Example A: Find the *center distance* for two 8 DP gears. One gear (N_1) has 30 teeth, and the other gear (N_2) has 42 teeth.

Rule: Find the total number of teeth in both gears and divide by two times the diametral pitch.

TABLE 11.4

Basic rules and formulas for diametral pitch spur gear computations

To Compute:	*Known Factors*	*Formula*	*Operation*
Addendum *(A)*	Diametral pitch (DP)	$A = \frac{1}{DP}$	Divide 1 by the diametral pitch
Center distance (CD)	Diametral pitch	$CD = \frac{N_1 + N_2}{DP \times 2}$	Divide the sum of the teeth in both gears by two times the diametral pitch
Chordal thickness (CT)	Pitch diameter	$CT = \frac{1.5708}{DP}$	Divide the constant 1.5708 by the diametral pitch
Clearance *(C)*	Diametral pitch	$C = \frac{0.157}{DP}$	Divide the constant 0.157 by the diametral pitch
Dedendum *(D)*	Diametral pitch	$D = \frac{1.157}{DP}$	Divide the constant 1.157 by the diametral pitch
Diametral pitch (DP)	Number of teeth and outside diameter (OD)	$DP = \frac{N + 2}{OD}$	Add 2 to the number of teeth and divide by the outside diameter
Diametral pitch	Number of teeth *(N)* and pitch diameter	$DP = \frac{N}{PD}$	Divide the number of teeth by the pitch diameter
Number of teeth	Outside diameter (OD)	$N = (OD \times DP) - 2$	Multiply the outside diameter by the diametral pitch and subtract 2
Outside diameter	Diametral pitch and number of teeth	$OD = \frac{N + 2}{DP}$	Add 2 to the number of teeth and divide by the diametral pitch
Outside diameter	Pitch diameter (PD)	$OD = PD + \frac{2}{DP}$	Divide 2 by the diametral pitch and add to the pitch diameter
Pitch diameter	Number of teeth and diametral pitch	$PD = \frac{N}{DP}$	Divide the number of teeth by the diametral pitch
Pitch diameter	Outside diameter and number of teeth	$PD = \frac{OD \times N}{N + 2}$	Multiply the outside diameter by the number of teeth and divide by the number of teeth plus 2
Whole depth (WD)	Diametral pitch	$WD = \frac{2.157}{DP}$	Divide the constant 2.157 by the diametral pitch

Formula

$$CD = \frac{N_1 + N_2}{2DP}$$
$$= \frac{30 + 42}{2 \times 8} = \frac{72}{16}$$
$$= 4.5 \text{ in.}$$

Example B: Find the *outside diameter* of the blank for a 42-tooth gear with 8 diametral pitch.

Rule: Add 2 to the number of teeth and divide by the diametral pitch.

Formula

$$OD = \frac{\text{number of teeth} + 2}{DP}$$
$$= \frac{42 + 2}{8}$$
$$= \frac{44}{8} = 5.5 \text{ in.}$$

Example C: Find the *pitch diameter* of a 30-tooth gear with 8 diametral pitch.

Rule: Divide the number of teeth by the diametral pitch.

Formula

$$PD = \frac{\text{number of teeth}}{DP}$$
$$= \frac{30}{8}$$
$$= 3.75 \text{ in.}$$

Example D: Find the *whole depth* for any gear with 8 diametral pitch.

Rule: Divide the constant 2.157 by the diametral pitch.

Formula

$$WD = \frac{2.157}{DP}$$
$$= \frac{2.157}{8}$$
$$= 0.2696 \text{ in.}$$

If the gear teeth are to be checked for thickness at the pitch line with a gear tooth caliper (see Fig. 11.117), the tooth thickness and the *corrected* or *chordal* addendum for a particular diametral pitch must be known. The chordal addendum is the distance from the top of the gear tooth to the intersection of the pitch circle and the tooth face. Do not use the true addendum when setting the tongue of the vernier gear calipers. The corrected addendum may be computed by finding the chordal addendum for a given number of teeth in a machinist's handbook and dividing by the diametral pitch. For example, the addendum for a 30-tooth gear with a diametral pitch of 8 is found as follows:

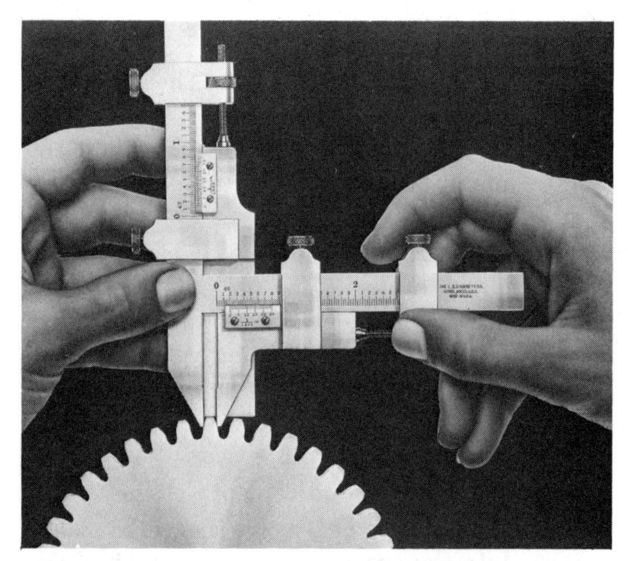

FIGURE 11.117 Gear tooth caliper. (*Courtesy L. S. Starrett Co.*)

$$\text{addendum} = \frac{1}{DP}$$
$$= \frac{1}{8} = 0.125 \text{ in.}$$

The *corrected* or *chordal* addendum for the same gear is found as follows:

$$\text{corrected addendum} = \frac{1.02055}{8} \quad \text{(from handbook)}$$
$$= 0.1279$$

Cutting a Spur Gear. After the necessary calculations have been completed, the proper cutter must be selected. Form cutters for milling involute gears usually come in sets of 8 for each diametral pitch. Sets of 15 cutters are also available, with the additional 7 cutters listed as *half-sizes.* For example, the 1-1/2 cutter (80 to 134 teeth) falls within the operating range of the 2 cutter (55 to 134 teeth). Involute gear cutters are correct in shape for only the *smallest* number of teeth in the range. Therefore, as the number of teeth to be cut increases, within the range the tooth shape becomes less accurate.

The profiles for a typical set of gear cutters are shown in Fig. 11.118. Notice that the 1 cutter in the set (135 teeth to rack) has relatively straight sides compared to the 8 cutter (12 to 13 teeth). On

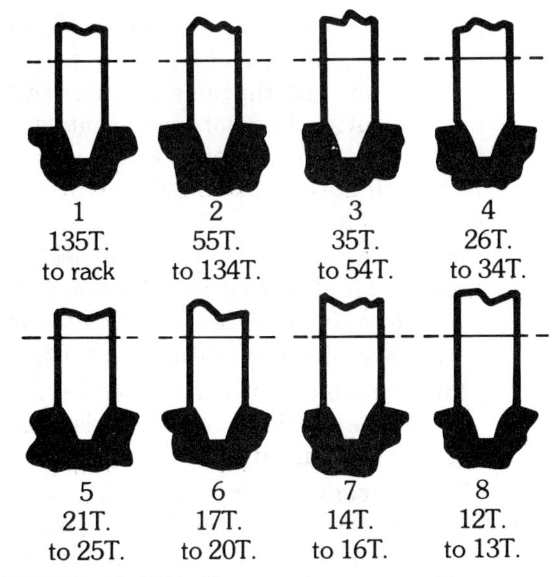

FIGURE 11.118 Gear tooth cutter profiles.

smaller gears of a given pitch diameter, the *face* of the tooth (portion above the pitch circle) is curved more to prevent interference between the gear teeth as they engage and disengage.

The following example explains the process for machining a 38-tooth spur gear of 14 diametral pitch:

1. Since 38 teeth fall within the range of the 3 cutter from the 14 DP set (see Table 11.5), it will be used.
2. Mount the cutter on the milling arbor so that the thrust during the cutting operation is directed *toward* the dividing head.

TABLE 11.5
Involute gear cutters

Standard Cutter No.	*Special Cutter No.*	*Range of Teeth Cut*
1		135 to rack
	1 ½	80–134
2		55–134
	2 ½	42–54
3		35–54
	3 ½	30–34
4		26–34
	4 ½	23–25
5		21–25
	5 ½	19–20
6		17–20
	6 ½	15–16
7		14–16
	7 ½	13
8		12–13

3. Mount the dividing head and footstock on the table of the milling machine. Check the footstock and dividing head centers for alignment.
4. Press the gear blank on a mandrel and mount it between the centers of the dividing head and the footstock (see Fig. 11.119).
5. Center the cutter over the gear blank. A precise method of doing this is shown in Fig. 11.120 and described below:

FIGURE 11.119 Machining spur gear teeth. *(Courtesy Cincinnati Milacron Inc.)*

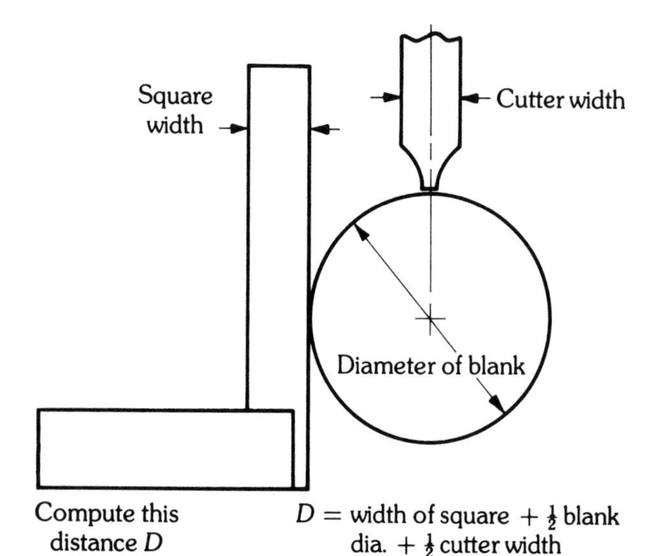

FIGURE 11.120 Centering the gear cutter.

(*a*) Place a toolmaker's square so that the blade is against the mandrel and extends above the center of the cutter.
(*b*) Add the width of the square blade, one-half the diameter of the mandrel, and one-half the thickness of the cutter.
(*c*) Using the proper micrometer, adjust the saddle so that the above measurement is achieved (see Fig. 11.120). The centerline of the cutter will then be above the centerline of the blank and mandrel.

6. Set the dividing head at the proper starting position and lock it.
7. Set the cutter to the whole depth of the gear tooth by:
   (*a*) Moving the table upward until the *slowly* rotating cutter touches a piece of cellophane tape stuck to the top of the gear blank.
   (*b*) Zero the dial on the vertical knee screw and feed upward the proper amount with the table placed so that the cutter does not make contact.
   (*Note:* A roughing cut at 75% of the working depth may be made before the final cut is taken.)
8. Partially machine the first tooth, feeding the gear blank so that its edge is 1/16 to 1/8 in. past the centerline of the cutter arbor, indexing and repeating the process.
9. Move the table away from the cutter and inspect the partial tooth for finish and possible irregularities.
10. If a gear tooth caliper is available, set the tongue to the corrected addendum and the horizontal slide to the tooth thickness and measure the tooth.
11. If the tooth is thicker or thinner than it should be, recheck the blank diameter and whole-depth calculations. Make any necessary adjustments.
12. Use the appropriate cutting fluid to extend cutter life and improve the finish of the gear teeth.
13. Remove the completed gear from the mandrel and deburr the teeth.

Cutting a Helical Gear. To machine a gear with helical teeth, the procedure differs in some respects from the procedure used in machining a spur gear. Since the teeth are at an angle to the axis of the gear blank, a milling machine with a universal table must be used and the table must be set at the angle specified as the helix angle of the gear. A gear train must also be set in position between the table feed screw and the dividing head. (Refer to 11.7.2 and Fig. 11.107). The gears in the gear train must be selected by referring to tables

FIGURE 11.121. Machining helical gear teeth. (*Courtesy Cincinnati Milacron Inc.*)

provided with the milling machine or by use of appropriate machinist's handbooks.

The proper cutter must also be selected. It will differ from the cutter specified for a comparable spur gear because the gear blank rotates while it is being fed past the cutter. A typical setup for machining a helical gear is shown in Fig. 11.121.

The *milling machine*, in its various forms, sizes, and applications, is found in almost every type of machine shop. In recent years, larger and more powerful milling machines have been developed to fully utilize the characteristics of new cutting tool materials. This, of course, has led to increased productivity per worker hour and machine hour.

The application of numerical control to milling operations has also increased the versatility and productivity of the milling machine and helped create areas of specialization. However, the machinist must still have an understanding of basic milling theory and practice. This basic knowledge, along with more specialized information about cutting tool materials, machining characteristics of metals, and other factors, allows the machinist to perform an increasing number of milling operations with experience.

REVIEW QUESTIONS

11.2 TYPES AND CONSTRUCTION FEATURES

1. Of what materials are the main castings of a milling machine usually made?
2. What are the functions of the *saddle* on a universal milling machine?
3. Describe the types of bearings used on a vertical milling machine spindle, and explain their functions.
4. Explain the purpose of a *backlash* eliminator, and indicate where it is mounted.
5. Explain the main difference between a *plain* and *universal* knee and column milling machine.
6. In what two directions can a fully universal head move on a ram-type vertical milling machine?
7. In what directions can the head on an overarm-type vertical milling machine be moved?
8. For what types of work are planer-type milling machines generally used?
9. What is the primary function of a jig boring machine?

11.3 MILLING MACHINE ATTACHMENTS AND ACCESSORIES

1. What two types of vertical milling heads can be used on horizontal milling machines?
2. Briefly explain how a reciprocating-type slotting attachment works.
3. Name the three basic types of milling machine vises.
4. Briefly describe the process of setting a swivel vise at a given angle to the face of the column.
5. What is the ratio of the worm and wheel in most indexing heads?
6. Name three work-holding devices that can be used on indexing heads.
7. What is the difference between direct indexing and simple indexing?
8. Name three types of arbors that can be used on horizontal milling machines.
9. What type of arbor is used for a heavy straddle milling operation?
10. Briefly explain how an offset boring head works.

11.4 MILLING CUTTERS

1. From what two materials can solid end mills be made?
2. On a wide plain milling cutter, what is the purpose of having helical teeth?
3. Why are some plain milling saws ground with side clearance?
4. Describe the difference between plain and side-milling cutters.
5. What are two advantages of stagger-tooth side-milling cutters?
6. By what two methods can the inserts be attached on inserted tooth cutters?
7. For what purpose are shell end mills with serrated peripheral teeth used?
8. What type of end mill is used to cut a keyseat with rounded ends in a shaft?
9. Why are form cutters ground with a zero face relief angle?
10. Name three types of form cutters.

11.5 ALIGNMENT AND SETUP PROCEDURES

1. Describe the process of aligning a semiuniversal head on a vertical milling machine.
2. What defect results from attempting to flycut a flat surface on a vertical milling machine with a semiuniversal head that is slightly tilted if the table feed is used?
3. Explain how to align the table on a universal milling machine.
4. Explain how the fixed jaw of a swiveling vise is aligned at a 90° angle to the face of the column.
5. What procedure is used to set a tilting vise at a 30° angle to the table using a universal bevel protractor?
6. List five accessories that can be used to help clamp work to a milling machine table.
7. When work is mounted in a milling machine vise, why should the thrust forces produced by the cutter be directed against the solid jaw?
8. What precautions should be taken when castings that have rough surfaces are held in a milling vise?
9. When a helical plain milling cutter is installed in a horizontal milling machine, how can you tell whether the thrust forces produced will be toward the column?

11.6 BASIC MILLING OPERATIONS

1. Why can carbide cutters be run at a higher cutting speed than high-speed-steel cutters?
2. Compute the spindle speed for a 4-in.-diameter cutter that is to be operated at a cutting speed of 320 ft/min.
3. What two limiting factors must the machinist consider when selecting cutting speeds and feed rates for a particular job?
4. What two major factors must be considered when determining the feed rate for a particular job?
5. Give two reasons why light chip loads should be selected when using thin slitting saws.
6. Explain why climb milling is generally not used when machining castings or forgings.
7. What two types of side-milling cutters can be used in straddle milling operations?
8. Explain the process of setting up cutters for a straddle milling operation and checking the distance between them.
9. Explain how a flycutter with two cutters can be set up to take a roughing and finishing cut at the same time.
10. Why are end mills with a helix angle that is opposite the direction of rotation often used for profiling operations?

11.7 ADVANCED MILLING OPERATIONS

1. Briefly explain the process of direct indexing.
2. What is the function of the sector arms on an indexing head?
3. Explain how to set up the indexing head to cut: a 32-tooth gear; a 72-tooth gear.
4. What two accessories can be used for angular indexing?
5. Explain how you would set up the indexing head to space holes 84° apart.
6. On a rotary table with a 72 : 1 worm ratio, how many degrees will the table move when the crank is turned: 1 revolution: 4 1/2 revolutions?
7. List two purposes for which rotary tables that are geared to the table feed screw can be used.
8. What two types of cutters can be used to mill a helical groove in a shaft?
9. Using a sketch, show how to find the developed length of a helix.
10. Explain how the lead of a milling machine is computed.
11. What factors about a helix must be known to compute the direction and the angle of universal table movement?
12. By what two processes can spur gears be cut?
13. How do spiral bevel and hypoid gears differ?
14. Using a sketch, show how to develop an involute curve.
15. Compute the blank diameter for a 48-tooth gear of 16 DP.
16. Briefly explain the process of centering a gear blank under the gear cutter.
17. Explain the difference between *working depth* and *whole depth* with reference to gear teeth.

12

Shapers and Operations

12.1 INTRODUCTION

The shaper is used primarily to machine flat surfaces in the horizontal, vertical, or angular planes. With practice, the skilled operator can manipulate the shaper for machining irregular and simple curved surfaces, slots, keyways, and simple internal machining operations.

Shapers employ a single-point cutting tool similar to the lathe cutting tool. The cutting tool is fastened in a toolpost attached to a *ram.* The ram reciprocates, pushing the cutting tool back and forth across the workpiece. The forward stroke of the ram produces the cutting stroke on the work. With the return stroke, the tool is returned to the starting position. As the tool is returned, the table and workpiece are advanced the desired amount for the next cut (Fig. 12.1).

In terms of the rate of metal removal, shapers are relatively slow machine tools and not generally considered production machines. In most larger shops the shapers are being replaced by vertical milling machines, especially the versatile NC mill (see Chap. 16). Many smaller shops and toolrooms still use the shaper since it is easy to set up and operate. Another advantage of the shaper is that the cutting tools are relatively inexpensive and easily reshaped.

Opening photo courtesy of El Camino College.

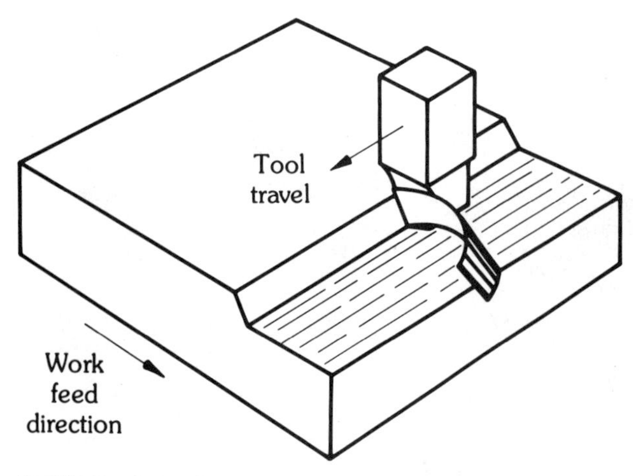

FIGURE 12.1 Cutting action of a horizontal shaper.

12.2 TYPES OF SHAPERS AND CONSTRUCTION FEATURES

Shapers are classified according to the position of the ram. There are two types. The most common is the horizontal; the other, the vertical shaper, is sometimes referred to as a *slotter*.

12.2.1 Horizontal Shapers

The principal parts of the horizontal shaper are shown in Fig. 12.2. The horizontal shaper is covered in detail because it is more common than the vertical shaper.

1. The *base* is a heavy casting that supports all the major component parts of the machine.
2. The *column* is a hollow casting mounted directly on the base, housing the operating mechanisms that drive the ram. The upper surface of the column has two precision machined ways on which the ram reciprocates. The front face of the column must be machined at right angles to the ram ways on the top of the column. The cross rail is mounted and moves on this front face.
3. The *cross rail*, also a casting, allows vertical and horizontal movement of the table. The cross-feed mechanism is attached to the cross rail.
4. The *table* is fastened to ways that move on the horizontal ways of the cross rail. The

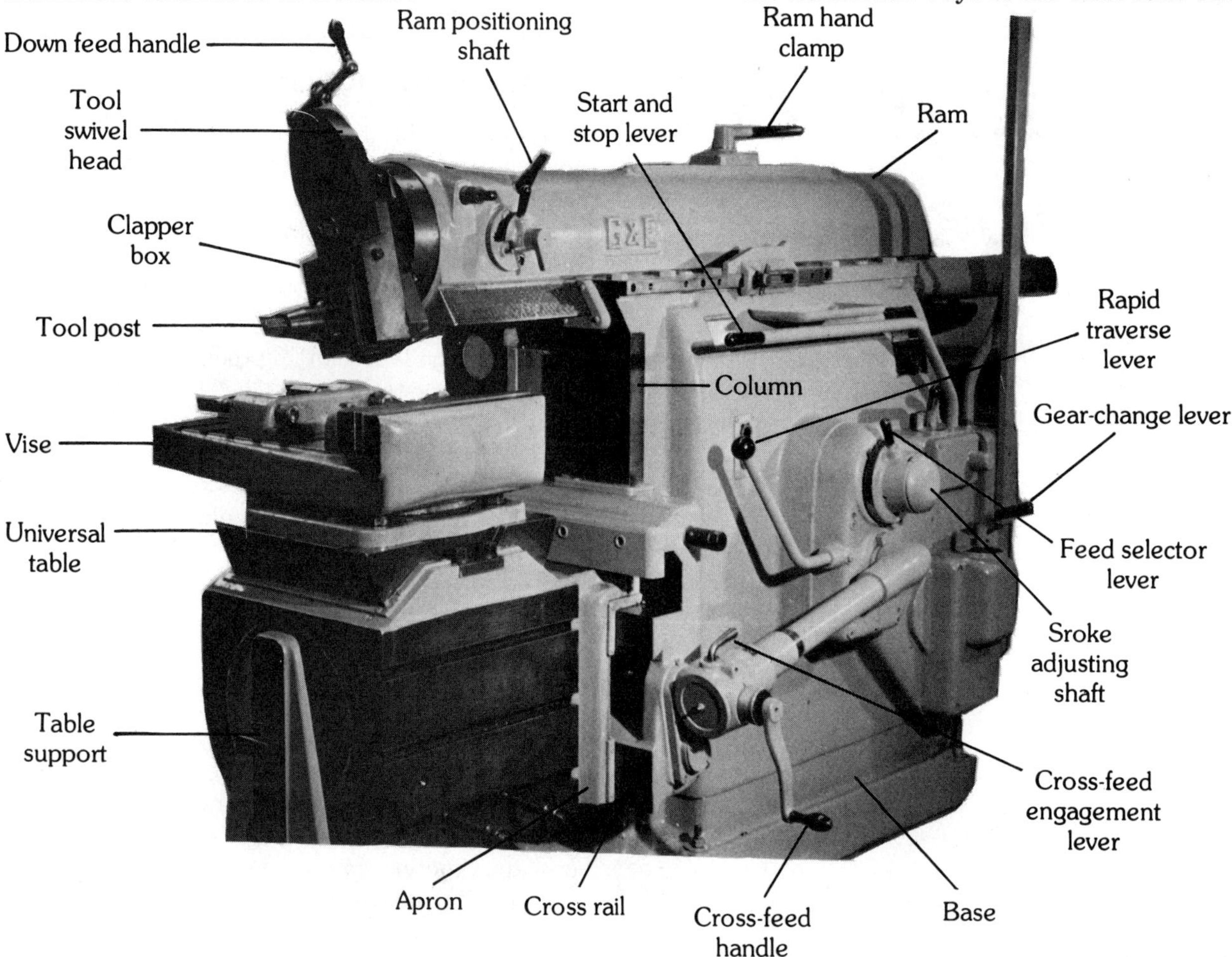

FIGURE 12.2 Principal parts of the horizontal shaper. (*Courtesy El Camino College*)

table is a hollow box casting with machined T slots on the top and sides. A vise is usually fastened to the top of the table to hold the workpiece. There are two types of tables: the *plain* and *universal*. The plain table is capable of horizontal and vertical movements only. The universal table, in addition to moving horizontally and vertically, can be swiveled to machine angles on the work.

5. The *table support* extends from the outer surface of the table to the base and is used on some larger shapers to support the weight of the table.
6. The *ram* is a heavy casting that reciprocates on the ways of the column. Ram movement is obtained either mechanically on crank-type shapers or hydraulically. Figure 12.3 illustrates the complete cycle of the crank drive mechanism. The ram is reciprocated by a pivoted rocker arm, which is fastened through linkage to the ram. The forward cutting stroke of the ram requires an arc of 220° and the return backstroke an arc of 140°. With these angles of arc, the ratio of cutting time to total cycle time is about 3:5, with the return time ratio about 2:5. In other words, the return stroke is faster. Hydraulically driven shapers are similar to the cranktype, except that the ram is driven by hydraulic oil pressure developed by a pump. A reversing valve changes the direction of the ram. Hydraulic shapers have the advantages of a wider range of speeds and feeds and uniform motion.
7. The *toolhead* is fastened to the front of the ram and can be swiveled for machining angular cuts. It contains the toolpost and toolholder that hold the cutting tool. The

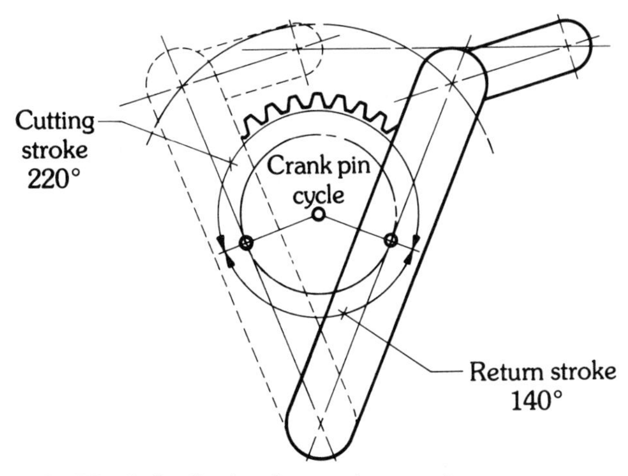

FIGURE 12.3 Cycle of a crank-type shaper.

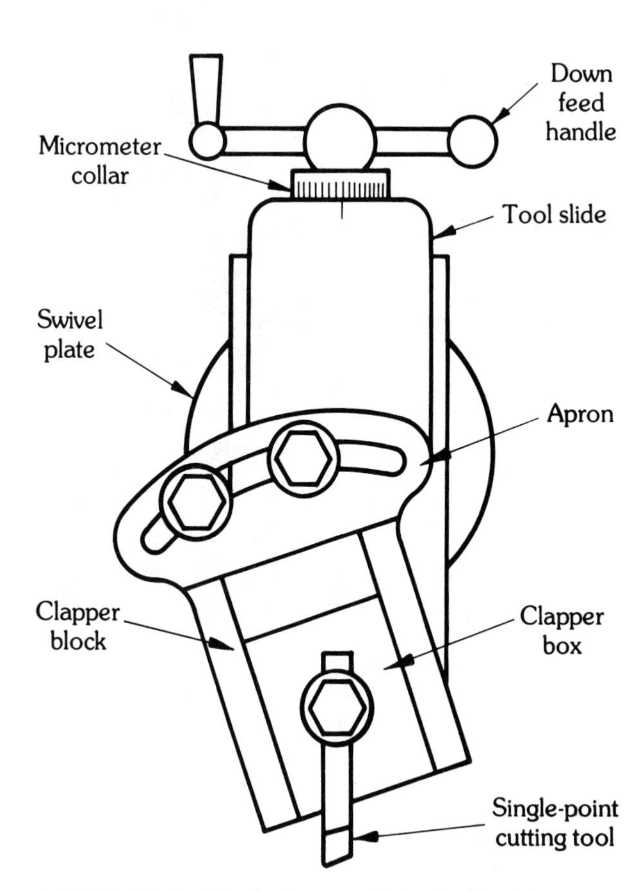

FIGURE 12.4 Principal parts of the toolhead.

toolslide is moved up or down by a *feed screw* to adjust for the proper depth of cut. A *micrometer collar* is attached to the feed screw (Fig. 12.4).

8. The *clapper box* is hinged to the toolslide. It can be swiveled a small amount in either direction (Fig. 12.4). The clapper box must be properly positioned when making vertical or angular cuts so the tool will swing out and away from the workpiece to provide clearance on the return stroke. For horizontal cuts, the clapper box is usually positioned vertically. The shaper cutting tool is held in the *toolpost* on the clapper box. On the cutting stroke, the pressure from the cutting tool holds the clapper box firmly against the toolhead.

12.2.2 Vertical Shapers

The vertical shaper is similar in operation to the horizontal crank-type shaper (Fig. 12.5). The obvious difference is the vertical position of the ram. A distinct advantage of the vertical shaper is that the cutting pressure or thrust is against a table bed, so there is no possibility of deflection of the worktable. Another advantage is that the table bed can be moved in three directions: longitudinally, laterally, and with a 360° rotary motion. The ver-

FIGURE 12.5 A vertical shaper attachment on a milling machine is used to cut slots, keyways, internal and external surfaces of various shapes and sizes. (*Courtesy Bridgeport Machines, Inc.*)

satility of the rotary table permits the vertical shaper to machine internal and external keyways, slots, splines, gears, and intricate die forms.

12.2.3 Size Classification of Shapers

The size of a shaper is determined by the maximum length of the stroke or ram movement. For example, a 17-in. shaper can machine a 17-in. cube; a 40-cm shaper will machine a 40-cm cube. Vertical and lateral table movement is also usually equal to the stroke.

12.3 CUTTING TOOLS FOR SHAPERS

Cutting tools used on shapers are similar to those used on lathes. Fig. 12.6 shows cutting tools for various machining operations performed on the shaper. Most shaper cutting tools require only a small amount of relief—generally, 3 to 5° of end and side relief. Side rake angles vary, depending on the material that is being machined. For steel, 10 to 15° is usually used. Cast iron requires 5 to 10° of side rake, aluminum 20 to 30°.

Toolholders used on the shaper are also similar to those used on lathes. However, the square hole that houses the toolbit is parallel to the base on shaper toolholders. Frequently, the universal or swivel-head toolholder is used on the shaper. As shown in Fig. 12.7, the universal toolholder can be positioned for five different types of cuts.

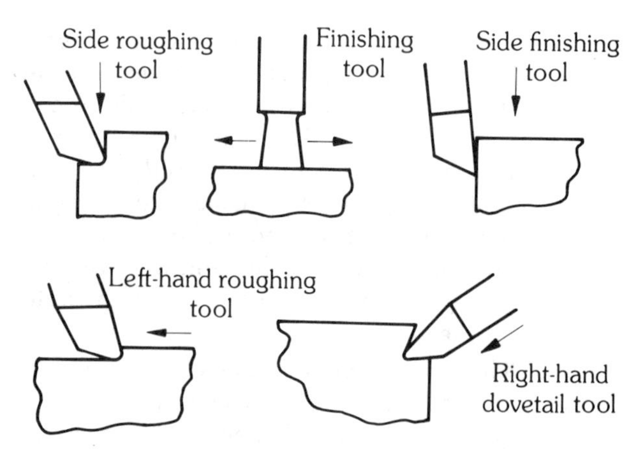

FIGURE 12.6 Shaper tools for common shaping operations.

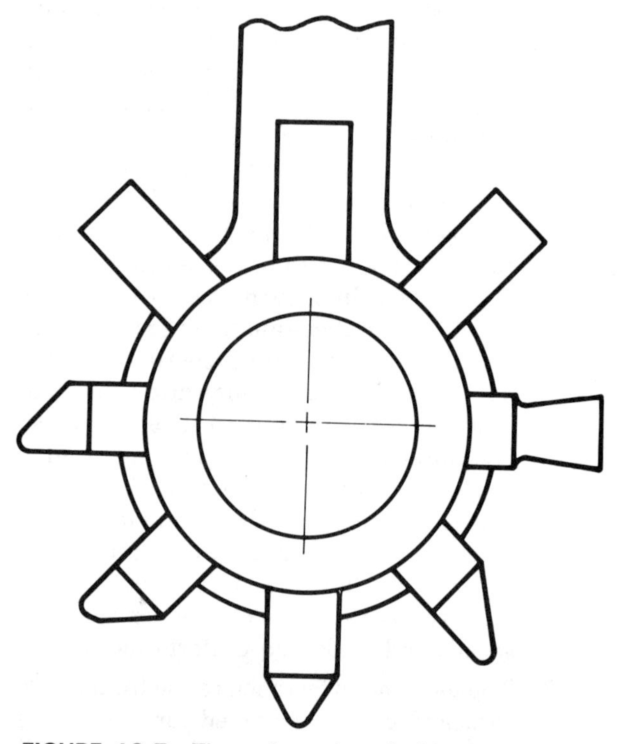

FIGURE 12.7 The universal toolholder is commonly used on horizontal shapers.

12.4 WORK-HOLDING DEVICES

Several types of work-holding devices are used on shapers. Each type requires the work to be clamped rigidly. If the work moves during an operation, it may cause serious damage to the shaper or injury to the operator.

12.4.1 Vises

Most parts to be machined on the shaper can be held in a *vise*. *Parallels* are used to support the work above the vise jaw, parallel to the table and vise bottom. Always make sure that the vise is clean and free of burrs before clamping it to the table.

12.4.2 Clamps, Straps, and T Bolts

When the workpiece is too large for a vise or irregular in shape, it is often clamped directly to the table.

12.5 SHAPER OPERATIONS

Most work machined on a shaper is held in a vise. Therefore, the following procedures, setups, and operations apply when the workpiece is mounted in a vise.

12.5.1 Shaper Adjustments

Before shaping a part, it is necessary to set up the shaper properly for the operation.

Ram Adjustments. Two adjustments must be made on the ram prior to machining the workpiece. First, the stroke length must be adjusted. This is accomplished by turning the *stroke-adjusting shaft* or stroke selector (Fig. 12.8). Most shapers have a scale on the ram with a pointer to indicate the length of stroke. The stroke length is adjusted when the ram is in its extreme return position. It is usually set approximately 1 in. (25 mm) longer than the length of the work to be machined.

The second adjustment is for positioning the cutter. The ram is adjusted so the ram travel machines the entire length of work. For the correct position of the ram to be set, the ram must be at the extreme return position of the stroke. Loosen the *hand clamp* and crank the ram position shaft (Fig. 12.9) until the cutting tool clears the beginning of the work by approximately 1/2 in. (12.7 mm). This procedure automatically allows a 1/2-in. (12.7-mm) clearance at the end of the workpiece. Next, tighten the ram hand clamp.

FIGURE 12.8 Adjusting the stroke length with the stroke-adjusting shaft. (*Courtesy El Camino College*)

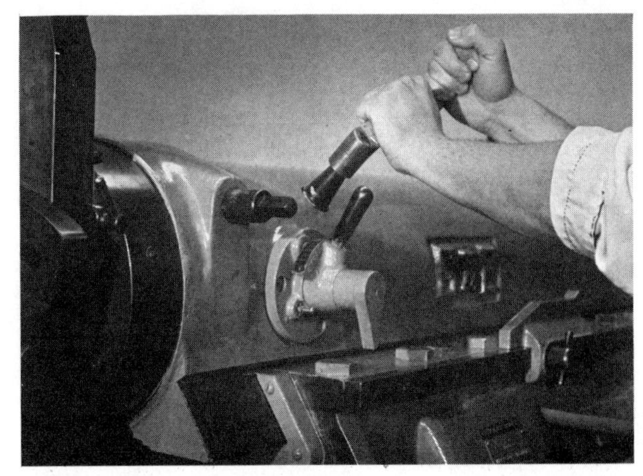

FIGURE 12.9 Adjusting the position of the ram is done by turning the ram position shaft. (*Courtesy El Camino College*)

Speed and Feed Adjustments. The speed of a shaper is the number of cutting strokes the ram makes per minute. The cutting speed selected for the shaper depends on the following:

1. Type of material being cut
2. Type of cutting tool
3. Rigidity of the setup and the machining tool
4. Depth of cut
5. Use of cutting fluids

The formula used to find the correct number of strokes per minute for a shaper is as follows:

$$\text{For inch speed: } N = \frac{\text{CS} \times 7}{L}$$

$$\text{For metric speed: } N = \frac{\text{CS}}{L} \times 0.6$$

where N = number of strokes
CS = cutting speed of material to be machined, in feet or meters per minute
L = length of stroke required, in inches or meters

Table 12.1 gives suggested cutting speeds and feeds for HSS cutting tools for the common types of metals.

TABLE 12.1
Suggested cutting speeds and feeds for a shaper using HSS cutting tools

	Cutting Speed		*Feed*	
Material	*ft/min*	*m/min*	*in.*	*mm*
Steel				
Mild	80–100	24–31	0.010	0.25
Tool and alloy	40–60	13–20	0.010	0.25
Cast iron				
Gray	100–120	31–40	0.015	0.37
Malleable	80–100	27–31	0.015	0.37
Aluminum	200–300	62–100	0.012	0.30
Brass	150–200	50–62	0.010	0.25

Example A: Find the number of strokes required to machine a 10-in. piece of mild steel with a cutting speed of 100 ft/min.

$$N = \frac{\text{CS} \times 7}{L}$$

$$= \frac{100 \times 7}{10}$$

$$= 70 \text{ strokes per minute}$$

Example B: Calculate the number of strokes per minute required to machine a piece of cast iron with a cutting speed of 20 m/min and a stroke length of 50 cm.

$$N = \frac{\text{CS}}{L} \times 0.6$$

$$= \frac{20}{0.5} \times 0.6$$

$$= 24 \text{ strokes per minute}$$

Feeds. Shaper feed is the distance the work moves after each cutting stroke. In general, the amount of feed required depends on the same variables that determine cutting speeds. The operator needs to use experience and judgment when selecting feed rates. The beginner should keep two things in mind: coarse feeds are used with heavy roughing cuts, and fine feeds are used with finishing cuts.

Setting the feed on hydraulic shapers is a matter of adjusting the feed control lever, located at the rear of the shaper on the operator's side. The crank-type shaper's feeds are regulated by a connecting feed rod. To vary the feed rate, move the connecting feed rod closer to the center of the driving mechanism for finer feeds, and away from the center for coarser feed rates (Fig. 12.10).

FIGURE 12.10 The amount of feed is regulated by the position of the feed rod. (*Courtesy El Camino College*)

12.5.2 Shaper Safety

1. Always wear eye protection.
2. Make sure that the work being machined is securely fastened.
3. Turn the machine ram *by hand* through a complete stroke to make sure that the tool and ram head clear the housing and workpiece. (*Note:* On some shapers this is not possible; therefore, put the machine through a complete cycle at a slow speed, and carefully observe the position of the ram head and tool.)
4. Always stop the machine before attempting to make any adjustments to the tool or ram.
5. Use a chip guard in front of the vise to catch flying chips coming from the work.
6. Stop the machine when taking any measurements on the work.
7. All machine guards should be securely in place before you operate a shaper.
8. Never use your hands to remove chips.
9. Allow approximately 18 in. (45 cm) of clearance behind the reciprocating ram.

12.5.3 Aligning the Shaper Vise

Shaper vises can be swiveled 360°. The jaws may be set either parallel to or at right angles to the ram stroke. For long, narrow workpieces, the vise jaws are placed parallel to the ram stroke to reduce machining time required. For work that is approximately square, the vise jaws are set perpendicular to the ram. This procedure allows greater depth of cuts and heavier feed rates because the vise jaw acts as a support or stop at the end of the workpiece. When extreme accuracy is required, a dial indicator is used to align the vise in either of the two positions.

To align the vise jaws *parallel* to the *ram*, use the following procedure:

1. Position the vise so that it is approximately parallel to the ram by the lines located on the base of the swivel vise.
2. Snug the T bolts lightly.
3. Adjust the length and position of the ram stroke slightly shorter than the length of the jaws.
4. Place a dial indicator in the toolpost and position the button on the gage against the solid jaw (Fig. 12.11).

FIGURE 12.11 Aligning the vise parallel to the ram travel. (*Courtesy El Camino College*)

5. Move the ram back and forth by hand or with the shaper set at its lowest speed, and adjust the vise until the dial (pointer) of the indicator does not move.
6. Tighten the vise securely and recheck the accuracy of the vise. Readjust if necessary.

To align vise jaws *perpendicular* to the *ram*, use the same procedure, but move the table back and forth using the longitudinal feed handle. The ram is stationary.

12.5.4 Shaping a Flat Horizontal Surface

Probably the most frequent operation performed on a shaper is machining a true flat surface on a part (Fig. 12.12). Here is the procedure:

FIGURE 12.12 Shaping a horizontal surface. (*Courtesy El Camino College*)

1. Mount the workpiece on parallels and place a round piece of stock between the movable jaw and workpiece if the work is not square. Secure the vise. Make sure that the vise is clean and remove all burrs from the workpiece. Seat the workpiece firmly against the parallels using a soft hammer.
2. Position the clapper box, toolholder, and cutting tool as shown in Fig. 12.13.
3. Set the ram to the proper length and position it over the workpiece.
4. Calculate and set the proper ram speed (number of strokes per minute) and feed rate.
5. Position the cutting tool over the highest spot of the workpiece. Using the feed handle on the tool slide, feed the tool until it lightly touches a paper feeler above the workpiece.
6. Moving the ram and table by hand, position the cutting tool at the beginning of the work.
7. Set the cutting tool to the proper depth of cut using the micrometer collar.
8. Start the ram and engage the table feed.
9. After the cut is complete, return the table and work to the starting position.
10. Continue to take successive roughing cuts leaving 0.010 to 0.015 in. (0.25 to 0.37 mm) for the finishing cut.

FIGURE 12.13 Correct position of the clapper box, toolholder, and cutting tool for a horizontal cut. (*Courtesy El Camino College*)

Hints on shaping a horizontal surface:

1. Carefully scribe a layout line indicating the finished depth on the outer end of the workpiece. This will help determine when you are getting close to the finish cut.
2. Use the proper cutting fluid on finish cuts to improve the surface finish.
3. Use a micrometer or vernier caliper to accurately measure the thickness of the workpiece.
4. When shaping cast iron, be sure the first cut is deep enough to cut under the scale completely (otherwise, the tool will rapidly become dull), or use a carbide cutting tool.
5. File or grind a 45° bevel on the outer end to the workpiece. This keeps the chips from flying out a great distance and prevents a ragged edge where the tool leaves the work.
6. If chatter marks appear on the machined surface, check the toolholder and/or the cutting tool for rigidity.
7. If a workpiece has irregular surfaces where it is to be clamped, place emery cloth (abrasive side toward work) between the vise jaws and workpiece. This technique makes the setup more secure.

FIGURE 12.14 Shaping a vertical surface. (*Courtesy El Camino College*)

12.5.5 Shaping a Vertical Surface

(*See* Fig. 12.14)

1. Mount the vise so that the solid jaw is perpendicular to the ram.
2. Mount the workpiece in a clean vise so that the edge to be machined extends approximately 1/2 in. (12.7 mm) beyond the vise jaws.
3. For vertical cuts, the clapper box must be swiveled *away* from the surface to be machined. This prevents the cutting tool from dragging and scoring the machined surface on the return stroke.
4. Set the ram for the correct stroke length and position for the work length.
5. Select the correct ram speed.
6. Carefully check that the ram and toolhead clear the workpiece.
7. Start the shaper and hand feed the cutting tool down the side of the workpiece approximately 0.010 in. (0.25 mm) at the end of each return stroke. (*Note:* Some shapers have an automatic vertical feed. If so, engage the appropriate lever or knob to use the power feed.)

12.5.6 Squaring a Block in a Shaper

One of the most common shaper jobs is squaring a piece of metal. The procedure for squaring a block of metal is similar to shaping a horizontal surface.

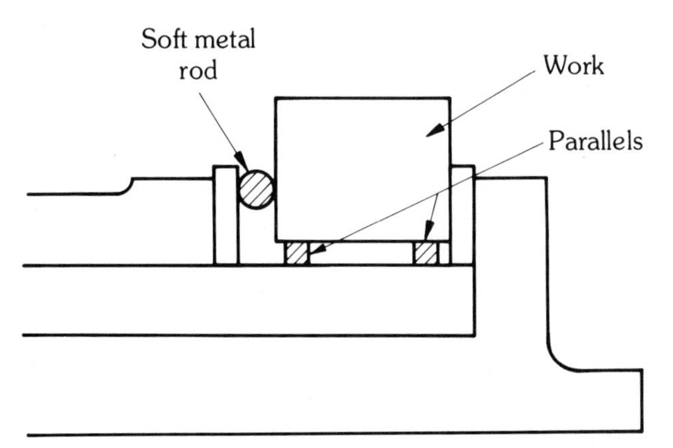

FIGURE 12.15 A soft metal rod is placed between the work and movable jaw. This holds the work flat against the solid jaw.

1. Mount the workpiece in a vise with the largest side up. Place a round rod between the movable jaw and the workpiece so that the work will be held flat against the solid jaw (Fig. 12.15).
2. Machine the first side, taking enough cuts to just clean up the surface.
3. Deburr the machined side and rotate the work so that the machined surface is against the solid jaw. (*Note:* Do not forget to place the round rod between the movable jaw and the work.)
4. Machine side 2 and check squareness using a machinist's square. Continue rotating the newly machined surface and using the round rod until the fourth side (last surface) is to be machined. On the last surface the round rod is no longer needed. (*Note:* Deburr each machine side before mounting it back in the vice and check the workpiece size with a micrometer.)
5. To square the ends of the block, use vertical cuts on each end as described in Sec. 12.5.5.

12.5.7 Other Shaping Operations

Some other operations peformed on the horizontal shaper are shaping angles, dovetails, grooves, and keyways.

Shaping Angles. Machining angles on the shaper can be accomplished in three ways:

1. The easiest method is to scribe a layout line on the workpiece at the desired angle. Mount the work in the vise so that layout line is parallel to the top surface of the vise jaw. Then machine the workpiece using conventional horizontal cuts (Fig. 12.16).

FIGURE 12.16 One method of shaping an angle on a part. Mount the workpiece at an angle so that the layout line is parallel to the top surface of the vise.

FIGURE 12.17 Shaping an angle on a workpiece by swiveling the vise to the desired angle. (*Courtesy El Camino College*)

2. A second method of shaping angles is to swivel the universal vise and workpiece to the desired angle (Fig. 12.17) and take conventional horizontal cuts to bring the workpiece to finished size.
3. A third method of shaping angles is to set the toolhead at a desired angle and feed the cutting tool as shown in Fig. 12.18. Note the position of the clapper box in this technique. This prevents the tool from dragging on the return stroke.

Shaping Dovetails. Dovetails can be machined on the shaper by angular shaping with the toolhead set at an angle (Fig. 12.18). Dovetails are made in sets of two, internal and external, and form an assembly (Fig. 12.19). Dovetail assemblies

FIGURE 12.18 Angles may be shaped on a workpiece by setting the toolhead at the desired angle. This is the operation of shaping a dovetail. (*Courtesy El Camino College*)

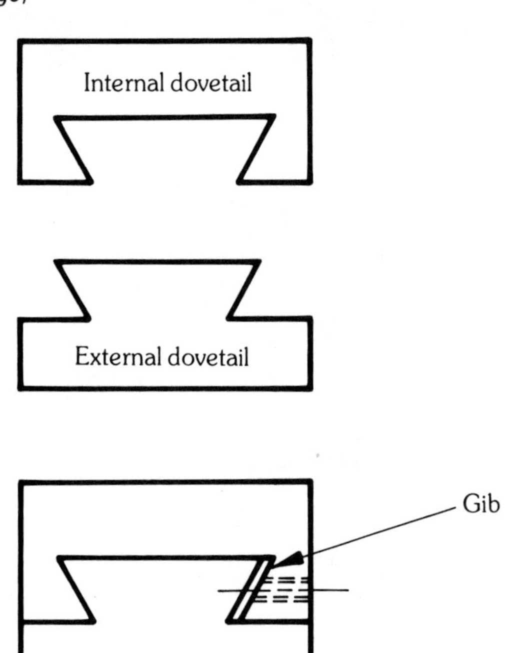

FIGURE 12.19 Dovetail assembly. A gib is used in the assembly to compensate for wear.

are used in toolsides and table assemblies for machine tools such as lathes, milling machines, and shapers.

Shaping Keyways and Grooves. A *groove* is a narrow channel or slot machined into a piece

FIGURE 12.20 Shaping a groove.

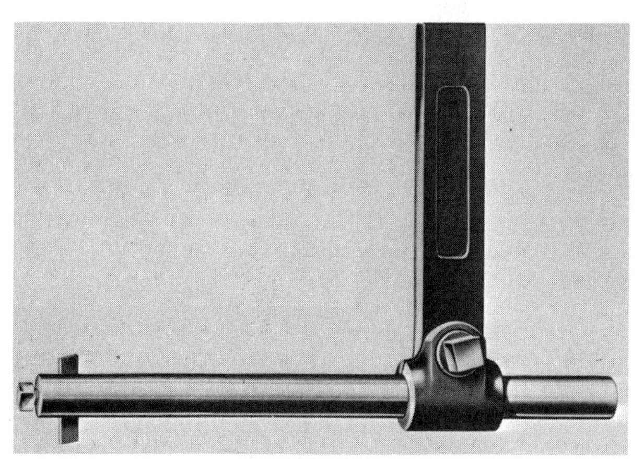

FIGURE 12.21 An extension bar is used to shape a keyway.

of metal. Figure 12.20 shows a groove being machined on the shaper. A *keyway* is the groove cut on the inside of a pulley, gear, or wheel. A *keyseat* is a keyway machined on a shaft. A *key*, usually a square piece of metal, is fitted into a keyway or keyseat to form a positive drive between a shaft and gears, pulleys, and so forth. Figure 12.21 shows a keyway being machined in a gear using an *extension bar* (also called a *poke bar*). Keyseats are machined the same way as a groove. In either case, the size of the keyway or keyseat being machined is determined by the width ground on the cutting tool. Wide slots may be machined by making two or more cuts.

REVIEW QUESTIONS

12.2 TYPES OF SHAPERS AND CONSTRUCTION FEATURES

1. Describe the operational differences between vertical and horizontal shapers.
2. Briefly describe the functions of the following horizontal shaper components:
   (*a*) Base
   (*b*) Column
   (*c*) Cross rail
   (*d*) Table
   (*e*) Table support
   (*f*) Ram
   (*g*) Toolhead
   (*h*) Clapper box
3. How are shapers classified as to size?

12.3 CUTTING TOOLS FOR SHAPERS

1. How many degrees of end and side relief are necessary for shaping tools?
2. How many degrees of rake angle are required for the following materials?
   (*a*) Steel
   (*b*) Cast iron
   (*c*) Aluminum
3. How do lathe toolholders differ from those used on shapers?
4. Describe the swivel toolholder used on shapers.

12.4 WORK-HOLDING DEVICES

1. What work-holding device is most commonly used on a shaper?
2. When are straps and T bolts used to secure the part?

12.5 SHAPER OPERATIONS

1. Briefly describe the two necessary ram adjustments prior to machining a part.
2. List five variables in selecting a particular cutting speed for a shaping job.
3. Find the correct number of strokes per minute for the following shaping jobs:
   (*a*) Shaping an 8-in-long piece of tool steel at 60 ft/min
   (*b*) Machining a 6 1/2-in. piece of aluminum at 300 ft/min
   (*c*) Machining a piece of cast iron 30 cm long at 28 m/min
   (*d*) Shaping a 38 cm-long piece of brass at 55 m/min
4. Explain how the feed mechanisms are set up for hydraulic and crank-type shapers.
5. Briefly describe the safety factors to consider when operating a shaper.
6. Explain how the shaper vise is properly aligned parallel to the ram; perpendicular to the ram.
7. Describe the setup procedures for shaping a part flat and parallel.
8. What precautions must be taken to avoid tool damage when shaping cast iron?
9. What is the purpose of filing a 45°bevel on the end of the part?
10. How should work with irregular surfaces be properly clamped in the shaper vise?
11. Briefly describe the setup procedures for shaping a vertical surface on a workpiece.
12. What is the correct procedure for squaring a block of material in a shaper?
13. Describe three methods of machining angles on the shaper.
14. Explain the difference between a keyseat and a keyway. How is each machined on a shaper?

13

Grinding Machines, Operations, and Abrasive Processes

13.1 INTRODUCTION

Grinding, or abrasive machining, is one of the most rapidly growing metal-removal processes in manufacturing. Many machining operations previously done on conventional milling machines, lathes, and shapers are now being performed on various types of grinding machines. Greater productivity, improved accuracy, reliability, and rigid construction characterize today's industrial grinding machines.

Grinding, or abrasive machining, is the process of removing metal in the form of minute chips by the action of irregularly shaped abrasive particles. These particles may be in bonded wheels, coated belts, stones, or simply loose.

13.2 TYPES OF GRINDING MACHINES

Grinding machines have advanced in design, construction, rigidity, and application far more in the last decade than any other standard machine tool in the manufacturing industry. Grinding machines (also called grinders) fall into four categories:

1. Surface grinders
2. Cylindrical grinders

Opening photo courtesy of Gallmeyer & Livingston Co.

3. Centerless grinders
4. Special types of grinders

13.2.1 Surface Grinders

Surface grinders are used to produce flat, angular, and irregular surfaces. In the surface grinding process, the grinding wheel revolves on a spindle and the workpiece, mounted on either a reciprocating or rotary table, is brought into contact with the grinding wheel.

Four types of surface grinders are commonly used in industry:

1. The *horizontal spindle/reciprocating table* surface grinder. This surface grinder is the most commonly used type in school shops and industry (Fig. 13.1). It is available in various sizes to accommodate large or small workpieces. With this type of surface grinder, the work moves back and forth under the grinding wheel. The grinding wheel is mounted on a horizontal spindle and cuts on its periphery as it contacts the workpiece. The worktable is mounted on a saddle which provides cross-feed movement of the workpiece (Fig. 13.2). The wheelhead assembly moves vertically on a column to control the depth of cut required. The construction of this type of surface grinder is explained in greater detail later in the chapter.

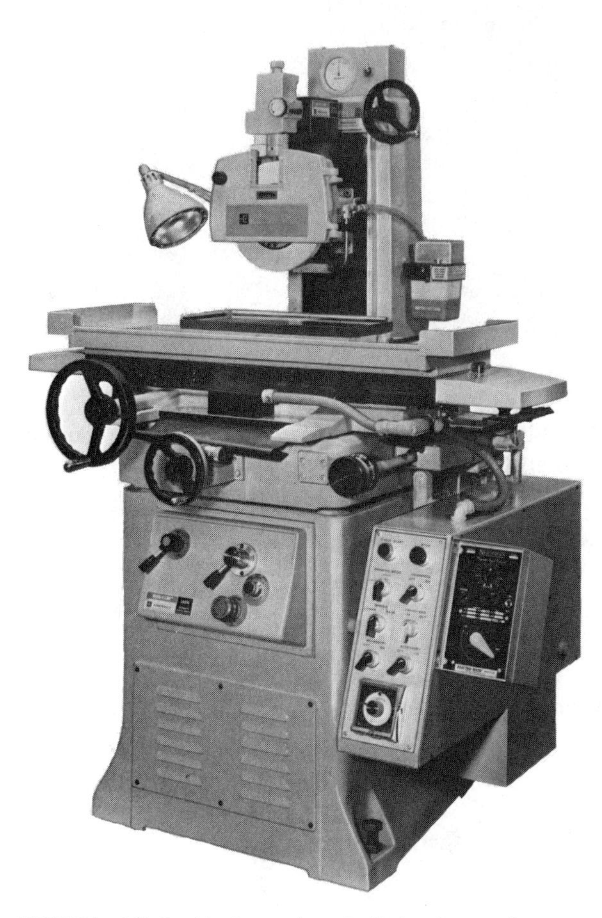

FIGURE 13.1 Horizontal spindle/reciprocating table surface grinder. (*Courtesy Boyar-Schultz Corp.*)

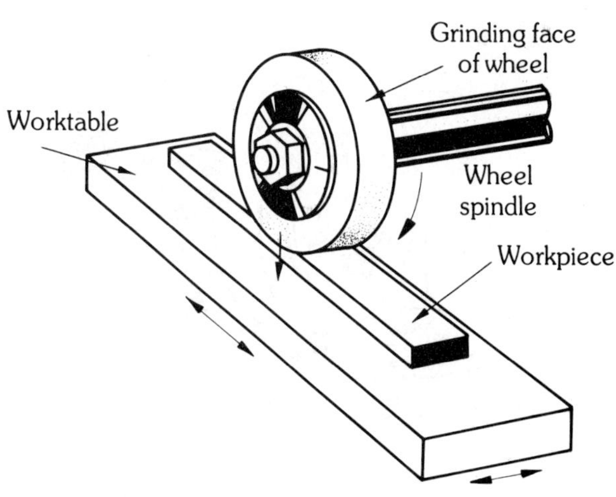

FIGURE 13.2 Operating principle of the horizontal spindle/reciprocating table surface grinder.

FIGURE 13.3 Horizontal spindle/rotary table surface grinder. (*Courtesy Stoffel Grinding Systems*)

2. The *horizontal spindle/rotary table* surface grinder (Fig. 13.3). This surface grinder also has a horizontally mounted grinding wheel that cuts on its periphery. The workpiece rotates 360° on a rotary table underneath the wheelhead. The wheelhead moves across the workpiece to provide the necessary cross-feed movement. The metal-removal rate is controlled by the amount of downfeed of the wheelhead assembly (Fig. 13.4).
3. The *vertical spindle/reciprocating table* surface grinder (Fig. 13.5). This type of grinding machine is particularly suited for

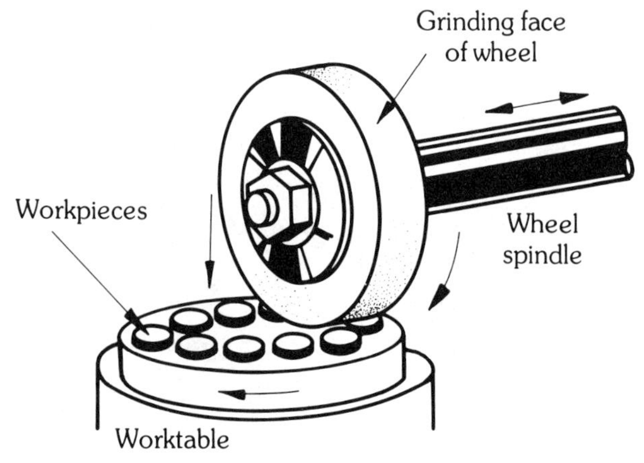

FIGURE 13.4 Operating principle of the horizontal spindle/rotary table surface grinder.

FIGURE 13.5 Vertical spindle/reciprocating table surface grinder. (*Courtesy Hill Acme Co.*)

grinding long and narrow castings like the bedways of an engine lathe. It removes metal with the face of the grinder wheel while the work reciprocates under the wheel. The wheelhead assembly, like on most other types of surface grinders, moves vertically to control the depth of cut. Cross feed is accomplished by the table moving laterally. The table is mounted on a saddle unit (Fig. 13.6).

Surface grinders of this type are generally larger machines capable of heavy metal-removal rates and large volume production.

4. The *vertical spindle/rotary table* surface grinder (Fig. 13.7). This type of grinding machine is also capable of heavy cuts and high metal-removal rates. Vertical spindle machines use cup, cylinder, or segmented wheels. Many are equipped with multiple spindles to successively rough, semifinish, and finish large castings, forgings, and

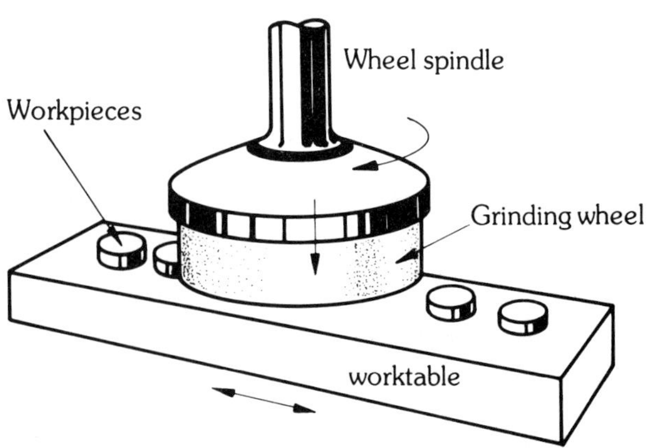

FIGURE 13.6 Operating principle of the vertical spindle/reciprocating table surface grinder.

FIGURE 13.7 Vertical spindle/rotary table surface grinder. (*Courtesy Blanchard Machine Tools*)

FIGURE 13.8 Operating principle of the vertical spindle/rotary table surface grinder.

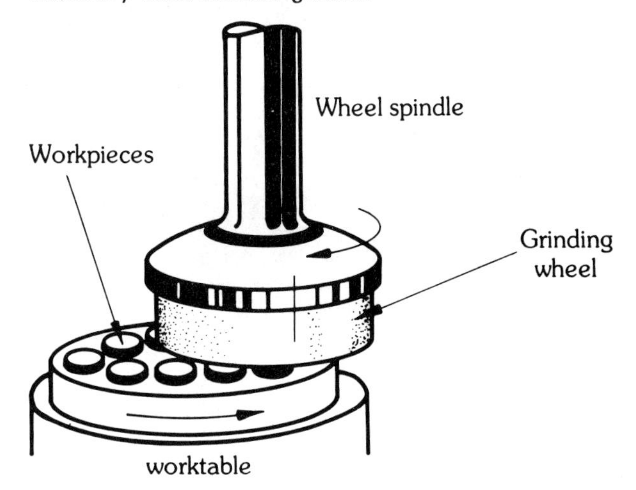

welded fabrications (Fig. 13.8). These grinding machines are available in various sizes and have up to 225-hp motors to drive the spindle.

13.2.2 Cylindrical Grinding Machines

Cylindrical grinding is the process of grinding the outside surfaces of a cylinder. These surfaces may be straight, tapered, or contoured. Cylindrical grinding operations resemble lathe-turning operations. They replace the lathe when the workpiece is hardened or when extreme accuracy and superior finish are required. Figure 13.9 illustrates the basic motions of the cylindrical grinding machine. As the workpiece revolves, the grinding wheel, rotating much faster in the opposite direction, is brought into contact with the part. The workpiece and table reciprocate while in contact with the grinding wheel to remove material.

The cylindrical grinder (sometimes called a center-type grinder) is equipped with a headstock, tailstock, table, and wheelhead (Fig. 13.10). The headstock and tailstock are equipped with centers that support the outer ends of the workpiece during the grinding operations. The tailstock center is a dead (nonrevolving) center while the headstock center may or may not revolve during grinding. When extreme accuracy is required, the two supporting centers must remain stationary as the workpiece revolves. When both centers are dead, precision sizes and good finishes can be obtained, because there is no possibility of runout from the headstock spindle. Cylindrical grinders may also hold a workpiece with work-holding devices such as universal, independent, or collet chucks.

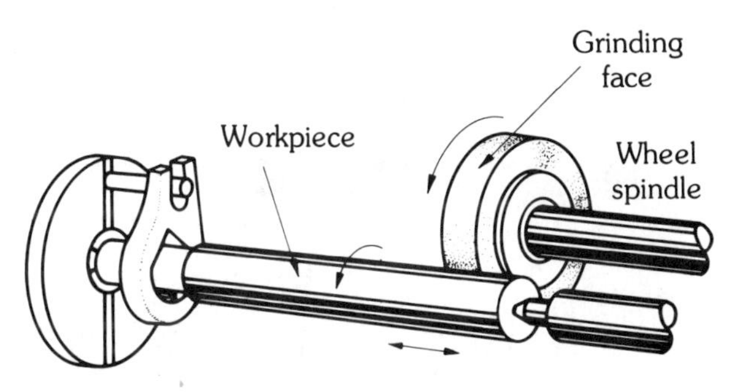

FIGURE 13.9 Operating principle of the cylindrical grinding machine.

There are two basic cylindrical grinders: the *plain* and *universal*. Plain cylindrical grinders are somewhat more rigid and are generally considered production machines. They are used primarily for grinding straight or slightly tapered surfaces on cylindrical parts. The plain type has a fixed wheelhead and headstock. The head and tailstock units are mounted on a table that swivels 10 or 12° for grinding shallow tapers on the work.

Universal grinding machines are toolroom machines capable of grinding parts with a great variety of configurations. They are more versatile because the wheelhead and headstock swivel.

Plain and universal grinders can be used for *form* grinding operations. Form grinding requires that the grinding wheel be dressed or shaped to the reverse of the form to be ground (Fig. 13.11). The

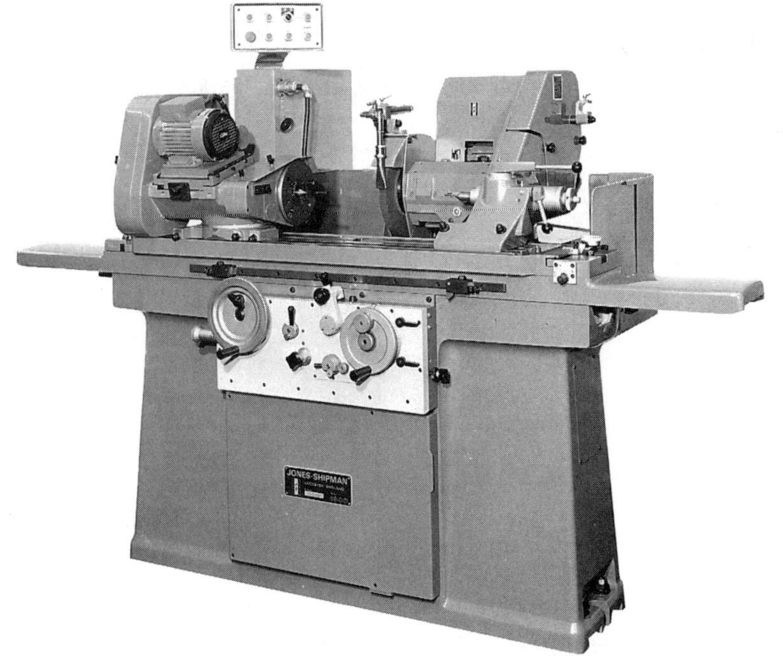

FIGURE 13.10 Universal cylindrical grinding machine. (*Courtesy Brown and Sharpe Grinding Machines, Inc.*)

FIGURE 13.11 Centerless grinding machine. (*Courtesy Cincinnati Milacron Inc.*)

part is ground by plunging the wheel directly into the workpiece to a predetermined depth without any movement.

Cylindrical grinder construction and operation are explained in detail later in the chapter.

13.2.3 Centerless Grinding Machines

Centerless grinding machines eliminate the need to have center holes for the work or to use work-holding devices. In centerless grinding, the workpiece rests on a workrest blade and is backed up by a second wheel, called the *regulating wheel* (Fig. 13.11). The rotation of the grinding wheel pushes the workpiece down on the workrest blade and against the regulating wheel. The regulating wheel, usually made of a rubber-bonded abrasive, rotates in the same direction as the grinding wheel and controls the longitudinal feed of the work when set at a slight angle (Fig. 13.12). By changing this angle and the speed of the wheel, the workpiece feed rate can be changed. The diameter of the workpiece is controlled by two factors: the distance between the grinding wheel and regulating wheel, and by changing the height of the workrest blade.

Centerless grinders are considered production grinding machines. They grind parts by three different methods:

1. *Through-feed centerless grinding*, as illustrated in Fig. 13.13, is used for grinding straight cylindrical parts.
2. *End-feed centerless grinding* is used only on tapered work (Fig. 13.14). The grinding wheel, regulating wheel, and workrest blade are set in a fixed position in relation

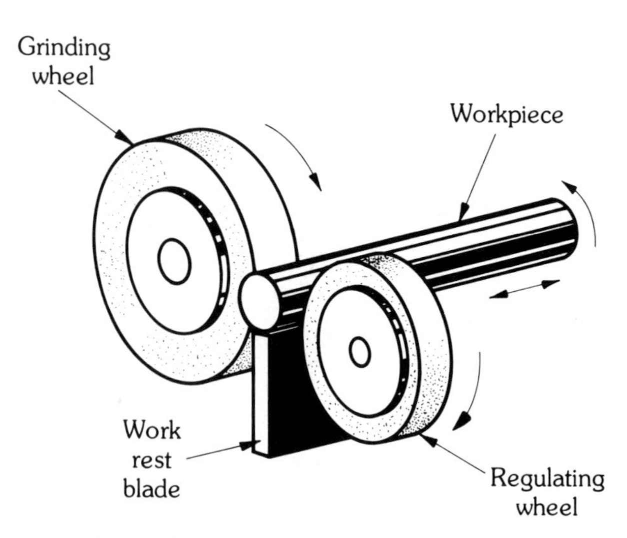

FIGURE 13.12 Operating principle of a centerless grinder.

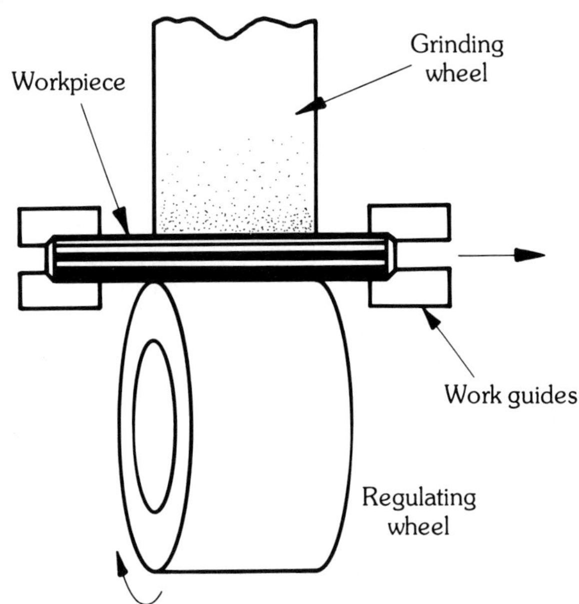

FIGURE 13.13 Through-feed centerless grinding.

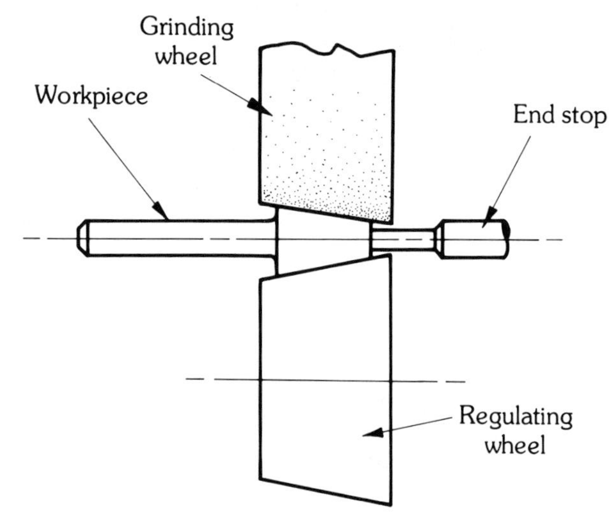

FIGURE 13.14 End-feed centerless grinding.

to each other. The part is fed in from the front and stops against a fixed, end stop. In this method, the grinding wheel or regulating wheel, or both, are dressed to the required taper.

3. *In-feed centerless grinding* is used when the part has a shoulder, head, or multiple diameters. It is similar to the plunge grinding method used on cylindrical grinding machines (Fig. 13.15). In this method there is no axial movement of the work and the length of the ground surface is generally limited to the width of the grinding wheel. When the ground length is longer than the width of the wheels, one end of the part is supported by the workrest blade. The other end rests on an outboard roller support.

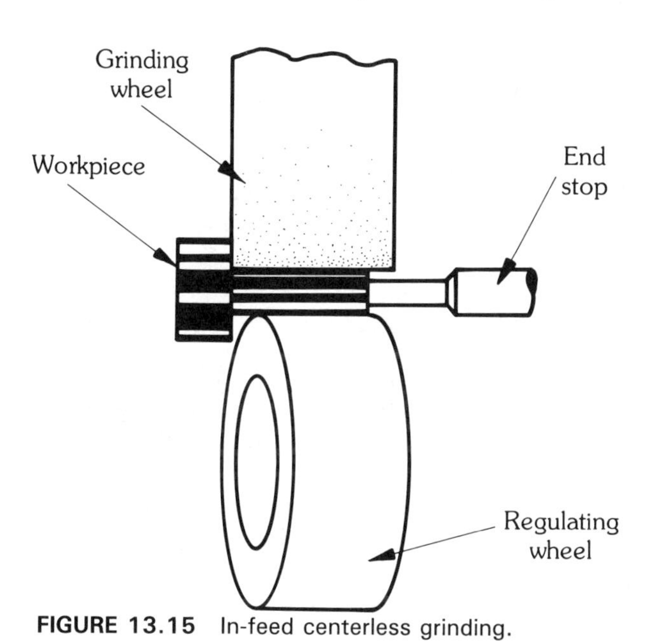

FIGURE 13.15 In-feed centerless grinding.

13.2.4 Special Types of Grinding Machines

Special types of grinders are grinding machines made for specific types of work and operations. A brief description of the more commonly used special types follows:

Internal Grinders. Internal grinders are used to accurately finish straight, tapered, or formed holes. The most popular internal grinder is similar in operation to a boring operation in a lathe. The workpiece is held by a work-holding device, usually a chuck, and revolved by a motorized headstock. The grinding wheel is revolved by a separate motor head in the same direction as the workpiece. It can be fed in and out of the work and also adjusted for depth of cut.

Crankshaft Grinders. Crankshaft or crankpin grinding machines are similar to cylindrical centertype grinders. They are used for grinding the crankpins and main bearing journals of automobile, aircraft, diesel, and other types of engines. Figure 13.16 illustrates the method used in crankpin grinding.

Tool and Cutter Grinding Machines. These grinding machines (Fig. 13.17) are designed to sharpen milling cutters, reamers, taps, and other machine tool cutters. The general-purpose tool and cutter grinder is the most popular and versatile tool grinding machine. Various attachments are

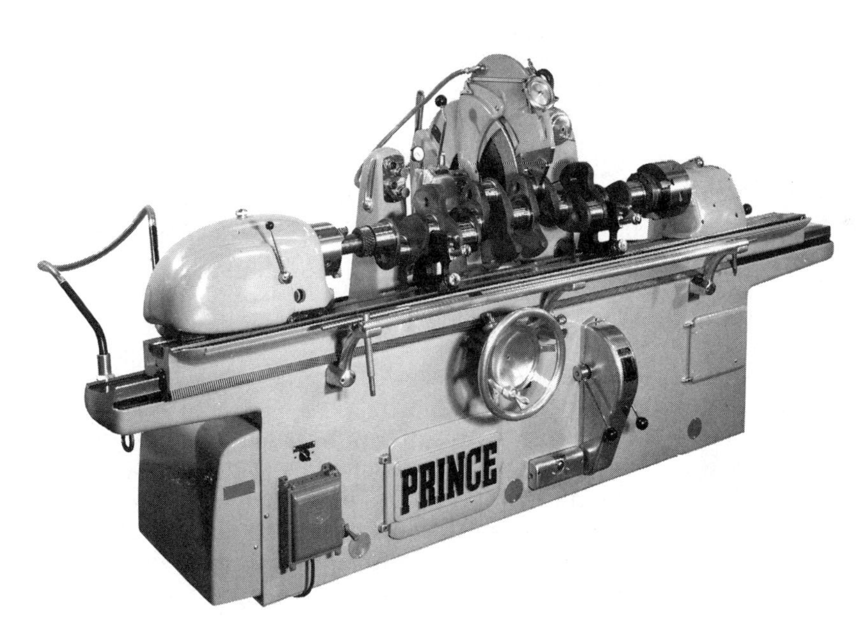

FIGURE 13.16 Crankshaft grinding machine. *(Courtesy Stoffel Grinding Systems)*

FIGURE 13.17 Tool and cutter grinding machine. *(Courtesy Gallmeyer & Livingston Co.)*

FIGURE 13.18 Jig grinding machine. *(Courtesy Moore Special Tool Co., Inc.)*

available for sharpening most types of cutting tools. It can also be used as a light-duty universal cylindrical grinder. Tool and cutter grinding operations and procedures are discussed in further detail later in this chapter.

Jig Grinding Machines. Jig grinders were developed to locate and accurately grind tapered or straight holes. Jig grinders are equipped with a high-speed vertical spindle for holding and driving the grinding wheel (Fig. 13.18). They utilize the same precision locating systems as do jig borers (see Chap. 10).

Thread Grinding Machines. These are special grinders that resemble the cylindrical grinder. They must have a precision lead screw to produce the correct pitch, or lead, on a threaded part. Thread grinding machines also have a means of dressing or truing the cutting periphery of the grinding wheel so that it will produce a precise thread form on the part. Thread grinding machines are discussed in greater detail in Chap. 6.

13.3 GRINDING WHEELS AND ABRASIVES

Grinding machines remove metal by means of a grinding wheel. Grinding wheels are composed of thousands of small abrasive grains held together by a bonding material. Each abrasive grain is a cutting edge. As the grain passes over the workpiece it cuts a small chip, leaving a smooth, accurate surface. As each abrasive grain becomes dull, it breaks away from the bonding material because of machining forces and exposes new, sharp grains.

There are five distinct elements to be considered when selecting a suitable grinding wheel for a specific application:

1. Type of *abrasive*
2. *Size* of the abrasive *grain*
3. *Grade* or hardness of the wheel
4. Structure or arrangement of the grains
5. *Bond:* the material holding the abrasive particles together

13.3.1 Types of Abrasives

Two types of abrasives are used in grinding wheels: *natural* and *manufactured.* Except for *diamonds,* manufactured abrasives have almost totally replaced natural abrasive materials. Even natural diamonds have been replaced in some cases by synthetic diamonds.

The manufactured abrasives most commonly

used in grinding wheels are aluminum oxide, silicon carbide, cubic boron nitride, and diamond.

Aluminum Oxide. Aluminum oxide is made by refining bauxite ore in an electric furnace. The bauxite ore is first heated to eliminate any moisture, then mixed with coke and iron to form a furnace charge. The mixture is then fused and cooled. The fused mixture resembles a rocklike mass. It is washed, crushed, and screened to separate the various grain sizes.

Aluminum oxide wheels are manufactured with abrasives of different degrees of purity (or different chemistries) to give them certain characteristics for different grinding operations and applications. The color and toughness of the wheel are influenced by the degree of purity.

General-purpose aluminum oxide wheels, usually gray and 95 percent pure, are the most popular abrasives used. They are used for grinding most steels and other ferrous alloys. White aluminum oxide wheels are nearly pure and are very *friable* (able to break away from the bonding material easily). They are used for grinding high-strength, heat-sensitive steels.

Silicon Carbide. Silicon carbide grinding wheels are made by mixing pure white quartz (sand), petroleum coke, and small amounts of sawdust and salt and firing the mixture in an electric furnace. This process is called *synthesizing* (combining) the coke and sand. As in the making of aluminum oxide abrasive, the resulting crystalline mass is crushed and graded by particle size.

Silicon carbide wheels are harder and more brittle than aluminum oxide wheels. There are two principal types of silicon carbide wheels: *black* and *green*. Black wheels are used for grinding cast irons; nonferrous metals like copper, brass, aluminum, and magnesium; and nonmetallics such as ceramics and gem stones. Green silicon carbide wheels are more friable than the black wheels and used for tool and cutter grinding of cemented carbide.

Cubic Boron Nitride. Cubic boron nitride (CBN) is an extremely hard, sharp, and cool cutting abrasive. It is one of the newest manufactured abrasives and two-and-a half times harder than aluminum oxide. It can withstand temperatures up to 2500°F (1370°C). CBN is produced by high-temperature, high-pressure processes similar to those used to produce manufactured diamond and is nearly as hard as diamond.

CBN is used for grinding superhard high-speed steels, tool- and die-steels, hardened cast irons, and stainless steels. Two types of cubic boron nitride wheels are used in industry today. One type is metal-coated to promote good bond adhesion and used in general-purpose grinding. The second type is an uncoated abrasive for use in electroplated metal and vitrified bond systems.

Diamond. Two types of diamond are used in the production of grinding wheels: *natural* and *manufactured*. Natural diamond is a crystalline form of carbon and very expensive. In the form of bonded wheels, natural diamonds are used for grinding very hard materials such as cemented carbides, marble, granite, and stone.

Recent developments in the production of manufactured diamonds have brought their cost down and led to expanded use in grinding applications. Manufactured diamonds are now used for grinding tough and very hard steels, cemented carbide, and aluminum oxide cutting tools.

13.3.2 Abrasive Grain Size

The *size* of an abrasive grain is important because it influences stock-removal rate, chip clearance in the wheel, and the surface finish obtained.

Abrasive grain size is determined by the size of the screen opening through which the abrasive grits pass. The number of the nominal size indicates the number of the openings per inch in the screen. For example, a 60-grit-sized grain will pass through a screen with 55 openings per inch, but it will not pass through a screen size of 65. A low grain size number indicates a large grit, and a high number indicates a small grain.

Grain sizes vary from 6 (very coarse) to 1000 (very fine). Grain sizes are broadly defined as *coarse* (6 to 24), *medium* (30 to 60), *fine* (70 to 180), and *very fine* (220 to 1000). Figure 13.19 shows a comparison of three different grain sizes and the screens used for sizing. Very fine grits are used for polishing and lapping operations, fine grains for fine-finish and small-diameter grinding operations. Medium grain sizes are used in high stock-removal operations where some control of surface finish is required. Coarse grain sizes are used for billet conditioning and snagging operations in steel mills and foundries, where stock-removal rates are important and there is little concern about surface finish.

13.3.3 Grade

The *grade* of a grinding wheel is a measure of the strength of the bonding material holding the individual grains in the wheel. It is used to indicate the relative *hardness* of a grinding wheel. Grade

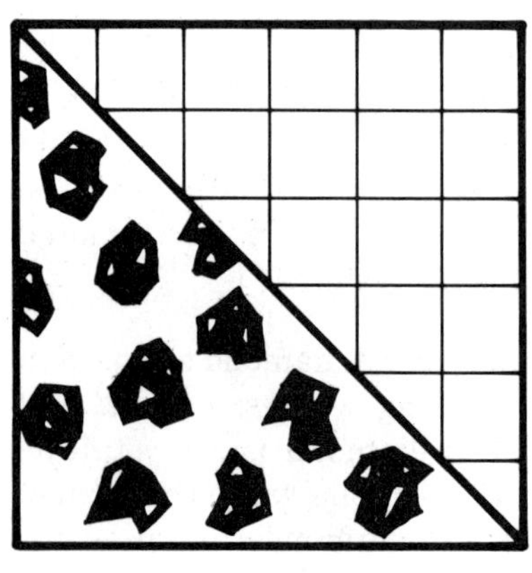
6–grain size

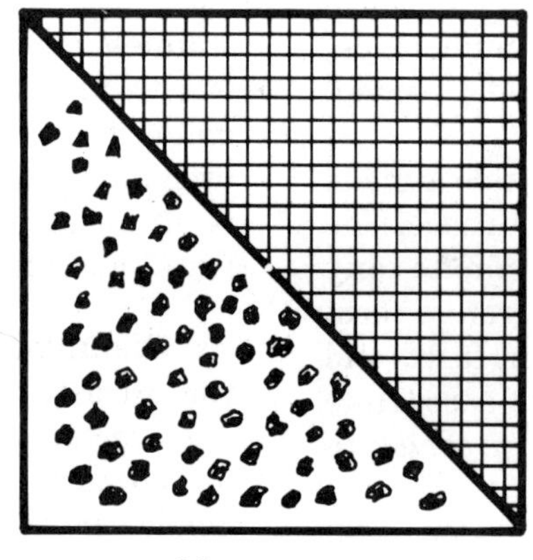
24–grain size

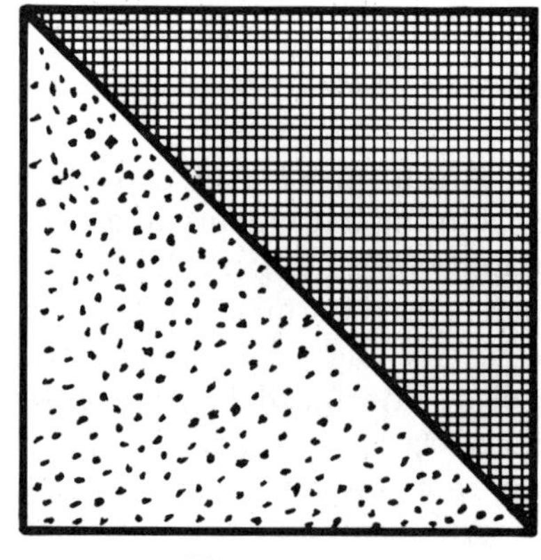
48–grain size

FIGURE 13.19 Comparison of three different grain sizes.

or hardness refers to the amount of bonding material used in the wheel, *not* to the hardness of the abrasive. A soft wheel has less bonding material than a hard wheel.

The range used to indicate grade is A to Z, with A representing maximum softness and Z maximum hardness. The selection of the proper grade of wheel is very important. Wheels that are too soft tend to release grains too rapidly and wheel wear is great. Wheels that are too hard do not release the abrasive grains fast enough and the dull grains remain bonded to the wheels causing a condition known as *glazing*. The grade selected for a grinding operation depends on four factors:

1. *Type of material.* Softer-grade wheels are generally used on hard material, harder-grade wheels on soft materials.
2. *Rigidity of machine.* Softer wheels are used on rigidly constructed machines. Less rigid machines that tend to vibrate require harder wheels.
3. *Wheel contact.* Harder wheels are used when the area of wheel contact on the work is small. When the area of contact is large, softer wheels are required.
4. *Nature of the workpiece.* Harder wheels are required for grinding forms or plunge grinding, to hold the contour. Harder wheels are also used when removing large amounts of materials. Softer wheels are generally used when continuous wheel wear is necessary to maintain sharp cutting edges and prevent damage to heat-sensitive materials.

The factors just described for proper wheel grade selection are general guidelines; they are not hard and fast rules to be followed in all applications.

13.3.4 Structure

The *structure* of a grinding wheel refers to the relative spacing of the abrasive grains; it is the wheel's density. There are fewer abrasive grains in an open-structure wheel than in a close-structure wheel. Figure 13.20 shows a comparison of different structures used in a grinding wheel.

Two general guidelines to use when selecting a structure for a grinding wheel are:

1. Closely spaced grains give finer finishes and hold form better in plunge or form grinding but generally do not release grains as readily. Close-grain structures tend to load up more easily because there is little room for chip clearance.
2. Open-structure wheels are generally used on larger wheels for heavy cuts and ductile materials.

The structure of a wheel is designated by a number from 1 to 15. The higher the number, the more open the structure; the lower the number, the more dense.

13.3.5 Bond

Abrasive grains are held together in a grinding wheel by a bonding material. The bonding material does not cut during a grinding operation; its main function is to hold the grains together with varying degrees of strength (grade). Standard grinding wheel bonds are vitrified, silicate, shellac, resinoid, rubber, and metal.

Vitrified Bond. Vitrified bonds are used on more than 75 percent of all grinding wheels. Vitrified bond material is comprised of finely ground clay and fluxes with which the abrasive is thoroughly mixed. The mixture of bonding agent and

An open-structure grinding wheel

A medium-structure grinding wheel

A dense- or closed-structure grinding wheel

FIGURE 13.20 Comparison of three different grain structures.

abrasive in the form of a wheel is then heated to 2400 °F (1315 °C) to fuse the materials.

Vitrified wheels are strong and rigid. They retain high strength at elevated temperatures and are practically unaffected by water, oils, or acids. One disadvantage of vitrified bond wheels is that they exhibit poor shock resistance. Therefore, their application is limited where impact and large temperature differentials occur. Newer types of vitrified bonded wheels are commonly used for speeds up to 12,000 surface feet per minute (sfpm) (3657 m/min). Older wheels having vitrified bonds were limited to 6500 sfpm (1981 m/min).

Silicate Bond. This bonding material is used when heat generated by grinding must be kept to a minimum. Silicate bonding material releases the abrasive grains more readily than other types of bonding agents. Speed is limited to below 4500 sfpm (1372 m/min).

Shellac Bond. Shellac is an organic bond used for grinding wheels that produce very smooth finishes on parts such as rolls, cutlery, camshafts, and crankpins. They are not generally used on heavy-duty grinding operations.

Resinoid Bond. Resinoid-bonded grinding wheels are second in popularity to vitrified wheels. Phenolic resin in powdered or liquid form is mixed with the abrasive grains in a form and cured at about 360 °F (197 °C). Resinoid wheels are used for grinding speeds up to 16,500 sfpm (5029 m/min). Their main use is in rough grinding and cutoff operations. Care must be taken with resinoid-bonded wheels since they will soften if they are exposed to water for extended periods of time.

Rubber Bond. Rubber-bonded wheels are extremely tough and strong. Their principal uses are as thin cutoff wheels and driving wheels in centerless grinding machines. They are also used when extremely fine finishes are required on bearing surfaces.

Metal Bond. Metal bonds are used primarily as bonding agents for diamond abrasives. They are also used in electrolytic grinding, where the bond must be electrically conductive.

13.3.6 Standard Grinding Wheel Marking

A standard method of marking grinding wheels has been adopted by grinding wheel manufacturers. This marking system is used to describe the wheel composition as to type of abrasive, grain size, grade, structure, and bond type. Figure 13.21 illustrates this standard marking system.

Guide to Markings

Position 1: type of abrasive. The type of abrasive is signified by a code as follows:

52 – A – 60 – H – 10 – V – 28

52 – manufacturer's symbol to identify the exact type of abriasive

A – type of abrasive

60 – grain size

H – grade

10 – structure

V – bond

28 – manufacturer's private marking to identify wheel

FIGURE 13.21 Standard bonded abrasive wheel marking system for aluminum oxide and silicon carbide wheels.

A—aluminum oxide
C—silicon carbide
BN—cubic boron nitride
D—natural diamond
MD—manufactured diamond

The type of abrasive may or may not be preceded by the manufacturer's prefix to exactly identify kind or class of a particular type of abrasive.

Position 2: grain size. A number is used to indicate the grit size of the abrasive. Most grinding wheels use grain sizes from 8 to 600.

Position 3: grade. Position 3 indicates the grade or hardness of the wheel. Letters are used in sequence from A to Z, with Z indicating maximum hardness.

Position 4: structure. The structure of a wheel is indicated by a number from 1 to 15, with 15 signifying an open structure with the widest grain spacing.

Position 5: bond type. The fifth position of the marking system is used to indicate the following types of bonding materials:

B—resinoid
BF—resinoid reinforced
E—shellac
R—rubber
RF—rubber reinforced
S—silicate
V—vitrified

Sometimes, manufacturers use additional symbols in this position to indicate a special variation of a particular bond type. There is no standardization in the use of these extra symbols.

Position 6: manufacturer's record. The last position is reserved for a manufacturer's record mark and other wheel characteristics.

13.3.7 Standard Grinding Wheel Shapes and Faces

Nine standard wheel *shapes* and 12 standard wheel *faces* have been adopted for general use by most grinding wheel manufacturers. Figure 13.22 shows the most common standard wheel shapes used on all types of grinding machines. Figure 13.23 illustrates the standard wheel faces used on most grinding wheel shapes.

Special grinding wheel shapes are used in rare or highly specialized grinding operations, but only the nine standard shapes are described here.

Straight Wheels. Types 1, 5, and 7 straight grinding wheels are the most frequently used for the majority of grinding operations. They are used for surface grinding, cylindrical and internal grinding operations, off-hand grinding, and tool and cutter grinding. Note that type 1 has no recessed sides, type 5 has one recessed side, and type 7 has two recessed sides. Recessed sides provide clearance for mounting flanges on certain machines.

Cylinder Wheels. Type 2 wheels are used on horizontal and vertical surface grinders. They are

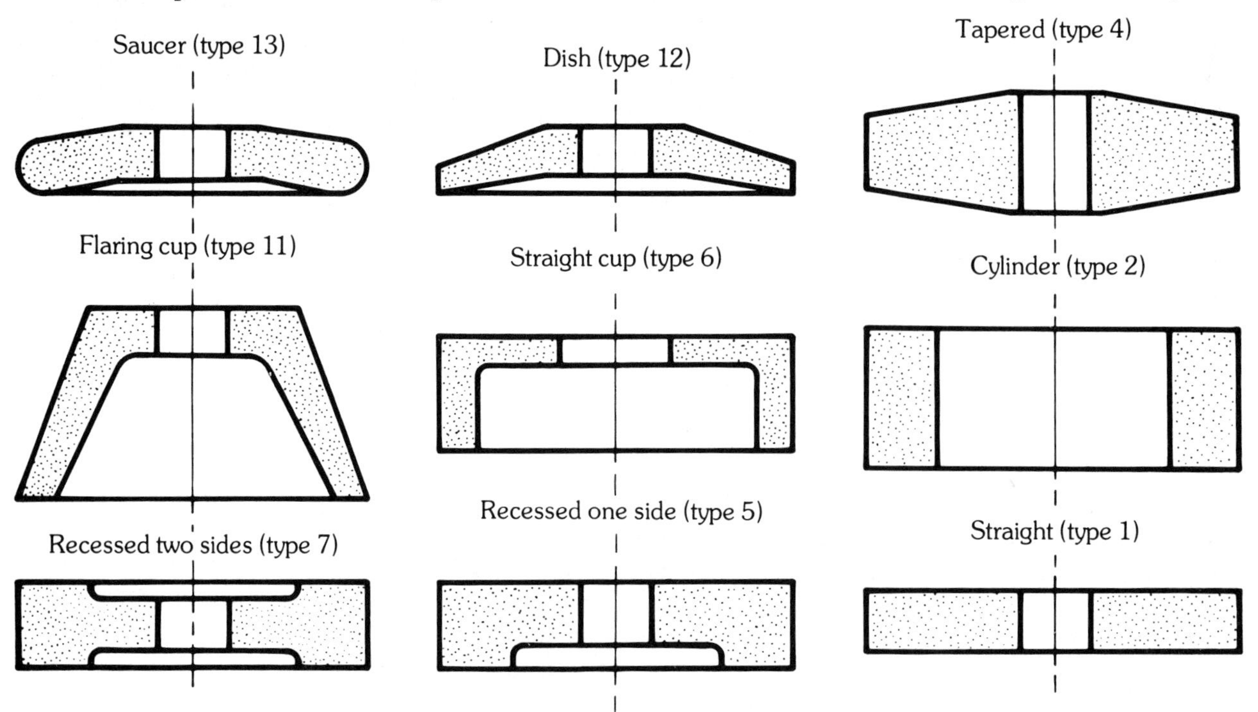

FIGURE 13.22 Common grinding wheel shapes used on all types of grinding machines.

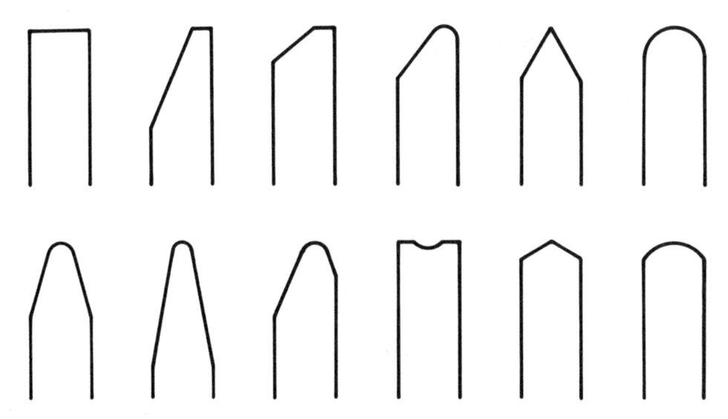

FIGURE 13.23 Common grinding wheel faces used on most grinding wheels.

designed to grind on either the face or periphery of the wheel.

Tapered Wheels. Type 4 is similar to type 1 except that both sides of the wheel are tapered to prevent the wheel from breaking. Type 4 wheels are used mainly for snagging operations where heavy pressures are applied.

Straight Cup Wheels. Type 6 wheels are used primarily for vertical and horizontal surface grinding machines. They are also used for some tool and cutter grinding operations.

Flaring Cup Wheels. Type 11 is a tool and cutter grinding wheel for sharpening milling cutters, reamers, form cutters, and other cutting tools.

Dish Wheels. Type 12 is also a tool and cutter grinding wheel used for most cutter sharpening operations. The thin edge at the periphery allows the edge of the wheel to enter thin and narrow places such as the flutes of form milling cutters.

Saucer Wheels. Type 13 wheels are used for a variety of tool and cutter grinding and some light snagging operations.

Mounted Wheels. Illustrated in Fig. 13.24 are small wheels of different sizes and shapes. These grinding wheels are mounted on steel shafts that allow them to be used in portable drill motors and hand grinders to deburr or break sharp edges on workpieces. Mounted wheels are also used in some internal grinding machines (Fig. 13.25) and jig grinding operations. They are also used in offhand grinding for difficult corners and small areas such as shaping and finishing die cavities.

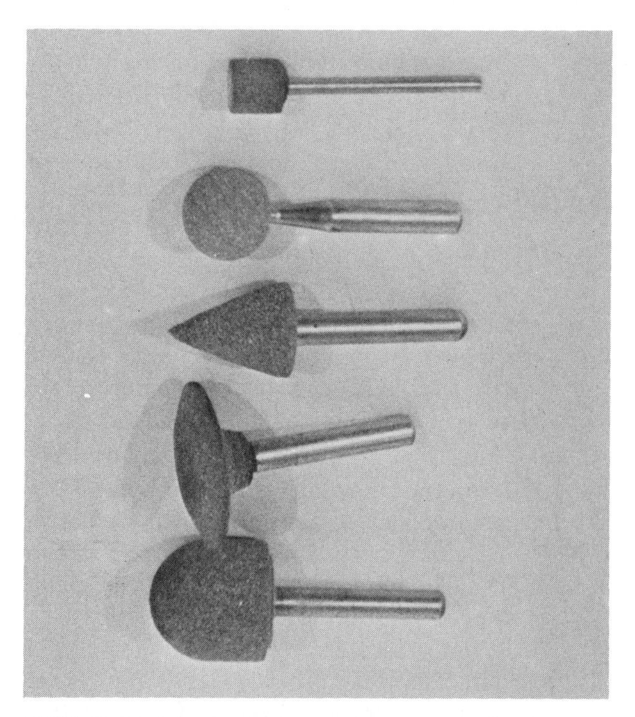

FIGURE 13.24 Assortment of mounted wheels.

FIGURE 13.25 Small mounted wheel being used for an internal grinding operation.

13.3.8 Grinding Wheel Selection

Before attempting to select a grinding wheel for a particular operation, the operator should consider the following six factors for *maximum* and *safe* results:

1. *Material to be ground.* If the material to be ground is carbon steel or alloy steel, *aluminum oxide* wheels are usually selected. Extremely hard steels and exotic alloys should be ground with *cubic boron nitride* or *diamond.* Nonferrous metals, most cast irons, nonmetallics, and cemented carbides require a *silicon carbide* wheel. A general rule on grain size is to use a *fine*-grain wheel for hard materials, and a *coarse*-grain wheel for soft and ductile materials. *Close*-grain spacing and *soft* wheels should be used on harder materials,

while *open* structures and *harder* wheels are preferable on soft materials.

2. *Nature of the grinding operation.* Finish required, accuracy, and amount of material to be removed must be considered when selecting a wheel. Fine and accurate finishes are best obtained with *small* grain size and grinding wheels with *resinoid, rubber,* or *shellac* bonds. Heavy metal removal is obtained with coarse wheels with *vitrified* bonds. Another factor to consider is the shape of the wheel.
3. *Area of contact.* The area of contact between the wheel and workpiece is also important. *Close*-grain spacing, *hard* wheels, and *small* grain sizes are used when the area of contact is small. Conversely, *open* structures, *softer* wheels, and *larger* grain sizes are recommended when the area of contact is large.
4. *Condition of the machine. Vibration* influences the finish obtained on the part as well as wheel performance. Vibration is generally due to loose or worn spindle bearings, worn parts, out-of-balance wheels, or insecure foundations. *Harder* wheels perform best when vibration is present.

   A second factor is the available *horsepower.* Heavy metal removal requires high horsepower. A general rule is that as more horsepower is available, *harder* wheels should be selected.

   A third factor relates to the *size* of the grinding machine. This may seem obvious, but is worth noting. Light-duty, toolroom grinders, for example, do not usually have adequate horsepower or rigid enough construction and frequently do not have coolant systems. *Softer* wheels should be selected for these machines. *Harder* wheels are used on heavy-duty production grinding machines, especially when coolant is used.
5. *Wheel speed.* Wheel speed affects the *bond* and *grade* selected for a given wheel. Wheel speeds are measured in surface feet per minute (sfpm) or meters per minute (m/min). Vitrified bonds are commonly used to 6,500 sfpm (1981 m/min) or in selected operations up to 12,000 sfpm (3657 m/min). Resinoid-bonded wheels may be used for speeds up to 16,500 sfpm (5029 m/min). Grinding wheels should never be operated at speeds higher than those for which they are rated. They may fracture and fly apart during operation, *causing severe injury to an operator and damage to the machine.*

   When wheel speed is higher, *softer*-grade wheels should be used. *Harder* wheels should be used at slower speeds because slower speeds tend to wear softer wheels more quickly.
6. *Grinding pressure.* Grinding pressure is the rate of in-feed used during a grinding operation; it affects the grade of wheel. A general rule to follow is that as grinding pressures increase, harder wheels must be used.

13.3.9 Storage, Inspection, Mounting, and Balancing Wheels

Wheel Storage. All grinding wheels should be properly stored in suitable shelves or bins. Wheel storage rooms should be dry and not subject to extreme temperature change. Wheels with organic bonds may become seriously damaged if subject to dampness. The following four guidelines will help ensure proper storage of grinding wheels:

1. Straight and tapered grinding wheels are best stored on edge. Make sure that suitable cradles have been constructed to prevent the wheels from rolling.
2. Cylinder and large straight wheels may be stacked on their sides provided corrugated cardboard or other cushioning material is placed between them.
3. Small cup, saucer, mounted wheels, and other shaped wheels should be stored in boxes, bins, or drawers.
4. Thin cutoff wheels may also be stacked provided suitable cushions are placed between each wheel.

Inspection of Grinding Wheels. All grinding wheels are breakable, and some are extremely fragile. Great care should be taken in handling grinding wheels. New wheels should be closely inspected immediately after receipt to make sure they were not damaged during transit. Grinding wheels should also be inspected prior to being mounted on a machine.

To test for damage, suspend the wheel with a finger and gently tap the side with a screwdriver handle for small wheels and a *wooden* mallet for larger wheels (Fig. 13.26). An undamaged wheel will produce a clear ringing sound; a cracked wheel

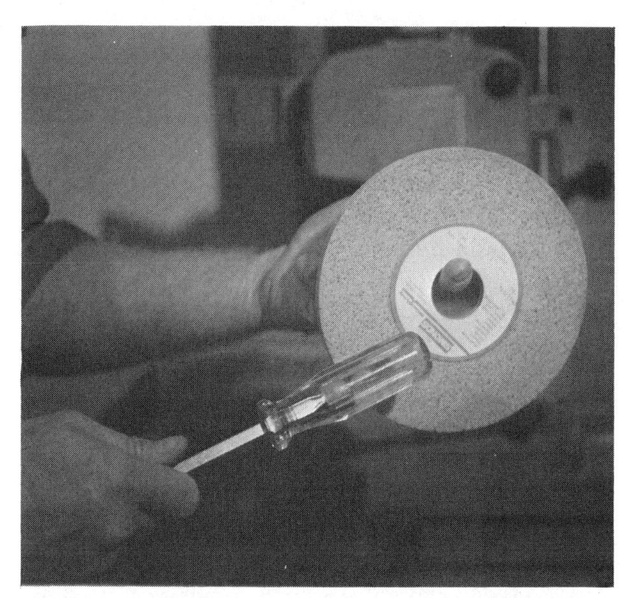

FIGURE 13.26 Testing a grinding wheel for internal cracks. (*Courtesy Orange Coast College*)

will not ring at all. Vitrified and silicate bonded wheels give a more distinct ring than organic bonded wheels.

Mounting Grinding Wheels. All wheels should be closely inspected before mounting:

1. Make sure that the bond of the grinding wheel is appropriate for the spindle speed of the grinding machine. Remember, operating a wheel at a speed faster than is recommended by the manufacturer can cause it to break.
2. Blotters or suitable compressible material must be used on each side of the grinding wheel. They should be larger in diameter than the flange washers used. This allows uniform distribution of the flange pressure (Fig. 13.27).
3. Grinding wheels should fit freely on the spindle of the machines. They should never be forced on the spindle because this may cause the wheel to fracture. The spindle must be clean and free of burrs before mounting the wheel.
4. Any bushings necessary to correct the hole size of the wheel should not extend beyond the sides of the wheel because they may interfere with the flange washers.
5. Make sure that the flange facings and any washers used are clean and free of burrs.
6. Care should be taken not to overtighten the spindle nut or the wheel may be damaged. Tighten the nut just enough to hold the wheel firmly.

Balancing Wheels. It is important to balance wheels over 10 in. (254 mm) before they are mounted on a machine. The larger the grinding wheel, the more critical balancing becomes. Grinding wheel balance also becomes more critical as speed is increased. Out-of-balance wheels cause excessive vibration, produce faster wheel wear, chatter, poor finishes, damage to spindle bearings, and can be dangerous.

The proper procedure for balancing wheels is to first statically balance the wheel. Next, mount the wheel on the grinding machine and dress. Then remove the wheel and rebalance it. Remount the wheel and dress slightly a second time.

Balancing wheels is done by shifting weights on the wheel mount (Fig. 13.28). The wheel is installed on a balancing arbor and placed on a balancing fixture (Fig. 13.29). The weights are then shifted in a position to remove all heavy points on the wheel assembly.

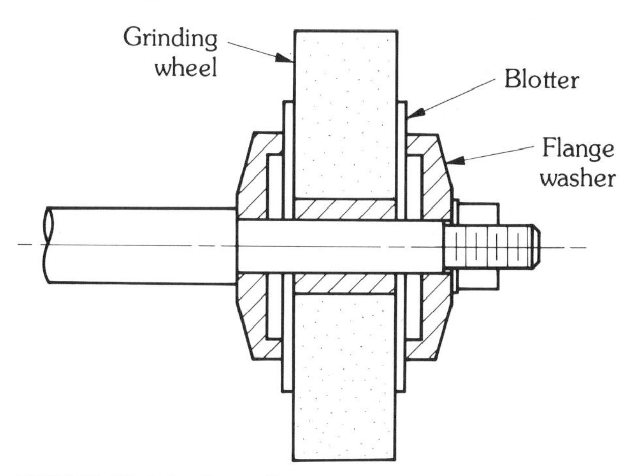

FIGURE 13.27 Properly mounted grinding wheel using blotters and recessed flanges.

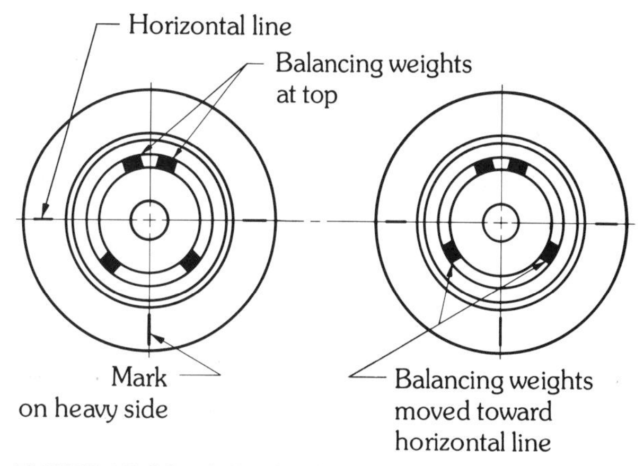

FIGURE 13.28 Adjusting balancing weights to balance a grinding wheel.

FIGURE 13.29 Grinding wheel balancing stand. (*Courtesy Brown and Sharpe Grinding Machines, Inc.*)

Today, many production grinders are equipped to balance the wheel while it is running on the machine.

13.4 SURFACE GRINDING OPERATIONS

The most frequent grinding operation performed is surface grinding. The horizontal spindle, reciprocating table surface grinding machine is the most commonly used and is discussed in detail.

13.4.1 Surface Grinder Construction

Before learning the various operations that can be performed on a surface grinder, it is important to become acquainted with the construction features and operating characteristics. Figure 13.30 illustrates the major components of a typical hydraulic surface grinder

Base. The base is generally a heavy iron casting. At the back of the base is a column supporting the wheelhead. On top of the base are machined ways where the saddle mounts. On hydraulic models the reservoir and pump are usually located inside the base.

Saddle. The saddle is mounted on the ways of the base and provides the cross-feed motion of the table, either manually or hydraulically.

Table. The table is fitted to the saddle and can be moved manually or hydraulically. The table moves to the left or right of the operator when he is facing the machine (traverse). It has T slots to accommodate mounting of vises and other work-holding devices. A magnetic chuck is used most often to hold parts.

Wheelhead. The wheelhead is mounted on the column of the machine. A handwheel with a micrometer collar moves the wheelhead up and down on accurately machined ways. This movement controls the depth of cut. The spindle on which the grinding wheel is mounted is located in the wheelhead in the horizontal position. Perhaps the most important elements in any grinding machine are the *spindle bearings*. Various types of spindle bearings are used for grinding wheel spindles. Regardless of the type of bearing, it is important that they are properly adjusted and exhibit minimum wear. Worn bearings permit the spindle and grinding wheel to vibrate, causing poor grinding results.

Coolant Systems. A cutting fluid, usually referred to as a *coolant*, is used in most grinding operations. Its main function is to keep temperatures generated by the grinding process to a minimum. Grinding coolants also provide a lubrication film between the work and wheel, control the grinding dust emitted, and remove the residue of abrasive grains and metal particles (called swarf) from the grinding area. Most grinding coolants are soluble oil and water or nonoily chemicals (see Chap. 15). The coolant is directed to the wheel and work by a pump from a reservoir (Fig. 13.31).

Size Capacity. The size of a surface grinder is classified according to its work capacity. Work capacity is based on the size of the magnetic chuck, the table size, and the table travel length.

13.4.2 Work-Holding Devices and Accessories for Surface Grinders

Almost any work-holding device used on a milling machine or drill press can be used on surface grinders. Vises, rotary tables, index centers, and other fixtures are used for special setups. However, the most common work-holding device on surface grinders is the *magnetic chuck*.

Magnetic Chucks. Magnetic chucks hold the workpiece by exerting a magnetic attraction

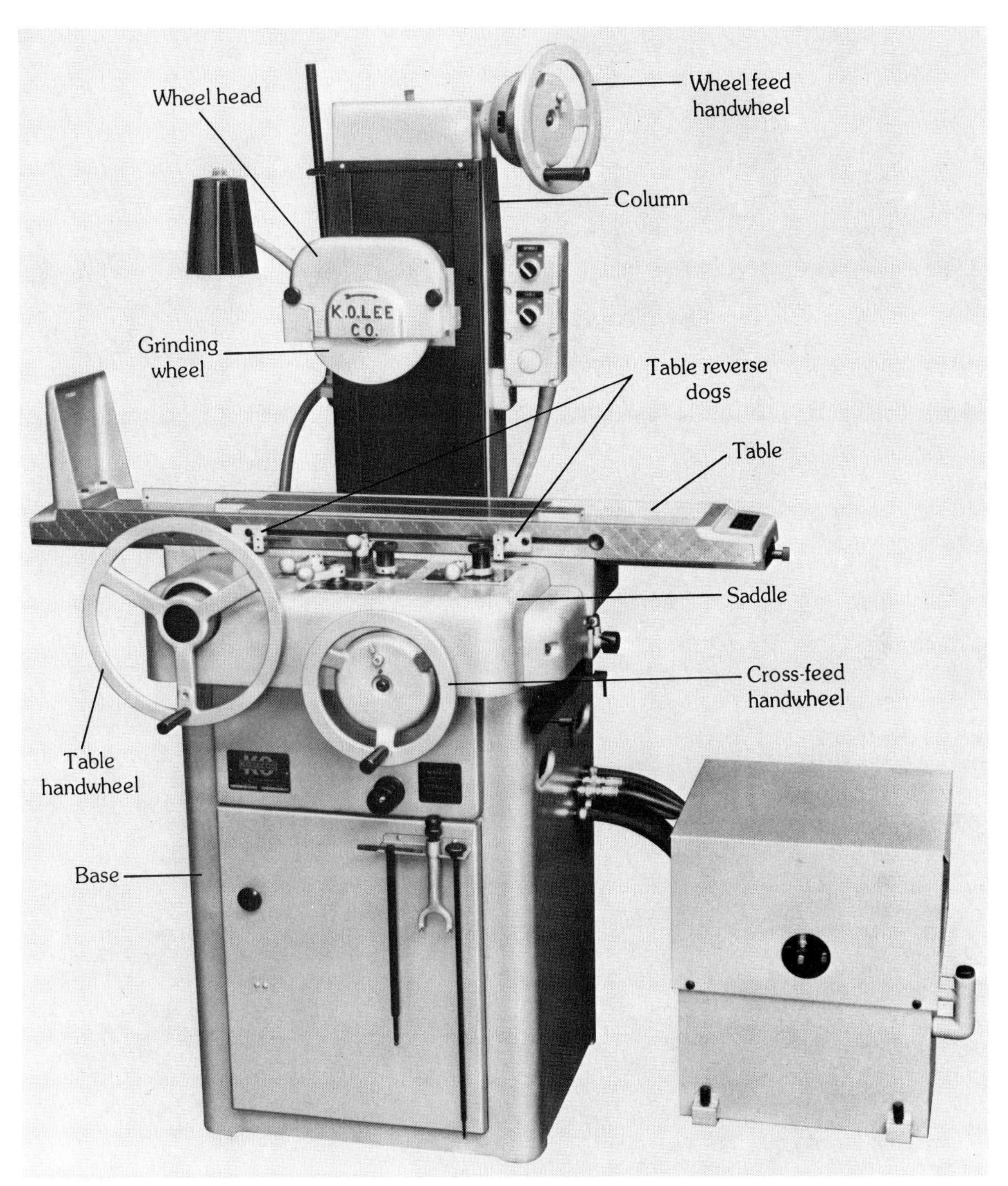

FIGURE 13.30 Principal parts of a modern hydraulic grinder. (*Courtesy K.O. Lee Co.*)

on the part (Fig. 13.32). Only magnetic materials such as iron and steels may be mounted directly on the chuck. Two types of magnetic chucks are available for surface grinders: the permanent magnet and the electromagnetic chucks.

On *permanent magnet* chucks, the holding power comes from permanent magnets. The work is placed onto the chuck and a hand lever is moved to energize the magnets. The *electromagnetic* chuck operates on 110 or 220 volts and is energized by

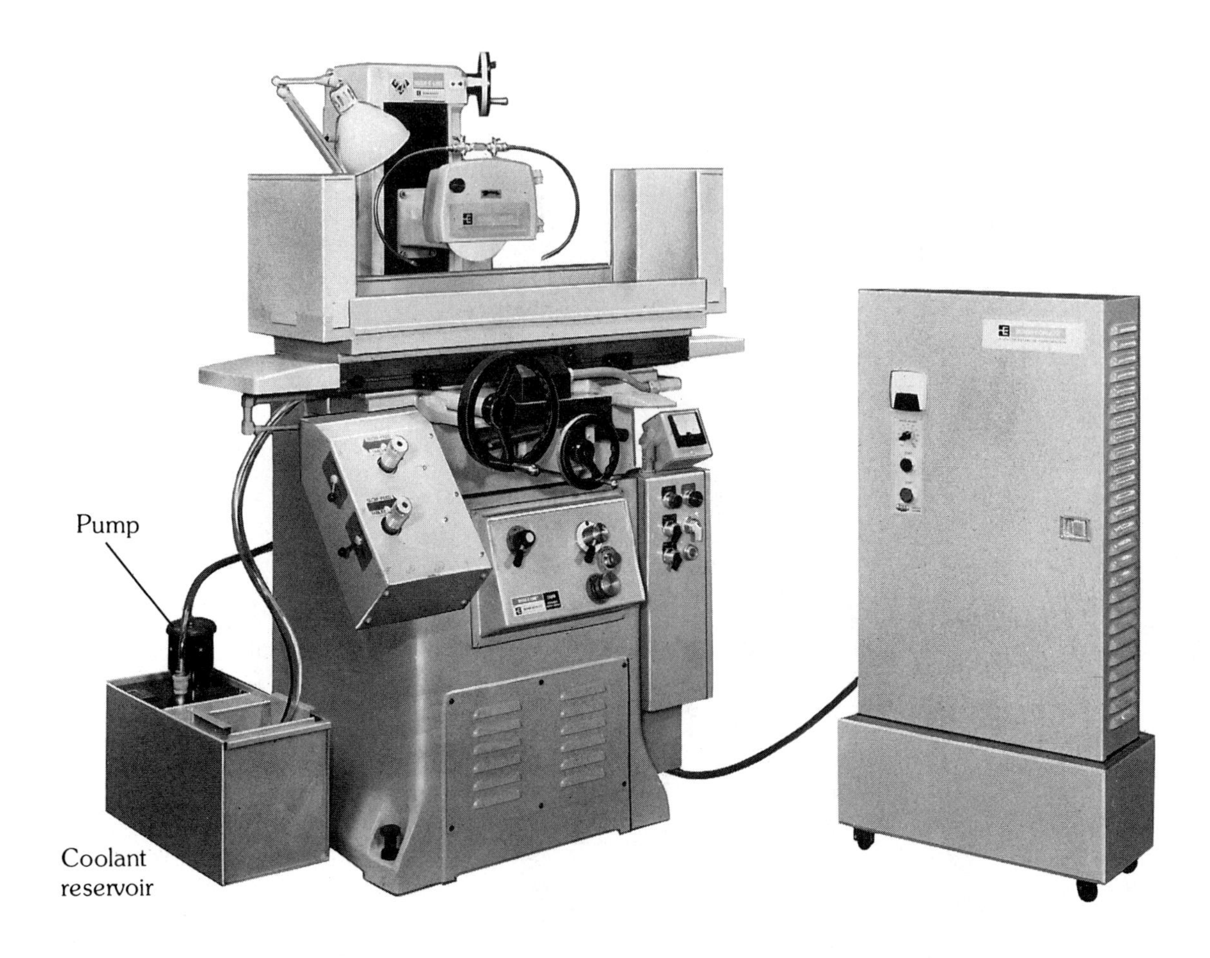

FIGURE 13.31 Grinding machine coolant reservoir and pump. (*Courtesy Boyar-Schultz Corp.*)

FIGURE 13.32 Permanent magnetic chuck used on reciprocating surface grinders.

a switch. This type of chuck has two advantages. First, the holding power may be adjusted to suit the area of contact of the workpiece; small amounts of current are used with smaller parts, large amounts with larger parts. A second advantage is the demagnetizer switch. It reverses the current flow momentarily and neutralizes the residual magnetism from the chuck and workpiece.

Two commonly used accessories are magnetic parallels (or blocks) and magnetic V blocks. These accessories provide an extension of the magnetism to the work. *Magnetic parallels* allow workpieces of different heights or projections to be held on magnetic chucks (Fig. 13.33). *Magnetic V blocks* allow round work or parts with 90° edges to be held (Fig. 13.34).

Angle Plates. Angle plates permit thin work to be held securely (Fig. 13.35). They are also used when grinding a completely square part. These procedures are discussed thoroughly later in the chapter.

13.4.3 Surface Grinding Safety

It is important that the operator learn the rules of safety before using a surface grinding machine.

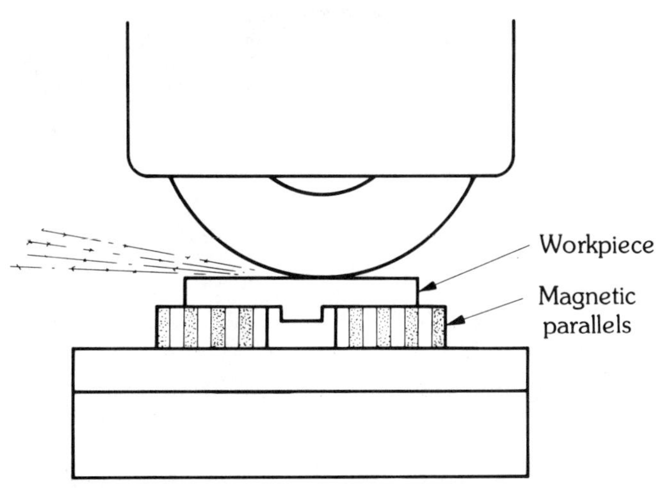

FIGURE 13.33 Magnetic parallels are used to support workpieces that have different heights.

FIGURE 13.34 Magnetic V block being used to grind a 45° angle. (*Courtesy Orange Coast College*)

FIGURE 13.35 An angle plate is used to support thin parts for surface grinding operations.

Grinding wheels are very fragile cutting tools. They cannot take rough use or they may break, causing serious injury to the operator.

1. *Always* wear safety goggles during any grinding operation.
2. *Always* inspect a grinding wheel prior to mounting it on a machine.
3. When first starting a grinder, stand aside in case the wheel is cracked or damaged.
4. Make sure that *all* wheel guards are in place and secure.
5. Make sure that the work is properly mounted and secure.
6. Make sure that the work clears the grinding wheel before starting the machine.
7. Check the grinding wheel to make sure that it is the correct type for the machine.
8. Never attempt to remove or install work until the wheel has completely stopped.
9. Do not force work against a wheel; to do so may cause it to break.
10. Never run coolant against a wheel that is not rotating.

13.4.4 Dressing and Truing the Wheel

Dressing a wheel refers to removing the glaze from a dull wheel, restoring it to its original shape, or removing loaded material from the wheel. Dressing the wheel exposes new cutting edges and is done as needed during the grinding operation.

Truing a wheel means removing enough abrasive material from the wheel face so the periphery is concentric with the spindle rotation and the wheel face is parallel to the table. A wheel must be true and sharp for best grinding results.

True and dress a wheel as follows:

1. Clean the magnetic chuck thoroughly with a rag. Check by running a hand over the chuck to see that there are no abrasive grains or material left on the surface.
2. Place a dressing diamond in a proper holder and tighten.
3. Set the diamond and holder on the chuck and make sure that the diamond is pointing in the same direction the wheel rotates (Fig. 13.36). (*Note:* The direction of wheel rotation is usually marked on the outer face of the wheel guard.)
4. Position the table so that the centerline of the wheel is slightly to the right of the diamond point.
5. Using the wheel feed crank handle, lower the wheel approximately 1/16 in. (1.5 mm) above the diamond point.

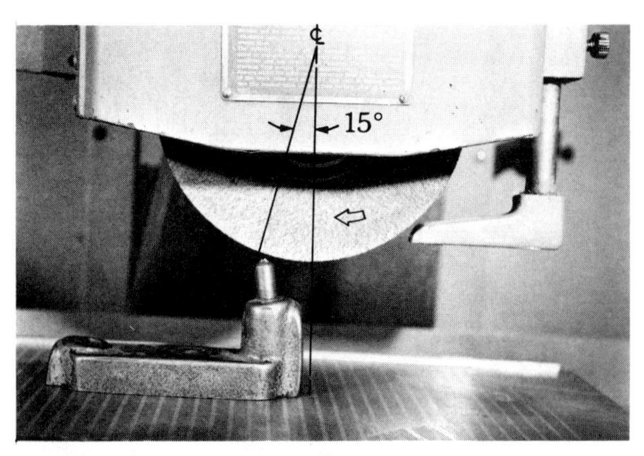

FIGURE 13.36 Correct position of the diamond for truing and dressing a grinding wheel. (*Courtesy Orange Coast College*)

6. Move the table in and out (laterally) until the diamond is under the high spot of the wheel face.
7. Start the grinding wheel and allow the spindle bearings to warm up (about 5 minutes).
8. Slowly lower the wheelhead until the diamond just touches the wheel.
9. Using the cross-feed handle, slowly feed the diamond across the face of the wheel.
10. Continue to feed the diamond across the face of the wheel taking 0.001 in. (0.025 mm) successive cuts off the face of the wheel.
11. Once the wheel has been trued or the loaded material removed, take several 0.0005-in. (0.01-mm) passes.

Hints on truing and dressing:

1. If the grinding operation to be performed will be done without coolant (dry), dress the wheel dry. If wet, dress the wheel wet.
2. If the wheel is being trued or dressed for finish operations, feed the diamond slowly. If heavy metal removal is the objective, use higher rates of feed during dressing. Rapid feed rates produce a coarser grain structure; slower feeds produce a finer grain structure.
3. The diamond is held in its holder at a 10 to 15° angle. It should be rotated periodically to ensure that the point of the diamond remains in a conical form. If held vertically, the diamond point will wear away quickly.

FIGURE 13.37 Surface grinding a part flat and parallel. (*Courtesy Gallmeyer & Livingston Co.*)

13.4.5 Grinding Flat and Parallel Surfaces

The most common operation performed on a surface grinder is grinding flat or parallel surfaces on a part (Fig. 13.37). Here is the procedure:

1. Make sure that the correct wheel is properly mounted on the spindle. Examine the wheel to see if it is sharp and not glazed or excessively loaded with material. If necessary, dress the wheel (see Sec. 13.4.4).
2. Thoroughly clean the face of the magnetic chuck and check for burrs or nicks. If burrs are present, carefully hone them away.
3. Remove any burrs from the work and place it on the center of the chuck.
4. Energize the magnetic force of the chuck to secure work.
5. On hydraulic surface grinders, set the table reversing dogs so that the wheel clears the end of the work approximately 1 in. (25.4 mm). Set the cross-feed reversing dogs so the wheel face extends beyond the edges of the workpiece.
6. Set the table speed to the fast position for longer work, slow position for shorter work.

7. Adjust the cross-feed rate from 0.020 to 0.100 in. (0.51 to 2.54 mm) for roughing operations and 0.005 to 0.015 in. (0.13 to 0.37 mm) for finishing cuts.
8. Using the table handwheel (usually the largest) and the cross-feed handwheel, position the work under the grinding wheel.
9. Next, lower the wheel to approximately 1/16 in. (1.5 mm) above the workpiece.
10. Start the machine and allow the spindle to warm up. (*Note:* On hydraulic grinders a separate switch is used to start the hydraulic pump.)
11. Slowly lower the wheel until it just touches the work (a few sparks will be generated). [*Caution:* The area of initial contact between the wheel and the work may or may not be the high spot of the part. Therefore, it is good practice to raise the wheel a few thousandths of an inch (0.09 to 0.1 mm) and slowly *hand feed* the entire surface of the work under the wheel.]
12. After the work has been checked for high spots, start the automatic table travel and cross-feed movements.
13. Turn on the coolant pump and adjust the flow valve so that a generous amount of coolant is applied to the work surface.
14. Lower the grinding wheel for successive cuts and continue to grind until the first surface has been completed. [*Note:* The depth of cut varies with the size of the machine and the amount of stock to be removed. As a general rule, roughing cuts are 0.002 to 0.005 in. (0.05 to 0.12 mm) and finish cuts between 0.0005 to 0.001 in. (0.012 to 0.025 mm).]
15. The shutdown procedure for surface grinding machines is, first, to stop the table and cross-feed movements, making sure that the wheel is located off the workpiece. Next, stop the coolant supply. Finally, shut down the wheel. Let large wheels run 3 to 5 minutes to thoroughly remove the coolant.
16. After the wheel has stopped *completely*, release the magnet, carefully remove workpiece from the chuck, and deburr any sharp edges.
17. To grind the second side of the part, repeat operational steps 2 through 16.

FIGURE 13.38 Using a depth mike to measure a part on a surface grinder. (*Courtesy Orange Coast College*)

[*Note:* If the workpiece requires an accurate thickness, it must be measured periodically while grinding the second side. Use a depth micrometer (Fig. 13.38), or remove the part and use an outside micrometer.]

13.4.6 Grinding a Part Square

Another operation frequently performed on the surface grinder is squaring a block of material. Squaring means that all six sides are square to one another. Following is the procedure for squaring a block of material:

1. Grind the two largest surface areas flat and parallel. This procedure is explained in detail in Sec. 13.4.5.
2. Mount one finished side to an angle plate, as shown in Fig. 13.39. Note the position of the parallel; it raises the workpiece higher than the top of the angle plate and gives an approximate parallelism between the edge to be ground and the chuck. Use C clamps to secure the work to the angle plate. Make sure they are in a position to clear the wheel and column of the machine.
3. After the work is properly mounted, turn on the magnetic chuck.
4. Adjust all reversing dogs and the table and cross-feed rates.
5. Position the work under the wheel, with the wheel approximately 1/16 in. (1.5 mm) above the top surface of the work.
6. Start the wheel. Slowly lower the wheelhead until the wheel just touches the work, then raise it slightly. As before, manually move the table and cross-feed

FIGURE 13.39 A precision angle plate is used when grinding a surface square to another. (*Courtesy Orange Coast College*)

controls and check for the high spot of the material.

7. Start the hydraulic table and cross-feed pump; turn on the coolant pump and adjust the flow.
8. Take successive cuts until the surface is completed. Side 3 is the first side to be ground square. Remove the part, and using a precision cylindrical square or other device, check the squareness of the side. This completes side 3.
9. Side 4 may be ground by placing side 3 (the side just completed) directly on paper and the magnetic chuck—if the area of contact is large enough to secure the part firmly. To check this, energize the magnet, grip the workpiece with your hands, and try to remove it. If the work can be removed, it is safer to mount it on the angle plate or surround the part with additional pieces of magnetic material (Fig. 13.40).
10. Grind side 4 to the required size using the same procedure as for side 3. Again, check for squareness and dimensional length.
11. With sides 1 to 4 complete, side 5 can now be ground. Side 5 will require two angle plates to mount the part correctly. Figure 13.41 shows the proper way to mount the part. The technique of using two angle plates ensures that side 5 will be square to the four surfaces finished previously.
12. After the part has been properly mounted, grind surface 5 using steps 3 through 8.

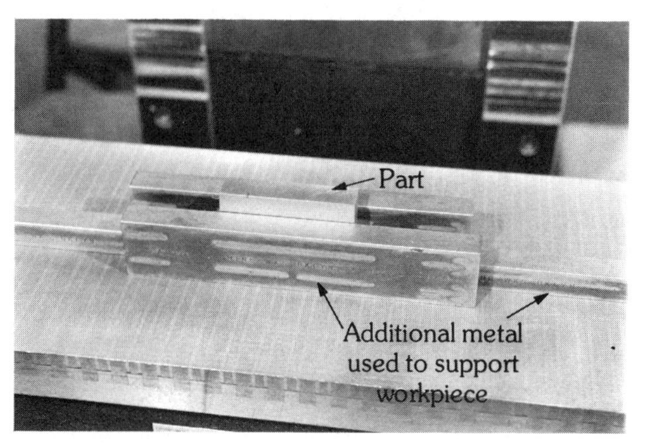

FIGURE 13.40 Additional pieces of metal are used to support parts that have smaller surface areas. (*Courtesy Orange Coast College*)

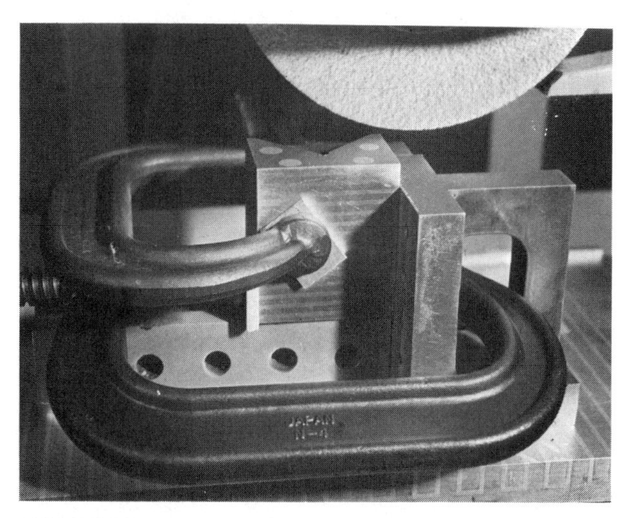

FIGURE 13.41 Two angle plates are used to grind a surface square to two other surfaces. (*Courtesy Orange Coast College*)

13. The procedure for grinding side 6 is similar to that for side 4. That is, it may be mounted directly to the magnetic chuck if the area of contact is large enough to hold the part firmly, or mounted using two angle plates as side 5. In any case, side 6 must be carefully ground to the finished size.

13.4.7 Grinding Angular Surfaces

It is sometimes necessary to grind a surface on a part at an angle. The procedures for grinding angular surfaces are essentially the same as for grinding flat surfaces. However, the part is generally mounted at an angle rather than flat on the magnetic chuck.

Figure 13.42 shows one method of mounting the work at an angle using an adjustable *universal vise.* This method is generally used when

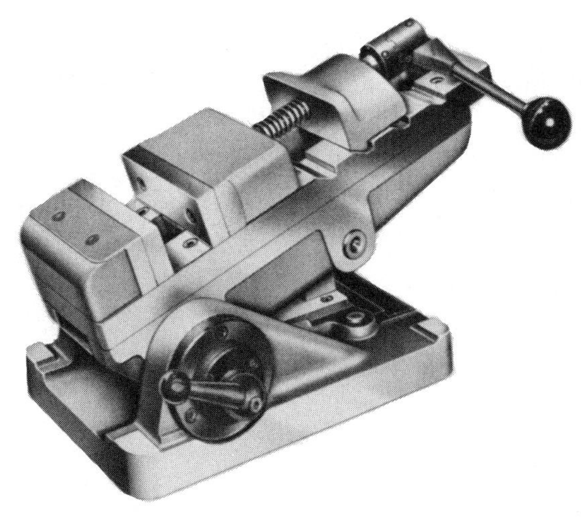

FIGURE 13.42 An adjustable angle sine vise is used to grind an angle on a part on the surface grinder. (*Courtesy Brown and Sharpe Grinding Machines, Inc.*)

FIGURE 13.43 An angle plate is often used when grinding angles on a part.

precise accuracy is not required. Another method is to mount the work against an angle plate (Fig. 13.43). With this method the part is tilted to the required angle by the use of the sine bar and gage blocks. This is a very accurate method. It is usually done on a surface plate first and then taken to the grinding machine for the machining operation.

Angular surfaces can also be ground by mounting the work directly on the magnetic chuck and using a wheel that has been dressed to the proper angle. Figure 13.44 shows how the wheel is dressed to the required angle. This technique is called *form grinding*.

13.4.8 Other Surface Grinding Operations

Form grinding. Form grinding is the process of producing angular or curved surfaces on a part. These special surfaces are produced by a grinding wheel dressed to the *reverse* of the shape

FIGURE 13.44 Grinding wheel dressed to a required angle is used to grind a part.

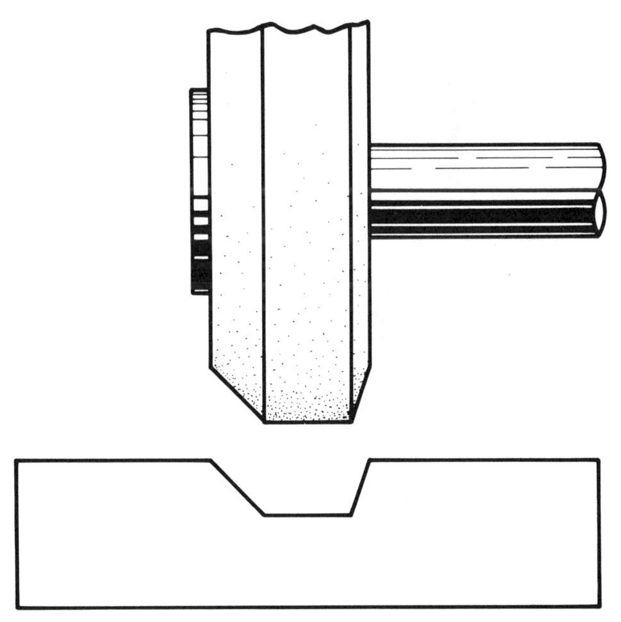

FIGURE 13.45 Contoured part and dressed grinding wheel.

of the finished surface. Figure 13.45 shows a typical contour part and properly dressed wheel.

End Mill Sharpening. The end cutting edges of an end mill may be sharpened using a surface grinder and special fixture (Fig. 13.46). Note how the fixture tilts the milling cutter so that the proper clearance is ground on the cutting edge.

Cutting Off in a Surface Grinder. Thin cutoff wheels may be mounted on the spindle of a surface grinder and used to cut off sections of hardened parts. The wheelhead is slowly lowered into the workpiece as the table reciprocates.

FIGURE 13.46 A special fixture is used when sharpening an end mill on a surface grinder.

13.4.9 Hints on Surface Grinding and Surface Grinding Problems and Solutions

Producing high-quality work on surface grinders requires considerable knowledge and practice. The beginner should gain as much knowledge as possible about the grinding machine, characteristics of the various grinding wheels and setups, and procedures for grinding operations. The following list of eight suggestions should aid the beginner in producing high-quality work on the surface grinder.

1. Like other cutting tools, abrasive wheels become dull and worn. Keep the wheel sharp and true by dressing as needed.
2. Thin, distorted, or warped material, when mounted on the magnetic chuck, may require shimming to ensure that the magnetic pull does not spring the work. This procedure is usually done on the first surface mounted. After this surface is ground flat, it is placed on the chuck, and side 2 is ground parallel to the first side.
3. Nonmagnetic materials may be ground using the magnetic chuck by completely surrounding the material with magnetic material, locking it in.
4. Surface grinding produces extremely sharp edges on parts. Always handle newly ground parts carefully. Round the sharp edges with an oilstone or fine abrasive cloth.
5. Small nicks, burrs, or scratches on the magnetic chuck can cause inaccurate results in surface grinding. Hand stoning with a medium-grit oilstone cleans off light burrs. More serious defects may require grinding the surface of the chuck to restore it to its original accuracy. Use a soft wheel with adequate coolant and light cuts for this procedure.
6. Grinding is *usually* (not always) a finishing operation on a part. The typical surface grinding job has been previously machined on a milling machine and hardened in a heat-treating furnace. It is time consuming

TABLE 13.1

Causes of and solutions for surface grinding problems

Problem	*Causes*	*Solutions*
Chatter on workpiece	Wheel out of balance	Rebalance. Make sure that wheel fits spindle. Tighhten wheel mounting flange.
	Wheelhead	Check spindle bearing clearance for balance of motor and pulleys.
	Wheel out of round	Redress.
	Faulty coolant	Clean coolant tank, replace with correct coolant mixture.
	Wheel too hard	Change to softer grade.
Burns or discoloration on work	Improper operation	Decrease the rate of in feed or cross feed. Increase work rate.
	Improper wheel	Use softer grade. Redress the wheel rougher. Use greater volume of coolant. Dress wheel frequently.
Burnished work surface (wheel rubbing, not cutting)	Wheel glazed or acting too hard	Dress wheel frequently. Use softer wheel. Use more open-structured wheel. Use coarser wheel. Take lighter cuts.
Irregular scratches on work surface	Dirty coolant	Clean coolant tank and replace fluid. Use finer grit
	Improper wheel	Use harder wheel. Clean wheel guard.
	Dirty table surface	Clean table before changing part.

to leave excessive material on the surfaces to be ground. The amount of stock left for the grinding operation depends on the nature of the part—that is, the length, width, warpage present, type of material, and whether it has been through-hardened or carburized. Consider these factors when machining the part prior to the grinding operation.

7. A surface grinding machine comes equipped with some form of coolant system. If not, adequate ventilation systems should be utilized to ensure proper grain and particle removal from the grinding area.
8. Sometimes a thin piece of paper is placed between the work and magnetic chuck. The paper will protect the chuck from scratches from rough or abrasive metals such as castings or forgings.

Table 13.1 shows various surface grinding problems, causes, and solutions. Use this table along with the operator's guide or handbook when encountering problems with surface grinding operations.

13.5 CYLINDRICAL GRINDING OPERATIONS

Cylindrical grinding is the process of grinding outside cylindrical surfaces. Center-type cylindrical grinders can grind straight, tapered, or multiple diameter parts. The most commonly used cylindrical grinder is the universal type. It is the most versatile cylindrical grinder and for that reason is discussed in detail.

13.5.1 Universal Cylindrical Grinder Construction

Figure 13.47 illustrates the major component parts of the universal cylindrical grinding machine. It is important that you become familiar with the machine before operating it.

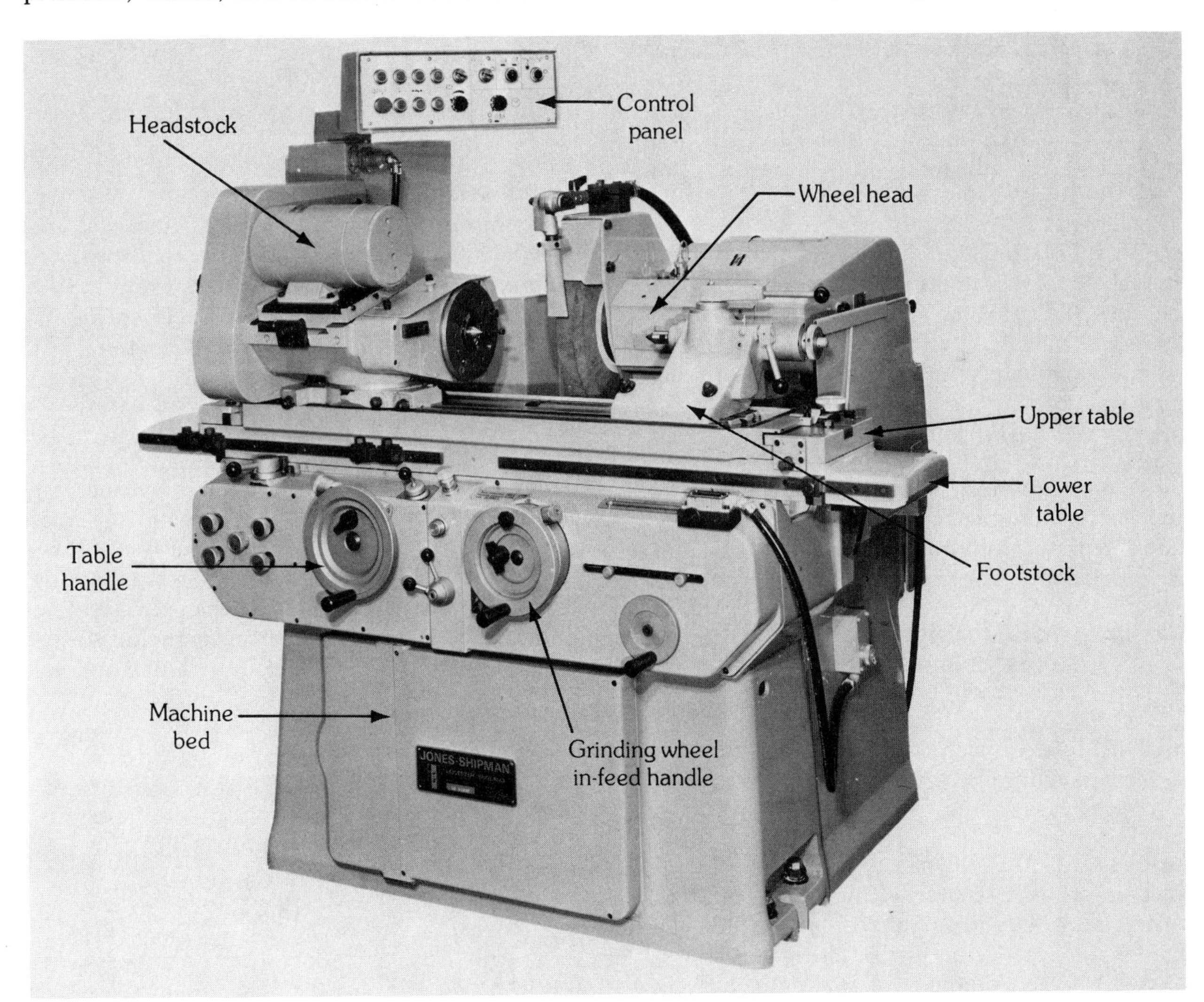

FIGURE 13.47 Major parts of a universal cylindrical grinder. (*Courtesy Brown and Sharpe Grinding Machines, Inc.*)

Machine Bed. This is the base of the machine. It is made of cast iron to provide strong and rigid support for the other parts. On the top surface are accurate machine ways for the sliding table.

Tables. There are two tables mounted to the machine bed. The lower or *sliding table* rests directly on the bed. It provides the reciprocating motion of the machine either hydraulically or mechanically. The upper or *swivel table* is mounted on top of the sliding table and may be swiveled for grinding tapers and angles. The sliding table supports the headstock and the footstock.

Headstock. The headstock is mounted on the left side of the swivel table and contains a motor to rotate the work. A *spindle nose* is provided to mount different types of work-holding devices such as chucks and faceplates. Most work, however, is held between centers. The center mounts directly into the hollow spindle of the headstock.

Footstock. The footstock supports and aligns the outer end of the work. It slides along the ways of the table to accommodate different lengths of work. A lever on the footstock is used to retract the spring-loaded spindle for loading the part and provides the proper tension on the part.

Wheelhead. The wheelhead is fastened to a cross slide, permitting the wheel to be fed toward the table and the work automatically or by hand. Universal machines allow the wheelhead to be swiveled for plunge grinding of angles. Wheelheads must be accurately balanced to prevent vibration and chatter marks on the part.

Coolant Systems. Coolants are used on cylindrical grinding operations to control abrasive dust and provide a better finish on the work. Complete coolant systems are built into most machines.

13.5.2 Work-Holding Devices and Accessories for Cylindrical Grinders

Work-holding devices and accessories used on center-type cylindrical grinders are similar to those used on engine lathes.

Centers. The primary method of holding work is between centers. The points on these centers may be high-speed steel or tungsten carbide. A lubricant is used with either type and is applied between the point of the center and the center hole on the work (Fig. 13.48).

FIGURE 13.48 Centers are the primary method of holding work on a cylindrical grinder. Always use a lubricant on dead centers.

Chucks. Independent, universal, and collet chucks can be used on cylindrical grinders when the work is odd-shaped or contains no center hole (Fig. 13.49). They are used also for internal grinding operations.

Steady-Rest. A steady-rest or backrest is used during grinding to support long, thin workpieces (Fig. 13.50).

Internal Grinding Attachments. An internal grinding attachment may be mounted on the wheelhead for internal grinding operations. It contains a separate high-speed motor to drive the grinding wheel. The attachment can be swung out of the way when not in use (Fig. 13.51).

13.5.3 Grinding a Straight and Parallel Outside Diameter

Before attempting to use a cylindrical grinder (see Fig. 13.52), review Sec. 13.4.3. These safety rules also apply to cylindrical grinding operations. However, special care must be exercised when working around the rotating headstock. The procedures for testing, balancing, and mounting grinding wheels are similar to those for surface grinding and are as follows:

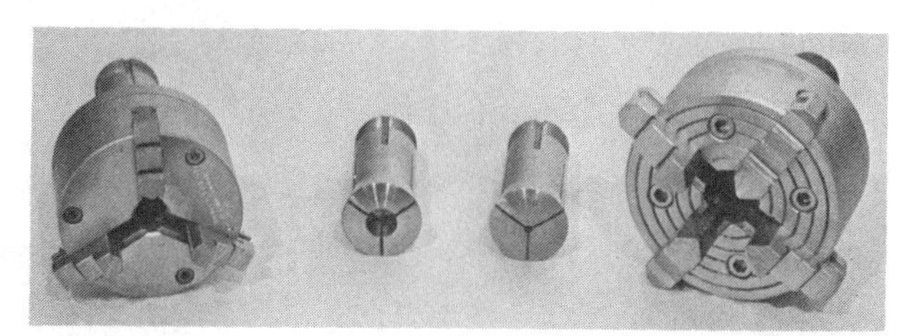

FIGURE 13.49 A variety of work-holding devices are used on cylindrical grinders.

FIGURE 13.50 The steady rest is used to support long workpieces during internal grinding operations. (*Courtesy Brown and Sharpe Grinding Machines, Inc.*)

1. Examine the grinding wheel to see if it is the proper type for the material to be ground. Check to see if it is clean and sharp and has been properly mounted.

FIGURE 13.51 High-speed internal grinding attachment used for grinding the inside diameter of a part.

2. Dress the wheel if necessary. Mount the diamond and holding fixture on the machine table (Fig. 13.53). Start the wheel and run it several minutes to warm up the spindle bearings. Adjust the wheel until the diamond just touches the wheel. Take consecutive passes of 0.001 in. (0.02 mm) with the diamond until the wheel is true and sharp. Use a coolant during the dressing operation.

FIGURE 13.52 Grinding a straight and parallel shaft.

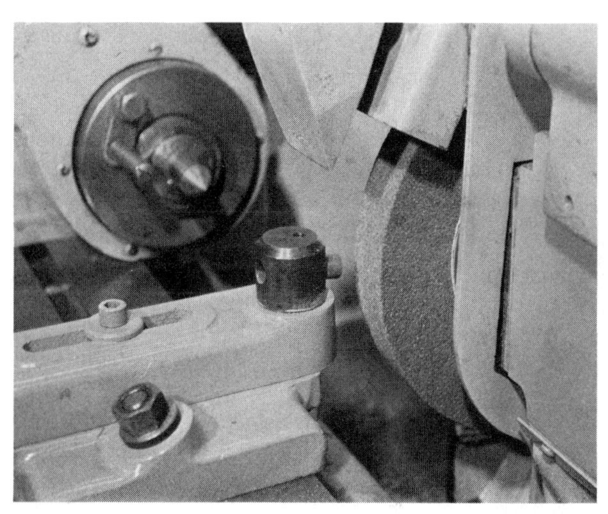

FIGURE 13.53 Dressing the wheel on a cylindrical grinder.

FIGURE 13.54 Checking the alignment of centers using a dial indicator.

3. Check the alignment of centers. Mount the test bar on clean centers and position a dial indicator on the center line of the bar (Fig. 13.54). Then, using the table handwheel, move the test bar past the dial indicator from one end to the other. Adjust the swivel table until the dial indicator reads zero for the entire length of the test bar.
4. Clean the center holes of the work and the centers on the machine. Lubricate the center holes with a heavy center point lubricant.
5. Position the footstock and clamp in place slightly closer to the headstock than the length of the workpiece. The spring inside the footstock helps keep the proper tension on the work.
6. Place a dog on the left end of the workpiece and hand tighten it.
7. Mount the workpiece between centers. Use the hand lever on the footstock to retract the center and let the workpiece fit snugly between centers.
8. Position the dog on the work so that the headstock drive pin fits in the slot on the dog (Fig. 13.55).
9. Set the table stops so that the wheel face will overrun each end of the workpiece about one-third the thickness of the wheel. (*Note:* The work is generally machined smaller in diameter where the dog is mounted.)
10. Adjust the work speed for the diameter and type of material to be ground. Also, make sure that the work is rotating in the same direction as the grinding wheel.
11. Cylindrical grinders are equipped with dwell or tarry timing devices. They allow the automatic table to pause a few seconds before changing direction and permit the wheel to clear itself at the end of the work. Set the timer to the proper cycle. Short cycles are used on smaller diameters, longer cycles on larger work.
12. Set the proper table traverse speed. For *roughing cuts* the traverse speed should be approximately one-half to two-thirds of the wheel width for each revolution of the work. For finish grinding operations, set the traverse speed for a slower rate.

FIGURE 13.55 The drive pin should fit freely in the dog.

13. Start the grinding wheel and let the spindle bearing warm up.
14. Start the workpiece revolving, and slowly bring the wheel into contact with the work. Back the wheel from the work a few thousandths of an inch (about 0.07 mm).
15. Manually move the table traverse and slowly feed the entire length of the work past the wheel. This checks for any high spots along the workpiece.
16. After the work has been checked for high spots, start the coolant, engage the automatic table feed, and take sufficient passes to clean up the work. In-feed the wheelhead approximately 0.002 to 0.003 in. (0.025 to 0.076 mm) per pass.
17. Stop the coolant, spindle, wheelhead, and table feed and check the part taper. Adjust table if necessary (Fig. 13.56).

FIGURE 13.56 Adjusting the swivel table.

18. Continue to take successive roughing and finish cuts until the part is to finish size. There should be no visible sparks generated on the last pass taken. This is called spark out and ensures that the work is round and concentric. Stop and check for size periodically.

13.5.4 Grinding Tapers and Angles on a Part

Work that requires tapers or angles may be ground by several methods on a universal cylindrical grinder. For grinding shallow tapers (usually up to 10°, *the table is swiveled* the required amount and the part is ground in the same manner as in parallel grinding. This is the only method that can be used on plain cylindrical grinders because the spindle and wheelheads are fixed.

Steep tapers can be ground by two methods. One method is to *swivel the wheelhead* and plunge grind the work surface. In this method the table remains stationary and the wheel is fed into the rotating work. A second method is to *swivel the workhead.* This method requires that the work be held with a chuck while the table is fed back and forth across the wheel face.

13.5.5 Cylindrical Grinding Problems and Solutions

Successful cylindrical grinding requires a thorough knowledge of the machine tool, the proper set-up procedures, and general grinding principles. Some problems encountered on cylindrical grinders are similar to those in surface grinding operations. Table 13.2 shows various cylindrical grinding problems, causes, and solutions. Use this table along with the operator's guide when encountering problems on the cylindrical grinder.

13.6 TOOL AND CUTTER GRINDING OPERATIONS

Tool and cutter grinding machines are used primarily to sharpen cutting tools such as reamers, taps, and milling cutters. There are several types of tool and cutter grinding machines; however, the *universal tool and cutter grinder* is the most popular because of its broad capabilities. With the proper attachments and/or accessories, the universal tool and cutter grinder may be used for cylindrical grinding operations, internal grinding, surface grinding, and cutoff grinding operations.

Because the number of different cutting tools that need sharpening is so large, only the basic principles of tool and cutter operation are described in this section.

13.6.1 Universal Tool and Cutter Grinder Construction

The universal tool and cutter grinder is similar in construction and operation to the universal cylindrical grinding machine. Become familiar with the major parts and accessories before attempting to operate it. Figure 13.57 illustrates the major parts of the universal tool and cutter grinder.

Base. The base is a heavy, boxlike casting. The top surface has machined ways on which the saddle is mounted.

TABLE 13.2

Causes of and solutions for cylindrical grinding problems

Problem	*Causes*	*Solutions*
Chatter or wavy marks on the workpiece	Wheel out of balance	Rebalance, make sure wheel fits spindle. Tighten wheel flange.
	Wheel too hard	Use softer wheel. Dress wheel coarser.
	Workpiece support inadequate	Use steady rest, adjust tension on workpiece.
	Machine vibrations	Check motor and machine for excessive vibrations. Check level of machine. Check pulleys and V belts.
Spiral marks on workpiece	Machine improperly operated	Slow work speed, decrease depth of cut.
	Work loose on centers	Adjust tension of centers.
	Centers loose in spindles	Make sure that spindle and centers are clean and free of nicks and burrs.
Burns or discoloration of work	Improper operation	Decrease speed of wheel. Increase work speed. Decrease depth of cut. Increase coolant supply.
	Improper wheel	Use a more open wheel. Use softer wheel. Increase speed of dressing cycle.
Work out of round	Misalignment of machine parts	Check head and footstocks for proper alignment and secure clamping. Check headstock bearings. Check swivel table setting. Check table gibs.
	Work centers	Adjust work on centers for proper tension. Keep work centers clean and oiled. Regrind centers if worn. Check work center holes.
	Improper machine operation	Insufficient dwell. Allow part to spark out. Decrease table feed rate.
Rough finish	Incorrect wheel	Use finer wheel. Use softer wheel.
	Improper dressing	Decrease wheel dressing cycle time.
	Improper machine operation	Decrease table travel feed rate. Increase work speed. Use steady rest.

Saddle. The saddle is mounted on the base and has machined ways for the table assembly. It moves in and out by means of a cross-feed handwheel located on the front of the machine.

Tables. Two tables are mounted on the saddle. The lower or *sliding table* is mounted directly on the saddle; it provides the reciprocating movement of the machine. The upper or *swivel table* is

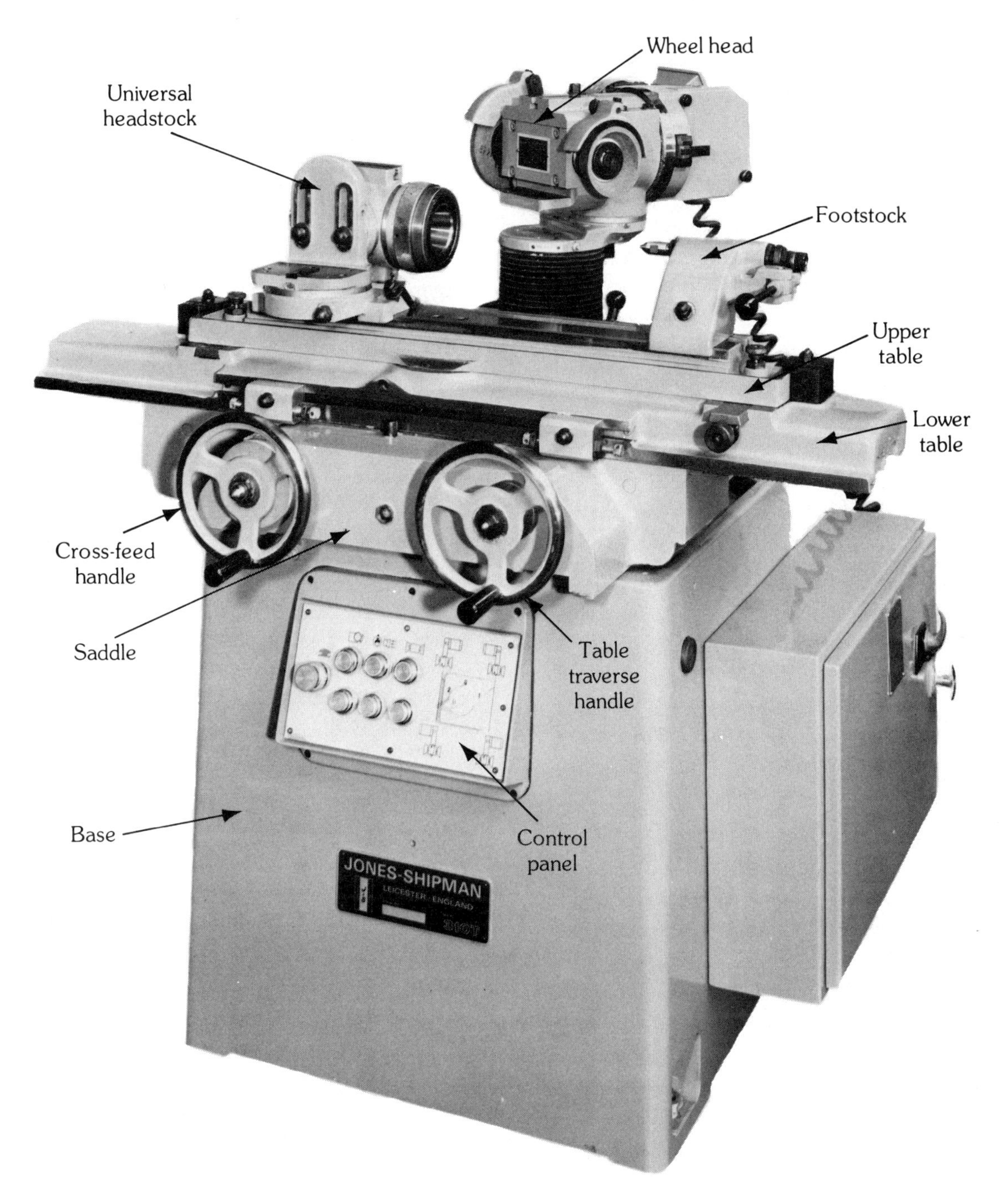

FIGURE 13.57 Major parts of a universal tool and cutter grinding machine. (*Courtesy Brown and Sharpe Grinding Machines, Inc.*)

mounted on the sliding table and may be swiveled for grinding tapers. It has T slots for mounting work and attachments for various grinding operations.

Universal Headstock. The headstock is equipped with a motor and mounted on the swivel table. It is used to support cutting tools in sharpening operations. The headstock has a *spindle nose* on which work-holding devices may be mounted. It can be swiveled vertically and horizontally.

Footstock. The footstock is mounted on the table and used to support the outer end of the work, mandrels, and arbors.

Wheelhead. The wheelhead is mounted on a column. It may be swiveled 360°, lowered or raised by handwheels, and has two arbors for mounting different types of grinding wheels.

Exhaust Sytems. Exhaust systems are required when grinding dry. These are used to prevent abrasive material from spreading through the shop environment and for eye protection.

13.6.2 Tool and Cutter Accessories

Centering Gage. The centering gage is an essential tool for cutter sharpening operations. It is used to align the center of the footstock, the headstock, and the toothrest with the center of the wheelhead spindle. Figure 13.58 shows the centering gage aligning the wheelhead spindle height.

Toothrests. A toothrest is used to rigidly support the tooth of the cutter while the cutter is being sharpened (Fig. 13.59). It may be fastened to the wheelhead or table, depending on the type of grinding wheel used or the type of cutter being sharpened. A toothrest may be *plain* or *micrometer.* The micrometer toothrest allows small vertical adjustments using a micrometer. Various types of *toothrest blades* are inserted into the toothrest holder. The shape of the toothrest blade depends upon the type of cutter being sharpened, the spacing of the teeth, and the rake angle. Figure 13.60 illustrates several types of toothrest blades.

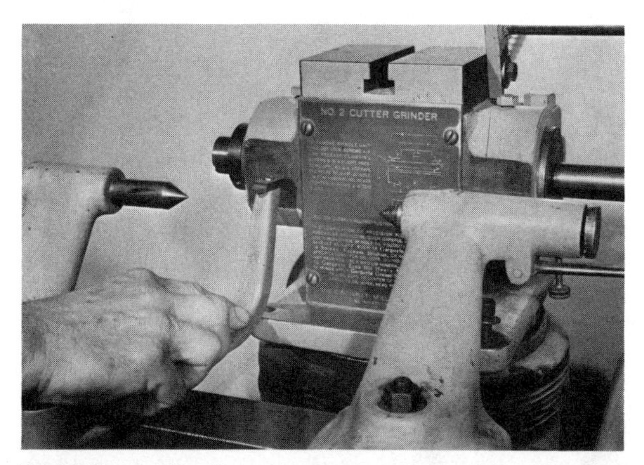

FIGURE 13.58 The center gage is used to align the wheelhead height. *(Courtesy El Camino College)*

Arbors and Mandrels. During a cutter sharpening operation, the cutting tool must be mounted properly to ensure accurate results. Precision arbors and mandrels are used for holding cutting tools (Fig. 13.61).

13.6.3 Cutting Tool Geometry

To achieve accurate cutter sharpening results, it is important to understand the geometry of cutting tools.

Relief Angles. Relief (sometimes referred to as clearance) is the portion of material ground away directly behind the cutting edge of the tool. It allows the cutting edge *and only the cutting edge* to enter the work during the cutting operations. It is measured from an angle formed by a line tangent to the periphery of the cutter and the slope of the land (the width of surface behind the cutting edge) (Fig. 13.62). Correct relief angles are essential for proper cutter performance. Excessive relief does not support the cutting edge properly, causes chatter, and the cutter soon becomes dull. Insufficient relief causes the cutting edge to rub on the work causing friction that will rapidly dull the cutter.

Two relief angles, primary and secondary, are found on most cutting tools. *Primary* relief is that portion ground away directly behind the cutting edge. The *secondary* relief angle gives additional clearance behind the primary relief angle and controls the width of the land.

FIGURE 13.59 The toothrest is used to support teeth of a milling cutter during a sharpening operation. *(Courtesy Brown and Sharpe Grinding Machines, Inc.)*

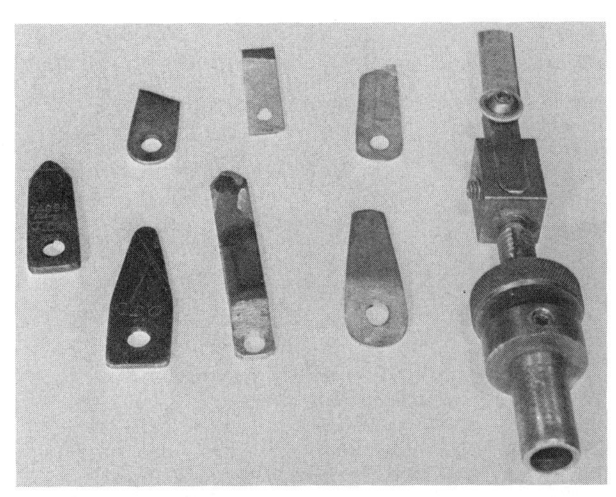

FIGURE 13.60 Several types of toothrest blades. (*Courtesy El Camino College*)

FIGURE 13.61 Arbors and mandrels are used for holding milling cutters on the tool and cutter grinder during sharpening operations.

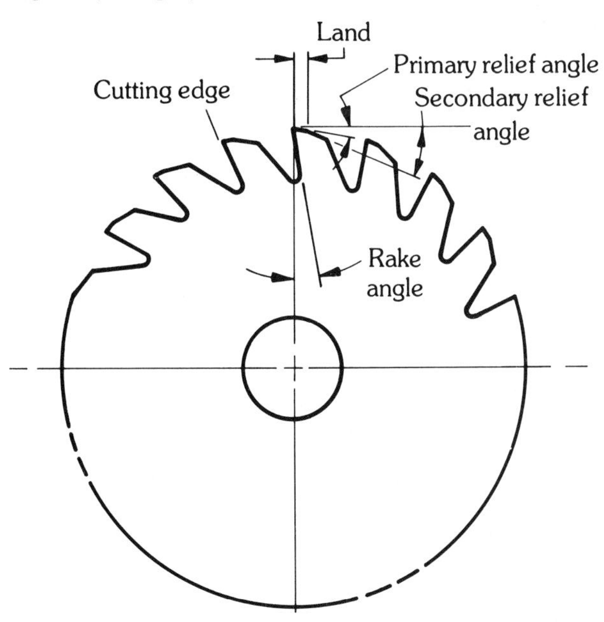

FIGURE 13.62 Relief and rake angles on a milling cutter.

The primary relief angle varies on milling cutters depending on the diameter of the cutter and the type of material to be ground. Small cutters require more relief than large cutters. Table 13.3 shows the approximate primary and secondary relief angles for average diameter, high-speed-steel cutters. Secondary relief angles are generally 3 to 5° larger than primary relief angles.

TABLE 13.3
Suggested relief angles using high-speed cutters

Material to Be Machined	*Primary Relief Degrees*	*Secondary Relief Degrees*
Carbon and alloy steels	3–5	6–8
Cast irons	4–7	7–10
Bronze (hard)	4–7	7–10
Brass and bronze (soft)	10–12	13–15
Aluminum	10–12	13–15
Magnesium	10–12	13–15
Plastic	10–12	13–15
Stainless steel	5–7	8–10
Titanium	7–11	10–14

Land. The land is the width of the surface directly behind the cutting edge (Fig. 13.62) and is ground to the relief angle. It varies in width from 1/64 to 1/16 in. (0.39 to 1.59 mm), depending on the diameter of the cutter. The land width is controlled by the secondary relief angle. Too large a land width, caused by repeated sharpening of the primary relief angle, makes the heel behind the cutting edge drag over the cutting surface.

Rake Angle. The rake angle on a milling cutter is the angle formed between the tooth face and a line passing through the axis of rotation of the cutter (Fig. 13.62). It is most important in the formation of the chips generated by the cutting operation. To ensure a smooth chip flow during the cutting operation, the rake angle must be ground smooth and free of burrs.

13.6.4 General Tool and Cutter Setups and Operations

Because of the large variety of cutting tool types and designs available for machining operations today, only the *basic* setup principles are discussed in this section. For detailed setup precedures for each type of cutting tool, the instruction manual of the tool and cutter grinder is a good source of information.

Wheel Selection. Cutting tools can be sharpened by using either the periphery of a *straight wheel* or the face of a *cup wheel.* The cup wheel grinds the lands of the teeth flat. The straight wheel leaves the land slightly concave (called hollow grinding) behind the cutting edge. Large-diameter straight wheels are preferred because they leave a smaller radius on the land.

Other considerations are type of abrasive, grain size, grade, structure, and bond of wheel to be used. In general, aluminum oxide wheels of medium grain size (60 to 80), with a softer grade (J or K), and vitrified bonds are used for most high-speed-steel cutter grinding operations. Silicon carbide wheels are generally used to rough grind tungsten carbide cutters, and diamond wheels (150 to 180 grain) are used to finish sharpening operations.

Excessive *wheel speeds* should be avoided because they tend to temper and heat check the surface of the cutting edges. The surface speed of the grinding wheel should be between 4000 to 5000 sfpm (1220 to 1525 m/min) for most cutter grinding operations.

General Tool and Cutter Setup Procedures. There are two widely accepted methods of sharpening cutters. One method is to mount the toothrest so that the grinding wheel rotates *off* the cutting edge of the cutter (Fig. 13.63). In this method the cutter is pushed against the toothrest during the grinding operation. The relief angle is produced by raising the wheelhead above the center of the cutter a certain amount depending on the diameter of the wheel and the desired relief. A burr is produced on the cutting edge, which must be hand-stoned to remove it.

The second method is to rotate the grinding wheel *onto* the cutting edge of the cutter (Fig. 13.64). This technique produces a keener, burr-free cutting edge. Another advantage of rotating the wheel onto the cutting edge is that the heat is directed away from the cutting edge and dissipated into the body of the cutter. With this method it is necessary to hand-hold the cutter against the toothrest, otherwise the rotating wheel also rotates the cutter. Therefore, the first method is considered safer. The second method requires the wheelhead to be *lowered* below the center of the cutter to obtain the necessary relief.

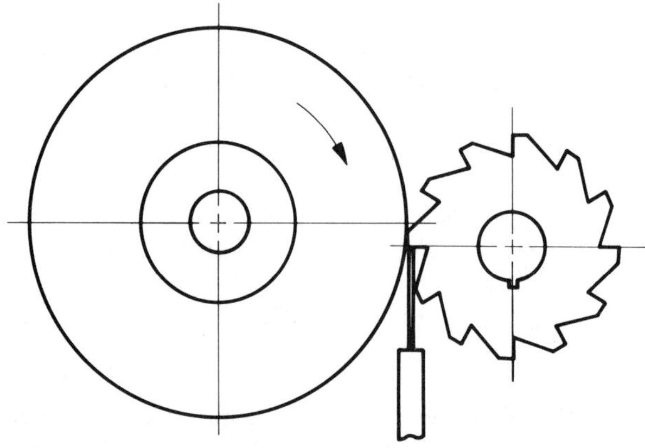

FIGURE 13.63 When the toothrest is mounted this way, the grinding wheel rotates off the cutting edge of the cutter.

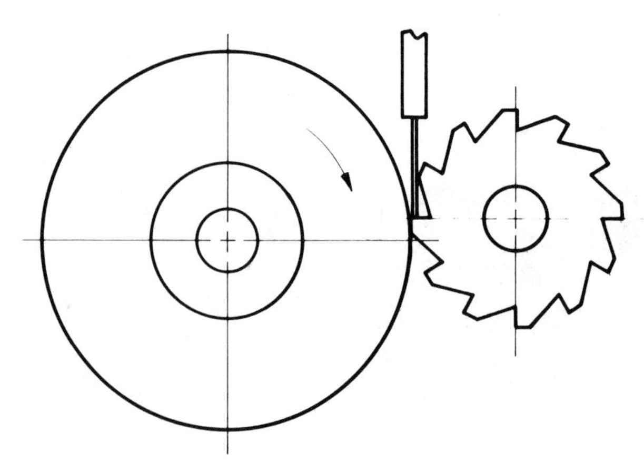

FIGURE 13.64 When the toothrest is mounted this way, the grinding wheel rotates onto the cutting edge of the cutter.

Producing the Relief Angle. The relief angle is produced by properly setting the grinding wheel, cutter, and toothrest. Different types of wheels generally use different methods. For *straight wheels* the procedure is as follows:

1. Using a center gage or height gage, bring the cutter and wheel centerlines into the same plane.
2. Mount a toothrest to the table and adjust the tip to the same center height.
3. Raise or lower (depending on the wheel rotation) the wheelhead the proper amount to obtain the required relief.

The distance *(D)* the wheelhead is raised or lowered is calculated as follows:

$$D = \text{relief angle} \times \text{wheel diameter} \times 0.0087 \quad \text{constant}$$

or

$$D = \theta \times d \times 0.0087$$

Example: Calculate the distance to lower or raise the wheelhead to obtain a 10° relief angle on a cutter using a 7-in. (177.8-mm) straight grinding wheel.

$$\begin{aligned} D &= \theta \times d \times 0.0087 \\ &= 10 \times 7 \times 0.0087 \\ &= 0.609 \quad \text{distance wheelhead is moved} \end{aligned}$$

The procedure for using *cup wheels* is the same as for straight wheels; however, the toothrest is mounted on the *wheelhead* rather than the table. Also, when calculating the distance to move the wheelhead, there is one exception: the *cutter* diameter is used in place of the wheel diameter. That is, the distance the wheelhead is raised or lowered is calculated as follows:

$$D = \text{relief angle} \times \text{cutter diameter} \times 0.0087 \quad \text{(constant)}$$

or

$$D = \theta \times d \times 0.0087$$

Types of Cutters and Their Setups. Cutting tools differ in design and method of sharpening. They may be classified into two general categories:

1. Cutters that require sharpening the *relief angle* behind the cutting edge. The cutting edge may be on the periphery, side, or end of the cutter, depending on the type. This category includes cutters such as plain milling cutters, side-milling cutters, end mills, face mills, reamers, and slitting saw.
2. Cutters that require sharpening on the *cutting face* of the teeth. On this type of cutter the face is ground so the profile of the tooth is not altered. This category includes form cutters, taps, hobs, involute gear cutters, and form tools.

The following discussion illustrates and gives a brief description of some typical tool and cutter setups.

Fig. 13.65 illustrates a *plain milling cutter* being sharpened using a cup wheel. The toothrest is mounted on the wheelhead, and the wheel and toothrest are lowered to obtain the necessary relief. *Side-milling cutters* require two setups. To sharpen the periphery, the setup is the same as described for plain milling cutters. Side grinding is done by mounting the cutter on a stub arbor and installing the arbor into the universal headstock (Fig. 13.66). The headstock is then tilted the required amount to achieve the relief angle on the tooth. Note the position of the toothrest.

End milling cutters also require two separate setups to completely sharpen the cutter. The end-cutting edges are ground while the shank of the tool is held in the universal headstock and the headstock tilted to the required relief angle (Fig. 13.67). To sharpen the periphery it may be supported either between centers or held in a special fixture.

FIGURE 13.65 Plain milling cutter being sharpened with a cup wheel. (*Courtesy Cincinnati Milacron Inc.*)

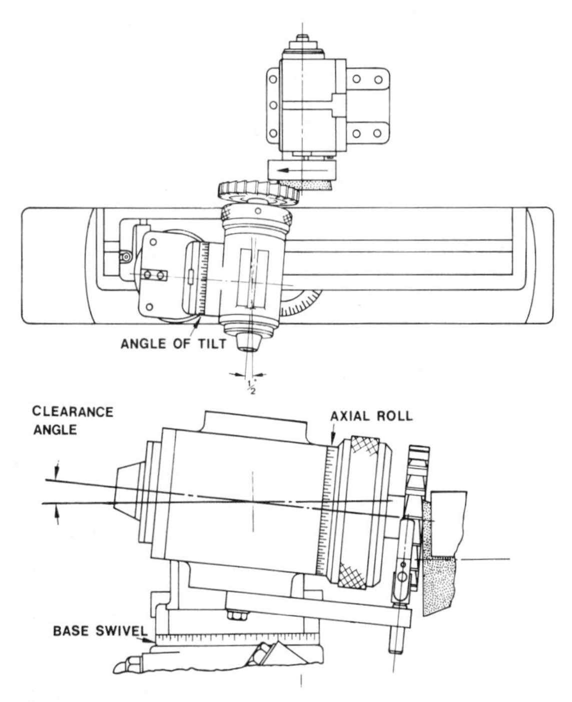

FIGURE 13.66 Sharpening the side of a milling cutter. The cutter is mounted on an arbor. (*Courtesy Brown and Sharpe Grinding Machines, Inc.*)

Formed milling cutters are sharpened radially on the cutting face, usually with a dish wheel. Fig. 13.68 shows the sharpening of an involute gear forming cutter.

There are many different types of *reamers*, each requiring a special setup procedure. Most machine *reamers* cut on a chamfered end, so most sharpening operations are done on the end of the

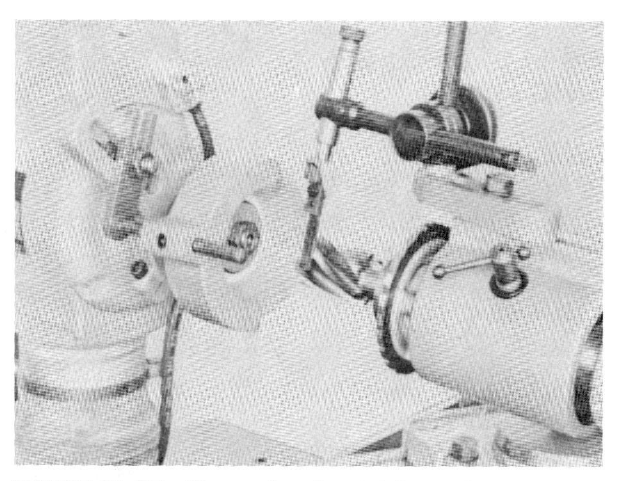

FIGURE 13.67 Sharpening the periphery of an end mill.

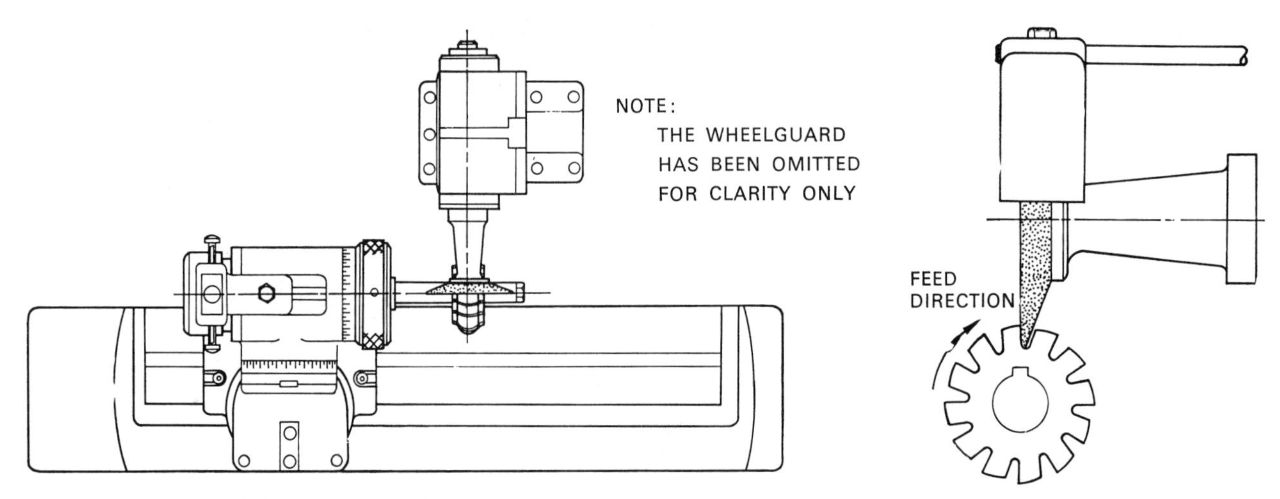

FIGURE 13.68 A dish wheel is used to sharpen the face of a gear cutter. (*Courtesy Brown and Sharpe Grinding Machines, Inc.*)

FIGURE 13.69 Sharpening the end of a reamer using a cup wheel. (*Courtesy El Camino College*)

reamer. Fig. 13.69 illustrates a machine reamer being ground on the end cutting edges. When the periphery of a reamer requires sharpening, it is generally mounted between centers and sharpened to the next smaller nominal size.

REVIEW QUESTIONS

13.2 TYPES OF GRINDING MACHINES

1. List and briefly describe the four types of surface grinders.
2. Compare the differences between plain and universal cylindrical grinders.
3. How is form grinding accomplished on cylindrical grinding machines?
4. How is the work fed on a centerless grinder?
5. Of what material is the regulating wheel usually made?
6. List and briefly describe the three types of centerless grinders.
7. Briefly describe the operating principle of internal grinding machines.

13.3 GRINDING WHEELS AND ABRASIVES

1. List the four kinds of abrasive materials used for grinding wheels, and briefly describe the applications of each.
2. How does abrasive grain size influence grinding operations?
3. What system is used to size abrasive grains?
4. Describe how the grade of an abrasive wheel affects its operation.
5. What is meant by *hard grinding wheel; soft grinding wheel*?
6. What does the structure of a grinding wheel indicate?
7. List and briefly describe four different types of bonding materials used for grinding wheels.
8. What do the numbers and letters of the following examples of wheel designations represent?
   (a) A–40–J–12–V
   (b) C–80–L–8–B
   (c) A–120–M–6–S
9. Sketch the nine standard wheel shapes, and give an application for each.
10. List six factors to consider when selecting a grinding wheel for a particular grinding application.
11. How should a grinding wheel be inspected prior to being mounted on a machine?
12. Briefly describe the procedure for mounting a grinding wheel on a machine spindle.
13. What may occur if a grinding wheel is not balanced properly?

13.4 SURFACE GRINDING OPERATIONS

1. Briefly describe the functions of the following major components of a surface grinder:
   (a) Base
   (b) Saddle
   (c) Table
   (d) Wheelhead
   (e) Coolant system
2. Describe the operating principles for permanent magnetic and electromagnetic chucks used on surface grinders.
3. What is the purpose of the demagnetizer switch on the electromagnetic chuck?
4. List 10 safety rules to observe during surface grinding operations.
5. What are the major differences between dressing and truing a grinding wheel?
6. Describe the procedures for dressing and truing a grinding wheel on a surface grinding machine.
7. Briefly describe the steps necessary to grind a part flat and parallel.
8. How are angle plates used when squaring a part on a surface grinder?
9. Describe three methods of grinding angular surfaces on a part using a surface grinder.
10. How can end mill cutters be sharpened on a surface grinder?
11. Describe the procedure for mounting warped or distorted parts on a magnetic chuck.
12. How may nonmagnetic materials be secured for surface grinding operations?

13.5 CYLINDRICAL GRINDING OPERATIONS

1. Briefly describe the following major component parts of the cylindrical grinding machine:

(a) Bed
(b) Sliding table
(c) Swivel table
(d) Headstock
(e) Footstock
(f) Wheelhead

2. How are centers used on the cylindrical grinder?
3. What three types of chucks are used on the cylindrical grinder?
4. When is a steady rest used on a cylindrical grinding machine?
5. Briefly describe the proper setup procedures for grinding a straight diameter on a part.
6. List the three methods used and describe how tapers and angles may be ground on a universal cylindrical grinder.
7. What are the four possible causes of chatter on the workpiece in cylindrical grinding?
8. Describe the possible solution for parts that are burned during cylindrical grinding operations.

13.6 TOOL AND CUTTER GRINDING OPERATIONS

1. Briefly describe the following major parts of a universal tool and cutter grinder:
   (a) Base
   (b) Saddle
   (c) Sliding table
   (d) Swivel table
   (e) Universal headstock
   (f) Footstock
   (g) Wheelhead
2. What is the function of the centering gage for tool and cutter grinding operations?
3. What are the differences between the plain and micrometer toothrests?
4. Describe the purposes of the toothrest blade.
5. Briefly describe the purposes of arbors and mandrels for cutter grinding operations.
6. What are the suggested land widths for most milling cutters?
7. What happens if the land width of a cutter is too large?
8. Sketch a typical side-milling cutter and label the following parts:
   (a) Rake angle
   (b) Primary relief angle
   (c) Secondary relief angle
   (d) Land
9. What two wheel shapes are used for most tool and cutter grinding operations?
10. Compare the two methods of placing the toothrest blade on the machine during cutter grinding operations.
11. How are relief angles produced on a milling cutter grinding operation for straight wheels? For cup wheels?

14

Basic Metallurgy and Heat Treatment

14.1 INTRODUCTION

Machine shop operations almost always involve metallic materials, so machinists must understand the physical characteristics of common metals. The machinist is usually responsible for choosing appropriate cutting tool materials, selecting speeds and feeds, and in some cases specifying or performing simple heat-treatment operations. The choices made will affect the efficiency of the machining operations and the quality of the products in terms of durability, appearance, and accuracy.

The main emphasis in this chapter is on *physical metallurgy,* which involves the chemical composition, heat treatment, forming, and machining of metals. There are relatively few basic metals that the machinist uses, but there are many variations, or *alloys,* of each metal. Alloys are formed by adding other metallic elements to change the strength, corrosion resistance, or other characteristics of the parent metal. In some cases, the added elements do not combine chemically but form a mechanical mixture, with particles of the added material dispersed in the original metal. An example is adding lead to steel to improve the steel's machineability.

Opening photo courtesy of Bethlehem Steel Corp.

14.2 PHYSICAL CHARACTERISTICS

14.2.1 Strength

One physical characteristic of concern to the machinist is the strength of the metal. Its strength will affect which cutting speed and cutting tool materials the machinist selects. The *tensile strength* of a metal is expressed in *pounds per square inch* (psi). For example, a low-carbon steel may have a tensile strength of 70,000 psi (Fig. 14.1). The tensile strength of a metal can be altered by heat treatment, the addition of alloying elements, cold-working, or any combination of these processes.

The terms *stress* and *strain* are used when discussing the strength of materials. The three types of stress of concern to the machinist are *tensile stress* (the forces that tend to pull materials apart), *compressive stress* (the forces that tend to squeeze materials together), and *shear stress* (the forces that tend to cut materials). Stress is measured in pounds or kilograms per unit area.

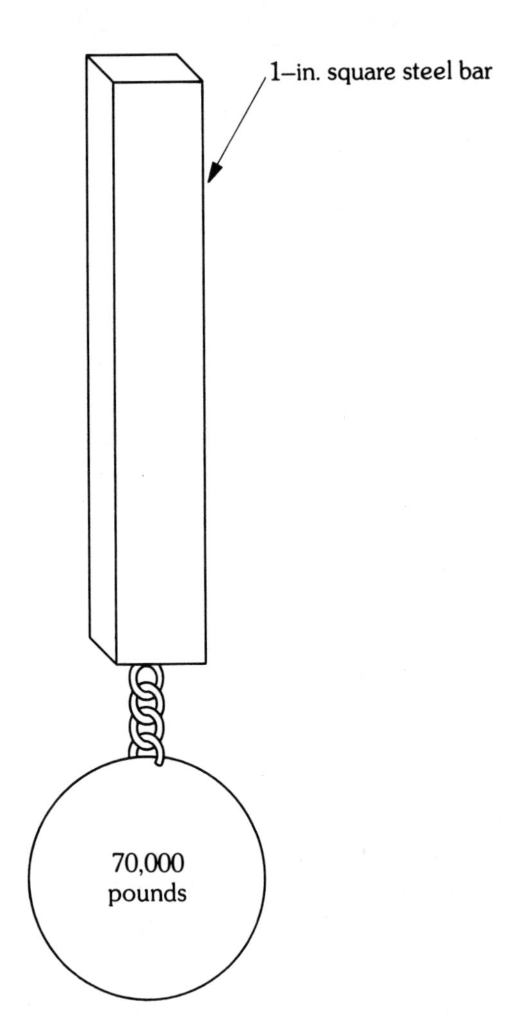

FIGURE 14.1 The weight is applying a tension load to the bar.

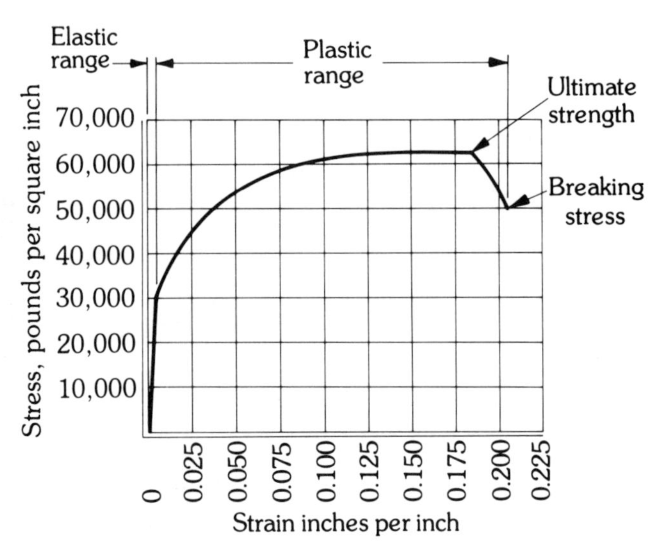

FIGURE 14.2 Stress–strain diagram for steel.

Strain is the distortion or deformation that results from the application of a load to an object. The amount of strain is usually expressed as a percentage of the length of the object or in inches of deformation per inch of length of the object. The deformation is temporary, and the object returns to its original size and shape if its elastic limit has not been exceeded (Fig. 14.2).

14.2.2 Elasticity

Elasticity is a characteristic of metals that affects the stress–strain relationship. It is a measure of the stiffness of any given metal. Metals have an *elastic limit,* and a metal specimen loaded to a point below the elastic limit will return to its original dimensions. For example, a spring can be compressed and released many times. If the elastic limit is not exceeded, the spring will return to its original length when unloaded. In any machining or forming operation in which metal is cut or bent, the elastic limit of the material has been exceeded.

14.2.3 Hardness

Hardness is a characteristic of metals that is important to the machinist. A hard metal resists machining operations such as drilling, turning, milling, and hand operations such as filing, center punching, and hacksawing. The hardness of metals can be tested with machines such as the Rockwell Hardness Tester (Fig. 14.3).

The hardness of most metals can be changed by heat treatment, cold-working, or both. Some austenitic materials, like certain stainless steels, cannot be hardened by heat treatment, but they can be rolled or hammered to increase strength and

FIGURE 14.3 Hardness tester being used to check the hardness of a steel part.

hardness. Carbon and alloy steels with more than 0.25 to 0.30 percent carbon can be hardened by heat treatment. As hardness increases, brittleness increases and ductility decreases. For example, a high-carbon steel such as AISI–C1095 is very brittle in its fully hardened condition and breaks suddenly if loaded beyond the elastic limit.

14.2.4 Ductility

Ductility is the characteristic that allows metals to be bent, shaped, and formed without breaking or tearing. Metals like low-carbon steels, alloy steels in fully annealed form, annealed copper, and aluminum are quite ductile and can be formed into complex shapes (Fig. 14.4). Ductility is usually expressed in terms of percentage of elongation before failure during a strength test. Metals such as gray cast iron and white iron are not ductile and break before they bend or stretch. The terms *plasticity* and *malleability* are also used to indicate the difficulty or ease of forming materials.

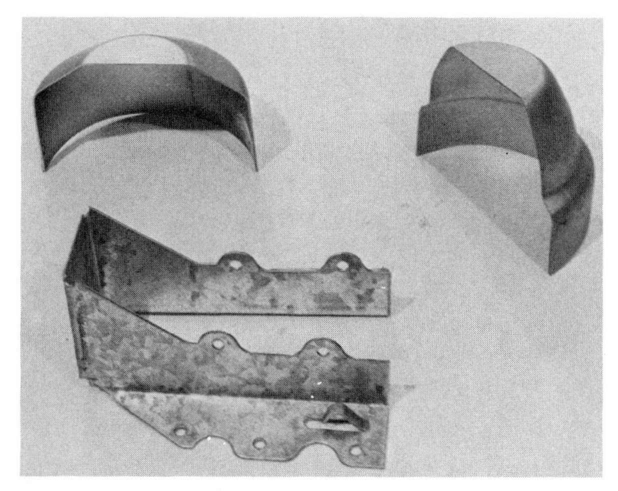

FIGURE 14.4 Complex stampings made from ductile metal.

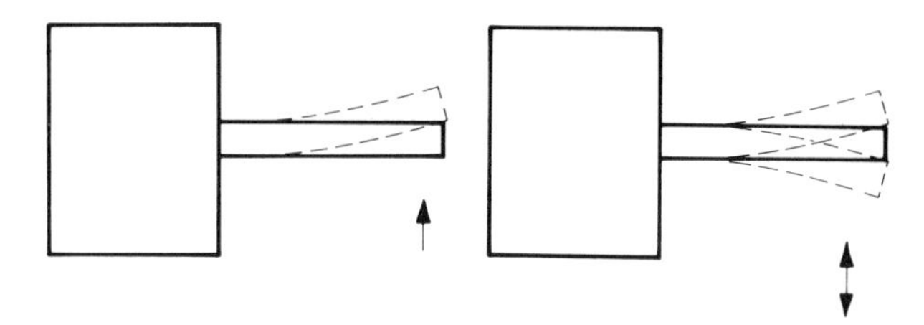

FIGURE 14.5 Fatigue testing determines the ability of a material to withstand repeated deflections.

14.2.5 Fatigue Resistance

Fatigue resistance is the ability of a metal to withstand repeated impacts or bending loads without fracturing. These loads are below the elastic limit of the material and are not intended to cause permanent deformation. As shown in Fig. 14.5, the loads may be applied in one or more directions.

One factor affecting fatigue resistance is the smoothness of the object. Objects with smooth surfaces and properly radiused corners can withstand more load cycles before failure. Therefore, the machinist who does smooth, well-finished work free from scratches contributes to the fatigue resistance of the objects being machined. *Shot peening*, in which the surface of a smooth part is bombarded with small, hard steel balls to produce a surface with a slight compressive stress, is another way to increase the fatigue life of metal parts.

14.2.6 Machineability

Machineability is a characteristic affecting the speed at which metal can be cut. The standard material for establishing machineability ratings is a freemachining steel, AISI–B1112, which has been given a machineability rating of 100. Metals such as copper, aluminum, brass, and magnesium

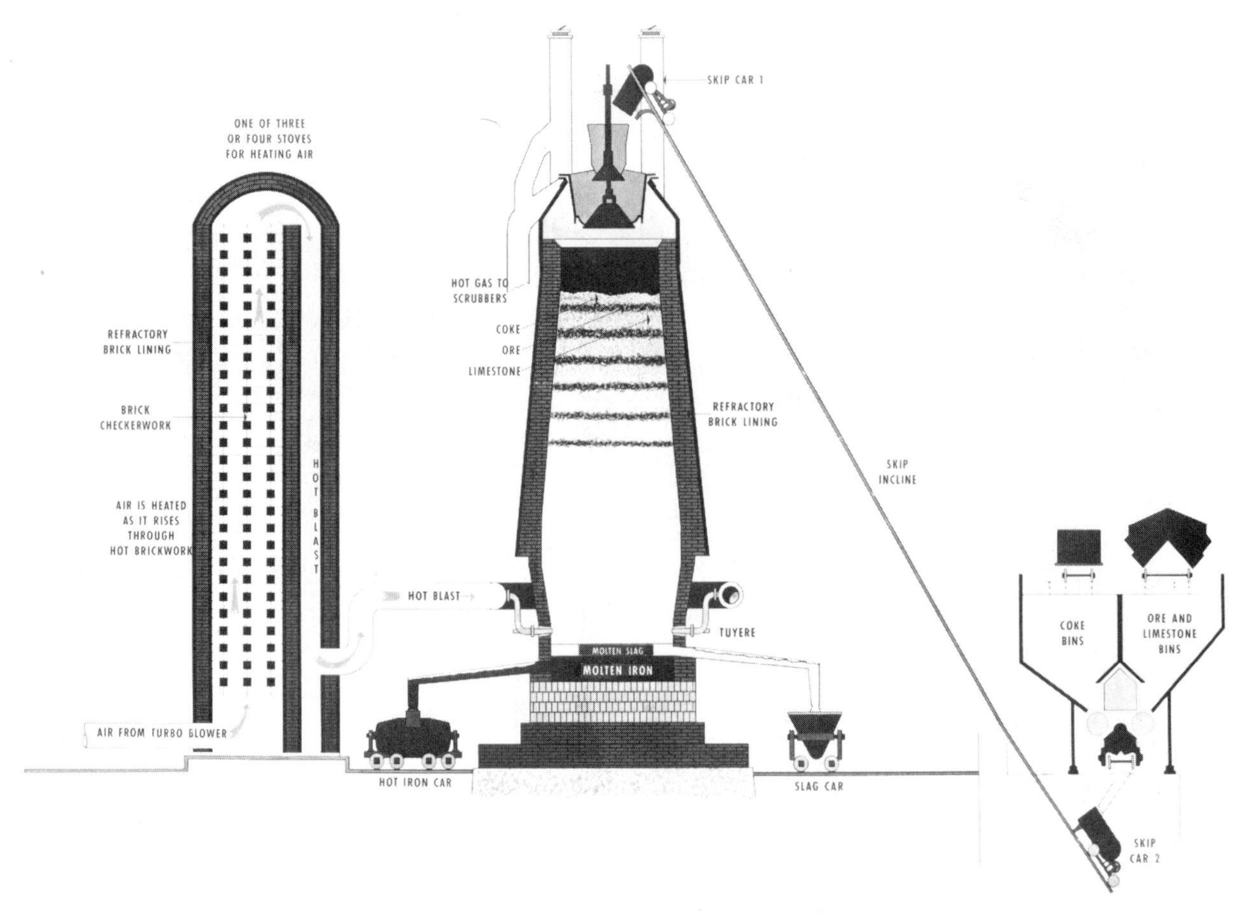

FIGURE 14.6 The blast furnace is used to reduce iron ore to pig iron. (*Courtesy Bethlehem Steel Corp.*)

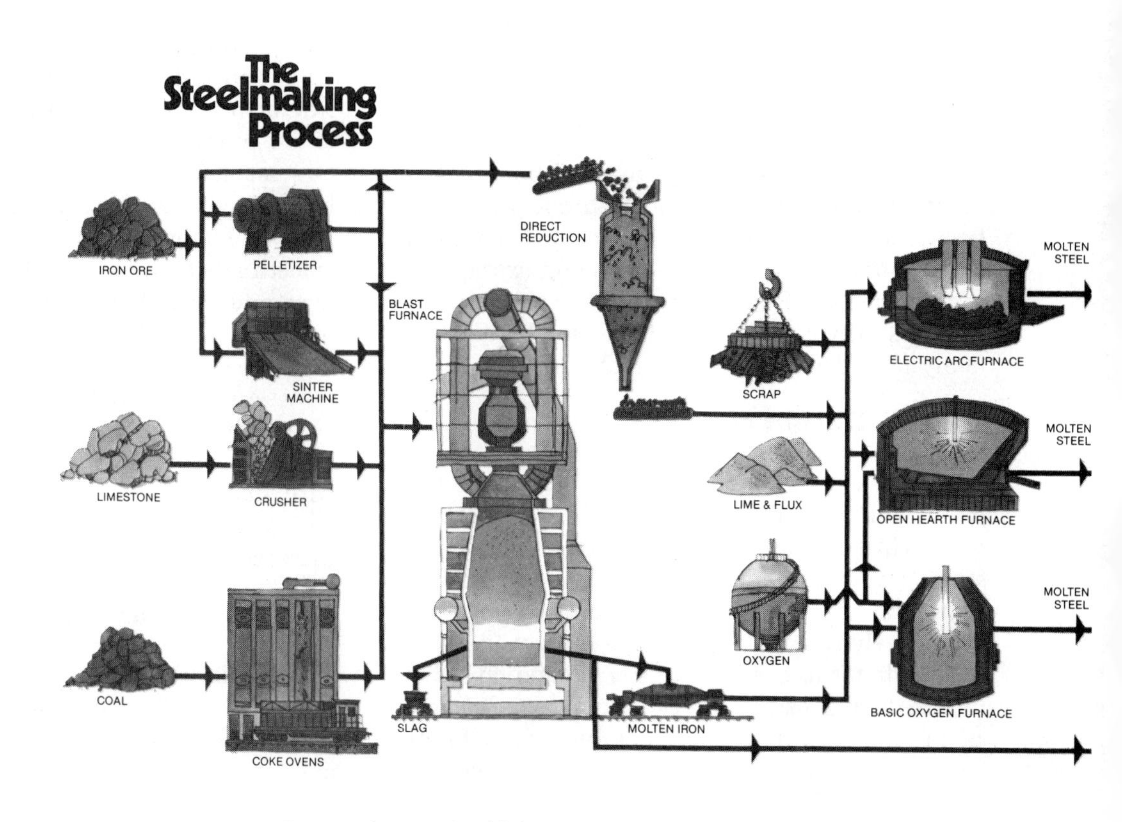

FIGURE 14.7 Flow diagram for steelmaking. (*Courtesy American Iron and Steel Institute*)

usually have machineability ratings above 100. Alloy steels, cast iron, and plain carbon steels with varying carbon content have machineability ratings below 100.

14.3 FERROUS METALS

The basic component of all ferrous metals is iron. Other materials, or *alloying elements,* may be added to the iron to alter its strength, corrosion resistance, and other characteristics. Iron ore is smelted in a blast furnace (Fig. 14.6) to produce pig iron. The blast furnace is charged from the top with a mixture of *iron ore, coke,* and *limestone.* Coke, which is made from metallurgical grade coal, is the fuel; limestone is the *flux*, or scavenger of impurities. As the result of heat and a chemical reaction, metallic iron (pig iron) is produced. At the bottom of the furnace the *slag,* which is a mixture of molten limestone and impurities, floats on the pool of molten iron.

The pig iron drawn from the bottom of the blast furnace can be cast into *pigs* or sent in molten form directly to the open hearth or basic oxygen furnace for making steel (Fig. 14.7). Since pig iron still contains some impurities, it must be processed further before it can be used.

14.3.1 Cast Iron

The four major types of cast iron are *gray iron, white iron, malleable iron,* and *ductile iron.* The carbon content of cast iron varies from about 1.7 to 4.5 percent, depending on the type of iron.

Gray Cast Iron. This is the most common form of cast iron and is used extensively in castings for machine tools, automotive parts, and other industrial products. It is relatively inexpensive, has high compressive strength, and is fairly easy to cast and machine. It contains up to 4 percent carbon and varying amounts of silicon and other elements. Typical gray iron castings are shown in Fig. 14.8.

Most of the carbon in gray cast iron is in the form of graphite flakes, which makes the iron difficult to weld and reduces the tensile strength. The tensile strength of the common classes of gray cast iron varies from about 20,000 to 60,000 psi. The presence of free graphite in gray cast iron improves its machineability since it acts as a dry lubricant.

White Cast Iron. The composition of white cast iron, sometimes called *chilled iron,* is similar to that of gray cast iron, except that in some cases the carbon content is lower. It is hard, brittle material with great wear resistance. It can be machined only by grinding. When fractured, the metal appears white, especially at the surface, since all or nearly all of the carbon is in the form of iron carbide. The term *chilled iron* refers to the rapid cooling of the metal that takes place when it is cast, caused by the use of metal or graphite *chills* in the mold (Fig. 14.9).

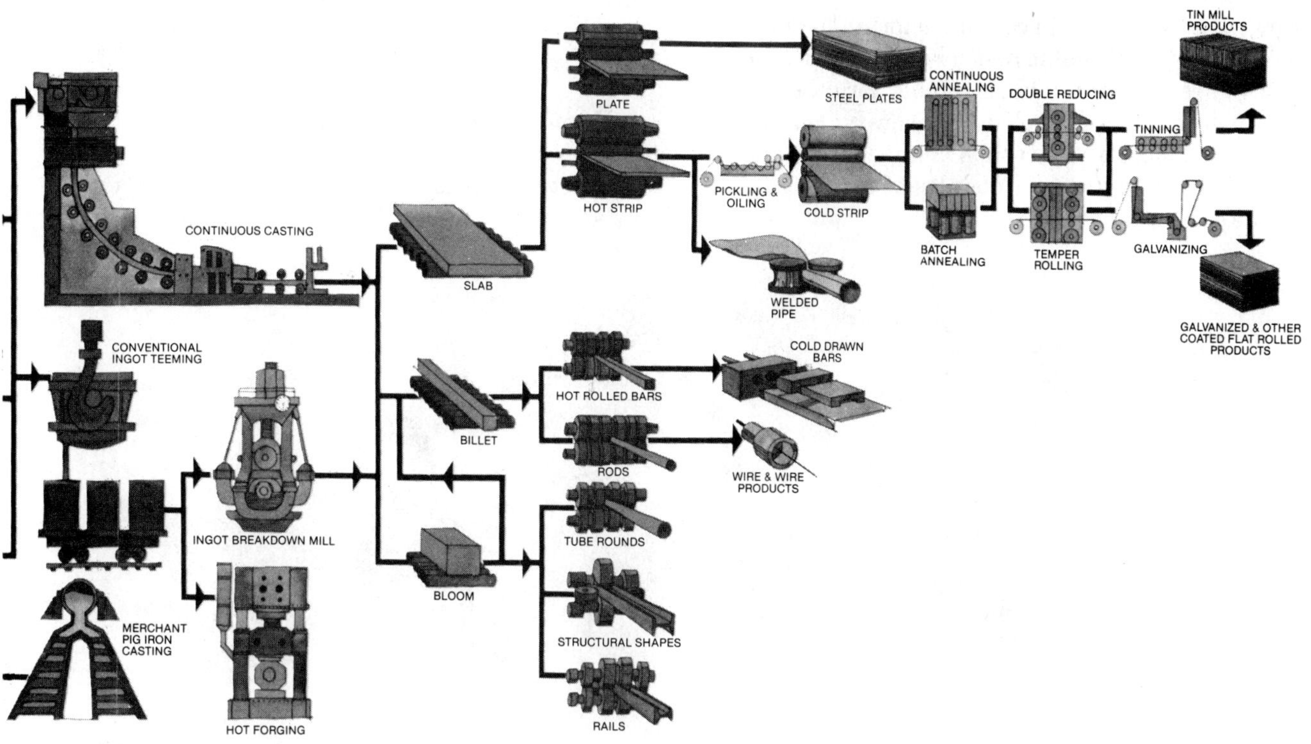

FIGURE 14.7 *(Continued)*

FIGURE 14.8 Gray cast iron castings.

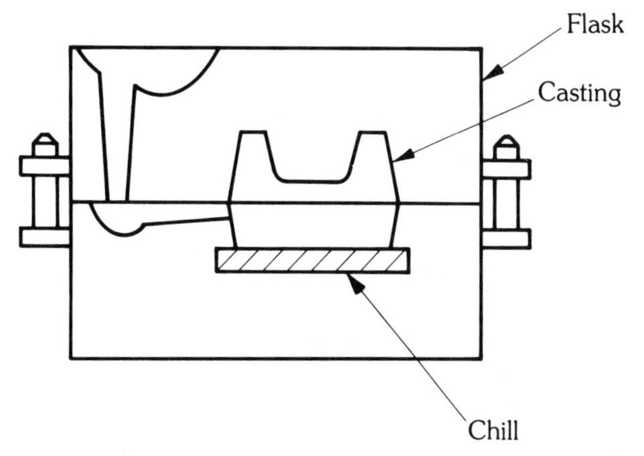

FIGURE 14.9 The *chill* in the sand mold is used to cool the casting rapidly.

Malleable Iron. Malleable iron castings begin with a white iron structure in the *as cast* condition with the carbon in combined form. To make the castings machineable and shock resistant, the castings are heated to about 1600°F for a period of up to 48 hours and cooled very slowly. During this process most of the carbon changes form and is dispersed as very small nodules of carbon in a steel-like matrix.

Malleable iron has a tensile strength of 50,000 psi or more and can be bent and stretched to a certain extent before it fractures. It is used in automotive and industrial components where a low-cost, easily machineable casting with good shock resistance is needed, such as in the cast crankshaft shown in Fig. 14.10.

Ductile Iron. This variation of gray cast iron is also known as *nodular iron* because the carbon is dispersed in the material in the form of small nodules rather than flakes. Ductile iron is a strong, shock-resistant material that can be readily welded, machined, and heat-treated to tensile strengths above 100,000 psi. To make ductile iron, a small amount of magnesium is injected into the molten cast iron just before it is poured into the sand molds.

14.3.2 Carbon Steels

Carbon steels are produced by removing all but the desired amounts of carbon, manganese, sulfur, phosphorous, and other elements from pig iron. In modern steelmaking practices, the basic oxygen and the open-hearth furnaces are used to produce most of the steel. Electric furnaces are mainly used to produce medium- and high-carbon steels and alloy steels.

All steels contain some carbon, and the carbon content determines the *hardenability* of the steel (Fig. 14.11) as well as other characteristics. The amount of carbon is therefore closely related to the strength and wear resistance that can be developed by heat treatment.

Classification. An understanding of the steel classification system initially developed by the Society of Automotive Engineers (SAE) and later adopted and slightly modified by the Amer-

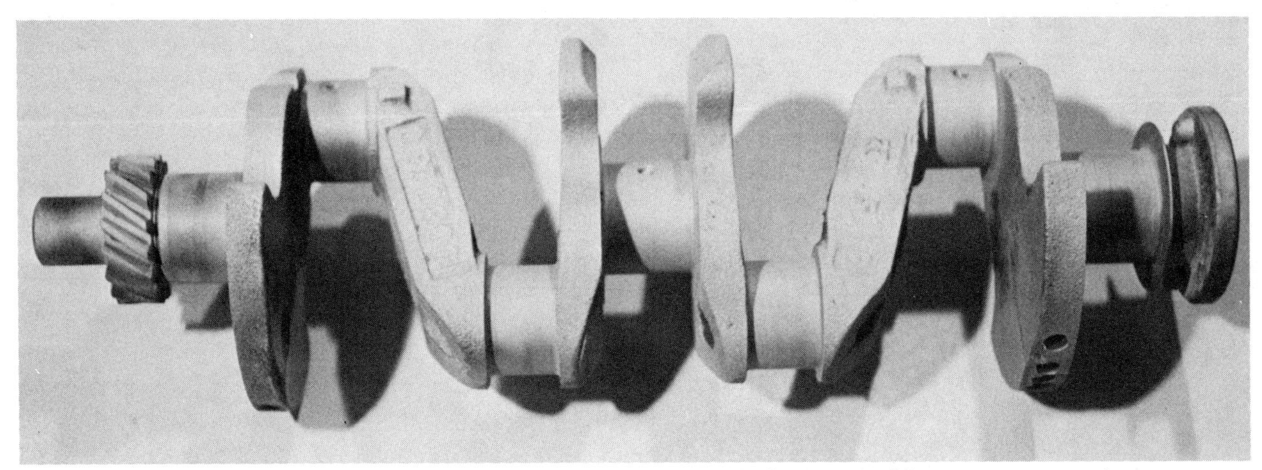

FIGURE 14.10 Malleable iron castings are shock resistant and may be heat-treated for greater wear resistance.

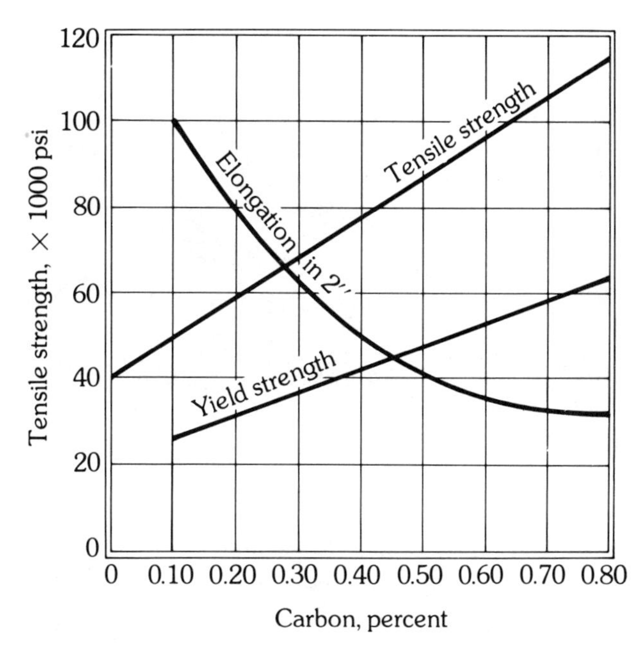

FIGURE 14.11 Effect of carbon content on tensile strength, yield strength, and elongation on annealed steel.

ican Iron and Steel Institute (AISI) is necessary for the machinist who has to select different steels for particular applications. The numbering system for identifying steels consists of the code AISI and the classification numbers and one or two letters. The letter B stands for acid Bessemer steel, C for basic open-hearth, D for acid open-hearth, and E for electric furnace steel.

The first digit of the classification number indicates the *series* or *class* of steel. The second digit indicates the presence and amount of the *alloying element* in increments of 1 percent. The last two digits in a four-digit number indicate the *carbon content* in hundredths of 1 percent. In a five-digit number, the last three digits indicate carbon content, and the amount of carbon in the steel is always 1 percent or more. The letter L inserted between the second and third digits means that lead has been added to improve machineability. The letter B indicates that at least 0.0005 percent boron has been added to improve the hardening characteristics. Major steel categories are shown in Table 14.1.

Two typical carbon steel designations are:

AISI–C1025

AISI—American Iron and Steel Institute

C—basic open-hearth steel
1—plain carbon
0—no major alloying element
25—0.25 percent carbon content

TABLE 14.1
Steel classification

Number	*Major Categories*
10 XX	Carbon steels with no other significant alloying elements
11 XX	Free machining, resulfurized carbon steels
12 XX	Similar to 11 XX, but on open-hearth steels
13 XX	Manganese steels
20 XX	Nickel steels
31 XX	Nickel-chrome steels
40 XX	Plain molybdenum steels
41 XX	Chrome-molybdenum steels
43 XX	Nickel-chrome molybdenum steels
46 XX	Nickel-molybdenum steels
48 XX	Nickel-molybdenum steels
50 XX, 51 XX, 52 XX	Plain chrome steels
61 XX	Chrome-vanadium steels
86 XX	Nickel-chrome-molybdenum steel, but lower alloying percentage than 43 XX
92 XX	Silicon steels

AISI–12L14

AISI—American Iron and Steel Institute
1—carbon steel
2—free-machining, resulfurized, rephosphorized
L—leaded
14—0.14 percent carbon content

Low-Carbon Steels. Steels with less than 0.30 percent carbon are classified as low-carbon steels and are not hardenable except by carburizing. Steels in this carbon content range are available *hot-rolled* or *cold-finished* in a variety of shapes ranging from sheet to round bar stock. Hot-rolled steels have some mill scale on the surface and are not as accurately sized as cold-finished materials. Steel may be cold-finished by *rolling* or *drawing*. Sheet and flat bar stock are cold-finished by rolling. Bar stock such as round or hexagonal is finished by drawing through a die. Bar stock can also be cold-finished by turning and/or grinding, as in the case of drill rod.

Cold-finished low-carbon steel bar is used extensively in making machined parts. When lead and sulfur are added, the machineability of the steel is greatly increased, but it becomes unsuitable for fusion welding.

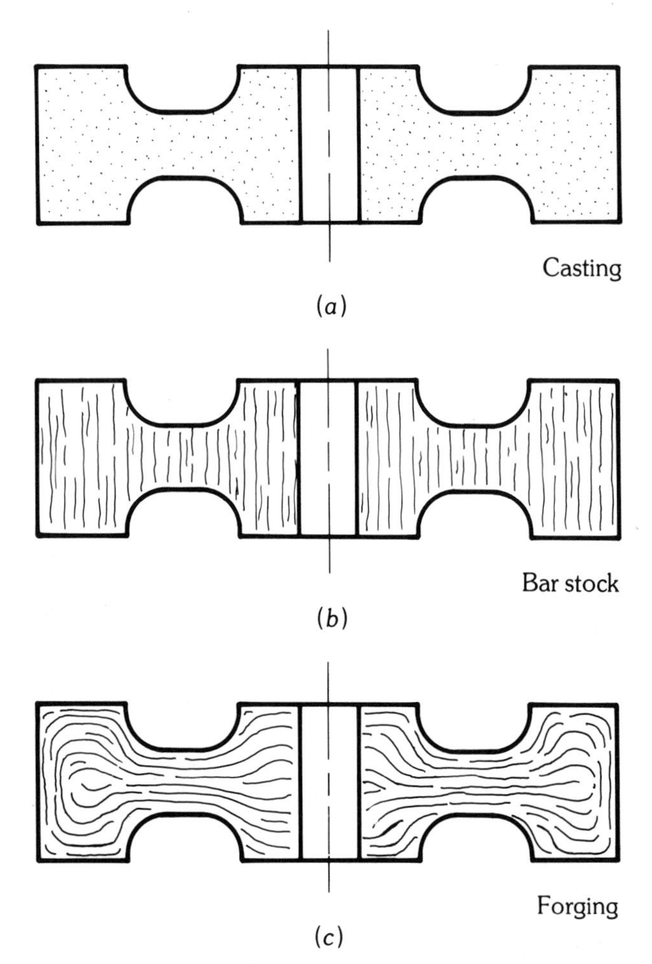

FIGURE 14.12 Differences in grain structure in (*a*) cast, (*b*) bar stock, and (*c*) forged parts.

Medium-Carbon Steels. The carbon content of medium-carbon steels is between 0.30 and 0.60 percent. Steels in the medium-carbon group are used extensively for machine parts since they can be machined readily and partially or fully hardened to improve tensile strength and wear resistance. Many industrial products that require high strength and shock resistance are forged to shape so that the grain flow pattern follows the contour of the part (see Fig. 14.12). Forgings of medium-carbon steel are used for hand tools such as wrenches and hammers and for automotive parts that must be strong and fatigue-resistant such as rear axle shafts. Medium-carbon steels are not as easily welded as low-carbon steels.

High-Carbon Steels. Steels in this category contain over 0.60 percent carbon, and no particular upper limit is specified. Most high-carbon steels used for making tools, dies, punches, and similar equipment have a carbon content of 0.009 to 1.10 percent. A few die steels (such as AISI–D3) have as much as 2.25 percent carbon.

High-carbon steels can be readily hardened and are used extensively for cutting tools for woodworking and some metalworking operations (Fig. 14.13). Steels of this type are available in bar and sheet form, either in hot-rolled or in cold-rolled sheets and strip. High-carbon steels are also available in precision-finished form such as drill rod or ground bars and strip.

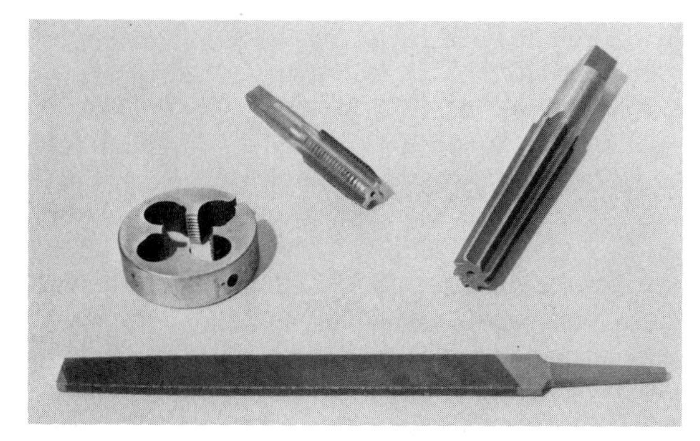

FIGURE 14.13 Many cutting tools are made of heat-treated high-carbon steel.

14.3.3 Alloy Steels

Alloy steels are a combination of iron, carbon, and one or more *alloying elements* in significant amounts. Of the approximately 12 commonly used alloying elements, manganese, chromium, and nickel are most often used. Alloy steels are more difficult to make than carbon steels and are therefore usually more expensive. The presence of alloying elements provides certain characteristics that cannot be obtained with carbon steels. The extra cost is justified when very high strength and fatigue resistance are desired.

Alloying Elements. Alloying elements may have differing effects on steel. In some cases a small amount of an element has one effect, such as increasing tensile strength; adding a large amount of the same element provides corrosion resistance or some other desired characteristic. The seven common alloying elements and their effects are as follows:

1. *Manganese* is a basic element in all alloy and carbon steels. When the manganese content exceeds 1.65 percent, the steel is classified as an alloy steel. It is an effective deoxidizer and combines with sulfur in steel to form manganese sulfide inclusions. These act as an internal lubricant to improve machineability. Manganese also adds strength and hardness, particularly in higher-carbon-content steels, but it tends to reduce weldability and ductility. High-

manganese steels are very tough and wear-resistant.

2. *Nickel* increases the toughness, fatigue resistance, wear resistance, and hardenability of most steels when used in small amounts. It lowers the critical temperature of steel and thus broadens the temperature range for heat treatment. In large amounts, in conjunction with chromium, it makes steels *stainless,* or corrosion resistant. Nickel steels are also used in very low-temperature applications.
3. *Chromium* is one of the most versatile alloying elements. It increases hardenability, toughness, resistance to wear, and tensile strength. Some die steels have as much as 12 percent chromium, which makes the steel air-hardening when the proper amount of carbon is present. All stainless steels contain chromium in amounts ranging from 12 to 26 percent.
4. *Molybdenum* is used in various alloy steels in conjunction with chromium, manganese, nickel, cobalt, and other elements. It increases hardenability of a steel for a given carbon content and helps make hardened steels tough and fine-grained. Molybdenum is used in many steels subjected to high temperatures and is also used in high-speed-steel cutting tools.
5. *Silicon,* a nonmetallic element, is a powerful deoxidizer present in all steels in amounts ranging from about 0.20 to 2.20 percent. It increases the strength of steel, and, when used with manganese, it helps produce strong alloy steels with high shock resistance.
6. *Vanadium,* which is used in relatively small amounts in alloy steels, prevents or reduces grain growth in steels as they are heat-treated. It is used as an alloying element in steels for springs, gears, axles, hand tools, and other objects requiring impact resistance, toughness, and strength.
7. *Tungsten* is used extensively in high-speed steels since it increases the strength and the high-temperature hardness of the steel. It is referred to as a refractory metal since it melts at high temperatures (6098 °F) and is so hard it can be machined only by grinding.

Parts such as the racing engine crankshaft and connecting rod shown in Fig. 14.14 and Fig. 14.15 are subjected to high mechanical stresses and repeated load reversals in use. Therefore they must be light as well as durable to be effective. The alloy steels usually selected for engine components of this type are compounded to provide the best possible balance of wear resistance, tensile strength, and resistance to fatigue fractures. A typical alloy steel might contain almost all of the alloying elements listed above, with one or more being present in relatively large amounts. For example, an alloy steel in which the dominant alloying elements are chromium and molybdenum (such as AISI–E4340) might be referred to as "chrome-moly." That term only names a general group of steels while the AISI numerical designation specifies and describes a particular steel alloy.

14.3.4 Stainless Steels

This unique group of steels is divided into three major categories: the 300 series, 400 series, and precipitation hardening group. The various stainless steel alloys have been assigned a three-digit designation in the 200, 300, 400, or 500 range. Only

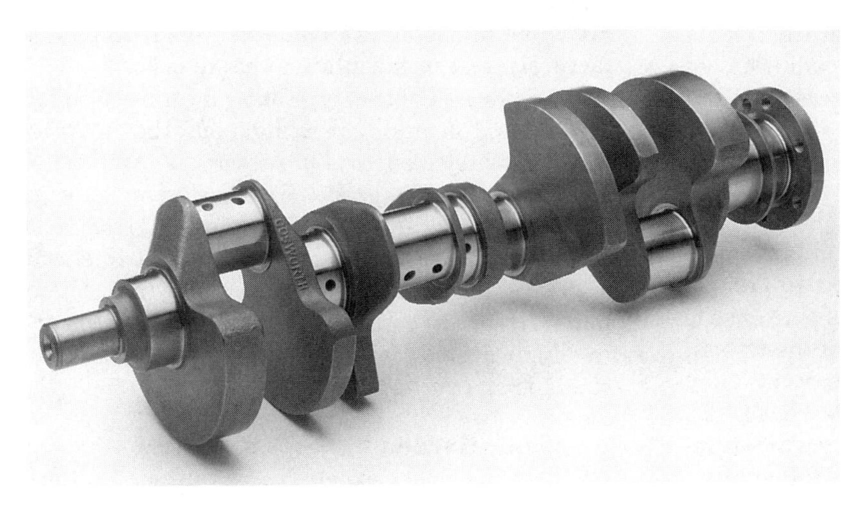

FIGURE 14.14 Racing engine crankshaft forged from alloy steel. (*Courtesy Cosworth Engineering, Inc.*)

FIGURE 14.15 Racing engine connecting rod machined from alloy steel forging and heat-treated for maximum strength. *(Courtesy Cosworth Engineering, Inc.)*

the 300 and 400 series are covered since those two categories include about 90 percent of the commonly used stainless steels.

300 Series. The steels in this group have a number of characteristics in common. They are low in carbon content, with a range of 0.03 to about 0.15 percent. When the letter L is added to the designation (*example:* 304L) the carbon content is usually held to 0.03 percent. The nickel content may vary from 6.5 to 22 percent, and the chromium content from 16 to 26 percent. Other elements such as manganese, silicon, and molybdenum are also added.

All 300-series stainless steels are *austenitic* and therefore cannot be hardened by heat treatment. They will work-harden, however. This characteristic, along with the low thermal conductivity (about one-third the heat-conducting ability of carbon steels), causes problems in machining these materials. To improve machineability, sulfur or selenium may be added in amounts up to 0.35 percent. Steels of this type have a suffix in the designation indicating the presence of these elements (*example:* 309S and 303Se). The addition of selenium and/or sulfur somewhat reduces corrosion resistance and weldability.

400 Series. The steels in this group are basically iron–carbon–chromium alloys, with minor amounts of other elements added. The 400-series steels are subdivided into two groups: the *martensitic* and *ferritic* steels. The martensitic steels can contain as much as 1.20 percent carbon, but the amount is usually about 0.15 percent. They are hardenable to varying degrees by heat treatment. The machineability of the martensitic stainless steels is better than that of the 300 series.

The ferritic stainless steels cannot be hardened by heat treatment. Their machineability rating is approximately 55 percent, using B-1112 screw stock as the standard at 100 percent. When about 0.15 percent sulfur is added, the machineability improves to about 90 percent, or about 150 sfpm using high-speed-steel cutters.

Stainless steels can be cast, forged, and welded and are available in sheet, plate, tube, and bar form.

14.3.5 Tool Steels

Traditionally, tool steels have been identified by brand names rather than by a number system. In recent years AISI has developed a uniform designation system in which a letter identifies each major tool steel group. There are a number of different steels in each group, and each steel is identified by a suffix number. For example, one molybdenum-based high-speed steel is known as M10 and is available from at least 11 sources. The characteristics of the more widely used tool and die steels are found in handbooks available from steel manufacturers and in machinists' handbooks.

Since steels of this type are generally used for cutting and forming operations on other metals, they are subjected to severe service conditions. Therefore, they must be wear-resistant, able to withstand heat, and must not deform during heat treatment. They must also be resistant to shock and fatigue because the tools made from these materials are often used in punch press and shearing applications. Typical tooling made from these steels is shown in Fig. 14.16.

Water-Hardening Steels. These are the simplest and least expensive of the tool and die

FIGURE 14.16 Punch and die set used to perforate and form metal parts.

steels. In most cases the carbon content ranges between 0.60 and 1.40 percent. Manganese, silicon, and other alloying elements are added in small amounts to increase wear resistance and toughness and to prevent or reduce grain growth during heat treatment. Water is a severe quenching medium since it very rapidly reduces the temperature of the object being quenched. In some cases, particularly when quenching thin or complex parts, cracking and/or distortion may occur.

Oil-Hardening Steels. This is the largest group of tool and die steels. It includes materials such as some types of drill rod, some shock-resisting steels, and several high-carbon, high-chromium steels. There is generally less chance of distortion or cracking when heat-treating steels of this type because oil is a milder quenching agent than water. The carbon content may be as high as 1.45 percent in some alloys, especially those containing a large amount of molybdenum.

Air-Hardening Steels. Steels that can be air-hardened have carbon contents ranging from 0.50 to 2.25 percent. In most cases the chromium content is high. A steel like AISI-D2 has a chromium content of 12 percent, for example. Air-hardening steels tend to be very stable dimensionally during heat treatment, and deformation is at a minimum. Generally, the hardening temperatures are higher for air-hardening steels than for other tool steels. In most cases the hardening temperatures are in the range of 1650 to 1875 °F, but steels such as AISI–H24 must be heated to about 2100 °F before quenching in still air or with compressed air.

FIGURE 14.17 Cutting tools made of high-speed steel.

Air-hardening steels are widely used for cutting tools for both metallic and nonmetallic materials and for forming and processing equipment where resistance to wear, shock, and the effects of heat are important.

High-Speed Steels. This type of steel is used primarily for cutting tools for metals and other materials (see Fig. 14.17). It can retain its hardness at temperatures up to about 1100 °F. There are several categories of high-speed steels, including the molybdenum, tungsten, tungsten–cobalt, and molybdenum–tungsten grades. Molybdenum and cobalt are sometimes added, to increase the hardness at high temperatures or *red hardness.*

The molybdenum high-speed steels contain a comparatively small amount of tungsten but have a molybdenum content of about 8 to 9.5 percent. In their fully annealed form, high-speed steels are machineable by normal machine shop methods. After they are hardened, they can be machined only with abrasives.

High-speed steels are hardened by heating to about 1650 °F, cooling to below the critical temperature and then heating quickly to the range 2150 to 2375 °F, depending on the type of steel. The steel is quenched in air or oil and then *tempered* by heating to the range 1000 to 1100 °F and cooling in air. In some cases the tempering is done in a bath of heated salt. The hardness of high-speed steel heat treated in this manner is about Rockwell C-65, making it suitable for the majority of metal-cutting operations.

14.4 NONFERROUS METALS

The term *nonferrous* refers to all metals that are not iron-based, such as copper, aluminum, and nickel. This includes a wide variety of materials, some of which are used as primary metals and some as alloying elements. They are generally more expensive than ferrous metals because they are more difficult to separate from their ores and are often scarce and found only in certain parts of the world. There are many nonferrous metals, but we shall discuss only those of primary importance in machine shop work.

Nonferrous metals often have features that make them desirable for certain industrial operations. In some cases they are more corrosion-resistant than ferrous metals, easier to machine and fabricate, have good heat and electrical conductivity, and are lighter in weight. With some exceptions, they are not as strong as steel but are generally quite ductile and have high machineability ratings.

14.4.1 Aluminum

Aluminum is a light metal derived from *bauxite,* an ore mined in various parts of the world. Incidentally, aluminum is the most abundant metal in the earth's crust. Recovering metallic aluminum from ore involves the use of electrolytic processes that are relatively expensive.

Aluminum weighs about 170 lb/ft^3, or approximately one-third the weight of steel. It is a good conductor of heat and has about 60 percent of the electrical conductivity of copper. Pure aluminum is very reactive and when exposed to the atmosphere, a thin, tenacious oxide layer forms almost immediately. This prevents further corrosion but also interferes with some welding processes.

Classification. A four-digit system for classifying aluminum alloys has been devised by the Aluminum Association, a trade association of aluminum producers and processors. The classification system, shown in Table 14.2, indicates the major alloying element or elements found in each series and certain other characteristics. The first digit denotes the major alloying element, and if the second digit is 0, no particular control has been applied to the amounts of alloying elements. If the second digit is from 1 through 9, the manufacturer has exerted varying degrees of control over the alloy content.

In the 1XXX-series aluminum alloys, the last two digits indicate the purity of the aluminum. For example, a common alloy such as 1100 is 99 percent pure aluminum, with some control exerted by the manufacturer over the remaining 1 percent of the metal.

In the 2XXX- through 7XXX-series alloys, the last two digits are assigned by the manufacturer to identify a particular alloy. For example, one alloy containing copper as its main alloying element is referred to as 2024. Aluminum alloys may also be classified as *heat-treatable* or *nonheat-treatable.* The 1XXX-, 3XXX-, and 5XXX-series alloys are nonheat-treatable, although heat is used to anneal these materials after they have been work-hardened. The 2XXX-, 6XXX-, and 7XXX-series alloys are heat-treatable, and their temper condition often affects their machineability. For example, in the fully soft, or *O* condition, they tend to be gummy and difficult to machine cleanly. The common temper designations are shown in Table 14.3.

TABLE 14.2
Aluminum classification

Number	*Alloyed with:*
1XXX	Commercially pure
2XXX	Copper
3XXX	Manganese
4XXX	Silicon
5XXX	Magnesium
6XXX	Magnesium and silicon
7XXX	Zinc

TABLE 14.3
Temper designations—aluminum

Temper Designation	*Heat Treatment*
O	Annealed wrought alloys
F	Fabricated or cast
T-2	Annealed cast alloys
T-3	Solution-heat-treated and cold-worked
T-4	Solution-heat-treated
T-5	Artificially aged only
T-6	Solution-heat-treated and artificially aged
T-7	Solution-heat-treated and stabilized
T-8	Solution-heat-treated, cold-worked, and artificially aged
T-9	Solution-heat-treated, artificially aged, and cold-worked

Wrought Aluminum Alloys. The stronger wrought alloys, especially the copper–aluminum (2XXX series), magnesium–silicon–aluminum (6XXX series), and zinc–aluminum (7XXX series) alloys, are used extensively in making machined products. These materials are generally machined in the heat-treated form and can be machined at relatively high cutting speeds with both high-speed-steel and carbide cutting tools.

Cast Aluminum Alloys. Aluminum can be readily cast by the sand casting, pressure die casting, and permanent mold processes. Many of the castings are finished by various machining processes, including turning, drilling, and milling. Some cast aluminum alloys that contain large amounts of silicon are quite abrasive and can be machined efficiently at high speeds only by using carbide and ceramic tools. High-speed-steel toolbits may be used at moderate cutting speeds (250 to 300 sfpm).

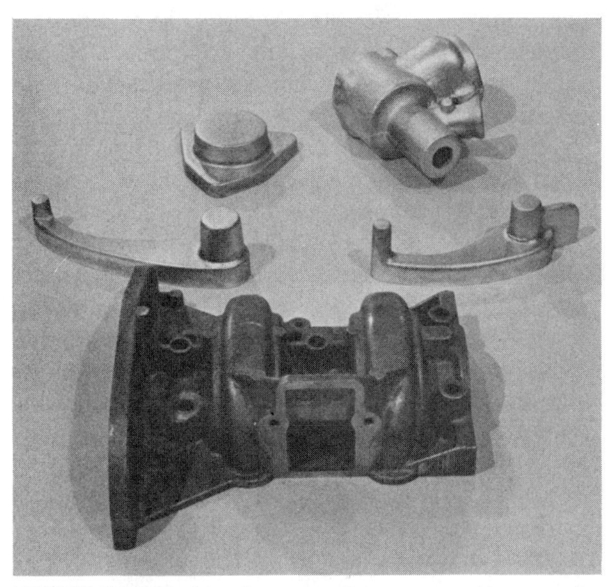

FIGURE 14.18 Aluminum castings are used in many industrial products.

Aluminum castings may be machined in the as-cast, or F condition, or they may be heat-treated to the T-4, T-5, or T-6 condition (see Table 14.3). Some zinc–aluminum alloys age-harden without heat treatment after being cast. As a rule, castings that have age-hardened or have been heat-treated machine more readily. They are also stronger. Some typical aluminum castings are shown in Fig. 14.18.

Many aluminum alloys can be formed by forging in closed dies either by impact or by press forging. The forging operation is usually done with the metal at a temperature well below the melting point. The resulting products, such as the pistons in Fig. 14.19, are almost always stronger than a casting of comparable weight and size and are more resistant to fracture. Forged aluminum parts are almost always heat-treated.

14.4.2 Magnesium

Magnesium has come into general use only in the last 50 to 60 years. It is the lightest of the commonly used metals, weighing about 106 lb/ft^3, approximately two-thirds the weight of aluminum and one-fifth the weight of steel. Magnesium bar and castings are readily machineable, but precautions must be taken to keep the cutting tools sharp and

FIGURE 14.19 Pistons machined from heat-treated aluminum alloy forgings. (*Courtesy Cosworth Engineering, Inc.*)

to avoid taking cuts that leave fine, powdered chips. If a dull tool is used when taking a fine cut, the chips may ignite, resulting in a fire very difficult to extinguish.

Classification System. The classification system for magnesium alloys consists of a prefix of one or two letters that indicates the major alloying elements and the number or numbers showing the percentage of each alloying element. The common alloying elements for magnesium are in Table 14.4. A suffix letter may be used to show changes in the original alloy. For example, alloy AZ63A is about 6.0 percent aluminum, 3.0 percent zinc, and the remainder magnesium.

TABLE 14.4
Major alloying elements in magnesium

Letter	*Alloying Element*
A	Aluminum
E	Rare earths
H	Thorium
K	Zirconium
M	Manganese
Z	Zinc

Characteristics. Magnesium alloys can be heat-treated or cold-worked to increase their tensile strength. The common temper designations for magnesium are:

F—as fabricated or as cast
T4—solution-heat-treated
T5—artificially aged
T6—solution-heat-treated and artificially aged

Magnesium can also be extruded into a variety of shapes or forged. The highest tensile strengths can generally be developed by heat-treating forged or extruded products to the T-6 condition.

14.4.3 Copper and Copper-Based Alloys

Copper has been used longer than any other metal. Bronze (copper and tin) alloys suitable for making tools, weapons, and other hardware were developed as long as 4500 years ago. Copper alloys readily with nickel and many other metals and is the main constituent of the various brass and bronze alloys. It can also be used for plating.

Some other useful properties of copper are its high heat and electrical conductivity and its corrosion resistance. It can be soldered, welded, and joined by silver brazing, and because of its ductility, it can be formed and forged readily.

Classification. Most copper alloys are known by descriptive names as well as by numbers in a classification system. In some cases, numbers are used to indicate content of the main constituents, such as in 70–30 brass (also known as cartridge brass), which is made up of 70 percent copper and 30 percent zinc. A confusing aspect of this classification system is that some materials that are made up of copper and zinc and by general definition would be classed as *brasses* are known as *bronzes.* Manganese bronze, for example, has 59 percent copper, 39 percent zinc, 0.7 percent tin, and traces of manganese.

Bronze. Most bronze alloys contain copper and tin as their main constituents, but there are some bronzes that contain very little tin, if any. Generally, bronzes may contain such alloying elements as aluminum, phosphorus, manganese, and silicon, all of which add strength and other desirable characteristics. When bronze is used as a bearing material, lead is often added in amounts up to about 10 percent. The addition of lead generally reduces tensile strength and improves machineability.

The strongest of the copper alloys is *beryllium bronze,* also known as beryllium copper. It usually contains up to 2.0 percent beryllium, which is a toxic material, and small amounts of nickel and cobalt. It is a heat-treatable material that can be hardened to around Rockwell C-40 scale and is difficult to machine. Sparkproof tools and some aerospace components are made from this material.

Brass. Brasses are copper–zinc alloys to which other metals can be added. When lead is added, the machineability rating of brass is raised substantially. The addition of small amounts of selenium also improves machineability. Brass alloys can be cast, forged, and formed by a variety of processes since in the annealed condition the material is generally quite ductile.

The brass bar stock that is used in general machine shop work and in automatic screw machine operations is cold-drawn and available in various tempers, ranging from soft to fully work-hardened. Seamless brass tubing in a variety of wall thicknesses is also used for machined parts. Some typical machined brass parts are shown in Fig. 14.20.

FIGURE 14.20 Typical parts machined from brass bar stock.

14.4.4 Titanium

Although titanium is used in industry in relatively small amounts, its high strength/weight ratio has made it very useful in the aircraft industry and similar applications. It is about 60 percent as heavy as steel and can be alloyed and heat-treated to achieve ultimate tensile strengths of 180,000 psi.

Titanium can be cast, forged, and formed with some difficulty. It can also be rolled and drawn into sheet, plate, and bar stock. An oxide film forms readily on the freshly machined titanium and prevents further oxidation at temperatures below 600°F.

The machineability of titanium varies from 20 to about 35 percent, using B-1112 steel as the standard. The scale on forgings and other hot-worked parts is highly abrasive, resulting in rapid tool wear. The tendency of the material to gall and adhere to tool bits, milling cutters, and drills requires constant attention to the condition of cutting tools.

14.4.5 Miscellaneous Materials

A number of specialized alloys have been developed to meet certain industrial needs, and these materials are usually processed and finished by machining. Only the more common ones are discussed. It is recommended that the manufacturer's machining recommendations be followed for the more unusual materials.

Nickel-Based Alloys. The most widely used nickel–copper alloy is *Monel,* a combination of about 66 percent nickel, 30 percent copper, and varying amounts of manganese, iron, and silicon. A letter prefix and numbers are used to identify the Monel alloys. Ordinary Monel (no letter designation) is hardened by cold-working, and tensile strengths up to 120,000 psi are attainable. R Monel with about 0.35 percent sulfur added is the most machineable of the alloys. K Monel contains 3.0 percent aluminum and can be precipitation-hardened to about 180,000 psi tensile strength. The H and S Monel alloys, which contain 3.0 percent and 4.0 percent silicon, respectively, are used for castings and can be hardened.

Inconel, in its various forms, is a nickel–chromium alloy that contains up to 6.75 percent iron and smaller amounts of aluminum and titanium. These alloys are strong, heat-resisting materials with a tensile strength of over 200,000 psi. They are very difficult to machine since they are abrasive and tough.

Bearing Metals. The term *babbitt* is used to describe some bearing metals that fall into two main categories: *lead* babbitts and *tin* babbitts. Lead babbitts are used for bearings in slow- and medium-speed industrial machinery and are usually composed of about 80 percent lead and tin, antimony, and copper in varying amounts.

Tin babbitts are much more suitable for high-speed bearings capable of carrying heavy loads. One common tin babbitt contains about 80 percent tin, 8 percent antimony, 10 percent lead, and 2 percent copper. All the babbitt alloys are easy to machine. In some cases they are fitted to the shaft they support by hand scraping and burnishing.

14.5 HEAT TREATMENT

Almost all metals react to some form of heat treatment. Heat may be used to either harden or soften a metal. With metals that can only be work-hardened, for example, heat must be used for *annealing,* or softening, purposes.

The demand for higher strength, more fatigue resistance, and greater wear resistance in metal products has made the heat treatment of metals a highly sophisticated process. Although the basic heat-treating operations consist of heating and cooling metals, the need for precise control of temperatures and cooling rates has led to the development of various furnaces and instruments to eliminate much of the guesswork.

14.5.1 Furnaces

Gas, oil, or electricity may be used as the source of heat in heat-treatment furnaces. Gas is widely used since it is a fairly economical and relatively clean fuel (Fig. 14.21). Oil is also used because of its low cost. Electric furnaces are more expensive to operate but provide a cleaner atmosphere and

FIGURE 14.21 Gas-fired furnace used for heat-treating metals. (*Courtesy El Camino College*)

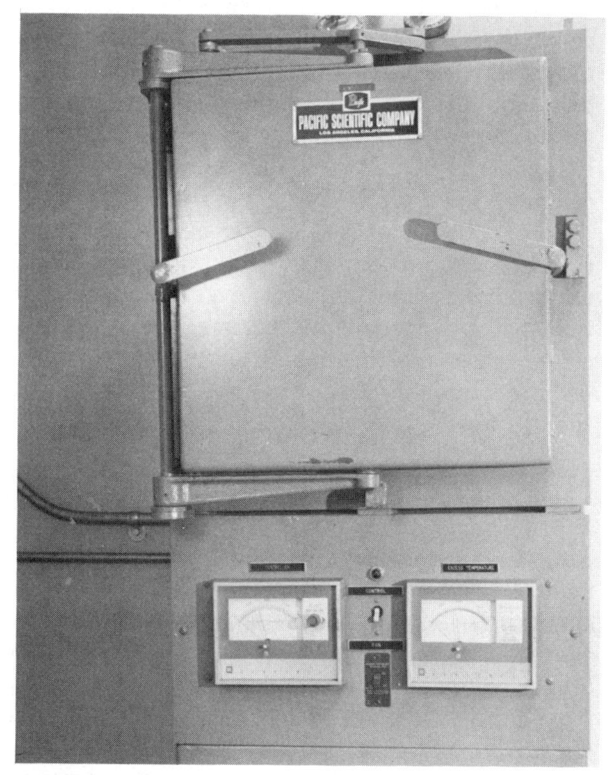

FIGURE 14.22 Electric furnace for heat-treating metals. (*Courtesy El Camino College*)

are easier to control (Fig. 14.22). An inert atmosphere, using argon or a similar gas, may be used in an electric furnace to prevent scaling of parts. The inert gas displaces the air in the furnace.

Hardening furnaces must be capable of producing temperatures in the 2300 to 2400 °F range. Gas furnaces usually have a forced-draft blower driven by an electric motor, which forces a mixture of gas and air into the combustion area. Once the furnace has reached its preset temperature, the control mechanism switches the heat source off or on to maintain the temperature within a ±20 °F range.

FIGURE 14.23 Salt bath type of furnace. (*Courtesy El Camino College*)

FIGURE 14.24 Thermocouple control to hold the furnace temperature at a given level. (*Courtesy El Camino College*)

Tempering furnaces usually operate in the 350 to 1200 °F range and are equipped with controls that maintain the preset heat level. A thermocouple is used as the temperature-sensing device in the furnace or in the pot, as shown in Fig. 14.23).

Pot-type furnaces are used for heat treating metal in salt or lead baths or for carburizing (Fig. 14.23). The advantages of using a salt or lead bath is that scaling is greatly reduced since the parts are not in contact with air while heated. Pot-type furnaces can be closely controlled by the use of immersion-type thermocouples and suitable electrical controls (Fig 14.24).

14.5.2 Heat-Treating Processes: Ferrous Metals

The heat treatment of ferrous metals consists of three basic steps: (1) heating the steel or iron to the proper temperature, (2) holding it at that tem-

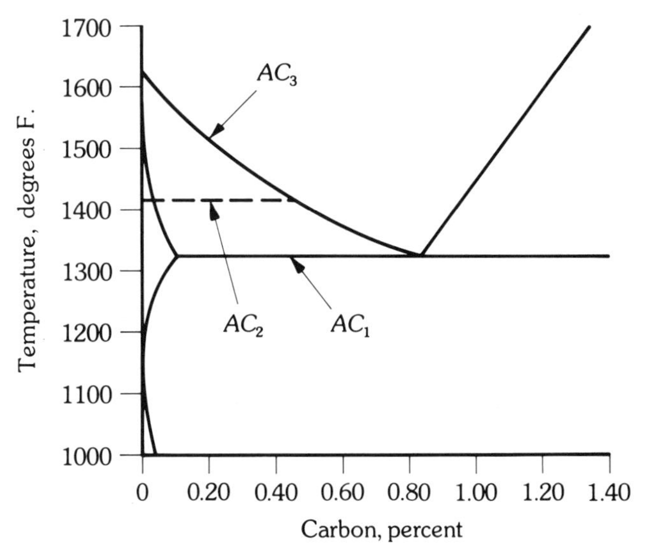

FIGURE 14.25 Iron–carbon diagram.

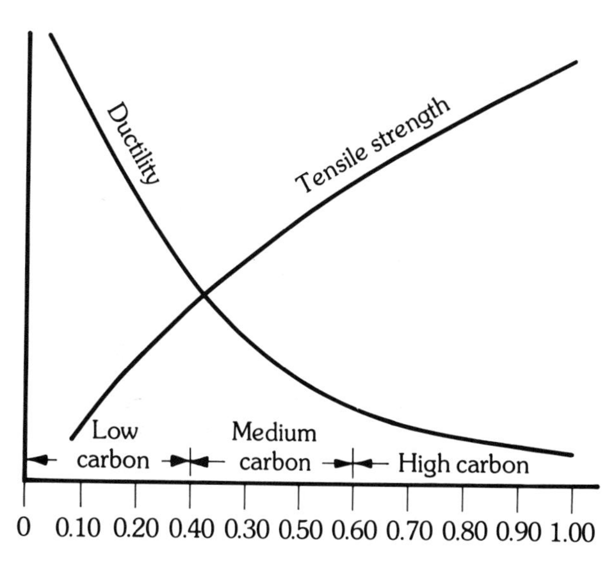

FIGURE 14.26 Relationship between carbon content, ductibility, and tensile strength.

perature (soaking it) until it is at a uniform temperature throughout, and (3) cooling it at the appropriate rate. The factors that differ among the various heat-treating processes are the temperature to which the material is heated and the cooling rate.

A knowledge of the relationship between carbon content and the hardenability of steel along with the use of the chart in Fig. 14.25 will help you understand the various heat-treating processes. Plain carbon steels with less than 0.25 percent carbon are not hardenable except by carburizing or case hardening. Steels in the medium-carbon range are hardenable to a certain extent, depending on carbon content. High-carbon steels (0.60 percent or more carbon content) can be hardened to the extent that they can be used as cutting tools. The addition of certain alloying elements such as manganese increases the depth of the hardened portion of carbon and alloy steels. The ductility and tensile strength of steel parts are also affected by the carbon content, as shown in Fig. 14.26.

Quenching Media. Since the *rate* at which steel is cooled is a major factor in heat treatment, selection of the appropriate quenching medium is of critical importance. The four common quenching media are *brine, water, oil,* and *air.* Brine is the most severe quenching medium, and gives the greatest temperature drop in a given time period. Water is less severe, and oil is the mildest of the liquids used for quenching. Tallow has also been used as a mild quenching material by blacksmiths.

Air-hardening steels may be quenched in still air or with compressed air to lower the temperature more rapidly. The manufacturer's or steel distributor's recommendations should be followed closely with regard to quenching high-carbon or alloy steels during heat treatment since cracks and distortion result from the use of a medium that is too severe.

Salt and Lead Baths. Molten salt at temperatures ranging from about 300 to 2000 °F may be used for quenching high-speed steels or for tempering. Since the temperature of the salt bath can be controlled very accurately, the tempering of complex tool steels is often done by this method. The salt can be easily removed from steel parts with hot water.

Molten lead may also be used for quenching, tempering, and stress-relieving steel parts. The operating range of lead baths is from about 700 to 1700 °F. One problem with the use of molten lead in heat treating is that toxic lead vapors may be released.

Safety Note: Metal parts must be *dry* or *warm* before they are placed in a salt or lead bath. Any moisture on the part turns to steam and causes an eruption of molten lead or salt. Salt and lead bath heat-treating processes must always be done in a well-ventilated area.

Annealing. A metal that has been *fully annealed* is in its softest possible condition and the internal stresses have been relieved. In the case of steel, the temperature is raised to about 50 °F above the upper critical temperature (line AC_3 in Fig. 14.25), at which point the carbon goes completely into solution with the iron. The proper annealing temperature is determined by the carbon content and the presence of other alloying elements. The part is held at this temperature, or *soaked,* about 1 hour for each inch of thickness. Cooling is accomplished by leaving the part in the furnace after it has been shut off or by packing it in lime to retard the cooling rate. A cooling rate of not more than 50 °F per hour is recommended. Most steels have the highest possible machineability rating and are formed most easily in the fully annealed condition.

Normalizing. After normalizing, steel is usually somwhat harder than after annealing and may have a slightly higher tensile strength. The steel is heated to a temperature of 50 to 100 °F above the upper critical temperature (line AC_3 in Fig. 14.25) and is then cooled in still air. It is held at the normalizing temperature only long enough for the entire object to heat evenly because either excessive heat or excessive time at the normalizing temperature results in undesirable grain growth and possible decarburization. Castings, forgings, and machined parts are normalized to relieve stresses caused by hot-working, cold-working, or machining and to produce a more uniform grain structure.

Hardening. The basic process of hardening steel without the addition of carbon and/or nitrogen during the hardening process consists of (1) selection of a plain carbon or alloy steel with 0.30 percent or more carbon content, (2) heating to about 50 to 100 °F above the critical temperature as determined by the carbon content, and (3) cooling the object at the proper rate in an appropriate quenching medium.

The hardening process produces a hard case of varying depth as shown in Fig. 14.27. The plain carbon steels have a relatively shallow hardened area. As certain alloys are added, notably manganese and chromium, the depth of the hardened area increases, leaving a much smaller soft core.

The *rate* at which the steel is cooled affects the hardness, with rapid cooling resulting in the greatest hardness. If the object is thin or complex, rapid cooling may distort or crack it, so the shape of the part must be considered in choosing a cooling medium. Steel parts should not be left in the fully hard condition since cracking may occur because of the stresses caused by quenching.

Tempering should be done as soon as possible after hardening to reduce the possibility of cracking.

Pack Carburizing. In this process carbon is added to the outer surface of low-carbon steel parts by heating the parts to the range 1650 to 1800 °F while in close contact with a carbonaceous material such as petroleum coke, charcoal, or peach pits. The steel parts are usually placed in a crucible with a lid or in an austenitic stainless steel box. Because low-carbon steels are austenitic (a solid solution of all its components) above the upper critical temperature, carbon is absorbed and the surface becomes high-carbon steel. The depth of carbon penetration is determined by the amount of carbon available and the length of time the parts are held at the carburizing temperature, usually 3 to 10 hours.

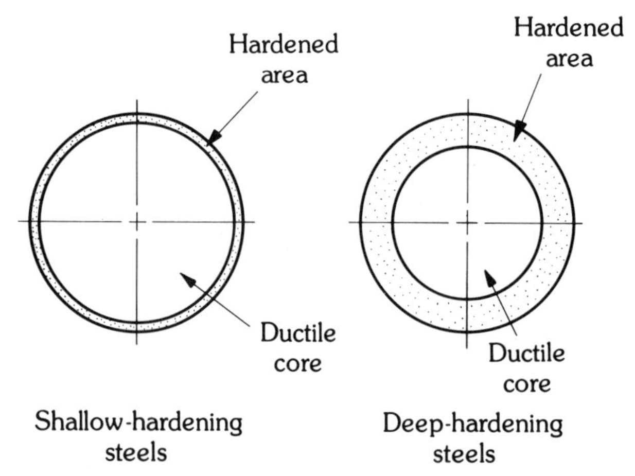

FIGURE 14.27 Variations in depth of hardened area.

The carburized part may be quenched in oil or water and finished by grinding, lapping, or honing. The resulting surface will be very hard and wear resistant, but it will be somewhat brittle unless tempered to the desired toughness. Carburized objects may be tempered in the range 300 to 400 °F to prevent chipping and cracking of the hardened surface.

When it is necessary to carburize only certain surfaces of a complex part, such as the small crankshaft in Fig. 14.28, the surface areas that are *not* to be carburized (such as threads) are plated with copper. The copper acts as a barrier, preventing carbon from entering the hot steel.

Cyanide Hardening. Sodium cyanide, potassium cyanide, and other salts are the hardening medium in which the parts to be hardened are immersed. The molten cyanide at about 1400 to 1650 °F releases carbon and nitrogen that are absorbed by the steel. The resulting case is relatively thin (0.002 to about 0.020 in.) but very hard. The parts are usually quenched directly after removal from the salt bath and may be tempered.

Safety Note: Cyanide salts and their fumes are *very* poisonous and must not be taken internally, allowed to enter cuts, or inhaled. Proper ventilation of the work area is *essential.*

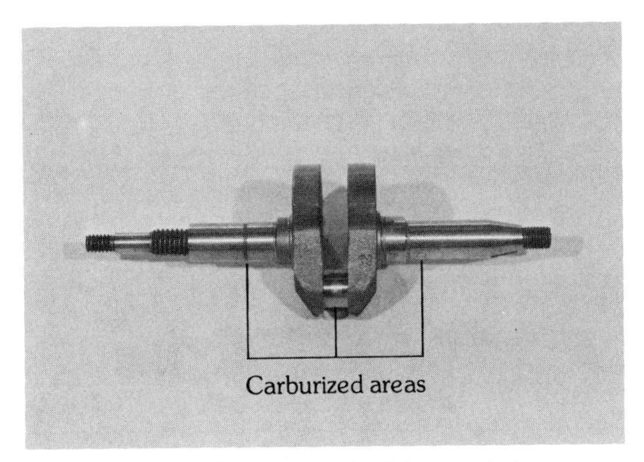

FIGURE 14.28 Selective carburizing—only bare areas absorb carbon.

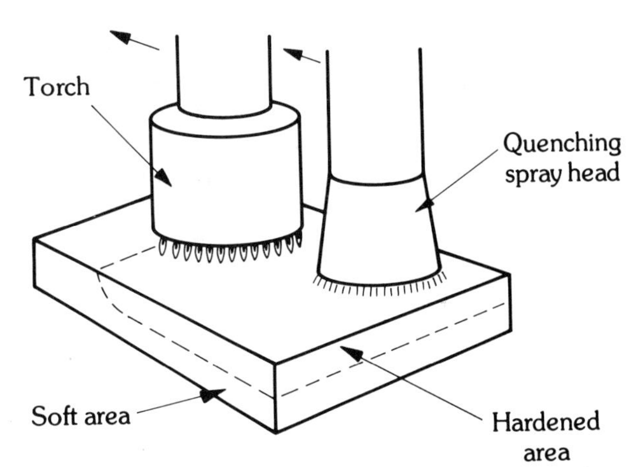

FIGURE 14.29 Flame-hardening the surface of an object.

Nitriding. In nitriding, nitrogen is added to the surface of some special steels by heating the part in an ammonia gas atmosphere or in a salt bath. The steel is usually heated to a temperature of 900 to 1000°F, and there is no distortion or scaling. The process is quite slow, and the hardened case on the part usually ranges from 0.001 to 0.005 in. in thickness. If a thicker case is desired (up to 0.015 to 0.020 in.), the process may take up to 96 hours, which makes it quite expensive. Nitriding produces a very hard surface in the Rockwell C-70 range that is highly wear-resistant. No quenching is required. Cutting tools made of high-speed steel and aircraft engine parts, such as cylinders and crankshafts, may be nitrided to extend their operating life.

Flame Hardening. This process can be used with carbon or alloy steels with more than 0.30 percent carbon and with most types of malleable or ductile cast irons. Basically, the process consists of rapidly heating the surface of the object to above the critical temperature with a gas or oxyacetylene flame and immediately quenching it by a water spray or air jet that closely follows the flame (Fig. 14.29). The part may then be tempered, if desired. Parts of many different shapes, including the ways of machine tools and gear teeth, can be hardened by this method. The surface may be hardened to depths ranging from 0.025 in. to as much as 0.200 in. Any subsequent machining must be done by grinding, honing, or lapping. The process is widely used in the machine tool, automotive, and farm equipment industries, among others.

Induction Hardening. Circular cross-sectional parts, such as shafts, gears, and similar objects, can be surface hardened by the induction process if they contain more than 0.30 percent carbon. The source of heat is a high-frequency induction coil placed around the part, as shown in Fig. 14.30. The entire surface of the part, or only certain portions, may be hardened. On an automotive engine camshaft, for example, only the cam lobes and the bearing surfaces are hardened, as shown in Fig. 14.31. These surfaces are then finished by grinding.

The rapid reversals of polarity that occur in the high-frequency alternating magnetic field cause

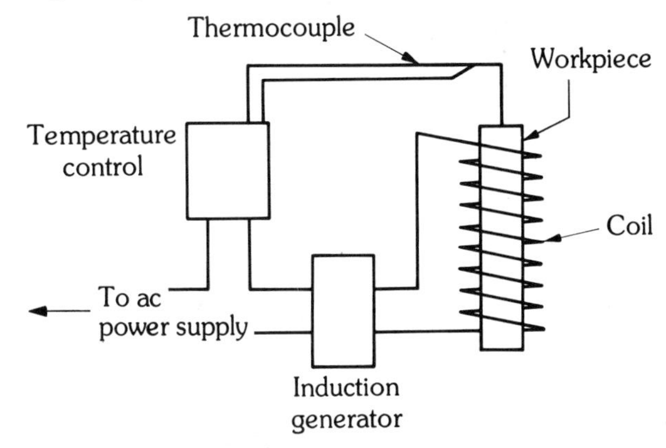

FIGURE 14.30 Schematic diagram of induction-hardening equipment.

FIGURE 14.31 Automobile engine camshaft that has been induction-hardened.

the surface of the object to become heated. The current frequency may range from about 1000 to more than 500,000 cycles per second. As the frequency increases, the heated area becomes shallower, and at the higher frequencies the hardened case may be only 0.010 to 0.015 in. thick. When lower frequencies are used, the case may be up to 0.200 in. thick.

Tempering. The tempering process may vary considerably, depending on the composition of the steel, the hardening process used, and the balance between hardness and shock resistance, or toughness, desired. Basically, tempering consists of heating a hardened steel part to a predetermined temperature below the critical temperature and cooling it in still air or quenching it in water, oil, or a salt or lead bath. The ultimate tensile strength drops as the tempering temperature is raised, as shown in the chart for AISI–4130 steel in Fig. 14.32, but the toughness of the material increases, as shown by the increase in elongation before failure.

Tempering temperatures range from 300 to 1200 °F, depending on the material being tempered and the toughness or hardness desired. Plain carbon steels may be tempered by using a simple furnace or an oxyacetylene or gas torch, and a means of polishing part of the surface of the object. The temperature of the steel can be estimated by the

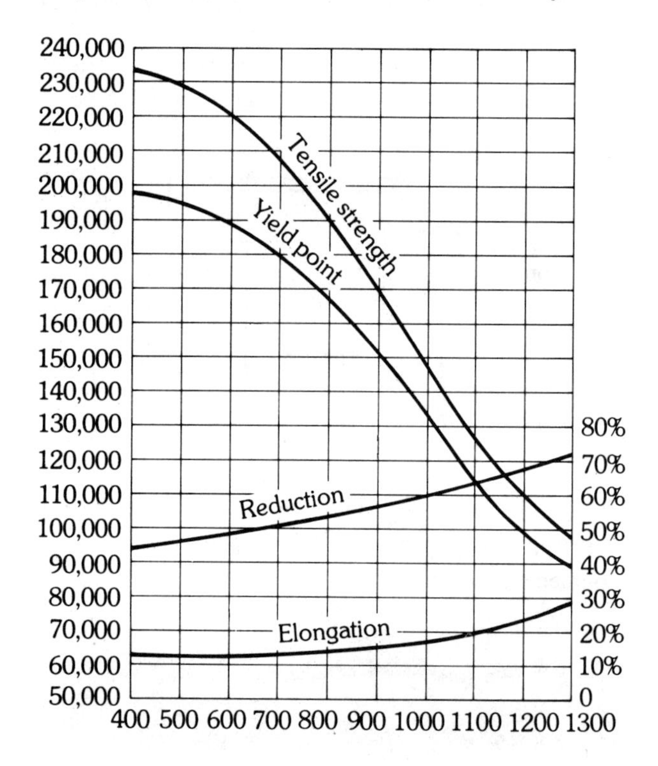

FIGURE 14.32 Effect of different tempering temperatures on AISI–4130 steel.

TABLE 14.5

Tempering color–temperature relationships

Color	*Temperature (°F)*	
Light blue	570–600	Soft
Violet	540–570	
Purple	510–540	
Brown	480–510	
Medium yellow	450–480	
Straw	420–450	
Light straw	390–420	Hard

temper colors that appear on the polished surface as heat is applied (see Table 14.5 for typical color–temperature relationships). After the steel has been slowly brought up to the proper temperature, as indicated by the oxide colors on the polished portion, it is allowed to cool in still air or quenched in water or oil.

14.6 METAL-TESTING PROCESSES

The testing of metal and metal parts during various stages is a necessary part of the quality assurance process in industry. The two general categories of tests are *destructive* and *nondestructive.* Destructive testing includes tensile, bending, torsion, compression, fatigue, and impact testing. It is designed to determine the failure point and other characteristics of materials such as bar stock and semifinished or finished parts that are intended to be tested to destruction. Nondestructive tests such as dye-penetrant, magnetic particle, ultrasonic, and radiography are used to inspect welds, castings, forgings, and machined parts. They are widely used in the aircraft industry and in the manufacture of other highly stressed machinery.

14.6.1 Hardness Testing

Although almost all hardness testing procedures leave a mark or identification on the object tested, determination of hardness is not considered a destructive testing process. The simplest way to check the hardness of a piece of material is to use a fine three-cornered file. This method has been used by machinists for many years, and it gives a good indication of the machineability and other characteristics of the material being tested. As shown in Fig. 14.33, the file method can be used to check the approximate hardness of objects that cannot be placed in a bench-type tester or are beyond the capacity of portable testers. As a general rule, any material that cannot be pene-

FIGURE 14.33 File being used to test the hardness of a steel object.

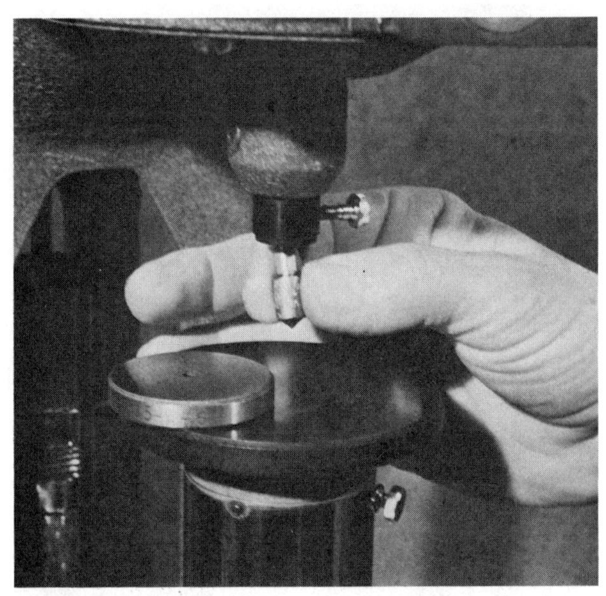

FIGURE 14.34 Installing a penetrator on a hardness tester.

trated by the teeth of a good file will be in the range Rockwell C-50 to C-60 and must be machined with abrasives or carbide tools.

Rockwell Hardness Tester. The Rockwell Hardness Tester determines the hardness of a material by applying a known pressure to a penetrator and indicating the depth of penetration as a scale valve. The Rockwell B and C scales are the most widely used in machine shop work. The B scale is used for measuring the hardness of most nonferrous metals, cast iron (except white iron), and unhardened steels. The penetrator for the B scale is a 1/16 in.-diameter hardened steel ball. The minor load is 10 kilograms (kg) and the major load is 100 kg.

The C scale is used to measure hardened steel and other metals whose hardness lies beyond the range of the B scale. A diamond penetrator with a conical 120° angle point is used. The minor load is 10 kg, and the major load is 150 kg.

A number of manufacturers make Rockwell Hardness Testers, and each has minor operating variations. The instructions for a particular machine should be read carefully and followed closely. However, general instructions for performing the Rockwell test are as follows:

1. Identify the material to be tested. Select the proper penetrator. Check it for nicks or cracks, and install it as shown in Fig. 14.34.
2. Install the proper anvil. (An anvil with a V slot is used when testing round specimens.)
3. Make sure that the proper major load weights are in place for the scale being used (100 kg for the B scale and 150 kg for the C scale).
4. Make sure that the specimen is free of scale and foreign matter, and place it on the anvil.
5. Bring the item to be tested into contact with the penetrator by turning the handwheel.
6. Apply the minor load and set the hardness scale to the set point.
7. Apply the major load and let the needle come to a stop.
8. Remove the major load and read the hardness on the appropriate scale.
9. Turn the handwheel to release the minor load and remove the specimen. It is recommended that the test be repeated several times and the readings be averaged.

Brinell Hardness Tester. The Brinell test is used on both ferrous and nonferrous metals, except for very hard steels. A steel or tungsten carbide ball 10 mm in diameter is forced into the surface of the material to be tested. The load can be applied hydraulically or by a handwheel, as shown in Fig. 14.35.

The operating principle of this machine is that a given load forces the ball deeper into soft materials than into hard materials. The diameter of the depression is measured with a Brinell Microscope, and a *Brinell number,* which is an indication of the hardness, is derived.

FIGURE 14.35 Brinell Hardness Tester. (*Courtesy Tinius Olsen Testing Machine Co., Inc.*)

For soft materials and almost all nonferrous metals, a 500-kg load is used; for steel the load is 3000 kg. A time element is involved. For steel the load must be applied for 15 seconds, and for nonferrous metals, 30 seconds.

Scleroscope Hardness Tester. The operating principle of the scleroscope tester is that a diamond tipped weight, when dropped onto a hardened steel surface from a given height, rebounds farther from harder surfaces. The basic scleroscope consists of frame, adjusting mechanisms, and a graduated glass tube through which the weight falls. The scale on the tube has 140 graduations. Very hard steels may cause the weight to rebound to the 110 mark; mild steel produces a reading of about 30. Very soft materials, such as lead and tin, make the weight rebound only to the 2 to 4 level. Scleroscope testers are bench-mounted or portable.

The approximate equivalents of Brinell numbers to the Rockwell C and other scales and to the Shore Scleroscope hardness number are shown in the table in the appendix. The approximate tensile strength of a material can be found in this table if the hardness is known. The ultimate tensile strength may also be determined by dividing the Brinell reading by 2 and multiplying by 1000. (*Example:* A steel with a Brinell hardness number of 200 has a tensile strength of about 100,000 psi.)

14.6.2 Spark Testing

The machinist often needs to identify ferrous metals. Some metals, such as gray cast iron, can be recognized easily, but others (those in bar form)

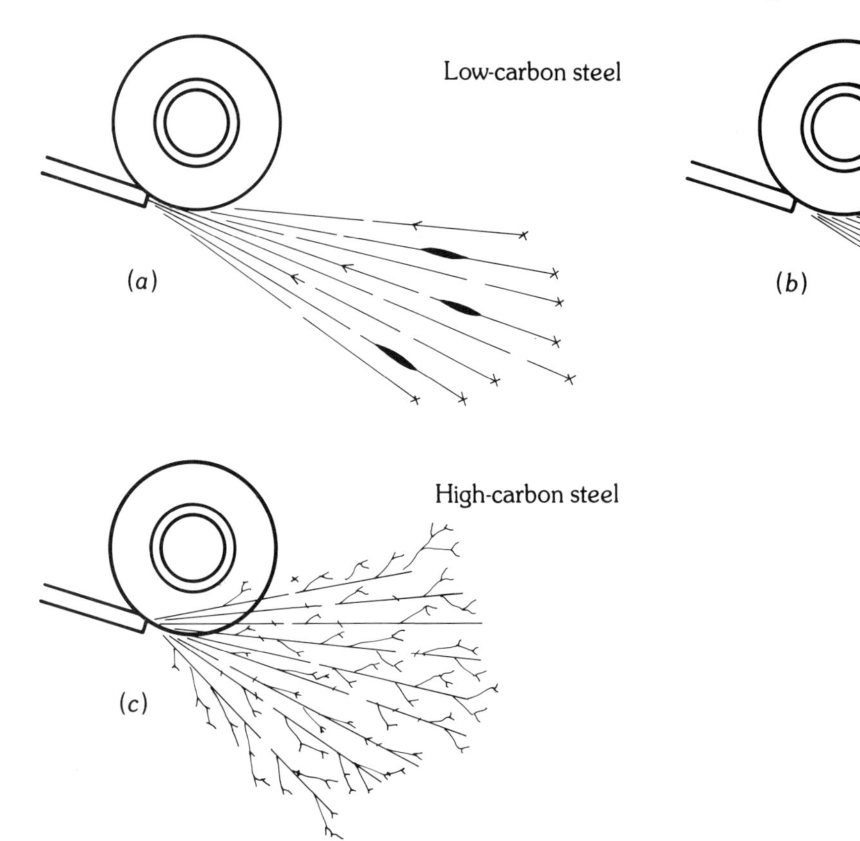

FIGURE 14.36 Spark testing with a grinding wheel may be used to identify certain types of steels.

must be spark tested. Although the spark testing procedure is not particularly accurate, the carbon content can be determined within certain limits. The steel bar being tested is brought into contact with a grinding wheel, and the sparks are examined.

As shown in Fig. 14.36, the sparks from low-, medium-, and high-carbon steels vary considerably. As the carbon content increases, the white stars, or bursts, at the end of the shaft of each spark are more numerous. High-speed steels produce short, red sparks that end in small, dark stars, or bursts.

The material in this chapter is not intended as a complete coverage of the extensive field of metallurgy. The topics introduced are of direct use to the machinist while selecting materials and processes that result in reliable, economical, and appropriate finished parts.

As machined products become more complex, more testing equipment of all types will be used in machine shops. The machinist who understands the operating principles of the equipment and the basic testing procedures will find advancement opportunties greatly increased.

REVIEW QUESTIONS

14.2 PHYSICAL CHARACTERISTICS

1. Briefly explain tensile stress, compressive stress, and shear stress.
2. What happens to a metallic material that is stressed beyond its *elastic limit*?
3. Explain how the fatigue resistance of a machine part may be improved.

14.3 FERROUS METALS

1. In what form is the major part of the carbon in gray cast iron, and how does this affect the tensile strength?
2. Briefly describe the structure of malleable iron after it has been heat-treated.
3. How does the addition of small amounts of lead to low-carbon steel affect the machineability?
4. Briefly explain the difference between cold-finished and hot-rolled steels of the same chemical analysis.
5. What effect does the addition of nickel have on the heat-treatment characteristics of alloy steels?
6. What are the two major alloying elements used in 300-series stainless steels?
7. By what means are *martensitic* stainless steels hardenable?
8. Identify the three major categories of tool steels in terms of the quenching medium used during hardening.
9. Why are high-speed steels more durable than high-carbon steels as cutting tool materials?

14.4 NONFERROUS METALS

1. How does the addition of silicon to aluminum alloys affect machineability?
2. What are the alloying elements used in the 2XXX-, 6XXX-, and 7XXX-series wrought alloys?
3. What precautions should be taken when machining magnesium?
4. What effect does the addition of lead have on the machineability of brass or bronze?
5. What alloying element is sometimes added to copper to make it very strong?
6. What characteristics of titanium make it relatively difficult to machine?
7. What are the main alloying elements in Monel metal?

14.5 HEAT TREATMENT

1. List the three steps in hardening medium- or high-carbon steel.
2. What four common quenching media may be used in heat treatment?
3. Describe the process of annealing a medium- or high-carbon steel object.
4. What materials can be used as sources of carbon in pack carburizing?
5. Briefly explain the process of nitriding.
6. Explain why hardened steel parts must almost always be tempered.
7. What ferrous metals can be flame hardened, and what is the minimum carbon content?

14.6 METAL-TESTING PROCESSES

1. What two major categories of tests can be used on metal objects?
2. How may the relative hardness of a metal object be tested with a file?
3. Briefly explain the operating principle of the Rockwell Hardness Tester.
4. What characteristics may be identified by spark-testing ferrous metals?

15

Metal-Cutting Theory and Cutting Fluids

15.1 INTRODUCTION

In metal-cutting operations, the objectives are to remove material rapidly and economically, produce the best surface finish possible, and get the longest possible cutting tool life. In actual practice, all of these objectives are seldom achieved.

Since the machining characteristics of metallic materials vary greatly, many different types of cutting tool materials are used. Cutting fluids also affect the metal cutting process drastically, a factor that the machinist must consider. Therefore, an understanding of the nature of cutting tool materials, their geometry, cutting fluids, and the machineability of at least the common metals is necessary for anyone wishing to become an expert machinist. The ability to select the proper processes and materials and to resolve problems as they occur will save time and money, and give better results.

15.2 CUTTING TOOL MATERIALS

Hard metals have been used to cut or deform other metals for many thousands of years. Before the development of powered machine tools, the lack of good cutting tool materials was not a serious problem. Within the last 150 years, however, better

Opening photo courtesy of The Pratt & Whitney Co., Inc.

cutting tool materials have been invented or developed. Generally, as better materials became available, larger and more powerful machine tools were built to produce machined parts more rapidly and economically.

15.2.1 High-Carbon Steel

High-carbon steels have been used longer than any of the other cutting tool materials. They are still used for low-speed machining operations and for some cutting tools for wood and plastics. They are relatively inexpensive and easily heat-treated, but they will not withstand severe use or temperatures above 350 to 400°F. Many taps, dies, hand reamers, files, inexpensive drills, and other similar tools are made of high-carbon steel (see Fig. 15.1); refer also to Table 15.1.

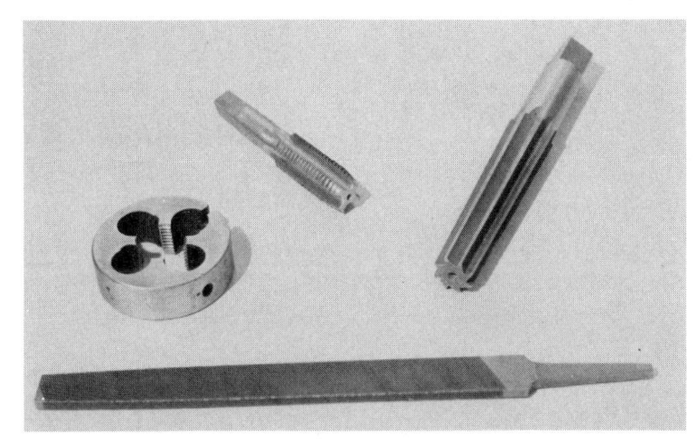

FIGURE 15.1 Cutting tools made of high-carbon steel.

Steels in this category are hardened by heating to above the critical temperature, quenching in water or oil, and tempering as desired. When tempered at 325°F, the hardness may be as high as Rockwell C-62 to C-65. High-carbon steel cutting tools are often nitrided at temperatures ranging from 930 to 1000°F to increase the wear resistance of the cutting surfaces and reduce galling and seizing. Some high-carbon steel taps are also treated in this manner.

Note that cutting tools of hardened high-carbon steel must be kept cool while being resharpened. If a blue color appears on the part being ground, it probably has been accidentally softened to some extent.

15.2.2 High-Speed Steel

The development of high-speed steel was a major advance in cutting tool technology. It was known as early as 1870 that the addition of large amounts of tungsten (up to 18 percent) to carbon steels allowed them to retain their hardness at much higher temperatures than plain high-carbon steels. A major breakthrough occurred when metallurgists developed the proper heat-treating processes to make high-tungsten-content steels into usable cutting tool materials.

The introduction of the high-speed-steel alloy known as 18–4–1 (also known as T-1), which retains its cutting edge at temperatures up to the range 1000 to 1100°F, allowed cutting speeds for high-carbon steel tools to be doubled in some cases. Tool durability and the time between resharpenings were also greatly increased. This also led to the development of more powerful and rigid machine tools and greater productivity.

The basic 18–4–1 (T-1) high-speed steel contains 18 percent tungsten, 4.1 percent chromium, 1.1 percent vanadium, about 0.70 to 0.80 percent carbon, 0.30 percent manganese, 0.30 percent silicon, and the remainder is iron. Many variations of this alloy have been developed, most of which have cobalt and 0.70 to 0.80 percent molybdenum added. Increasing the vanadium content to 5 percent (T-15) improves the wear resistance. Tungsten-type high-speed steels may have as much as 12.0 percent cobalt added and are then known as either *cobalt-type* or *super-high-speed steels* since the resistance to heat is increased.

Molybdenum-type high-speed steels contain only about 1.5 to 6.5 percent tungsten but have 8.0 to 9.0 percent molybdenum, 4.0 percent chromium, and 1.10 percent vanadium, along with 0.30 percent each of silicon and manganese and 0.80 percent carbon. The *molybdenum–tungsten* high-speed steels, also known as 6–6–2, 6–6–3, and 6–6–4 steels, consist of approximately 6 percent molybdenum, 6 percent tungsten, and vanadium in amounts ranging from about 2 to 4 percent. The carbon content ranges up to 1.30 percent, and is higher than in other common high-speed steels.

High-speed steels are used for cutting tools for both metallic and nonmetallic materials. In some cases, complex cutters are cast to shape, annealed, machined, and heat-treated. Some typical high-speed-steel cutting tools are shown in Fig. 15.2.

FIGURE 15.2 Typical cutting tools made of high-speed steel.

TABLE 15.1

Classification and characteristics of tool steels

Type	*Group*	*Machin-ability*	*Principal Alloying Elements*	*Carbon Content (%)*	*Characteristics*	*Applications (Tools)*
Water-hardening tool steels	W	100	Carbon, chromium, vanadium	0.60–1.40	Easy to machine; hard case, tough core; inexpensive	Light- or medium-impact tooling; cold heading; knurling, coining
Shock-resistant tool steels	S	85	Silicon, chromium, tungsten, molybdenum (increased hardness, heat and wear resistance)	0.30–0.70	Oil and water hardening, shock resistant at normal temperatures	Chisels, shear blades, punches
Oil-hardening cold-worked steels	O	80	Silicon, manganese, chromium, tungsten	0.90–1.45	Safer hardening, less dimensional change; inexpensive, readily available	Blanking, bending, shearing, gages
Air hardening	A	85	Manganese, chromium, tungsten, vanadium	0.70–2.25 (A7)	Excellent dimensional stability; good wear resistance; harder to machine	Intricate shapes; long slender broaches, delicate dies, etc.
High-carbon, high-chromium cold-worked steels	D	40–50	12% Chromium, silicon, tungsten, vanadium	1.00–2.35	High wear resistance; deep hardening properties, low dimensional change	Dies, master gages; long-wearing tools
Hot-worked tool steels	H	75 50 60	Chromium Tungsten Molybdenum	0.25–0.60	Deep hardening Resists high temperatures Same as for tungsten, less expensive	High work temperatures; hot forge dies, hot shears, molding dies
High-speed tool steels	T M	40–55 45–60	Tungsten, cobalt Molybdenum	0.70–1.00 0.80–1.50	High red hardness High abrasion resistance	Cutting tools, dies
Special-purpose tool steels	L	90	Chromium, molybdenum, vanadium	0.50–1.00	Good wear resistance	Bearings, rollers, form tools, dies, molds
	F	75	Tungsten, chromium	1.00–1.25	Resistant to wear and low-temperature shock loadings	
(Low carbon)	P	75–100	Chromium, molybdenum, nickel	0.03–0.07	May be carburized, easy to machine	

15.2.3 Cast Alloys

The term *cast alloys* refers to materials made up of about 50 percent cobalt, 30 percent chromium, 18 percent tungsten, and 2 percent carbon. The alloy content of these nonferrous metals varies, but cobalt is the dominant material, and cutting tools made of the cast alloys (commonly called Stellite) remain hard up to 1500 °F. The hardness is approximately Rockwell C-60 to C-62. This material cannot be annealed and always air-hardens when it cools. Therefore, cast-alloy tools are cast to shape and finished by grinding.

Because of their ability to resist heat and abrasion, the cast alloys are used for some engine and gas turbine parts and for cutting tools. They are also very corrosion-resistant and remain tough at temperatures up to 1500 °F, but are somewhat more brittle than high-speed steels.

15.2.4 Cemented Carbides

Carbide tools, also referred to as *sintered* carbide tools, are capable of operating at cutting speeds about three times as fast as high-speed steel. They came into common use about 50 years ago and were responsible for a large increase in machine productivity. Machines with higher horsepower and greater rigidity were developed to take advantage of the carbide tools.

The major constituent of tungsten carbide cutting tools is tungsten carbide powder which is composed of 95 percent powdered tungsten and 5 percent pure carbon in finely powdered form. These two materials are heated and combined, forming extremely hard tungsten carbide particles. This is mixed with 5 to 10 percent powdered cobalt, which serves as the binder, and a small amount of paraffin wax. This mixture, to which titanium carbide or tantalum carbide may be added in varying amounts to change the characteristics of the tool, is compressed into a form slightly larger than its finished shape. The part is presintered by heating it to about 1500 °F to burn out the wax. It is then reheated to the range 2500 to 2600 °F to complete the sintering process. At this point, the cobalt melts and forms a binder, or *matrix,* around the carbide particles, which have not melted. The hardness, which is partly dependent on the amount of pressure used to form the presintered parts, ranges from 85 to 90 on the Rockwell C scale.

The amount of cobalt used to bind the carbides together affects the toughness and shock resistance of the tool. Tools with more cobalt binder are more shock-resistant, but not as hard.

Carbide tools are classified into two major categories. One category is composed of the straight tungsten carbides (classes 1 to 4), which are hard and have very good wear resistance. They are best suited to machining cast iron, nonferrous metals, and some nonmetallic materials that are abrasive. The harder grades of carbides may also be used for drawing dies, or other applications where wear resistance is important and the shock loadings are slight.

The second category (classes 5 to 8) includes the combinations of tungsten and tantalum or titanium carbide. In some cases all three are used. These carbides are generally used for machining steel. They are resistant to cratering, which is a serious problem when tungsten carbide is used to machine steel. (Refer to Table 15.2 for additional information.)

Carbide inserts can be clamped in position or silver-brazed to the toolholder for turning operations (Fig. 15.3). Milling cutters may have carbide inserts that are clamped in place or brazed (Fig. 15.4). Some mills are made from solid carbide of the appropriate grade.

15.2.5 Ceramic Tools

Ceramic cutting tools are made of aluminum oxide powder that is compacted and sintered into triangular, square, or rectangular inserts, as shown in Fig. 15.5. They may be sintered without a binder or with a small amount of some type of glass. They have been in use for only 30 to 35 years and cannot be used effectively on low-powered machine tools. Very rigid and powerful machine tools are necessary to make full use of these tools' heat

TABLE 15.2

Cemented carbide classification

Class	*Type*	*Application*
For Machining Cast Iron, Nonferrous Metals, and Nonmetallics		
1	Tungsten	Roughing cuts
2	Tungsten	General-purpose
3, 4	Tungsten	Light cuts and finishing operations
For Machining Steel		
5	Tungsten	Roughing cuts
6	Titanium	General-purpose
7, 8	Tungsten, titanium, and tantalum in varying amounts	Light cuts and finishing operations

Note: Other grades of carbides are available for special applications, such as interrupted cuts and jobs requiring extremely long tool life. Manufacturers' handbooks should be used to help select cutting tool materials for unusual operations.

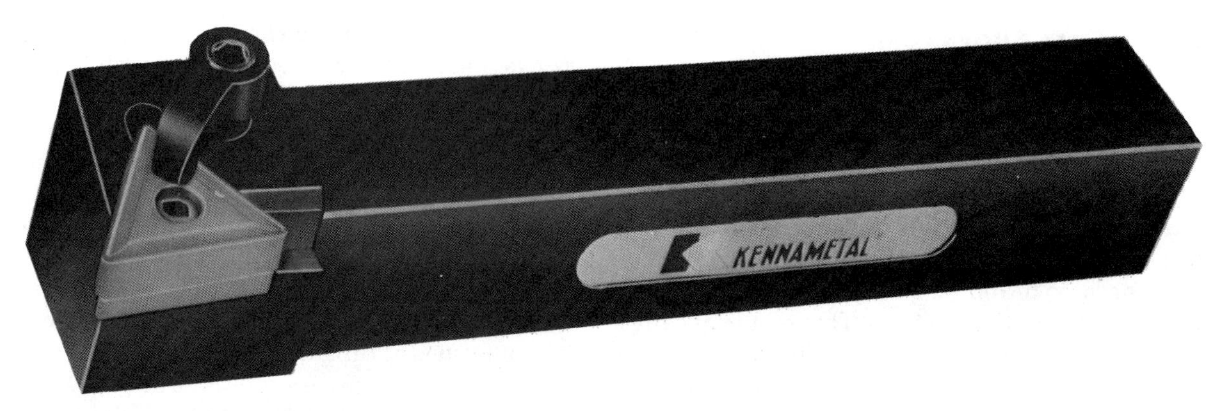

FIGURE 15.3 Lathe cutting tool with carbide insert. (*Courtesy Kennametal Inc.*)

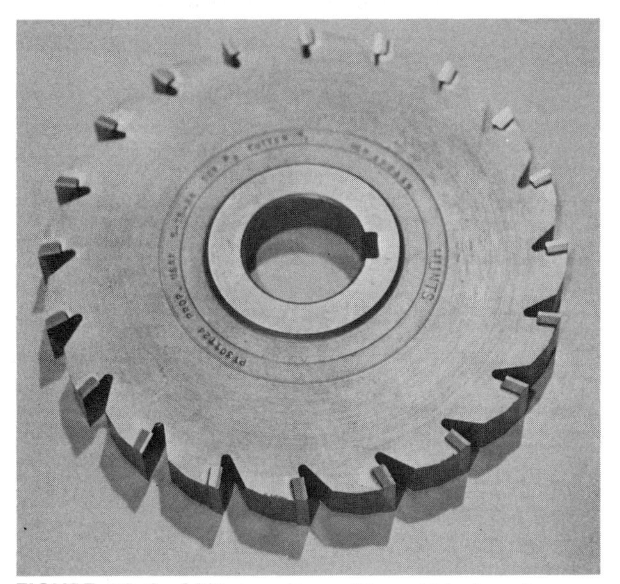

FIGURE 15.4 Milling cutter with carbide inserts.

FIGURE 15.5 Ceramic inserts for cutting tools.

resistance and hardness. Ceramic tools are very hard and chemically inert but are more brittle than carbides and other cutting tools.

Ceramic tool inserts can be made by the *cold-pressed* or *hot-pressed* method. Cold-pressed tools are compacted at a pressure of about 40,000 to 50,000 psi and are then sintered at temperatures of 2000 to 3000 °F. Hot-pressed ceramic inserts are sintered while under pressure and are denser.

The compressive strength of ceramic tools is very high, and they have low heat conductivity. Since they are quite brittle, they must be very well supported in the toolholder or they can be easily damaged or broken if the machine vibrates or chatters. Ceramic tools are very wear-resistant and in the proper machine can be operated at about twice the cutting speed of carbide cutting tools. In some cases, they can be operated at even higher speeds. Ceramic tools should not be used for interrupted cuts.

In recent years, diamonds have become more widely used as single-point cutting tools. They are particularly effective when used to cut highly refractory and abrasive metals such as high-silicon content aluminum alloys. The mass production of automotive pistons is an example of the effective use of diamonds to increase production rates and produce better finishes on machined parts. Although diamond tools are expensive, the increase in the number of parts produced between tool changes makes them cost-effective.

15.3 METAL CUTTING: THEORY AND PRACTICE

The manner in which a single-point tool actually cuts has been the subject of much speculation and discussion. It was first thought that the tool peeled off metal in the form of chips in the same manner that a knife is used to cut shavings from a piece of wood.

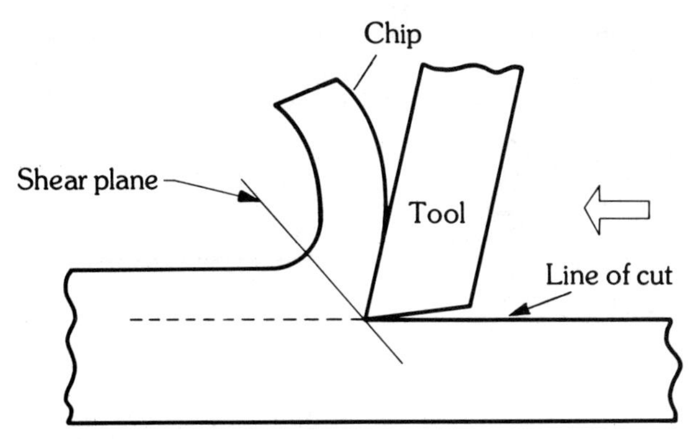

FIGURE 15.6 Metal-cutting operation.

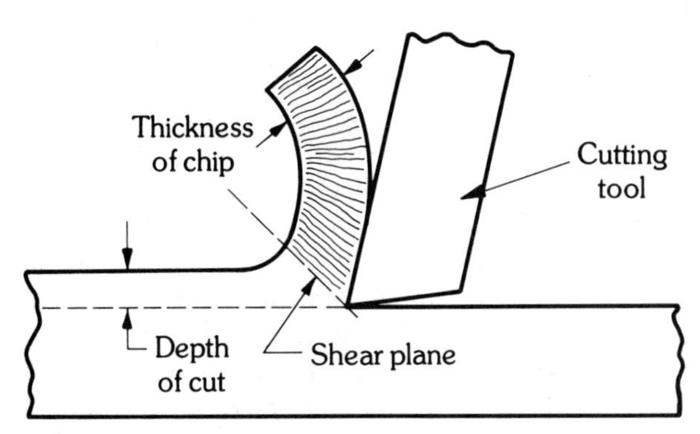

FIGURE 15.7 Metal chip showing deformation.

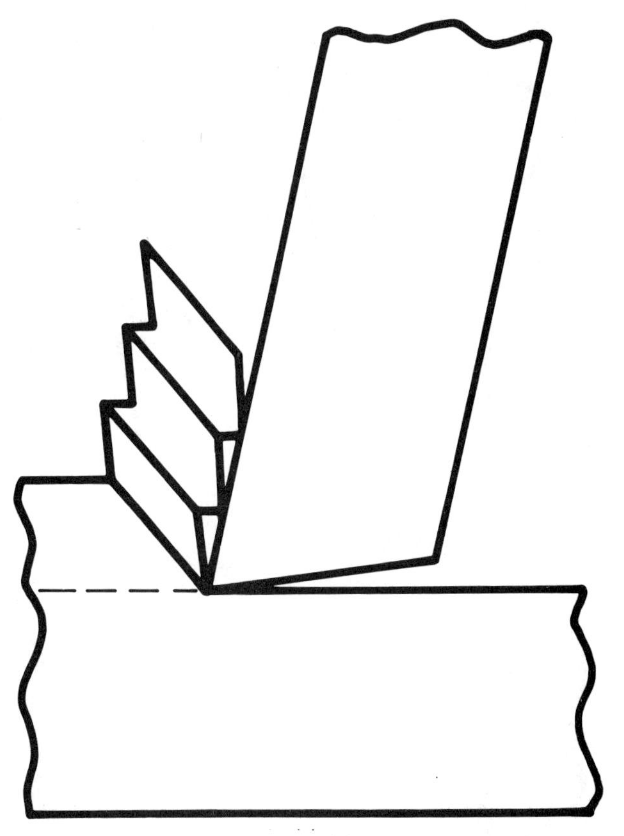

FIGURE 15.8 Segmented chip.

Actually, the chip is formed by deformation and failure of the metal immediately ahead of the cutting tool, as shown in Fig. 15.6. The location and angle of the *shear zone* and the nature of the chip are determined by the geometry of the tool and the physical characteristics of the material being cut. All chips are formed as a result of the shearing action that takes place when the tool moves metal on the surface of the workpiece. The deformation that takes place in shown in Fig. 15.7.

15.3.1 Chip Types and Characteristics

Segmented Chips. Materials that fracture easily, such as gray cast iron, produce a segmented chip, as shown in Fig. 15.8. As the tool moves along the workpiece, metal is displaced ahead of it. Since the metal is brittle rather than ductile, it fails, or breaks, after being only slightly deformed and displaced. Chips of this type are easily removed from the machine, so a chip breaker is usually not required on the tool.

Continuous Chips. When soft or medium hardness materials that are ductile and have a low coefficient of friction are machined, the chip is usually long and continuous. Depending on the material being machined and the rake angles on the top of the tool, the chip curls, or is straight and stringy. A chip breaker is used to fracture the chip into small bits so it does not get tangled in the machine and become a safety hazard. The formation of a continuous chip is shown in Fig. 15.9.

If the material being cut has a low coefficent of friction and the top of the toolbit is well polished, the material will not adhere to the toolbit. For example, some end milling cutters and drills are made with polished or hard-chrome plated flutes for machining soft, ductile materials.

Built-Up Edge. Soft materials that have a high coefficient of friction tend to stick to the top of the entering edge of the toolbit, causing what is known as a built-up edge (see Fig. 15.10). This is caused by the heat and pressure of the cutting operation, which in turn causes the material being machined to temporarily weld to the tip of the cutting tool. This change in the geometry of the tip of the cutting tool tends to generate more heat and usually results in a rougher surface finish. Tool life is also shortened because of excess heat and cratering (Fig. 15.11).

There is no single solution to the problem of built-up edges on drills, lathe cutting tools, and milling cutters. Changes in tool geometry, and the

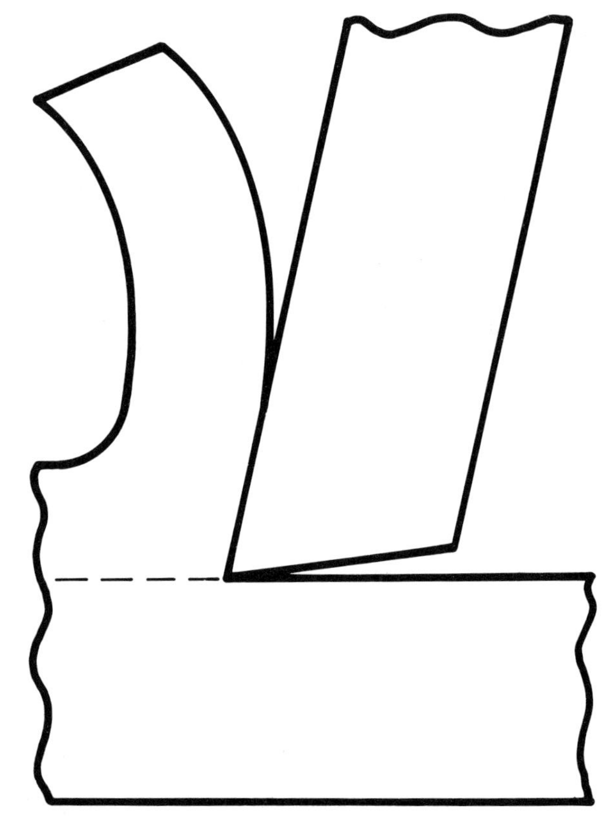

FIGURE 15.9 Continuous chip.

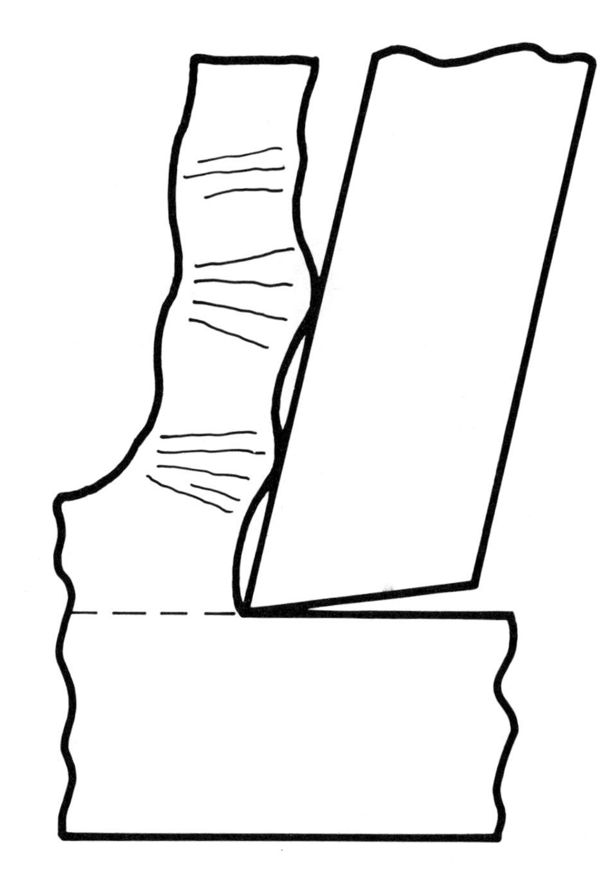

FIGURE 15.10 Built-up edge on toolbit.

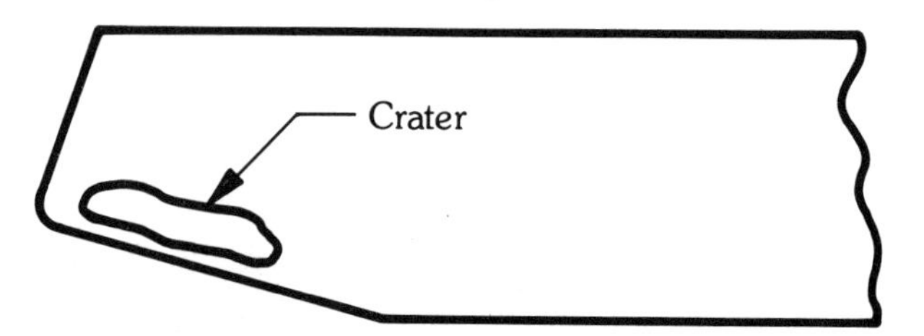

FIGURE 15.11 Cratered tool.

use of chip breakers, cutting fluids, and the best combination of cutting speed and feed all help to reduce the severity of the problem. The use of nitrided and polished cutting tools also helps.

15.3.2 Cutting Tool Geometry

The machinist must have an understanding of the theory and practice of selecting and preparing cutting tools. The performance of metal-cutting machines is closely related to tool shape and the manner in which the tool contacts the workpiece. In the case of milling cutters, the machinist usually cannot easily alter the geometry or shape of the teeth on the cutter. In lathe, shaper, and planer operations, however, the toolbit can be ground and set at the discretion of the machinist or set-up person.

The terms used to describe the various angles and forms to be considered in selecting or grinding a toolbit are often confusing. All the commonly used terms are identified in Fig. 15.12, using three views of a toolbit.

1. Back-rake angle
2. Side-rake angle
3. Front relief angle
4. Side relief angle
5. Side-cutting angle
6. Lip angle
7. Nose angle
8. Nose radius

Note that the position of the toolbit in the machine affects almost all these angles. Before selecting or grinding a lathe tool, for example, the machinist must know if the tool will be held horizontally or in a 16° holder (Figs. 15.13 and 15.14).

Rake Angles. On drills, as shown in Fig. 15.15, the rake angle is determined by the helix angle of the drill. For drilling hard materials, the rake angle may be excessive, resulting in insufficient strength and heat removal capability in the cutting edge. The machinist may then reduce the rake angle, as shown in Fig. 15.16. This also keeps

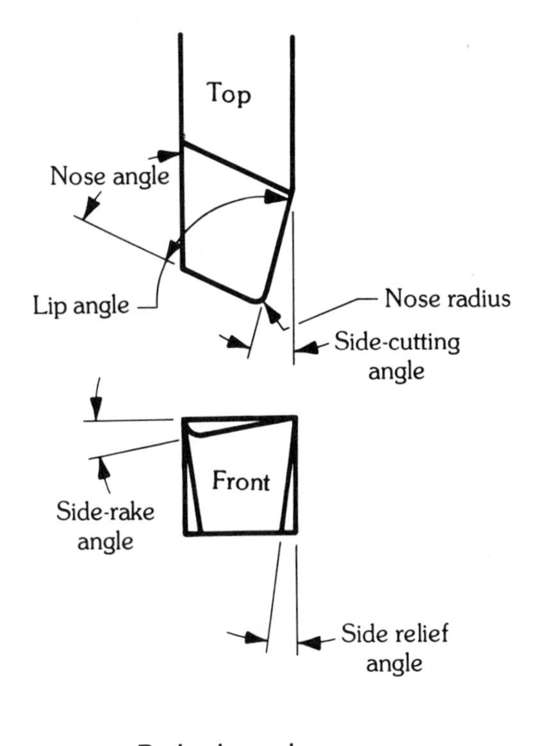

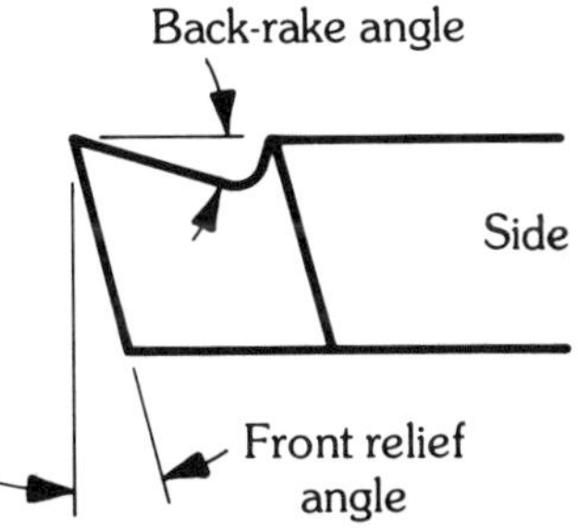

FIGURE 15.12 Cutting tool geometry.

FIGURE 15.14 Toolbit in a 16½° toolholder.

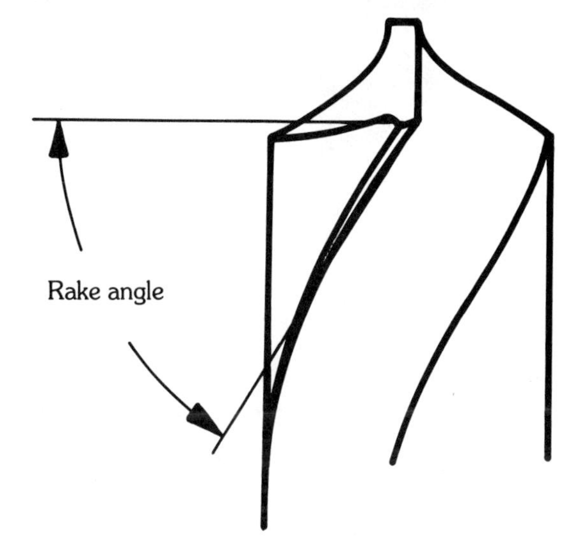

FIGURE 15.15 Rake angle on a drill.

FIGURE 15.13 Toolbit held horizontally.

FIGURE 15.16 Drill rake angle reground.

the drill from grabbing in soft materials such as some brass and bronze alloys.

Toolbits and cutters of all types may have positive, negative, or zero rake angles. This variation in geometry is shown with the carbide-tipped shell end mills in Fig. 15.17. Since all the cutters are right-hand cut (right-hand rotation as viewed from the end opposite the cutting edges), the back-rake angle is changed by machining the flutes of the cutter with a right-hand helix (positive rake), no helix (zero rake), or left-hand helix (negative rake), as shown in Fig. 15.18. The side-rake angle may also

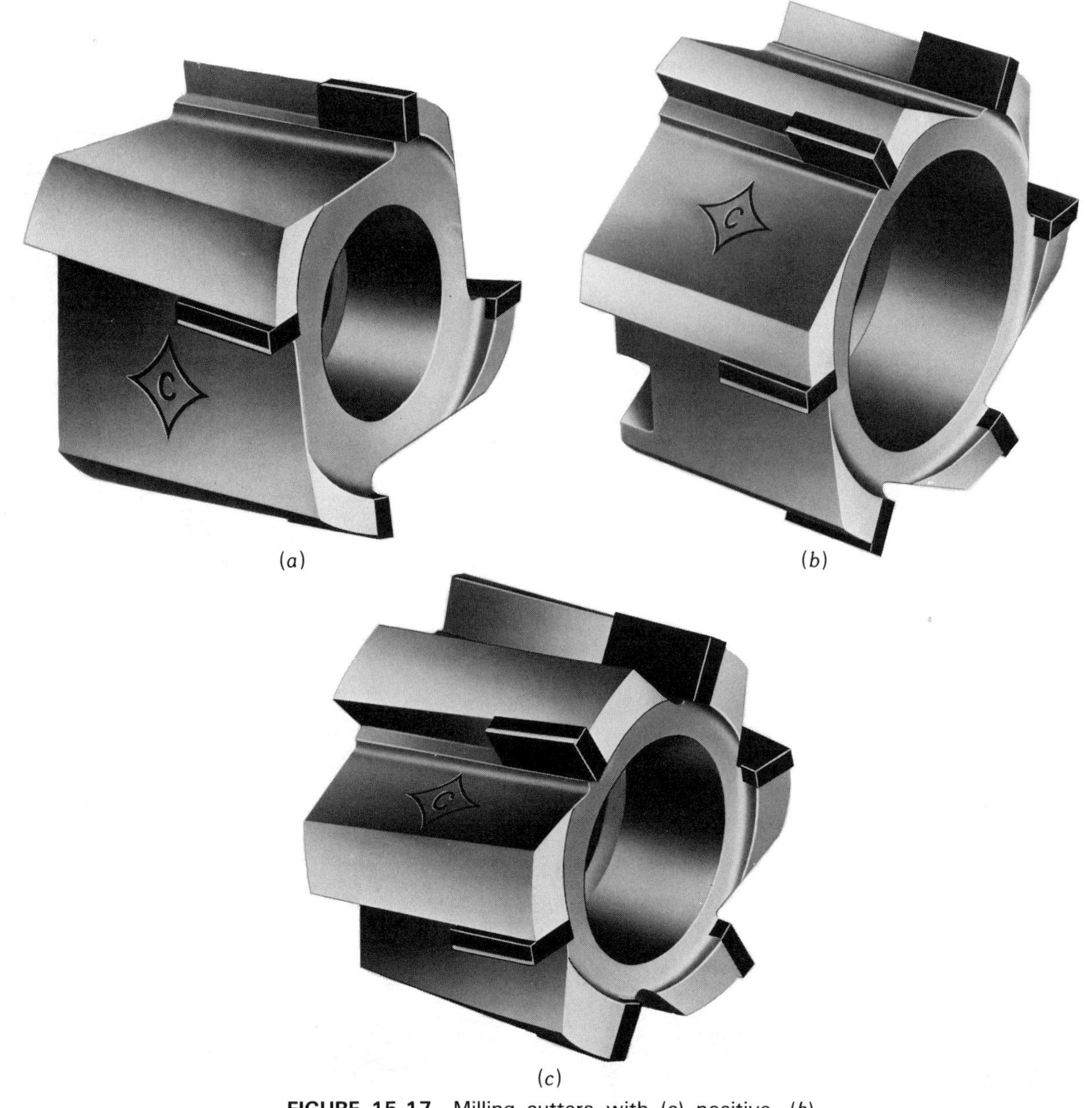

FIGURE 15.17 Milling cutters with (*a*) positive, (*b*) neutral, and (*c*) negative back-rake angles.

be positive, neutral, or negative, depending on the relationship of the face of the cutter to the center point. Form cutters such as gear tooth or sprocket cutters and threading tools usually have zero rake angle, as shown in Fig. 15.19, since the exact opposite of their shape must be produced on the part being machined.

On the lathe toolbits, the machinist must decide whether positive, neutral, or negative back- and side-rake angles are needed, depending on the material to be cut, machine power available, and other factors. Negative back rake produces a pressure pattern that tends to push the tool away from the work, as shown in Fig. 15.20. Since the *lip angle* of the tool (Fig. 15.12) is about 90°, the tool has greater strength, but more power is required to remove metal at a given rate. The lip angle must be reduced if a positive rake angle is desired. This results in a weaker tool point but provides for greater freedom of cutting action.

Relief Angles. All cutting tools must have side relief, front relief, or both (Fig. 15.12.) The relief angle will vary, depending on the material being machined. Side and front relief angles for all cutting tools should be kept to a minimum to increase tool life and reduce the possibility of tool breakage.

When setting the cutting tool on the lathe, the tool tip must be set exactly on center, especially

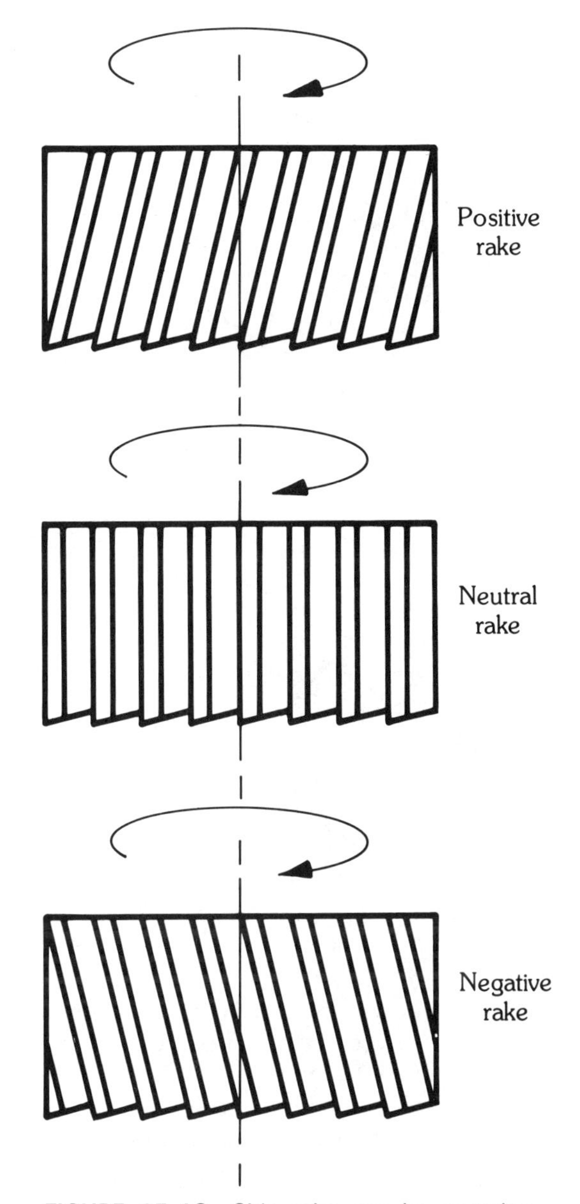

FIGURE 15.18 Side rake can be negative, neutral, or positive.

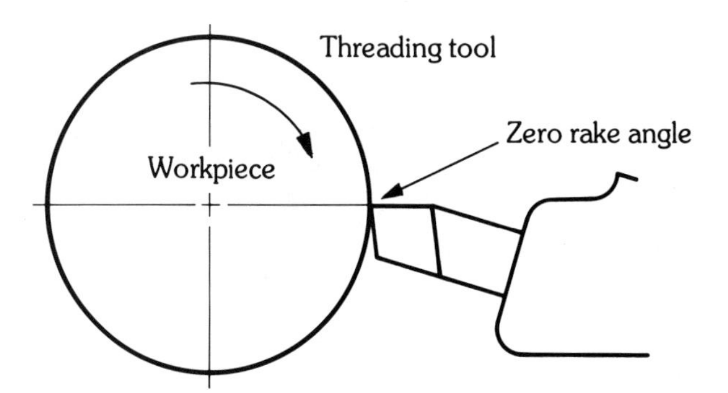

FIGURE 15.19 Form cutter with zero rake angle.

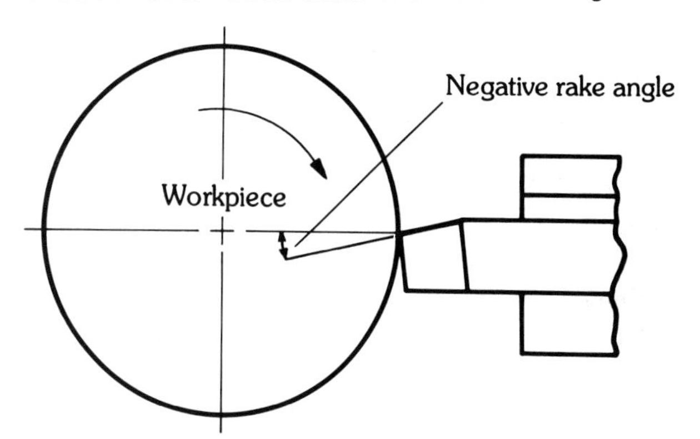

FIGURE 15.20 Lathe tool with negative back rake.

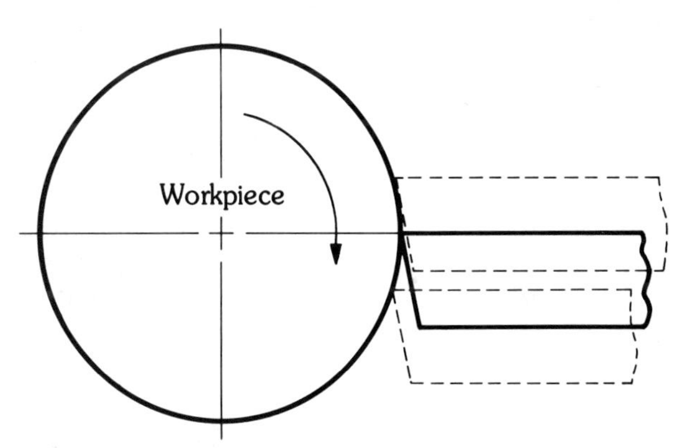

FIGURE 15.21 Above-center, on-center, and below-center settings. Note effect on relief angle.

when minimum front relief angles are being used. If the tool is set above center, the tip rubs rather than cuts. If the tool is set below center, the effective relief angle will be increased, thus weakening the tool (Fig. 15.21). When using carbide cutters, primary and secondary relief may be used, as shown in Fig. 15.22. The use of secondary relief allows the cutting edge to be resharpened with minimum metal removal.

Plain milling cutters are ground with relief on the periphery only; side-cutting milling cutters are ground with relief on the sides as well as the periphery of the teeth. The primary relief (Fig. 15.23) is measured from an imaginary line tangent to the periphery of the cutter at the tip of the tooth. Excessive relief angles weaken the teeth of milling cutters and cause premature failure. [*Note:* The

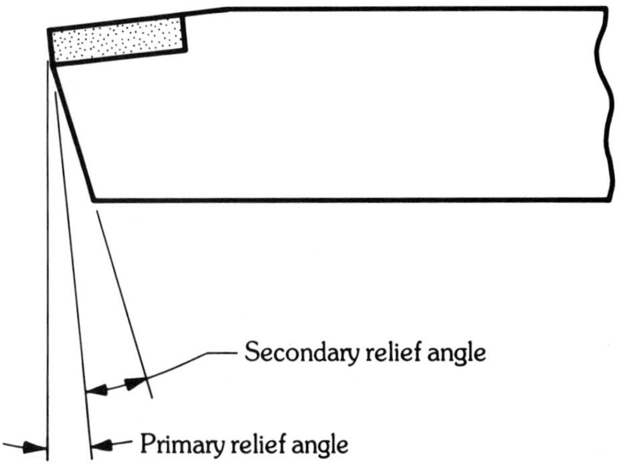

FIGURE 15.22 Primary and secondary relief angles on a carbide insert.

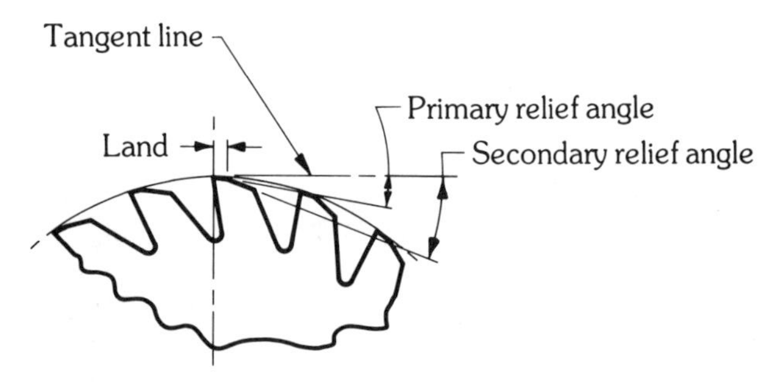

FIGURE 15.23 Primary and secondary relief angles on a milling cutter tooth.

secondary relief angle (Fig. 15.23) is sometimes referred to as the *clearance angle* or the *peripheral clearance angle.*]

Nose Radius. The *nose* of the tool is the rounded part that contacts the workpiece. The radius of the nose for tools used to take light cuts at fine feed rates is generally small (0.010 to 0.030 in.), particularly when machining delicate parts of a small diameter. The use of a large nose radius (0.060 in. or more) increases the strength of the tool and produces a good finish at high feed rates. However, there is also more possibility of chatter if the machine and tool setup are not rigid, particularly when turning small-diameter workpieces that are relatively long.

Chip Breakers. A chip breaker can be ground as an integral part of the toolbit (Fig. 15.24) or be of the clamp-on type (Fig. 15.25). The chip breaker prevents the formation of long chips that may become tangled in the machine or work. These chips are a safety hazard and also very inconvenient, particularly on turret lathes and automatic screw machines.

Chip breakers are not needed when machining materials such as gray cast iron, hard bronze alloys, and most aluminum castings since the chips fracture into small pieces during the cutting operation. Materials such as annealed low-carbon steel, wrought aluminum alloys, and most stainless steels machine with a long, stringy chip. A chip breaker must be used, particularly on tools for automatic lathes, turret lathes, and screw machines.

Surface Finish. The surface finish of a machined part is affected by factors such as tool geometry, feed rate, cutting speed, the machineability of the material, and the use of cutting fluids. In lathe operation the nose radius of the tool is of particular importance. It must be smooth and the right size for the feed rate and depth of cut. Honing the nose of the toolbit with a small oilstone

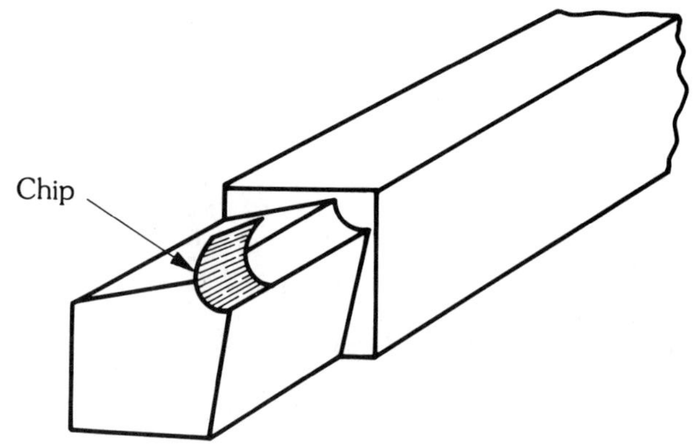

FIGURE 15.24 Chip breaker ground in a toolbit insert.

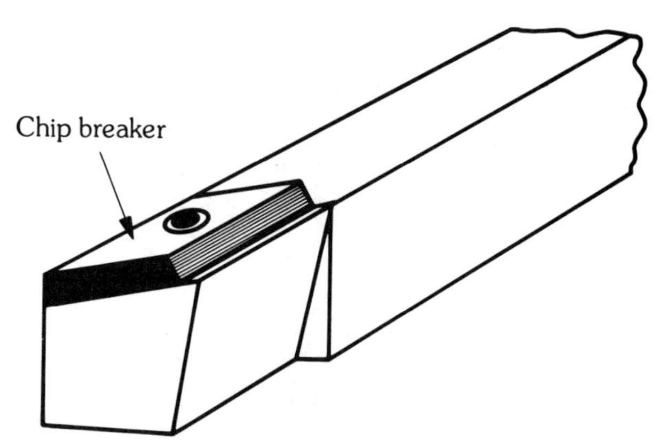

FIGURE 15.25 Clamp-on chip breaker.

after it has been ground is a highly recommended way to improve the surface finish of the part.

Some materials, particularly medium- and high-carbon steels and alloy steels without lead or other machining additives, tend to tear as they are cut. If the temperature at the toolbit tip is high because of high cutting speed, high feed rates, or poor tool geometry, a built-up edge is likely to form on the toolbit. As the built-up edge tears off and a new one is formed, there is a momentary change in the cutting action of the tool, causing a rough spot on the surface being machined.

Reducing friction at the point of chip formation is the most effective way of improving surface finish. This can be done by using lead or manganese sulfide in the steel, improving the finish on the tool, or establishing the best possible tool geometry. Finding the best feed rate and cutting speed combination, using cutting fluids, or a combination of all these factors may help.

15.4 CUTTING FLUIDS

Cutting fluids of various compositions have always been used extensively in machining operations. Since some of the highest specific pressures (up to 300,000 psi) in any type of mechanical work occur in metal cutting, fluids are used to reduce heat buildup and therefore extend tool life. Usually, the proper cutting fluid also allows higher cutting speeds and feeds and helps produce a better finish on the machined part.

The chips of some metals have a tendency to temporarily weld onto the top surface of the toolbit. This *built-up edge* is an unstable condition since it is continually formed and released. This is generally undesirable because it results in greater power consumption, poorer finish, and possible cratering of the toolbit. Although cutting fluids are only one remedy for this problem, they are widely used to reduce friction and heat, which in turn reduces the formation of a built-up edge on the toolbit.

FIGURE 15.26 Solid cutting lubricant used on a bandsaw blade.

15.4.1 Desirable Cutting Fluid Characteristics

An effective cutting fluid has the following six characteristics. Some are more effective lubricants; others are more effective coolants. The machinist must understand cutting fluid characteristics and be able to choose the one best suited to a particular job.

1. *Lubrication.* The cutting fluid must prevent or reduce formation of built-up edge on tools or cratering and lubricate some of the working parts on automatic screw machines and similar tools.
2. *Cooling capability.* The cutting fluid must be able to reduce the temperature of the tool and the workpiece to extend tool life and allow higher cutting speeds and feed rates.
3. *Rust and residue prevention.* The workpiece and the machine must be protected from rust and the formation of scum.
4. *Safety.* The cutting fluid must be nontoxic and nonflammable under normal working conditions. It should not cause skin irritation.
5. *Stability.* The cutting fluid should not support bacterial growth (become rancid) in use or in storage, and its characteristics should not change with age.
6. *Compatibility.* The cutting fluid must be compatible with other lubricants used on the machine.

15.4.2 Types of Cutting Fluids

There are four major categories of cutting fluids: straight oils, chemically treated oils, soluble oils, and chemical–water compounds. Compressed air or solid lubricants can also be used in certain applications. For example, solid lubricants in stick form are often used in tapping, broaching, and bandsawing operations (Fig. 15.26). Compressed air can be used alone or in conjunction with a mist coolant.

Straight Oils. Straight mineral oils are used extensively in machining nonferrous metals and free-machining steels. They are not chemically active and do not stain the machined parts. Straight mineral oils have relatively good lubricating qualities, but they are not very effective coolants. They are very stable cutting fluids and can be reused many times if properly cleaned and filtered. A typical application of straight mineral cutting oils is on the machine shown in Fig. 15.27.

When tougher nonferrous metals such as some bronzes and Monel metals are being cut, mineral oils blended with fatty animal (lard) or vegetable oils (usually castor oil) are sometimes used. The addition of animal or vegetable oils, however, can cause disagreeable odors or the formation of gummy substances on the machine. Small amounts of sulfur are sometimes added to increase tool life without causing discoloration of the machined parts.

The addition of animal or vegetable oils apparently increases the film strength and penetrating ability of mineral oils, thus reducing the formation of built-up edges on cutting tools and improving the surface finish of machined parts. Lightly sulfurized fatty oil–mineral blends are usually used on metals with a machineability rating of 70 percent or higher.

FIGURE 15.27 Mineral cutting oil in use on a gear cutting machine. (*Courtesy The Gleason Works*)

Chemically Treated Oils. These cutting oils are chemically active at relatively low temperatures (200 to 300 °F) and have very good lubricating properties, particularly when chlorine and sulfur are added. There are three general categories of chemically treated oils:

1. *Sulfurized mineral oils.* Oils in this group are used almost exclusively with tough low-carbon and low-alloy steels. The oil becomes chemically active at relatively low temperatures and stains any copper base alloys. Usually the sulfur content is below 1.0 percent. The sulfur in the oil reacts with ferrous metals when the cutting oil is exposed to the temperatures and pressures produced by the metal-cutting process. An iron sulfide film that resists high pressures and prevents galling is formed between the tool and the chip.
2. *Sulfurized and chlorinated mineral oils.* The addition of about 1 percent chlorine to sulfurized oils containing as much as 3 percent sulfur results in a cutting oil that is chemically active over a wider temperature range. Chlorine becomes active in the presence of clean iron at a lower temperature than sulfur. Oils of this type are extensively used in broaching, threading, hobbing, and general machining of tough low-carbon steels and alloy steels. They may cause corrosion of steel parts when moisture is present and are not used when machining copper base alloys.
3. *Fatty oil blends.* These cutting oils contain mineral oil, animal or vegetable fatty oils, and varying amounts of sulfur and chlorine. They may contain as much as 6 percent each of chlorine and sulfur. This combination produces an oil that withstands extreme pressures and is particularly good when very tough materials must be cut at slow speeds. They become active at low temperatures and therefore are very useful for broaching and similar processes.

Soluble Oils. The oil in soluble, or emulsifiable, oils is treated with soaps of various types so that it mixes readily with water. The mixture is white, and since it is mostly water, it is an excellent coolant. The oil provides some lubrication and prevents corrosion of both the workpiece and the machine.

The ratio of water to oil may vary from 1 part oil to 10 parts water, to 1 part oil to 100 parts water. When more oil is used, the mixture is suitable for heavier machining operations and provides better corrosion protection. When the water content is high, such as in a 1 : 100 mixture, the cooling effect is greatest and the mixture is used mostly for grinding operations. Soluble oils are much less expensive than mineral oils, and when protected from contamination they last a long time.

Fatty oils of either the animal or vegetable type can be added to soluble oils to increase their lubricating properties. Phosphorus, sulfur, or chlorine may also be added to soluble oils to increase their ability to protect the cutting tool during heavy machining operations. For use with hard-to-cut materials, the oil to water ratio usually ranges from 1 : 8 to 1 : 20. Soft water is usually best for mixing soluble oils.

Chemical–Water Compounds. During the past 40 years, chemical–water cutting fluids have become popular because of their excellent cooling and rust-inhibiting properties. They are widely used in grinding machines and other applications where effective cooling is necessary. For grinding applications, the solution is usually 1 part of chemical concentrate to 200 parts of water.

For other machining applications, wetting agents can be added to improve heat dissipation. If additional lubrication is required to prevent built-up edges on cutting tools, sulfur or chlorine may be added. For heavier machining, particularly when working with tough, tenacious materials, ratios of 1 : 5 parts of concentrate to water to about 1 : 30 parts may be used.

Chemical–water compounds work best when soft water is used. If hard water must be used, more chemical concentrate must be used for a given amount of water.

Also note that some concentrates cause the paint on some machine tools to deteriorate, particularly when strong solutions are used.

15.4.3 Cutting Fluid Application

On most simple machine tools, such as the vertical milling machine shown in Fig. 15.28, there are no provisions for pumping cutting fluid to the cutting tool. Hand application using a squeeze bottle or oil can is sufficient in such cases. On larger machines, a cutting fluid sump, a pump, lines, and a return system are usually provided, as shown in Fig. 15.29. As the cutting fluid returns to the sump by gravity, it is filtered, and solids are allowed to settle out before it is reused.

Flood Application. For most machining applications, the cutting fluid is directed to where the cutting tool contacts the work in a steady stream at low pressure. When the work or cutter rotates slowly, the cutting fluid wets the work and tool thoroughly enough to do an effective job of cooling and lubricating. When the work or cutter rotates rapidly, the cutting fluid may be thrown off. Experimenting with the placement of the cutting fluid nozzle may help, but in some cases other methods of delivering cutting fluid must be used. In some operations, two or more nozzles are used, especially when cutting fluid must be directed to both sides of a tool as shown in Fig. 15.30.

FIGURE 15.28 Hand application of cutting fluid.

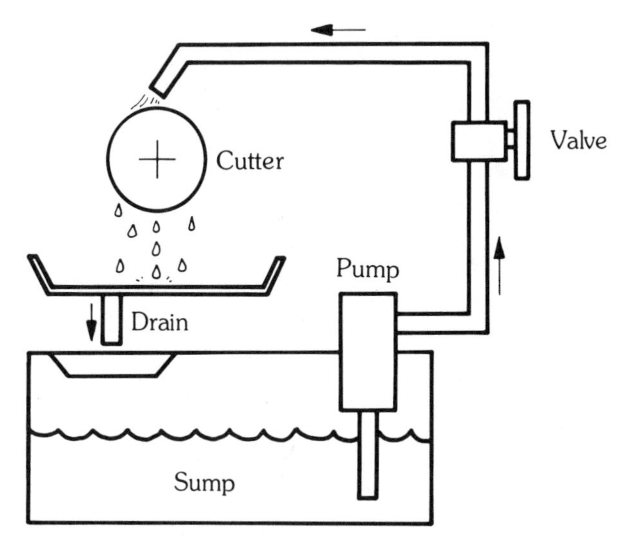

FIGURE 15.29 Cutting fluid system.

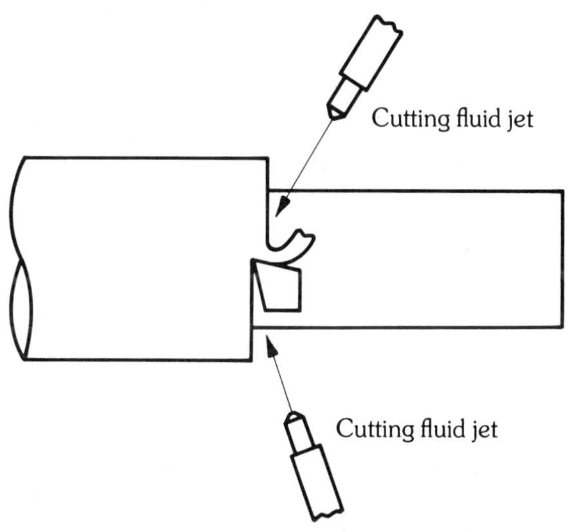

FIGURE 15.30 Cutting fluid delivery above and below the cut.

Mist Application. For certain light drilling, milling, and turning applications, the cutting fluid is delivered in mist form, as shown in Fig. 15.31. The cutting fluid is converted to mist form with compressed air and can be easily directed into hard-to-reach places. The amount of cutting fluid in the airstream is controlled with a needle valve in the line from the container, as shown in Fig. 15.32.

The cutting fluids used in mist coolant systems may be light cutting oils, soluble oils, chemical-water compounds, or wax and water mixtures. In comparison to other methods of application, the amount of cutting fluid used is quite small. In some cases the cutting fluid to air ratio may be 1 : 4000. The combination of compressed air, which cools as it expands, and finely divided cutting fluid results in very effective cooling. The amount of lubrication provided can be changed as needed by selecting cut-

FIGURE 15.31 Mist coolant system provides both lubricating and cooling action.

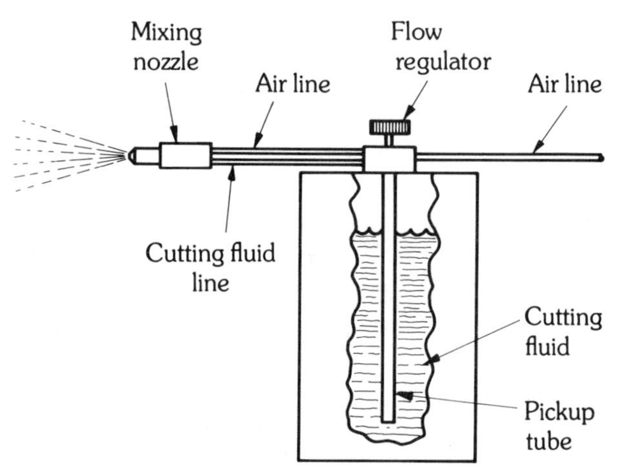

FIGURE 15.32 Schematic of mist coolant system.

ting fluids with more or less lubricating capability.

15.4.4 Cutting Fluid Selection and Use: General Recommendations

The selection of the most effective cutting fluid for any machining operation always involves several factors that must be well understood by the machinist. The major factors follow, but this is by no means a complete list. The machinist may also refer to Table 15.3 or to the recommendations of cutting fluid manufacturers for specific applications.

1. *Workpiece material.* Careful identification of the material being machined and the nature of any heat treatment that affects hardness is necessary. The machineability rating is a good indication of what cutting fluid should be used.
2. *Cutting tool material.* The type of cutting tool material used—high-speed steel, carbides, or ceramics—often determines whether a cutting fluid is used at all.
3. *Tool life.* In production operations, tool life is an important factor since the time spent changing and sharpening tools is nonproductive. Use of cutting fluids with the correct combination of lubrication and cooling properties maximizes tool life.

TABLE 15-3
Cutting fluid applications

	Machining Operation[a]			
Material	*Drilling*	*Milling*	*Turning*	*Threading*
Aluminum	A, C, D	A, C, D	A, C, D	A
Brass	Dry A, C, D	A, C, D	Dry A, C, D	A,C
Bronze	A, C, D	A, C, D	A, C, D	A, C
Ductile iron	A, B, C, D	A, B, C, D	A, B, C, D	A, B
Gray cast iron	Dry D	Dry C, D	Dry C, D	Dry A, B
Steel				
Low carbon	A, B, C, D	A, B, C, D	A, B, C, D	A, B
Medium carbon	A, B, C, D	B, C, D	B, C, D	B
High carbon and alloy	A, B, C, D	B, C, D	B, C, D	B

Note: The final choice of cutting fluid should be made only after considering such factors as cutting speed, feed rate, finish desired, and expected tool life.

[a]A, straight mineral oils; B, chemically treated oils; C, soluble oils; D, chemical–water compounds.

4. *Cutting speed.* Higher cutting speeds are economically desirable but may result in shorter tool life. The proper cutting fluid allows a sizable increase in cutting speed.
5. *Type of machining operation.* Metal-cutting operations vary in severity because of tool geometry and other factors. Broaching, threading with a die or single-point tool, tapping, and operations involving large form tools are regarded as the most severe. Light milling, turning, and drilling operations do not produce heavy pressures or high temperatures. Cutting fluid selection for severe machining processes is naturally more critical.
6. *Method of application.* How the cutting fluid is applied affects tool life, cutting speed, and the amount of metal removed in a given time period.
7. *Finish desired.* Prevention of a built-up edge on the tool is necessary in producing a good finish. Cutting fluids that generate a low shear strength film on the tool face and chip and provide effective cooling must be used when finish is important.

The information, procedures, and recommendations presented in this chapter have a direct effect on the quality and cost of machined parts produced in all types of machine shops. Those who are skilled machinists, toolmakers, prototype machinists, and setup people must constantly make decisions based on this information.

REVIEW QUESTIONS

15.2 CUTTING TOOL MATERIALS

1. Name three common cutting tools that can be made of high-carbon steel.
2. At what temperature does high-carbon steel start to soften?
3. What precaution should be taken when sharpening a cutting tool made of high-carbon steel?
4. What effect does tungsten have when added to high-carbon steels?
5. Name the two major types of high-speed steels.
6. What is the *binder*, or sintering material, that holds the tungsten carbide particles together in a finished cutting tool?
7. Briefly describe the process of sintering carbide tools.
8. Explain how the amount of binder in a carbide tool affects its hardness and toughness.
9. By what two means are carbide inserts attached to toolholders?
10. What is the main material used in ceramic cutting tools?
11. Name one disadvantage of ceramic cutting tools.

15.3 METAL CUTTING: THEORY AND PRACTICE

1. Sketch a side view of a cutting tool, workpiece, and chip and show the *shear zone.*
2. Name two materials that produce a continuous chip when machined.
3. Explain why gray cast iron machines with a segmented chip.
4. What is a *built-up edge* on a toolbit, and how does it affect surface finish?
5. How may the *rake angle* of a drill bit be altered?
6. Why should the *rake angle* of a cutting tool be reduced when cutting soft brass?
7. What is the effect of excessive front and side relief on a cutting tool?
8. How will the size of the nose radius on a cutting tool affect surface finish?
9. Briefly describe two different types of chip breakers.
10. Why should friction be kept at a minimum at the point of the tool where the chip is formed?

15.4 CUTTING FLUIDS

1. What are the two main functions of a cutting fluid?
2. What are the four main categories of cutting fluids?
3. Why are lard oil and castor oil sometimes added to straight mineral cutting oils?
4. What two chemicals may be used to increase the effectiveness of mineral cutting oils?
5. What type of cutting oil should be used for a slow speed, high-pressure cutting opera-

tion such as tapping or broaching a low-carbon steel part?

6. What type of cutting fluid should be used for maximum cooling effect on a precision grinding machine?
7. Name two applications where a cutting lubricant in stick form could be used.
8. Explain the process of mist application of cutting fluids.

16

Numerical Control

16.1 INTRODUCTION

Numerical control (commonly called NC) has given the manufacturing industry new and greater controls in the design and manufacturing of products. Today, many thousands of NC machines are in use in large and small machine shops. In these machine shops, NC may be used to control a simple drilling machine or to contour mill a part too complex to machine economically by conventional methods.

Numerical control of machine tools is simply *the control of machine tool functions by means of coded instructions.* Examples of *machine tool functions* are: moving the table, turning the spindle on or off, changing the cutting tool, indexing a part, or turning the cutting fluid on or off. *Coded instructions* are *alphanumeric* expressions used for controlling a particular machine tool function (Fig. 16.1).

16.2 NUMERICAL CONTROL DEVELOPMENT

Automation is nothing new. The word *automation* is derived from the word *automatic,* which means self-regulating, self-moving, or self-controlling. *Automation* is a term often applied to manufacturing and today's machine tools. One of the earliest attempts at automation was in the control of sound. Many cathedrals in Europe used rotating drums with pins to automatically ring chimes. This same concept was later used in music boxes. In 1863,

Opening photo courtesy of The Pratt & Whitney Co., Inc.

FIGURE 16.1 A modern NC machining center. *(Courtesy Pratt & Whitney Machine Co., Inc.)*

M. Fourneaux patented the first automatic player piano. Fourneaux used a 12-in.-wide paper strip that was perforated so air could be drawn through the perforations and actuate different piano keys.

In the clothing industry, attempts at automation began in the early 1700s when Falcon used punched cards to control knitting machines in England. Each hole in a particular card controlled mechanical linkage, which in turn controlled the different color of patterns woven into a piece of cloth.

The first computer was designed by Charles Babbage in the mid-1800s; however, it was never finished. Not until 1945 was the first accurate computer built for the Army Ordnance Department. Computers were one of the first tools of what many call the second industrial revolution. It was the computer that led to the development of numerical control.

Research and development on NC in industry was started in 1947 by John C. Parsons, owner of a corporation that manufactured helicopter rotor blades. Parsons coupled a jig boring machine with a computer to aid machining operations.

By 1949, the U.S. Air Force became aware of the increasing problems in machining complex shapes needed for modern airplanes and missiles. In 1951, the Air Materiel Command of the Air Force awarded a contract to Parsons and the Massachusetts Institute of Technology (MIT) for further research and development of an experimental NC machine that would manufacture contoured parts. The prototype machine was displayed at MIT in 1953.

Soon after, Giddings and Lewis, a leading machine tool building firm, came out with an NC profiling mill. The Air Force quickly placed orders for the new machine. General Electric and Burgmaster were also early pioneers in NC machine tools. By 1960, more than 100 NC machines were exhibited at a machine tool show in Chicago.

NC is definitely here to stay. There have been many changes in numerical control since the early efforts of John Parsons and MIT. Computers are now playing an important role in the NC industry. They are used to control one or more machine tools from a central location, commonly referred to as *direct numerical control* (DNC). Minicomputers and microprocessors are now being built and incorporated in many machines to direct machine tool functions. This *computer numerical control* (CNC) has the benefit of storing many different programs that can later be retrieved. DNC and CNC are discussed in greater detail later in the chapter.

16.3 INDUSTRIES USING NC

Many different varieties and sizes of NC equipment are available. Almost all industries utilize or are affected by NC, and there appears to be *no end* to its application. The following is only a partial list of the industries using NC equipment today.

The *aerospace* industry is by far the greatest user of NC equipment. NC really evolved to the state of the art it is in today because of the complex parts needed by the aerospace industry. The *electronics* industry uses NC to manufacture and assemble component parts on circuit boards; the *electrical* industry to wind coils for different types of motors. Another big industry to utilize NC is the *automotive* industry. Machine tools, welding and forming equipment, assembly lines, and drafting machines all use NC equipment. *Quality control* in larger manufacturing facilities utilizes NC to measure and inspect all types of parts.

16.4 TYPES OF NC MACHINE TOOLS

16.4.1 Drilling Machines

Drilling machines are the least expensive and easiest to operate of all NC machine tools. Most drilling machines have a worktable that can be controlled in two directions. The first NC drilling machine was a simple single-spindle, two-axis drilling machine. The spindle on this type of drilling machine needed an operator to feed the cutting tool into the workpiece. The worktable movements were the only portion of this machine tool controlled by NC.

Figure 16.2 shows a more modern type of drill-

FIGURE 16.2 An eight-spindle NC drilling machine. (*Courtesy Tapmatic Corp.*)

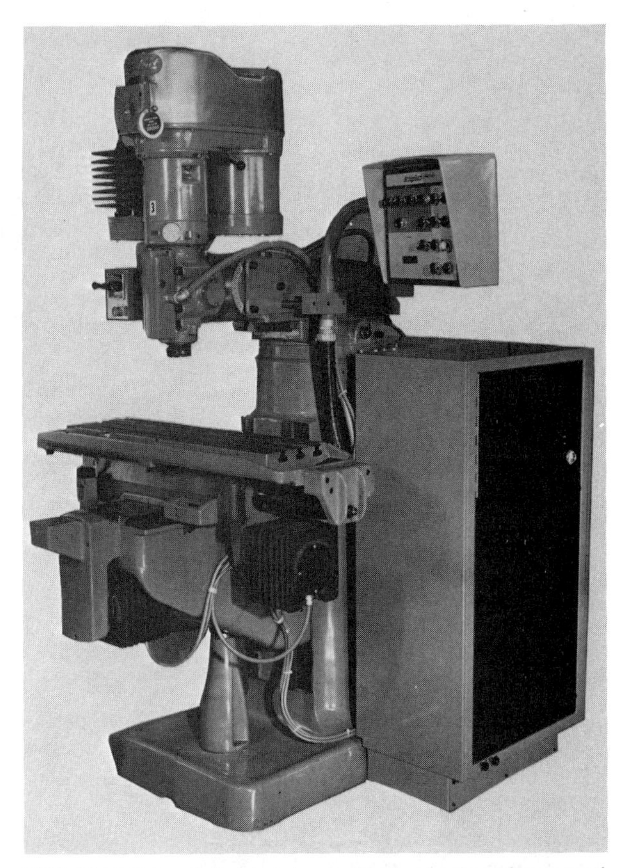

FIGURE 16.3 A light-duty vertical spindle column and knee NC mill. (*Courtesy Bridgeport Machines, Inc.*)

ing machine. This machine has eight spindles mounted in a turret. Eight different cutting tools may be installed in the turret and used when different sized holes are required. Turret drilling machines have a third axis, the *spindle* axis, and each spindle can be programmed to drill holes of different depths.

16.4.2 Milling Machines

Milling machines are the largest selling category of NC machine tools. They may have single or multiple spindles mounted either vertically or horizontally. NC milling machines have at least two axes of movement; some may have as many as five or six. Figure 16.3 shows a simple, single-spindle, ram and turret milling machine. It is a light-duty mill used for drilling operations and point-to-point (straight-line) milling. It is the easiest to program and operate. This type of machine is also available with continuous path (the ability to mill in two or more directions simultaneously), additional axes, and mechanisms to change cutting tools (tool changers).

A large, multiaxis milling machine is shown in Fig. 16.4. This milling machine is commonly referred to as a *machining center*. Machining centers can perform a variety of machining operations with only one setup.

16.4.3 Lathes

There are two major types of NC lathes: the *engine* and the *turret*. The engine lathe is generally a two-axis machine having longitudinal and transverse motions. Most NC engine lathes have the ability to turn, face, bore, machine external or internal tapers, and machine threads (Fig. 16.5).

NC turret lathes generally have three axes that include carriage, cross-slide, and turret ram motions [Fig. 16.6*(a)*]. The turret ram employs six or eight positions for a variety of cutting tools [Fig. 16.6*(b)*].

16.4.4 Boring Machines

NC boring machines are generally large machine tools that fall in two categories: those that revolve

FIGURE 16.4 Large, multiaxis horizontal spindle NC milling machine.

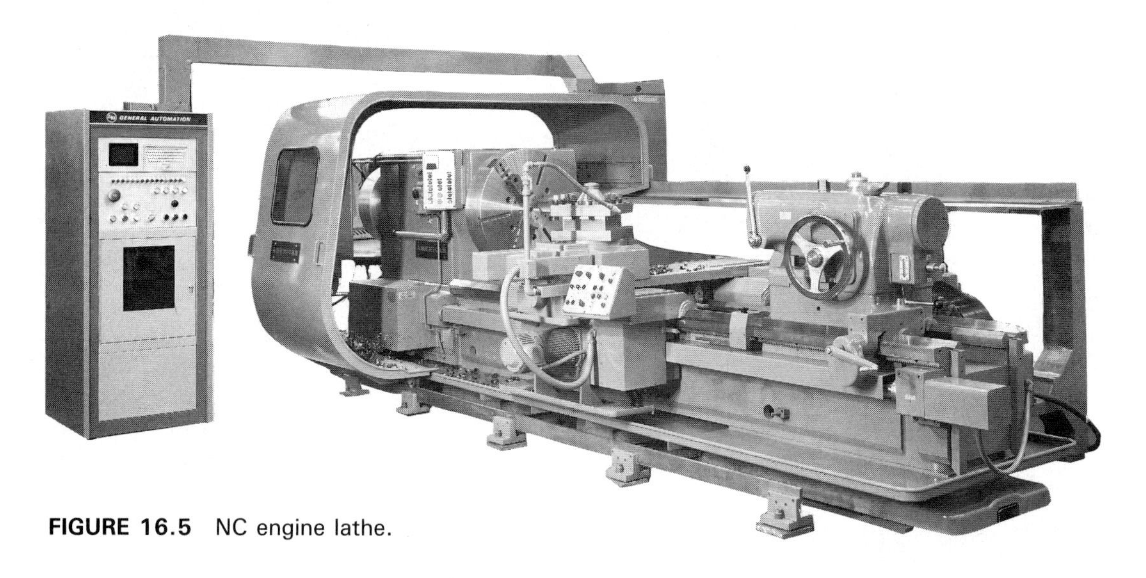

FIGURE 16.5 NC engine lathe.

the workpiece vertically and machine the part with a single-point tool, as shown in Fig. 16.7, and those in which the work is stationary and a horizontal revolving cutting tool is used (Fig. 16.8).

16.4.5 Grinding Machines

NC has been adapted to surface as well as cylindrical grinding operations. A continuous-path NC surface grinder can grind irregular shapes and contours necessary for complex parts such as cams, sprockets, or gears.

The foregoing list of NC applications to machine tools illustrates only those machine tools commonly found in the majority of shops. However, as machines and machine tools become more complex and the cost of microprocessors drops, NC will undoubtedly expand.

16.5 ADVANTAGES AND DISADVANTAGES OF NC

16.5.1 Advantages

1. By using the NC, human error is greatly reduced. The NC machine produces accurate reproductions of engineering data and the consistency of parts is improved, resulting in less scrap.
2. NC machines can produce complex parts that could not be made by traditional methods or conventional machines.
3. NC reduces tooling and fixture costs. Part of these savings are the result of the elimination of elaborate and specially designed tools and fixtures. Storage problems for tools and fixtures are also reduced.

(*a*)

(*b*)

FIGURE 16.6 (*a*) NC turret lathe; (*b*) lathe turret and cutting tools. (*Courtesy Pratt & Whitney Machine Co., Inc.*)

FIGURE 16.7 Vertical spindle NC boring machine.

FIGURE 16.8 Horizontal spindle NC boring machine. (*Courtesy Kanematsu-Goshu, Inc.*)

4. NC increases flexibility. Engineering changes in production parts are less costly and more rapid, since changes with NC are quickly accomplished by changing a tape rather than building new jigs and fixtures.
5. Production planning is easier and more effective with NC equipment because manufacturing capacity is more constant, predictable, and efficient. Cost estimates are improved because of the reliability and efficiency of NC.
6. NC reduces floor space requirements. Most NC machine tools will do many different machining operations on a part in one setup compared with traditional methods that include routing the part through

several conventional machines. With a single NC machine capable of a variety of operations in a single setup, fewer fixtures are needed, smaller lots can be produced economically, and less storage space is needed.

16.5.2 Disadvantages

1. Equipment costs are high. NC machines require high initial investments.
2. NC requires retraining existing personnel to become skilled programmers and operators.
3. Mechanical, electrical, and electronic maintenance personnel are required when working with NC machines.
4. NC is not suitable to long-run application.

16.6 NC MACHINE TOOL ELEMENTS

NC machine tool elements consist of axis nomenclature, dimensioning systems, control systems, servomechanisms, and open- or closed-loop systems. It is important to understand each element prior to actual programming of a numerically controlled part.

16.6.1 Axis Nomenclature

NC machine tools base their axis nomenclature on the standard Cartesian coordinate system. Figure 16.9 shows the standard Cartesian coordinates with *X*, *Y*, and *Z* axes. The *X*, *Y*, and *Z* labels for

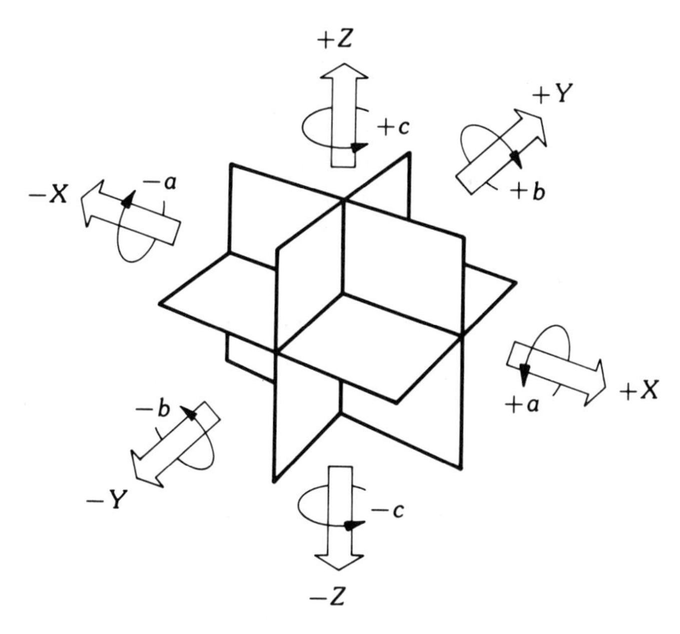

FIGURE 16.9 Standard Cartesian coordinate system.

the three major axes are standard with most NC machine tools. Since different types of NC machine tools use different table motions, specific definitions relating an axis to a table motion are not possible. However, standard definitions for *X*, *Y*, and *Z* axes relating to most machines are as follows:

X axis:

1. Must be horizontal
2. Must be perpendicular to the *Z* axis
3. Is generally the longest axis of movement

Y axis:

1. Perpendicular to *X* and *Z* axes

Z axis:

1. Is always parallel to the spindle and perpendicular to a plane established by *X* and *Y*.

In addition to the *X*, *Y*, and *Z* axes, several rotation movements can be accomplished on NC machine tools around each of the axes. *A*, *B*, and *C* axes can provide rotation about the *X*, *Y*, and *Z*. The *A* axis is rotary motion around the *X*, the *B* axis is rotary motion around the *Y*, and the *C* axis provides rotary movements around the *Z* axis (Fig. 16.9). Notice that Fig. 16.9 also shows plus and minus signs for the *X*, *Y*, and *Z* axes. The signs indicate a direction from the zero point (also known as the origin) along the three major axes. These plus and minus signs on most NC machines are used to indicate the direction of travel between two points. This is discussed further in the section on programming.

16.6.2 NC Measuring Systems

The term *measuring system* in NC refers to the method a machine tool uses to move a part from a reference point to a target point. A target point may be a certain location for drilling a hole, milling a slot, or other machining operation. The two measuring systems used on NC machines are the *absolute* and *incremental.*

Absolute System. The absolute (also called coordinate) measuring system uses a *fixed* reference point (origin). It is on this point that all positional information is based. In other words, all the locations to which a part will be moved must be given dimensions relating to that original fixed reference point. Figure 16.10 shows an absolute measuring system with *X* and *Y* dimensions, each based on the origin.

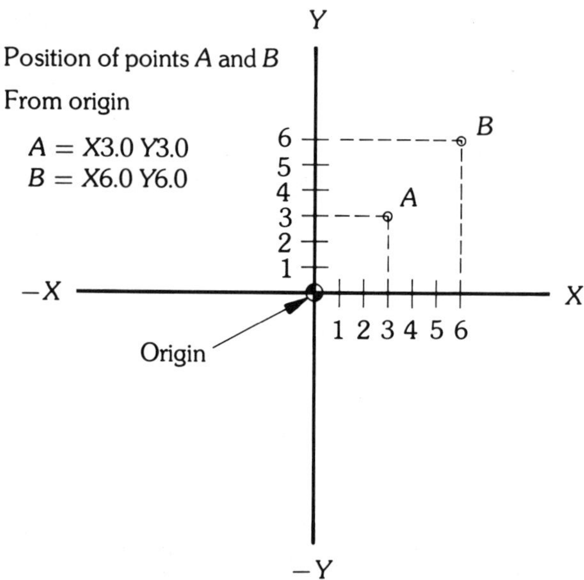

FIGURE 16.10 The absolute measuring system bases all of its locations from a fixed origin.

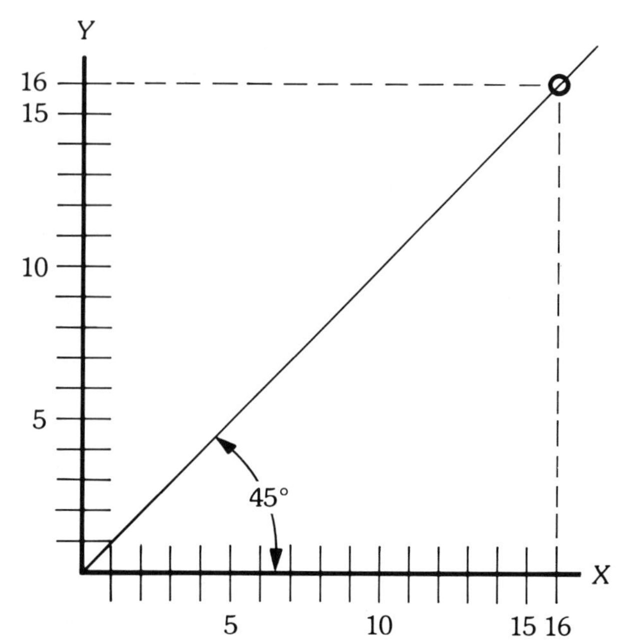

FIGURE 16.12 On point-to-point systems a 45° path is generated when equal values of X and Y are programmed.

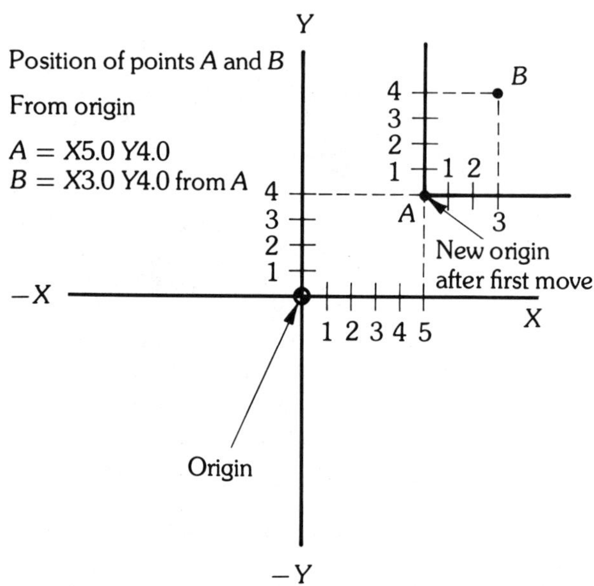

FIGURE 16.11 The incremental measuring system establishes a new origin after each move.

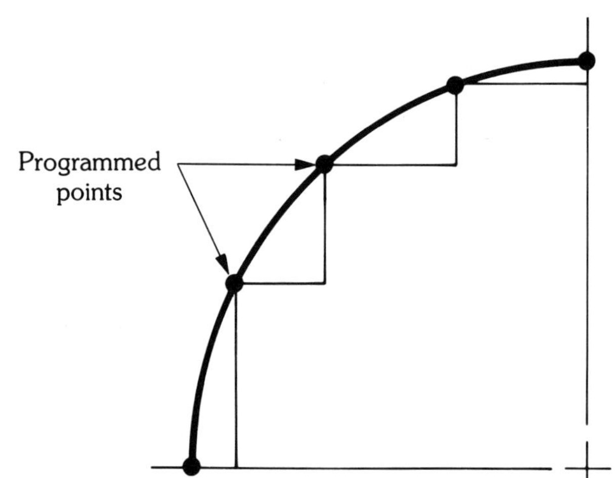

FIGURE 16.13 Arcs and angles may be generated on point-to-point systems by programming a series of small steps.

Incremental System. The incremental measuring system (also called delta) has a floating coordinating system. With the incremental system, the machine establishes a new origin or reference point each time the part is moved. Figure 16.11 shows X and Y values using an incremental measuring system. Notice that with this system, each new location bases its values in X and Y from the preceding location. One disadvantage to this system is that any errors made will be repeated throughout the entire program, if not detected and corrected.

16.6.3 NC Control Systems

There are two types of control systems commonly used on NC equipment: *point-to-point* and *continuous path.*

Point-to-Point Systems. A point-to-point controlled NC machine tool, sometimes referred to as a positioning control type, has the capability of moving only along a straight line. However, when two axes are programmed simultaneously with equal values (X2.000 in., Y2.000 in.) a 45° angle

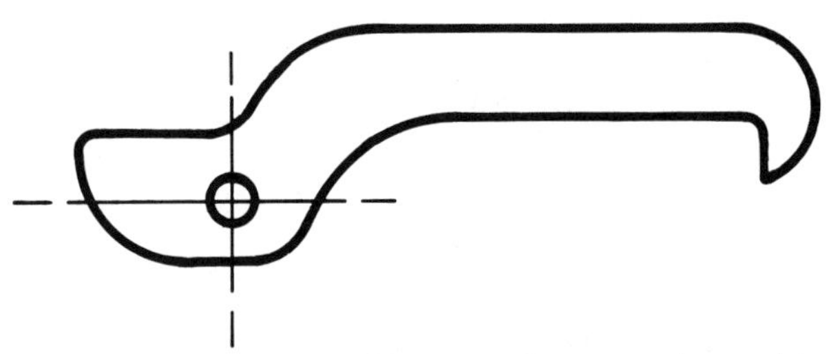

FIGURE 16.14 A continuous-path system is used to machine complex parts with many contours and radii.

will be generated (Fig. 16.12). Point-to-point systems are generally found on drilling and simple milling machines where hole location and straight milling jobs are performed. Point-to-point systems can be utilized to generate arcs and angles by programming the machine to move in a series of small steps (Fig. 16.13). Using this technique, however, the actual path machined is slightly different from the cutting path specified.

Continuous-Path Systems. Machine tools that have the capability of moving simultaneously in two or more axes are classified as continuous-path or contouring. These machines are used for machining arcs, radii, circles, and angles of any size in two or three dimensions. Continuous-path machines are more expensive than point-to-point systems and generally require a computer to aid programming when machining complex contours (Fig. 16.14).

16.6.4 NC Servomechanisms

NC servomechanisms are devices used for producing accurate movement of a table or slide along an axis. Two types of servos are commonly used on NC equipment: *electric stepping motors* and *hydraulic motors.*

Stepping Motors. Stepping motor servos are frequently used on less expensive NC equipment. These motors are generally high-torque power servos and mounted directly to a lead screw of a table or tool slide (Fig. 16.15). Most stepping motors are actuated by magnetic pulses from the stator and rotor assemblies. The net result of this action is that one rotation of the motor shaft produces 200 steps. Connecting the motor shaft to a 10-pitch lead screw allows 0.0005-in. movements to be made ($1/200 \times 1/10 = 0.0005$ in.).

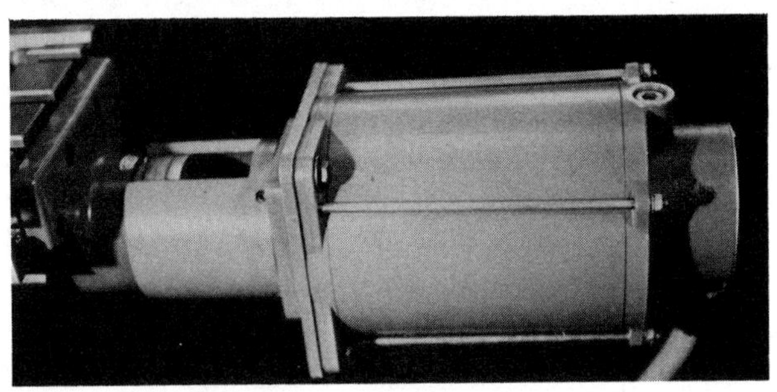

FIGURE 16.15 Stepping-motor servo drive.

Hydraulic Motors. Hydraulic servos produce a fluid pressure that flows through gears or pistons to effect shaft rotation. Mechanical motion of lead screws and slides is accomplished through various valves and controls from these hydraulic motors. Hydraulic servos produce more torque than stepping motors. However, they are more expensive and very noisy. Most large NC machines use hydraulic servos.

16.6.5 Open- and Closed-Loop Systems

Open-Loop. NC machines that use an open-loop system contain no *feedback* signal to ensure that a machine axis has traveled the required distance. That is, if the input received was to move a particular table axis 1.000 in., the servo unit generally moves the table 1.000 in. There is no means for comparing the actual table movement with the input signal, however. The only assurance that the table has actually moved 1.000 in. is the reliability of the servo system used. Open-loop systems are, of course, less expensive than closed-loop systems.

Closed-Loop. A closed-loop system compares the actual output (the table movement of 1.000 in.) with the input signal and compensates for any errors. A *feedback* unit actually compares the amount the table has been moved with the input signal. Some feedback units used on closed-loop systems are transducers, electrical or magnetic scales, and synchros. Closed-loop systems greatly increase the reliability of NC machines.

16.7 INPUT MEDIA AND TAPE STANDARDS FOR NC

As NC has developed over the past years, various input media have been used to present the information through a machine control unit (MCU) to the machine tool. *Most* NC machines today receive their signals from a punched 1-in. wide, 8-channel tape. Other forms of input media used are punched cards and magnetic tape.

16.7.1 Punched Tape Materials

Punched tapes are available in paper, paper–Mylar, solid Mylar, and other laminates. *Paper* is by far the least expensive. Paper tapes can be purchased oiled or nonoiled and come in a variety of colors. Oiled tapes are generally used to help lubricate the punches on the perforating equipment.

Paper–Mylar tape is a laminate of a thin piece of plastic between two strips of paper. It is more expensive than paper tape, but it is much more durable and hard to tear.

Solid–Mylar is the strongest material available and is almost indestructible. It is the most expensive tape material available but is not affected by oil, water, or other substances.

16.7.2 Tape Specifications and Standards

All NC equipment uses the same coding standards and specifications for tape size developed by the Electronic Industries Association (EIA) and the Aerospace Industries Association (AIA). Figure 16.16 shows a standard 1-in., 8-channel tape with complete tape dimensions.

The coding system adopted by the EIA is the *binary coded decimal* (BCD). Figure 16.17 shows a standard 1-in. EIA NC tape. The numbers 1 through 8 represent the channel number (also called track) assigned to the tape.

1. Channels 1 to 4 are used for numerical data for numbers 1 through 9.
2. Between channels 3 and 4 are sprocket holes used to advance or rewind the tape through the reader.
3. Channel 5 (marked CH) is a parity channel. With the EIA system an odd number of holes *must* be punched in each row of the tape, or the tape reader does not recognize the code and stops the machine. For example, notice the number 3 in Fig. 16.17. The number 3 in the BCD system consists of holes punched in channels 1 and 2 (2 + 1 = 3). With holes in only channels 1 and 2 the

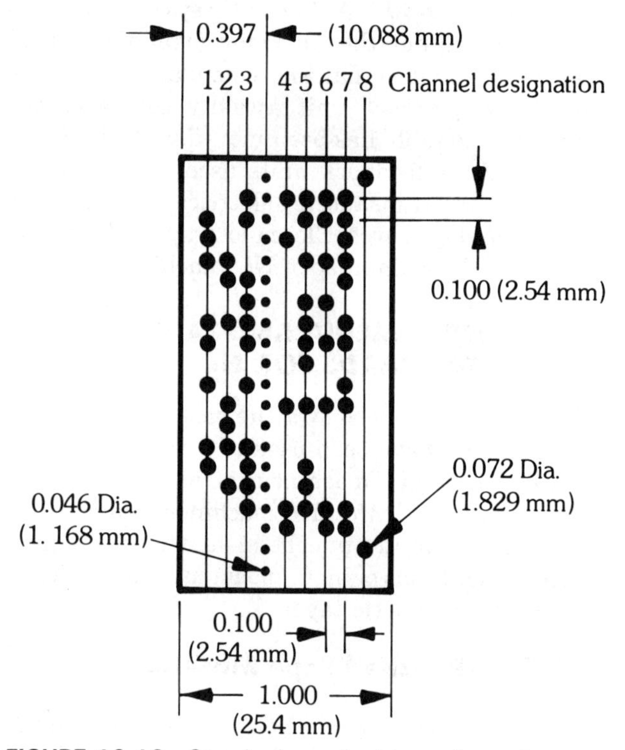

FIGURE 16.16 Standard punched tape dimensions.

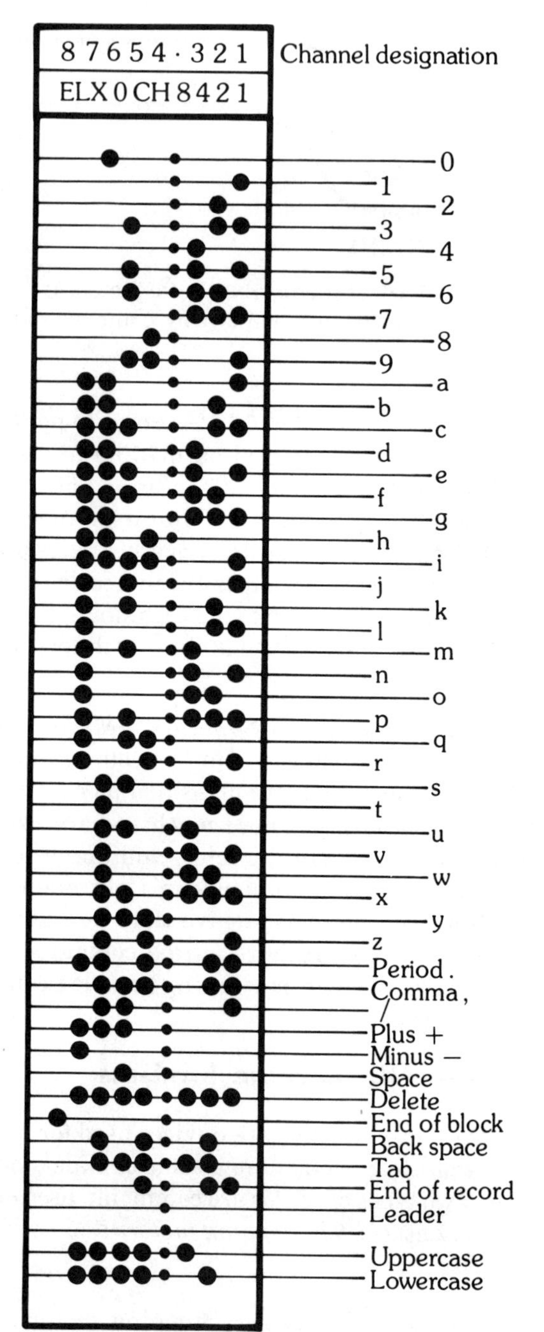

FIGURE 16.17 Electronics Industries Association standard coding for 1-in.-wide, 8-channel punched tape.

number 3 has an even number of holes in the row; therefore, a third hole is punched in the parity channel.

4. A single hole punched in channel 6 always represents a zero. Some *letters* of the alphabet also use channel 6 for their coding.
5. Channel 7, marked X, is used to code some letters in the alphabet and other symbols such as periods and minus signs.
6. Channel 8 is generally marked EL meaning end of line or EOB, end of block. An EOB is always used to complete a particular block of information used to direct table movements, miscellaneous functions, and so forth.

Tape Definitions. Certain terms are used to describe the NC tapes. A list of definitions for these terms follows:

Bit is a single character. It is either a punched hole or the absence of a punched hole on the tape.

Character is a series of punched holes to signify a letter, number, or symbol.

Row is a line of holes perpendicular to the edge of the tape. Each row represents a certain character.

Word is a set of characters that gives the NC machine a *single* instruction. A word may consist of one or more rows of punched holes. *Example:* An EOB is a word that requires a single row, while X10750 is a word that contains six rows.

Block is a group of words that gives the NC machine a complete statement to act upon. A block may contain one or more words in its statement (*example:* N1G81X2000Y1000M03). This particular block of information tells the NC machine: this is line 1 (N1); move the table 2 in. in the *X* direction and 1 in. in the *Y* direction; turn the spindle on (clockwise) M03; and finally, drill a hole (G81) at that location.

Channel or track is a line of holes or spaces *parallel* to the edge of the tape.

16.7.3 Tape Readers

Most NC machines use either *mechanical* or *photoelectric* tape readers to interpret the data on the punched tape.

Mechanical readers are generally used on most inexpensive point-to-point machines. They contain small mechanical fingers on a roll that detect the presence or absence of a punched hole on the tape. Mechanical readers are comparatively slow, reading up to 120 rows per second, and therefore are not used on continuous path machines.

FIGURE 16.18 Photoelectric tape reader. (*Courtesy El Camino College*)

FIGURE 16.19 Tape-perforating typewriter used for punching NC tapes.

Photoelectric readers (Fig. 16.18) can read up to 1000 rows per second. They are more expensive than mechanical tape readers, but since they are much faster, they are used on most NC machines. Photoelectric tape readers transmit a beam of light onto the punched tape, which is then read to detect the absence or presence of a hole.

16.7.4 Tape Punching Equipment

Numerical control tapes are punched by special typewriters. These machines have almost the same keyboard as a regular typewriter and are also equipped with mechanical punching devices to perforate the tape at the same time the program is typed (Fig. 16.19). Most tape-perforating equipment has tape readers that duplicate existing tapes and allow editing of incorrectly punched tapes.

16.7.5 Tape Formats

When the tape reader in the MCU reads the paper tape, it must interpret the bits of information placed on the tape, and then complete the required functions. The arrangement of these bits on the tape is called the *tape format.* There are several

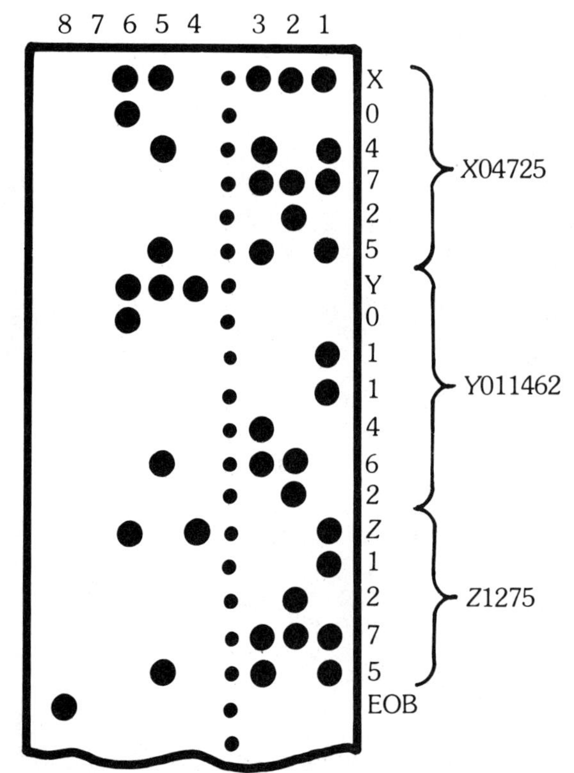

FIGURE 16.20 Word address tape format.

tape formats used, but most NC equipment uses either the *word address* or *tab sequential* format.

Word Address Tape Format. This type of tape format uses *alpha characters* (letters) and *numerical data* (Fig. 16.20). An alpha character, called an address, precedes each piece of numerical data or word to describe or identify the function of the numerical data. With the word address format, the blocks of information may be of any length, and the words used need not be in any particular sequence. In the following paragraphs, the most commonly used alpha characters are described.

Alpha characters *X, Y, Z, a, b, c,* and so on, are used to indicate a particular axis or axes of the machine tool.

The *G* address is used to signify a preparatory function on the machine tool. Preparatory functions direct a machine tool through given machining operations or manipulation of an axis. Examples of *G* addresses standardized by the EIA are as follows:

CODE	*FUNCTION*
G00–G03	Reserved for contouring only
G04	Dwell
G05	Hold
G06–G07	Unassigned
G08–G12	Reserved for contouring only
G13–G16	Axis selection
G17–G21	Reserved for contouring only
G22–G24	Unassigned
G25–G29	Permanently unassigned
G30–G32	Reserved for contouring only
G33	Thread cutting, constant lead
G34–G35	Reserved for contouring only
G36–G39	Reserved for control use only
G40–G49	Reserved for contouring only
G50–G79	Unassigned
G80	Fixed cycle cancel
G81	Fixed cycle 1
G82	Fixed cycle 2
G83	Fixed cycle 3
G84	Fixed cycle 4
G85	Fixed cycle 5
G86	Fixed cycle 6
G87	Fixed cycle 7
G88	Fixed cycle 8
G89	Fixed cycle 9
G90–G99	Unassigned

An *M* character is called a *miscellaneous* function. It precedes numerical data to prepare the NC system for operations such as turning the spindle or coolant on or off, rewinding the tape, and so forth. The following list indicates the various EIA standardized *M* functions.

CODE	*FUNCTION*
M00	Program stop
M01	Optional (planned) stop
M02	End of program
M03	Spindle CW
M04	Spindle CCW
M05	Spindle OFF
M06	Tool change
M07	Coolant 2 ON
M08	Coolant 1 ON
M09	Coolant OFF
M10	Clamp
M11	Unclamp
M12	Unassigned
M13	Spindle CW and coolant ON
M14	Spindle CCW and coolant ON
M15	Motion +
M16	Motion −
M17–M24	Unassigned
M25–M29	Permanently unassigned
M30	End of tape
M31	Interlock bypass
M32–M35	Constant cutting speed
M38–M39	Unassigned
M40–M45	Gear changes if used; otherwise, unassigned
M46–M49	Reserved for control use only
M50–M99	Unassigned

An *F character* or command is used to program the proper feed rate in in./min.

An *S character* is used to program correct spindle speeds. Some NC equipment has a manual adjustment for changing spindle speeds and on those machines the *S* command is not used.

The *letter T* is used to signify a spindle position on turret drilling and milling machines. For example, *T*4 orders spindle number four to be put in the correct position for a machining operation.

The word address tape format also uses the alpha character *N* to designate the *sequence number* or block number during the actual programming process.

The word address format is the easiest to learn and the most flexible to use for beginning programmers. Most NC equipment uses this format.

Tab Sequential Tape Format. This type of tape format uses no alpha characters to describe numerical data or words. However, words or data must appear in a sequential order and be preceded by the *tab* code. The tab code is punched on the paper tape to separate words and numerical data on each block of information throughout the entire program. An example of the tab sequential format is shown in Fig. 16.21. The usual sequence used in a block of information is sequence number, *X* value, *Y* value, *Z* value, feed rate, spindle speed, miscellaneous function, and finally an EOB. In this format, each sequence must be preceded by the tab code, otherwise the data will be interpreted incorrectly.

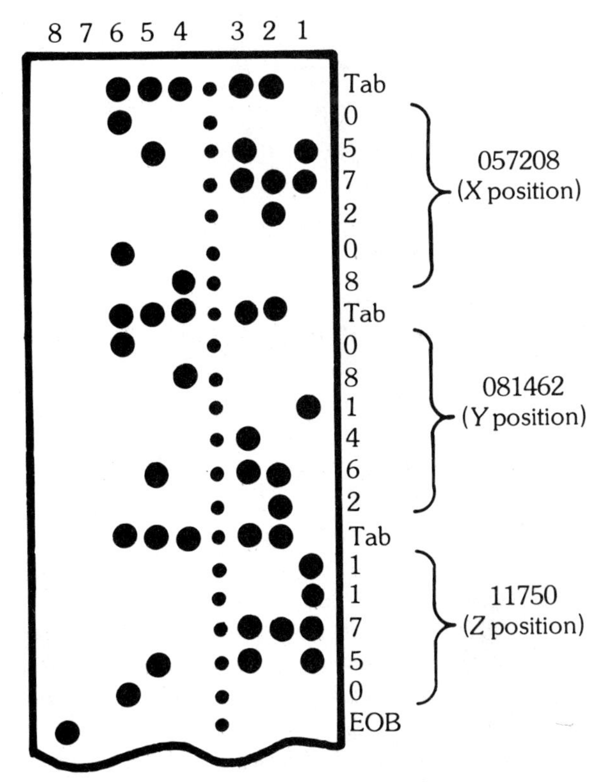

FIGURE 16.21 Tab sequential tape format.

16.8 PROGRAMMING

There are two basic types of NC programming: manual and computer-assisted. Each type of programming requires the NC programmer (often referred to as a parts programmer) to consider some fundamental elements prior to the actual programming steps.

16.8.1 Factors to Consider Prior to Programming a Part

1. *Type of dimensioning system.* The first step is to select the best NC machine to complete the job. Consider what NC machines are available at the time the job is to be machined and which machine is best suited for the particular job. After selecting the appropriate machine tool, determine what type of dimensioning system the machine uses—either an absolute or incremental system.
2. *Types of tape format.* The majority of NC machine tools use either word address or tab sequential as the tape format. The machine tool selected determines which tape format to use.
3. *Axis designation.* Another consideration is the axis designation of the machine tool. In most cases, the programmer has already considered this fundamental element when she or he selected the best NC machine tool to complete the job. The most important factor in axis designation is the location and position of the spindle. At this time the parts programmer also determines how many axes are available on the machine tool, that is, *X, Y, Z, a, b, c,* and so on, and whether the machine tool has a continuous path or point-to-point control system.
4. *Miscellaneous functions available.* With the wide variety of NC equipment available, there are also a number of options and miscellaneous (*M* functions) available. The parts programmer must determine what options and *M* functions are available on the selected machine tool.
5. *Machine tool zero-point system.* A *zero point* is a reference on the machine tool with which the mounted workpiece must be in a correct relationship, so all machining operations can be completed accurately. NC machines use either a fixed- or floating-zero system.

A *fixed zero* is a location on the machine table,

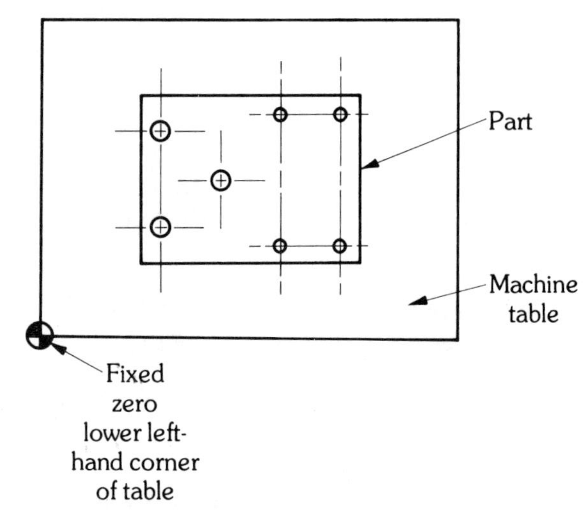

FIGURE 16.22 Fixed-zero system.

usually the center or lower left-hand corner as the operator faces the machine. If the machine uses an *absolute* system, all the dimensions used to carry out the machining operations are based from this fixed zero (Fig. 16.22). When the machine is equipped with an incremental dimensioning system, the first dimension is located from the fixed zero.

Floating zeros are more convenient to use and, as the name implies, the zero point can be established anywhere on the machine table. With the floating zero, the operator sets the part to be machined in any location on the table and locates the tool at the most convenient place to begin the program.

16.8.2 Manual Programming Steps

With the large number of different NC machine tools available today, it would be impossible to show a sample program for each type of machine. Most NC programmers program a machine by completing a series of procedures or steps. These six general programming steps are explained as follows:

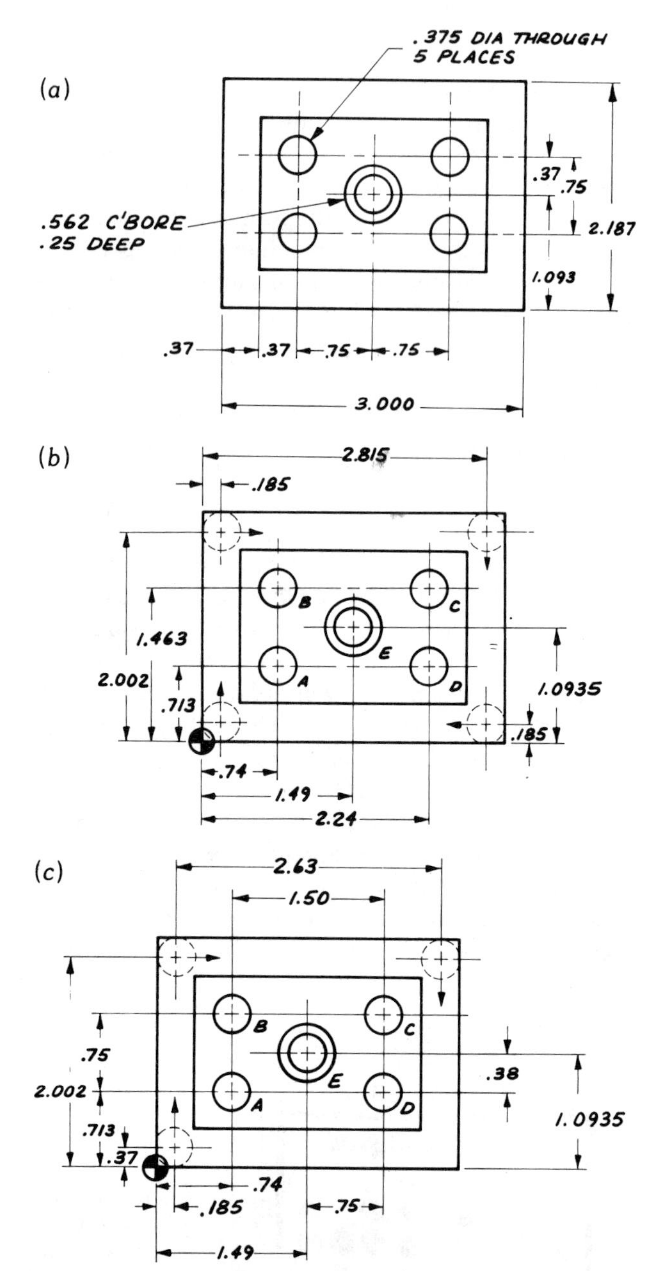

FIGURE 16.23 (*a*) Original shop drawing; (*b*) absolute format; (*c*) incremental format.

1. *Prepare an NC coordinate drawing.* The first step in making an NC program is to convert the existing blueprint or shop drawing into an NC coordinate drawing. This is done according to the type of dimensioning system the NC machine has—absolute or incremental. Figure 16.23 shows the original shop drawing converted to an NC coordinate drawing. Figure 16.23*(b)* shows the drawing converted to an absolute format, and Fig. 16.23*(c)* is an incremental format. The dashed lines indicate the cutter tool path for milling the perimeter of the part. Notice that each hole to be drilled has a letter. If each hole is labeled with a letter, the sequence of machining operations and locations of each hole is easier to keep track of. This concept will be clearer later when a sample part is programmed.
2. *Plan operations sequence.* The second step is to determine a sequence of operations. If the part requires a variety of machining operations, the programmer must decide when each is to take place. Should the part first be milled and then the holes drilled? Each NC part to be programmed is unique and the NC programmer must decide the shortest and most efficient machining time when he plans the operations sequence.

3. *Prepare program manuscript.* From the operations sequence a program manuscript is prepared. At this point all preparatory addresses are added to the sequence (milling, drilling cycles) along with feed rates, spindle speeds, and miscellaneous functions. The manuscript should also specify all machining instructions to the machine operator so that he may successfully complete the part.
4. *Punch tape.* After the manuscript is completed, a punched tape is prepared. The punched tape is the input for the NC machine tool.
5. *Verify tape.* The punched tape and program must then be verified. There are several methods of verification. Perhaps the most convenient is to install the punched tapes into the MCU of the NC machine and make a dry run. Another technique commonly used when a variety of operations are to be performed is to machine a wooden block or other test material. The sample part is then measured to see that all dimensions are correct and the machined part is acceptable.

   Modern shops often have a *plotter* to verify punched tapes (Fig. 16.24). An NC plotter uses a pen to draw the path of the cutter and locates the center of the holes to be drilled. The drawing the plotter produces is compared to the original shop drawing for accuracy.
6. *Make machine run.* The last step in programming a part is the machining of an actual part. After the part has been completed, all dimensions are carefully checked and any errors corrected. Periodic inspection is made on parts during the machine run to minimize reject parts.

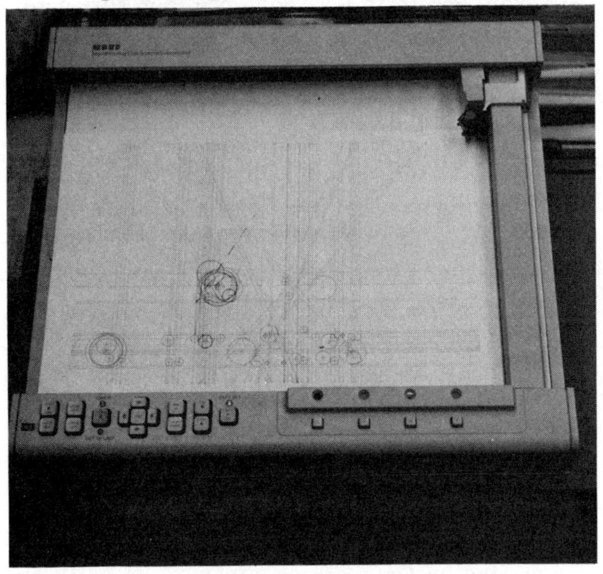

FIGURE 16.24 The NC plotter is used to draw the part of the cutting tool and plot all points for drilled holes. (*Courtesy Schlumberger Technologies, Inc.*)

Part Program Example. It would be difficult to illustrate example programs of each NC machine tool, function, tape code, and dimensioning system. Therefore, one sample program—a *word address* format, *incremental positioning* system—has been selected because this is a common NC system on machine tools.

Figure 16.25 shows a part drawing of an adapter plate. It requires four holes of 1/4-in. diameter and two holes of 3/4 in. diameter. In addition, a 0.375-in. slot is to be milled 0.250 in. deep.

Figure 16.26 shows a detailed manuscript and program for the adapter plate. The manuscript format is typical of the manuscripts used by parts programmers in industry. At the top of the manuscript are special headings to describe the part, machine tool, and cutting tools needed to complete the part. Directly below the heading are vertical columns. The first column, N, is the sequence number for each block of information of the program. The second column, G, is the word address code for preparatory commands. The next three columns are the machine axes *X, Y,* and *Z.* Column F is used to list the feed rates used, and the *M* column is used for coding miscellaneous functions. The last column is used for listing any special instructions to the operator.

Manuscript formats vary with the different types of NC machines; however, most are similar to the one illustrated. Here is an explanation of program example 1:

Line *N*1: starts the drilling sequence from the origin to the first hole A1. The G81 code puts

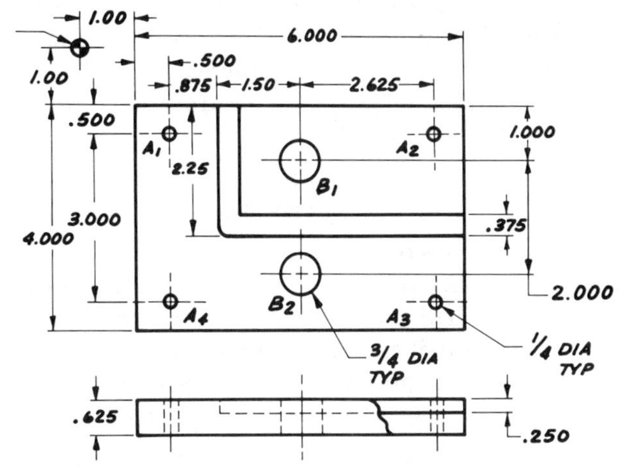

FIGURE 16.25 Adapter plate.

DATE	MACHINE TOOL INCREMENTAL 3 AXIS VERTICAL MILL	PAGE 1 of 1
PROGRAMMER L. BEAUTY	TOOLING Center Drill; $\frac{1}{4}$, $\frac{3}{4}$ drill; $\frac{3}{8}$ End Mill	
Part name Adaptor plate	Part number WYV5353	

N	G	X	Y	Z	F	M	Special instructions
1	81	1500	−1500		200	03	Center drill all holes
2		2375	−500				1000 RPM A_1–A_4
3		2625	500				B_1–B_2
4			−3000				
5		−2625	500				
6		2375	−500				
7						06	$\frac{1}{4}$ drill thru 6 holes
8			3000			57	1500 RPM A_1–A_4
9		2375	−500				B_1–B_4
10		2625	500				
11			−3000				
12		−2625	500				
13		−2375	−500				
14						06	$\frac{3}{4}$ drill thru 2 holes
15		2375	500			57	500 RPM B_2, B_1
16			2000				
17						06	$\frac{3}{8}$ End Mill Set
18	80	1125	1375			57	Stop for 0.250 inch
19			−2250		10	52	Depth of cut
20		4625					
21		−7125	3125		200	53	Return to origin
22						02	Stop spindle, rewind tape

FIGURE 16.26 Manuscript for adapter plate.

the machine in a drilling mode. The machine table moves at (F) 200 in./min between all hole locations. The 03 (M code) is used to start the spindle running. (*Note:* The depth of the center-drilled holes is usually adjusted at the machine, as are all further cutting tools. There are different devices used for setting spindle depths, usually cams or turret stops. These vary with different manufacturers of NC milling machines.)

Lines 2 to 6: locate the part for the center-drilling operation of each hole.

Line 7: stops the machine and spindle for a tool change. In this case, a 1/4-in. drill is installed into the drill chuck. The cycle start is actuated after the tool change. When the cycle start is pressed, the spindle automatically turns on and the next block of information is read.

Line 8: moves the table to the first hole to be drilled at 1/4 in. (A1). An M57 code is programmed to index the depth stop turret to a new preset depth for the 1/4-in. drill.

Lines 9 to 13: move the part to the remaining locations for the 1/4-in. drilling operation.

Line 14: another M06 code used to stop the machine and spindle. This time a 3/4-in. drill is installed for drilling holes B1 and B2. Again, the cycle start button must be depressed to start the spindle and machine functions.

Line 15: locates the table to the first 3/4-in. hole to be drilled (B2). The M57 code is used to index the depth stop turret for the spindle to the correct depth for the 3/4-in. drill.

Line 16: locates the part to the second 3/4-in. hole B1.

Line 17: M06, tool change, function used to stop the spindle and machine. A 3/8-in. mill is installed by the operator.

Line 18: begins with a preparatory function, G80 (sometimes other G functions are used), which changes the drilling mode to a milling mode of operation. This line also directs the part to the correct location to begin the milling operation. The M57 code is used to index the depth stop turret to a new preset depth for the 3/8-in. end mill.

Line 19: begins the milling operation by lowering the spindle M52. A feed rate of 10 in./min is programmed to feed the work and table during the milling operation. (*Note:* The use of miscellaneous function 52 may vary with different NC machines. Other numbers could be used to lower the spindle for the milling operation.

Line 20: completes the milling operation by moving the part to its final location.

Line 21: an M53 code is programmed to raise the spindle. *X* and *Y* dimensions are listed to locate the spindle back to the origin. Notice a feed rate F200 is used to move the table at 200 in./min quickly back to the origin.

Line 22: the last block of information programmed, the miscellaneous function M02. The M02 code stops the spindle and rewinds the punched tape back to the start of the program.

Summary. It is very important that the NC programmer carefully goes through a series of fundamental elements and programming steps. Each NC machine tool has its own unique characteristics, its own coded language (format) that cause the tool motions and functions to occur. The parts programmer must write a manuscript in the format for the specific machine tool on which the part is to be machined. The manuscript is next transferred to punched tape by tape perforation equipment, and then verified either by a plotter or on the actual machine tool.

16.8.3 Computer-Assisted Programming

Along with the rapid advances in numerical control are those of computer technology and computer languages. *Computer-assisted* or *computer-aided* parts programming has been a major factor in the rapid growth of NC. Without computers to aid in NC programming, machining complex parts that require contour milling and/or many hole locations would be tedious and lengthy. The parts shown in Fig. 16.27 have a variety of hole sizes and a large number of milling cuts. These parts would require many hours, if not several days, to conventionally machine or hand program. Utilizing a computer to assist the programming, it should take only several hours.

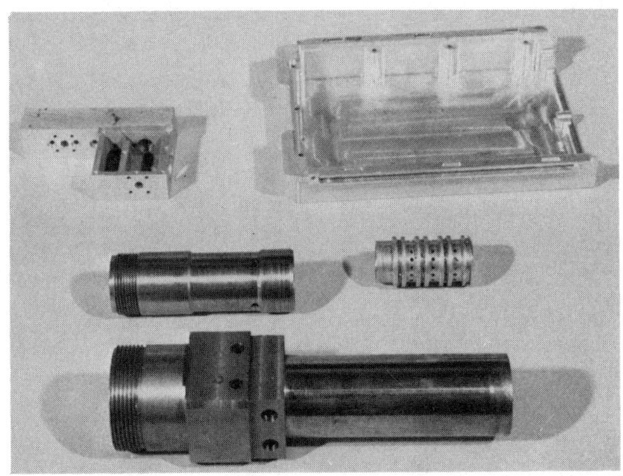

FIGURE 16.27 Complex parts are easily programmed using computer-assisted programming.

Another advantage of the computer is that it reduces the chance of human error. The computer is actually a high-speed calculator that can handle thousands of complicated mathematical problems correctly in a short period of time.

There are many computer languages used in parts programming, a few of which are:

Compact II—computer-assisted NC part programming

APT—automatically programmed tool

CAMP 1—compiler for automatic machine programming

SNAP—simplified numerical automatic programmer

PRONTO—program for numerical tools

ADAPT—Air Materiel Command developed apt

Many of these languages use similar words. They all use a pidgin English-type of vocabulary that helps the new programmer considerably when first learning a computer language.

The procedure used when utilizing the computer to aid programming is similar to manual programming. After receiving a drawing of the part, the programmer defines the part geometry with reference to the standard Cartesian coordinate (*X*, *Y*, and *Z*) system. After defining all the geometry of the part (part lines and planes, location of holes, and so on), the programmer describes the desired cutter path to, and/or around these lines, planes, and locations. The programmer also specifies all tool changes, feed rates, spindle speeds, and all other functions necessary to completely machine the part.

The next stage consists of having the completed program keypunched onto computer cards, and then submitted to the computer. A time-sharing system is often used in place of keypunch cards. Time-sharing systems allow the programmer direct contact with the computer by telephone. Along with the parts program a machine *postprocessor* program must be prepared. The function of the postprocessor is to translate the input data (original parts program) into a suitable format for a specific machine tool model and make. In other words, there would be a postprocessor for each type, make, and model number of NC machine. These programs are processed by the computer which produces a punched tape and a print out of the program manuscript listing all coordinates, machine functions, and so on.

The foregoing explanation of computer-assisted programming has necessarily been oversimplified. For further information, see your instructor for additional references and literature.

16.9 FURTHER DEVELOPMENTS IN NC

16.9.1 Computer Numerical Control

Computer numerical control (CNC) is the next step in numerical control. (*Note:* CNCs are NC machines. NC is a general term applied to machine tools as defined previously. The differences are explained below.)

Today, most CNC machine tools utilize a microprocessor or minicomputer built into the machine control unit of the machine. A CNC is a *soft-wired* system, meaning that once the program is loaded into the memory of the computer, no hardware is necessary to transfer the numerical codes into the controller.

Advantages of CNC systems are increased flexibility and increased capabilities. Programs can be created or edited and new information added (input) directly at the machine tool. In addition, programs may be entered, by means of a program tape, into the computer memory. Thus it is possible to run each workpiece from the stored data, eliminating the controller having to read the tape for each part.

CRTs (cathode ray tubes) are available for most CNC machines. CRTs resemble small television screens. (Fig. 16.28). Their purpose is to display words and numbers that are part of the program. The keyboard next to the screen permits the operator/programmer to enter data and visually edit any programs that may be in error. Optional tape punches may be connected to this unit so that once edited information is used to produce correct parts,

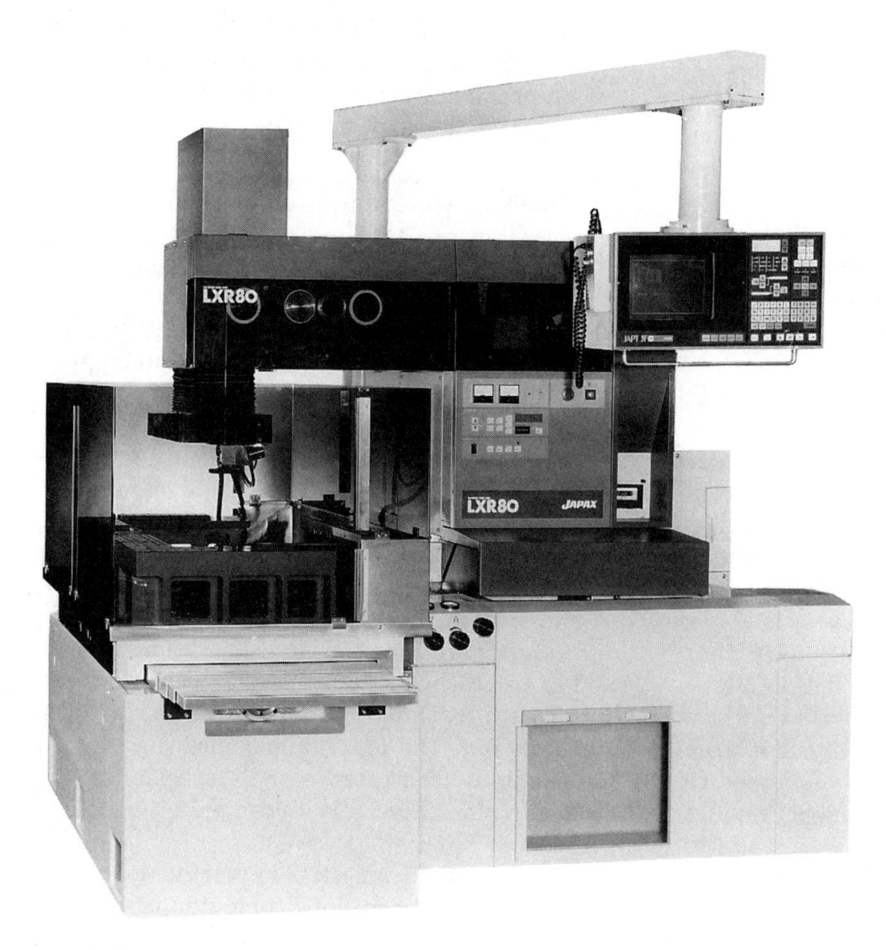

FIGURE 16.28 Modern CNC traveling-wire EDM. (*Courtesy Japax, Inc., a division of Bridgeport Machines, Inc.*)

a new tape can be generated at the machine for future use.

The majority of CNC machines will compute and execute special programmed routines called canned cycles. Canned cycles permit common machining operations, such as pocket milling and bolt hole circle drilling, to be performed by a single descriptive statement. Finally, CNC machines have the ability to read either EIA or ASCII codes, handle either incremental or absolute positioning systems, and react with either inch or metric input.

16.9.2 Direct Numerical Control

Direct numerical control (DNC) results when several NC or CNC machines are directly connected to and controlled by a large computer or several smaller computers. The computer eliminates the need for a machine control unit, a punched tape and tape perforation equipment, thereby decreasing downtime and maintenance of individual machines. Another advantage of DNC is that easy modification and editing of programs can be accomplished at the central computer.

REVIEW QUESTIONS

16.2 NUMERICAL CONTROL DEVELOPMENT

1. Define numerical control.
2. What is the role of NC in today's technology and automation?
3. Briefly explain how work by the following people led to the development of NC:
   (a) Fourneaux
   (b) Falcon
   (c) Charles Babbage
   (d) John C. Parsons

16.3 INDUSTRIES USING NC

1. Briefly explain how the following industries use NC:
   (a) Aerospace
   (b) Electronic and electrical
   (c) Automotive
   (d) Quality control
   (e) Fabrication

16.4 TYPES OF NC MACHINE TOOLS

1. Name and briefly describe two types of NC drilling machines.
2. Why are NC milling machines the largest selling NC machine tools?
3. Explain the operations an NC mill can perform.
4. List and describe the two types of NC lathes used in industry.

16.5 ADVANTAGES AND DISADVANTAGES OF NC

1. List and explain seven advantages of NC.
2. List and explain four disadvantages of NC.

16.6 NC MACHINE TOOL ELEMENTS

1. Briefly describe the X, Y, and Z axes.
2. Sketch a vertical and horizontal mill, engine lathe, and turret drill press. Label all table and slide motions with the correct axis.
3. If an NC rotary table is placed on a vertical milling machine worktable, what axis would it be on?
4. Compare absolute and incremental NC measuring systems.
5. Explain the point-to-point control system used on NC machine tools.
6. Compare electric stepping motors to hydraulic systems when used on NC machine tools to move tables and slides.
7. Explain the differences between an open- and closed-loop system. Also, give an example of each system outside the area of NC.

16.7 INPUT MEDIA AND TAPE STANDARDS FOR NC

1. List three types of input media used on NC machine tools. Which is the most commonly used?
2. Name the association that sets the coding standards and specifications for NC.
3. Sketch an 8-channel punched tape. Label all the channels and briefly describe the function of each.
4. Briefly explain the following terms that describe a punched tape:
   (a) Bit
   (b) Character
   (c) Row
   (d) Word
   (e) Block
   (f) Channel
5. Describe the operating principles of mechanical and photoelectric tape readers.
6. How are NC tapes perforated for tape readers?

7. Explain the differences between the word address and tab sequential tape formats. Which term is the more commonly used?
8. How are G addresses used on the word address formats? M functions?

16.8 PROGRAMMING

1. Describe five factors the parts programmer considers before preparing a program manuscript for a job.
2. Explain the necessary steps a programmer must complete to successfully prepare a program manuscript for a given part.
3. Prepare an absolute-zero, tab sequential program for the part shown by Fig. 16.25. Use the manuscript (Fig. 16.26) as an aid to compute the problem.
4. Briefly explain the process of parts programming using the computer as an aid.
5. What is a postprocessor and how is it utilized in computer-aided programming?

16.9 FURTHER DEVELOPMENTS IN NC

1. Explain the difference between CNC and DNC.
2. Which system will be developed more in the future?

17

Electrical Machining Processes

17.1 INTRODUCTION

Electrical discharge machining, electrical discharge grinding, and electrochemical machining are newer machining processes developed to efficiently remove metal without mechanical forces. These newer methods were developed primarily to machine extremely hard and tough materials and to produce complex-shaped parts. They are generally used when conventional machining methods are either impractical or impossible.

17.2 ELECTRICAL DISCHARGE MACHINING

Electrical discharge machining, generally referred to as EDM, is a process used to remove metal from any material, hard or soft, that conducts electricity. The dc power supply provides the energy for electrical discharges between the electrode and workpiece. The power supply is designed to control the amount of energy and the number of discharges per second. These discharges are actually intermittent sparks from the electrode (cutting tool) that erode the metal away from the workpiece. The cavity thus produced conforms to the shape of the electrode (Fig. 17.1). During the EDM process, the cut-

Opening photo courtesy Charmilles Technological Corp.

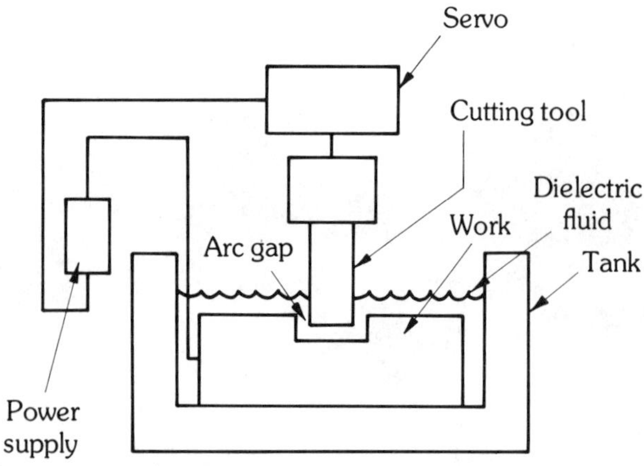

FIGURE 17.1 Basic components of an electrical discharge machine.

ting tool and the workpiece must never touch each other. If this should happen there would be an electrical short circuit and no metal removal would take place.

Application of the EDM process requires thorough knowledge of the fundamental principles of the EDM machine, EDM power supplies, cutting tools, coolant, and flushing techniques.

17.2.1 The EDM Machine

Electrical discharge machining requires a machine that has provisions to hold the electrode, a mounting surface for the workpiece, a power supply to provide the energy for the cut, and the necessary controls.

Most EDM machines resemble vertical milling machines. They have a base, column, head, and tank. The *base* generally serves as a reservoir to hold the dielectric fluid. The *column* supports the *head* assembly directly over the workpiece. The workpiece is mounted in a *tank* and positioned by a table generally mounted on the base (Fig. 17.2). The tank contains a dielectric fluid that surrounds the part during the EDM process.

The cutting tool is advanced and retracted by a servo system by means of a quill or slide. The servo system can be hydraulically or electrically controlled. The function of the control unit in the servo system is to maintain a constant gap between the electrode and the workpiece. It does this by comparing the *voltage* between the workpiece and the electrode to a reference voltage in the power supply. The servo feeds the electrode toward the workpiece until the reference voltage is obtained. At this time, the EDM process begins. However, if the reference voltage drops, the servo unit retracts the electrode from the workpiece until the reference voltage is again obtained. Generally, the voltage drop occurs when metal particles between

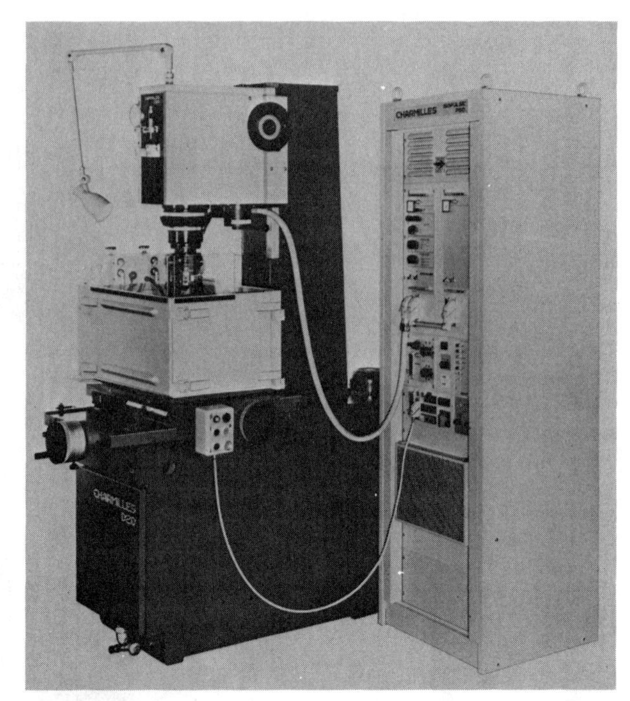

FIGURE 17.2 Electrical discharge machine. (*Courtesy Charmilles Technological Corp.*)

the electrode and workpiece short-circuit the system. Once the metal particles are flushed away by the dielectric fluid, the servo unit again feeds the electrode toward the workpiece.

17.2.2 EDM Power Supplies

The power supply of an EDM provides the energy for electrical discharges between the cutting tool and workpiece. This energy is supplied in the form of dc pulses. Several types of power supply have been used for EDM. Although there are many differences among them, each design is used for the same basic functions:

1. Supply dc voltage between the tool and workpiece
2. Control this energy so it produces the desired results

Resistance–capacitance and *pulse*-type power supplies are the two most commonly used today in industry. Resistance–capacitance, or simply the *RC* circuit, was widely used on the first EDM machines. It is also known as the *relaxation oscillator* type and commonly used on less expensive EDM circuits. To fully understand the basic principle of the *RC* circuit, one should clearly understand the operating principles of a capacitor. When a capacitor is connected to a dc source, it is charged to full voltage almost instantaneously. The capacitor is charged through a resistor from the dc source

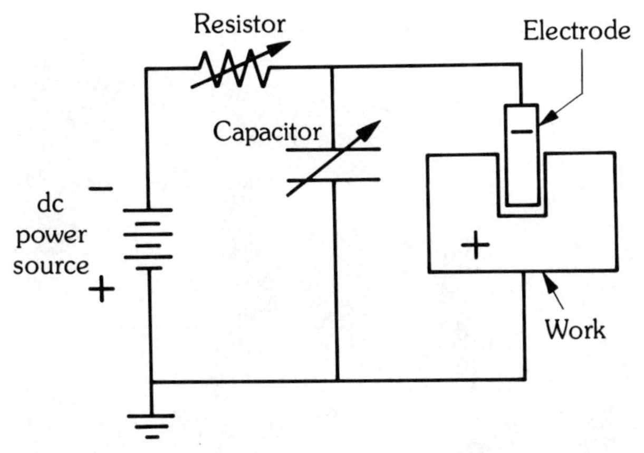

FIGURE 17.3 Resistance-capacitance (RC) circuit EDM power supply.

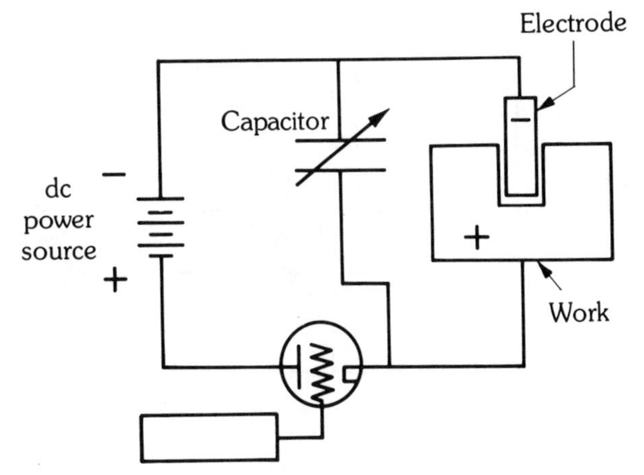

FIGURE 17.4 Pulse circuit EDM power supply.

to a level sufficient to break down the dielectric fluid at the machining gap (Fig. 17.3). When the dielectric fluid breaks down, a spark jumps from the tool to the workpiece and erodes the workpiece.

Advantages of the *RC* circuit power supply are:

1. It is simple and reliable.
2. It is less expensive.
3. It works well on small intricate jobs.

Disadvantages of the *RC* circuit are:

1. Coarser finishes result.
2. Overcuts occur around the tools, making it difficult to maintain close tolerances.
3. Large metal particles require constant flushing to prevent short circuits.
4. Supplying energy to the capacitor through a charging resistor consumes time and limits the number of possible discharges in a given unit of time.

The pulse-type power supply has become more popular in recent years. It has several distinct advantages over the *RC* circuit, although their basic circuitry is similar. Vacuum tubes or solid-state devices have been added in the system to achieve a rapid pulsing switch effect (Fig. 17.4). This pulse switching effect increases the frequency of discharges at least 10 times greater than with the basic *RC* circuit. This increase in frequency results in lower operating voltages but as many as 500,000 sparks or discharges per second.

The main advantages of the pulse-type circuit are:

1. Accurate control of roughing and finishing cuts
2. Better surface finishes due to increase in the number of discharges
3. Less overcut around the cutting tool

Rate. The amount of metal removed in a given unit of time during the EDM process is called the *removal rate.* The primary factor involved in metal-removal rate is the amount of *current* or *energy* (measured in amperes) applied to the workpiece. As the current increases, metal-removal rate increases as shown in Fig. 17.5. A spark of 1 ampere (A) erodes a certain amount of metal. When the current is doubled, the energy in the discharge is also doubled and approximately twice the amount of metal is removed. Modern EDM machines have power supplies capable of generating more than 100 A to the workpiece.

Frequency. The surface finish is controlled by the number of discharges per second, more often referred to as the *frequency* of sparks. The greater the amount of energy applied, the greater amount of material removed. However, when greater amounts of current are used, larger craters are eroded from the work, causing a rougher surface

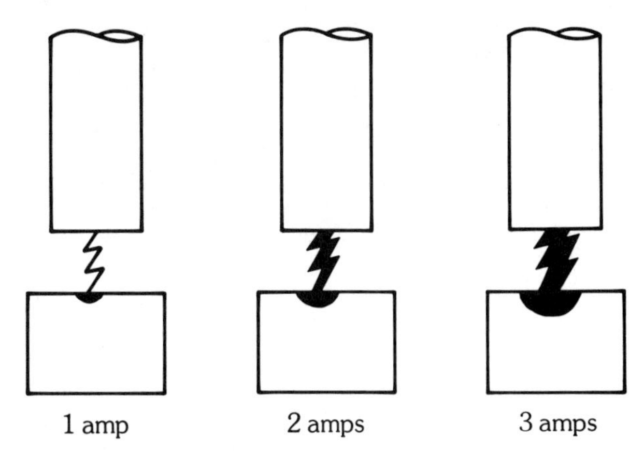

FIGURE 17.5 Effect of current on metal removal rate. As the amperage increases, the removal rate increases.

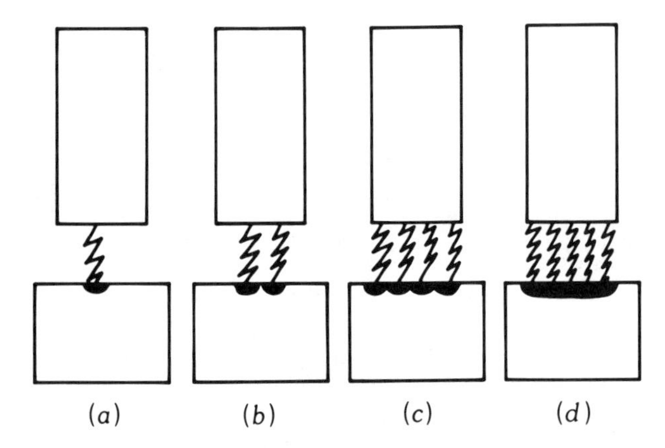

FIGURE 17.6 Effects of surface finish with the increase of frequency and constant current.

finish. To maintain increased metal-removal rates and at the same time improve the surface finish, it is necessary to increase the frequency of the discharges.

Figure 17.6 illustrates four machining circumstances. Each cutting tool has 5 A of current. Figure 17.6*(b)* shows the frequency per second doubled. The total volume of metal removed is the same as *(a)*, but each crater is one-half the diameter of the craters in the first workpiece, resulting in a smoother surface. When the frequency is increased even more, as shown in *c* and *d*, the surface finish is even finer. Remember, if the *frequency* of discharge increases and the current remains constant, the surface finish becomes finer. Typical frequencies used in modern EDM machines range from 300,000 to 500,000 cycles per second.

17.2.3 The Electrode

The electrode is the *cutting tool* in the EDM process. It is the means by which the electric current is carried to the workpiece. With normal EDM machining applications, the workpiece is the positive terminal of the power supply and called the *anode.* The electrode, called the *cathode,* is the negative terminal. This setup is called *standard polarity* and is used in the majority of EDM operations.

The size and shape of the electrode determines the size and shape of the cavity machined. The shape machined on the workpiece is the opposite of the shape of the electrode. In other words, the electrode can be thought of as the reverse of the shape produced on the workpiece (Fig. 17.7).

Electrode Materials. Many different materials can be used for the electrode in EDM. The only physical property an electrode *must* have is to be electrically conductive. However, other considerations when selecting a material to be used for the electrode are (1) it can be easily machined to shape, and (2) it can produce rapid metal-removal rates from the workpiece. The most widely used electrode materials are graphite, yellow brass, copper, copper and tungsten, silver and tungsten, and zinc alloys.

FIGURE 17.7 The electrode on an EDM produces a cavity opposite in shape. The electrode in the case produces an internal gear in the workpiece. (*Courtesy Charmilles Technological Corp.*)

Much experimentation and research have been done to find the best and most economical material for electrodes. The characterics of each type of machining application usually dictate the selection of the best electrode material. The most efficient electrode materials are those that exhibit the highest metal-removal rates from the workpiece combined with lowest wear rate from the electrode. Electrode wear results from the spark erosion process because not all the energy is dissipated in the workpiece. This is an important consideration since excessive wear of the electrode material is responsible for gradual loss of the original tool configuration. Usually, before excessive tool wear occurs, the tool is removed and resurfaced.

The best all-around electrode material is

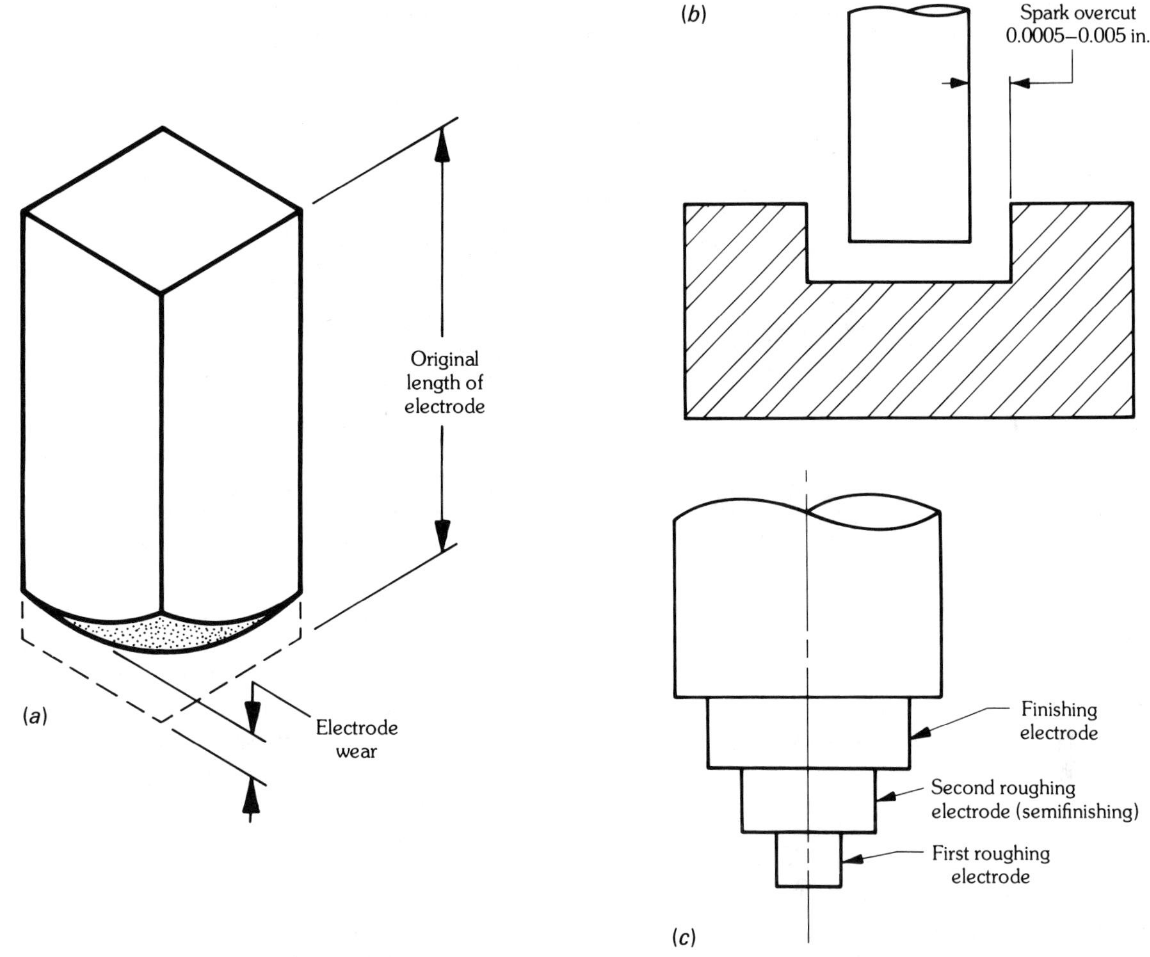

FIGURE 17.8 (*a*) Effects of electrode wear during the EDM process. (*b*) Spark overcut that occurs during the EDM process. As the current increases, the overcut increases. (*c*) A stepped electrode is often used so that an accurate finish size will be obtained.

graphite. It is medium-priced and available in various sizes and shapes. The tool wear of graphite is minimal. Graphite can be easily machined by turning, milling, boring, and grinding. Extremely high metal-removal rates and good surface finishes can be achieved with graphite.

Electrode Design and Construction. The design and construction of the electrode are very important to efficient electrical discharge machining. Electrodes can be fabricated from commercial stock, cast, or machined. Precision electrodes are generally machined.

Electrodes must be manufactured slightly smaller than the cavity or hole to be machined because of the *overcut* that occurs during the spark erosion process. The amount of overcut varies from 0.0001 to 0.008 in. and depends on the type of electrode. Because of the overcut characteristic, manufacturers of EDM machines provide spark overcut and wear-ratio charts that serve as guidelines. Figure 17.8*(a)* shows the cutting tool wear that occurs on an electrode; spark overcut characteristics are shown in Fig. 17.8*(b)*.

There are many different configurations and designs of electrodes. Because of electrode wear, a common technique used in cutting tool design is the *stepped electrode* [Fig. 17.8*(c)*]. Two or more steps are used for successive roughing, semifinishing, and finishing cuts so that the finished hole is accurate in size and shape. This same technique is employed when machining cavities with intricate detail by using several different electrodes.

Another design in construction of electrodes is the *cemented electrode.* The electrode is cemented to a shank with an electrically conductive, quick-curing epoxy adhesive (Fig. 17.9). The main benefit of this technique is the use of a small amount of electrode material, resulting in material cost and machining time savings.

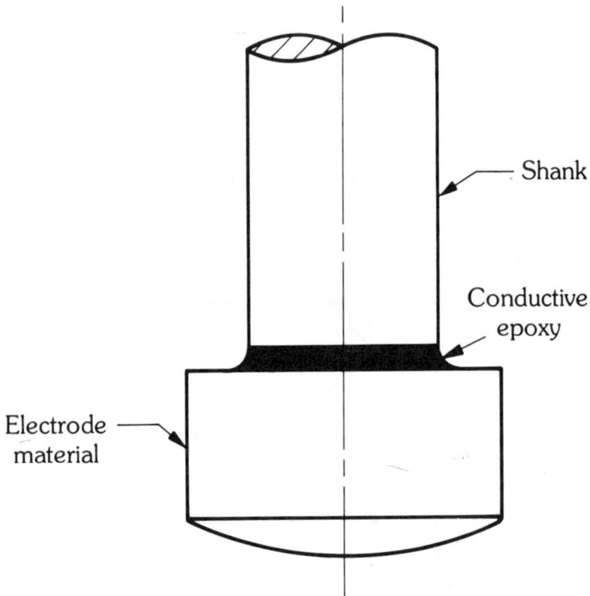

FIGURE 17.9 A cemented electrode is used to economize on electrode material.

17.2.4 Dielectric Fluid and Flushing Techniques

During the EDM process the workpiece and electrode are submerged in a dielectric fluid. The dielectric fluid, commonly called the *coolant,* has three basic functions. First, the fluid is present to *cool* the work and cutting tool. Without coolant, the metal particles stick to either the cutting tool or workpiece, and the heat from this transfer destroys the workpiece. Second, the fluid forms a *dielectric barrier* between the electrode and workpiece. This means the fluid is an insulator until enough voltage is reached to cause it to break down and allow the spark to occur. Dielectric fluids are similar to light-weight oils or kerosene and usually have a dielectric strength of 250 volts (V) per mil (0.001 in. or 0.025 mm). Therefore, a dielectric fluid of 250 V/mil with the electrode 0.001 in. (0.025 mm) from the work needs 250 V to cause a spark. Usually these high voltages are not common in the EDM process. Voltages of 80 to 100 are utilized that require maintaining a very close arc gap between the tool and work. With a close gap any minute metal particles from the machining process may short-circuit the system.

To keep the metal particles from short-circuiting the system, the coolant *flushes the particles* away from the tool and workpiece. This is the third function of the dielectric fluid. It is accomplished by forcing the fluid through the arc gap. Removing the metal particles from the arc gap is perhaps the most important single factor for efficient electrical discharge machining.

There are four basic methods by which the coolant can be forced through the arc gap. These flushing techniques (Fig. 17.10) are the *pressure through, reverse flow, vacuum,* and *vibration* methods.

The *pressure through* method [Fig. 17.10*(a)*] forces coolant down through a hollow electrode. In this method metal particles are rapidly forced away from the machining area. The hollow electrode also has the advantage of decreasing machining time because less metal is removed and a core drops out when the electrode goes through the work.

The *reverse flow* flushing method shown in Fig. 17.10*(b)*] is utilized when the workpiece has a previously machined hole. The fluid is forced under pressure through a manifold and the workpiece. The main advantage of this flushing technique is that the metal particles are forced from the sides of the workpiece and electrode, decreasing the chance of short circuits.

The *vacuum flow* method is also utilized when the workpiece has a previously machined hole. A partial vacuum is used to draw the dielectric fluid past the electrode, through the arc gap and workpiece [Fig. 17.10*(c)*]. The vacuum flow technique improves machining efficiency, reduces smoke and fumes, and helps reduce taper in the workpiece.

In some cases it may not be possible to have a coolant hole in either the workpiece or the electrode. When cavities, blind holes, or deep slots are to be machined, the *vibration method* is generally used. With this method, the electrode is vibrated, causing a pumping action that agitates the metal particles out of the arc gap [Fig. 17.10*(d)*]. This requires the EDM to have special attachments to vibrate the tool.

Each flushing technique described serves the same purpose: removing the metal particles from the arc gap and thus improving the efficiency of the EDM process.

17.2.5 EDM Applications

Electrical discharge machining is widely used by small and large manufacturers for production *dies, molds,* and *complex parts.* EDM applications are particularly suitable for production of parts that:

1. Are difficult to machine because of toughness or hardness of material
2. Have thin, intricate shapes or contours too difficult to machine by conventional techniques
3. Contain odd-shaped cavities or holes

The increased use of EDM for production work has been due largely to improved machines, elec-

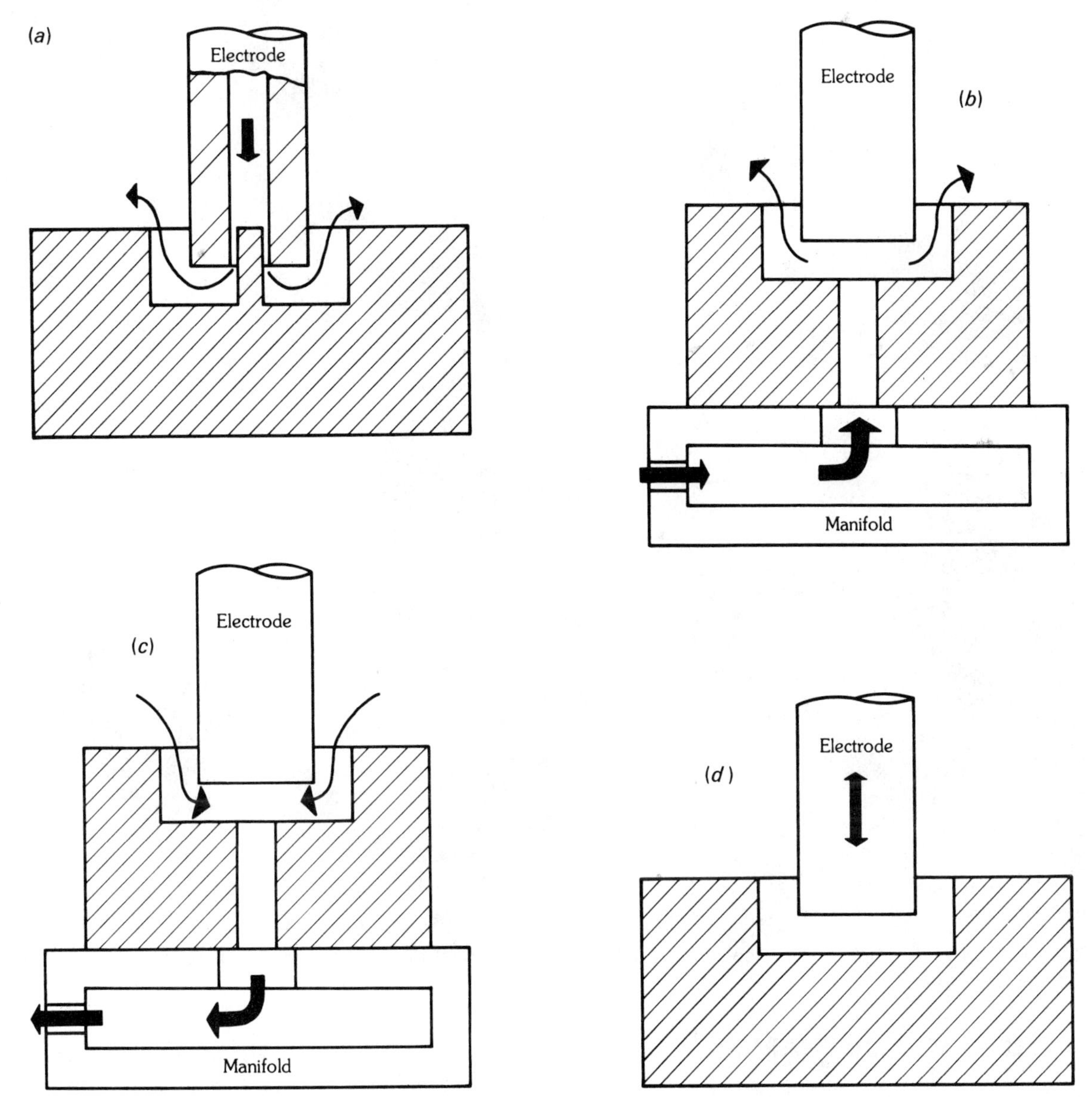

FIGURE 17.10 (*a*) The pressure-through flushing method forces coolant down through a hollow electrode. (*b*) The reverse-flow flushing method forces coolant through a manifold and a previously machined hole in the work. (*c*) With the vacuum-flow flushing method, a partial vacuum is used to draw the coolant past the electrode. (*d*) The vibration-flushing method uses vibration of the electrode to circulate the coolant.

trode materials, power supplies, improved techniques, and research on EDM. Examples of typical work produced by EDM are shown in Fig. 17.11.

17.2.6 Traveling-Wire EDM

Numerically controlled traveling-wire EDM has become increasingly popular in the past few years. The principal difference between conventional EDM and traveling-wire EDM is that the electrode is a fine wire that passes through the workpiece (Fig. 17.12). The wire feeds from a supply spool, then travels over rollers, through the workpiece, and finally, recoils onto a rewind spool (Fig. 17.13).

The wire can be brass, copper, tungsten, or molybdenum and is generally 0.003 to 0.012 in. (0.08 to 2.59 mm) in diameter. The workpiece is clamped to the machine table, which moves in X and Y axes by NC or CNC. NC and CNC controls with proper voltages, amperages, and frequencies allow for precise cutting patterns of virtually any profile or complex geometric shape. Tolerances of 0.0002 in. (0.013 mm) can generally be held in any electrically conductive material.

FIGURE 17.11 Examples of typical work produced by electrical discharge machining. (*Courtesy Charmilles Technological Corp.*)

17.3 ELECTRICAL DISCHARGE GRINDING

Electrical discharge grinding (EDG) is similar to EDM except that the electrode is a rotating wheel. EDG removes materials by rapid repetitive electrical discharges between the rotating wheel and the workpiece (Fig. 17.14). The wheel, usually graphite, supplies dc power at 80 to 100 V, 1/2 to 200 A, and rotates at 100 to 150 rpm. As in the EDM process, a dielectric fluid flushes the minute metal particles away from the work and wheel, preventing short circuits.

EDG is used in grinding hard cutting tool materials such as carbide and high-speed steel for form tools. Brittle or fragile parts may be machined by the EDG process, whereas abrasive grinding may cause fracturing because of the cutting forces involved. Another application of EDG is the grinding of thin sections on parts. Traditional abrasive grinding can often cause severe distortion on parts with thin sections or walls. Many jet engine parts are manufactured and reconditioned by use of EDG.

17.4 ELECTROCHEMICAL MACHINING

Electrochemical machining (ECM) is a method of removing material without the use of mechanical or thermal energy. In ECM, electric energy is combined with a chemical process to remove material. An electrolyte, generally a solution of water and inorganic salts, is pumped under pressure to control the conductivity and temperatures between the tool (electrode) and workpiece (anode) (Fig. 17.15).

With the ECM process, the material is flushed away from the gap between the workpiece and electrode by the flow of electrolyte. The dissolved material is subsequently removed from the electrolyte by a series of filters.

ECM has many of the same advantages as the

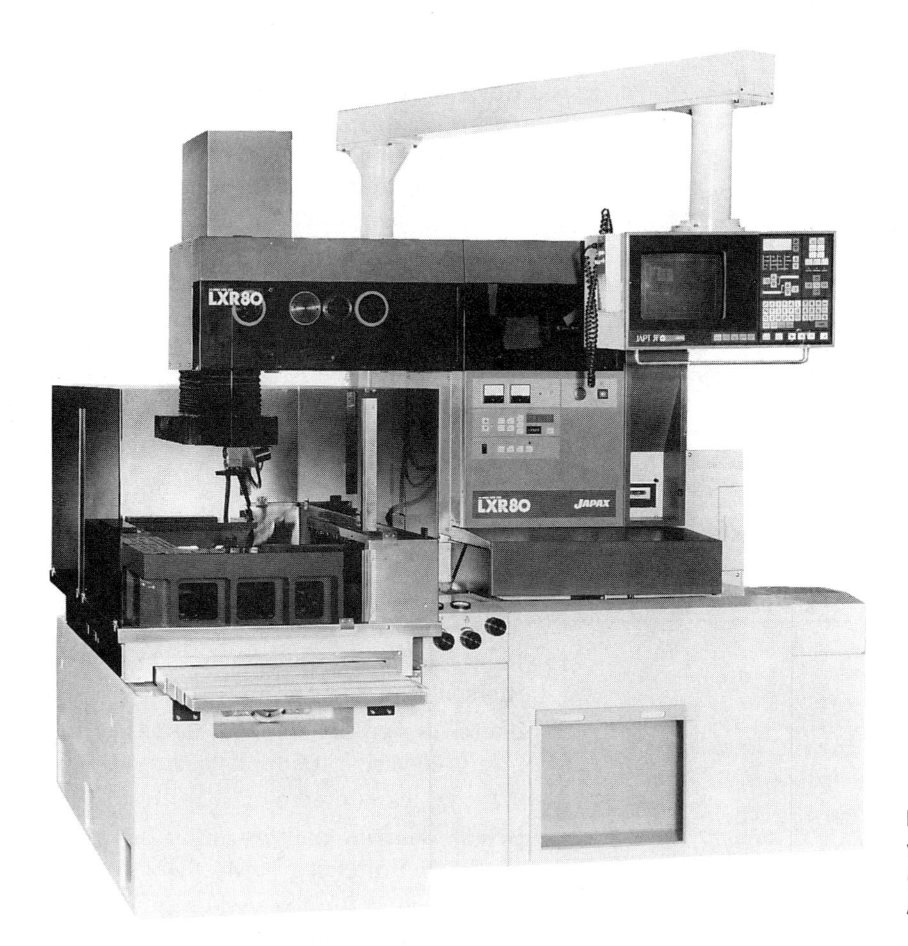

FIGURE 17.12 The CRT permits visual display of the program. (*Courtesy Japax, Inc., a division of Bridgeport Machines, Inc.*)

FIGURE 17.13 A traveling-wire EDM machining stacked parts. (*Courtesy Japax, Inc., a division of Bridgeport Machines, Inc.*)

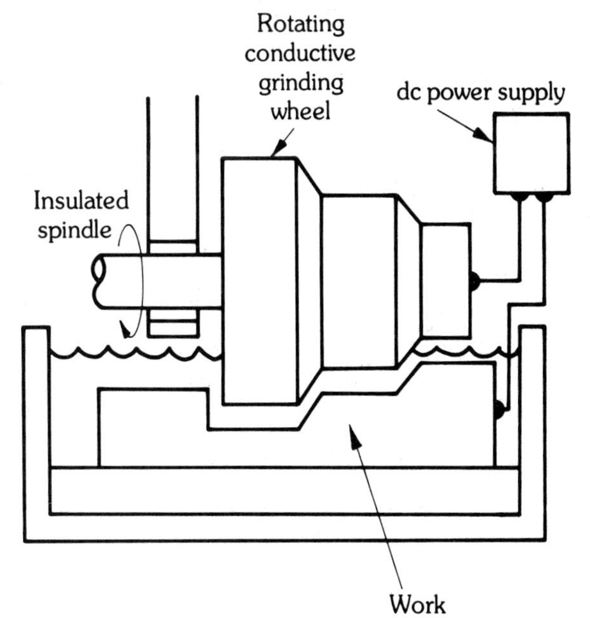

FIGURE 17.14 Basic components of electrical discharge grinding.

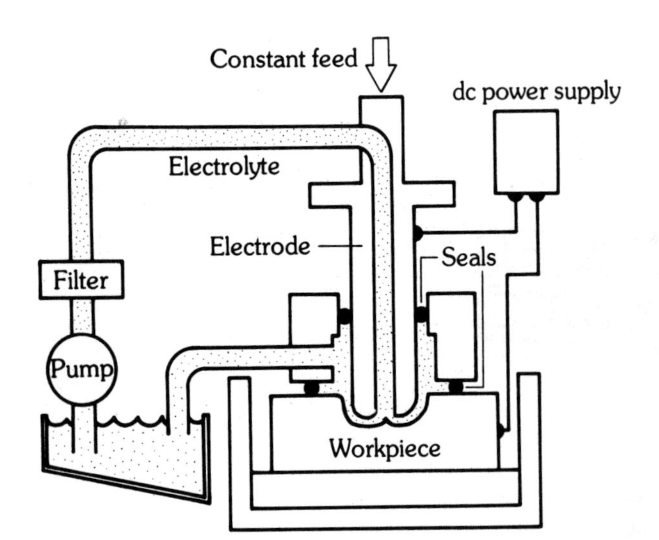

EDM process. High metal-removal rates in tough or hard alloys are its chief characteristics. It can machine odd-shaped, intricate, thin, and fragile sections in metals of any hardness. Another advantage of ECM is that during the machining operation burr-free parts are produced. This is due to the concentration of current at the edges of the workpiece that provides automatic rounding or absolutely burr-free parts. ECM machines are expensive, and the average machine shop is not generally equipped with them.

FIGURE 17.15 Basic components of electrochemical machining.

REVIEW QUESTIONS

17.2 ELECTRICAL DISCHARGE MACHINING

1. Give a brief description of the electrical discharge machining (EDM) process.
2. What special properties must the workpiece have to be machined by the EDM process?
3. Name and briefly describe the four *major* component parts of a typical EDM machine.
4. Describe the two purposes of the EDM power supplies.
5. Compare the operational differences between the resistance–capacitance and the pulse-type power supplies used on EDM machines.
6. Describe the concept of metal-removal rate with the EDM process. How is it measured?
7. What is frequency and how does it control the surface finish of the part in EDM?
8. Describe how the electrode (cutting tool) works in the EDM process.
9. Name the six common electrode materials. Which is most commonly used, and why?
10. How are electrodes manufactured for EDM applications?
11. Describe the concept of overcut caused by the spark erosion process.
12. Sketch a typical stepped electrode and briefly describe its application.
13. Why are cemented electrodes used?
14. Briefly describe the three purposes of the dielectric fluid used in the EDM process.
15. What type of fluid is generally used for the dielectric fluid?
16. Briefly describe and give an application of the four basic flushing techniques used on EDM machines.
17. Describe three characteristics that give the EDM process an advantage over conventional machining processes.

17.3 ELECTRICAL DISCHARGE GRINDING

1. Describe the electrical discharge grinding (EDG) process.
2. List four typical applications of EDG.

17.4 ELECTROCHEMICAL MACHINING

1. Describe the electrochemical machining (ECM) process.
2. What advantages does the ECM process offer over EDM?

Appendices

Appendix A: Using a Pocket Calculator to Solve Shop Mathematical Problems

Modern electronic technology has provided a new tool, the pocket calculator, which is an excellent tool for use in school, industry, and home. Today, electronic calculators are relatively inexpensive, compact, portable, and accurate. They can take much of the task out of solving tedious mathematical problems. Even the least expensive calculators are programmed to handle most calculations the machinist may have to do.

The calculator should not be used in place of mastering basic arithmetic skills, nor should it replace the concept of numbers and their relationships. The wise student will consider these suggestions without relying on the calculator as the *only* device for solving mathematical problems.

There are many calculators on the market, costing from under $10 to several hundreds of dollars. Only the basic four-function (+, −, ×, ÷) calculator is discussed in this appendix.

1. KEYS AND SWITCHES

Power switch. This turns the calculator on or off.

Numeric keys. Most calculators have a standard 1 to 9 numeric keyboard as well as 0 and decimal point (.) keys.

Clear key. On most calculators this is a multifunction key (CE). On the first push it clears the display of the last entry. A second push clears the calculator of all previous calculations.

Result key (=). This key immediately places the answer to any calculation on the display.

Operate keys (+, −, ×, ÷). These keys perform any operation and instruct the calculator regarding the next operation to be performed.

2. BASIC OPERATING INSTRUCTIONS

1. ***Number entry***: to enter a number, press the numeric key(s) in normal *sequence.*

Example: Enter 10.125.

Key Sequence	***Display***
Press the clear key twice	0
Press 1	1
Press 0	10
Press (.)	10.
Press 1	10.1
Press 2	10.12
Press 5	10.125

(Note: To clear an incorrect entry press the clear key once.)

2. ***Addition and subtraction***

Example A: Add 1.062 + 3.5.

Enter	***Press***	***Display***
	Clear key twice	0
1.062	+	1.062
3.5	=	*Answer* 4.562

Example B: Subtract 4.125 − 1.687.

Enter	***Press***	***Display***
	Clear key twice	0
4.125	−	4.125
1.687	=	*Answer* 2.438

3. Multiplication and division

Example A: Multiply 1.2 × 5.125.

Enter	***Press***	***Display***
	Clear key twice	0
1.2	×	1.2
5.125	=	*Answer* 6.15

Example B: Divide 6.875 ÷ 4.120.

Enter	***Press***	***Display***
	Clear key twice	0
6.875	÷	6.875
4.12	=	1.6686893

(Note: Most calculators display only the highest eight digits. The last number in the display may or may not be rounded off; therefore, never trust the last number in the *display* if you want a rounded answer, or always work with one more digit of accuracy than you need. This allows you to round off to the proper degree of accuracy.)

4. Chains: Adding and subtracting fractions and whole numbers

Example A: Add 17/8 + 31/8 + 53/16.

With the simple four-function calculator we must first solve (or convert) each fraction into a decimal value. Using the calculator, solve the fractions first using simple division:

7/8 = 0.875

1/8 = 1.125

3/16 = 0.1875

Next, add the decimal (converted fraction) to the whole number in the original number. (This procedure can be done relatively easily.)

1 + 0.875   3 + 0.125   5 + 0.1875

Finally, add each number using the same procedure for ordinary addition of numbers.

(Note: The same procedure is used when subtracting or multiplying fractions. Remember first to convert the fraction into a *decimal* and then perform any additional arithmetical operations.)

Example B: Calculate $\frac{2 \times 6 + 1.2}{1.125}$

Since all the factors in the problem are decimals, the solution is simple. First, multiply (×) 2 times 6, add (+) 1.2, then divide (÷) by 1.125, and, finally, press the (=) key. The display should read 11.733333.

Example C: Calculate the correct rpm rate for a 2.75-in.-diameter workpiece to be machined at 110 spfm.

First, set up the problem:

$$\text{rpm} = \frac{4 \times \text{CS}}{\text{diameter}} = \frac{4 \times 110}{2.75} = 160$$

or

4 (×) 110 (÷) 2.75 (=) 160   displayed answer

Appendix B: Decimal Equivalent Charts for Number- and Letter-Size Drills

Number-size drills

No.	Size of Drill (in.)	No.	Size of Drill (in.)	No.	Size of Drill (in.)	No.	Size of Drill (in.)
1	0.2280	21	0.1590	41	0.0960	61	0.0390
2	0.2210	22	0.1570	42	0.0935	62	0.0380
3	0.2130	23	0.1540	43	0.0890	63	0.0370
4	0.2090	24	0.1520	44	0.0860	64	0.0360
5	0.2055	25	0.1495	45	0.0820	65	0.0350
6	0.2040	26	0.1470	46	0.0810	66	0.0330
7	0.2010	27	0.1440	47	0.0785	67	0.0320
8	0.1990	28	0.1405	48	0.0760	68	0.0310
9	0.1960	29	0.1360	49	0.0730	69	0.0292
10	0.1935	30	0.1285	50	0.0700	70	0.0280
11	0.1910	31	0.1200	51	0.0670	71	0.0260
12	0.1890	32	0.1160	52	0.0635	72	0.0250
13	0.1850	33	0.1130	53	0.0595	73	0.0240
14	0.1820	34	0.1110	54	0.0550	74	0.0225
15	0.1800	35	0.1100	55	0.0520	75	0.0210
16	0.1770	36	0.1065	56	0.0465	76	0.0200
17	0.1730	37	0.1040	57	0.0430	77	0.0180
18	0.1695	38	0.1015	58	0.0420	78	0.0160
19	0.1660	39	0.0995	59	0.0410	79	0.0145
20	0.1610	40	0.0980	60	0.0400	80	0.0135

Letter-size drills

A	0.234	J	0.277	S	0.348
B	0.238	K	0.281	T	0.358
C	0.242	L	0.290	U	0.368
D	0.246	M	0.295	V	0.377
E	0.250	N	0.302	W	0.386
F	0.257	O	0.306	X	0.397
G	0.261	P	0.323	Y	0.404
H	0.266	Q	0.332	Z	0.413
I	0.272	R	0.339		

Appendix C: Physical Properties of Common Metals

Metal	Symbol	Specific Gravity	Melting Point °C	Melting Point °F	Lb/in.³
Aluminum					
Cast	Al	2.56	658	1217	0.0924
Rolled	Al	2.71	—	—	0.0978
Antimony	Sb	6.71	630	1166	0.2424
Bismuth	Bi	9.80	271	520	0.3540
Boron	B	2.30	2300	4172	0.0831
Brass	—	8.51	—	—	0.3075
Cadmium	Cd	8.60	321	610	0.3107
Carbon	C	2.22	—	—	0.0802
Chromium	Cr	6.80	1510	2750	0.2457
Cobalt	Co	8.50	1490	2714	0.3071
Copper	Cu	8.89	1083	1982	0.3212
Gold	Au	19.32	1063	1945	0.6979
Iron	Fe	7.86	1520	2768	0.2634
Cast	Fe	7.218	1375	2507	0.2605
Wrought	Fe	7.70	1500–1600	2732–2912	0.2779
Lead	Pb	11.37	327	621	0.4108
Magnesium	Mg	1.74	651	1204	0.0629
Manganese	Mn	8.00	1225	2237	0.2890
Mercury	Hg	13.59	−38.8	−37.7	0.4909
Molybdenum	Mo	10.2	2620	4748	0.368
Monel metal	—	8.87	1360	2480	0.320
Nickel	Ni	8.80	1452	2646	0.319
Selenium	Se	4.81	220	428	0.174
Silicon	Si	2.40	1427	2600	0.087
Silver	Ag	10.53	961	1761	0.3805
Sodium	Na	0.97	97	207	0.0350
Steel	—	7.858	1300–1378	2372–2532	0.2839
Tantalum	Ta	10.80	2850	5160	0.3902
Tin	Sn	7.29	232	450	0.2634
Titanium	Ti	5.3	1900	3450	0.1915
Tungsten	W	19.10	3000	5432	0.6900
Uranium	U	18.70	—	—	0.6755
Vanadium	V	5.50	1730	3146	0.1987
Zinc	Zn	7.19	419	786	0.2598

Appendix D: Hardness Conversion Table: Rockwell and Brinell

Brinell Hardness Number, 3000-kg, 10-mm Ball		Rockwell Hardness Number		Rockwell Superficial Hardness Number Superficial Diamond Penetrator			
Standard Ball	*Tungsten-Carbide Ball*	*1/16 Diamond Ball, 100 kg, B Scale*	*Diamond Ball, 150 kg, C Scale*	*15 N Scale*	*30 N Scale*	*45 N Scale*	*Tensile Strength (Approximate; 100 psi)*
248	248	–	24.2	71.7	45.1	24.5	118
241	241	100.0	22.8	70.9	43.9	22.8	114
235	235	99.0	21.7	70.3	42.9	21.5	111
229	229	98.2	20.5	69.7	41.9	20.1	109
223	223	97.3	–	–	–	–	104
217	217	96.4	–	–	–	–	103
212	212	95.5	–	–	–	–	100
207	207	94.6	–	–	–	–	99
201	201	93.8	–	–	–	–	97
197	197	92.8	–	–	–	–	94
192	192	91.9	–	–	–	–	92
187	187	90.7	–	–	–	–	90
183	183	90.0	–	–	–	–	89
179	179	89.0	–	–	–	–	88
174	174	87.8	–	–	–	–	86
170	170	86.8	–	–	–	–	84
167	167	86.0	–	–	–	–	83
163	163	85.0	–	–	–	–	82
156	156	82.9	–	–	–	–	80
149	149	80.8	–	–	–	–	73
143	143	78.7	–	–	–	–	71
137	137	76.4	–	–	–	–	67
131	131	74.0	–	–	–	–	65
126	126	72.0	–	–	–	–	63
121	121	69.0	–	–	–	–	60
116	116	67.6	–	–	–	–	58
111	111	65.7	–	–	–	–	56

Appendix E: Basic Formulas and Solutions for Right Triangles

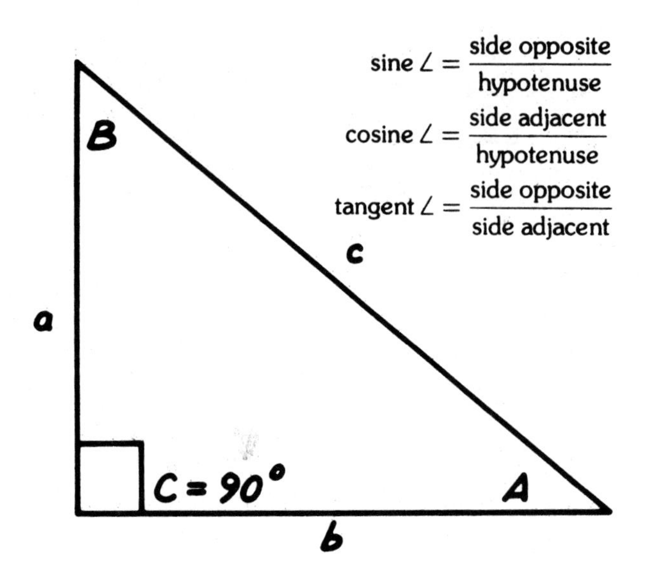

Known	Solution	
Sides *a* and *b*	$c = \sqrt{a^2 + b^2}$	$\tan B = \frac{b}{a}$
Side *c* and angle *B*	$b = \sin B \times c$	$a = \cos B \times c$
Sides *a* and *c*	$b = \sqrt{c^2 - a^2}$	$\cos B = \frac{a}{c}$
Side *a* and angle *B*	$b = \tan B \times a$	$c = \frac{a}{\cos B}$
Sides *b* and *c*	$a = \sqrt{c^2 - b^2}$	$\sin B = \frac{b}{c}$
Side *b* and angle *B*	$a = \frac{b}{\tan B}$	$c = \frac{b}{\sin B}$
Side *b* and angle *A*	$a = \tan A \times b$	$c = \frac{b}{\cos A}$
Side *c* and angle *A*	$a = \sin A \times c$	$b = \cos A \times c$
Side *a* and angle *A*	$b = \frac{a}{\tan A}$	$c = \frac{a}{\sin A}$

Appendix F: Morse Tapers: Sizes and Specifications

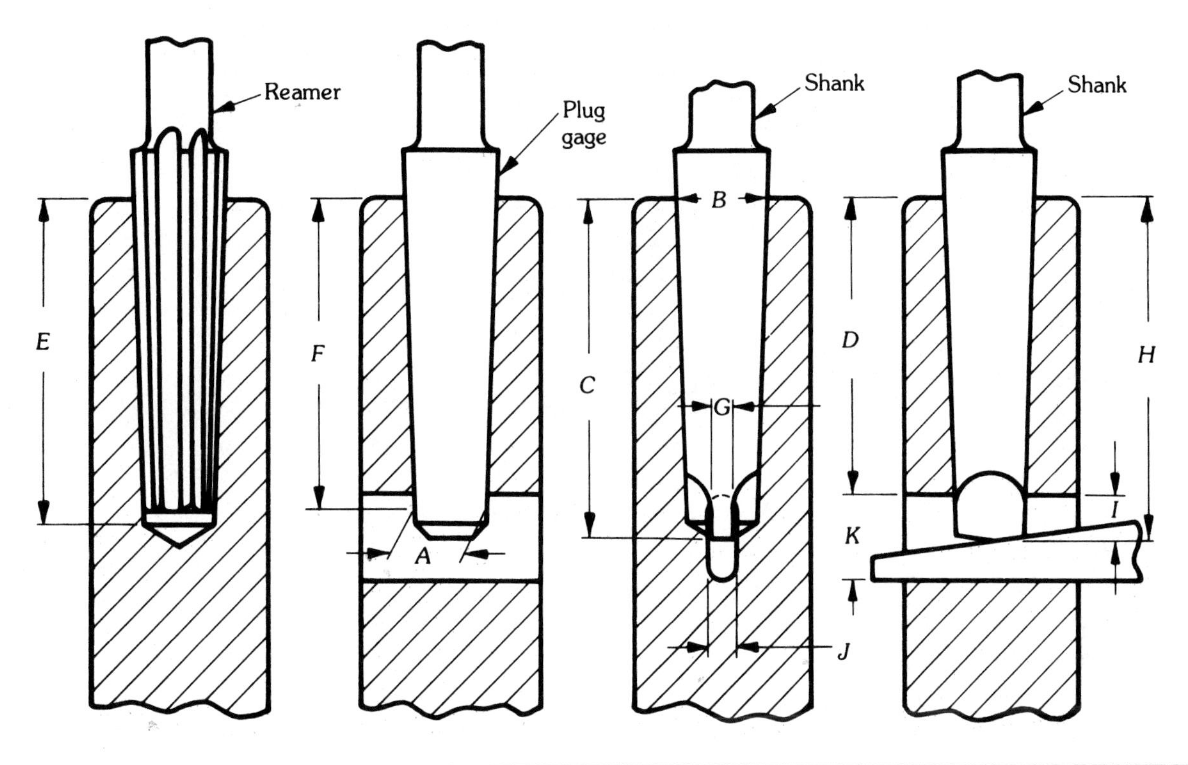

Number of Taper	Diameter of Plug at Small End	Diameter at End of Socket	Shank		End of Socket to Tang Slot	Socket		Standard Plug Depth	Thickness	Tang			Radius	Tang Slot		Taper per in.	Taper per Foot
			Whole Length	Depth		Depth of Drilled Hole	Depth of Reamed Hole			Length	Radius	Diameter		Width	Length		
	P	A	B	C	L	M	N	O	D	E	F	G	H	J	K		
0	0.25200	0.35610	2 11/32	2 7/32	1 15/16	2 1/16	2 1/32	2	0.156	1/4	5/32	15/64	3/64	0.172	9/16	0.052050	0.62460
1	0.36900	0.47500	2 9/16	2 7/16	2 1/16	2 3/16	2 5/32	2 1/8	0.203	3/8	3/16	11/32	3/64	0.218	3/4	0.049882	0.59858
2	0.57200	0.70000	3 1/8	2 15/16	2 1/2	2 21/32	2 39/64	2 9/16	0.250	7/16	1/4	17/32	1/16	0.266	7/8	0.049951	0.59941
3	0.77800	0.93800	3 7/8	3 11/16	3 1/16	3 5/16	3 1/4	3 3/16	0.312	9/16	9/32	23/32	5/64	0.328	1 3/16	0.050196	0.60235
4	1.02000	1.23100	4 7/8	4 5/8	3 7/8	4 3/16	4 1/8	4 1/16	0.469	5/8	5/16	31/32	3/32	0.484	1 1/4	0.051938	0.62326
4 1/2	1.26600	1.50000	5 3/8	5 1/8	4 5/16	4 5/8	4 9/16	4 1/2	0.562	11/16	3/8	1 13/64	1/8	0.578	1 3/8	0.052000	0.62400
5	1.47500	1.74800	6 1/8	5 7/8	4 15/16	5 5/16	5 1/4	5 3/16	0.625	3/4	3/8	1 13/32	1/8	0.656	1 1/2	0.052626	0.63151
6	2.11600	2.49400	8 9/16	8 1/4	7	7 13/32	7 21/16	7 1/4	0.750	1 1/8	1/2	2	5/32	0.781	1 3/4	0.052138	0.62565
7	2.75000	3.27000	11 5/8	11 1/4	9 1/2	10 5/32	10 5/64	10	1.125	1 3/8	3/4	2 5/8	3/16	1.156	2 5/8	0.052000	0.62400

Appendix G: English–Metric Conversion Factors

Multiply:	*By:*	*To Find:*
Centimeters	3.28083×10^{-2}	Feet
Centimeters	0.3937	Inches
Cubic centimeters	3.53145×10^{-3}	Cubic feet
Cubic centimeters	6.102×10^{-2}	Cubic inches
Cubic feet	2.8317×10^{-4}	Cubic centimeters
Cubic feet	2.8317×10^{-2}	Cubic meters
Cubic feet	6.22905	Gallons, British Imperial
Cubic inches	4.329×10^{-3}	Gallons, U.S. liquid
Cubic inches	3.72×10^{-3}	Gallons, U.S. dry
Cubic inches	3.604×10^{-3}	Gallons, British Imperial
Cubic inches	16.38716	Cubic centimeters
Cubic meters	35.3145	Cubic feet
Degrees, angular	0.0174533	Radians
Foot pounds	0.13826	Kilogram meters
Feet	30.48	Centimeters
Gallons, British Imperial	1.20094	Gallons, U.S.
Gallons, British Imperial	277.42	Cubic inches
Gallons, U.S.	0.832702	Gallons, British Imperial
Gallons, U.S. dry	268.8	Cubic inches
Gallons, U.S. liquid	231	Cubic inches
Gallons, U.S. liquid	3.78543	Liters
Grams, metric	2.20464×10^{-2}	Pounds, avoirdupois
Horsepower, metric	0.98632	Horsepower, U.S.
Horsepower, U.S.	1.01387	Horsepower, metric
Inches	2.54	Centimeters
Kilograms	2.20462	Pounds
Kilogram meters	7.233	Foot pounds
Kilograms per square centimeter	14.2234	Pounds per square inch
Liters	3.53145×10^{-2}	Cubic feet
Liters	0.26417	Gallons, U.S. liquid
Pounds, avoirdupois	453.592	Grams, metric
Pounds per square inch	7.031×10^{-2}	Kilograms per square centimeter
Radians	57.29578	Degrees, angular
Square centimeter	0.1550	Square inches
Square feet	0.0929034	Square meters
Square inches	6.45163	Square centimeters
Square meters	10.7639	Square feet

Appendix H: Decimal Equivalents for Metric Drills

mm	in.	mm	in.	mm	in.	mm	in.	mm	in.	mm	in.
0.04	0.0016	0.92	0.0362	2.65	0.1043	4.75	0.1870	7.75	0.3051	12.00	0.4727
0.06	0.0024	0.95	0.0374	2.70	0.1063	4.80	0.1890	7.80	0.3071	12.25	0.4823
0.08	0.0032	0.98	0.0386	2.75	0.1083	4.90	0.1929	7.90	0.3110	12.50	0.4921
0.10	0.0039	1.00	0.0394	2.80	0.1102	5.00	0.1969	8.00	0.3150	12.75	0.5020
0.12	0.0047	1.05	0.0413	2.85	0.1122	5.10	0.2008	8.10	0.3189	13.00	0.5122
0.15	0.0059	1.10	0.0433	2.90	0.1142	5.20	0.2047	8.20	0.3228	13.25	0.5217
0.18	0.0071	1.15	0.0453	2.95	0.1161	5.25	0.2067	8.25	0.3248	13.50	0.5315
0.20	0.0079	1.20	0.0473	3.00	0.1181	5.30	0.2087	8.30	0.3268	13.75	0.5413
0.22	0.0087	1.25	0.0492	3.10	0.1221	5.40	0.2126	8.40	0.3307	14.00	0.5512
0.25	0.0098	1.30	0.0512	3.15	0.1240	5.50	0.2165	8.50	0.3346	14.25	0.5610
0.28	0.0110	1.35	0.0532	3.20	0.1260	5.60	0.2205	8.60	0.3386	14.50	0.5709
0.30	0.0118	1.40	0.0551	3.25	0.1280	5.70	0.2244	8.70	0.3425	14.75	0.5807
0.32	0.0126	1.45	0.0571	3.30	0.1299	5.75	0.2264	8.75	0.3445	15.00	0.5906
0.35	0.0138	1.50	0.0591	3.35	0.1319	5.80	0.2284	8.80	0.3465	15.25	0.6004
0.38	0.0150	1.55	0.0611	3.40	0.1339	5.90	0.2323	8.90	0.3504	15.50	0.6102
0.40	0.0157	1.60	0.0630	3.45	0.1358	6.00	0.2362	9.00	0.3543	15.75	0.6201
0.42	0.0165	1.65	0.0650	3.50	0.1378	6.10	0.2402	9.10	0.3583	16.00	0.6229
0.45	0.0177	1.70	0.0669	3.55	0.1398	6.20	0.2441	9.20	0.3622	16.25	0.6398
0.48	0.0189	1.75	0.0689	3.60	0.1417	6.25	0.2461	9.25	0.3642	16.50	0.6496
0.50	0.0197	1.80	0.0709	3.65	0.1437	6.30	0.2480	9.30	0.3661	16.75	0.6595
0.52	0.0205	1.85	0.0728	3.70	0.1457	6.40	0.2520	9.40	0.3701	17.00	0.6693
0.55	0.0217	1.90	0.0748	3.75	0.1476	6.50	0.2559	9.50	0.3740	17.25	0.6791
0.58	0.0288	1.95	0.0768	3.80	0.1496	6.60	0.2598	9.60	0.3780	17.50	0.6890
0.60	0.0232	2.00	0.0787	3.90	0.1535	6.70	0.2638	9.70	0.3819	17.75	0.6988
0.62	0.0244	2.05	0.0807	3.95	0.1555	6.75	0.2658	9.75	0.3839	18.00	0.7087
0.65	0.0265	2.10	0.0827	4.00	0.1575	6.80	0.2677	9.80	0.3858	18.25	0.7185
0.68	0.0268	2.15	0.0847	4.10	0.1614	6.90	0.2717	9.90	0.3898	18.50	0.7283
0.70	0.0276	2.20	0.0866	4.15	0.1634	7.00	0.2756	10.00	0.3939	18.75	0.7382
0.72	0.0284	2.25	0.0886	4.20	0.1654	7.10	0.2795	10.10	0.3976	19.00	0.7489
0.75	0.0295	2.30	0.0906	4.25	0.1673	7.20	0.2835	10.25	0.4035	19.25	0.7579
0.78	0.0307	2.35	0.0925	4.30	0.1693	7.25	0.2854	10.50	0.4134	19.50	0.7577
0.80	0.0315	2.40	0.0945	4.40	0.1732	7.30	0.2874	10.75	0.4232	19.75	0.7776
0.82	0.0323	2.45	0.0965	4.50	0.1772	7.40	0.2913	11.00	0.4331	20.00	0.7874
0.85	0.0335	2.50	0.0984	4.60	0.1811	7.50	0.2953	11.25	0.4429		
0.88	0.0346	2.55	0.1004	4.65	0.1831	7.60	0.2992	11.50	0.4528		
0.90	0.0354	2.60	0.1024	4.70	0.1850	7.70	0.3032	11.75	0.4626		

Appendix I: Temperature Conversion

Celsius to fahrenheit

°C	°F	°C	°F	°C	°F	°C	°F	°C	°F	°C	°F	°C	°F
0	32	330	626	480	896	630	1166	780	1436	930	1706	1080	1976
100	212	340	644	490	914	640	1184	790	1454	940	1724	1090	1994
200	392	350	662	500	932	650	1202	800	1472	950	1742	1100	2012
210	410	360	680	510	950	660	1220	810	1490	960	1760	1110	2030
220	428	370	698	520	968	670	1238	820	1508	970	1778	1120	2048
230	446	380	716	530	986	680	1256	830	1526	980	1796	1130	2066
240	464	390	734	540	1004	690	1274	840	1544	990	1814	1140	2084
250	482	400	752	550	1022	700	1292	850	1562	1000	1832	1150	2102
260	500	410	770	560	1040	710	1310	860	1580	1010	1850	1160	2120
270	518	420	788	570	1058	720	1328	870	1598	1020	1868	1170	2138
280	536	430	806	580	1076	730	1346	880	1616	1030	1886	1180	2156
290	554	440	824	590	1094	740	1364	890	1634	1040	1904	1190	2174
300	572	450	842	600	1112	750	1382	900	1652	1050	1922		
310	590	460	860	610	1130	760	1400	910	1670	1060	1940		
320	608	470	878	620	1148	770	1418	920	1688	1070	1958		

Fahrenheit to celsius

°F	°C	°F	°C	°F	°C	°F	°C	°F	°C	°F	°C	°F	°C
32	0	660	349	960	516	1260	682	1560	849	1880	1026	2180	1193
212	100	680	360	980	527	1280	693	1580	860	1900	1038	2200	1204
400	204	700	371	1000	538	1300	704	1600	871	1920	1049	2220	1216
420	216	720	382	1020	549	1320	716	1620	882	1940	1060	2240	1227
440	227	740	393	1040	560	1340	727	1640	893	1960	1071	2260	1238
460	238	760	404	1060	571	1360	738	1660	904	1980	1082	2280	1249
480	249	780	416	1080	582	1380	749	1680	916	2000	1093	2300	1260
500	260	800	427	1100	593	1400	760	1700	927	2020	1105	2320	1271
520	271	820	438	1120	604	1420	771	1740	949	2040	1116	2340	1284
540	282	840	449	1140	616	1440	782	1760	960	2060	1127	2360	1293
560	293	860	460	1160	627	1460	793	1780	971	2080	1138	2380	1305
580	304	880	471	1180	638	1480	804	1800	982	2100	1149	2400	1316
600	316	900	482	1200	649	1500	816	1820	993	2120	1160		
620	327	920	493	1220	660	1520	827	1840	1004	2140	1171		
640	338	940	504	1240	671	1540	838	1860	1015	2160	1182		

Appendix J: Inch–Millimeter Equivalents

Millimeters to Inches						Inches to Millimeters					
mm	*in.*	*mm*	*in.*	*mm*	*in.*	*in.*	*mm*	*in.*	*mm*	*in.*	*mm*
0.01	0.0004	0.35	0.0138	0.68	0.0268	0.001	0.025	0.290	7.37	0.660	16.76
0.02	0.0008	0.36	0.0142	0.69	0.0272	0.002	0.051	0.300	7.62	0.670	17.02
0.03	0.0012	0.37	0.0146	0.70	0.0276	0.003	0.076	0.310	7.87	0.680	17.27
0.04	0.0016	0.38	0.0150	0.71	0.0280	0.004	0.102	0.320	8.13	0.690	17.53
0.05	0.0020	0.39	0.0154	0.72	0.0283	0.005	0.127	0.330	8.38	0.700	17.78
0.06	0.0024	0.40	0.0157	0.73	0.0287	0.006	0.152	0.340	8.64	0.710	18.03
0.07	0.0028	0.41	0.0161	0.74	0.0291	0.007	0.178	0.350	8.89	0.720	18.29
0.08	0.0031	0.42	0.0165	0.75	0.0295	0.008	0.203	0.360	9.14	0.730	18.54
0.09	0.0035	0.43	0.0169	0.76	0.0299	0.009	0.229	0.370	9.40	0.740	18.80
0.10	0.0039	0.44	0.0173	0.77	0.0303	0.010	0.254	0.380	9.65	0.750	19.05
0.11	0.0043	0.45	0.0177	0.78	0.0307	0.020	0.508	0.390	9.91	0.760	19.30
0.12	0.0047	0.46	0.0181	0.79	0.0311	0.030	0.762	0.400	10.16	0.770	19.56
0.13	0.0051	0.47	0.0185	0.80	0.0315	0.040	1.016	0.410	10.41	0.780	19.81
0.14	0.0055	0.48	0.0189	0.81	0.0319	0.050	1.270	0.420	10.67	0.790	20.07
0.15	0.0059	0.49	0.0193	0.82	0.0323	0.060	1.524	0.430	10.92	0.800	20.32
0.16	0.0063	0.50	0.0197	0.83	0.0327	0.070	1.778	0.440	11.18	0.810	20.57
0.17	0.0067	0.51	0.0201	0.84	0.0331	0.080	2.032	0.450	11.43	0.820	20.83
0.18	0.0071	0.52	0.0205	0.85	0.0335	0.090	2.286	0.460	11.68	0.830	21.08
0.19	0.0075	0.53	0.0209	0.86	0.0339	0.100	2.540	0.470	11.94	0.840	21.34
0.20	0.0079	0.54	0.0213	0.87	0.0343	0.110	2.794	0.480	12.19	0.850	21.59
0.21	0.0083	0.55	0.0217	0.88	0.0346	0.120	3.048	0.490	12.45	0.860	21.84
0.22	0.0087	0.56	0.0220	0.89	0.0350	0.130	3.302	0.500	12.70	0.870	22.10
0.23	0.0091	0.57	0.0224	0.90	0.0354	0.140	3.56	0.510	12.95	0.880	22.35
0.24	0.0094	0.58	0.0228	0.91	0.0358	0.150	3.81	0.520	13.21	0.890	22.61
0.25	0.0098	0.59	0.0232	0.92	0.0362	0.160	4.06	0.530	13.46	0.900	22.86
0.26	0.0102	0.60	0.0236	0.93	0.0366	0.170	4.32	0.540	13.72	0.910	23.11
0.27	0.0106	0.61	0.0240	0.94	0.0370	0.180	4.57	0.550	13.97	0.920	23.37
0.28	0.0110	0.62	0.0244	0.95	0.0374	0.190	4.83	0.560	14.22	0.930	23.62
0.29	0.0114	0.63	0.0248	0.96	0.0378	0.200	5.08	0.570	14.48	0.940	23.88
0.30	0.0118	0.64	0.0252	0.97	0.0382	0.210	5.33	0.580	14.73	0.950	24.13
0.31	0.0122	0.65	0.0256	0.98	0.0386	0.220	5.59	0.590	14.99	0.960	24.38
0.32	0.0126	0.66	0.0260	0.99	0.0390	0.230	5.84	0.600	15.24	0.970	24.64
0.33	0.0130	0.67	0.0264	1.00	0.0394	0.240	6.10	0.610	15.49	0.980	24.89
0.34	0.0134	—	—	—	—	0.250	6.35	0.620	15.75	0.990	25.15
						0.260	6.60	0.630	16.00	1.000	25.40
						0.270	6.86	0.640	16.26	—	—
						0.280	7.11	0.650	16.51		

Appendix K: Decimal Equivalents of an Inch by 1/64s

Fraction	Decimal	Fraction	Decimal
1/64	0.015625	33/64	0.515625
1/32	0.03125	17/32	0.53125
3/64	0.046875	35/64	0.546875
1/16	0.0625	9/16	0.5625
5/64	0.078125	37/64	0.578125
3/32	0.09375	19/32	0.59375
7/64	0.109375	39/64	0.609375
1/8	0.125	5/8	0.625
9/64	0.140625	41/64	0.640625
5/32	0.15625	21/32	0.65625
11/64	0.171875	43/64	0.671875
3/16	0.1875	11/16	0.6875
13/64	0.203125	45/64	0.703125
7/32	0.21875	23/32	0.71875
15/64	0.234375	47/64	0.734375
1/4	0.25	3/4	0.75
17/64	0.265625	49/64	0.765625
9/32	0.28125	25/32	0.78125
19/64	0.296875	51/64	0.796875
5/16	0.3125	13/16	0.8125
21/64	0.328125	53/64	0.828125
11/32	0.34375	27/32	0.84375
23/64	0.359375	55/64	0.859375
3/8	0.375	7/8	0.875
25/64	0.390625	57/64	0.890625
13/32	0.40625	29/32	0.90625
27/64	0.421875	59/64	0.921875
7/16	0.4375	15/16	0.9375
29/64	0.453125	61/64	0.953125
15/32	0.46875	31/32	0.96875
31/64	0.484375	63/64	0.984375
1/2	0.5	1	1

Appendix L: Tap Drill Sizes for Unified Screw Threads with Approximately 75% Depth

Unified Coarse			*Unified Fine*		
Tap Size	*Threads per inch*	*Tap Drill Size*	*Tap Size*	*Threads per inch*	*Tap Drill Size*
No. 5	40	No. 38	No. 5	44	No. 37
No. 6	32	No. 36	No. 6	40	No. 33
No. 8	32	No. 29	No. 8	36	No. 29
No. 10	24	No. 25	No. 10	32	No. 21
No. 12	24	No. 16	No. 12	28	No. 14
1/4	20	No. 7	1/4	28	No. 3
5/16	18	F	5/16	24	I
3/8	16	5/16	3/8	24	Q
7/16	14	U	7/16	20	25/64
1/2	13	27/64	1/2	20	29/64
9/16	12	31/64	9/16	18	33/64
5/8	11	17/32	5/8	18	37/64
3/4	10	21/32	3/4	16	11/16
7/8	9	49/64	7/8	14	13/16
1	8	7/8	1	14	15/16
1 1/8	7	63/64	1 1/8	12	1 3/64
1 1/4	7	1 7/64	1 1/4	12	1 11/64
1 3/8	6	1 7/32	1 3/8	12	1 19/64
1 1/2	6	1 11/32	1 1/2	12	1 27/64
1 3/4	5	1 9/16			
2	4 1/2	1 25/32			
NPT National Pipe Thread					
1/8	27	11/32	1	11 1/2	1 5/32
1/4	18	7/16	1 1/4	11 1/2	1 1/2
3/8	18	19/32	1 1/2	11 1/2	1 23/32
1/2	14	23/32	2	11 1/2	2 3/16
3/4	14	15/16	2 1/2	8	2 5/8

Note: Outside diameter (OD) of a number size tap or machine screw may be found as follows:

OD = ($N \times 0.013$) + 0.060

where N = screw size number)

Example: Find the diameter of a No. 6 screw:

OD = ($N \times 0.013$) + 0.060)
= (6×0.013) + 0.060
= 0.138 in. diameter

Appendix M: Tap Drill Sizes for ISO Metric Screw Theads

Thread Size and Pitch	Inch Equivalent		Recommended Tap Drill Size		Probable Hole Size (in.)	Probable Percent Thread
	Size	TPI	Drill	Decimal		
M1.6 × 0.35	0.0630	72.5689	1.25 M/M	0.0492	0.0507	69
M1.8 × 0.35	0.0709	72.5689	1.45 M/M	0.0571	0.0586	69
M2 × 0.4	0.0788	63.4921	1.60 M/M	0.0630	0.0647	69
M2.2 × 0.45	0.0867	56.4334	1.75 M/M	0.0689	0.0706	70
M2.5 × 0.45	0.0985	56.4334	2.05 M/M	0.0807	0.0826	69
M3 × 0.5	0.1182	50.7872	2.5 M/M	0.0984	0.1007	68
M3.5 × 0.6	0.1378	42.3370	2.9 M/M	0.1142	0.1168	68
M4 × 0.7	0.1575	36.2845	3.3 M/M	0.1299	0.1328	69
M4.5 × 0.75	0.1772	33.8639	3.75 M/M	0.1476	0.1508	69
M5 × 0.8	0.1969	31.7460	No. 19	0.1660	0.1692	68
M6 × 1	0.2363	25.4001	5 M/M	0.1968	0.2006	70
M7 × 1	0.2756	25.4001	6 M/M	0.2362	0.2400	70
M8 × 1.25	0.3150	20.3211	H	0.2660	0.2701	70
M8 × 1	0.3150	25.4001	J	0.2770	0.2811	66
M10 × 1.5	0.3937	16.9319	8.5 M/M	0.3346	0.3390	71
M10 × 1.25	0.3937	20.3211	8.75 M/M	0.3445	0.3491	70
M12 × 1.75	0.4725	14.5138	13/32 in.	0.4062	0.4109	69
M12 × 1.25	0.4725	20.3211	10.75 M/M	0.4232	0.4279	70
M14 × 2	0.5512	12.7000	12 M/M	0.4724	0.4772	72
M14 × 1.5	0.5512	16.9319	12.5 M/M	0.4921	0.4969	71
M16 × 2	0.6300	12.7000	14 M/M	0.5512	0.5561	72
M16 × 1.5	0.6300	16.9319	14.5 M/M	0.5709	0.5758	71
M18 × 2.5	0.7087	10.1595	15.5 M/M	0.6102	0.6152	73
M18 × 1.5	0.7087	16.9319	16.5 M/M	0.6496	0.6546	70
M20 × 2.5	0.7875	10.1595	17.5 M/M	0.6890	0.6942	73
M20 × 1.5	0.7875	16.9319	18.5 M/M	0.7283	0.7335	70
M22 × 2.5	0.8662	10.1595	19.5 M/M	0.7677	0.7729	73
M22 × 1.5	0.8662	16.9319	20.5 M/M	0.8071	0.8123	70
M24 × 3	0.9449	8.4667	53/64 in.	0.8281	0.8340	72
M24 × 2	0.9449	12.7000	22 M/M	0.8661	0.8720	71
M27 × 3	1.0630	8.4667	24 M/M	0.9449	0.9511	73
M27 × 2	1.0630	12.7000	63/64 in.	0.9844	0.9914	70

Glossary

abrasive, artificial silicon carbide, aluminum oxide, and other materials.

abrasive, natural diamond, emery, garnet, flint, pumice, and other materials.

absolute location exact distance from a fixed reference point.

absolute reference point point established as zero on both axes for absolute dimensioning.

AC common term for alternating current.

acceleration increasing rate of change in speed.

acetone flammable liquid used as a solvent and cleaner.

acid compound that when dissolved in water produces hydrogen ions.

acid bath used to clean metal objects.

Acme thread screw thread with an included angle of 29°.

active block numerical control block of information in control of machine functions.

actuator device that moves a machine part or other object.

adapter device that allows objects of different shapes or sizes to be fitted together.

addendum part of a gear tooth that lies outside the pitch circle.

air gages comparison devices that sense variations in airflow rates and convert them into dimensional scale or gage readings.

allowance minimum clearance or maximum interference permitted between mating parts.

alloy homogenous mixture of two or more metals.

aluminum silvery white metal that is one-third the weight of steel.

amalgam alloy of mercury and one or more metals.

ampere unit of measurement of electrical current produced by 1 volt acting through a resistance of 1 ohm.

angle difference in direction between two straight lines.

annealing softening of a substance, usually a metal, by heating it above the critical temperature and cooling it slowly.

anode positive pole or electrode of an electrolytic cell.

anvil steel or hard iron block on which metal is shaped or forged.

apprentice learner of a trade or occupation who has contracted to work as a learner for a specified term.

arbor shaft on which an object being machined or a cutter may be mounted. *See also* Mandrel.

arc portion of the circumference of a circle.

arc welding fusion welding of metals by use of an electric arc as a heat source.

argon inert gas used in certain welding and heat-treating operations.

asbestos fibrous mineral that is noncombustible, nonconducting, acid resistant, and hazardous.

axis line that is a reference point for mechanical or geometrical relationships.

backgear compound gear train used on some machine tools to reduce speed and multiply torque.

ball bearing bearing consisting of an inner race, an outer race, and balls that roll when the inner or outer race turn.

band filing continuous filing operation done on a bandsaw with a blade composed of a band and file segments.

band polishing continuous polishing operation done with an abrasive strip in place of a bandsaw blade.

baseline line or surface from which dimensions are shown in baseline dimensioning on a drawing.

bed main horizontal part of a lathe that carries the headstock, carriage, and tailstock.

bench work operations done with hand and measuring tools and in which machine tools are not used.

bevel gears gears that transmit power between two shafts which meet at an angle.

bill of material list of parts and the quantities of each required for an object or machine.

bore to machine a hole with a true cylindrical internal surface.

boring mill upright machine with a revolving table that carries the work.

boron carbide fine powder that is almost as hard as diamond.

brass alloy of copper and zinc; in some cases other materials are added.

brazing joining two pieces of metal without melting either one by using a brazing alloy that melts at a lower temperature than the materials being joined.

brine water that has been saturated or nearly saturated with salt.

broach long tool with cutting teeth that is pushed or pulled through a hole in a workpiece to reshape or enlarge the hole.

bronze alloy of copper and tin; in some cases other materials are added.

burnish to polish a metallic surface by rubbing it with another metallic, harder surface.

burr sharp or ragged edge on a machined metal object.

bushing sleeve, usually of bronze, that is fitted into a machine part to serve as a bearing.

buttress thread screw thread that is triangular in cross section but has one face at 90° to the longitudinal axis of the screw.

cadmium white, ductile, metallic element used to plate steel and as an alloying element.

calibrate to check and set the accuracy of index marks on measuring instruments.

caliper tool that measures the inside or outside diameter or other dimensions of a workpiece.

cam device mounted on a rotating shaft that transposes rotary motion into uniform or nonuniform reciprocating motion.

carbon nonmetallic element found in all organic substances that is used as an alloying element in ferrous metals.

carburizing heating low-carbon steel in a closed container with carbon to form a high-carbon-content skin (case) on the steel part.

carriage part of a lathe that moves along the ways and includes the apron, cross slide, and other parts.

case hardening forming a thin, hard skin (case) on iron and steel alloys.

castings parts made by pouring molten metal into sand, plaster, or metal molds.

cast iron alloys of iron, carbon, and other materials that are primarily used for making castings (also known as gray iron, malleable iron, ductile iron, etc.).

center drill short drill combined with a countersink, used to prepare the end of a shaft for a center.

center gage gage that sets the threading tool for cutting V threads.

centerless grinding grinding operation in which the work is supported by a blade or rest between the grinding and the regulating wheels.

centerline line on a drawing or layout indicating the center of the object.

centimeter one one-hundredth of a meter, or 0.3937 in.

chisel, cold chisel, such as a flat or a cape chisel, for cutting cold metal.

chromium grayish-white metallic element used in alloying steels and for plating.

chuck device that holds workpieces or cutting tools in drill presses, lathes, and other machine tools.

climb milling metal-cutting operation in which the table moves in the same direction as the part of the cutter below the arbor.

cold drawing finishing of metal bars by drawing them through a die while they are cold.

cold rolling finishing metal sheets or flat bars by rolling them between polished rolls while they are cold.

cold sawing any sawing process in which the chips are not heated to the softened state.

collet clamping or holding device for cutters or workpieces.

combination set group of measuring and layout tools containing a rule, square head, center head, and protractor.

Compact II language for computer-assisted part programming of NC tools.

comparator measuring device that may be set with gage blocks and measures deviations from a standard setting.

copper reddish, soft, ductile metal with very good heat and electrical conductivity and is the basic element in brass and bronze.

counterbore cutting tool fitted with a pilot that enlarges a hole for parts of its length.

datum line any base or reference line from which dimensions are taken or calculations are made.

DC common expression for direct electrical current.

decarburization removal of carbon from metals by heating.

degree 1/360 of the circumference of a circle.

diamond crystalline form of carbon (the hardest

known mineral) used as a cutting tool and a grinding tool and to dress grinding wheels.

die tool that cuts an external thread.

dimension line line on a drawing that indicates to what portion of the drawing the dimension refers.

dividers tool with pointed legs for measuring or setting off distances.

dividing head mechanical device that spaces the perimeter of a workpiece into equal parts (also called an indexing head).

draw filing metal-finishing operation done with a single-cut file held across the long axis of the workpiece.

drill jig device that accurately guides one or more drills into an object by means of accurately positioned hard bushings.

ductile iron high-strength type of cast iron that will bend without fracturing.

ductility characteristic of metals that allows drastic changes in their shape without breaking.

elasticity tendency of a hard, strong metal to return to its original shape when force is applied below the elastic limit.

elastic limit point beyond which a force can cause permanent deformation.

electromagnet magnet of variable strength produced by passing current through conductors around a soft iron core.

eutectic alloy that has the lowest possible melting point with its components.

exploded view view in which the parts of an assembly are shown separately and in the order in which they are assembled.

extrusion forming metal bars, structural shapes, and tubes by forcing the metal through a die.

faceplate circular plate for holding workpieces that may be attached to the nose of a lathe spindle.

feedscrew any screw used to move a machine tool table or carriage in relation to a cutter or toolbit.

ferrous metals all metals that are alloys of iron, carbon, and other materials.

fillet concave curve connecting two surfaces that meet at an angle.

finish mark mark used on a drawing or sketch to indicate that a surface will be machined.

fit relationship in terms of size between two parts of an assembly.

fixed gages nonadjustable gages such as taper, plug, ring, and form gages.

flute part of a reamer, tap, or similar tool on which the cutting edge is ground or machined.

fly cutter cutter body for facing operations on milling machines into which one or more cutter bits may be placed.

forging forming hot metal into a particular shape by pressure or impact.

furnace device for heating and heat treating metals, using electrical current, oil, or gas for heating.

gage blocks precisely dimensioned and hardened blocks that are available in sets and used to set or calibrate other measuring tools.

gear toothed wheel that transmits motion positively.

gear train two or more gears that drive machine parts at a specified ratio of speeds.

gib strip of steel or bronze used to adjust the fit between mating machine parts.

granite rock composed of quartz, feldspar, and mica from which dimensionally stable surface plates and angle plates are made.

grinding machining process for removing metal by the cutting action of grains of abrasive in a wheel or disk.

gun drill single fluted drill for drilling deep, accurate holes.

half nut nut that is split lengthwise and used to engage a lead screw for threading on a lathe.

half section section view of a drawing that shows an internal view of only one-half of a drawing.

hardening developing maximum strength and wear resistance in medium- or high-carbon steel by heating and quenching rapidly.

heat treatment process of altering the physical conditions of metals by heating and cooling.

helical gear gear in which the teeth follow a helical path at a specified angle to the center line of the gear.

helix angle angle of a thread, worm, or gear tooth with a line perpendicular to the longitudinal axis.

high-carbon steel steel that has more than 0.60 percent carbon.

high-speed steel steel that contains tungsten or molybdenum as the major alloying element and is a heat-resistant cutting tool material.

hob cutter for cutting worms, gears, and some types of threads.

honing finishing process that uses abrasives to smooth and straighten holes.

hook-tooth blade type of saw blade with a tooth profile that has positive rake.

hot-rolled steel steel rolled to shape while heated to the plastic condition.

idler gear intermediate gear in a gear train.

impact test used to test materials for resistance to fracturing when a sudden load is applied.

indenture legal document binding an apprentice to an employer for a period of time.

independent jaw chuck lathe chuck whose jaws may be moved independently of one another.

indexing dividing a circle or a portion of a circle into regular or irregular spaces.

inert gas gas that may be used as a shield in welding or heat treating to prevent oxidation or scaling.

inserted tooth cutter milling cutter with separate cutting teeth held in position by screws or locking devices.

inspection measuring and evaluating materials, components, and completed objects for compliance with specifications.

interference fit size relationship between two parts that require the use of force for assembly.

internal gear gear with the teeth on the internal circumference of a ring.

internal thread thread cut on the inside of a hole.

involute gear teeth gear teeth whose cross-sectional shape is based on an involute curve.

ipm inches per minute, a term used to express the feed rate on milling machines.

ipr inches per revolution, a term used to express the feed rate on lathes.

iron metallic element that is the basic ingredient in ferrous metals.

isometric form of perspective drawing done without reference to vanishing points.

jig device that locates and holds a workpiece and guides the cutting tools.

jig boring machine machine somewhat similar to a vertical milling machine that accurately locates and bores holes.

job shop machine shop that specializes in repair work and making single items rather than manufacturing.

journeyman skilled worker who has served an apprenticeship.

kerf slot cut by a saw blade.

keyseat recessed groove in a shaft into which a key is set.

key way groove in a shaft or hub in which a key may move or be attached.

kilogram 1000 grams in metric measure, equal to 2.204 lb.

knurling forming a raised diamond pattern on a metal surface by cutting or displacement action.

land space between the flutes or grooves on reamers, drills, taps, counterbores, and similar tools.

lapping finishing external or internal surfaces accurately by using very fine abrasives.

lathe machine tool for turning, facing, boring, and similar operations.

laying out marking surfaces to be machined to specified dimensions.

lead soft, ductile, blue-gray, heavy metal that melts at 327°C.

lead axial distance that a nut moves in one revolution on a thread.

lead screw screw on a lathe that is used to move the carriage for thread cutting.

liter metric measure of volume, equal to 1.057 qt.

low-carbon steel steel containing less than 0.30 percent carbon.

magnesium very light metal (about 106 lb/ft^3) that alloys readily with aluminum and other metals.

magnetic chuck device that uses permanent magnets or electromagnets to hold iron or steel workpieces.

major diameter outside diameter of a screw thread.

malleable iron cast iron that has been partially decarburized by prolonged heating and is less brittle than gray cast iron.

mandrel shaft onto which an object to be machined may be pressed or held.

medium-carbon steel steel with a carbon content of 0.30 to 0.60 percent.

metallurgy science of separating metals from their ores, alloying and heat treating them, and determining their physical and chemical characteristics.

meter fundamental unit of length in the metric system: 39.37 in. in length.

metrology science of measurement.

micrometer caliper measuring tool based on an accurate lead screw and both true and representative scales.

micron one millionth of a meter.

mil one thousandth of an inch.

milling machining surfaces by using a rotating cutter with one or more teeth.

millwright mechanic who installs and maintains machinery.

minor diameter diameter of a screw thread measured at the bottom of the threads.

minute in the measure of angles, 1/60 of a degree.

miter gears bevel gears that transmit power at a 90° angle and are always of equal diameter, pitch, and number of teeth.

modulus of elasticity ratio of stress per unit of area to the strain per unit of length, with the strain being below the elastic limit of the material.

molybdenum metal used as an alloying element in steel, including high-speed steel.

nickel strong, noncorrosive, white metal used as an alloying element in steels and for plating.

nickel silver alloy of copper, nickel, and zinc (also known as German silver).

nitriding hardening the surface of steel objects by heating them in the presence of ammonia gas or other nitrogen-bearing compounds.

nonferrous metals any metal in which the main constituent is not iron.

normalizing heating steel or iron objects above the critical temperature and cooling them in the open air.

numerical control (NC) control of a machine or a process by command instructions in symbolic form.

obtuse angle angle greater than 90°.

ohm basic unit of electrical resistance.

oilstone abrasive stone that is oiled and used to sharpen cutting tools.

open shop business in which both union and nonunion workers are employed.

optical comparator machine that enlarges the profile view of a part and compares it to a standard profile.

orthographic projection graphical presentation of an object with each view, such as front, top, or side, showing a face of the object.

outside caliper tool that measures the outside dimensions of an object.

patternmaker craftsperson who makes the wooden or metal patterns used in casting metal objects.

permanent magnet special magnet steel that retains its magnetic power indefinitely.

pickling removing scale and oxides from the surface of metal objects with an acid solution.

pictorial sketch freehand drawing that shows more than one surface of an object.

pinion gear smaller gear of set of two gears.

pipe thread sharp V thread cut on a taper of 3/4 in./ft.

pitch distance between the crests of adjacent threads or a given point on adjacent gear teeth.

planer machine tool with reciprocating table, used to machine flat and angular surfaces.

plug gage accurate fixed gage used to check the size of holes.

powder metallurgy forming parts out of powdered metal by compacting the powder into a mold under great pressure and heating it.

protractor tool that measures and lays out angles.

punch press machine tool that cuts and forms metal parts.

pyrometer instrument that measures temperatures over a wide range.

quadrant quarter circle, usually identified by a number.

quenching cooling a metal object at a particular rate to change its physical characteristics.

quill hollow shaft or spindle.

rack gear with a straight pitch line.

radial drilling machine drilling machine with an arm that moves up or down and around a vertical column that has a head which moves fore and aft on the arm.

radius gage device that measures the radius of a fillet or a rounded corner.

rake angle angle between the top face of a cutting tool and a line at a 90° angle to the workpiece.

reamer tool with cutting edges used to finish the interior of holes.

riffler small file, usually curved, used to finish the inside surface of cavities.

ring gage gage with an accurately finished hole, used to check the outside diameter of machined parts.

rolled threads threads formed by hardened rolls or dies without the removal of material.

rpm abbreviation for revolutions per minute.

saddle assembly on a lathe that moves fore and aft on the ways.

screw grooved, helical path around a cylinder.

screw pitch gage device that checks the pitch of a thread.

scroll thread cut on the face of a flat plate and follows a spiral path outward from the center.

sector part of a circle bounded by two radii and the arc that subtends them.

sensitive drill press drilling machine that has only a hand feed for the quill assembly.

serrations toothlike projections on the edge or surface of a part.

servo device that provides the power to move machine parts at the command of the operator or numerical control system.

setup person skilled worker who prepares machines for use by less-skilled machine operators.

shim thin metal spacer used in the adjustment of machine parts.

silicon carbide refractory and abrasive material made by sand, coke, and sawdust in an electric arc furnace.

silver white, ductile metal that is an excellent conductor of heat and electricity.

silver brazing brazing similar or dissimilar materials by using an alloy of silver and other metals.

sine bar tool that may be set at a precise angle to the surface of a surface plate by use of gage blocks.

slide caliper caliper that consists of a graduated fixed portion and a movable portion with an index or reference mark.

snap gage nonadjustable double-ended gage that may be used for checking internal or external sizes.

spindle part of a machine that rotates and carries either the workpiece or cutters.

spur gear gear with the centerline of the teeth parallel to the centerline of the shaft on which the gear is mounted.

stainless steel alloy that contains iron, carbon, chromium, and sometimes nickel.

strain to stretch a material past its elastic limit.

surface grinding grinding flat surfaces with the workpiece held on a reciprocating or rotating table.

surface speed rate of movement of the work relative to the tool, or vice versa, usually expressed in feet per minute or meters per minute.

tachometer device that indicates the rotational speed of a shaft or other object.

tap tool that cuts threads in a hole by means of cutting edges on the flutes.

taper attachment attachment on a lathe that allows simultaneous longitudinal and cross movement of the saddle and cross slide, thus cutting a taper.

thermocouple simple electric generator made by welding together two dissimilar materials at one end. When the welded end is heated, an electrical potential proportional to the heat is produced.

thread miller machine that cuts threads by use of a rotating milling cutter.

tin silvery white, soft metal used in solders and as a plating material.

titanium strong, grayish metal that weighs less than steel.

tolerance allowable variation in the dimension specified for a machined part.

toolmaker skilled machinist who specializes in the making of jigs, gages, fixtures, and other tooling.

trammel beam compass in which movable heads are attached to a rectangular or round bar.

trepanning removal of a circular piece of material from inside a steel bar, plate, or billet.

tumbler gear intermediate gear in a gear train that reverses the direction of rotation of the driven gear.

tungsten heat-resistant metal used as an alloying element in high-speed steel.

tungsten carbide iron-gray powder composed of carbon and tungsten and used in sintered form as a cutting tool material.

turret lathe lathe with a revolving turret head mounted on either the saddle or a special tailstock.

ultimate strength highest unit stress that a material can withstand prior to failure.

unit stress stress, or load, on a unit of area, usually expressed in pounds per square inch.

universal grinding machine grinding machine on which the table, wheelhead, and work head swivel.

universal milling machine milling machine that has a graduated two-piece saddle, thus allowing the table to be swiveled.

vanadium metal used as an alloying element in steel to improve shock resistance and forgeability.

vernier small auxiliary scale on a micrometer or other precision measuring tool, used to subdivide increments on the true scale or other scales.

vertical boring mill large machine tool that carries the work on a rotating table with a vertical spindle.

vertical turret lathe lathe with a vertical spindle and ways that carries the work in a chuck and has the cutting tools mounted in a five-position turret and side heads.

V ways ways on machine tools in the shape of an inverted V with the top flattened.

waviness wavelike irregularities in a surface.

ways longitudinal surfaces on a machine on which components such as carriages and tables move.

wheel truing cutting off the irregularities on a rotating grinding wheel with a diamond dresser.

white iron extremely hard cast iron that results from pouring the hot metal into a mold with a chill plate in it.

whole depth total depth of a gear tooth from the top to the root line.

work hardening hardening that occurs when certain metals are hammered or rolled at room temperatures.

working drawing drawing that contains all the information for making an object.

worm gear set gear reduction composed of a worm and wheel with nonintersecting axes.

yellow brass alloy of about 70 percent copper and 30 percent zinc.

yield point point at which an object under load starts to deform without the addition of more load.

zinc bluish-white metal used in a variety of alloys and as a coating for steel.

Index

C

D

E

F

G

H

I

J

K

L

M

S

T

U

V

W